LEAST-COST ELECTRIC UTILITY PLANNING

HARRY G. STOLL

With Contributing Authors:

LEONARD J. GARVER
GARY A. JORDAN
WILLIAM W. PRICE, JR.
RICHARD M. SIGLEY, JR.
RICHARD S. SZCZEPANSKI
JAMES B. TICE

Systems Development & Engineering Department
General Electric Company
Schenectady, New York

WILEY

A WILEY-INTERSCIENCE PUBLICATION

JOHN WILEY & SONS

New York • Chichester • Brisbane • Toronto • Singapore

Library of Congress Cataloging-in-Publication Data:

Stoll, Harry G.
Least-cost electric utility planning / Harry G. Stoll, with
contributing authors, Leonard L. Garver . . . [et al.].
p. cm.
"A Wiley-Interscience publication."
Bibliography: p.
Includes index.
ISBN 0-471-63614-2
1. Electric power systems—Planning. 2. Electric utilities—
Planning. I. Title.
TK153.S834 1989 88-21980
621.3′1′068—dc19 CIP

Printed in the United States of America

10 9 8 7 6 5 4 3 2

CONTENTS

PREFACE

Today's utility power system planner is responsible for planning the reliable and efficient operation of a multi-billion-dollar, high-technology power system enterprise. This book describes the key elements and tools necessary to carry out the planning responsibilities for this system.

The utility power system business is undergoing major changes. Fuel price excursions of the 1970s have created uncertainty in future fuel prices. The cost of fuel accounts for 40–60% of the cost of electricity.

Since the 1970s, electricity load growth has experienced a dramatic slowdown from prior historical levels. Conservation, change in industrial composition, and change in economic growth are creating new patterns in load growth.

The Public Utility Regulatory Policies Act of 1978 requires that utilities purchase power from industrial cogenerators and small power producers at rates up to the utility avoided cost. Cogenerators are the first nonutility, nonregulated independent power producers. Future legislation and policies can lead to further deregulation of the electric power industry. Utilities are developing planning strategies to address these outcomes.

The utility industry is becoming more cost-conscious and cost-competitive, as well as more innovative in its methods of addressing the issues. For example, there has been increased use of transmission interconnections to low-cost generation sources. New technologies, including high-voltage direct-current transmission, series compensation, and shunt compensation, will contribute toward achieving reliable least-cost transmission.

The overall cost of designing, constructing, and operating a new-generation plant has increased significantly over the last two decades in response to higher construction costs and increased environmental control equipment. New-generation technology is being developed to meet environmental requirements at lower cost.

Never before has the goal of "getting more out of what is there" been more of a reality in the utility industry. Generating plants are being extended well beyond their normal 40-year life. Transmission systems are being upgraded and uprated. New technology is being retrofitted on existing equipment to assure reliable, low-cost performance.

Sweeping changes under way in the utility industry are continually presenting new challenges to utility power system planning. This book presents the key principles necessary to develop planning strategies that will meet these changes.

The book is divided into seven areas with one to three chapters that address each topic. The material covered includes:

• Economics, finance, and regulation	Chapters 2–5
• Industrial power economics	Chapter 6
• Load demand and management	Chapters 7–8
• Generation system reliability	Chapters 9–11
• Generation system production cost	Chapters 12–13
• Capacity planning	Chapters 14–15
• Transmission planning	Chapters 16–17

Chapters 2 through 5, in the areas of economics, finance, and regulation, provide the necessary foundation for evaluating economic decisions, determining electricity price, evaluating financing requirements, and understanding utility business incentives.

A review of industrial power economics, specifically as it relates to industrial cogeneration of process steam and electricity, is presented in Chapter 6. Cogeneration and independent power production have become a major source of electrical power. This chapter presents examples of strategies utilities use for working with industrial power projects.

Methods for understanding and forecasting load growth while recognizing the impact of conservation, price elasticity, and economic growth are presented in Chapters 7 and 8. Peak-load forecasting and load management economic evaluation methods are also presented. The load demand forecast is one of the most important elements of planning since it forms the basis for a capacity addition strategy.

Chapters 9 through 11 present the principles of generation system reliability. Chapter 9 discusses the sources and interpretation of generating unit reliability data. The methods and examples presented illustrate how to project future generating unit reliability performance. Chapters 10 and 11 present the theory and examples of single-area and multiarea power system reliability evaluations. Topics include maintenance scheduling, loss-of-load probability

(LOLP), emergency operating procedures, optimal cost-justified reliability levels, and multiarea reliability with transmission limitations.

The principles for simulating the hourly operation of a power system are presented in Chapters 12 and 13. The topics covered include hydroelectric plant dispatch, thermal commitment and dispatch, and probabilistic simulation to account for random generating unit outages and load uncertainty. Chapter 13 focuses on multiarea production simulation with transmission network restrictions.

Chapters 14 and 15 present the integration of reliability evaluation, production simulation, and investment cost analysis as they affect least-cost capacity planning. Several capacity planning analysis approaches are discussed. These chapters include discussions of the economic basis for a balanced mix of generating unit types and short- and long-term marginal costs as well as a methodology for integrated end-use supply-and-demand planning. A decision tree methodology and examples are also presented for developing capacity plans with minimum business risk in an uncertain business environment.

Chapters 16 and 17 present procedures and analysis for bulk power transmission planning. Chapter 16 focuses on the static (nontransient) issues of transmission planning, including line loadability guidelines, AC and DC power flows, and horizon-year planning. Chapter 17 focuses on power system stability methods and issues of modern interconnected power systems.

Each subject area addresses power system theory and principles and applies them to realistic utility examples. Results from solved examples are expanded to illustrate the sensitivity and direction of key parameters.

HARRY G. STOLL

ACKNOWLEDGMENTS

This book had its origins with the course in Economics and Planning that I cotaught for several years in General Electric's Power Systems Engineering Course (PSEC). Each year, I added one or two chapters and tried them out on the PSEC students who demonstrated great powers in uncovering errors and omissions; thanks to all for being so kind.

In March 1987, I agreed to refine these chapters into a book for John Wiley & Sons. With excellent council from Dr. Fred Ellert, my Department General Manager, I supplemented my own experience with that of the many experts within the department.

It is with pleasure that I acknowledge my contributing authors for their untiring energy in working with me to bring this volume to fruition. Their sage council, developed from years of electric utility experience, provided a wealth of practical insights. I am truly honored to have had the opportunity to work with each of them on this book. Dick Sigley's expertise contributed to the economics, finance, and regulatory chapters. Rit Szczepanski was invaluable in refining the chapters on load forecasting and generation planning. Gary Jordan and Len Garver provided immeasurable assistance on the generation reliability and production simulation chapters. Bill Price and Jim Tice developed excellent material for the transmission planning and stability chapters.

I also want to thank many others in the department for their council and suggestions. Dr. Leon Kirchmayer provided early inspiration for this work. I am grateful to Glenn Haringa, Harry Heiges, and Marv Schorr for their comments on the recent finance, tax law, and environmental law revisions. Bob Gentner, Bob Van Housen, John Kovacik, and Jim Oplinger's review of the cogeneration chapter and the preparation of the cycle performance parameters for the examples was greatly appreciated. Valuable suggestions were offered by Paul Albrecht on the reliability chapters, Kim Wirgau on power-flow calculations, Mel Crenshaw on excitation systems, Tom Younkins on prime mover dynamics, Farhad Nozari and Glenn Breuer on HVDC transmission, Nick Miller on the interregional stability example, and Charlie Concordia on system stability.

I am very grateful for the overall support that I received from GE management during the production of the book. Bjorn Kaupang's review and Nelson Simons's untiring dedication to helpful recommendations greatly enhanced the material. I especially thank Dr. Fred Ellert for his constant encouragement, advice, and support for this project.

Special appreciation is given to Paula Rosenberg for the unsparingly-employed-red-ink-pencil-edit corrections to the book, along with hyphenation and grammar tutorials. Perhaps she rewrote as much as we wrote. Lucienne Walker provided editorial coordination. I am grateful to Carol Bradshaw, Jackie Brumley, and Holly Powers for their dedication in word processing the text. Special thanks to Jan Nolan for triumphing cheerfully over the seemingly never-ending train of revisions, format changes, and deadlines.

Finally, I want to thank the families of the contributing authors for the time the authors spent on the book that they would have rather spent with you. I especially thank my own family, Patricia, Heather, Kristen, and Alison for their understanding and love.

H. G. S.

1

THE UTILITY PERSPECTIVE

As a first step in planning the future utility power system, it is beneficial to review the path that has led to the current situation of the utility industry. This section presents an historical overview of many of the changes experienced by the electric utility industry since the 1970s.

A major changing force affecting the utility industry began in 1973 with the fuel price excursions as shown in Figure 1.1. Prior to the 1970s, oil, natural gas, and coal were low-priced, stable sources of energy. Oil prices increased dramatically in 1973 and again in 1979. Aided by natural gas deregulation, natural gas prices followed oil prices. The rise in oil prices removed a competitive ceiling on coal prices, which then also increased.

However, oil prices began to decrease in the 1980s with natural gas following suit, as a result of price competition. In fact, real oil prices in the late 1980s were lower than the prices in the middle and late 1970s.

The energy price excursions of the 1970s had a major impact on energy demand. Figure 1.2 illustrates the U.S. natural gas consumption for every unit of real U.S. Gross National Product (GNP). During the 1960s and early 1970s, real natural gas prices declined in real terms. When prices decline, economic theory suggests that demand should increase, which was the case during that period. At the same time, gas consumption per unit of GNP increased.

In the early 1970s, gas prices quadrupled for \$5/bbl equivalent to \$23/bbl equivalent by the early 1980s. Consumption per unit of GNP decreased by about one-half. Economists refer to this as "price elasticity," or the change in demand per change in price. For natural gas, the price elasticity is approximately −.50 (a fourfold price increase changing demand twofold). The price elasticity phenomenon for oil was nearly identical.

Electricity demand is governed by the same economic principles as other

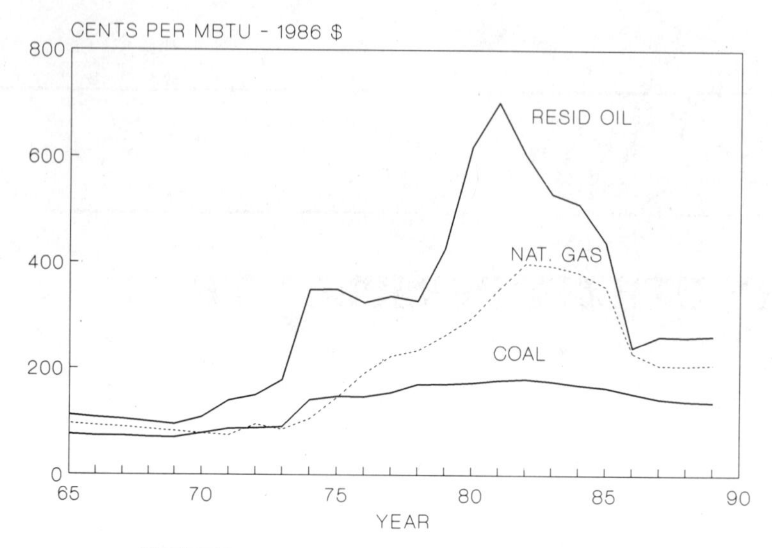

FIGURE 1.1 U.S. utility fuel cost price trend.

energy forms as illustrated in Figure 1.3. Since the price excursions were not as large as those for oil and gas, the demand did not change as drastically. An examination reveals that when electricity prices rise, however, electricity demand per unit of GNP output decreases. Overall, the electricity price elasticity lies in the $-.25$ to $-.50$ range.

Coal, petroleum, and natural gas (and nuclear energy) are the primary U.S. sources of energy, as shown in Figure 1.4. During the late 1960s and early

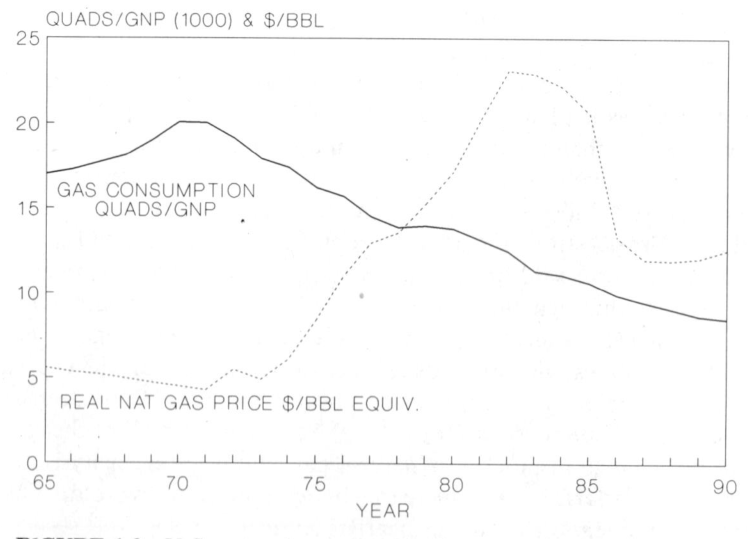

FIGURE 1.2 U.S. natural gas consumption per unit of GNP output.

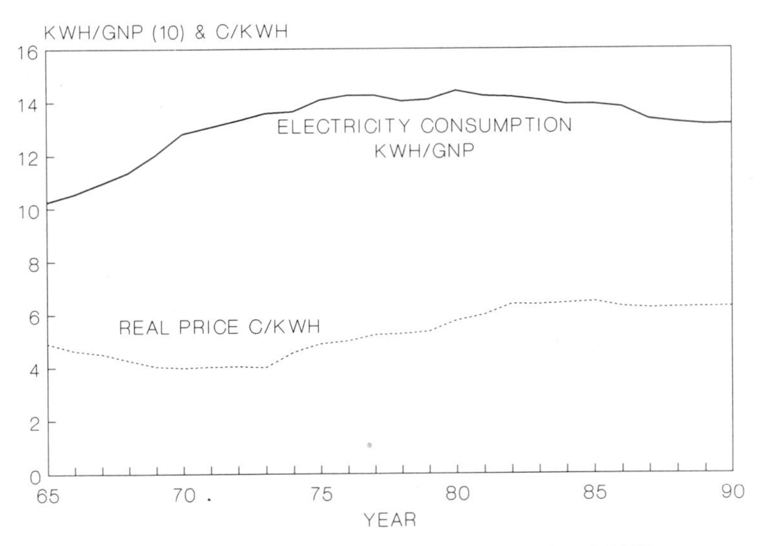

FIGURE 1.3 U.S. electricity consumption per unit of GNP output.

1970s, oil and natural gas gained market share, largely at the expense of coal. In the late 1970s, coal and nuclear energy increased in primary market share as a result of the relative price increases.

Primary energy sources may be utilized directly to produce end-use work (i.e., in transportation) or produce process heat and space-conditioning heat. Alternatively, the primary energy sources may be converted into an end-use electricity product. Conversion to electricity has been the preferred means of

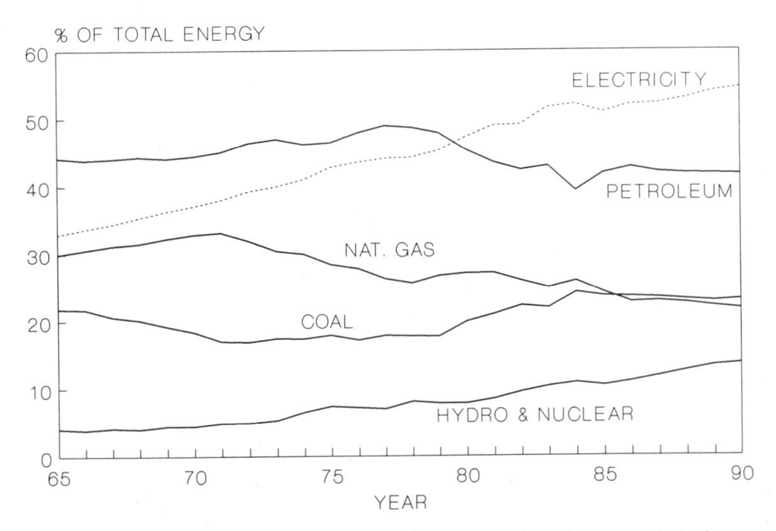

FIGURE 1.4 U.S. primary energy share and electricity market share.

energy utilization as historically shown by the increasing electricity end-use market share. Electricity increased in end-use market share from 32% in 1965 to more than 50% in the late 1980s.

The rate of electricity growth slowed significantly in the late 1970s and 1980s from the historical 7%/year growth rate of the 1950s and 1960s. This trend is illustrated in the 5-year rolling average rates in Figure 1.5. Contributing to this reduced growth rate was slower economic growth, lower household formation growth, customer conservation, and the decline of the electricity intensive "smokestack" industries. Load growths in the late 1980s were in the 2–4%/year range.

As electricity energy consumption was transitioning to a new growth rate plateau, peak demand growth was also undergoing transition, as implied by the load factor history shown in Figure 1.6. "Load factor" is defined as the ratio of the average demand divided by the peak demand. With increased use of air conditioning, the national load factor increased during the 1950s as utilities switched from a strong winter peak to a biseasonal peak of equal summer and winter load. During the 1960s, the load factor declined as the summer peak became dominant as a result of the significant increase in the penetration of air conditioning.

In the late 1970s and early 1980s, conservation contributed to further decreasing the load factor. This is because average demand is typically reduced more than peak demand when electricity is conserved. Deindustrialization of high-electric-intensity industries also contributed to decreasing load factor with the loss of these high-load-factor consumers. The small amount of load management installed during the 1970s and 1980s (approximately 1% of residential consumers) provided a small increasing load factor influence.

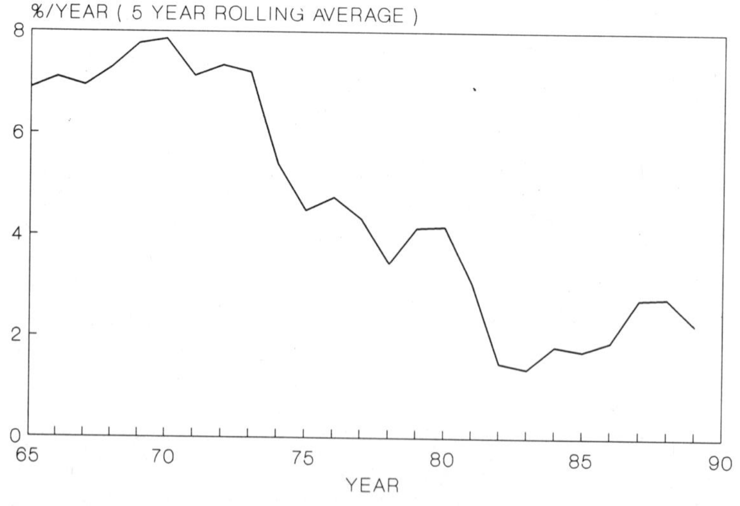

FIGURE 1.5 U.S. national electricity growth rate—5-year rolling average.

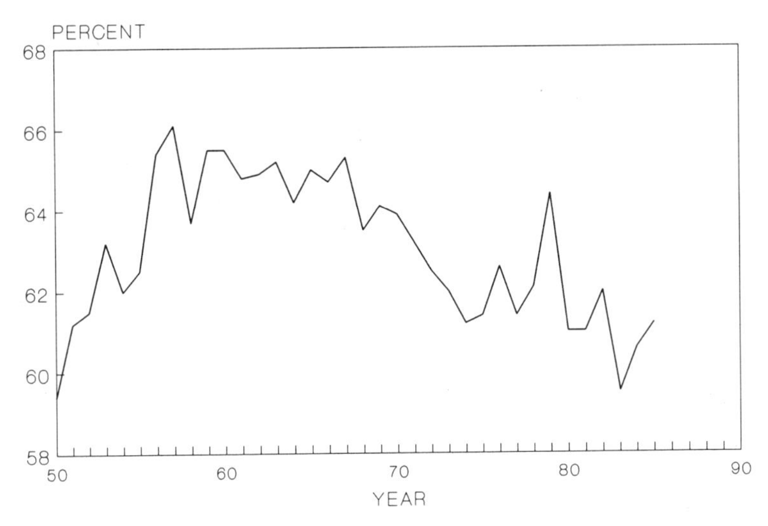

FIGURE 1.6 U.S. annual load factor trend.

Even within the utility industry, technological and conservation changes have had an impact on load levels. Losses in electricity transmission and distribution have been reduced from the 11% level in 1960 to the 8% level in the 1980s as presented in Figure 1.7.

During the 1970s, energy issues became prominent. One issue was to improve the efficiency of energy utilization. The major piece of legislation to focus on this issue was the 1979 Public Utilities Regulatory Policies Act (PURPA). This act required that utilities purchase power and pay as much

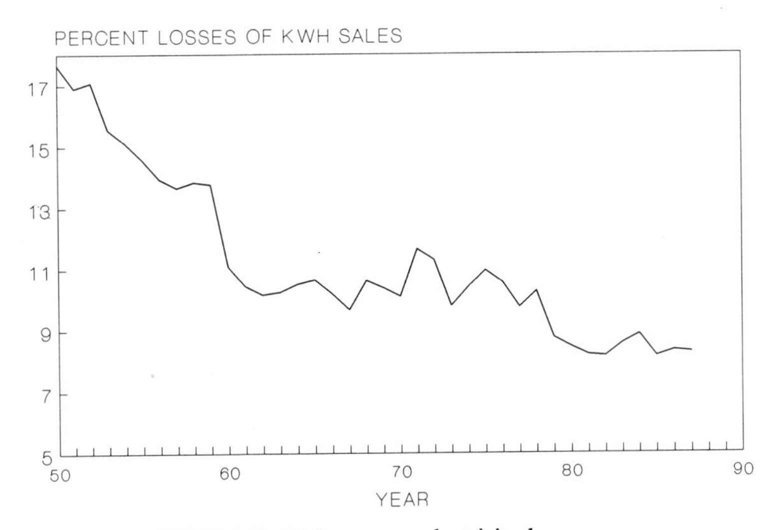

FIGURE 1.7 U.S. average electricity losses.

as the utility's avoided cost per kilowatt-hour for the cogenerated electricity. (Cogeneration involves the simultaneous production of process steam and electricity.) This legislation and associated energy price increases led to the quadrupling of new cogeneration capacity additions, as shown in Figure 1.8.

During the 1960s and 1970s, nonutility generated power from industrial companies (cogeneration and small power producers) averaged 500 MW of additions per year. These capacity additions were very small in comparison to the 15,000 MW/year of utility additions during the same time period. After 1980, however, nonutility capacity additions rose significantly and continued into the late 1980s as a major source of electric capacity additions. The key incentive for an industrial company to begin cogeneration is that it reduces overall industrial process steam and electricity costs.

Prior to the 1970s, electric utilities were constructing new plants at an average of 15,000 MW/year. Since base-load power plants require 5–10 years to construct, utilities had many thousand megawatts of generating capacity in construction during the mid-1970s, when load growth transitioned from 7%/year growth to the average 3%/year level of the 1980s. Since it was often economical to finish the construction on those plants that were already in the construction process, more capacity was placed in service than was necessary to meet the peak-load demands on the domestic utility power system. Consequently, reserve margins increased as shown in Figure 1.9. (Reserve margin is the percentage amount of capacity in excess of the peak-load demand. A 20–25% value is typically required to reliably serve the customer load demands.)

Throughout the 1980s, electric utilities had little need for additional capacity because of the high reserve margin. Consequently, new capacity construction programs were reduced. While the national reserve margin value in 1990

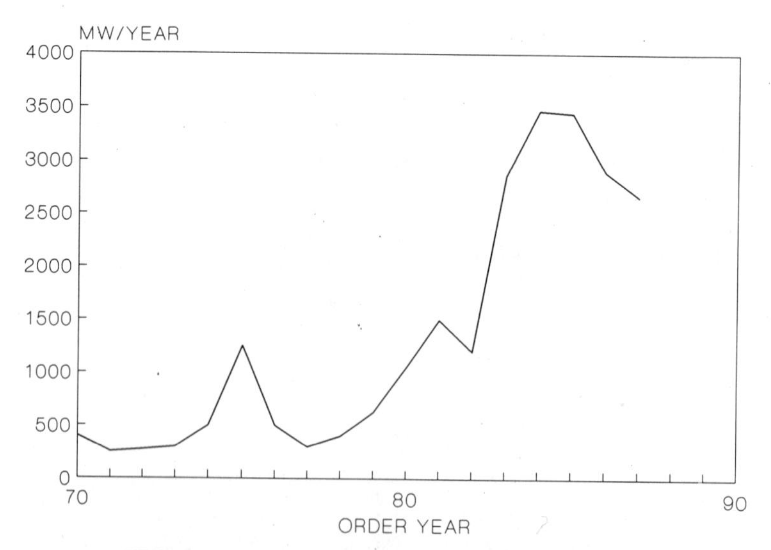

FIGURE 1.8 U.S. non-utility capacity additions.

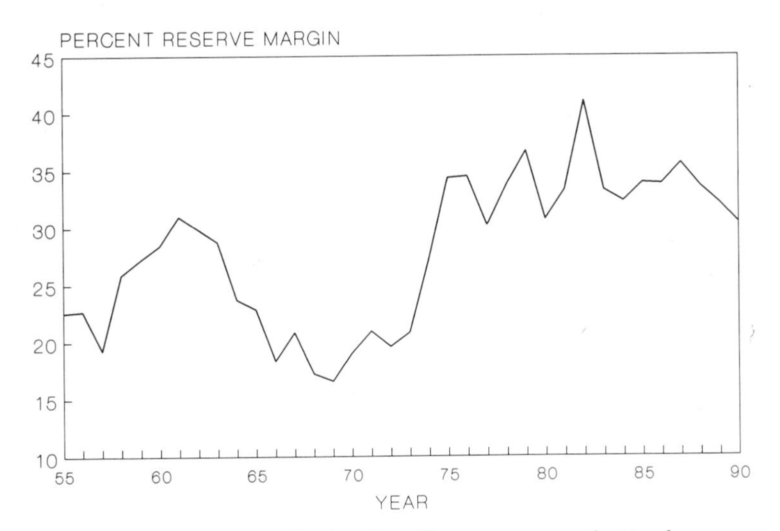

FIGURE 1.9 U.S. electric utility reserve margin trend.

was still above the 20–25% target, individual utilities and regions of the country have lower values and may need to add capacity in the early 1990s.

A utility industry issue of the major importance is its aging fleet of generating capacity, as illustrated in Figure 1.10. Because of the decreased load growth of the post-1970 time period and decreased new capacity addition requirements, the average fleet age is increasing. By the year 2000, almost 30% of fossil steam capacity in the United States will be 30 years or older. This is

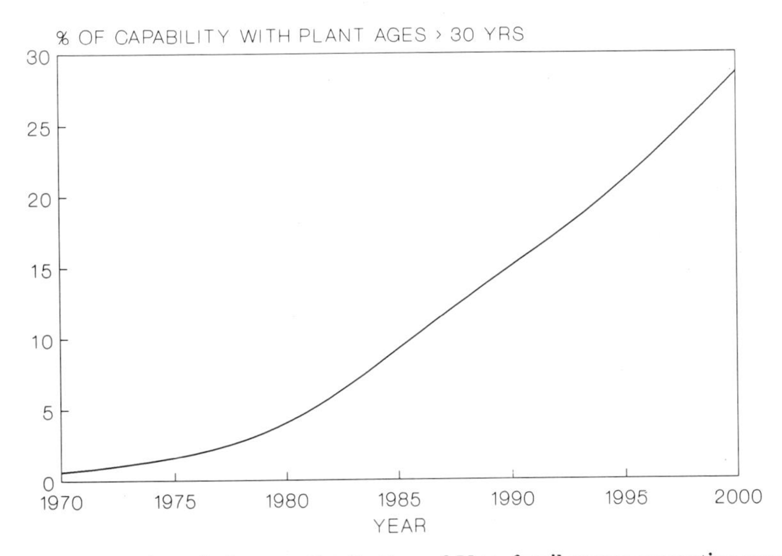

FIGURE 1.10 Cumulative age distribution of U.S. fossil-steam generating capacity.

an important issue for the industry to recognize because, as generating units age, reliability and efficiency degrade.

Generating units have traditionally been retired at 40–45 years of age. Because of the high cost of new generating units, however, utilities have turned to upgrade and life-extension projects for generating units. This reduces the need for new generating units by reinvigorating the generating unit with improved reliability and efficiency. In many cases, new technology enhancements to existing plants are made to further enhance performance.

Bulk power transmission has been generally driven by the same issues as generation. With significant price differential between coal and oil–gas fuels, it has become economical to transport power over hundreds of miles from low-cost coal and hydroelectric generating plants to higher fuel cost regions of the country. This has led to several large high-voltage direct-current (HVDC) and high-voltage ac (HVAC) transmission installations.

The subsequent chapters present the key principles necessary to develop reliable, least-cost planning strategies to successfully guide the power system from today into the future.

For the utility industry to remain healthy and reliable, least-cost power systems are required for the future. Integrated end-use and supply planning is of great importance and involves many technologies, including load conservation, load management, cogeneration, independent (nonutility) power production, utility generation capacity additions, transmission system access, and interutility resource integration. This book presents the principles necessary for understanding and developing a reliable, least-cost planning strategy.

2

INTRODUCTION TO UTILITY FINANCIAL ACCOUNTING

Electric utilities are business enterprises that provide electric power service. The measurement of the utility's success in providing reliable, low-cost electric power service is its financial performance.

This chapter presents the basic accounting rules that guide the financial reporting of electric utilities. Subsequent chapters develop engineering economics in terms of these utility financial accounting procedures. The bottom-line measure of an electric utility's operations and planning, however, is its financial performance.

2.1 BASIC FINANCIAL REPORTS

There are two basic financial reports provided by any business enterprise. The first one is the balance sheet, a picture of the financial status at a given point in time. The balance sheet contains a summary listing of the assets (items *owned* by the utility) and the liabilities (debts *owed* by the utility) on a particular date, usually December 31 of each year. The second financial report is the income statement, a cumulative record of the revenues and expenses incurred by the firm during a stated period of time. The fiscal year can be any 12-month term, but is usually January 1 through December 31 of a calendar year.

These two financial reports are linked to each other as shown in Figure 2.1. Assume that the balance sheet in 1993 is the starting point. This record is modified by the operational results reported on the income statement during 1994 to produce a new balance sheet as of December 31, 1994. Similarly, the balance sheet for 1994 will be subsequently modified through time by the 1995

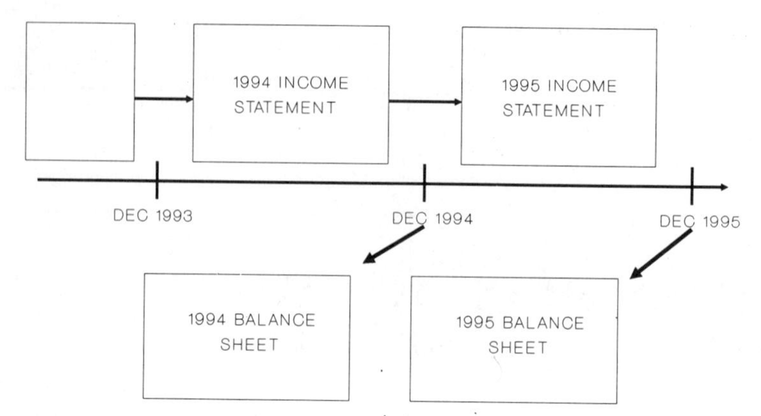

FIGURE 2.1 The two primary financial reports, the balance sheet and income statement, must be coordinated.

income statement, which would then produce a balance sheet at the end of 1995.

One key balance sheet accounting equation is Equation (2.1), which states that the assets of the corporation equal the liabilities of the corporation plus the owner's equity (or owner's ownership) in the business. The other important equation is the income equation, which states that the revenues minus all expenses equals income or profit from operating the business.

Accounting equation:

$$\text{Assets} = \text{liabilities} + \text{owner's equity} \tag{2.1}$$

Income equation:

$$\text{Income} = \text{revenues} - \text{expenses} \tag{2.2}$$

2.1.1 The Balance Sheet

A typical electric utility balance sheet is shown in Table 2.1. The assets are shown in the left-hand column and the liabilities plus the owner's equity are shown on the right.

2.1.1.1 Plant at Original Cost. The most important category for an electric utility is the investment in utility plant, typically 90% of the utility's total assets. Utility plant, the first item under assets, is the total utility plant in service based on the original cost of installing that plant. For example, a plant may have been placed in service in 1960 at a cost of $100 million ($100M). Another plant placed in service in 1985 may have cost $200M. These two items, if they were the only plants in the utility system, would be reported together in the original cost category as a $300M asset.

TABLE 2.1 The Balance Sheet

Assets	Liabilities and Owner's Equity
Plant at original cost	Common stock
	Paid in capital
	Retained earnings
	Total common equity
Less depreciation reserve	Preferred stock
Net plant	Total equity
Construction work in progress	Long-term debt
	Total capitalization
Total plant	Current liabilities
Current assets	Short-term debt
Cash	Accounts payable
Accounts receivable	Customer deposits
Materials and supplies	
Other investments	
Deferred debits	Deferred credits
Total assets	Total liabilities

2.1.1.2 Depreciation Reserve. Although the plant is recorded on the basis of the total plant in service at original cost, it is expected to provide useful service for only 30–40 years. Therefore, it is necessary to provide for the replacement (or preserve the original value) of this plant when the plant is retired (worn out) in 30–40 years. Depreciation reserve can be viewed as an account that provides an accumulation of funds that could be used to replace original equipment when it is retired (assuming no inflation) or to pay back the original investors. The accumulated depreciation can also be viewed as the economic life that has been consumed from the assets.

2.1.1.3 Net Plant. The accumulated depreciation reserve, which is funded every year out of operating revenues, is shown as a deduction from the plant at original cost, leaving net plant-in-service value. Net plant-in-service is the original worth of the plant that is in service less the accumulated depreciation since the plant was originally placed in service. This is also called the "book value of plant," which may or may not equal the plant's market value.

2.1.1.4 Construction Work in Progress. The construction-work-in progress (CWIP) account reports how much the utility has spent on power plants that are under construction and not yet in service. For example, if the utility spent $200M through 1990 on constructing a plant that would be placed in service in 1992, the CWIP account in 1990 would report a balance of $200M. If another $100M were spent to complete the plant, then when it would be placed in service in 1992 the CWIP account would be decreased by $300M and the plant-in-service account would be increased by $300M.

2.1.1.5 Total Plant. The total plant assets of the utility are the sum of the net plant in service plus the construction work in progress on new plants. For electric utilities, the total net plant is typically 90% of the total assets of the utility.

2.1.1.6 Current Assets. Other utility assets are current assets, which generally include four accounts. The first is cash. A utility typically has a cash balance in a checking account at a bank in order to provide for the funds needed to conduct day-to-day business. The cash balance can include not only cash and bank accounts but also temporary or marketable securities that the utility can easily convert into cash on short notice.

The second account is accounts receivable, which is money that is owed to the utility but has not yet been paid. The largest example is electric bills that have been sent to customers but have not yet been paid. When the utility sends out an electric bill in the mail, the unpaid bills immediately fall into a category called "accounts receivable." As these bills are paid, they will be deducted from the accounts receivable category and placed into the cash category. Normally, utilities will have a 30–45-day accounts receivable, depending on the frequency of its bills and the payment terms for customers. Many utilities try to encourage prompt payment of bills by providing a credit discount for early payment or a late payment fee.

The third account, which may constitute 2–4% of the total assets, is materials and supplies on hand. A large fraction of the materials, supplies, and inventories is the fuel supply at power plants. Normal practice for utilities is to have a 60–90-day inventory of fuel. A fuel inventory of 60–90 days for all the power plants can be a very significant asset, since that is roughly one-quarter of a year's fuel expense. Another significant component is spare and renewal parts for power plants, transmission, and distribution equipment.

The last current asset is other utility investments. These may be either short or intermediate-term investments that are not immediately convertible to cash. A utility may also have longer-term investments in real estate, fuels, consulting, or other energy-related businesses.

2.1.1.7 Deferred Debits. Finally, deferred debits include unamortized debt discount and expense and other work in progress.

The sum of the asset items (Sections 2.1.1.1–2.1.1.7) equals the total assets of the utility.

2.1.1.8 Equity. On the liability side of the balance sheet, the first major category is owner's equity. The largest type of equity for investor-owned utilities is in the form of common equity. The utilities issue stock on the stock exchanges, which permits public ownership of the electric utility. Each common stockholder helps to elect the board of directors by casting a vote according to the number of shares owned.

The proceeds from issuing common stock to the public are recorded under the equity section entitled common stock and paid in capital. Under "common stock'' is listed the par value times the number of shares. Under ''paid in capital'' is the excess between market value and par value. Par value is a nominal value per share, usually equal to the original stock price at first issue, while the market value is the price on the stock exchange. The utility issues common stock at the market value on the financial markets at the date of issue. The total cumulative proceeds from the sale of all the stock over the entire history of the corporation is recorded under these two categories.

The last item constituting common equity is retained earnings. Retained earnings are the cumulative profits of the utility enterprise over its history that have been reinvested back into the company. When a utility declares a profit, for example $100 million in a year, this profit can be divided in two ways. Part of it, typically about 65%, will be mailed to the stockholders, proportional to the number of shares owned, in the form of a common dividend payment, normally on a quarterly basis. The balance of the $100M or profit, typically $35M, is retained earnings and will be reinvested in the utility. The $35M that is retained during the year is added to all the previously accumulated retained earnings and recorded under ''retained earnings'' on the balance sheet. These retained earnings, or retained profits, permit the utility to have an increased infusion of capital that permits the utility to expand its capital plant to meet the needs of its customers. Utilities may also need to supplement retained earnings by additional new stock offerings and other types of financing.

2.1.1.9 Preferred Stock. The other form of equity is called ''preferred stock.'' Preferred stock differs from equity in that preferred stock has a fixed yield associated with each offering and usually no vote in company business. For example, a utility may issue preferred stock bearing a dividend of 12%/year. In contrast, common stock has no fixed yield, and dividends and profits on common stock are entirely dependent on the financial success of the utility. Preferred stock, however, has a guaranteed yield associated with it. If the utility does very well in a year, the preferred stockholder still receives only the stipulated dividend.

2.1.1.10 Long-term Debt. Long-term debt is a major source of funds for electric utilities. It represents the amount of money that the utility has borrowed from debt holders on a long-term basis. Traditionally, electric utilities have offered bonds with a stipulated interest rate (e.g., 10%) and a stipulated maturity (typically 30 years). A utility issues these bonds as the need arises to meet the expansion needs of the enterprise. All the long-term debt (or bonds) that the utility has issued and not retired (i.e., not paid back to the borrowers) is recorded under the category of ''long-term debt.''

Long-term debt holders have a call or priority on the assets of the corporation in the event that it fails to pay interest on the debt and be required to enter into a bankruptcy proceeding. If a business should fail, the preferred stockholders have the second claim to the business assets and the equity shareholders would receive the remaining balance of the assets. Consequently, long-term debt holders have an advantage of additional security over equity holders in the business. This is why bonds typically may be issued at an interest rate somewhat less than the preferred stock dividend rate.

2.1.1.11 ***Total Capitalization.*** The sum of the equity and the long-term debt equals the total capitalization of the utility. Capitalization is the quantity of money the utility has raised in external financial markets to create and maintain its business enterprise.

2.1.1.12 ***Current Liabilities.*** The utility also has current liabilities, just as it has current assets. Current liabilities are of a short-term nature and are involved in operating the business on a daily basis.

"Current liabilities" includes the short-term debt and any long-term debt that is payable within 1 year. For example, a utility may need $100M for a construction program in this year with $10M required each month for the first 10 months. The utility can elect to borrow $10M each month from a bank or line of credit and then issue a long-term offering of $100M when the utility believes the financial market conditions are appropriate for large long-term financing. In this way, a utility can typically save a considerable sum of money by judiciously choosing its timing for long-term financing.

"Accounts payable" exist when a utility has been billed for services or materials from suppliers but has not paid these bills. If the utility has received $10M of materials and supplies from a manufacturer, $10M would first be entered into the "materials and supplies" account and the "accounts payable" account. After the bill is paid, the accounts payable would be decreased by $10M and the cash account would be decreased by $10M. Customer deposits are self-explanatory.

2.1.1.13 ***Deferred Credits.*** The utility may also have additional entries called "deferred credits," such as a federal tax liability incurred but deferred because of the tax laws. A utility may be able to defer payment of taxes by liberalized tax depreciation of assets (e.g., a sum of $40M). The taxes are not forgiven, only deferred. This future liability of $40M of additional taxes will be recorded as a deferred credit. Essentially, the utility is recording a future obligation that is a liability to the utility.

The sum of the equity, long-term debt, current liability, and deferred credits equals (Sections 2.1.1.8–2.1.1.13) the total liabilities (and owner's equity) of the utility.

2.1.1.14 ***Balance Sheet—Summary.*** Total assets and total liabilities must be equal; hence the "balance" sheet.

Whenever a transaction is made, two entries must be made in order to maintain a balance of assets equaling liabilities. Either (1) an asset account and a liability account must be changed together, (2) an asset account must be increased and other asset account decreased, or (3) a liability account must be increased and another liability account must be decreased. This is referred to as the "double-entry bookkeeping system."

Example. Suppose that the utility has just purchased a new substation transformer that will be placed in service to replace one that has failed. The first activity that occurs is that the utility receives the transformer from the manufacturer. When the transformer is delivered and the utility receives the bill, two things will happen. First, the plant-in-service account is increased by $10 million ($10M), the cost of the new transformer, and the accounts payable entry is increased by $10M, the invoice amount for the transformer. Both the assets and the liabilities have been modified but still equal (balance) each other. The utility may wait 30 days and then pay the $10M bill. Then, the accounts payable is decreased by $10M and the utility cash account is decreased by $10M. Again, total assets will be equal to total liabilities. As an option, the utility may elect to borrow $10M to pay for this transformer. It would issue $10M from long-term debt, in which case the long-term debt entry is increased by $10M and the cash account is increased by $10M, the amount of cash that the utility receives from its bondholder.

In summary, the final balance sheet changes are that both plant-in-service and total assets are each increased by $10M. On the liability side, the long-term debt is increased by $10M and, consequently, the total liabilities will be increased by $10M.

2.1.2 The Income Statement.

The income statement is the second key financial report that a utility or business enterprise produces. A typical utility income statement is illustrated in Table 2.2. The income statement is a cumulative record of all the business transactions for a year of operation.

2.1.2.1 Revenues. At the top of the income statement are revenues from utility sales of electricity, natural gas, steam, and other sources. Shown on the next several lines as deductions from the revenues are the expenses and costs of doing business.

2.1.2.2 Expenses. The largest expense for a utility is its operation expense for producing electricity. The largest component of operation expense is fuel cost, which typically represents approximately 25–50% of the total cost of producing electricity. The other operation costs include staffing: engineers, administrative staff, meter readers, plant attendants, and others, including the company president.

Maintenance expense is another major expense item. This covers the cost of maintaining power plants, transmission lines, and distribution networks to perform reliably and economically. Maintenance expenses are a significant item in the utility's total cost of operations, ranging from 5 to 10% of the cost of electricity.

Depreciation expense accounts for the annual charges associated with economic wear-out of utility plant equipment. Power plants, transmission lines, and distribution equipment have very long lives, typically 30–40 years. There-

TABLE 2.2 Income Statement

Revenues
Less expenses
Operating
Maintenance
Depreciation
Income taxes
Property taxes
Operating income
Other nonoperating income (net)
Allowance for funds used during construction
Total income
Less interest
Earnings
Less preferred stock dividends
Available for common stockholders
Less common stock dividends
Annual retained earnings (profits)

fore, the depreciation that is accrued is based on that life expectancy, and annual depreciation expense is one over life times the original plant in service. This is quite modest in comparison to other business enterprises where depreciation may occur over 3–5 years.

Income taxes and property taxes are expenses that the utility pays to operate a business. Federal taxes are proportional to the taxable income. Property taxes are proportional to the value of the property that is in service.

2.1.2.3 ***Operating Income.*** If expenses are subtracted from revenues, the result is operating income. Operating income provides for interest charges and dividend requirements that the utility has to provide to its debt-holders and equity participants.

2.1.2.4 ***Nonoperating Income.*** Total income that the utility receives may include two other components. The first is nonoperating income, which is typically small in the utility industry. Nonoperating income may be, for example, the rent that the utility would charge to a telephone company for access to power distribution poles to attach telephone cables. It can also include rents on utility-owned properties and other miscellaneous sources of income that are not directly associated with the production of electricity.

2.1.2.5 ***Allowance for Funds Used During Construction.*** The second source of other income is called "allowance for funds used during construction." When a utility is building a power plant, the utility finances its construction through bonds and equity offerings in the financial market. The utility incurs interest charges and dividend requirements on the money borrowed to build the new plant. The accumulated interest charges and dividends can be a significant item. Because of their significance, an accounting procedure was developed called "allowance for funds used during construction," a noncash offset to the increased financing charges associated with the construction of a power plant. After 1978, the allowance for funds used during construction was divided into two segments, "equity funds used during construction" and "borrowed funds used during construction." This procedure will be discussed later on in the chapter. It is a standard accounting practice used principally by the electric utility industry.

2.1.2.6 Total Income. The total of operating income and the two sources of nonoperating income yields total income. From total income, the interest charges are deducted. Interest charges include both long-term interest that is paid on bonds and other financing vehicles, including short-term interest charges.

2.1.2.7 Earnings. The total income less the interest charges equals the earnings or profits to the corporation. The objective of any business enterprise is to generate the earnings necessary to provide adequate return on investments to its equity participants. If earnings fall short, the business enterprise will not have reasonable access to the financial markets and will not be able to expand its system to meet consumer demands. A sufficient level of earnings must be generated to permit the utility to have access to capital markets at *reasonable* financing rates.

The earnings or profits that the utility generates must provide for the preferred stock dividends for the preferred stockholders as well as common stock dividends for the common stock owner. After preferred and common stock dividends are subtracted from earnings, the remainder is retained earnings or retained profits for the year. Retained earnings are reinvested into the business.

2.1.2.8 Available for Common Stockholders. "Available for common stockholders" is an item that is also usually itemized on an income statement. This item reflects the net earnings of the utility less the preferred stock dividends. "Available for common stockholders" is exactly that: the total amount of money available for use by the common stockholders. This money may be completely paid out in dividends to the common stockholders, or it may be completely retained by the utility and reinvested. Typically, 60–70% of the funds that are available for common stockholders are paid out in common stock dividends by the utility industry. This varies, of course, from utility to utility, and depends on the company's financial health. On the average, the utility industry reinvests 30–40% of funds available to common stockholders. These are credited to retained earnings.

2.1.2.9 Retained Earnings. Retained earnings on the income statement are added to the beginning of year accumulated retained earnings on the balance sheet to compute the retained earnings at year end.

2.1.2.10 Income Statement—Example. The example data presented in Table 2.3 illustrate the relationship between the income statement and the balance sheet. In order to focus on the account changes, this example will show only changes in accounts arising from an additional fuel purchase.

TABLE 2.3 Double-Entry Bookkeeping Example

Part A	Journal entries	
	Date	*Item*
	February 26	Utility purchases fuel = $50M
	March 26	Utility pays the fuel bill
	April 1–31	Utility burns the fuel
Part B	Balance-sheet change February 28	
	Assets	*Liabilities*
	Materials and supplies +$50M	Accounts payable +$50M
Part C	Balance sheet change March 31	
	Assets	*Liabilities*
	Cash −$50M	Accounts payable −$50M
Part D	Income statement change April 1–30	
	Revenues	$0
	Less expenses (operation fuel)	$50M
	Income	−$50M
	Annual retained earnings	−$50M
Part E	Balance-sheet change April 30	
	Assets	*Liabilities*
	Materials and supplies −$50M	Retained earnings −$50M
Part F	Income statement change May 1–31	
	Revenues	$50M
	Less expenses (operation fuel)	$0
	Income	$50M
	Annual retained earnings	$50M
Part G	Balance sheet change May 31	
	Assets	*Liabilities*
	Cash +$50M	Retained earnings +$50M

As shown in the transaction record or journal (part A), on February 26, the utility purchases additional fuel for $50M. On March 26, the utility pays the additional fuel bill of $50M. During the month of April, the utility burns the fuel that it purchased back in February.

Part B illustrates the change in the balance sheet as a result of the transaction that occurred during February. As a result of purchasing the fuel, the utility shows a $50M liability in accounts payable. Since it has received the fuel, the utility has increased its material and supply inventory and reports that as a $50M asset under a current asset account.

Part C presents the cumulative balance sheet changes as of March 31. The utility pays the fuel bill and decreases its cash asset account by $50 million. At the same time, the accounts payable account is decreased by $50M because the $50M fuel bill has been paid. The utility still has the fuel in its materials and supplies inventory recorded as $50M.

In April, the utility burns the fuel and produces electricity. The change in the income statement is noted in part D. The income statement prior to April 1 has not been impacted because the utility has not used the fuel. In the income statement, the utility incurs an operation cost of $50M worth of fuel.

Without changes in any of the other income statement accounts, this results in the reduction of income by $50M and a decrease in retained earnings by $50M. Because the retained earnings were changed by $50M, this must be reflected in the balance sheet on April 30 as shown in part E. Thus, the retained earnings are reduced by $50M.

Since the fuel was consumed during the month of April, it is no longer in the materials and supplies inventory, which has been reduced by $50M. Consequently, the materials and supplies inventory shows a cumulative change of zero. Thus, the net balance sheet change at the end of April shows a cash decrease of $50M to pay for the fuel and a decrease of $50M in retained earnings.

Suppose that the utility recovers the cost of fuel in its electric rates. To reflect that scenario, the May electric revenue is increased by $50M. The income statement change in May is shown in part F. The utility receives $50M more in revenues (cash), which leads to $50M more in retained earnings. The balance-sheet change on May 31 is shown in part G. The cash account has been increased by $50M in May, which leads to a cumulative cash change of zero. The same is true for the retained-earnings account.

In this example, the utility purchased fuel, burned the fuel, and received additional revenue to cover the cost of the fuel. The final balance sheet showed no final asset or liability change by the end of May. This example illustrated how the income statement and balance sheet work together in the double-entry accounting process. An expense item will be processed through the income statement and will essentially appear as net income or retained earnings of the utility. The accounting has been simplified by assuming that there were no interest charges on the utility investments or borrowings for the $50M worth of cash, or that there were any income tax effects.

2.2 INTRODUCTION TO TAX ACCOUNTING

This section introduces the principles and impacts of tax accounting on the income statement and profitability of an electric utility.

2.2.1 Depreciation

One key consideration in electric utility accounting is depreciation. Since electric utilities are capital-intensive industries, the recovery of the investment in plant and equipment is essential to their continued financial health. Depreciation is an accounting procedure in which the expensing of the initial investment in a power plant is distributed over the expected operating life of the plant. For example, a $400M power plant will be charged to a utility's customers over the 40-year economic life of the plant.

The standard or book approach, termed "straight-line depreciation," begins with the $400M plant investment and divides it by 40 years of active service life. This yields a depreciation expense of $10 million per year and the annual depreciation charge that is recorded on the income statement as the depreciation cost of this power plant.

Table 2.4 shows three depreciation schedules for an asset with a 10-year life. The straight-line procedure yields a depreciation rate for a 10-year plant investment of 10% per year. The sum-of-years digits (SYD) procedure provides a more rapid depreciation rate in the first years and a lower rate during the later years. The accelerated cost recovery system (ACRS) has a depreciation rate that is significantly larger in the first several years and zero in the later years. With ACRS, an asset has a shorter life than with other rules where SYD was used. The straight-line depreciation method is always used, by ac-

TABLE 2.4 Sample Depreciation Annual Rates

Year	Remaining Life	Straight Line	Sum-of-Years Digits (SYD)	Accelerated Cost Recovery System (ACRS)[a]
1	10	.10	.182	.15
2	9	.10	.164	.22
3	8	.10	.145	.21
4	7	.10	.127	.21
5	6	.10	.109	.21
6	5	.10	.091	0
7	4	.10	.073	0
8	3	.10	.055	0
9	2	.10	.036	0
10	1	.10	.018	0

[a]Assets with a 10-year life for straight-line or SYD methods, have only a 5 year life under ACRS (Tax Reform Act of 1986).

counting convention, to calculate book depreciation expense in the income statement. The straight-line method or faster methods are used for *tax* depreciation calculations as permitted by the Internal Revenue Service (IRS) tax code.

The depreciation method of SYD can easily be calculated for any asset life. The calculation is shown in Equation (2.3), where the numerator is the remaining life of equipment at the beginning of the year, or the total life minus equipment age as of the beginning of that year. The denominator is the sum of all the ages from age one to life. However, the sum of all the ages from age equal one to life is merely the SYD and can be expressed as life times life plus 1 divided by 2.

$$\text{Depreciation SYD (age)} = \frac{\text{remaining life}}{\text{SYD}} \tag{2.3}$$

but

$$\text{SYD} = \sum_{\text{age}=1}^{\text{life}} (\text{age}) = \frac{\text{life}(\text{life} + 1)}{2} \tag{2.4}$$

For example, a piece of equipment with a 10-year life has an SYD equal to 55. The depreciation rate in year 6 is 5 divided by 55, or .091, as follows:

$$\text{Life} = 10 \text{ years}$$

$$\sum_{\text{Age}=1}^{\text{Life}} (\text{age}) = \frac{10(11)}{2} = 55$$

$$\text{Depreciation (age at beginning of year 6)} = \frac{10 - 5}{55} = \frac{5}{55} = .091$$

2.2.2 Income Taxes

A simplified example illustrates the income tax calculation. Suppose that the utility has a total revenue of $1000 and total expenses of $600. The utility has recently bought new equipment costing $1000, with a life of 10 years. These data are shown in Table 2.5.

The objective is to calculate the income tax and income statement in the first year based on two methods. In method 1, straight-line book depreciation will also be used for tax purposes. In method 2, the tax depreciation rate will be based on the ACRS tax depreciation schedule of Table 2.4.

Method 1, Book depreciation is used. The taxable income is $1000 in revenue less $600 of expenses and less $100 tax depreciation. This results in $300 of taxable income when computed using Equation (2.5). Income tax is as-

TABLE 2.5 Simplified Example (First Year)

Revenue	\$1000
Expenses	\$ 600
Asset with initial cost = \$1000	
Life = 10 years	
Depreciation (First year)	
Book (straight line)	\$ 100
Tax (ACRS)	\$ 150

sumed to be 34% of taxable income, or \$102 as shown in the following calculation:

$$\text{Taxable income} = \text{revenue} - \text{expenses} - \text{tax depreciation}$$

$$\text{Taxable income} = 1000 - 600 - 100 = \$300 \tag{2.5}$$

$$\text{Income tax} = .34 \cdot \text{Taxable Income}$$

$$\text{Income tax} = .34 \cdot 300 = \$102 \tag{2.6}$$

Method 2, ACRS depreciation is used. The taxable income is \$250, and the income tax is shown in the following calculation:

$$\text{Taxable income} = 1000 - 600 - 150 = \$250$$

$$\text{Income tax} = .34 \cdot \$250 = \$85$$

2.2.3 Income Statements

The income statement for the first year is shown in Table 2.6. The income statement for each method begins with a revenue of \$1000 and subtracts expenses, book depreciation, and taxes. The result is operating income.

$$\begin{aligned}\text{Operating income} &= \text{revenue} - \text{expenses} \\ &\quad - \text{book depreciation} - \text{taxes}\end{aligned} \tag{2.7}$$

In method 2, where ACRS tax depreciation is used, the income taxes of \$85 yield an operating income of \$215. Note in both examples that book depreciation is the same.

On the basis of the results in Table 2.6, it can be seen that method 2 is the best tax depreciation method to use. In method 2, an accelerated tax depreciation is used because it yields a lower tax and a larger operating income or profit.

Table 2.7 completes the results by showing not only the first year income statement but also the income statement in the tenth year, assuming that costs

TABLE 2.6 Simplified Example Results Income Statement

	Method 1 Straight-Line Tax Depreciation	Method 2 ACRS Tax Depreciation
Revenue	$1000	$1000
Less		
Expenses	600	600
Depreciation	100	100
Taxes	102	85
Operating income	198	215

and revenues are constant. In method 1, the operating income and income statement are the same not only in the first and tenth years but also in all the intervening years because tax depreciation is the same each year.

2.2.4 Flow-Through Accounting

Completing the calculations through the tenth year for the ACRS case (case 2 in Table 2.7) yields an operating income of $215 in the first year but only $164 in the tenth year. The accelerated tax depreciation method using ACRS provides a higher operating income than straight-line depreciation in the first year but, of course, provides a lower operating income in the final year. If the calculations for all the intervening years were shown, the cumulative operating

TABLE 2.7 Income Statements (Constant Revenue)

	(1) Straight-Line Tax Depreciation		(2) "Flow-Through" ACRS Tax Depreciation		(3) "Normalized" ACRS Tax Depreciation	
	1st Year	10th Year	1st Year	10th Year	1st Year	10th Year
Revenue	1000	1000	1000	1000	1000	1000
Less						
Expenses	600	600	600	600	600	600
Taxes paid	102	102	85	136	85	136
Taxes deferred					17	−34
Depreciation	100	100	100	100	100	100
Operating income	198	198	215	164	198	198

income over the 10-year period would be the same regardless of what tax depreciation schedule is used. Only the yearly distribution of the operating income changes with the depreciation method. In the ACRS tax depreciation example, more operating income is generated in the first year because the taxes are lower. In the later years, however, less operating income is produced because the taxes are higher.

This example illustrated an enterprise that had a constant revenue and expenses over 10 years. One might have expected that the operating income, or profits, should have been constant over this period. However, using ACRS tax depreciation, the operating income is higher than the straight-line depreciation example during the first year. If the performance of this business were examined, one might believe that the operating income would remain at the $215 level. However, we know that the operating income will be decreasing over the remaining 9 years. Therefore, the operating income has been overstated in the first year and does not present an objective financial report.

An accounting method called "normalizing" has been established to report the financial performance on an average or normal year basis. The normalization procedure is achieved by inserting another expense on the income statement called "deferred taxes," as shown in column 3 of Table 2.7.

2.2.5 Normalized Accounting

In the normalized ACRS tax depreciation treatment, recognition is made of the fact that taxes are lower in the first year than they would be during the 10-year period. The taxes paid in the first year were only $85, and accelerated depreciation saved $17 over the straight-line method.

Because the cumulative taxes paid over the 10-year period are independent of the depreciation schedule, the $17 saving in the first year represents a future tax liability. Consequently, the $17 is a deferred tax to be paid in the future, is recorded in this method as a future expense that is reported in year 1.

The results presented in method 3 of Table 2.7 are revenue of $1000, expenses of $600, taxes paid of $85, and provision for the payment of future taxes (or deferred taxes) of $17. The depreciation for book purposes is still $100, yielding an operating income of $198.

Similarly, in the tenth year under a normalized ACRS depreciation technique, the taxes paid are $136. The deferred taxes are actually negative $34, indicating that money has already been accumulated in prior years to permit these taxes to be paid at this time. The result is that the operating income is a constant $198 for all years. Thus, the income statement now presents a "normal" financial outlook. This accounting treatment is referred to as "normalized," while the nonnormalized treatment is referred to as "flow-through."

If the income statement were computed under the normalized treatment for all the intervening years between the first and tenth years, the deferred taxes would sum to zero over this 10-year period.

Normalized accounting is used widely in the utility industry to account for unusual circumstances or delayed period expenses. In the utility industry, under most recent law, all investor-owned electric utilities must use normalized tax depreciation methods in their income statement reporting.

Continue with this example and compute the change in the balance sheet associated with an income statement based on an ACRS tax depreciation method normalized for the first year.

2.2.6 Cash Report

To compute the change in balance sheet, a third key utility financial statement, the cash report, must be introduced. The cash report is the same as a sources-and-uses-of-funds statement. The objective of the cash report is to show the sources and disposition of cash throughout the year.

The cash report begins with the cash account balance at the beginning of the year. All the sources of cash are presented, as well as all the applications of cash. The difference between sources of cash and applications of cash is the computed net change in cash. The beginning cash balance added to the change in cash equals the cash at the end of the year. Table 2.8 shows a typical cash report in the first year based on the normalized ACRS tax depreciation example and assuming a zero cash balance at the beginning of the year.

Examine the cash that is generated from operations, the income statement for the normalized ACRS example in Table 2.7. In that example, the cash would have been obtained from the sale of electricity. When expenses to pro-

TABLE 2.8 Cash Report Funds Statement

Normalized ACRS tax depreciation example (first year) (all costs in $)			
Cash at beginning of the year			0
Cash from operations (sources)			
Internal			
Net income	198		
Depreciation	100		
Deferred taxes	17		
Cash from external sources			
New financing			
Long-term debt	0		
Preferred stock	0		
Common stock	0		
Total sources of cash		315	
Application of cash			
Capital expenditures	0		
Dividends	100		
Total funds dispensed		100	
Cash at end of the year			215

duce the revenue are subtracted from revenue, the net cash that is generated is the operating income (net profits).

However, some of the expenses listed in Table 2.7 do not represent a "cash" expense. Depreciation is an expense that involves no cash transaction; neither do deferred taxes. Thus, sources of cash are from operating income and the noncash expenses, depreciation, and deferred taxes.

Return to Table 2.8. Note that the places where cash has been generated from operating the utility are from net profits, the charge for depreciation and deferred taxes. The total cash generated by operating the utility in the first year is $198 plus $100 plus $17. Cash could also be derived by borrowing from the financial markets through stock and debt financing. In this example, these are zero. Thus, the total sources of cash are $315.

Cash is also applied to operate the utility business. For example, if the utility had a construction program to build more plants, then cash would be needed. This transaction would be recorded under capital expenditures. In this example, there is no construction program and the capital expenditures are zero. Since the utility has to pay dividends to its stockholders, assume that $100 is distributed in dividends. Thus, the total cash spent is $100.

The cash at the end of the year can now be computed as the cash at the beginning of the year plus the total cash from operations and external sources ($315) less the dispersed cash ($100), or $215.

2.2.7 Retained-Earnings Statement

The retained-earnings statement also plays an important role. This statement shows how the accumulated retained-earnings change from the beginning of the year to the end of the year.

For this example, if the retained earnings in the beginning of the year are $800, then the retained earnings at the end of the year are increased by the net income or profits generated by the utility ($198) less the amount that the utility pays out in dividends ($100). Thus, the accumulated retained-earnings balance at the end of the year is $898, as shown in Table 2.9.

2.2.8 Balance Sheet

At this point, the balance sheet can be computed for the first year, as shown in Table 2.10. The balance sheet is shown as of January 1 or December 31 of the previous year. The objective is to modify the balance sheet from January 1 to reflect the operation of the entire year through December 31, year end.

Begin with examining the fixed assets, or utility plant. In the example, the utility plant in service in the example changes as $1000 in "construction work in progress" is assumed to go into service on January 1, with "utility-plant-in-service" now at $2000.

The next item is the accumulated depreciation. The accumulated depreciation to date is assumed to be $1000 as the old plant is fully depreciated while

TABLE 2.9 Retained-Earnings Statement

Retained earnings	
Balance (beginning of year)	800
Add: net income	198
Less: dividends	100
Retained-earnings balance (end of year)	898

the new plant has an annual book depreciation of $100. This results in an accumulated depreciation at the end of the year of $1100. Since the utility is not building any more new power plants, "construction work in progress" at year end is zero. Current assets, cash-on-hand, began at zero and increased by $215, based on Table 2.8. Thus, the cash at year end is $215. The total assets at the beginning of the year are $1000, $1115 at year end.

On the liabilities side, suppose that the long-term debt began with $100 and was not changed during the year. Common stock began at $100 and also wasn't changed during the year. However, retained earnings were $800 and, as a result of the retained earnings report in Table 2.9, grew to $898.

The provision for accumulated deferred taxes also must be accounted for. The income statement reported $17 as taxes deferred. These taxes are a liability that is incurred today but will be paid in some future year. Consequently, they must be reported on the liability side of the balance sheet as a future liability,

TABLE 2.10 Balance Sheet
NORMALIZED ACRS TAX DEPRECIATION EXAMPLE (FIRST YEAR)

	January 1 (Beginning of Year)		December 31 (End of Year)
Assets			
Fixed assets			
Utility plant in service	1000		2000
Less accumulated depreciation	1000	(Δ = 100)	1100
Construction work in progress	1000		0
Current assets			
Cash, receivables	0	(Δ = 215)	215
Total assets	1000		1115
Liabilities			
Long-term debt	100		100
Preferred stock	0		0
Equity			
Common stock	100		100
Retained earnings	800		898
Current liabilities	0		0
Provision for deferred taxes	0		17
Total liabilities	1000		1115

and are reported as "provision for deferred taxes." There were zero deferred taxes at the beginning of the year, and $17 of deferred taxes were reported as an expense. Thus, the "provision for accumulated deferred taxes" is $17 at year end.

The sum of the liabilities at the end of the year is $1115, which also equals the assets.

2.2.9 Flow-Through versus Normalized Tax Accounting

Now examine the impact of flow-through versus normalized tax accounting on the electric utility industry. The industry has its electric rates (or revenue) regulated so that it achieves a specified net income or return on investment. Thus, the objective of the regulated utility is to operate the utility with the lowest possible costs, thereby achieving the specified net income. Assume that the utility is regulated so that the net income must be $215, as shown in Table 2.7 with method 2, "flow-through" accounting. The income statement for the regulated utility for the first year is shown in Table 2.11.

The values in Table 2.11 were calculated using the results from Equations (2.8) through (2.12). Equation (2.8), the revenue equation, shows revenue equal to expenses plus taxes plus book depreciation plus net income.

$$\begin{aligned} \text{Revenue} &= \text{expenses} + \text{taxes paid} + \text{taxes deferred} \\ &\quad + \text{book depreciation} + \text{net income} \end{aligned} \tag{2.8}$$

TABLE 2.11 Income Statement
REGULATED UTILITY

	(1) Book	(2) Flow-Through	(3) Normalized
Revenue	1025	1000	1025
Less			
Expenses	600	600	600
Taxes paid	110	85	93
Taxes deferred	—	—	17
Depreciation	100	100	100
Net income	215	215	215
	Cash Report		
Funds from operations			
Net income	215	215	215
Depreciation	100	100	100
Deferred taxes	0		17
Total	315	315	332

Equation (2.9) expresses taxable income. Equation (2.10) is the equation for taxes paid.

$$\text{Taxable income} = \text{revenue} - \text{expenses} - \text{tax depreciation} \tag{2.9}$$

$$\text{Taxes paid} = \text{tax rate} \cdot \text{taxable income} \tag{2.10}$$

Equation (2.11) is the equation for deferred taxes that are equal to tax rate times book depreciation minus tax depreciation, if normalized. If not normalized, deferred taxes are zero.

$$\begin{aligned}\text{Deferred taxes} &= \text{tax rate} \cdot (\text{tax depreciation} - \text{book depreciation}) \\ &\qquad \text{with normalized accounting} \\ &= 0 \text{ with flow-through accounting}\end{aligned} \tag{2.11}$$

Substitution of Equations (2.9) and (2.10) into Equation (2.8) leads to Equation (2.12).

$$\begin{aligned}\text{Revenue} &= \text{expenses} + \text{tax rate} \cdot (\text{revenue} - \text{expenses} \\ &\qquad - \text{tax depreciation}) \\ &\quad + \text{taxes deferred} + \text{book depreciation} + \text{net income}\end{aligned} \tag{2.12}$$

$$\begin{aligned}\text{Revenue} &= \text{expenses} + \text{book depreciation} \\ &\quad + \frac{\text{tax rate}}{(1 - \text{tax rate})}(\text{book depreciation} - \text{tax depreciation}) \\ &\quad + \frac{\text{taxes deferred} + \text{net income}}{1 - \text{tax rate}}\end{aligned} \tag{2.13}$$

For normalized accounting, substitute Equation (2.11) and

$$\text{Revenue} = \text{expenses} + \text{book depreciation} + \frac{\text{net income}}{1 - \text{tax rate}} \tag{2.14}$$

Using data from Table 2.11 and Equation (2.14) for the normalized case, revenue is \$1025.

$$\text{Revenue} = 600 + 100 + \left(\frac{215}{1 - .34}\right) = \$1025$$

Using the revenue calculated in Equation (2.14), the taxable income can be computed using Equation (2.9), the taxes paid using Equation (2.10), and deferred taxes using Equation (2.11).

$$\text{Taxable income} = 1025 - 600 - 150 = \$275$$

$$\text{Taxes paid} = .34(275) = \$93$$

$$\text{Deferred tax} = .34(150 - 100) = \$17$$

The cash report for funds from operations shows that the total funds from operations is $315 when straight-line tax depreciation is used, $315 based on the flow-through method with ACRS depreciation, and $332 when the normalized method is used with ACRS depreciation.

Which accounting procedure in Table 2.11 would the utility investor prefer? Similarly, which accounting procedure would the rate payer who paid the revenue for this utility through its electric rates use? The answers to these questions point out a dichotomy of the electric utility industry. The rate payer would prefer the flow-through ACRS accounting method because it leads to the lowest revenue requirements, or lower electric rates, consistent with providing the utility a fair return on its investment, or fair net income. On the other hand, the utility investor would prefer the normalized treatment because it provides a fair rate of return on net income, as well as a larger cash flow from operations. Larger operations cash flow means that the utility will need less external financing to complete its construction projects and expansion plan. Less external financing benefits current investors. Further discussion of this concept is found in Chapter 5.

2.3 ACCOUNTING FOR INTEREST CHARGES DURING CONSTRUCTION

Major utility power plant projects typically require 4 or more years to complete. Because of the length of construction, interest charges on the money borrowed to finance the project constitute a significant cost. Public service commissions have traditionally adopted the criteria that charges for power plants cannot be passed onto the customer until the plant is used and useful. Therefore, all the charges incurred during the plant construction period, including cash disbursements and interest charges, must be accrued in the construction-work-in-progress account. These changes are *not* included in the rate base (or earning base) of the utility until the plant comes into service. Typically, 20–25% of the cost of a coal-fired power plant is interest charges during construction. For a nuclear plant, the percentage is much higher.

Interest charges and dividends paid on new stock issued during construction are reported on the utility income statement financial report as part of the total interest charges and dividend payments. Interest charges during construction have a negative effect on the utility income statement because they subtract from the gross income and, consequently, lower the net income. Net income is generally lower for a utility that is constructing a new power plant than for a utility that is not. As a result of this effect, an accounting procedure called "allowance for funds used during construction" (AFDC) has been adopted by the utility industry to better represent the actual current performance as reported on the income statement. The following examples illustrate the influence of construction on the utility financial report and further explain industry use of AFDC. Also discussed are the accounting principles used to properly report AFDC.

"Allowance for funds used during construction" is a normalized accounting treatment of the interest charges and dividend payments of a power plant during a construction period. The AFDC normalization procedure is similar to that used for income tax purposes.

Examine a utility income statement in which the expenses do not inflate over a five-year period, depreciation is constant, tax depreciation equals book depreciation, and the income tax rate is 50%. Table 2.12 illustrates this example for a utility that has no construction over a 5-year period. The utility has operating expenses of $600, book depreciation of $100, a long-term interest expense of $100, and a return on equity of $100. These data lead to federal taxes of $100 and, consequently, a revenue of $1000 using Equation (2.12). Because utility expenses are constant, the revenue requirements are $1000 each year.

Table 2.13 provides a base for understanding how the income statement would appear if the utility had a construction program. The data are the same as in Table 2.12, except that the utility builds a power plant and spends $1000 in year 1 and another $1000 in year 2. Thus, the new plant costs $2000. In the process of building the power plant, the utility sells only bonds, at a 10% interest rate to finance the project. Long-term interest expense is then increased in year 1 by $100 because $1000 is spent in year 1 for the plant. Interest expense is increased in year 2 to $200 because the company sells another $1000

TABLE 2.12 Allowance for Funds Used During Construction (AFDC) Example
NO CONSTRUCTION PROJECT
TAX DEPRECIATION = BOOK DEPRECIATION
TAX RATE = 50%

	Year 1	Year 2	Year 3	Year 4	Year 5
Revenue	$1000	$1000	$1000	$1000	$1000
Operating expenses	600	600	600	600	600
Depreciation	100	100	100	100	100
Taxes	100	100	100	100	100
Long-term debt Interest expense	100	100	100	100	100
Return on equity	100	100	100	100	100

where Tax = tax rate (revenue − operating expense − tax depreciation − interest expense)

Tax = .5(1000 − 600 − 100 − 100) = $100

TABLE 2.13 Construction of New Plant—No AFDC Treatment (CWIP in Rate Base)

Assume:

Table 2.12 data

The utility builds a power plant spending $1000 in year 1 and another $1000 in year 2

The power plant is installed in year 3 at a total cost of $2000

The utility sells bonds at 10% interest during construction to finance the power plant, thus increasing its long-term interest expense in year 1 by $100 and in year 2 by $200

The power plant is retired in year 5

The utility uses 3-year straight-line tax depreciation.

	Year 1	Year 2	Installation of Power Plant Year 3	Year 4	Year 5
Revenue	1100	1200	1866	1800	1733
Operating expenses	600	600	600	600	600
Depreciation	100	100	766	767	767
Taxes	100	100	100	100	100
Long-term debt interest expense	200	300	300	233	166
Return on equity	100	100	100	100	100

Where The utility must maintain $100 return on equity
Depreciation on the new power plant = 2000/3 = 666.67 $/year

in bonds for a total of $2000, in order to pay for the new plant. For simplicity, assume the plant has a life of three years and will be retired at the end of year 5.

In addition, assume that the utility is regulated and receives a $100 return on equity as its fair rate of return. Table 2.13 illustrates the revised income statement. Note in year 1 that the utility has a higher long-term debt interest than was shown in Table 2.12. Revenue requirements are higher in year 1 in order to maintain the return on equity requirement because the long-term debt interest expense is increased by $100. Similarly, in year 2 the revenue requirements are higher by $200 because the interest expense has increased. Finally, in year 3 the revenue is further increased because the new power plant has entered service and will be depreciated. The depreciation increases by $667 per year while the plant is depreciated. At the end of years 3, 4, and 5, the bond holders are paid their interest and one-third of the outstanding bonds are bought back (recalled) each year.

When examining Table 2.13, several things are apparent. First, the revenue requirement has increased in years 1 and 2 because of the new power plant construction. Since the new power plant is not in service, the utility customers are paying for electricity service based on a utility plant that is not used and

TABLE 2.14 Construction of New Plant—No AFDC Treatment (No Rate Increase During Construction)

			Installation of Power Plant		
	Year 1	Year 2	Year 3	Year 4	Year 5
Revenue	1800	1000	1866	1800	1733
Expenses	600	600	600	600	600
Depreciation	100	100	766	767	767
Taxes	50	0	100	100	100
Long-term debt interest	200	300	300	233	166
Return on equity[a]	50	0	100	100	100

[a]Regulate return on equity to achieve $100 after plant installation.

useful. This is unacceptable in traditional electric utility rate proceedings. Consequently, something must be done in years 1 and 2.

Table 2.14 adjusts the results of Table 2.13 so that there will be no rate increase during the construction period. The revenue in years 1 and 2 is required to be $1000, as in the case of the utility that had no construction. After the plant goes into service (in years 3, 4, and 5), the revenues are adjusted to permit the utility to have a return on equity of $100. In this case, it is noted that while the revenue in years 1 and 2 is $1000, the return on equity that the utility receives is much lower, $50 in year 1 and $0 in year 2.

This earnings statement is distressing to the electric utility. When a utility is constructing a power plant, it must support its external financing by demonstrating sound credit worthiness. In this example, however, one measure of credit worthiness, return on equity, is very low ($50) in year 1 and $0 in year 2. Therefore, the utility is in the poor financial position of requiring external financing at a time when its financial credit worthiness is weak.

Poor financial performance during the construction period can be normalized to better reflect the utility's true income statement. Table 2.15 presents the general principle of AFDC accounting. The first step is to capitalize or incorporate the interest charges during construction into the final cost of the power plant. If the power plant costs $2000 and if there are $300 of accumulated interest charges, then plant cost is recorded as $2300. AFDC during the

TABLE 2.15 AFDC Accounting Principles

"Capitalize" or incorporate the interest charges during construction (debt and equity) into the final cost of the plant; interest costs during construction are depreciated (or normalized) over the life of the plant
Offset the increased interest costs due to a construction project by a counter entry on the income statement; this counter entry statement appears as an "other income" item

construction period is then depreciated over the life of the plant as if these interest charges (and dividends) were actual expenditures for the plant. The AFDC that is imputed (added) to the final cost of the power plant is based on the AFDC rate set by the regulatory agency.

The second accounting treatment is the offset of AFDC by a counterentry on the income statement. This counterentry appears as another income item and is called "allowance for funds used during construction" (AFDC).

In the past, AFDC was not a taxable item, nor was the depreciation associated with AFDC a depreciable expense for tax purposes. Since the Tax Reform Act of 1986 does not allow the interest portion of AFDC as a tax deduction, the interest expense is now capitalized for tax depreciation purposes.

In the example, the AFDC in the first year is $100 associated with a $100 AFDC interest expense. In the second year, AFDC is $200 associated with a $200 interest expense increase. Thus, the interest expense for tax purposes is $100 in year 1 and $200 in year 2.

With these principles in mind, an income statement based on AFDC accounting can be constructed for a utility building a new power plant. An example is shown in Table 2.16. The results of Table 2.16 in years 1 and 2 are similar to the results in Table 2.14, except for the other income called "AFDC." When that is applied to the income statement, it leads to a return on equity of $100 in year 1 and year 2. Thus, this accounting treatment meets one of the principle guides for AFDC normalization: it assures that return on equity is acceptable and equal to that which the utility would have received if it had no construction program. In addition, the revenue is independent of the construction program.

For years 3, 4, and 5, when the plant becomes used and useful, two points are noted. First, the book depreciation base of the new plant is no longer $2000 but is $2300, to be depreciated over 3 years. Second, the tax depreciation base on the new plant is also $2300, or $766 per year. Consequently, when the tax report is computed, it leads to a tax requirement of $100 per year as shown in Table 2.16. The revenue is computed as $1966 for year 3. Note that the total revenue over the 5-year period is the same whether there is AFDC accounting, as shown in Table 2.16, or no AFDC accounting, as shown in Table 2.13. Similarly, the tax collected is the same.

The funds report is summarized in Table 2.17 for the first and fifth years both with and without AFDC accounting. The principal difference is in the sources of internal funding. Note that AFDC is reported as income on the income statement, but that there actually is no cash charge associated with that income. Consequently, it needs to be subtracted form the internal sources of funds. The conclusion may be drawn that no AFDC accounting [also called "construction work in progress" (CWIP) in the rate base] leads to a higher cash flow to the utility during the construction period. However, higher electric rates are required during a construction period. AFDC accounting is the accepted vehicle for income statement accounting for power plants under construction.

Some public service commissions have recognized the cash flow merits of

TABLE 2.16 AFDC Accounting Example[a]

	Year 1	Year 2	Year 3	Year 4	Year 5
Revenue	1000	1000	1966	1900	1833
Expenses	600	600	600	600	600
Depreciation	100	100	866	867	867
Taxes	100	100	100	100	100
Other income (AFDC)	100	200	0	0	0
Long-term debt interest	200	300	300	233	166
Return on equity	100	100	100	100	100

[a]*Note:*

Book depreciation on the new plant $= \frac{2000 + 300}{3} = 766.67$

Tax depreciation on the new plant $= \frac{2000 + 300}{3} = 766.67$

non-AFDC accounting or CWIP in rate base and have permitted utilities to partially or fully employ non-AFDC accounting of construction projects.

When a large or expensive power plant enters commercial service, electric rates generally increase. Consumer groups have argued that a several-year phase-in of these costs should be implemented to avoid a step change in electric rates, or "rate shock." Phase-in is a second type of normalization, which is essentially an extension of the AFDC process after the power plant enters service.

In summary, AFDC is an accounting treatment used within the utility industry to normalize interest costs during construction over the operating life of a power plant. With AFDC accounting, electric rates are the same as if the utility had no construction program. On the other hand, with CWIP in the rate base, electric rates increase during construction periods and provide a higher cash flow to the utility.

2.4 OTHER UTILITY TAXES

At various times, federal tax law has had investment tax credit provisions that were eliminated by the Tax Reform Act of 1986. The objective of the invest-

TABLE 2.17 Funds Report

	No AFDC (CWIP in Rate Base)		AFDC Accounting	
Sources of Internal Funds	Year 1	Year 5	Year 1	Year 5
Return on equity	100	100	100	100
Depreciation	100	767	100	867
Less AFDC	0	0	100	0
Total sources	200	867	100	967

ment tax credit legislation has been to stimulate new plant investment by industry, including electric utilities.

Investment tax credit is a one-time credit against taxes that is deducted from the federal income taxes paid. In previous years, the rate was 4% for utilities and most recently was 10%. For example, if a $1000 piece of equipment is purchased, the purchaser receives a $100 credit that can be used to reduce taxes in the current year. Utility power plants typically are purchased over a period of 1–6 years. In each year, the investment tax credit (ITC) is applied to that portion of the plant that is purchased in that year, commonly called "ITC on progress payments."

Prior to 1986, there was a maximum limit, typically 85%, on the amount by which taxes could be reduced. For example, if the utility were to purchase a piece of equipment for $1000, the investment tax credit would be $100. If the utility's income taxes for that year were $50, then the utility could apply only 85% of $50 to be a tax credit. The utility would then pay a tax of $7.50 and, therefore, would have used $42.50 of that $100 tax credit. The balance of tax credit, $57.50, could be carried back or could be applied against the next year's taxes. This process of carrying forward unused investment tax credit is called "carry forward." The investment tax credit was allowed to be carried backward for 3 years to restate prior income taxes, or it was to be carried forward and applied toward future tax reductions.

Utility industry investment tax credit benefits were typically normalized. When normalized, a $100 tax credit could be amortized over the 40-year life of the equipment and investment tax credit was shown as a credit of $100 divided by 40 years, or $2.50 per year.

A utility also pays local or property taxes on its plant investments. Property taxes are typically assessed on the basis of first cost or on the real value of assets. Real value can include the net replacement costs or the first-cost value, but does not include the effects of inflation or depreciation. These rates are typically set by local government jurisdictions and can range from 1 to 6%, but generally average around 2.5%.

2.5 SUMMARY

Through simplified examples, this chapter has introduced basic utility financial accounting issues which are discussed in greater detail in subsequent chapters.

BIBLIOGRAPHY

Brigham, E. F., and T. J. Natell, "Normalization Versus Flow Through for Utility Companies Using Liberalized Tax Depreciation," *Accounting Review,* Vol. 49, No. 3, July 1974, pp. 436–447.

Lamp, G. E., and J. C. Hempstead, "Treatment of Deferred Taxes and Unamortized Investment Tax Credit in the Revenue Requirement Equation," *Engineering Economist,* Vol. 21, No. 2, Winter 1976, pp. 79–88.

Naylor, T. H., *Corporate Planning Models,* Addison-Wesley, Reading, MA, 1979.

Suelflow, J. E., *Public Utility Accounting—Theory and Application,* MSU Public Utility Studies, 1973.

PROBLEM. The Desert Island Power Company has been created. A record of the first year's operations includes:

Thousands of Dollars	
Sale of stock	1000
Sale of bonds ($i = 5\%$)	1000
Purchase of generation–transmission–distribution system	2000
Purchase of fuel (on credit)	200
Fuel inventory consumed	170
Billing for sale of electricity	400
Revenue from sale of electricity	370
Wages	120
Payment for fuel	180
Interest payment to bondholder	50

Construct the income statement and balance sheet assuming no taxes.

3

TIME VALUE OF MONEY

If a depositor places $100 in a bank account that pays 7% interest per year, $197 could be expected in that account after 10 years as a result of interest paid and compounded over the 10-year period. When choosing to put the money in the bank for 10 years, the depositor essentially indicated indifference about having $100 today or $197 10 years from now. Thus, 7%/year is this depositor's time value of money.

Others may have different perspectives of the time value of money and may invest their funds in other ways. Some may elect not to deposit any money into the bank at 7% interest but rather to consume their discretionary income today rather than save it for a future time. In this case, the time value of money for this potential depositor is higher than 7%; otherwise, the discretionary income would have been saved rather than spent.

Everyone has one's own personal time value of money. People in their twenties and thirties typically have a very high time value of money and, because of their limited discretionary incomes, prefer to consume goods and services rather than to save. On the other hand, people in the 50-year age bracket save a great deal more and typically use such conservative investments as banks.

3.1 BUSINESS CONCEPT

Business owners also have a time value of money. A business may have profits this year and the owners must decide what to do with these profits. They can be invested in expanding the business, placed in a bank account and have interest earned, distributed to the owners of the business, or used to buy back existing loans or debts. The merits of each of these alternatives rest on how the business views its time value (or cost) of money. Time value of money is the opportunity cost of capital for the business.

In utility terminology, the time value of money is also referred to as the "discount or present worth rate." For electric utilities, the time value of money can be computed by examining the cost of investment capital or simply the cost of money. The cost of money is the weighted average composite interest cost of bonds, the yield rate of new preferred stock, and the rate of return on common stock offerings by the utility. These components can be combined into a term called the "discount or present worth rate."

For example, a utility could finance its new offerings as follows:

- 50% with bonds at an interest of 7%
- 10% with preferred stock and a dividend yield of 11%
- 40% with common equity offerings that are expected to achieve a 15% return

The composite cost of capital for new financing would be 10.6% per year (.05 • 7 + .1 • 11 + .4 • 15). In economic analysis studies, this composite cost of new capital would be the discount rate or time value of money to the utility.

Some utilities use an after tax discount rate such as the following:

$$\text{Discount rate} = (1 - t) \cdot b \cdot B + p \cdot P + c \cdot C \tag{3.1}$$

where t = incremental income tax rate
b = bond (or debt) interest rate (%/year)
B = fraction that is debt financed
p = preferred stock dividend rate (%/year)
P = fraction that is preferred stock financed
c = common stock equity rate (%/year)
C = fraction that is financed by common stock

The after-tax rate accounts for the fact that bond (debt) interest is a tax deductible item. Therefore, $(1 - t)$ is the net cost of debt to the utility.

There are six compound interest formulas that find application in engineering economics. The nomenclature that will be used in the formulas is presented in Table 3.1.

TABLE 3.1 Nomenclature

Symbol	Definition
P	= Present spot cash equivalent
F	= Future spot cash equivalent
R	= Recurring annual cash stream
n	= Number of years
i	= Present worth discount rate

3.2 SINGLE-PAYMENT INTEREST FACTORS

The first interest formulas involve a single payment that earns interest over a period of years. If the interest is allowed to accumulate with the principal over several years (in a bank account, for example), it is referred to as "compound interest." Conceptually, \$1 in principal may be deposited which, after one year, has achieved \$1 plus interest. This is then the beginning principal for the next year. This concept is illustrated in Figure 3.1, in which a principal amount at year zero is known and the objective is to compute the future amount F, which would be in the account at a future time n years later. This is simply the single-payment compound-amount factor, or compound interest factor (CIF) and is equal to the quantity one plus the interest rate raised to the nth power. Thus, if the future cash equivalent at year end is to be determined for a deposit of principal P today, then the CIF is used. The relationships are presented as Equations (3.2) and (3.3).

Compound-Interest Factor (CIF). Given P, find F (at n years in the future):

$$\text{CIF} = (1 + i)^n \tag{3.2}$$

$$F = \text{CIF} \cdot P \tag{3.3}$$

As an example of single-payment compounding, suppose that an investor has \$1000 in hand today and wants to know how much can be had in 10 years if the money was invested in a bank at 7% annual interest. For this example:

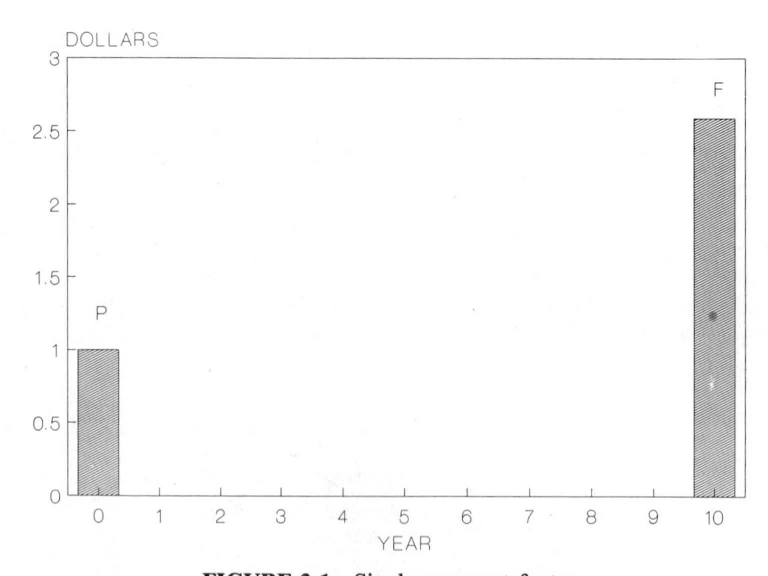

FIGURE 3.1 Single payment factor.

$$F = (1.96715)\ 1000 = \$1967.15$$

To roughly determine how long it takes to double your money, the number of years equals $72/i$, where i is in percent. At 7%, the number of years is 10.28.

In engineering economic analysis, the inverse of this factor finds extensive application and is called the "single-payment present-worth factor or present-value factor (PVF). For this factor, a future amount in year n of value F is known, and the objective is to find the value of the amount at time zero, P. The present value factor is the inverse of the compound-interest factor, and is shown as Equations (3.4) and (3.5).

Present-Value Factor (PVF). Given F (at n years in the future), find p:

$$\text{PVF} = \frac{1}{(1 + i)^n} \tag{3.4}$$

$$P = \text{PVF} \cdot F = \frac{1}{\text{CIF}} \cdot F \tag{3.5}$$

As an example of single-payment discounting, suppose that an investor wants to have \$1000 in a savings account 5 years from now and the bank is paying 7% compound interest. To calculate the single payment,

$$P = (.71299)\ 1000 = \$712.99$$

That is, \$712.99 has to be deposited for 5 years in order to have \$1000 when the interest rate is 7%.

3.3 UNIFORM SERIES FACTORS

3.3.1 Uniform Series (Present Worth) Factor

In economic analysis, there often exists uniform series of annual payments (an annuity) that extend from today through n years. To compute the present worth for the uniform annual series of payments, a convenient formula can be derived as illustrated in Figure 3.2.

The formula is computed by calculating the present worth of each one of the annual payments by discounting it back to time zero, as shown in Equation (3.6). The first annual payment is discounted by the quantity one plus the

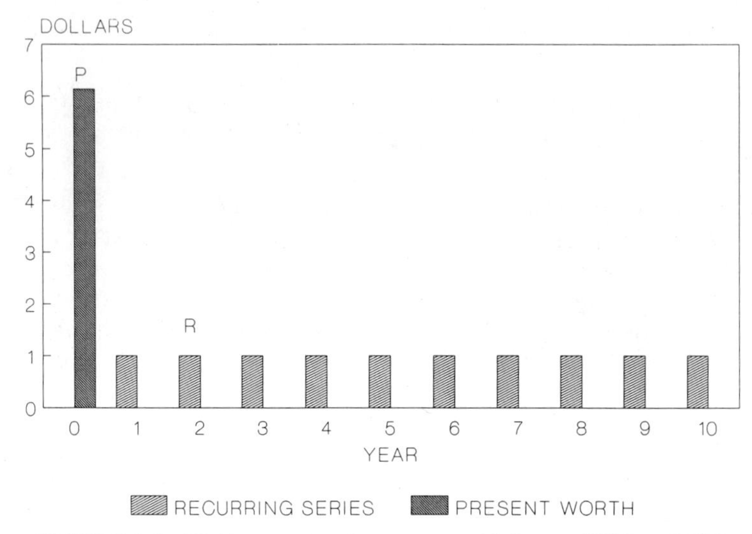

FIGURE 3.2 Uniform series (present-worth) factor (USF or PWF)

interest rate, and the last payment is discounted by one plus the interest rate raised to the nth power. The next step is to multiply this equation by one plus the interest rate, giving Equation (3.7). Then, subtraction of Equation (3.6) from Equation (3.7), noting that most terms cancel, results in Equation (3.8). This equation can be rearranged as shown in Equations (3.9) and (3.10).

Thus, the uniform series present-worth factor is shown in Equation (3.9) and is based only on the present-worth discount rate and the number of years. The uniform series present worth factor multiplied by the dollar amount of the annuity yields the immediate present cash equivalent of that future annuity.

$$P = \frac{R}{(1+i)} + \frac{R}{(1+i)^2} + \frac{R}{(1+i)^3} + \cdots + \frac{R}{(1+i)^n} \tag{3.6}$$

Multiply Equation (3.6) by $(1 + i)$:

$$P(1+i) = R + \frac{R}{1+i} + \frac{R}{(1+i)^2} + \cdots + \frac{R}{(1+i)^{n-1}} \tag{3.7}$$

Subtract (3.6) from (3.7):

$$Pi = R - \frac{R}{(1+i)^n} \tag{3.8}$$

Rearranging:

$$P = [\frac{(1 + i)^n - 1}{i(1 + i)^n}] \cdot R = \text{PWF} \cdot R \tag{3.9}$$

$$\begin{aligned} \text{PWF} &= \text{(uniform series) present-worth factor} \\ &= [\frac{(1 + i)^n - 1}{i(1 + i)^n}] \end{aligned} \tag{3.10}$$

As an example, an investor might want to know the equivalent payment, or one-time payment, that is the same as making equal payments into a savings account on a regular basis. Suppose that \$100 is saved on the last day of the year for 10 years in a savings account that pays 6% interest. What is the present worth equivalent amount? The answer is:

$$P = (7.36009)\ 100 = \$736.01$$

3.3.2 Capital Recovery Factor

Another useful formula is the capital recovery factor (CRF), which finds the equivalent value of a future annuity given the present cash equivalent. This is noted in Equation (3.11), where the CRF is merely the inverse of the (uniform series) present worth factor as shown in Equations (3.11) and (3.12).

As an example of a capital recovery factor calculation, assume that the interest rate is 10% for 30 years. The CRF of Equation (3.12) can then be calculated, and the result is .10608. Thus, \$1 today is equivalent to .10608 dollars every year for the next 30 years, assuming a discount rate of 10% per year.

Given P, find R for n years:

$$R = \frac{i(1 + i)^n}{(1 + i)^n - 1} \cdot P = \text{CRF} \cdot P \tag{3.11}$$

$$\text{CRF} = \frac{1}{\text{PWF}} = \frac{i(1 + i)^n}{(1 + i)^n - 1} \tag{3.12}$$

EXAMPLE

$$n = 30; \qquad i = 10\%; \qquad \text{CRF} = \frac{(.1)(1.1)^{30}}{(1.1)^{30} - 1} = .10608$$

The capital recovery factor helps to determine what equal regular payments are equivalent to a present amount of money. The CRF is the indicator for prospective home buyers to determine what their equal monthly payments will be when they borrow to make the biggest purchase in their lifetimes. For example, suppose that a mortgage of $100,000 is required and the interest rate is 10%. The end-of-year payment to pay interest and principal on the $100,000 over 30 years is:

$$R = (.10608)\ 100{,}000 = \$10{,}608$$

Of course, home mortgages are usually paid off monthly in which case an equivalent monthly rate would be used over 30 • 12, or 360, payment periods.

3.3.3 Compound-Amount Factor

The third uniform series factor is the compound-amount factor (CAF) that calculates the future worth of a uniform series. The CAF helps determine what future amount an investor would have if equal amounts R were placed in a savings account at the end of each year for N years. The CAF is the product of

$$F = \text{CIF} \cdot P = \text{CIF} \cdot \text{PWF} \cdot R$$

$$= (1 + i)^n \cdot \frac{(1 + i)^n - 1}{i\,(1 + i)^n} = \frac{(1 + i)^n - 1}{i} \qquad (3.13)$$

$$\text{CAF} = \text{compound amount factor} = \frac{(1 + \mathrm{i})^n - 1}{i} \qquad (3.14)$$

The compound-amount factor helps one determine how much money can be saved by regularly putting equal amounts of money into a savings account. Suppose that $100 is saved at the end of each year for 10 years in a savings account that pays 6%. On the same day the last payment is made, the saver will have:

$$F = (13.18079)\ 100 = \$1{,}318.08$$

3.3.4 Sinking-Fund Factor

The final uniform series interest factor is called the "sinking-fund factor" (SFF). In this case, a future payment is given and the objective is to calculate the value of an annuity in order to accumulate a future cash equivalent. This is shown in Figure 3.3.

The formula for the sinking-fund factor can be calculated using the compound-amount factor (CAF) or the results of the capital recovery factor (CRF) formula. A recurring annuity has a present worth equal to P, the present cash equivalent, times the CRF; P is equal to the present-value factor (PVF) times the future value as shown in Equation (3.15). Thus, the sinking fund

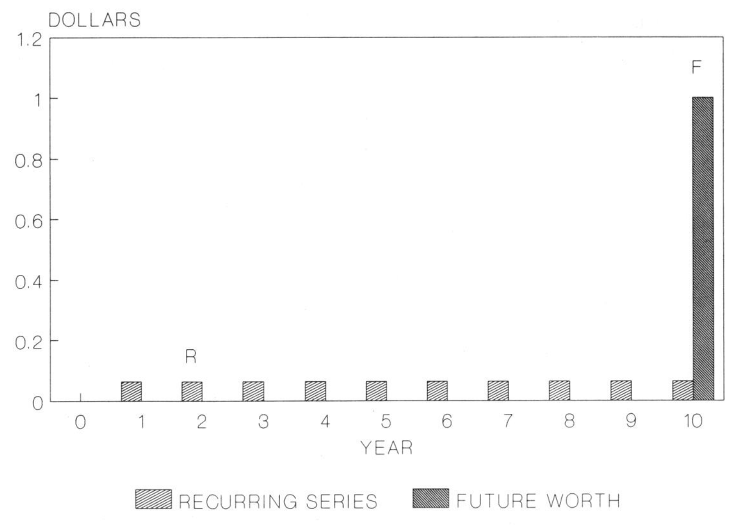

FIGURE 3.3 Sinking-fund factor (SFF).

factor is merely equal to the present-value factor times the capital recovery factor, as shown in Equation (3.16).

$$
\begin{aligned}
R = P \cdot \text{CRF} &= (\text{PVF} \cdot F) \cdot \text{CRF} \\
&= (\text{PVF} \cdot \text{CRF}) \cdot F \\
&= \frac{1}{(1+i)^n} \cdot \frac{i(1+i)^n}{(1+i)^n - 1} \cdot F \\
&= \frac{i}{(1+i)^n - 1} \cdot F \\
&= \frac{1}{\text{CAF}} \cdot F \\
&= \text{SFF} \cdot F
\end{aligned}
\tag{3.15}
$$

$$
\text{SFF} = \frac{i}{(1+i)^n - 1} \tag{3.16}
$$

The sinking-fund factor tells the saver how much money must be put into a savings account each year (a sinking fund) in order to have a given amount at the time the last payment is made. For example, mortgage bonds with a face value of \$1000 must be redeemed in 20 years. How much money should be put into a savings account at the end of each year to meet that bond commitment if the savings account pays 7%? The answer is:

$$R = (0.02439)\ \$1000 = \$24.39$$

The capital recovery factor and sinking fund factors are mathematically and physically related to each other (see Figure 3.4). The mathematical formula for the capital recovery factor is rewritten as shown in Equation (3.17), in which i is added and subtracted from the numerator. Factoring the common term, $(1 + i)^n - 1$, in the numerator and denominator leads to Equation (3.18). However, the second term in Equation (3.18) is recognized as the sinking fund factor, which leads to Equation (3.19). Thus, the capital recovery factor is equal to the discount rate plus the sinking-fund factor.

$$\text{CRF} = \frac{i\,(1 + i)^n}{(1 + i)^n - 1} = \frac{i\,(1 + i)^n - i + i}{(1 + i)^n - 1} \tag{3.17}$$

$$\text{CRF} = i + \frac{i}{(1 + i)^n - 1} \tag{3.18}$$

$$\text{CRF} = i + \text{SFF} \tag{3.19}$$

Now examine this relationship in light of a physical interpretation. Suppose that a $1000 investment in a piece of equipment was made, the equipment has a useful life of 10 years, and the discount rate is 10%. Assume that a loan was

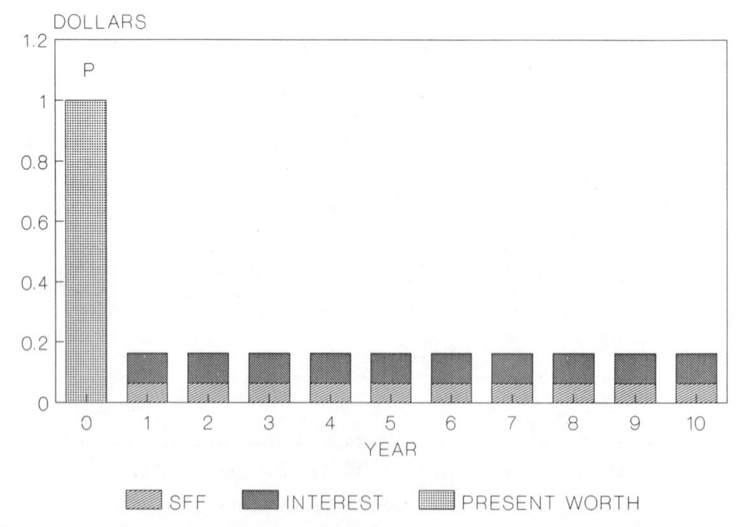

FIGURE 3.4 Capital recovery factor (CRF—provides return of investment + return on investment; SFF—provides return of investment).

taken out from the bank worth $1000 at 10% interest rate. Further assume that the bank is very flexible and several alternative repayment plans can be arranged.

The first alternative assumes the establishment of a payment schedule with the bank so that over the next 10 years both principal and interest are paid in uniform annual payments, similar to those for a house mortgage. In this case, the capital recovery factor could easily be applied as in Equation (3.11). The CRF would provide for interest and principal repayment. After 10 years, the loan to the bank would be paid. The CRF is calculated at 10% for 10 years to be .16275, for a bank payment of $162.75 per year.

Suppose, however, that a second payment plan pays the bank back in a different fashion. Suppose that the bank is paid its interest every year and the loan of $1000 is paid in year 10. In this case, the annual expenditures would be 10% of $1000, or $100 per year plus a big balloon payment of $1000 in year 10. This balloon payment can be obtained from a sinking fund account. In a sinking fund, an amount is set aside each year to accumulate to $1000 in 10 years. The sinking fund factor is calculated to be .06275 or $62.75 per year to obtain that payment goal. The combination of the $100 in interest payments plus the $62.75 sinking fund to provide for the balloon payment is $162.75. Not surprisingly, this is equal to the annual cost of the first alternative. In effect, the capital recovery factor has two pieces; one is the payment of interest or return on the investment, and the other provides for the return of the $1000 investment. The sinking fund factor, however, establishes a fund that provides only for the return of the investment.

In summary, the "time value of money" equations are widely used in planning because planning analysis generally compares alternatives that have different future costs. The present worth mathematics approach permits dollars in any year to be translated into a common-year dollar basis.

3.4 UNIFORM ANNUAL EQUIVALENT OF AN INFLATION SERIES

The previous section discussed uniform annual series and showed how to compute both present worth and future worth cash equivalents. In many planning problems, there can be a series of payments that increase in proportion to an inflation index called "uniform annual inflation series." In this section, a formula will be derived for calculating the present worth of a uniform annual inflation series.

The inflation series is:

$$A[1,\ 1 + a,\ (1 + a)^2,\ \ldots,\ (1 + a)^{n-1}] \tag{3.20}$$

where A = cost in the first year
a = inflation rate, per unit
n = number of years considered

The present worth P of this series at interest rate i is:

$$P = A \left[\frac{1}{1+i} + \frac{1+a}{(1+i)^2} + \frac{(1+a)^2}{(1+i)^3} + \cdots + \frac{(1+a)^{n-1}}{(1+i)^n} \right] \quad (3.21)$$

Multiplying (3.21) by $(1 + i)$ yields:

$$P(1+i) = A \left[1 + \frac{1+a}{1+i} + \left(\frac{1+a}{1+i}\right)^2 + \cdots + \left(\frac{1+a}{1+i}\right)^{n-1} \right] \quad (3.22)$$

Multiplying (3.21) by $(1 + a)$ yields:

$$P(1+a) = A \left[\frac{1+a}{1+i} + \left(\frac{1+a}{1+i}\right)^2 + \cdots + \left(\frac{1+a}{1+i}\right)^n \right] \quad (3.23)$$

Subtracting Equation (3.23) from Equation (3.22) yields:

$$P(i-a) = A \left[1 - \left(\frac{1+a}{1+i}\right)^n \right] \quad (3.24)$$

or

$$P = \frac{A\left[1 - \left(\frac{1+a}{1+i}\right)^n\right]}{i-a} \quad (3.25)$$

The present worth cash equivalent of a uniform annual inflation series has a simple formula, as presented in Equation (3.25).

The uniform levelized annual equivalent U of an inflation series can be computed as the present-worth cash equivalent times the capital recovery factor:

$$U = P \cdot \text{CRF} = \frac{A\left[1 - \left(\frac{1+a}{1+i}\right)^n\right]}{i-a} \cdot \text{CRF}$$

or

$$U = \frac{A\left[1 - \left(\frac{1+a}{1+i}\right)^n\right]}{i-a} \cdot \frac{i(1+i)^n}{(1+i)^n - 1} \quad \text{(levelized value)} \quad (3.26)$$

where A = cost in the first year
a = inflation rate, per unit

i = discount rate, per unit
n = number of years considered

The uniform levelized annual equivalent of the annual inflation series is that uniform value that has the same present-worth value as the series. A levelized inflation series factor is defined as the uniform levelized annual equivalent divided by the first-year cost. The levelizing factor is widely used in many studies because this formula permits the effects of uniform inflation to be conveniently expressed in terms of a single number.

Figure 3.5 presents the levelized inflation series factor in terms of the average inflation rate for the case of a 10-year analysis. If the annual inflation rate is 6% and the discount rate is 12%, then the levelized inflation series factor is approximately 1.25. This means that an inflating series that begins with one dollar and escalates at 6%/year for 10 years can be replaced by a uniform series that has a value of 1.25 over the 10-year life. The uniform series of 1.25 is much easier to use in economic studies and provides a more visual interpretation of the problem and a simpler solution technique than by treating each year individually.

3.5 SUMMARY

The time value of money formulas are repeated as follows:

Nomenclature:

P = Present spot cash equivalent
F = Future spot cash equivalent

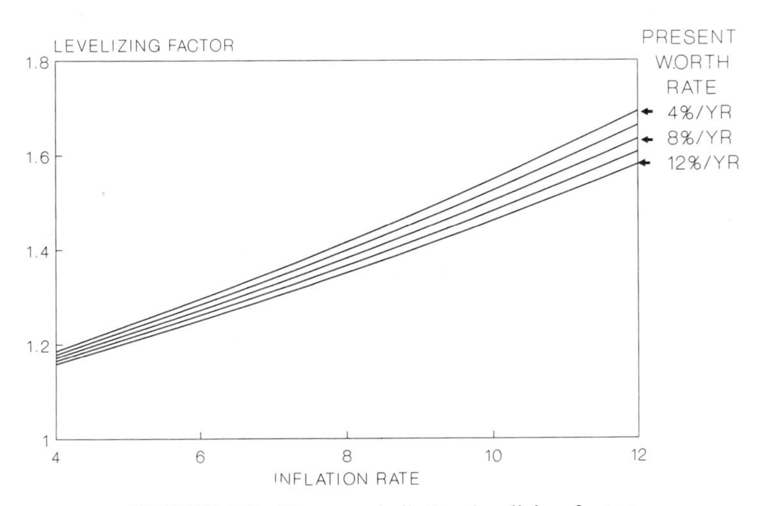

FIGURE 3.5 Ten-year inflation levelizing factor.

R = Recurring annual cash stream
n = Number of years
i = Present-worth rate (%/year)
a = Inflation rate (%/year)

	Abbreviation	Given	Find	Formula
Compound-interest factor	CIF	P	F	$(1 + i)^n$
Present-value factor	PVF	F	P	$\dfrac{1}{(1 + i)^n}$
(Uniform series) present worth factor	PWF	R	P	$\dfrac{(1 + i)^n - 1}{i\,(1 + i)^n}$
Capital recovery factor	CRF	P	R	$\dfrac{i(1 + i)^n}{(1 + i)^n - 1}$
Compound-amount factor	CAF	F	R	$\dfrac{(1 + i)^n - 1}{i}$
Sinking-fund factor	SFF	R	F	$\dfrac{i}{(1 + i)^n - 1}$
Levelizing factor for uniform inflation	LF	a	U	$\left(\dfrac{1 - [(1 + a) / (1 + i)]^n}{i - a}\right) \cdot \text{CRF}$

BIBLIOGRAPHY

American Telephone and Telegraph Company, *Engineering Economy,* 3d ed., McGraw-Hill, New York, 1977.

Grant, E. L., W. G. Ireson, and R. S. Leavenworth, *Principles of Engineering Economy,* 6th ed. Ronald Press, New York, 1976.

Newman, D. G., *Engineering Economic Analysis,* rev. ed., Engineering Press, San Jose, Cal., 1977.

Thuesen, G. J., and W. J. Fabrycky, *Engineering Economy,* 6th ed., Prentice-Hall, Englewood Cliffs, N.J., 1984.

White, J. A., M. H. Agee, and K. E. Case, *Principles of Engineering Economic Analysis,* Wiley, New York, 1977.

PROBLEMS

1 Plot a curve of percent compound interest (ordinate) versus years (abscissa) of the amount of time it takes to double your money if you invest it at a given percent interest. Use percent values of 2, 4, 6, 8, and 10%.

2 Someone owes you $500 but offers instead to give you $1000 in 10 years. If that person is sure to pay, and the going interest rate is 4%, what would you do?

3 An individual will save $1000 at the end of each year for 30 years. The bank will pay 6% compounded annually. How much will be available at the end of 30 years? What is the present value of this amount?

4 You borrow $10,000 from the bank at 8%. What are the equal annual payments including interest and principal that will ensure that the loan is repaid at the end of 10 years?

5 You have purchased some machinery for your company and have the choice of making payments as follows:

a $10,000 now.

b $6,000 now and $6,000 at the end of 5 years.

c $2,500 now and $1,000 at the end of each year for 10 years.

If the interest rate is 8%, which plan would you select?

6 Suppose that municipal bonds of $1000 par value pay 4% annual interest. You could invest your money at 7% in home mortgages. How much would you be willing to pay for the bonds? Assume that the bonds mature in 10 years. (Interpolate between 6 and 8 percent tables.)

7 A college plans to build a laboratory in 1990. The Ford Foundation granted them $840,000 in 1985. At what interest rate must the gift have been invested to have $1,000,000 for that purpose?

8 Customers A and B each deposit $1,000 in their respective savings accounts on the first of each year. At the end of each year, A withdraws an amount equal to the interest. Customer B leaves the accumulated amount deposited. At the end of year 10, A has $10,450 in the account, while B has $13,206.79. Which account has a higher interest rate?

9 Suppose that you have a contract to be paid the following amounts:

$1000 on January 1, 1992

$2000 on January 1, 1993

$3000 on January 1, 1994

$4000 on January 1, 1995

$5000 on January 1, 2004

If your i is 6%, what level of annuity would you settle for (payable annually, first payment January 1, 1993) if you were to receive it for 12 years? If you were to receive it for an eternity?

4

ECONOMIC EVALUATION

A business enterprise participating in a free-market system attempts to maximize its profits over a long-term period. In most cases, this method of conducting business encourages competition and assures that consumers and society receive the greatest benefit in terms of the lowest acceptable price for standards of product quality. An electric utility, however, is granted a monopoly franchise within a service territory subject to the condition that its rates be regulated in a manner that allows a fair and reasonable return on its investment. The electric utility must charge the lowest electric rates possible consistent with providing an acceptable rate of return on its investment and an acceptable quality of electric service. A business enterprise in a free-market system and an electric utility have distinctly different business objectives, and their economic evaluation methods are markedly different.

4.1 UTILITY ECONOMIC EVALUATION METHODS

For a competitive business enterprise, the widely used economic evaluation method is called a "discounted cash-flow rate-of-return method." In this method, all of the cash flows are examined for each alternative through the time horizon of the evaluation. For each alternative, the cash flows are discounted at several different present-worth rates. That discount rate which results in the future cash flows equaling the initial investment is called the "discounted cash-flow rate of return." The project with the highest discounted cash-flow rate of return is considered the best choice. All projects that have a discounted cash-flow rate of return that exceeds the cost of money are worthwhile projects (ignoring investment risk).

For regulated utilities, the economic evaluation method most widely used is called the "minimum-revenue-requirements method." Because the utility is regulated, the rate of return on any investment is determined based on the allowed regulated return on investment. That return is a weighted average return on bonds, where interest return is based on bond ratings and the equity return allowed by the regulating commission. Therefore, the return to bondholders, return on equity, depreciation, tax charges, and revenues can all be calculated. The alternative that provides the lowest revenue requirements is considered the best choice.

Another evaluation method, the "investment-pay-back method," is used by both utilities and free-market enterprises for scoping analysis. The payback method is simply calculated as the number of years required for the net benefits to equal the initial investment. Business enterprises in a free-market system use this as a screening tool to examine a variety of alternatives. After narrowing down the pool of alternatives to a manageable size (typically five 5–10), the enterprise may then conduct discounted cash-flow rate-of-return analyses on the most promising candidates. In the regulated utility industry, the payback method is widely used on small discretionary investments, particularly for spare parts or retrofit activities of a utility.

The revenue requirements method, used predominately by the utility industry, will be examined in detail. Revenue requirements are the sum of two items: the annual fixed charges on a new investment and the annual expenses for fuel, operation, and maintenance. First examine the annual fixed charges on investment.

4.2 FIXED-CHARGE RATE

The concept of fixed-charge rates is widely used in the utility industry. "Fixed-charge rate" is defined as the annual owning costs of an investment as a percent of the investment. A typical value might be 20%/year. When an investment in utility plant is made and placed into service, the owning cost to the utility includes the following:

- Interest on bonds used to partially finance the project
- Equity return requirements of the stockholders who helped to finance the project
- Income taxes to be paid to state and local government
- Ad valorem (property) taxes and insurances to be paid
- Depreciation charge on this investment.

The fixed-charge rate has a yearly variation as shown in Figure 4.1. The fixed-charge rate might begin at a value of 28%/year and decrease to a value of 14%/year at age 30 years. Levelized or average value over the 30- or 40-

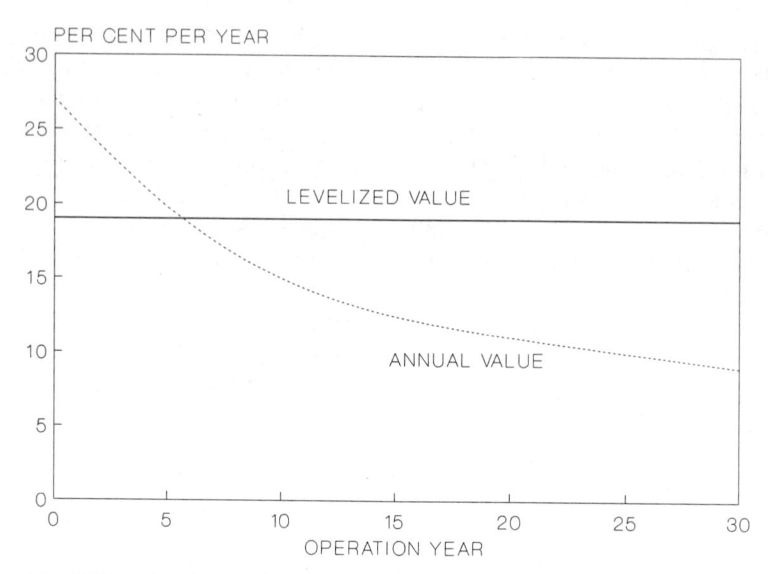

FIGURE 4.1 Comparison of yearly and levelized fixed-charged rates.

year plant life is also used. This can range from 18 to 22% on a levelized basis for a typical investor-owned utility, depending on the specific utility. The next section illustrates the calculation of the levelized and nonlevelized annual fixed-charge rate.

4.2.1 Capital Structure

To derive equations for the fixed charge rate, the utility income statement, income tax statement, and cash report must be examined in more detail. Table 4.1 presents a utility income statement, Table 4.2 shows a tax statement and Table 4.3 a cash report.

Income taxes paid are calculated in the income tax statement. The taxes paid may be reduced by a one-time investment tax credit on qualifying new plant investment made during the year. The investment tax credit was eliminated by the Tax Reform Act of 1986, except on transition plant where construction was begun before January 1, 1986 and will be in operation by December 31, 1991. The tax credit was 10% of the cost of the new investment. Thus, a new $100M plant expenditure would receive a $10M investment tax credit. If the tax liability is $54M, then the actual tax paid in this year would be $44M as a result of the $10M tax credit. Even though the investment tax credit has been eliminated, the quations will be developed with its inclusion, since a credit could be reinstituted at a future time.

The investment tax credit may have a normalized or flow-through accounting treatment. (In actuality, only partial flow-through was allowed by the IRS.) In the flow-through treatment, the income statement contains no adjustments for the one-time nature of the investment tax credit. In the normalized

TABLE 4.1 Income Statement

Revenue	REV_i
Less expenses	
Production cost	PC_i
Depreciation (book)	DB_i
Income taxes paid	$IT_i - ITC_i$
Deferred investment tax credit	$ITC_i \cdot NORMITC$
Amortization of investment tax credit	$AITC_i \cdot NORMITC$
Deferred income taxes	$DFIT_i$
Ad valorem taxes and insurance	AV_i

Operating income:

$$OPIN_i = REV_i - PC_i - DB_i - (IT_i - ITC_i) - DFIT_i - AV_i - ITC_i \cdot NORMITC + AITC_i \cdot NORMITC \quad (4.1)$$

Allowance for funds used during construction	$AFDC_i$
Less interest expense	INT_i

Net income (equity return):

$$NI_i = REV_i - PC_i - DB_i - (IT_i - ITC_i) - DFIT_i - AV_i - ITC_i \cdot NORMITC + AITC_i \cdot NORMITC + AFDC_i - INT_i \quad (4.2)$$

where

DB_i = book depreciation in year i

NORMITC = 1 if the investment tax credit is normalized;
0 if the investment tax credit is flow-through;
X where $0 < X < 1$ if a portion is normalized

DT_i = tax depreciation in year i

DBT_i = the book depreciation rate used in tax calculations for normalized accounting in year i

and

$$DFIT_i = t \cdot (DT_i - DBT_i) \cdot NORMDEPR = \text{deferred income taxes} \quad (4.3)$$

where

t = income tax rate

NORMDEPR = 1 if tax depreciation savings are normalized;
0 if tax depreciation savings are flow-through;
X where $0 < X < 1$ if a portion is normalized

and

$$AITC_i = (ITC/YRITC) \text{ if } i \leq YRITC = \text{amortization of investment tax credit}$$
$$= 0 \text{ if } i > YRITC \quad (4.4)$$

where YRITC = years to amortize the investment tax credit (YRITC = 1 is flow-through)

TABLE 4.2 Income Tax Statement

Revenue	REV_i	
Less deductible expenses		
Production cost	PC_i	
Depreciation (tax)	DT_i	
Ad valorem taxes & insurance	AV_i	
Interest expenses	INT_i	
Taxable income:		
	$TI_i = REV_i - PC_i - DT_i - AV_i - INT_i$	(4.5)
Income tax (tax rate = t):		
	$IT_i = t \cdot TI_i$	(4.6)
Less investment tax credit	ITC_i	
Income taxes paid:		
	$FIT_i = IT - ITC_i$	(4.7)

accounting treatment, the income statement adjusts or normalizes the effect of this one-time credit over several years (the investment tax credit amortization period). This adjustment is accounted for by establishing two additional lines on the income statement, "deferred investment tax credit," and "amortization of investment tax credit." The "deferred investment tax credit" line merely reverses the income tax credit to zero the effect of the credit on the "income

TABLE 4.3 Cash Report

Cash (start of year)	$CASH_i$
Cash from operations	
Net income	NI_i
Adjustments for noncash items	
Depreciation (book)	DB_i
Deferred income taxes	$DFIT_i$
Deferred investment tax credit	$ITC_i \cdot NORMITC$
Less amortization of investment tax credit	$AITC_i \cdot NORMITC$
Cash from external sources	
New financing	$FINANCE_i$
Less application of funds	
Capital expenditures (including AFDC)	$CAPEXP_i$
Refunding of financing	$REFUND_i$
Dividends	$DIVIDEND_i$
Decrease in working capital	$\Delta WKCAP_i$
Cash at year - end	$CASH_{i+1}$

$$CASH_{i+1} - CASH_i = (NI_i - DIVIDEND_i) + (FINANCE_i - CAPEXP_i - REFUND_i) - \Delta WKCAP_i + DB_i + DFIT_i + ITC_i \cdot NORMITC - AITC_i \cdot NORMITC \quad (4.8)$$

tax paid." The "amortization of investment tax credit" is shown as a credit to the expenses and is equal to the investment tax credit divided by the amortization period (typically 10–30 years) to yield an average tax benefit.

Deferred income taxes also require additional explanation. If flow-through accounting is required by the regulatory agency, then deferred income taxes are zero. If normalized accounting is allowed, then the tax benefits from the difference between tax and book depreciation is adjusted or normalized. The tax benefit from using tax depreciation rather than book depreciation was shown in Chapter 2, Section 2.2 as $t \cdot (\mathrm{DT}_i - \mathrm{DBT}_i)$, where t is the average income tax rate, DT is the tax depreciation, and DBT is the book depreciation for tax purposes. The deferred tax depreciation benefit will sum to zero over the life of the project because the total depreciation for book or taxes is the same; only the timing is different. The normalization procedure reverses the tax benefit by reporting an expense on the income statement that is equal to the benefit. In this way, the net income effects of accelerated tax depreciation are adjusted to present a normal (or average) income statement.

"Operating income" is the difference between "revenue" and "expenses." An additional nonoperating income is AFDC, which was discussed earlier in Chapter 2, Section 2.2.

Net income contains the dividends of the preferred stockholders and the earnings available to common stockholders.

The income tax statement presented in Table 4.2 is based on the formula of revenue less deductible expenses equals taxable income. Income tax is computed as the tax rate (typically, 34% for federal taxes) times the taxable income. The taxes paid are reduced by any applicable investment tax credit (currently zero).

The cash report (Table 4.3), also called a "source and application of funds" statement, shows how the transfer of funds impacts on cash. The first category is cash from operations. The first item is net income, which is a source of cash. Several items appear on the income statement as expenses but are not cash-related transactions. They would lead to more income flow. These items include book depreciation, deferred income tax, and investment tax credit accounts. Another source of funds is from external sources of financing including banks, bondholders, and new stockholders.

Funds are spent in several ways, including capital expenditures for new plant. Note that capital expenditures are usually entered to include AFDC. However, since AFDC appears as a noncash income on the income statement, it must appear as a negative source of funds on the funds report. Including AFDC with the capital expenditures as an application of funds achieves the accounting requirement.

Utility funds may be used to refund outstanding bonds as they mature and become due, or to buy back the utility's own stock. Dividends to preferred and common stockholders are also a use of utility funds.

Working capital encompasses the net current assets required to operate the business enterprise on a day-to-day basis. It includes inventories plus accounts receivable less accounts payable. Items classified on the income statement as cash expenses may have a timing difference between when the expense is reported on the income statement and when the cash is spent. The working capital entry accounts for this difference. For example, suppose that $40 of fuel was purchased in January and the fuel was burned in February and paid for in March. At the end of January, the balance sheet would show an increase in fuel inventory of $40 and an accounts payable of $40.

At the end of February, assuming $100 of revenue and no other expenses other than fuel, the income statement would be:

February	
Revenues	$100
Fuel	$40
Net income	$60

The balance sheet would show a decrease in fuel inventory by $40. The cash report for February would show:

Cash at beginning of February	$0
Cash from operations	
Net income	$60
Less application of funds	
Decrease in working capital	$40
Cash at end of February	$100

The cash at the end of Febrary is equal to the cash from the revenues since no expenses are paid until March.

4.3 FIXED-CHARGE RATE COMPONENTS

Begin the analysis of the fixed-charge rate by examining the incremental changes to utility financial accounts arising from a new utility plant being placed into service. For simplicity, assume that the plant cost is $1, including AFDC, and that the utility financed the plant from external sources. The mix of financing is:

Financing Mix Type	% of Each Type	Typical Value (%)	Interest Rate or Earnings Requirement (%)	Typical Value (%)
Long-term bonds	b	50	B	10
Preferred stock	p	10	P	11
Common stock	c	40	C	14.75
Total		100	Average = i	12
				= Bb + pP + cC

The interest on long-term bonds is tax-deductible; stock earnings are not.

The fixed-charge rate is defined as the annual cost of owning new utility plant in per unit of the initial installed capital cost. The amount that the revenue requirements in the income statement (Table 4.1) increase as a result of a $1 plant addition is equal to the annual fixed-charge rate. The fixed-charge rate varies yearly. It is largest when the plant is first installed and decreases as the plant ages. A uniform annual levelized fixed-charge rate is often calculated, which is the present-worth levelized average value of the fixed-charge rate.

The revenue requirements that increase when a new $1 plant addition is installed are (from Table 4.1):

	Category
Depreciation	2
Income taxes	3
Deferred income taxes	3
Investment tax credit normalizations	3
Ad valorem taxes and insurance	4
Interest expense	1
Net income	1

The items in the revenue requirements are grouped into four categories, which are discussed in the following sections.

4.3.1 Return on Investment

The interest expense and net income requirements represent the investors' financial returns on their investment. The interest rate and preferred stock dividend rate are established at the time of the financial offering. The common stock earnings rate is based on the utility's allowed rate of return on common stock.

Common stock earnings provide a common stock dividend, typically 60% of the available earnings; the remainder is reinvested in the utility business as retained earnings. Retained earnings are used to finance other capital projects.

For this fixed charge rate analysis, assume that all of the common stock earnings are paid as dividends. This assumption simplifies the fixed-charge rate derivation but does not influence the fixed-charge rate value.

Examine the cash report of Table 4.3 for any year after the plant has entered service. Assume that the utility will not need to change its working cash and working capital as a result of this $1 new plant addition. There are no new capital expenditures or new financing necessary after the plant is placed into service.

With these ground rules, the cash report [Equation (4.8)] reduces when:

$$\begin{aligned} \mathrm{NI}_i &= \mathrm{DIVIDEND}_i \\ \mathrm{Cash}_{i+1} &= \mathrm{CASH}_i \end{aligned}$$

$$\Delta WKCAP_i = 0$$
$$CAPEXP_i = 0$$
$$FINANCE_i = 0$$

Then:

$$Refund_i = DB_i + DFIT_i + (ITC_i - AITC_i) \cdot NORMITC \tag{4.9}$$

For the moment, consider the special case in which tax accounting is flow-through. Then:

$$DFIT_i = 0; \qquad ITC_i = AITC_i$$

and for flow-through:

$$Refund = DB_i \tag{4.10}$$

Equation (4.10) shows that each year, the business can refund part of the plant financing by using the funds derived from the book depreciation expense. Since the sum of the book depreciation expense is equal to $1, the entire amount of plant financing will be refunded over the life of the plant.

The outstanding financing at any year j is then the total less the accumulated depreciation reserve to date.

$$\begin{aligned} \text{Financing outstanding}_j \ &= 1 - \text{depreciation reserve}_j \\ \text{(flow-through)} \qquad & \\ &= 1 - \sum_{i=1}^{j-1} DB_i \end{aligned} \tag{4.11}$$

The bond interest expense at year j can be computed as the bond interest rate times the percentage in bonds times the financing outstanding.

$$\begin{aligned} \text{Interest expense}_j &= bB\left[1 - \sum_{i=1}^{j-1} DB_i\right] \\ \text{(flow-through)} \qquad & \end{aligned} \tag{4.12}$$

In the more general case of normalization [Equation (4.9)], additional funds from tax normalization can be used to refund the outstanding financing. Thus, the financing outstanding is:

$$\text{Financing outstanding}_j =$$

$$1 - \sum_{i=1}^{j-1} [DB_i + DFIT_i + (ITC_i - AITC_i) \cdot NORMITC] \tag{4.13}$$

At year j, the bond interest, preferred dividend and net income requirements are then:

Interest expense$_j$ =

$$bB \left[1 - \sum_{i=1}^{j-1} [DB_i + DFIT_i + (ITC_i - AITC_i) \cdot NORMITC]\right] \tag{4.14}$$

Preferred dividend$_j$ =

$$pP \left\{1 - \sum_{i=1}^{j-1} [DB_i + DFIT_i + (ITC_i - AITC_i) \cdot NORMITC]\right\} \tag{4.15}$$

Common stock earnings$_j$ =

$$cC \left[1 - \sum_{i=1}^{j-1} [DB_i + DFIT_i + (ITC_i - AITC_i) \cdot NORMITC]\right] \tag{4.16}$$

$$\begin{aligned} \text{Net income}_j &= \text{preferred dividends} + \text{common stock earnings} \\ &= (pP + cC) \cdot \left[1 - \sum_{i=1}^{j-1} [DB_i + DFIT_i \right. \\ &\quad \left. + (ITC_i - AITC_i) \cdot NORMITC]\right] \end{aligned} \tag{4.17}$$

4.3.2 Depreciation

The depreciation for book purposes DB_i can vary annually if installed plant varies, but if $CAPEXD_i = 0$, depreciation is a constant amount. For utility plant, generation, transmission, or distribution, the annual book depreciation rate is a uniform value equal to one divided by the book life. The book life varies for typical plant and is generally 30–40 years. For utility plant, the depreciation rate may range from 0.0333 to 0.025.

$$\text{Book depreciation} = DB_i = \frac{1}{\text{book life}} \tag{4.18}$$

4.3.3 Taxes

The income taxes are calculated from the income tax statement (Table 4.2).

Note that the depreciation used for tax purposes is usually *not* the same as that used for book purposes. There are several reasons for this.

Book depreciation is based on total cost of the plant, including AFDC. (Tax depreciation in the past has not included AFDC. The Tax Reform Act of 1986 excludes the AFDC debt portion as a current tax deduction and allows it to be capitalized for tax book purposes.) For coal-fired power plants, AFDC can represent 15–25% of the total capitalized cost of the plant; for nuclear plants, an even larger percentage. Tax depreciation cannot include nondepreciable assets such as land.

Generally, allowable tax depreciation lives are shorter than the book life of utility equipment. Under the ACRS (accelerated cost recovery system of the IRS Tax Law), utility plant may be depreciated over a 15-year period, whereas the book life may be 30–40 years.

Tax depreciation rates are sometimes higher than book rates. Typical depreciation rate schedules include the declining balance method, sum-of-years digits method, and the ACRS method. Under current law, the ACRS rates based on 1.5 declining balance are those typically used by taxpayers. Table 2.4 in Chapter 2 presents a comparison of these rates for a 10-year depreciation life.

Equation (4.19), based on Equation (4.5) and (4.6), presents the income tax equation.

$$\mathrm{IT}_i = t(\mathrm{REV}_i - \mathrm{PC}_i - \mathrm{DT}_i - \mathrm{AV}_i - \mathrm{INT}_i) \tag{4.19}$$

Substituting for the revenue from Equation (4.2), based on Table 4.1, yields:

$$\begin{aligned}\mathrm{IT}_i = t\,(&\mathrm{PC}_i + \mathrm{DB}_i + (\mathrm{IT}_i - \mathrm{ITC}_i) + \mathrm{DFIT}_i + \mathrm{AV}_i \\ &+ \mathrm{ITC}_i \cdot \mathrm{NORMITC} - \mathrm{AITC}_i \cdot \mathrm{NORMITC} - \mathrm{AFDC}_i \\ &+ \mathrm{INT}_i + \mathrm{NI}_i - \mathrm{PC}_i - \mathrm{DT}_i - \mathrm{AV}_i - \mathrm{INT}_i)\end{aligned} \tag{4.20}$$

Canceling similar terms yields:

$$\begin{aligned}\mathrm{IT}_i = t\,(\mathrm{IT}_i) + t\,(&\mathrm{DB}_i - \mathrm{DT}_i - \mathrm{ITC}_i + \mathrm{DFIT}_i \\ &+ \mathrm{ITC}_i \cdot \mathrm{NORMITC} - \mathrm{AITC}_i \cdot \mathrm{NORMITC} - \mathrm{AFDC}_i + \mathrm{NI}_i)\end{aligned} \tag{4.21}$$

Solving for the income tax yields:

$$\begin{aligned}\mathrm{IT}_i = \frac{t}{1-t}(&\mathrm{DB}_i - \mathrm{DT}_i + \mathrm{DFIT}_i - \mathrm{ITC}_i\,(1 - \mathrm{NORMITC}) \\ &- \mathrm{AITC}_i \cdot \mathrm{NORMITC} - \mathrm{AFDC}_i + \mathrm{NI}_i)\end{aligned} \tag{4.22}$$

Substituting the results from Equations (4.17) and (4.3) into Equation (4.22) yields the final result for income taxes, Equation (4.23).

$$\begin{aligned}\mathrm{IT}_i = \frac{t}{1-t}[&\mathrm{DB}_i - \mathrm{DT}_i + t \cdot (\mathrm{DT}_i - \mathrm{DBT}_i) \cdot \mathrm{NORMDEPR} \\ &- \mathrm{ITC}_i\,(1 - \mathrm{NORMITC}) - \mathrm{AITC}_i \cdot \mathrm{NORMITC} - \mathrm{AFDC}_i\end{aligned}$$

$$+ (pP + cC) \cdot \{1 - \sum_{j=1}^{i-1} [DB_j + t \cdot (DT_j - DBT_j) \cdot NORMDEPR$$

$$+ (ITC_j - AITC) \cdot NORMITC]\} \tag{4.23}$$

where $AFDC_i$ = zero after a generating unit enters commercial service (if there is no "phase-in")

IT_i = income tax liability for year i

DB_i = book depreciation expense in year i

DT_i = tax depreciation in year i

DBT_i = book depreciation for normalized tax accounting

t = income tax rate

ITC_i = investment tax credit in year i

$AITC_i$ = amortization for investment tax credit in year i

NORMDEPR = 1, if normalized tax depreciation accounting is used; 0, if flow-through tax depreciation accounting is used

NORMITC = 1, if normalized investment tax credit accounting is used; 0, if flow-through investment tax credit accounting is used

p = fraction of preferred stock issued (preferred stock ratio)

P = preferred dividend rate

c = fraction of common stock issued (common equity ratio)

C = earnings requirement for common stock

The income taxes paid are the tax liability less the investment tax credit:

$$\text{Income taxes paid} = IT_i - ITC_i \tag{4.24}$$

The deferred income tax is:

$$DFIT_i = t \cdot (DT_i - DBT_i) \cdot NORMDEPR \tag{4.25}$$

The deferred investment tax credit is:

$$\text{Deferred investment tax credit} = ITC_i \cdot NORMITC \tag{4.26}$$

The amortization of investment tax credit is:

$$\text{Amount of ITC} = AITC_i \cdot NORMITC \tag{4.27}$$

4.3.4 Ad Valorem (Property) Taxes and Insurance

The ad valorem (or property) taxes typically are based on the plant's initial capital cost including AFDC. A typical ad valorem tax may be 2–4% per year.

In some states, the ad valorem tax is based on the replacement cost or fair-market value of the plant. In this case, the ad valorem tax may increase as a result of general price inflation. Hence, the ad valorem tax rate based on the initial plant cost basis may increase (or decrease) through time.

$$\text{Ad valorem tax} = \text{AV}_i \tag{4.28}$$

4.4 TOTAL ANNUAL FIXED-CHARGE RATE

The total annual fixed-charge rate is the sum of the following components:

Annual fixed charges = interest expense [Equation (4.14)]
+ net income [Equation (4.17)]
+ book depreciation [Equation (4.18)]
+ ad valorem tax [Equation (4.28)]
+ income tax [Equation (4.24)]
+ deferred income tax [Equation (4.25)]
+ deferred investment tax credit [Equation (4.26)]
+ amortization of investment tax credit [Equation (4.27)] (4.29)

If each of these components is per unitized, the resulting value is the fixed-charge rate.

4.5 LEVELIZED ANNUAL FIXED-CHARGE RATE

The annual fixed-charge rate [Equation (4.29)] is powerful but cumbersome to apply because it changes every year. For a 20- or 30-year analysis, this can be a computational burden. For most economic analysis, the levelized annual (or average) fixed-charge rate is much easier to apply because only one number is carried in the calculations. The levelized fixed-charge rate will provide the same answer as the varying annual fixed-charge rate.

The general expression for levelizing any variable V_i is given in Equation (4.30):

$$\overline{V} = \text{CRF}_N \cdot \sum_{j=1}^{N} \frac{V_j}{(1.0 + r)^j} \tag{4.30}$$

where V_j = dollar amount of variable V in year j
r = present worth discount rate
CRF_N = capital recovery factor for N years at rate r
N = number of years to be levelized
$\overline{V}$ = levelized value of variable V

Note that a superscript — (vinculum, denoting "average") will denote a levelized value.

Equation (4.30) computes the levelized value by first computing the present worth of variable V over the N-year sum. The present worth is then converted into an annuity by multiplying by the capital recovery factor.

The levelized fixed charge rate can be computed by applying Equation (4.30) to each component of the annual fixed-charge rate of Equation (4.29). However, prior to applying Equation (4.30), some mathematical preliminaries are derived that will simplify the levelized fixed-charge rate calculations.

1. **Levelizing Formulas**

a. *Constant.* The levelized value of a constant, such as the book depreciation rate, has a form of:

$$d_i = \begin{cases} \dfrac{1}{\mathrm{M}} & \text{for} \quad 1 \le i \le M \\ 0 & \text{for} \quad i > M \end{cases} \tag{4.31}$$

where M = depreciation life in years. If the levelizing period is N years where $N \ge M$, then the application of Equation (4.30) to Equation (4.31) simplifies to:

$$\bar{d} = \frac{1}{M} \mathrm{CRF}_N \sum_{i=1}^{M} \left(\frac{1}{1+r}\right)^i$$

$$\bar{d} = \frac{1}{M} \frac{\mathrm{CRF}_N}{\mathrm{CRF}_M} \qquad \text{if } M \le N \tag{4.32}$$

If the depreciation life M is greater than the levelizing period (which rarely occurs in practice), then:

$$\bar{d} = \frac{1}{M} \mathrm{CRF}_N \sum_{i=1}^{N} \left(\frac{1}{1+r}\right)^i = \frac{1}{M} \tag{4.33}$$

b. *Accumulated Depreciation.* Another example is the accumulated depreciation component. Using tax depreciation as an example, we obtain

$$\mathrm{CUM}dt_j = \sum_{i=1}^{j-1} dt_i \tag{4.34}$$

where dt_i = tax depreciation rate in year i

$\mathrm{CUM}dt_j$ = cumulative tax depreciation up to year J

j = year

This can be simplified by noting that the summation:

$$\text{CUM}dt_j = \text{CUM}dt_{j-1} + dt_{j-1}; \qquad \text{CUM}dt_{j=0} = 0 \tag{4.35}$$

Applying the levelizing Equation (4.30), to Equation (4.35) results in

$$\begin{aligned}\overline{\text{CUM}dt} &= \text{CRF}_N \sum_{j=1}^{N} \frac{\text{CUM}dt_j}{(1+r)^j} \\ &= \text{CRF}_N \sum_{j=1}^{N} \frac{\text{CUM}dt_{j-1}}{(1+r)^j} + \text{CRF}_N \sum_{j=1}^{N} \frac{dt_{j-1}}{(1+r)^j}\end{aligned} \tag{4.36}$$

Rearranging the first term of Equation (4.36) yields:

$$\begin{aligned}\sum_{j=1}^{N} \frac{\text{CUM}dt_{j-1}}{(1+r)^j} &= \sum_{j=1}^{N} \frac{\text{CUM}dt_{j-1}}{(1+r)^{j-1}(1+r)} \\ &= \frac{1}{1+r} \sum_{j=1}^{N} \frac{\text{CUM}dt_{j-1}}{(1+r)^{j-1}}\end{aligned} \tag{4.37}$$

Further note that the $\text{CUM}dt_{j=0} = 0$, and that the cumulative tax depreciation at year N is unity, $\text{CUM}dt_{j=N} = 1.0$. Thus, Equation (4.37) can be rewritten as:

$$\begin{aligned}\frac{1}{1+r} \sum_{j=1}^{N} \frac{\text{CUM}dt_{j-1}}{(1+r)^{j-1}} &= \frac{1}{1+r} \sum_{j=1}^{N} \frac{\text{CUM}dt_j}{(1+r)^j} - \left(\frac{1}{1+r}\right) \frac{\text{CUM}dt_{j=N}}{(1+r)^N} \\ &= \frac{1}{1+r} \sum_{j=1}^{N} \frac{\text{CUM}dt_j}{(1+r)^j} - \left(\frac{1}{1+r}\right)\left(\frac{1}{1+r}\right)^N\end{aligned} \tag{4.38}$$

which, times CRF_N, is the first term in Equation (4.36). The second term in Equation (4.36) can be rewritten as (note that $dt_{j=0} = 0$):

$$\begin{aligned}\text{CRF}_N \sum_{j=1}^{N} \frac{dt_{j-1}}{(1+r)^j} &= \text{CRF}_N \sum_{j=2}^{N} \frac{dt_{j-1}}{(1+r)^j} \\ &= \frac{\text{CRF}_N}{1+r} \sum_{j=2}^{N} \frac{dt_{j-1}}{(1+r)^{j-1}} \\ &= \frac{\text{CRF}_N}{1+r} \sum_{j=1}^{N-1} \frac{dt_j}{(1+r)^j}\end{aligned} \tag{4.39}$$

In most levelized fixed-charge rate analyses, the levelizing period is longer than the tax depreciation life. If this is true, then $dt_{\text{year} = N} = 0$, and Equation (4.39) simplifies to:

$$\text{CRF}_N \sum_{j=1}^{N} \frac{dt_{j-1}}{(1+r)^j} = \frac{\text{CRF}_N}{1+r} \sum_{j=1}^{N} \frac{dt_j}{(1+r)^j} = \frac{\overline{dt}}{1+r} \tag{4.40}$$

where $\overline{dt}$ = levelized tax depreciation rate. Equations (4.38) and (4.40) can be substituted into Equation (4.36), yielding Equation (4.41):

$$\overline{\text{CUM}dt} = \frac{1}{1+r}\,\overline{\text{CUM}dt} - \frac{\text{CRF}_N}{(1+r)(1+r)^N} + \overline{dt}\left(\frac{1}{1+r}\right) \tag{4.41}$$

Rearranging and noting that $\text{CRF}_N/(1+r)^N = \text{SFF}_N$ yields

$$\overline{\text{CUM}dt} = \frac{\overline{dt} - \text{SFF}_N}{r} \tag{4.42}$$

The levelized cumulative depreciation rate is equal to the levelized depreciation rate minus the sinking-fund factor divided by the present-worth discount rate. This formula holds true regardless of the type of depreciation.

c. *Single Payment.* Investment tax credit is a one-time credit typically during the first service year. The levelized value of this credit is:

$$\overline{\text{ITC}} = \text{CRF}_N \cdot \frac{\text{ITC}_{\text{year}=1}}{(1+r)} = \text{ITC}_{\text{year}=1} \cdot \frac{\text{CRF}_N}{1+r} \tag{4.43}$$

d. *Accumulated Single Payment.* The accumulated investment tax credit form is:

$$\text{CUMITC}_j = \sum_{i=1}^{j-1} \text{ITC} \tag{4.44}$$

Note that the investment tax credit is typically applied when the plant enters commercial service in year 1. In that case, $\text{CUMITC}_{j=\text{year }1} = 0$, and for any year after commercial service $\text{CUMITC}_j = \text{ITC}$. Thus, the levelized accumulated investment tax credit is:

$$\overline{\text{CUMITC}} = \text{CRF}_N \sum_{i=1}^{N} \frac{1}{(1+r)^i} \sum_{j=1}^{i-1} \text{ITC}_j$$

$$= \mathrm{CRF}_N \sum_{i=2}^{N} \frac{1}{(1+r)^i} \mathrm{ITC}$$

$$= \mathrm{ITC} \cdot \mathrm{CRF}_N \left[\sum_{i=1}^{N} \left(\frac{1}{1+r}\right)^i - \frac{1}{1+r} \right]$$

$$= \mathrm{ITC} \cdot \mathrm{CRF}_N \left(\frac{1}{\mathrm{CRF}_N} - \frac{1}{1+r} \right)$$

$$= \mathrm{ITC}\left(1 - \frac{\mathrm{CRF}_N}{1+r} \right) \tag{4.45}$$

2. **Levelized Fixed-Charge Rate Formula.** The levelized fixed-charge rate components can be computed using the levelizing relationship formulas of Equations (4.33), (4.42), and (4.45).

   The first two components in Equation (4.29) are interest expense and net income. Inserting the levelized relationships in (4.14) and (4.17) yields:

Levelized interest and net income =

$$\underset{\text{all } j \text{ years}}{\text{Levelized}} \left\{ (bB + pP + cC) \cdot \left[1 - \sum_{i=1}^{j-1} \mathrm{DB}_i + t(\mathrm{DT}_i - \mathrm{DBT}_j) \right.\right.$$

$$\left.\left. \cdot \,\mathrm{NORMDEPR} + \mathrm{ITC}_i \cdot \mathrm{NORMITC} - \mathrm{ITC}_i \cdot \mathrm{DITC}_i \cdot \mathrm{NORMITC} \right] \right\}$$

since from Equation (4.25) $\mathrm{DFIT}_i = t(\mathrm{DT}_i - \mathrm{DBT}_i) \cdot \mathrm{NORMDEPR}$ and by definition $\mathrm{AITC}_i = \mathrm{ITC}_i \cdot \mathrm{DITC}_i$.

By substituting Equations (4.42), (4.42), (4.45) and (4.42), respectively, we obtain:

Levelized interest and net income =

$$(bB + pP + cC) \cdot \left(1 - \left\{ \left(\frac{\overline{\mathrm{DB}}}{r} - \frac{\mathrm{SFF}_N}{r} \right) \right.\right.$$

$$+ \left[t\left(\frac{\overline{\mathrm{DT}}}{r} - \frac{\mathrm{SFF}_N}{r} \right) - t\left(\frac{\overline{\mathrm{DBT}}}{r} + \frac{\mathrm{SFF}_N}{r} \right) \right] \cdot \mathrm{NORMDEPR}$$

$$+ \mathrm{ITC}\left(1 - \frac{\mathrm{CRF}_N}{1+r}\right) \cdot \mathrm{NORMITC}$$

$$\left.\left. - \,\mathrm{ITC} \cdot \mathrm{NORMITC}\left(\frac{\overline{\mathrm{DITC}}}{r} - \frac{\mathrm{SFF}_N}{r} \right) \right\} \right)$$

Simplifying yields:

Levelized interest and net income =

$$(bB + cC + pP)\ \{1 - [\frac{\overline{DB}}{r} - \frac{SFF_N}{r} + \frac{t}{r}(\overline{DT} - \overline{DBT}) \cdot NORMDEPR$$

$$+ ITC\ (1 - \frac{CRF_N}{1 + r}) \cdot NORMITC - NORMITC$$

$$\cdot \frac{ITC}{r} \cdot (\overline{DITC} - SFF_N)]\} \tag{4.46}$$

where $\overline{DITC}$ = levelized amortization rate of the investment tax credit
$ITC \cdot \overline{DITC}$ = levelized amortization cost of the investment tax credit
$\overline{DB}$ = levelized book depreciation
$\overline{DBT}$ = levelized book depreciation for normalized tax calculations
$\overline{DT}$ = levelized tax depreciation

Levelizing Equation (4.23) yields the income taxes:

Levelized income taxes =

$$\overline{IT} = \frac{t}{1 - t}[\overline{DB} - \overline{DT} + t \cdot (\overline{DT} - \overline{DBT}) \cdot NORMDEPR$$

$$- ITC \cdot (1 - NORMITC) \cdot \frac{CRF}{1 + r} - ITC \cdot \overline{DITC} \cdot NORMITC$$

$$+ (\frac{pP + cC}{pP + cC + bB}) \cdot (\text{levelized interest and net income})] \tag{4.47}$$

The levelized income taxes paid are computed by subtracting the levelized investment tax credit:

$$\text{Levelized income taxes paid} = \overline{IT} - ITC \cdot \frac{CRF_N}{1 + r} \tag{4.48}$$

$$\text{Deferred income taxes} = t \cdot (\overline{DT} - \overline{DBT}) \tag{4.49}$$

$$\text{Deferred investment tax credit} = (ITC \cdot \frac{CRF}{1 + r} \cdot NORMITC \tag{4.50}$$

$$\text{Amortization of investment tax credit} = ITC \cdot \overline{DITC} \cdot NORMITC \tag{4.51}$$

$$\text{Ad valorem taxes and insurance} = \overline{AV} \tag{4.52}$$

Example 4.1. Calculate the 30-year levelized annual fixed-charge rate for a plant investment with the following characteristics:

Bond interest rate	10%/year
Bond capitalization fraction (debt ratio)	50%
Preferred stock dividend rate	12%/year
Preferred stock capitalization fraction	10%
Common stock earnings rate	14.5%/year
Common stock capitalization fraction	40%
Composite cost of capital	12%/year
Investment tax credit	0%
Investment tax credit amortization period	15 years (or .06667/year)
Combined federal and state income tax rate	38%
Tax depreciation method (1986 Tax Law)	20-year ACRS
Normalization	Partial
Book depreciation life (for tax calculations)	20 years
Book depreciation life	30 years (or .0333/year)
AFDC (as % of installed plant investment)	20%
AFDC borrowed funds (as % of plant)	10%
Present-worth discount rate	12%/year
Ad valorem property tax rate and insurance (levelized)	3%/year

$$\text{Levelized ACRS depreciation} = .4066 \cdot \text{CRF} = .4066 \cdot .12414 = .0505$$

The levelized book depreciation rate for tax calculations (partial normalization) can be computed as the present worth of a 20-year uniform series levelized to 30 years.

$$\overline{\text{DBT}} = \left(\frac{1}{20 \text{ years}}\right) \cdot \text{USF}_{20 \text{ years}} \cdot \text{CRF}_{30 \text{ years}}$$

$$\overline{\text{DBT}} = (.05) \cdot (7.469) \cdot (1.2414) = .0464$$

Because AFDC is 20% of the installed capital cost, the depreciable tax cost prior to 1986 was only 80% of the installed plant. (Prior to 1986, only the non-AFDC portion of book cost was capitalized for tax purposes, but following the Tax Reform Act of 1986, the interest portion of AFDC is no longer allowed as a current deduction for tax purposes but may be capitalized for tax purposes. Since half of the AFDC is interest, the tax basis is now reduced by

TABLE 4.4 ACRS 20-Year Tax Depreciation Rate

Year	ACRS Rate %/Year	12% Present Worth Factor	Present Worth of ACRS Depreciation
1	.04	.893	.0357
2	.08	.797	.0638
3	.07	.712	.0498
4	.06	.636	.0382
5	.06	.567	.0340
6	.05	.507	.0254
7	.05	.452	.0226
8	.05	.404	.0202
9	.05	.361	.0181
10	.05	.322	.0161
11	.05	.287	.0144
12	.05	.257	.0129
13	.05	.229	.0115
14	.05	.205	.0103
15	.04	.183	.0073
16	.04	.163	.0065
17	.04	.146	.0058
18	.04	.130	.0052
19	.04	.116	.0046
20	.04	.104	.0042
21	0		0
22	0		0
23	0		0
24	0		0
25	0		0
26	0		0
27	0		0
28	0		0
29	0		0
30	0		0
			.4066

90% from the book basis.) The tax depreciation is 90% of that calculated above.

$$\overline{DT} = .9 \cdot .0505 = .0455; \qquad \overline{DBT} = .9 \cdot .0464 = .0418$$

The net investment tax credit would be .09 of the installed plant book cost, but ITC is now zero due to the 1986 tax law.

The levelized investment tax amortization rate is the present worth of a 20-year uniform series levelized over 30 years.

$$\overline{\text{DITC}} = \left(\frac{1}{20 \text{ years}}\right) \cdot \text{USF}_{20 \text{ years}} \cdot \text{CRF}_{30 \text{ years}}$$

$$= .05 \cdot (7.469) \cdot (.1241) = .0464$$

The fixed-charge rate components are:

a. Levelized interest and net income [Equation (4.46)]. Note that $(bB + pP + cC) = .12$.

$$= .12 \left[1 - \left(\frac{.0333}{.12} - \frac{.00414}{.12} + \frac{.38}{.12} (.0455 - .0418)(1) + 0 - 0\right)\right]$$

$$= .12 \left[1 - (.2775 - .0345 + .0117 + 0 - 0)\right]$$

$$= .0894$$

b. Levelized book depreciation = .0333.

c. Levelized income taxes [Equation (4.47)]:

$$= \left(\frac{.38}{1 - .38}\right) [.0333 - .0455 + .38(.0455 - .0418)(1)$$

$$- 0 - 0 + \frac{.07}{.12}(.0894)]$$

$$= .613 (.0333 - .0455 + .0014 - 0 - 0 + 0.522)$$

$$= .0254$$

The levelized income taxes paid are the tax liability less the investment tax credit = .0254 − (0) • (.1241/1.12) = .0254.

d. Deferred income taxes = .38 • (.0455 − .0418) = .0014.

e. Deferred investment tax credit = 0 • (.1241/1.12) (1) = 0.

f. Amortization of investment tax credit = (0) • (.0464) (1) = 0.

g. Ad valorem taxes and insurance = .03.

The total levelized fixed-charge rate is summarized as:

Levelized interest and net income	.0894
Levelized book depreciation	.0333
Levelized income tax paid	.0254
Deferred income taxes	.0014
Deferred investment tax credit	0
Amortization of investment tax credit	0
Ad valorem taxes and insurance	.03
Total	.1795

The levelized annual fixed-charge rate for this case is 18%/year.

Example 4.2. Compute the levelized fixed-charge rate using the data of Example 4.1. Use flow-through accounting for both the accelerated tax depreciation and the investment tax credit.

	Summary
Levelized interest and net income	
$= .12\ [1 - (\frac{.0333}{.12} - \frac{.00414}{.12} + 0 + 0 + 0)]$	.0908
Levelized book depreciation	.0333
Levelized income taxes paid =	
$.613\ [.0333 - .0455 + 0 - 0 - 0 + \frac{.07}{.12}(.0908)]$	
$= .613\ [.0333 - .0455 + .0482]$	.0250
Ad valorem taxes and insurance	.03
Total	.1791

Flow-through accounting under the new tax law results in a slightly lower fixed-charge rate, as might be expected from the earlier accounting examples of a regulated utility.

The fixed-charge rate for a government-owned utility (municipal, state, or federal) is lower than that of an investor-owned utility. The government-owned (or "public" utility) pays no income taxes, and the new capitalization of a public utility is typically debt-financed. In addition, the interest of public utility financings (bonds) are usually exempt from federal (and in some cases state) income taxes. Because interest income is tax-exempt, public utility bonds can be sold in the bond market with yields of 1–3%/year less than the non-tax-exempt bonds that investor-owned utilities issue.

Public utilities are usually exempt from paying property and ad valorem taxes as well. However, many public utilities provide a contribution in lieu of taxes as compensation to local governments.

Example 4.3. Using the same data as for Example 4.1, compute the levelized fixed-charge rate for a municipal public utility that is tax exempt, issues tax-exempt bonds at 8%/year, and provides 3% levelized contributions in lieu of taxes. Use a present-worth rate of 8%/year.

	Summary
Levelized interest and net income $= .08 \cdot [1 - (\frac{.0333}{.08} - \frac{.00414}{.08})]$	.0508
Levelized book depreciation	.0333
Contributions in lieu of taxes	.03
Total	.114

As shown, a public utility will have a levelized fixed-charge rate that is much lower than an investor-owned utility.

4.6 REVENUE REQUIREMENTS—EXAMPLE

The revenue requirements method is the economic evaluation method used predominately by the utility industry. Revenue requirements consist of two items: the annual fixed charges on a new investment and the annual expenses for fuel, operation, and maintenance. The decision criteria among alternatives is to implement projects that have the least present-worth revenue requirements.

Present-worth revenue requirements may be calculated and presented in three formats: the cumulative present-worth method, the levelized annual cost method, and the equivalent capitalized cost method. The cumulative present worth method is based on comparison of the cumulative present worth costs of alternative projects and is widely used.

The levelized annual cost method compares the levelized annual costs of alternative projects. Levelized annual costs may be computed by multiplying the cumulative present-worth costs by the capital recovery factor, which uniformly spreads the costs over the evaluation period.

The equivalent capitalized cost method compares alternatives by translating cumulative present-worth costs into an equivalent capitalized cost. The equivalent capitalized cost is defined as the initial equivalent investment cost that has the same cumulative present-worth cost as the actual cost stream. The equivalent capitalized cost may be calculated by dividing the cumulative present-worth cost by the product of the levelized annual fixed-charge rate times the uniform series factor. The equivalent capitalized cost method is also widely used.

These three revenue requirement methods are all equivalent. The minimum cost project selection is, therefore, the same regardless of the method because they differ only in the way in which the results are presented.

Example 4.4 illustrates these alternative methods for a project with a 5-year life, which is not typical of a major utility project.

Example 4.4. Two plans perform the same productive function but have different costs, as shown in Table 4.5.

TABLE 4.5 Revenue Requirements Example

	Plan A	Plan B
Life	5 years	5 years
Investment	1000	1200
Operating cost (year 1)	139	52.8
Inflation 8.0%/year		
Present worth rate = .097		
Levelizing factor = 1.156		
Uniform series factor = 3.82		
Levelized fixed-charge rate = .339		

Compute the 5-year cumulative present worth cost of each.

Present-worth revenue requirements (PWRR) of plan A:

$$\begin{aligned}\text{PWRR} &= \text{5-year present worth of fixed charges} + \text{operating cost}\\ &= .339 \cdot 1000 \cdot \text{(5-year uniform series factor)}\\ &\quad + 139 \cdot 1.156 \cdot \text{(5-year uniform series factor)}\\ &= .339 \cdot 1000 \cdot 3.82 + 139 \cdot 1.156 \cdot 3.82\\ &= \$1908\end{aligned}$$

Present-worth revenue requirements of plan B:

$$\begin{aligned}\text{PWRR} &= .339 \cdot 1200 \cdot 3.82 + 52.8 \cdot 1.156 \cdot 3.82\\ &= \$1787\end{aligned}$$

Therefore, select alternative B, since it has the least cumulative present worth cost, as shown in Table 4.5.

Compute the levelized annual cost for the same alternatives. Levelized annual revenue requirements of plan A:

$$1000 \cdot .339 + 139 \cdot 1.156 = \$500$$

Levelized annual revenue requirements of plan B:

$$1200 \cdot .339 + 52.8 \cdot 1.156 = \$468$$

Therefore, select plan B, which is the same conclusion reached earlier.

Compute the equivalent capitalized cost: The equivalent capitalized cost can be found by equating the cumulative present-worth fixed charges of the equivalent investment to that of the cumulative present-worth cost of the original cost stream.

$$\begin{aligned}&\text{Cumulative present-worth fixed charges}\\ &\quad = \text{cumulative present-worth cost}\end{aligned} \tag{4.53}$$

or

$$\begin{aligned}&\text{Equivalent capitalized cost} \cdot \text{levelized annual fixed-charge rate}\\ &\quad \cdot \text{uniform series factor} = \text{cumulative present-worth cost}\end{aligned} \tag{4.54}$$

The equivalent capitalized cost is solved as:

$$\begin{matrix}\text{Equivalent}\\ \text{Capitalized Cost}\end{matrix} = \frac{\text{Cumulative present-worth cost}}{\text{(levelized annual fixed-charge rate} \cdot \text{USF)}} \tag{4.55}$$

Total equivalent capitalized cost of plan A = $1000 + 139 • 1.156 • 3.82/(.339 • 3.82) = $1474.

Total equivalent capitalized cost of plan B = 1200 + 52.8 . 1.156/.339 = $1380.

Thus, as in the earlier methods, alternative B is the best choice. Note that the equivalent capitalized cost of the investment cost is the actual investment cost.

The results of this example are shown in Figure 4.2. The equivalent capitalized cost method has a presentation advantage. In many cases, the capital investment is of keen interest. The equivalent capitalized cost method provides an evaluation procedure in which the investment cost is not numerically changed. All the operating costs are translated into equivalent investment costs. Thus, the total evaluation is performed in terms relative to the investment. This can have a clearer interpretation than a cumulative present-worth cost or levelized annual cost method.

4.7 PAY-BACK METHOD

In cases of small discretionary investments, an investment pay-back method may be used for screening analysis. The pay-back period is defined as the number of years of incremental operating cost savings required to equal the incremental investment. Operating cost savings may be assumed to be constant with time, escalated as a result of inflation, or escalated and discounted with time. Acceptable pay-back periods for utility projects range from 3 to 5 years.

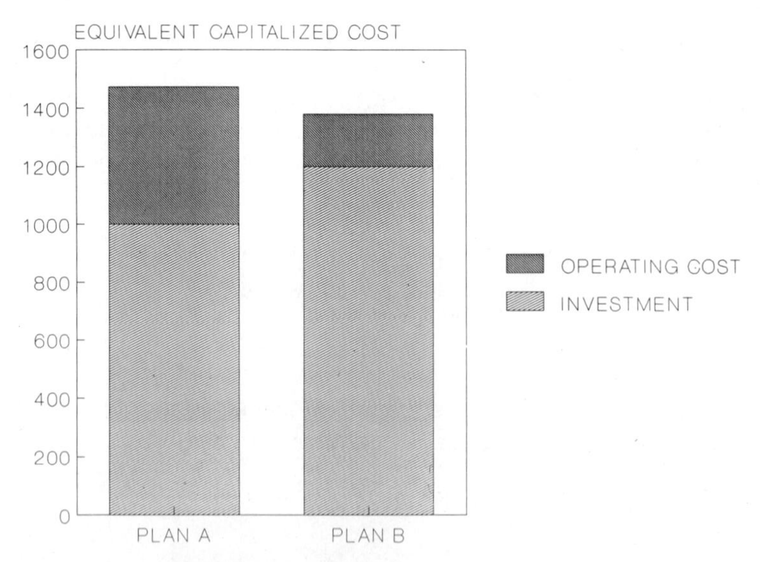

FIGURE 4.2 Equivalent capitalized cost.

Example 4.5. Using the following data, compute the investment pay-back benefit of plan B relative to plan A, assuming operating cost escalation

Year	Inflation Factor	Operating Cost Savings	Cumulative Operating Cost Savings	Investment
1	1.00	86.2	86.2	
2	1.08	93.1	179.29	$200
3	1.166	100.5	279.79	

SOLUTION

$$\text{Payback period} = \frac{\text{incremental investment}}{\text{incremental operating cost savings}}$$

$$= \frac{\$200}{86.2\ \$/\text{year}} = 2.32 \text{ years}$$

4.8 SUMMARY

This chapter introduced basic accounting principles for utility economic evaluation methods. It also discussed the utility income statement, balance sheet, and cash report. The concept of the annual fixed-charge rate was introduced and the equations for it were derived. The equations for the levelized annual fixed-charge rate were also calculated. The chapter concluded with utility revenue requirement and investment pay-back examples.

BIBLIOGRAPHY

Bary, C. W., and W. T. Brown, "Some New Mathematical Aspects of Fixed Charges," *AIEE Transactions,* Vol. 76, Part III, 230 (1957).

Baumol, W. J., *Economic Theory and Operations Analysis,* 3rd ed., Prentice-Hall, Englewood Cliffs, NJ, 1972, Chapters 18 and 19.

Heck, F. M., Jr., "The Cost-of-Capital in Economics Studies," *AIEE Transactions, Power Apparatus and Systems,* December 1961, p. 775.

Jeynes, P. H., "Annual Carrying Charges in Economic Comparisons of Alternate Facilities," *AIEE Transactions,* Vol. 70, 1951.

Jeynes, P. H., "Common Sense in Utility Economics," *Electrical World,* New York, November 23, 1959.

Jeynes, P. H., *Profitability and Economic Choice,* The Iowa State University Press, 1968.

Marsh, W. D., *Economics of Electric Utility Power Generation,* Oxford Engineering Science Series, 1980.

Michaelson, W. G., "Income Taxes and the Engineering Decision," *IEEE Transactions on Power Apparatus and Systems,* October 1967, p. 1250.

Neverman, A. W., "The Effect of an Earning Constraint on the Discount Rate," *Public Utilities Fortnightly,* February 16, 1978.

PROBLEMS

1 Your company's capital structure is 50% debt and 10% preferred stock, and the remainder is common equity. Debt can be issued at 11%, preferred at 12%, and common stock has a yield of 15.75%. Property tax is 3% of first cost. Flow-through accounting is used for tax savings. Compute the fixed-charge rate and annual revenue requirement for a $1,000,000 investment with an economic life of 30 years. The combined federal and state income tax rate is 46%, and the levelized tax depreciation rate is 5%. Assume that there is no investment tax credit.

2 Two different transformers are being considered for a substation. Transformer A costs $250,000 and has annual losses of $50,000 and has a life of 35 years. Transformer B costs $200,000 and has annual losses of $100,000 and a life of 35 years. The company's capital structure is 50% debt, at a cost of 10% (interest) and 50% common equity at a cost of 12%. Property tax rates are 2.5%, and the levelized income tax rate is calculated to be 2.5% of first cost. Calculate the fixed charge rate and levelized annual costs, and recommend which transformer should be purchased.

3 Given a 50,000-kW gas turbine with installed cost at $150/kW for a total cost of $7,500,000. The fixed-charge rate is 18%; cost of capital is 10%; and fuel, operation, and maintenance costs are $200,000 per year. Find

a. Annual levelized revenue requirements.

b. Present worth for a 30-year period.

4 Suppose that an electric system expansion for 5 years is proposed. A production simulation has calculated the fuel costs. The cost of purchased power or sale of power has been fixed by contractual agreement as shown below.

Choose 1995 to be the reference year for present worth. The average cost of capital is 10%; the fixed-charge rate for generation is 16% and for transmission 18%; generation costs are $400/kW with no inflation. Find the present worth for the study.

5 Two types of equipment perform the same job. Equipment A costs $10,000 to install and $500/year to operate and has an average life of 10 years.

Year	Add Generation (MW)	Millions of dollars		
		Add Transmission	Production Cost	Purchase (Sale) Power
1998	800	$48	$43.414	—
1999	—	40	47.898	$ 8.640
2000	800	62	49.922	—
2001	—	54	50.158	10.466
2002	1000	66	52.284	(5.220)

Equipment B costs $8,500 to install and $750/year to operate and has an average life of 8 years. The company's cost of money is 9%. Income tax for equipment A is 1.5% (of original cost) and for equipment B is 2.5%. Property tax and insurance are 2.5% of original cost. Assume that equipment is part of continuing plant (no inflation) and no salvage. Compute levelized annual cost for each alternative. At what purchase price for equipment B would the two alternatives be equally attractive?

5

FINANCIAL AND REGULATORY ANALYSIS

Engineering economics analysis methods presented in the previous chapters are extremely useful and widely used for economic decisionmaking. Corporate financial simulation provides a broader analysis methodology for decision-making by accounting for many "real-world" issues including electric rate regulation, fuel pass-through clauses, earnings per share growth, external financing requirements, and the stock market environment. These company and industrywide issues have an influence on utility decisions while expanding the scope of the economic model and minimizing the assumptions required for analysis.

5.1 CORPORATE FINANCIAL SIMULATION METHOD

The corporate financial simulation method block diagram is presented in Figure 5.1. Financial simulation prepares future financial performance forecasts of major utility financial reports: balance sheet, income statement, funds report, and retained-earnings report. This is done by simulating the financial transactions in the same manner as a utility would conduct business.

The overall flow diagram of Figure 5.1 can be segmented into nine key elements, which are detailed in Sections 5.1.1 through 5.1.9. The simulation process may be conducted on an annual basis or, for detailed analysis, on a monthly basis. If the financial analysis of the utility income statement determines that operating income or earnings fall below a target value, then a rate change proceeding is requested to adjust the rates and obtain satisfactory income or earnings.

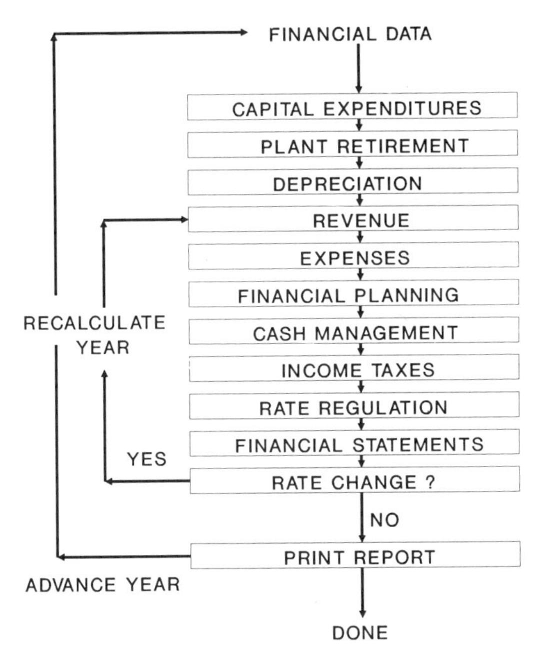

FIGURE 5.1 Corporate financial simulation method block diagram.

5.1.1 Capital Expenditures

Capital expenditures are comprised of money spent for utility plant additions and/or depreciable replacements and upgrades of existing plant facilities. Utility practice is to include allowance for funds used during construction (AFDC) in the capitalized expenditures.

Example 5.1. A utility plant addition is estimated to cost \$1 million (\$1M) in 1990 dollars. The expenditures will be over a 4-year period from 1991 through 1994. Assuming a cost escalation of 5%/year and a 10% AFDC rate, calculate the yearly expenditures and the total installed cost of the plant. The plant expenditure pattern is given as the first column in Table 5.1.

The expenditure is first escalated to the current-year dollars. Then AFDC is calculated on the first year's expenditure (where the current year's AFDC assumes a uniform expenditure throughout the year, or a yearly average of one-half the yearly expenditure). The cumulative expenditure is then calculated including the prior year's AFDC, thereby compounding the AFDC "interest." This is referred to as the "construction work in progress" (CWIP) account. Then AFDC is calculated based on the cumulative expenditures, including AFDC. The annual capital expenditure is the sum of the AFDC and the escalated expenditure.

Based on these calculations, the total installed plant cost in 1994, including

TABLE 5.1 Plant Expenditures Based on Cost Escalation

Year	Per Unit 1990 Expenditure	Escalated Expenditure	AFDC	Total Annual Expenditure Including AFDC
1991	0.100	0.105	0.005	0.110
1992	0.250	0.276	0.025[a]	0.301
1993	0.350	0.405	0.061[b]	0.466
1994	0.300	0.365	0.106[c]	0.471
Cumulative	1.000	1.151	0.197	1.348

[a] (0.276)(.1)(.5) + (.110)(.1) = 0.025.
[b] (0.405)(.1)(.5) + (.110 + .301)(.1) = 0.061.
[c] (0.365)(.1)(.5) + (.110 + .301 + .466)(.1) = 0.106.

AFDC, is now \$1.348M. Note that the AFDC, formerly referred to as "interest during construction," is 17.1% (.197/1.348) of the total cost of the project.

If all the plant construction expenditures took place in 1991 and the plant goes on-line in 1994, the total plant cost in 1994 would be \$1.467M, with AFDC constituting 28.5% of the total plant cost. If all of the construction expenditures took place in 1994, the total plant cost would be \$1.276M with AFDC as 4.7% of the plant cost. Thus, there is a strong economic incentive to have a short construction cycle at the time of plant installation.

The AFDC account is divided into two segments, borrowed funds and equity funds. The borrowed fund portion is that share of AFDC that is "financed" using nonequity funds (bonds, notes, short-term debt). The other portion is "financed" using equity funds (preferred stock, common stock). This segmentation of AFDC is required for accounting purposes on the utility income statement and for income tax purposes.

In some cases, utilities are permitted to place some of the construction expenditures for a project into the rate base during the construction period. In this case, AFDC is calculated only on the CWIP account balance not included in the rate base.

After a project is completed and the plant enters commercial service, the CWIP charges for a project are transferred to the appropriate plant-in-service account.

Periodically, an investment tax credit has been part of the U.S. Federal Tax Law. When in effect, the investment tax credit has been calculated as a percentage of the annual plant expenditure (excluding all AFDC). The tax credit may be taken on progress payments during construction or accumulated and taken when the plant enters commercial service. The Tax Reform Act of 1986 eliminated the investment tax credit on all plants except certain "transition" plants that began construction by December 31, 1985.

The plant cost of a project is needed for tax calculations. The tax cost of a project excludes the equity portion of AFDC. Prior to 1986, the tax cost of a project excluded all AFDC. The Tax Reform Act of 1986 requires that AFDC on borrowed funds be included in the tax cost. In addition, the tax base on a project was reduced by 5% when an investment tax credit was taken.

5.1.2 Plant Retirement

Plant retirements subtract from the plant-in-service account. Fossil fueled plants are usually fully depreciated, and demolition and plant removal costs are small. Decommissioning costs for nuclear units are more substantial.

5.1.3 Depreciation

Depreciation is calculated for all plants in service. Book depreciation is generally based on a straight-line depreciation method using a 30- to 40-year book life.

Tax depreciation, which is required for tax computation, is calculated on the basis of the appropriate tax depreciation method applicable to a specific plant based on its vintage. Sum- of-year-digits (SYD) methods are used for plants added after 1954 using tax depreciation lives from IRS guidelines. Guideline tax lives are 30 years for transmission and distribution projects, 28 years for steam electric plants, 20 years for nuclear and combustion turbine plants, and 50 years for hydroelectric plants.

Plants installed from 1971 to 1980 may use shorter tax life guidelines under the Asset Depreciation Range (ADR) Law (80% of guideline tax life). These plants are depreciated according to a double-declining balance method or SYD. Equipment installed after 1981 is treated according to the Economic Recovery Tax Act of 1981 and the Tax Equity and Fiscal Responsibility Act (TEFRA) of 1982. The Tax Reform Act of 1986 revises the tax lives for plant additions after 1986. Table 5.2 presents several tax depreciation schedules.

Prior to 1986, the tax depreciation rate was applied to the tax cost of the plant excluding AFDC less half of the investment tax credit. Since 1986, the tax depreciation rate must be applied to the cost of the plant, including the borrowed funds (bond interest, short-term debt) component of AFDC. The tax cost of the plant also excludes one-half of the received investment tax credit (ITC). ITC is allowed only for a "transition" plant and is completely phased out by 1991.

Example 5.2. Calculate the total book and tax depreciation in 1990 for a utility with four plants whose characteristics are:

Plant number	1	2	3	4
Installation year	1960	1975	1985	1989
Installed cost (excluding AFDC) $M	150	280	800	950
(Including borrowed AFDC)	165	310	900	1100
(Including all AFDC)	180	350	1000	1200

TABLE 5.2 Annual Tax Depreciation Rate Schedules for Utility Steam Plant

Tax Life in Years	SYD 28	ADR 22.5	TEFRA 15	1986 Tax Reform Act 20
1	0.07	0.09	0.05	0.04
2	0.07	0.08	0.10	0.07
3	0.06	0.07	0.09	0.07
4	0.06	0.07	0.08	0.06
5	0.06	0.06	0.07	0.06
6	0.06	0.06	0.07	0.05
7	0.05	0.05	0.06	0.05
8	0.05	0.05	0.06	0.05
9	0.05	0.04	0.06	0.05
10	0.05	0.04	0.06	0.05
11	0.04	0.03	0.06	0.05
12	0.04	0.03	0.06	0.05
13	0.04	0.03	0.06	0.05
14	0.04	0.03	0.06	0.05
15	0.03	0.03	0.06	0.05
16	0.03	0.03		0.04
17	0.03	0.03		0.04
18	0.03	0.03		0.04
19	0.02	0.03		0.04
20	0.02	0.03		0.04
21	0.02	0.03		
22	0.02	0.03		
23	0.01	0.03		
24	0.01			
25	0.01			
26	0.01			
27	0.01			
28	0.01			

Tax life	28	22.5	15	20
Book life	35	35	35	35
Investment tax credit taken ($M)	0	28	80	0

SOLUTION. Using Table 5.2, the tax depreciation is

Plant number	1	2	3	4
Tax base	150	280	760	1100
Depreciation rate	0	.03	.07	.07

Total tax depreciation in year 1990 = $138.6M

$$\text{Book depreciation} = \frac{180}{35} + \frac{350}{35} + \frac{1000}{35} + \frac{1200}{35} = \$78\text{M}$$

5.1.4 Revenue

Revenues are calculated on the basis of the electric rate tariff in effect. The rate tariff for each utility service class can have an energy component, demand component, and fixed component. The components may be dependent on the season and quantity of use. The revenue is computed by multiplying each rate component times the electricity consumption (energy and demand) and accumulating the results for all service classes.

In many utilities, an "automatic" fuel adjustment clause is present. The objective of the fuel adjustment is to immediately adjust electric revenue to compensate for any changes in fuel costs without having to prepare a full electric rate case proceeding. The fuel adjustment is typically based on actual fuel costs per kilowatt-hour compared to a reference fuel cost base established at the time of the last regulatory rate proceeding. Fuel adjustments may be full or partial.

Table 5.3 presents power system data for a 1990 base year and a 1992 current year. Most fuel adjustment revenue clauses are based on the difference in dollars per kilowatt-hour ($/kWh) fuel cost of sales. In this case, the fuel adjustment is:

$$\begin{aligned}\text{Fuel cost}\\ \text{revenue adjustment} &= 63{,}000{,}000 \cdot (23.13\text{–}21.85\,\$/\text{MWh})\\ &= \$80.6\text{M} \qquad (5.1)\end{aligned}$$

However, this $/kWh fuel cost adjustment inaccurately accounts for increased fuel costs from the base year. In Table 5.3, the utility installed 800 MW of additional coal-fired capacity in 1992. This plant is not in the rate base because the last regulatory proceeding was in 1990. The costs of this plant are not factored into the electric rate tariff. If they were, the carrying charges on the new plant would have increased rates slightly.

The automatic fuel cost adjustment reflects a fuel cost contribution from the new plant that lowers the automatic fuel adjustment. This creates an inconsistency that leads to revenues not in synchronism with utility costs. This may be compensated for by using a fixed fuel mix composition based on the reference year megawatt-hour generation mix instead of the actual year. Table 5.4 presents this calculation using the reference year sales to compute the required fuel adjustment revenues. The 1992 fuel adjustment is now based on a 23.31 $/MWh cost, and the fuel adjustment now is:

$$\begin{aligned}\text{Fuel adjustment} &= \$63{,}000{,}000\ \text{MWh} \cdot (23.31 - 21.85\ \$/\text{MWh})\\ &= \$91.98\text{M} \qquad (5.2)\end{aligned}$$

This fuel adjustment provides $11M more in revenue than does the simple $/kWh adjustment clause.

The fuel adjustment clause can account for an additional change. Note that

TABLE 5.3 Fuel Adjustment Clause Example

	Reference Year 1990					
Fuel Type	Capacity (MW)	Fuel Cost ($/MBtu)	Heat Rate (MBtu/MWh)	Operating Cost ($/MWh)	Sales (Millions MWh)	Total Cost (Millions $)
Hydroelectric	1,100	—	—	—	5	—
Nuclear	1,300	0.80	10.5	8.4	8	67.2
Coal	5,700	2.00	10.0	20.0	30	600.0
Oil & gas	5,700	4.00	10.0	40.0	15	600.0
Total	13,800				58	1,267.2
Average ($/MWh)			(1267.2/58) = 21.85			

the utility invested in capital improvements to improve the efficiency of oil- and gas-steam units, thus reducing heat rate from 10.0 to 9.8 MBtu/MWh in 1992 (refer to Table 5.3).

The fuel adjustment computations of Equations (5.1) and (5.2) are based on $/KWh and show that the efficiency benefit reduces revenues. The increased cost of capital improvement to achieve these efficiency benefits has not been factored into the analysis.

One way to overcome this situation is to make use of a $/MBtu-based fuel adjustment clause. In this method, both the fuel mix and the efficiency of the reference year are used to compute the fuel adjustment for the current year. Table 5.5 presents these calculations. The fuel adjustment now is:

$$\begin{aligned}\text{Fuel adjustment} &= (63{,}000{,}000 \text{ MWh})\,(23.54 - 21.85 \text{ \$/MWh}) \\ &= \$106.47\text{M} \end{aligned} \tag{5.3}$$

The $/MBtu-based fuel adjustment clause provides a revenue adjustment of $14.5M more, which contributes to paying for the cost of the capital improvement.

Other types of fuel adjustment provisions may be made. Some states have instituted special fuel adjustment clauses pertaining to efficiency and conservation measures that permit the fuel savings to be used to accelerate the cost recovery of the capital investment. The influence of fuel adjustment revenue clauses on utility decisionmaking incentives will be further discussed later in this chapter.

5.1.5 Operating Expenses

Operating expenses consist of the following expenses: fuel, nonfuel generation operation and maintenance (O&M), transmission and distribution O&M, cus-

Table 5.3 (continued)

		Current Year 1992			
Capacity (MW)	Fuel Cost ($/MBtu)	Heat Rate (MBtu/MWh)	Operating Cost ($/MWh)	Sales (Millions MWh)	Total Cost (Millions $)
1,100	—	—	—	5	—
1,300	.90	10.5	9.45	8	75.6
6,500	2.10	10.0	21.0	35	735.0
5,700	4.40	9.8	43.12	15	646.8
14,600				63	1,457.4
		(1457.4/63) = 23.13			

tomer account, and sales and general and administrative (G&A), including all overheads. Table 5.6 presents the national average composition of these expenses for 1985. Fuel and purchased power expenses comprise the largest share.

Comprehensive computer software programs are available to compute fuel and purchased power expenses for forecasting future costs of an electric utility. The remainder of the expenses can be modeled as functions of capital expense or as numbers of customers.

Note that wages and salaries represent only 15% of the total operating expenses. Thus, 85% of operating expenses are purchased quantities, fuel, and manufactured goods.

5.1.6 Financial Planning

External financing is required if the application of funds (principally capital expenditures and dividends) exceeds the internal sources of funds (principally net income, depreciation, and deferred taxes). This often happens when a utility is involved in a large construction program. Financial planning schedules the timing and types of external financing (stocks, bonds, preferred stock, short-term debt, etc.) according to financial constraints and guidelines.

When the internal sources of funds exceed the application of funds, then financial planning is needed to invest the "excess cash." This often occurs when a utility has only a small construction program or when no generation capacity is under construction.

The requirements for external financing can be computed from the cash report (or funds report), which lists the sources and application of funds. Table 5.7 presents a sample table. Consider the year 1990, when the total application of funds is $683M. Since the ending cash balance ($62M) is scheduled to be $2M more than the starting cash balance, the total sources of funds must

TABLE 5.4 Fixed Fuel Base Fuel Adjustment Clause Example

Fuel Type	Reference Year Capacity	Fuel Cost 1992 ($/MBtu)	1992 Heat Rate (MBtu/MWh)	Operating Cost 1992 ($/MWh)	Reference Sales Total (Millions MWh)	Cost (Millions $)
Hydroelectric	1100	—	—	—	5	—
Nuclear	1300	0.9	10.5	9.45	8	75.6
Coal	5700	2.10	10.0	21.0	30	630.0
Oil & Gas	5700	4.40	9.8	43.12	15	646.8
Total	13800				58	1352.4
Average ($/MWh)			(1352.4/58) = 23.31			

TABLE 5.5 $/MBtu-Based Fuel Adjustment Clause Example

Fuel Type	Reference Year Capacity	1992 Fuel Cost ($/MBtu)	Reference Year Heat Rate (MBtu/MWh)	($/MWh)	Reference Year Sales (Millions MWh)	Total Cost (Millions $)
Hydroelectric	1100	—	—	—	5	—
Nuclear	1300	0.9	10.5	9.45	8	75.6
Coal	5700	2.10	10.0	21.0	30	630.0
Oil & gas	5700	4.40	10.0	44.0	15	660.0
Total	13800				58	1365.6
Average ($/MWh)			$\frac{1365.6}{58} = 23.54$			

TABLE 5.6 Composition of Operating Expenses (1985)[a]

Type of Expense	Contribution (%)
Fuel and purchased power	64
Generation O&M	14
Transmission O&M	2
Distribution O&M	6
Customer account	4
Sales	0.1
G&A	10

[a]Wages and salaries constitute 15% of the total utility operating expenses.

TABLE 5.7 Sample Cash Report for 1990

	1990 (Millions $)	
Beginning cash balance		60
Funds from operations		
Net income	187	
Net cash expenses	63	
Depreciation and amortization	118	
Total funds from operations		368
Funds from outside sources		
Long-term debt	211	
Preferred stock	25	
Common stock	25	
Short-term debt	3	
Total funds from outside sources		264
Other sources of funds (rents, other businesses, etc.)		53
Total sources of funds		685
Construction and plan expenditures	341	
Dividends on preferred stock	27	
Dividends on common stock	115	
Retirement of securities and short-term debt		
Long-term debt	89	
Preferred stock	9	
Common stock	3	
Short-term debt	22	
Change in working capital	20	
Other applications of funds	57	
Total applications of funds		683
Ending cash balance		62

be $685M. Since the internal sources of funds total $368M, the required external sources are compiled as:

$$\text{Total funds from outside sources} = 685 - 53 - 368 = \$264\text{M}$$

The distribution of external financing among the various financing types is based on the utility's corporate policy. Traditionally, average utility total capitalization is 50% long-term debt, 10% preferred stock, and 40% common stock. If the existing long-term debt on the balance sheet is $1530M and the total capitalization is $3141M, then the fraction of the funds to be raised by long-term debt financing to achieve a 50% debt ratio is:

$$\frac{1530 - 89 + f_B \cdot 264}{3141 - 89 - 9 - 3 + 264} = \frac{1441 + f_B \cdot 264}{3304} = .5$$

where f_B is the fraction of new bonds to be sold. Note that the retirement of securities is considered in this calculation.

Performing the calculations yields: $f_B = .799$ or

$$\text{New long-term debt financing} = .799 \cdot \$264\text{M}$$
$$= \$211\text{M}$$

Similar calculations may be performed for the other financing types.

The selection of the financing mix between long-term debt, preferred stock, common stock, and short-term debt is influenced by many factors. Short-term debt financings (financings with maturities within approximately 1 year such as 90-day notes) are generally used between major financial offerings. Since the financial markets experience cycles during which interest costs and stock prices can fluctuate, short-term financing provides flexibility to execute a major financial offering at the proper opportunity, which allows capital costs to be "accumulated" until major financing is needed.

Long-term debt (bond) financing is timed to financial market cycles and is the lowest-cost financial vehicle. A high debt ratio leads to large fixed interest charges and significantly leverages earnings. A small percentage decrease in revenue can have a large percentage decrease in earnings, since interest charges are fixed at the time of borrowing. While the business practice of A-grade manufacturing and service corporations is a debt ratio of 20% or less, utilities can have similar ratings at 50% debt ratio due to the stability of the business and state and federal regulation of earnings. In situations where energy projects have take-or-pay contracts for the electricity and energy output, debt ratios of 80% have been justified.

Many bond offerings carry indenture constraints. A typical indenture limit is that no new bonds may be issued if the total interest coverage ratio falls below a specified value, typically 2.0 or more after tax. Interest coverage is defined as the income before interest expense divided by the interest expense. Interest coverage measures the margin by which bondholders have their interest requirements covered. If a utility interest coverage ratio falls below the specified value, the utility may not issue new bonds but must issue other, less secure, types of financing such as preferred or common stock.

Long-term debt offerings of bonds are rated by investment services such as Moody's and Standard and Poors. The bond rating has a key influence on the interest cost of the bond when issued. A highly rated bond has a lower interest cost. Key factors that influence the bond rating are interest coverage (higher coverage leads to higher bond rating), AFDC percent of earnings (higher AFDC leads to lower bond rating), and business risk. Table 5.8 presents financial rating criteria guidelines for utility bonds.

Equity financial instruments include preferred and common stocks. Preferred stocks are used by utilities to provide an equity financing vehicle with a lower cost than common stock. Common stock equity financing represents 35–45% of utility external financing. The common stockholders are the real owners of the utility. The worth of their financial holdings is dependent on the utility's financial performance. Thus, the stockholders bear the risks of the utility business. This risk is manifest in the yield (or interest cost) of stock (earnings-to-stock-price ratio) as shown in Figure 5.2. Lower-quality common stocks require significantly higher "interest cost" because they have higher risks than high-quality stocks.

One measure of a utility stock is its market-to-book ratio. This measures the current market price of the stock divided by the balance-sheet worth of the stock. Stock that is selling above book price means that new stockholders are paying a premium to own stock. This is indicative of a healthy, growing corporation. Stock that is selling below book means that new stockholders are purchasing stock for a price less than that of the original stockholders. Thus, the equity position of existing stockholders is "diluted" by new stockholders who are buying stock below book since both old and new stockholders have equal claims per share to company assets and earnings. Consequently, existing shareholders prefer that the company not issue any new stock if the stock is selling below book value.

TABLE 5.8 Financial Rating Criteria Guidelines for Electric Utilities

	AAA	Aa	A	Baa
Interest coverage pretax including AFDC	4.0 +	3.25–4.25	2.5–3.5	3.0 –
Debt ratio	45% –	42–47%	45–55%	53% +
Cash-flow adequacy (cash flow less AFDC and dividends as % of cash capital outlays)	...	40% +	20–50%	30% –

FIGURE 5.2 Influence of security rating on interest-rate costs.

The market-to-book ratio can be computed as:

$$\text{Market} : \text{book ratio} = (\text{price} : \text{earnings ratio}) \cdot (\text{earnings per Share}) / (\text{total balance-sheet equity/number of shares}) \tag{5.4}$$

$$\text{Market} : \text{book ratio} = (\text{price} : \text{earnings ratio}) \cdot (\text{return- on- equity}) \tag{5.5}$$

The price-to-earnings ratio is dependent on the specific utility financial performance and business risks. "Return on equity" is the ratio of earnings divided by the common stock balance sheet equity. For example, a 7.5 price-to-earnings ratio and a 14% return on equity results in a 1.05 market-to-book ratio.

5.1.7 Cash Management and Accounting

Bookkeeping for a utility company applies standard double-entry bookkeeping procedures (discussed in Chapter 2). Cash management involves adjusting the financing so that adequate cash is on hand monthly, weekly, daily, or even hourly. Excess cash is invested while short-term shortages are managed by bank loans or lines of credit.

5.1.8 Income Taxes

The Federal Income Tax law has been revised several times during the 10 years preceding 1988 but the overall concept of an income-based tax has remained.

The provisions of the Tax Reform Act of 1986 added an alternative minimum tax option, and that legislation has changed the tax accounting of AFDC.

The first step in determining income tax is the calculation of the taxable income.

$$\text{Taxable income} = \text{revenue} - \text{production expenses} - \text{tax depreciation} - (\text{interest expense} - \text{AFDC borrowed funds}) \quad (5.6)$$

The Tax Reform Act of 1986 modified the taxable income equation to exclude the borrowed portion of AFDC. Income tax is calculated as:

$$\text{Gross federal income tax} = \text{income tax rate} \cdot \text{taxable income} \quad (5.7)$$

where the income tax rate was set at 34% in 1987 (down from 46% in previous years). The rate is graduated, but a moderate sized company will have most of its income taxed at the top rate.

The Tax Reform Act of 1986 eliminated investment tax credit on new projects. Projects initiated before 1986 are governed by transition rules that permit tax credit to be applied. "Regular" federal income tax is adjusted by these credits as:

$$\text{Regular federal income tax} = \text{gross federal income tax} - \text{investment tax credit} \quad (5.8)$$

The Tax Reform Act of 1986 added an alternative minimum tax. In this case, the minimum tax is calculated by adjusting the taxable income and applying a minimum tax rate (20% in 1987). The actual federal tax paid is the larger of either the regular federal income tax or the alternative minimum tax.

The alternative minimum taxable income is adjusted by the difference between ACRS tax depreciation and alternative minimum tax depreciation.

$$\text{AMTI} = \text{alternative minimum taxable income} = \text{taxable income} - (\text{ACRSTD} - \text{AMTD}) \quad (5.9)$$

where

ACRSTD = ACRS tax depreciation

AMTD = alternative minimum tax depreciation based on straight-line depreciation tax life:

Plant Service Year	Minimum Tax Straight Line Tax Life (Steam Plant)
1981–1983	15 years
1984	18 years
1985–1986	19 years
1987 +	20 years

An adjusted net book income is then computed as:

$$\begin{aligned}\text{Adjusted net book income} &= \text{revenue} - \text{production expense} \\ &\quad - \text{book depreciation} - \text{interest} \\ &\quad + \text{nonoperating income} + \text{AFDC}\end{aligned} \tag{5.10}$$

The alternative minimum taxable income is adjusted by 75% (50% for the years 1987–1989) of the amount by which adjusted net book income exceeds alternative minimum taxable income, as follows:

$$\begin{aligned}\text{FAMTI} &= \text{final alternative minimum taxable income} \\ &= \text{AMTI} + .75 \cdot (\text{adjusted net book income} - \text{AMTI})\end{aligned} \tag{5.11}$$

The alternative minimum tax is calculated as:

$$\begin{aligned}&\text{Alternative minimum tax} \\ &\quad = (\text{FAMTI}) \cdot \text{minimum tax rate} - \text{investment tax credit}\end{aligned} \tag{5.12}$$

where the minimum tax rate is 20% in 1987 and investment tax credits on previous projects may be used to reduce the minimum tax up to 25% of the tax.

The regular federal income tax is compared to the alternative minimum tax. If the alternative minimum tax is larger, then the alternative minimum tax is paid and the difference between regular federal income tax and the alternative minimum tax is accrued and added to the value of any prior year. This accumulated value may be used to decrease taxes in future year when the regular tax exceeds the alternative minimum tax. If:

$$\text{Alternative minimum tax} > \text{regular income tax}$$

then:

$$\begin{aligned}&\text{Accumulated income tax difference} \\ &\quad = \text{accumulated income tax difference from prior years} \\ &\quad + (\text{alternative minimum tax} - \text{regular income tax})\end{aligned} \tag{5.13}$$

$$\text{Income tax paid} = \text{alternative minimum tax}$$

If:

$$\text{Alternative minimum tax} < \text{regular income tax}$$

Then:

$$\begin{aligned}\text{Income tax paid} &= \text{regular income tax} \\ &\quad - \text{accumulated income tax difference}\end{aligned} \tag{5.14}$$

However, the income tax paid cannot be reduced to less than the alternative minimum tax.

5.1.9 Rate Regulation

Electric rates are established during a regulatory rate proceeding. When changes to the utility system occur that lead to changes in utility cost or income, then a new regulatory rate proceeding is undertaken to determine new rates. The regulatory proceeding can take from 6 months to 2 years to complete.

The regulatory process begins with the determination of the acceptable return on equity. An acceptable return is based on several considerations including: what is required to instill financial confidence in the utility, what comparable utilities are allowed when operating in the same business risk environment, and what is fair and reasonable.

Revenues are then calculated to yield the acceptable return on equity. The revenues are allocated to various customer service classifications based on the costs of serving those classes. A rate tariff structure is designed for each class based on the functional costs (energy, demand, and customer costs) of service. Electric revenue regulation is performed on either of two criteria: the return on equity or the return on rate base.

Return on common equity involves calculating the net income available (income statement) for common stock divided by the average or end-of-year common equity (balance sheet). Revenue is increased or decreased so that the return-on-equity target is achieved.

Regulation on rate base is the more traditional index. The rate base is defined as:

$$\begin{aligned}\text{Rate base} = {} & \text{total plant in service} \\ & - \text{accumulated depreciation reserve} \\ & + \text{materials and supplies (optional)} \\ & + \text{fossil fuel inventory (optional)} \\ & + \text{working capital allowance (optional)} \\ & - \text{deferred income taxes (optional)} \\ & - \text{deferred investment tax credit (optional)} \\ & + \text{construction work in progress (optional)} \end{aligned} \tag{5.15}$$

The rate base largely includes net plant in service plus adjustments that vary with regulatory commissions. The rate base consists of the investment costs that a utility is paying as return to its investors. The utility must, therefore, earn that return. The regulated return on rate base is calculated by computing the composite cost of capital to the utility, including the regulated return on equity.

Example 5.3. A utility has the following capitalization:

	\$ Millions
Long-term debt	1000
Preferred stock	100
Common equity	900
Short-term debt	15
	2015

The utility has paid \$109M in long-term debt interest, \$2M in short-term debt interest, and \$11M in preferred stock dividends. The utility should have a regulated return on equity of 14%. Calculate the composite cost of capital for use in a return-on-rate-base evaluation.

SOLUTION. Total interest and equity costs are:

	\$Millions
Long-term debt	109
Preferred stock	11
Common equity	$.14 \cdot 900 = 126$
Short-term debt	2
	248

The composite cost of capital is $248/2015 = .123$ or 12.3%/year. The regulated return on rate base should then be set at 12.3%/year by the commission.

Return-on-rate-base regulation involves calculating the utility operating income (revenue less operating expenses for fuel, operation and maintenance, book depreciation, and taxes) divided by the rate base.

Rate regulation proceedings may require 6 months to 2 years to complete. Utilities require several months of preparation prior to a rate request change submittal. The regulating agency (public utility commission) staff studies the submitted documentation and may request additional supporting materials from the utility. Public hearings are conducted under an administrative law judge. Finally, the public utility commissioners review the facts and issues of the case and render a rate request decision.

Rate regulation can be based on a historical test-year period. In this case, the latest historical financial performance of the utility is examined, and the required revenue is determined so as to provide the regulated return on equity or return on rate base. Electric rates that provide the required revenue over the historical test-year period are then determined. These rates are incorporated into the electric rate tariff to be applied in the future.

The historical test-year procedure is based on factual utility operating costs (depreciation, fuel, interest, etc.), that is, historical statistics. The disadvantage of this procedure is that the electric rates to be applied in the future are based on utility costs over a prior historical period and do not reflect changes in utility costs due to inflation, load demand and plant additions. Consequently, in an inflationary economy, the achieved rate of return is usually less than the value set by the commission.

Another method of rate regulation is based on a forecasted future test-year period. In this method, utility operating costs are forecast for the future year during which the new electric rate tariff is to apply. The revenues and rates are calculated to provide the regulated return on equity or rate base. With this method, known future changes can be incorporated and the effect of inflation can be approximately included. In addition, the actual achieved rate of return closely matches the regulated value. The disadvantage of this method is that the utility costs used in the revenue determination are not historically based, but on projections that are subject to prolonged debate.

Between these two extremes (historical test year and future test year) is a continuum of practical approaches. Some proceedings are based on a historical test year, but modified by known future changes such as a generating unit being placed into commercial service. Another approach is a hybrid method in which a 1-year forward future test-year method is used. The rate request proceeding is conducted and the issues of the case are evaluated. Because the proceeding consumes approximately one year (on the average), the projected future operating cost data becomes historical fact by the conclusion of the proceeding. A final adjustment is then made to the electric rates that reflects any changes from the projected future test year to the actual historical test year.

Example 5.4 (Financial Simulation). Assume that the current date is January 1, 1990. The example utility projects a peak load in 1990 of 6000 MW with 7500 MW of capacity. The utility is constructing a new coal-fired steam unit that is to be completed in 1993. The plant has \$905M in CWIP as of 1989, and another \$575M is forecast to be spent to complete the project. In the latter 1990s the utility is planning to add combustion turbines. During the 1990s, the utility plans to spend \$40M–\$50M/year on transmission and distribution plant additions.

The historical balance sheet and income statement for 1989 are shown in Tables 5.9 and 5.10.

The public service commission uses a hybrid historical–future test-year electric-rate proceeding method that leads to a 1-year lag between the actual historical data year used in the rate proceeding and the application year of the new rate. Thus, actual 1989 data are used to establish 1990 electric rates. The public service commission also establishes a 14% rate of return on common equity criteria for rate regulation.

Part A. Compute the electric revenues and average electric rate that the utility will be permitted to charge. The utility has a 90% automatic fuel cost, ¢/kWh adjustment. Assume a combined federal and state income tax rate of 40% and a tax depreciation of \$152,336. The 1989 energy sales were 34,410,000 MWh.

TABLE 5.9 Sample Balance Sheet 1989
(ALL VALUES IN THOUSANDS OF $)

Assets		Liabilities	
Utility Plant		Capitalization	
Electric production	2,100,000	Common stock outstanding	1,000,000
Other plant	1,609,200	Retained earnings	354,651
Gas plant	0	Common stock equity	1,354,651
Construction work in progress	767,425	Preferred stock	400,000
Total nuclear fuel	0	Long-term debt + current maturities	2,115,000
Total utility plant	4,476,626		
Less depreciation reserve electricity	980,555	Total capitalization	3,869,651
Less depreciation reserve gas	0		
Less amount of nuclear fuel	0	Current + accrued liabilities	
Net utility plant	3,496,070	Short-term debt	10,661
		Accounts payable + miscellaneous	87,789
Other property & investments	0		
		Total current + accrued liabilities	98,450
Current accrued assets			
Cash + equivalent	9,932	Deferred credits	
Accounts received + deferred debits	101,087	Accumulated deferred FIT	1,961
Materials and supplies	362,973	Accumulated deferred ITC	0
Total current assets	473,992	Total deferred credits	1,961
Total assets	3,970,063	Total liability + capital	3,970,063

TABLE 5.10 Sample Income Statement 1989
(ALL VALUES IN THOUSANDS OF $)

Income Statement		
Electric revenue	1,370,506	
Gas revenue	0	
Electricity, fuel adjustment revenue	89,635	
Other revenue	0	
Total		1,460,141
Electric operating expense		
Fuel and O&M	777,000	
Net purchased power	0	
Other production expenses	49,489	
Total electric production	826,489	
Gas production expenses	0	
Depreciation expense	92,819	
Federal income tax liability (FIT)	71,988	
Other taxes	148,368	
Deferred + adjustments	1,961	
Total operating expenses		1,141,625
Operating income	318,515	
AFDC–equity funds	43,804	
Net other nonoperating income	0	
Income before interest		362,319
Long-term interest	188,105	
Short-term + other interest	2,236	
AFDC–borrowed funds (credit)	37,364	
Net income	209,342	
Preferred dividends	45,000	
Available to common equity	164,342	
Common dividends	139,690	
Net income after dividends		24,651
Common shares year average	50,000,000	
Earnings per share	3.2868	
Dividends per share	2.7938	
Payout ratio	.8500	
Stock book value-/share	27.0930	
Stock market price-/share	29.5815	
Market-to-book ratio	1.0918	
Price-to-earnings ratio	9.000	
Return on common equity	164,342/1,354,651 = .1213	

SOLUTION. The revenue that the utility should collect for the historical 1989 test year is calculated based on a 14% return on equity (ROE). The regulated available income to common stockholders for the historical year is (in thousands of dollars):

$$\text{Available to common} = .14(\text{equity base}) = .14(1{,}354{,}651) = \$189{,}651$$

The revenue may be calculated as:

$$\begin{aligned}\text{Revenue} = {} & \text{electric production} + \text{book depreciation} + \text{other taxes} \\ & + \text{income taxes} - \text{AFDC}_{\text{equity}} + \text{interest} - \text{AFDC}_{\text{borrowed}} \\ & + \text{preferred dividends} + \text{available to common equity} \qquad (5.16)\end{aligned}$$

Income taxes are dependent on equity earnings. Substituting historical year results yields:

$$\begin{aligned}\text{Revenue} &= ((777{,}000 + 49{,}489) + 92{,}819 + 148{,}368 + 1961) \\ &\quad + \text{federal income tax (FIT) liability} \\ &\quad - 43{,}804 + (188{,}105 + 2{,}236) - 37{,}364 \\ &\quad + 45{,}000 + 186{,}651 \\ &= \$1{,}413{,}461 + \text{FIT liability}\end{aligned}$$

Taxes are calculated as:

$$\begin{aligned}\text{Taxable income} = {} & \text{revenue} - (\text{electric production} + \text{other taxes} \\ & + \text{interest} - \text{AFDC}_{\text{borrowed}} + \text{tax depreciation}) \qquad (5.17)\end{aligned}$$

$$\begin{aligned}\text{Taxable income} &= \text{revenue} - (777{,}000 + 49{,}489 + 148{,}368 \\ &\quad + 188{,}105 + 2{,}236 - 37{,}364 + 152{,}336) \\ &= \text{revenue} - 1{,}280{,}170\end{aligned}$$

The federal and state income tax rate is a composite 40%. Thus, the revenue may be calculated as:

$$\text{Revenue} = \$1{,}413{,}461 + .40\,(\text{revenue} - 1{,}280{,}170)$$

$$\text{Revenue} = \$1{,}502{,}322$$

The average electric rate based on 34,410,000 MWh sales is:

$$\begin{aligned}\text{Electric rate} &= \$1{,}502{,}322{,}000/34{,}410{,}000\ \text{MWh} = 43.66\ \$/\text{MWh} \\ &= 4.366\ ¢/\text{kWh}\end{aligned}$$

Part B. The electric rate determined in 1989 is applied to year 1990. The 1990 MWh sales are 35,440 GWh. The 1990 cost elements are (in thousands of dollars):

Fuel and O&M	\$850,000
Other production expenses	\$ 53,013
Depreciation expense	\$ 92,702
Deferred and adjustments of Taxes	\$ 2,335
Other taxes (property)	\$149,541
AFDC equity	\$ 53,338
AFDC borrowed	\$ 44,737
Capital expenditures	\$241,703
Change in working capital	\$ 33,473
Automatic fuel adjustment Revenue	\$ 34,907
Tax depreciation	\$155,897
Common stock dividend payout ratio	85%

Compute the achieved return on equity in year 1990.

SOLUTION. This solution requires an iteration because capital expenditures may need to be financed. These influence the interest and dividend charges, which then influence the amount of internal cash generation.

The first step is to prepare an income statement. The electric revenue from the rate tariff is 43.66 \$/MWh • (35,440,000 MWh) = \$1,547,310 (thousand). Fuel adjustment revenues of \$34,907 (thousand) are added to this to obtain total revenues of \$1,582,217 (thousand). Preliminary taxable income is:

$$
\begin{aligned}
\text{Income tax} &= .40\,(\text{revenue} - \text{deductible expenses})\\
&= .40\,(1{,}582{,}217 - (850{,}000 + 53{,}013 + 149{,}541\\
&\quad + 188{,}105 + 2{,}236 - 44{,}737 + 155{,}897)\\
&= \$91{,}265
\end{aligned}
$$

A preliminary income statement is shown in Table 5.11.

The funds report can then be written (as shown in Table 5.12) and the total funds from outside sources calculated.

Because \$149,260 (thousands) of outside funds are required to be financed, the annual financing costs (approximately \$15,000 thousands) reduces the net income. The income statement must be re-evaluated. After several iterations, the income statement, cash report and balance sheet are finalized as shown in Tables 5.13, 5.14, and 5.15.

Return-on-equity is calculated as the available to common equity divided by the end-of-year equity as 201,170/1,416,827, or 14.2%. Note that the achieved return on equity is slightly higher than the regulated value. This can happen even in an inflationary period. In this example, the growth in kilowatt-hour

TABLE 5.11 Preliminary Income Statement
(ALL VALUES IN THOUSANDS OF $)

Revenue total	$1,582,217
Electric operating expense	
Fuel and O&M	850,000
Other production expense	53,013
Depreciation expense	92,702
FIT liability	91,265
Other taxes	149,541
Deferred and adjustments	2,335
Operating income	343,361
AFDC–equity funds	53,338
Income before interest	396,699
Long-term interest	188,105
Short-term and other interest	2,236
AFDC–borrowed funds	44,737
Net income	251,095
Preferred dividends	45,000
Available to common equity	206,095
Common dividends	175,216
Net income after dividends	30,899

sales, 3%/year, causes the revenues to increase by 3%. Because major capital additions are not being placed into service, utility costs of service are increasing slightly less than 3%/year. Revenues growth slightly outpaces the utility cost growth, and the return on equity is slightly higher than the regulated or allowed value.

This case is reversed in 1993 when the large coal-fired generating plant is placed into commercial service, causing utility costs to increase. Depreciation

TABLE 5.12 Preliminary Cash Report
(ALL VALUES IN THOUSANDS OF $)

Funds from operations	
Net income after dividend	30,879
Depreciation	92,702
Deferred and adjustments	2,335
Total funds from operations	125,916
Total funds from outside sources	149,260
Total sources of funds	275,176
Application of funds	
Capital expenditures	241,703
Change in working capital	33,473
Total application of funds	275,176

TABLE 5.13 Income Statement
(ALL VALUES IN THOUSANDS OF $)

	1989	1990
Electric revenue	1,370,506	1,547,381
Gas revenue	0	0
Electricity, fuel adjustment revenue	89,635	34,907
Other revenue	0	0
Total	1,460,141	1,582,287
Electric operating expense		
Fuel and O&M	777,000	850,000
Net purchased power	0	0
Other production expenses	49,489	53,013
Total electric production	826,489	903,013
Gas production expenses	0	0
Depreciation expense	92,819	92,702
FIT liability	71,988	88,156
Other taxes	148,368	149,541
Deferred + adjustments	1,961	2,335
Total operating expenses	1,141,625	1,235,746
Operating income	318,515	346,541
AFDC-equity funds	43,804	53,338
Net other nonoperating income	0	0
Income before interest	362,319	399,879
Long-term interest	188,105	196,955
Short-term + other interest	2,236	1,228
AFDC-borrowed funds (credit)	37,364	44,737
Net income	209,342	246,432
Preferred dividends	45,000	45,263
Available to common equity	164,342	201,170
Common dividends	139,690	170,994
Net income after dividends	24,651	30,175
Common shares year average	50,000,000	50,445,800
Earnings per share	3.3868	3.9878
Dividends per share	2.7938	3.3897
Payout ratio	.8500	.8500
Stock book value-/share	27.0930	28.0861
Stock market price-/share	29.5815	35.0906
Market-to-book ratio	1.0918	1.2779
Price-to-earnings ratio	9.000	9.0000

expense is increased by the plant entering service thus increasing operating costs. AFDC has increased earnings during construction by acting as other income (equity funds) and as a deduction to interest expense. When the plant is placed in service, AFDC is no longer applied, which contributes to an earnings decrease. If the plant is efficient and reduces the overall system fuel costs, then the automatic fuel adjustment clause decreases the revenues. Thus, earnings

TABLE 5.14 Cash Report
(ALL VALUES IN THOUSANDS OF $)

	1989	1990
Beginning cash balance	10,000	9,932
Funds from operations		
Net income after dividend	24,651	30,175
Noncash expenses		
Depreciation	92,819	92,702
Amortization of nuclear fuel	0	0
Provisions for taxes	222,317	240,031
Total funds from operations	339,787	362,909
Funds from outside sources		
Equity securities	0	37,000
Long-term debt	65,000	112,000
Short-term instruments	− 19,338	1,014
Total outside funds	45,661	150,014
Provision for interest + dividends		
Interest	190,342	198,184
Dividends	184,690	216,257
Total sources of funds	760,480	927,363
Application of funds		
Capital expenses—general	156,425	199,239
Capital expenses—other	42,464	42,464
Capital expenses—gas	0	0
Capital expenses—nuclear fuel	0	0
Interest paid	190,342	198,184
Taxes paid	220,356	237,697
Dividends paid	184,690	216,257
Repayment of long-term debt	0	0
Change in working capital	− 33,729	33,473
Change in other property and investments	0	0
Total applied funds	760,548	927,314
Ending cash balance	9,932	9,981

experience a sharp drop in a year in which a capital intensive plant enters commercial service. Figure 5.3 illustrates the achieved return on equity over the 10-year period. Note that with a 1-year regulatory lag, the achieved return on equity falls to 3% during the year when the coal plant enters commercial service.

The earnings per share for the cases of a 1-year regulatory lag and no regulatory lag are presented in Figure 5.4. With no regulatory lag, the earnings per share exhibit a smooth growth trend. In the 1-year regulatory lag case, the earnings per share have a fluctuating trend and markedly decrease when the new coal plant enters commercial service.

Figure 5.5 presents the average electricity price for the case of 1-year regula-

TABLE 5.15 Balance Sheet as of December 31

(ALL VALUES IN THOUSANDS OF $)

Assets			Liabilities		
Utility plant			Capitalization		
Electric production	2,100,000	2,100,000	Common stock outstanding	1,000,000	1,032,000
Other plant	1,609,200	1,638,519	Retained earnings	354,651	384,827
Gas plant	0	0	Common stock equity	1,354,651	1,416,827
Construction work in progress	767,425	966,665	Preferred stock	400,000	405,000
Total nuclear fuel	0	0	Long-term debt + current maturities	2,115,000	2,227,000
Total utility plant	4,476,626	4,705,183			
Less depreciation reserve (electricity)	980,555	1,060,111	Total capitalization	3,869,651	4,048,827
Less depreciation reserve (gas)	0	0			
Less amount of nuclear fuel	0	0	Current + accrued liabilities		
Net utility plant	3,496,070	3,645,072	Short-term debt	10,661	11,675
			Accounts payable + miscellaneous	87,789	95,134
Other property & investments	0	0			
			Total current + accrued liabilities	98,450	106,809
Current + accrued assets					
Cash + equivalent	9,932	9,981	Deferred credits		
Accounts received + deferred debits	101,087	109,543	Accumulated deferred FIT	1,961	4,296
Materials and supplies	362,973	395,335	Accumulated deferred ITC	0	0
Total current assets	473,992	514,859	Total deferred credits	1,961	4,296
Total assets	3,970,063	4,159,932	Total liabilities + capital	3,970,063	4,159,932

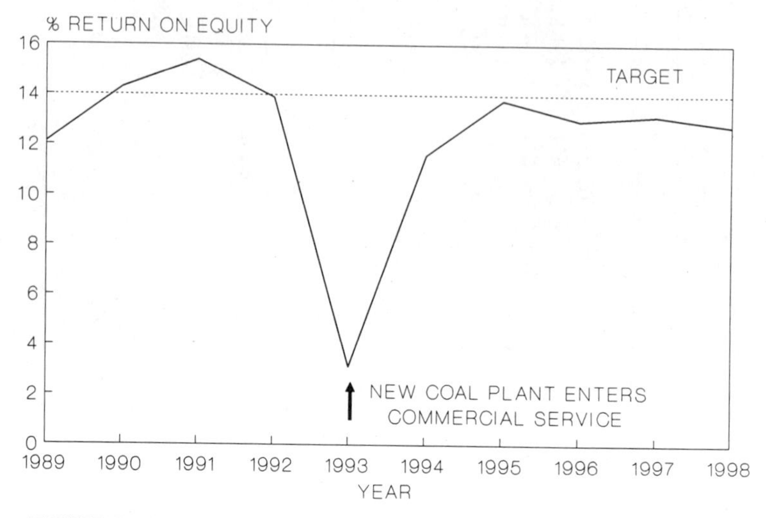

FIGURE 5.3 Sample utility return on equity over a 10 year period.

tory lag and no regulatory lag. In 1993, the coal plant enters commercial service. With no regulatory lag, the electric rates increase due to higher costs of service. With a 1-year regulatory lag, the rate increase is delayed by a year.

Since the utility financial performance markedly decreases when a capital-intensive plant enters commercial service, utilities have an incentive to reduce the time lag. This may be accomplished by a rate request immediately after the plant enters service, requesting immediate interim rate relief, or requesting

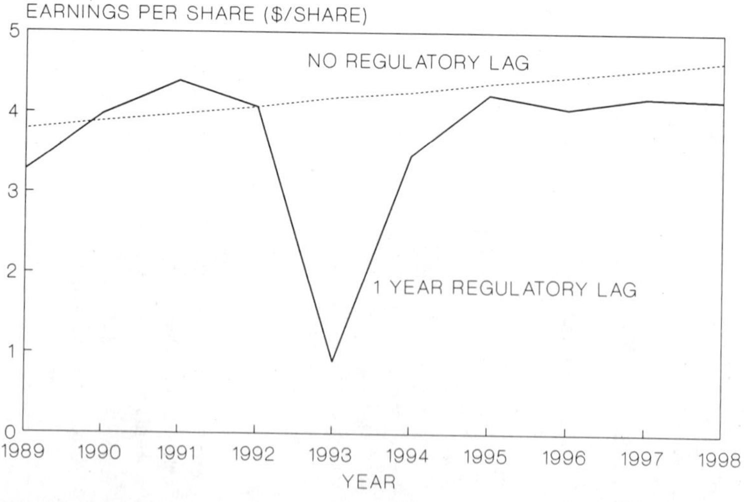

FIGURE 5.4 Sample utility earnings per share over a 10 year period (for a one-year regulatory lag and no regulatory lag).

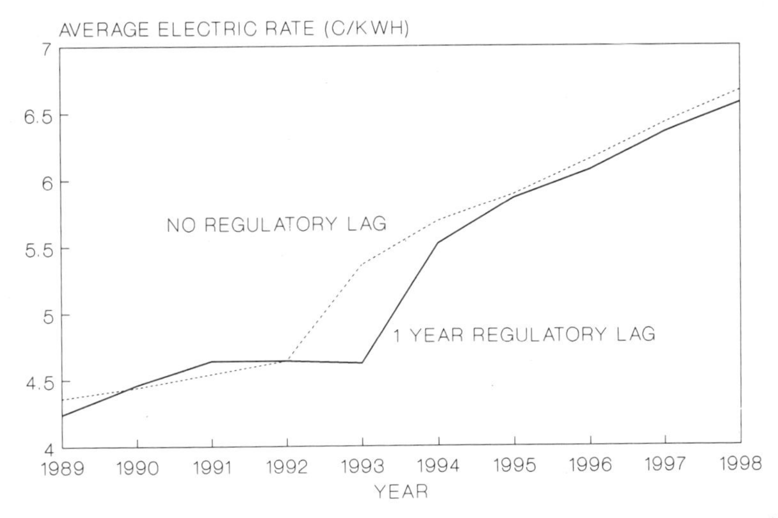

FIGURE 5.5 Average electric rate over a 10-year period (for a one-year regulatory lag and no regulatory lag).

a "pro-forma" adjustment to account for a plant entering commercial service in the near future.

Financial simulation is a very useful tool for planning. It provides information on future electric rates, external financing needs, and utility financial performance. For long-range planning, simplified financial models have proved useful. For short-range planning and budgeting, more detailed corporate financial models are usually employed.

5.2 AVERCH–JOHNSON BIAS

Regulated investor-owned utilities serve two masters, the electric consumer, whom the utility is chartered to serve at least cost, and the utility stockholders, who invest funds and who are entitled to a fair and reasonable return on their investment. Regulation by federal, state and local authorities provides a mechanism by which the utility can serve these two interests. However, regulation is an imperfect mechanism and can create decision biases.

One theoretical bias is the "Averch–Johnson bias." This theory suggests that utilities have a bias toward capital investment when the market-to-book ratio is greater than unity.

A simple example will illustrate this bias. An unregulated business is considering a $4M investment that produces a $1M/year savings in expenses. This investment has a 4-year investment pay-back. For illustration, consider that the investment is financed by sales of common stock. The simplified income statement and balance sheet for this unregulated business is shown in Table 5.16 both with and without the equity investment.

TABLE 5.16 Financial Performance of an Unregulated Business

	Income Statement		
	No Investment	With Equity Investment	Difference
Revenue	40	40	0
Expenses	28	27	−1
Interest	3	3	0
Earnings	9	10	+1

	Balance Sheet		
	No Investment	With Equity Investment	Difference
Plant and equipment	110	114	4
Current assets	10	10	0
Total assets	120	124	0
Equity	80	84	4
Debt	30	30	0
Current liabilities	10	10	0
Total liabilities	120	124	0

Assume that there are four million shares of common stock. The book value of common stock with no investment is $80M/4,000,000 shares, or $20/share. Assume that the stock is selling on the market at $16/share, or 80% of book value.

In the case of the $4M investment, the new stock shares are financed at $16/share. The total stock shares are then:

$$\text{Stock shares} = 4 \text{ million} + (\$4\text{M investment})/(\$16/\text{share})$$
$$= 4.25 \text{ million}$$

The earnings per share and return on equity after the investment are:

	No Investment	With Investment
Earnings per share	$9M/4 million = 2.25 $/share	$10M/4.25 million = 2.35 $/share
Return on equity	$9M/80 million = 11.25%	$10M/84 million = 11.90%

The earnings per share and return to equity improve with the investment. For a nonregulated business, this project is a strong candidate for implementation because it leads to improved earnings per share and increased stockholder net worth.

Suppose the market price of stock is other than 80% of book value. The earnings per share are plotted in Figure 5.6 as a function of the market-to-book ratio (4-year investment payback case label). A higher market-to-book ratio means fewer new shares need to be sold to raise the same amount of equity, and earnings per share increase with increasing market-to-book ratio.

Suppose, however, that the annual expense saving benefits are only $.3M/year rather than $1M/year. In this case (a 13-year investment pay-back period), the earnings per share decrease if the investment is made. The 13-year investment pay-back is a bad investment for the business.

Consider the same project for a regulated utility. Assume that the utility is regulated to a return on equity of 11.25% with no regulatory lag. In this case, the income statement and balance sheet are presented in Table 5.17. Note that the earnings are established based on the total utility equity times the 11.25% return on investment. The investment results in a revenue (or rate) reduction as the $1M expense saving is shared between the rate payer (in reduced rates) and the stockholder (in increased earnings).

The earnings per share and return on equity are computed as:

	No Investment	With Equity Investment
Earnings per share	$2.25	$2.22
Return on equity	11.25%	11.25%

Note that the earnings per share have *decreased* with the equity investment. A utility may be biased against this investment because it yields less earnings per share, which are a key stockholder financial indicator.

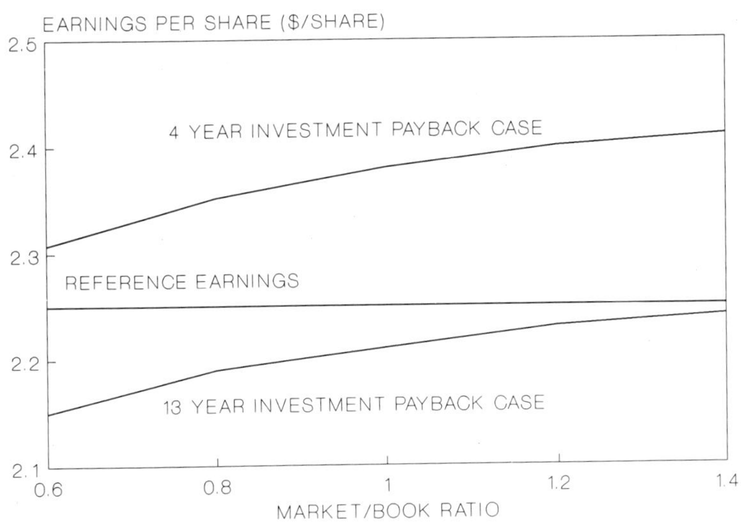

FIGURE 5.6 Unregulated business earnings-per-share results.

TABLE 5.17 Financial Performance of a Regulated Business

	Income Statement		
	No Investment	With Equity Investment	Difference
Revenue	40	39.45	− .55
Expenses	28	27	−1.00
Interest	3	3	0
Earnings	9	9.45 (.1125 • 84)	.45
	Balance Sheet		
	No Investment	With Equity Investment	Difference
Plant and equipment	110	114	4
Current assets	10	10	0
Total assets	120	124	0
Equity	80	84	4
Debt	30	30	0
Current liabilities	10	10	0
Total liabilities	120	124	0

The earnings per share results for other market-to-book stock ratios are presented in Figure 5.7. Note that the earnings per share increase with an increasing market-to-book ratio that is greater than unity. This affect arises due to the regulation of utility earnings. Note that the earnings-per-share curve of Figure 5.7 is the same for both a 4-year investment payback case or a 13-year investment payback case because of the regulation of utility earnings.

Figure 5.8 presents the earnings per share for a regulated utility for a $4M and $40M investment, both with a 4-year investment payback period. The larger the investment, the larger the earnings increase when the market-to-book ratio is greater than unity.

In this simplified characterization of a regulated utility (with no regulatory time lag), the earnings-per-share growth is independent of the overall economic benefits of the investment but is dependent on the magnitude of the investment. When the market-to-book ratio is greater than unity, the Averch–Johnson bias theory suggests that the utilities increase capital investment because it improves earnings per share. Similarly, when the market-to-book ratio is less than unity, utilities have an incentive to reduce capital investment.

5.3 REGULATORY INCENTIVES

The introduction of regulation creates other biases in addition to the Averch–Johnson bias.

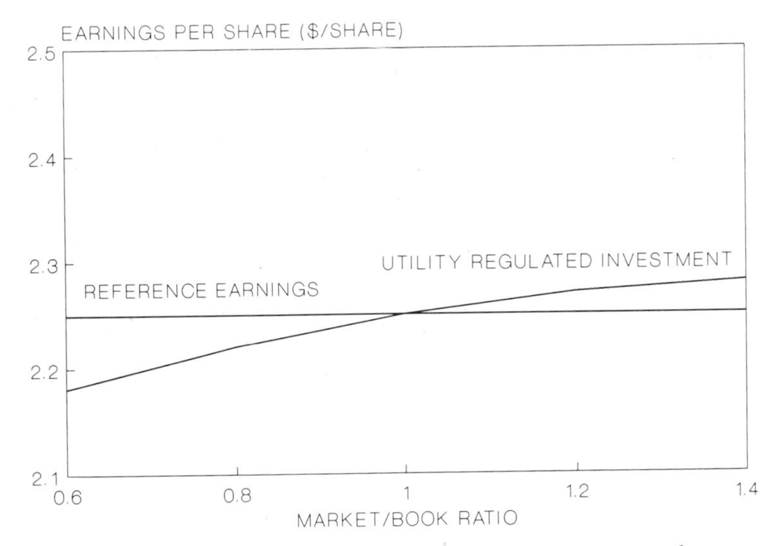

FIGURE 5.7 Regulated business earnings-per-share results.

One bias is evident with utility efficiency improvements when the utility has an automatic fuel cost pass-through and regulatory lag. Consider an example of improving the generation unit heat rate (efficiency) by 5% for the utility example of Section 5.1. Assume the utility implements the improvement in 1991 at a cost of $157M. The fuel cost saving in 1991 is $35M, and the investment payback is 157/35, or 4.48 years, an acceptable value for the electric

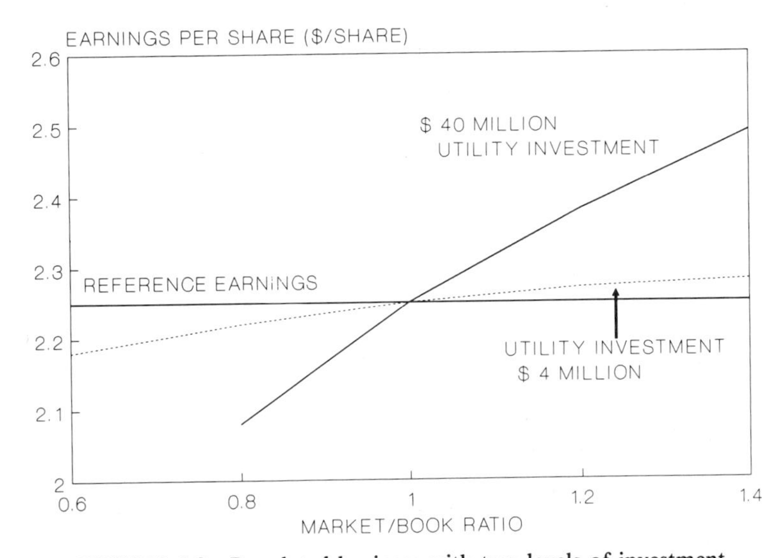

FIGURE 5.8 Regulated business with two levels of investment.

utility industry. In this example, the utility has an automatic 90% fuel cost (¢/kWh) revenue adjustment and a 1-year regulatory lag.

In 1991, when the project is implemented, the fuel cost decreases by $35M, relative to the reference case of no heat rate improvement, and the revenues decrease by $32M (90% fuel pass-through). The fixed costs of the $157M improvement (depreciation, taxes, interest) begin in 1991, and utility earnings significantly decrease in 1991. Figure 5.9 illustrates the earnings per share performance and contrasts the 1-year regulatory lag case with a case with no regulatory lag. In the long run, the earnings increase approaches $8M/year, which is the regulated 14% return-on-equity investment of $57M (36% of the $157M improvement is equity-financed). Figure 5.10 illustrates the cumulative present worth average return on equity for the case of no regulatory lag and for a 1-year regulatory lag. In the no regulatory lag case, the return on equity is the 14% regulated value. In the 1-year lag case, the cumulative achieved return on equity begins at −5% in 1991 and increases to 10% after 10 years.

Regulatory lag and the fuel-cost adjustment pass-through can result in projects receiving less than the regulated return on equity and in lower earnings per share. This effect is largely independent of the economic merits of the efficiency improvement. Doubling the efficiency benefits to a 10% efficiency improvement (a 2.3-year pay-back) improves the earnings in 1991 by only $3.5M.

With earnings erosion associated with this regulatory climate example, there is a utility bias against making efficiency improvements. Several state public utility commissions have instituted supplemental efficiency improvement incentive programs that permit accelerated utility cost recovery and improved earnings flow when efficiency improvements are made to the system.

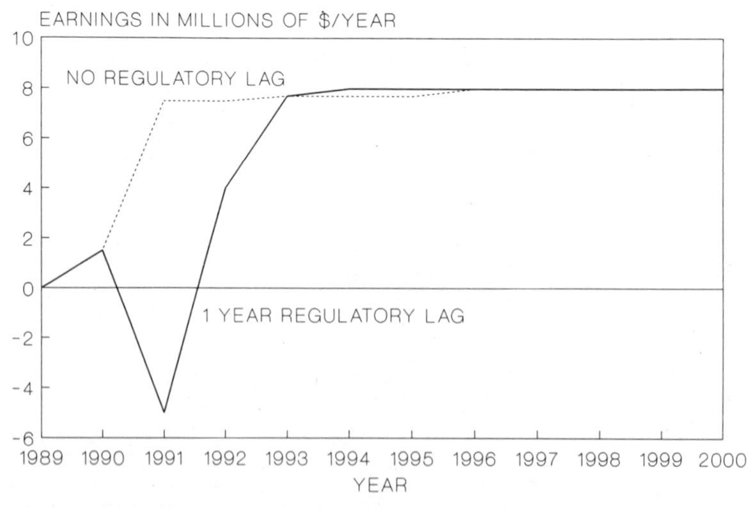

FIGURE 5.9 Effect of regulation and fuel pass-through on utility earnings from an efficiency investment.

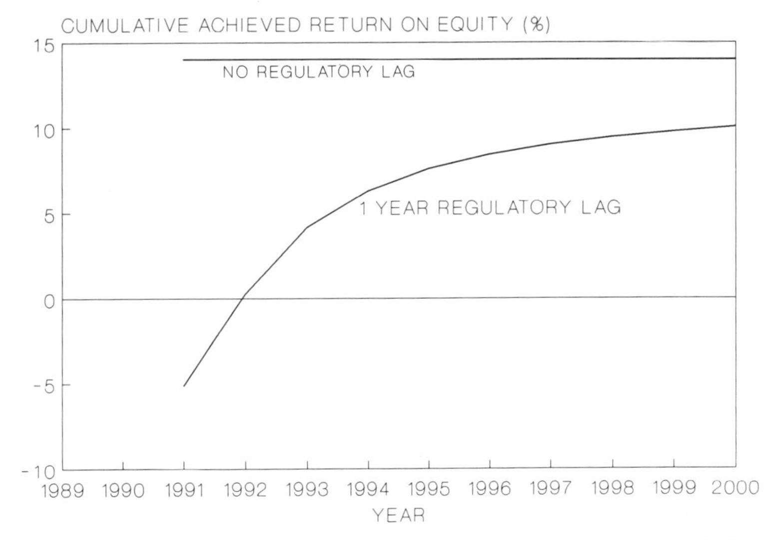

FIGURE 5.10 Effect of regulation and fuel pass-through on utility cumulative return on investment from an efficiency investment.

5.4 UTILITY INCENTIVES

Electric utilities are regulated to provide the lowest electric prices subject to a regulated return on investment. The price of electricity, ¢/kWh, may be higher or lower than the marginal cost of electricity. The marginal cost of electricity is the cost of providing one additional megawatt-hour of electricity to the consumer. The marginal cost, ¢/kWh, may include a capacity charge contribution. In periods of high generation reserve margins, the marginal cost of electricity is typically less than the average electricity price. In periods of low generation reserves, requiring the utility to install additional new capacity, marginal cost is typically greater than the average electricity price. Chapter 15 discusses the calculation of marginal costs.

Whenever the utility loses money (costs exceed revenue) in serving one segment of customers, the utility has three options: (1) lose money overall, (2) shift the cost burden of the losses to other customer segments, or (3) apply the additional cost burden of the losses back to the customer segment that is causing the losses. In the absence of a prompt rate regulation proceeding, the utility first encounters option 1. A rate proceeding may occur after a time lag. If the regulatory proceeding adjusts only the average electric rates, then option 2 results, with all the customer segments being affected. If the regulatory proceeding examines the cost of service of each customer segment, then option 3 results. Because cost-of-service accounting procedures are generally based on embedded costs rather than marginal costs, a combination of options 2 and 3 usually results. The same options are available when the utility makes additional earnings in serving one segment of customers.

The difference between the average electric price and marginal cost creates some utility incentives. In competitive nonregulated businesses, the average cost and marginal cost are numerically equal or nearly equal. In these businesses, biases are not present.

5.4.1 Conservation

As an example, consider the overall incentives to the utility and consumer with end-user (consumer) electric energy conservation. The costs are assumed to be:

Marginal cost of electricity	5¢/kWh
Price of electricity	6¢/kWh
Marginal cost of conservation	4¢/kWh

In the first case, the price of electricity is greater than the marginal cost. The marginal cost of conserving a kilowatt-hour is assumed to be less than the marginal cost of producing electricity.

Table 5.18 illustrates the incentives when the utility purchases the conservation equipment for the consumer. For every 5¢/kWh saved at the marginal cost, the utility loses 6¢/kWh that it would have collected by selling electricity through the rate tariff. The utility loses 5¢/kWh but the consumer gains 6¢/kWh. The overall cost to society (utility plus consumer) is a 1¢/kWh benefit. Therefore, the utility has a strong disincentive to purchase end-use conservation.

Consider the case when the consumer purchases the conservation equipment as shown in Table 5.19. In this case, the utility has a 1¢/kWh disincentive while the consumer has a 2¢/kWh positive incentive. Since the end-use consumer is permitted to implement conservation, the conservation will naturally take place, despite the detriment to the utility.

When the marginal cost of electricity is greater than the average price of electricity, the incentives change. Consider the case when the marginal cost of electricity is 7¢/kWh. Table 5.20 presents the conservation incentives. The utility has a 1¢/kWh incentive, the consumer has a 2¢/kWh incentive, and overall society has the sum of 3¢/kWh. In this case, the utility has an incentive to promote end-use conservation and is in an economic position to provide additional economic incentives.

It seems to be "fair" that the utility should return the benefits it receives to those customers that produce the benefits. In the example of Table 5.20, the customer should receive an additional 1¢/kWh. This could take the form of an incentive payment to the consumer, free financing, and so forth. With an incentive program, the consumer has a 3¢/kWh positive overall incentive to pursue end-use conservation. This further encourages other conservation projects that may be cost-justified up to 7¢/kWh.

While the concept of the utility providing the consumer with incentive payments is considered "fair" in the case of positive utility net benefits, it is not

TABLE 5.18 Sample Incentives for Conservation—Utility Purchased

	Utility		Consumer		Overall Society	
Gross benefits	Save marginal cost	5¢/kWh	Save electric rate	6¢/kWh	Save marginal cost	5¢/kWh
Costs	Cost of conservation	4¢/kWh		0	Cost of conservation	4¢/kWh
	Lost revenue	6¢/kWh				
Net benefits		−5¢/kWh		6¢/kWh		1¢/kWh

TABLE 5.19 Sample Incentives for Conservation—Consumer Purchased

	Utility		Consumer		Overall Society	
Gross benefits	Save marginal cost	5¢/kWh	Save electric rate	6¢/kWh	Save marginal cost	5¢/kWh
Costs	Lost revenue	6¢/kWh	Cost of conservation	4¢/kWh	Cost of conservation	4¢/kWh
Net benefits		−1¢/kWh		2¢/kWh		1¢/kWh

TABLE 5.20 Sample Incentives for Conservation—Consumer Purchased

	Utility		Consumer		Overall Society	
Gross benefits	Marginal cost	7¢/kWh	Electric rate	6¢/kWh	Marginal cost	7¢/kWh
Costs	Lost sales	6¢/kWh	Cost of conservation	4¢/kWh	Cost of conservation	4¢/kWh
Net benefits		1¢/kWh		2¢/kWh		3¢/kWh

necessarily considered "fair" when utility economic incentives are negative. Consider the case depicted in Table 5.19. The utility loss of 1¢/kWh should be "fairly" borne by the end-use conserving consumer. Otherwise two actions may occur: either the utility economic penalty is borne by all consumers or the consumer may pursue other conservation projects that incur costs of up to 6¢/kWh, which would yield negative overall societal benefit.

5.4.2 Incentive Sales

When the marginal costs are less than the electricity price, the utility has an incentive to promote additional sales. The net economic benefits from the additional sales (the difference between the price of electricity and marginal costs) can be "fairly" passed onto the purchaser of the additional electricity in the form of lower electricity rates.

5.4.3 Industrial Cogeneration

Industrial cogeneration of electricity had a renewed interest during the 1980s. Prior to this period, domestic U.S. cogeneration additions averaged 500 MW/year. In the 1980s, cogeneration additions have increased five times to approximately 2500 MW/year.

Special incentives result from cogeneration. Consider an example in which the marginal cost is less than the electricity price.

Marginal cost of electricity	5¢/kWh
Average price of electricity	6¢/kWh
Marginal cost of cogeneration	5.5¢/kWh

The marginal cost of cogeneration is expressed in ¢/kWh of electricity produced, assuming that a credit for any process steam has been included in the ¢/kWh value. (Chapter 6 discusses cogeneration economic analysis.)

Consider a situation in which the cogeneration facility displaces electricity that the industrial facility is using. Table 5.21 presents the incentives. Since the economic incentive is 0.5¢/kWh, the overall economic incentive for the industrial facility is to install cogeneration. The utility incentive, however, is −1¢/kWh and the overall societal benefit is −0.5¢/kWh. Since the cogenerator is permitted to build the facility, the natural economic outcome is for the cogeneration facility to be built and the utility to lose 1¢/kWh.

Recognizing this probable economic outcome with its inherent 1¢/kWh loss, utilities may reduce their electric rates to discourage the construction of a cogeneration facility. If the utility decreased its electric rates to 5.5¢/kWh, then the industrial facility's incentive to cogenerate decreases to zero, and the utility retains the electricity customer and loses only 0.5¢/kWh from the original 6¢/kWh price. From the utility perspective, a loss of 0.5¢/kWh (which is really a 50% reduction in "profit") by decreasing its rates to be competitive with a cogeneration project is better than a loss of 1¢/kWh if the project proceeds.

When the marginal cost is greater than the electric rate, the cogeneration incentive increases. Suppose that the marginal cost of electricity is 7¢/kWh. Table 5.22 presents the incentives. In this case, there are two options: the cogenerator can reduce in-plant electricity needs, or the cogenerator, under the Public Utility Regulatory Policy Act (PURPA), can sell the cogenerated power to the utility at marginal cost and simultaneously buy the power from the utility at the industrial rate tariff (6¢/kWh in this example). The overall societal net gain is the same regardless of which option is chosen. Since the PURPA option provides the greatest incentive to the cogenerator, the economic incentive to the utility for cogeneration is zero.

5.4.4 Incentives Conclusions

The difference between marginal costs and average electric rates creates differing incentives for the utility and the end-use electricity consumer. If electricity were priced on the basis of marginal costs, then these differing incentives would disappear.

However, electricity is priced on the basis of total embedded costs, not marginal cost. This assures that the regulated return on investment is achieved. Electric rate tariff structures are developed that provide "nearly" marginal cost price signals to end-use consumers and provide for full-cost recovery of the total embedded costs. As one example, the residential bill is comprised of a fixed monthly cost component and an energy (kWh) cost component. One "nearly" marginal cost residential rate includes a marginal cost-based energy (kWh) cost component and a fixed monthly cost component. The fixed component "adjusts" the total residential class revenues to equal the amount required for total embedded cost recovery of residential class costs.

The utility serves many interests, including those of the utility investor, the general electric consumer and individual electric consumers. Each of these interests has a stake in the outcome of a utility business decision. Depending on the issue being studied, each group may have a different perception on the merits of a utility decision. A "stakeholder analysis" is an analysis procedure in which the merits of the decision are examined from the perspective of each of the "stakeholders."

This analysis procedure permits an assessment of which stakeholders benefit from a utility decision and which receive no benefits or negative benefits. If a significantly impacted stakeholder receives negative benefits, then the utility decision may need to be modified to remedy this condition.

The advantage of a stakeholder analysis is that it permits concerns for each party to be identified and remedied early, thereby increasing the utility business decision success. Financial simulation and marginal cost analysis are key tools in stakeholder analysis.

TABLE 5.21 Cogeneration Incentives (In-Plant Needs Only)

	Utility		Consumer		Overall Society	
Gross benefits	Marginal cost	5¢/kWh	Electric rate	6¢/kWh	Marginal cost	5¢/kWh
Costs	Lost sales	6¢/kWh	Cost of conservation	5.5¢/kWh	Cost of conservation	5.5¢/kWh
Net benefits		−1¢/kWh		0.5¢/kWh		−0.5¢/kWh

TABLE 5.22 Sample Cogeneration Incentives

	Utility	Industrial Rate	PURPA	Cogenerator	Industrial Rate	PURPA	Overall Society	
Gross benefits	Marginal cost	7¢/kWh	7¢/kWh	Electric rates	6¢/kWh	7¢/kWh	Marginal cost	7¢/kWh
Costs	Lost sales buy-back	6¢/kWh	7¢/kWh	Cost of cogeneration	5.5¢/kWh	5.5¢/kWh	Cost of cogeneration	5.5¢/kWh
Net benefits		1¢/kWh	0		0.5¢/kWh	1.5¢/kWh		1.5¢/kWh

BIBLIOGRAPHY

Cicchetti, C. J., and J. L. Jurewitz. *Studies in Electric Utility Regulation*, Ballinger Publishing Company, Cambridge, MA, 1975.

Standard and Poor's, *Credit Overview, Corporate and International Ratings*, Standard and Poor's, New York, 1986.

PROBLEMS

1 The projected cost in 1990 dollars of a 500-MW combined-cycle power plant is \$650/kW. Assume that inflation averages 5%/year; the 5-year construction period has a per unit expenditure pattern of .10, .20, .30, .20, .20; construction starts in 1990; and the AFDC rate is 10%. Compute each year's annual expenditure including AFDC.

2 A utility steam-generating plant begins commercial operation in 1992. The capitalized cost for book purposes is \$1,000,000,000 and for tax purposes is \$900,000,000. Book life is 35 years and tax life is 20 years. What are both book and tax depreciation for the first 5 years of operation with tax depreciation calculated using the Tax Reform Act of 1986?

3 Accelerated tax depreciation reduces taxes during the early years of an investment. Taxes are higher in later years. For a book life of 35 years and a tax life of 20 years, calculate the per unit present worth of tax savings with a tax rate of 34% and a present-worth interest rate of 10%.

4 To demonstrate the application of fuel adjustment clauses, the reference year data for a sample utility system are given as follows:

Fuel Type	Capacity (MW)	Fuel Cost (\$/MBtu)	Heat Rate (MBtu/ MWh)	Operating Cost (\$/MWh)	Sales (Million MWH)	Total Cost (Million \$)
Hydro	500	—	—		2.0	
Nuclear	600	0.80	10.5		3.5	
Coal	2000	2.00	9.5		13.5	
Gas	1500	4.00	12.0		4.0	
Total	4600				22.5	

The empty columns can be quickly calculated. The changes for the current year are fuel costs of \$0.90, \$2.10, and \$4.40 for nuclear, coal and gas, respectively, and an increased total sales of 24 million MWh. Compute a \$/MBtu-based fuel adjustment and the dollar change to total revenues.

5 The solution of Equation (5.3) calculates the long-term debt percentage of total funds from outside sources. New long-term debt financing is \$211M.

Using Table 5.7, show the calculation for new preferred and common stock when the balance sheet shows \$321.4M of preferred stock and \$1318.6M of common stock.

6 A utility has the following capitalization structure in millions of dollars from its balance sheet:

Long-term debt	950
Preferred stock	200
Common equity	800
Short-term debt	50
Total	2000

The utility has paid \$85.5M in long-term debt interest, \$5M in short-term debt interest, and \$22M in preferred stock dividends. The utility wants to have a regulated return on equity of 13.5%. Calculate the corporate cost of capital for use in a return-on-rate-base evaluation.

6

INDUSTRIAL POWER GENERATION ECONOMICS

Industrial power generation has been an important factor in the history of electricity from the early beginnings of the power industry. As integrated electric utility companies grew throughout the twentieth century and provided reliable, low-cost power, industrial generation was largely replaced by power generated by electric utilities.

Today, industrial power generation competes with utility power generation when an industrial company needs both electricity and process heat. This coincident production of both electricity and process heat is termed cogeneration. Industrial cogeneration has long been used in the pulp and paper, textile, petroleum, chemical, and food industries (Gentner, 1988). Cogeneration can also be applied in the commercial sector, including hospitals and large office building complexes.

This chapter reviews cogeneration application considerations, the impact of the Public Utility Regulatory Policies Act (PURPA), and the economic evaluation methods applicable to nonregulated enterprises. The focus of this chapter is on nonutility ownership of industrial cogeneration, which is the dominant cogeneration market affecting the utility industry. This chapter also discusses the impact of cogeneration on the utility industry.

6.1 COGENERATION OVERVIEW

Cogeneration is defined as the coincident production of process heat and shaft power. Shaft power can be used to drive a pump, compressor, or, most often, an electric generator. Cogeneration is usually applied to topping cycles, where high-temperature energy is first used to produce shaft power. The exhaust energy that remains after the production of shaft power is of sufficiently high quality [typically 50 pounds per square inch absolute (psia) to 600 psia saturated, or superheated, steam] to be useful in thermal heat processes.

Figure 6.1 contrasts the fuel utilization effectiveness of a cogeneration cycle with that of a utility power generation cycle. The utility steam turbine power cycle, shown on the left, produces only shaft power. Based on the Second Law of Thermodynamics, the utility cycle must operate at a high temperature (boiler steam conditions), produce work, and reject heat to the surroundings (condenser). Overall utility steam turbine cycle efficiency is in the range of 35%. Utility combined cycle plants, which utilize gas turbine and steam turbine cycle (refer to Chapter 14 and Section 6.3), can achieve efficiencies in the range of 50%. The topping cogeneration cycle, shown on the right, generates power using high-temperature steam or gas ("topping cycle") and rejects its lower-temperature heat to the industrial process instead of a condenser. In the cogeneration case, as much as 85% of the fuel energy can be applied to producing useful process heat and shaft power.

One measure of the effectiveness of a cogeneration system in power production is the fuel chargeable to power. Consider two process heat cycles both supplying the same process steam as shown in Figure 6.2. The conventional boiler produces process steam only, and the system has auxiliary power requirements (AUX kW1), which are supplied by external power purchases. The cogeneration cycle produces both power (kW2) and process steam and also requires supplemental auxiliary power (AUX kW2).

The fuel chargeable to power is defined as the additional fuel output for cogeneration divided by the net incremental power output. It is measured in

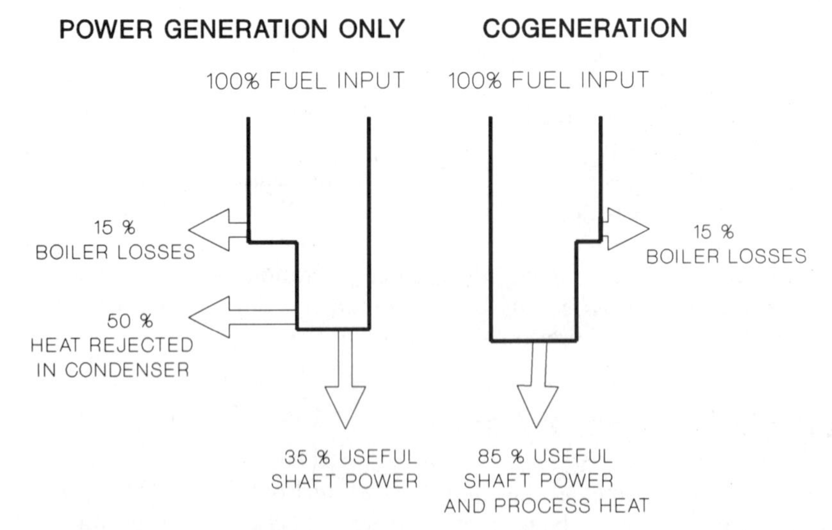

FIGURE 6.1 Contrast of power generation and cogeneration cycles.

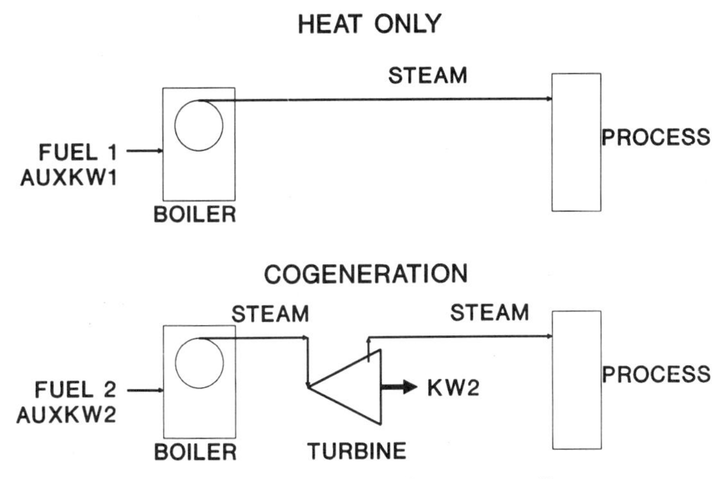

FIGURE 6.2 Fuel chargeable to power diagram.

Btu/kWh (British thermal units per kilowatt-hour), the same as a generating unit heat rate.

$$\text{Fuel chargeable to power} = \frac{\text{FUEL2} - \text{FUEL1}}{\text{kW2} - (\text{AUX kW2} - \text{AUX kW1})} \quad (6.1)$$

The fuel chargeable to power for topping steam turbine cogeneration cycles ranges from 4,500 to 10,000 Btu/kWh. For a condensing steam turbine in a utility generation plant, the fuel chargeable to power (heat rate) ranges from 9500 to 12,500 Btu/kWh. The cogeneration power cycle, therefore, can be a very fuel efficient method of producing electric power.

Cogeneration cycles can be segmented into two types: those involving steam-turbine cycles and those involving combustion-turbine cycles. Steam-turbine cycles use a steam boiler that can burn many different fuels including oil, gas, coal, wood, and refuse. The steam turbine is designed to extract steam to the desired process pressure levels, as illustrated in Figure 6.3.

Combustion-turbine (gas-turbine) cycles use a gas turbine that burns oil or gas fuel to produce power. The gas turbine exhaust, in the temperature range of 1000°F, is input to a heat-recovery steam generator (HRSG) that uses the hot gas turbine exhaust to produce steam. This steam passes directly to a process, or through a steam turbine with exhaust for a process steam purpose.

The power-to-heat ratio is a measure of the amount of shaft power produced per MBtu/hour of process steam. If an industrial process requires 500 MBtu/hour of heat and 20,000 kW of power, then the power-to-heat ratio is 20,000 kW/500 MBtu/hour, or 40. Steam-turbine cycles typically produce less power than do gas-turbine cycles per unit of heat supplied to the process. Top-

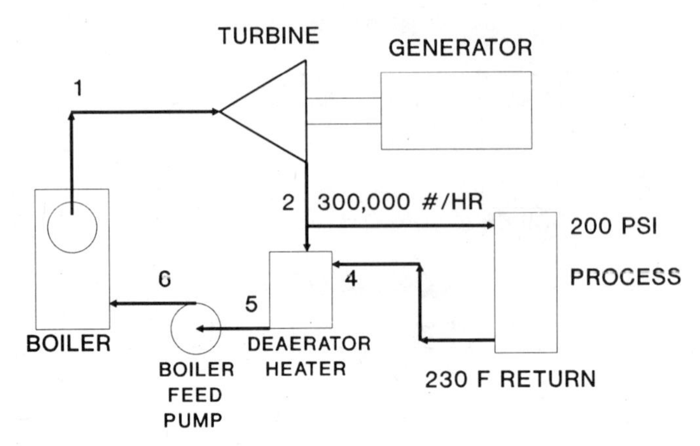

FIGURE 6.3 Flow diagram for a simplified cogeneration facility.

ping cogenerated steam-turbine cycles can effectively produce up to 85 kW/MBtu/hour; gas-turbine cycles can produce up to 300 kW/MBtu/hour.

Typical steam-turbine cogeneration cycles are applied when a low-cost fuel is available such as coal, wood chips, refuse, and so on, and the power-to-heat ratio is less than 85. Gas-turbine cycles are typically applied when the power-to-heat ratio is greater than 85, and when natural gas or oil is an available and economical fuel. Gas-turbine cycles typically have a lower capital cost than do steam-turbine cycles on a $/kW basis.

Process steam heat may also be required at several pressures and mass flow requirements. Process steam is generally needed as saturated steam to permit extraction by the industrial process for heat in the form of the heat of vaporization at constant temperature. Constant temperature heat extraction is important for process control. When significant superheat is present, the steam may be attemperated (mixed with water spray) to achieve a saturated condition. In some cases, the steam may be consumed in the industrial process. When this occurs, makeup water is returned to the system.

6.2 STEAM-TURBINE COGENERATION CYCLES

Steam-turbine cogeneration cycles provide lower power-to-heat ratios and burn lower cost fuels. The thermodynamic performance of steam turbine cogeneration cycles is calculated based on the Rankine heat cycle. Refer to the Appendix for a brief review of thermodynamics as applied to power-turbine cycles.

6.2.1 Initial Steam Conditions

Cogeneration power generation plants typically use lower inlet steam turbine steam conditions than do utility power generation plants. Typical steam conditions for new cogeneration plants are in the range of 800 psi to 1450 psi, in contrast to 2400 and 3600 psi for utility power generation plants.

The cogeneration system turbine size is typically much smaller than utility turbine sizes because the kilowatt power requirements of an industrial facility usually range from 10 to 100 MW. Because utilities serve the demands of thousands of customers, it is economical to size utility plants in the 300–800-MW range.

6.2.1.1 Simplified Steam Plant Example. Cogeneration plants are generally designed to meet specific industrial process heat and electricity needs. Detailed methods are used to design the plant and evaluate the thermodynamic performance. In this chapter simplified methods and designs are presented to illustrate the key concepts.

The next example is presented to illustrate a simplified steam-turbine cogeneration performance calculation. An actual plant design would use additional plant equipment not included in this example.

Example 6.1. An industrial facility requires 300,000/hour of steam at 200 psi (pounds per square inch), which is returned as process condensate at 230°F. The boiler is 85% efficient, and is to operate at 1400 psia (pounds per square inch absolute), 900°F. The turbine is 80% efficient and is to exhaust directly to the process. An open (deaerator) feedwater heater, shown in Figure 6.3, uses 200 psi of steam from the turbine exhaust. Feedwater heating improves the cycle efficiency. (In an actual case, the deaerator feedwater heater would not have this high a pressure.) Calculate the turbine kilowatt output, the Btu/hour of fuel consumed, the fuel chargeable to power, and the power-to-heat ratio.

SOLUTION. The first step is to calculate the turbine expansion line. Enthalpy at the turbine inlet (1400 psia, 900°F) is 1433 Btu/lb and entropy is 1.56 Btu/lb/°F. A theoretical isentropic expansion down to 200 psia leads to an enthalpy value of 1220 Btu/lb and a temperature of 420°F. The enthalpy change is 213 Btu/lb. An 80% efficient turbine achieves only 80% of that change, or 170.4 Btu/lb. Thus the enthalpy of the turbine exhaust is 1433–170.4 = 1,262.6 Btu/lb. This results in an entropy of 1.62 Btu/lb/°F and a temperature of 490°F.

The second step is to calculate the open feedwater heater heat balance. In this heater, the flow from the turbine extraction and from the process return are mixed together. Let M_T be the turbine flow. The feedwater heater energy balance is:

$$h_5\,(M_T) = 300{,}000\,h_4 + (M_T - 300{,}000)\,h_2$$

where h_5 is the enthalpy of state 5 in Figure 6.3. The heater output is designed to be a saturated liquid at 200 psi.

The enthalpy of point 5 on Figure 6.3 is 355 Btu/lb. The enthalpy of the returned saturated liquid at 230°F (point 4) is 198 Btu/lb.

The heat balance on the heater yields:

$$355\ M_T = 300{,}000\ (198) + (M_T - 300{,}000)\ 1262.6$$

$$M_T = 351{,}895\ \text{1b/hour}$$

The boiler feed pump increases the pressure from 200 to 1400 psia. The enthalpy change is a function of the specific volume, pump efficiency (75%), and the pressure change. The enthalpy change is approximated as the product of the specific volume times the pressure change times a 144/778 conversion factor.

$$\triangle h_{56} = .02 \cdot (1400 - 200) \cdot 144/778/.75 = 6\ \text{Btu/lb}$$

The enthalpy into the boiler is:

$$h_6 = 355 + 6 = \text{Btu/lb}$$

The boiler fuel input is:

$$\text{Fuel input} = (1433 - 361)\ \text{Btu/lb} \cdot \frac{351{,}895\ \text{lb/hour}}{.85\ \text{efficiency}}$$

$$= 443.8\ \text{MBtu/hour}$$

The turbine power output is the turbine enthalpy change times the mass flow divided by the Btu/hour-to-kilowatt conversion factor, 3412 Btu/kWh.

Gross turbine power output =

$$(170.4\ \text{Btu/lb}) \cdot \frac{351{,}895\ \text{lb/hour}}{3{,}412\ \text{Btu/kWh}} = 17{,}574\ \text{kW}$$

The generator electrical output is assumed to be 96% of the turbine mechanical shaft output, or 16,871 kW.

The boiler feed pump requires electric power to operate. Assuming a 96% efficient motor, the kW requirements are:

$$6\ \text{Btu/lb} \cdot (351{,}895\ \text{lb/hour})/3{,}412/.96,\ \text{or}\ 636\ \text{kW}$$

The net power output of the plant is the generator output less the boiler feed pump input, 16,871 − 636, or 16,234 kW.

The heat absorbed by the industrial process is the enthalpy change times the mass flow:

$$\text{Net heat to process} = (1{,}262.6 - 198) \cdot (300{,}000) = 319.4\ \text{MBtu/hour}$$

The fuel chargeable to power can be calculated by first calculating the boiler fuel input for a conventional boiler of 85% efficiency supplying 319.4 MBtu/hour to the industrial process.

$$\text{HEATI} = 375.8 \text{ MBtu/hour}$$

The fuel chargeable to power is calculated using Equation (6.1) as:

Fuel chargeable to power =

$$\frac{(443.8 - 375.8) \cdot \text{E06 Btu/hour}}{16{,}234} = 4194 \text{ Btu/kWh}$$

The power to heat ratio is:

$$\text{Power to heat ratio} = \frac{16{,}234 \text{ kW}}{319.4 \text{ MBtu}} = 50.8 \text{ kW/MBtu}$$

The power-to-heat ratio is shown in Figure 6.4 for other steam turbine throttle conditions. This graph is prepared by fixing the turbine expansion line and end-point conditions. In this way, the heat to process is held fixed. The throttle temperature corresponding to the desired throttle pressure is determined so as to lie on the expansion line. The power-to-heat ratio increases with increasing steam conditions.

Figure 6.5 presents the results of the previous example for a range of process

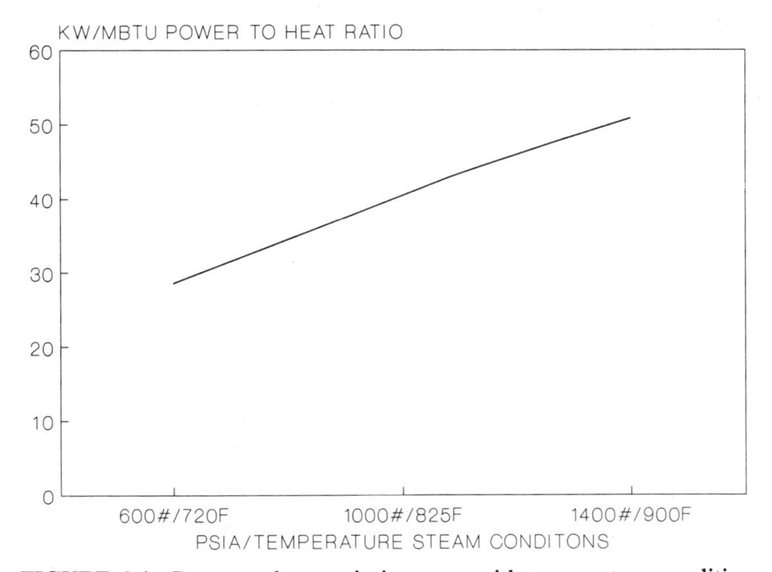

FIGURE 6.4 Power-to-heat ratio increases with temperature conditions.

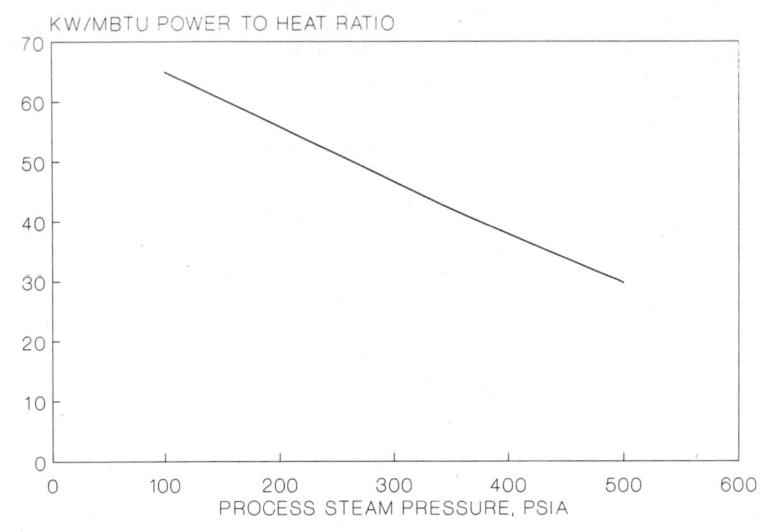

FIGURE 6.5 Power-to-heat ratio decreases with increasing process pressure.

steam conditions with 1400 psia, 900°F turbine conditions. The power-to-heat ratio decreases as process steam pressure conditions increase because at higher pressures there is less available enthalpy drop.

6.2.2 Reheat and Feedwater Heating

Cogeneration steam-turbine cycles seldom include a reheat section as do typical utility plant designs. If the cogeneration process pressure is 50 psi or greater, usually there is not enough available enthalpy after reheat to justify additional capital expense of the reheat section. In utility applications in which the turbine exhausts to 1 or 2 psia, there is a large enough enthalpy drop after the reheat to justify additional expense.

Only one feedwater heater was included in Example 6.1. In many industrial facilities, several process pressure levels are present and are the logical values to use in feedwater heating. Typically, three stages of feedwater heating are economical for steam-turbine cogeneration cycles (Kovacik, 1984).

6.2.3 Power Augmentation

In steam-turbine cogeneration cycles with no condensers, the power-to-heat ratio is limited to approximately 85 kW/MBtu net heat to process. This is dependent on process steam conditions, turbine inlet steam conditions and feedwater cycle, as illustrated in Figures 6.4 and 6.5. The power output may be less than the electrical needs of the industrial facility.

Electrical power generation can be increased by expanding part of the tur-

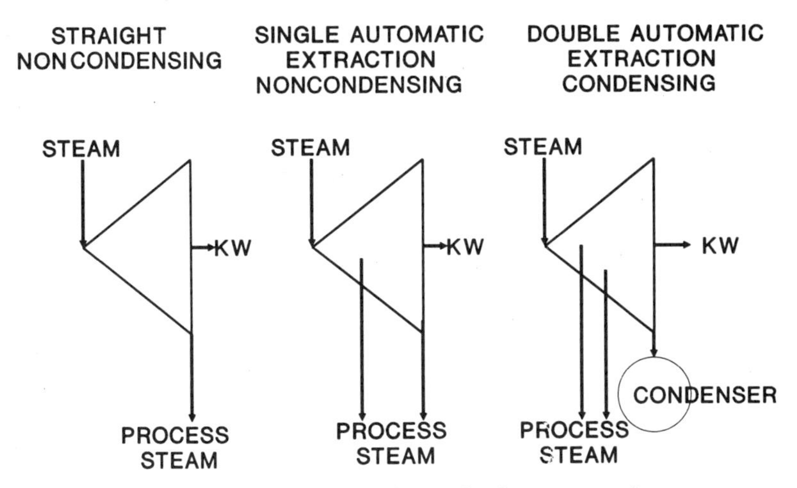

FIGURE 6.6 Steam-turbine cycles for cogeneration.

bine flow to a condenser (at 1 or 2 psia). Figure 6.6 illustrates three steam-turbine cogeneration cycles. Turbine 1 provides a single pressure level to the industrial process. Turbine 2 provides two pressure levels to the industrial process by extracting steam from the turbine. Turbine 3 provides two process pressure levels with part of the steam further expanded to a condenser to provide additional power output.

The fuel chargeable to power for a noncondensing turbine cycle is approximately equal to the 3412-Btu/kWh conversion factor divided by the boiler efficiency. This is because the heat rejected by the turbine is used in the industrial process.

For a condensing turbine cycle, the fuel chargeable to power for that fraction of steam expanded to the condenser is typical of a utility plant heat rate of 11,000–13,000 (for typical cogeneration plant steam conditions). To increase the power output from a cogeneration cycle (process heat demands are fixed), increased mass flow is introduced into the turbine and expanded to a condenser. The result is an increase in power-to-heat ratio but at the expense of a higher fuel chargeable to power and lower thermal efficiency.

Although condensing power is not as energy efficient as power from a noncondensing cycle, the economics can be favorable because of:

- Reduced utility demand charges for electricity.
- The availability of low cost fuels such as wood, coal, refuse, or process by-products.
- Excess thermal energy available from an industrial process can be used to generate steam.
- Reliability of other sources of power is in question.

6.2.4 Typical Steam Cycle Costs

Table 6.1 presents typical total installation costs for a noncondensing steam-cycle cogeneration plant with 150 psig (pounds per square inch gravity) process heat requirements and 200,000-lb/hour process steam requirement. These costs include "allowance for funds used during construction" (AFDC). If the plant size were to double, plant costs increase by approximately 75%.

Annual operating labor and maintenance cost estimates are also presented in Table 6.1. Doubling the plant size increases the operating labor by approximately 25%. The average annual maintenance cost is estimated at 3% of the installed cost per year.

6.3 GAS-TURBINE CYCLES

Gas-turbine cycles provide higher power-to-heat ratios than do steam-turbine cycles at favorable fuel chargeable to power. Figure 6.7 illustrates four typical gas-turbine cycles. Figure 6.7*A* is the simplest gas-turbine cycle configuration. The turbine exhaust at approximately 1000°F is input into a heat-recovery steam generator (HRSG) to produce steam at the required industrial process steam pressure. The gas-turbine shaft power drives an electrical generator.

In Figure 6.7*B*, the HRSG is designed to produce steam at elevated pressure and temperature conditions (typically 750–950°F). The steam may be expanded through a steam turbine that exhausts to the process steam header.

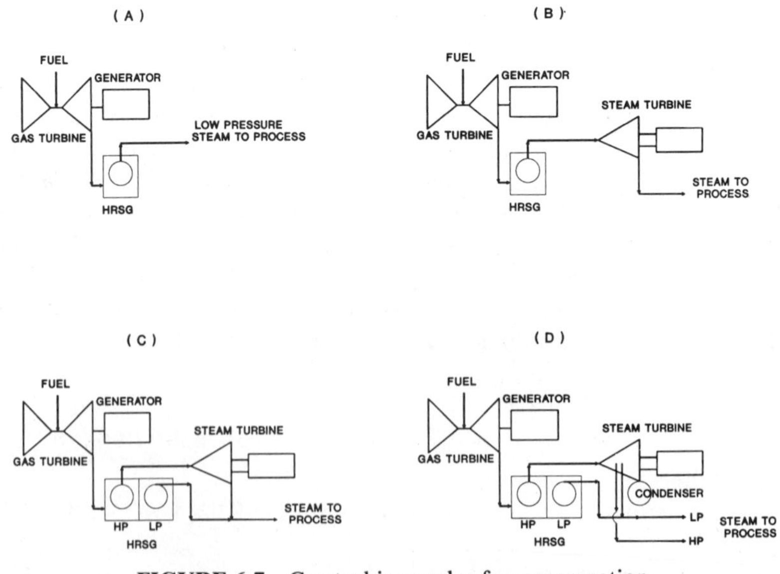

FIGURE 6.7 Gas-turbine cycles for cogeneration.

TABLE 6.1 Steam-Turbine Cycle: Typical Industrial Cogeneration—Total Installed Costs[a]
(MILLIONS 1990 \$; PROCESS HEAT AT 150 PSIG)

System	Stoker Coal No. SO_2 Scrubber	Stoker Coal SO_2 Scrubber	Gas-Fired Packaged Boiler	Municipal Refuse With Stack Cleanup	Gas-Fired Package Boiler No Generation
Fuel input (MBtu/hour)	289	289	289	300	245
Output (kW)	9,350	9,150	9,350	9050	0
Steam production @ 150 psi/kW/hour	200	200	200	200	200
Total installed cost (M $)	29.6	36.7	19.6	40.0	12.3
Operating labor costs (M $/year)	1.9	1.9	1.2	1.9	.8
Maintenance cost % of installed cost/year	3%	3%	3%	3%	3%

[a] Includes allowance for funds used during construction (AFDC), which is 15% of plant costs. Site location is in U.S. Gulf Coast labor area.

The combined use of gas-turbine and steam-turbine cycles is referred to as a combined cycle.

Figure 6.7*C* illustrates the configuration of a multiple-pressure HRSG. Multiple-pressure HRSGs provide increased recovery of the gas-turbine exhaust energy compared to a single-pressure unfired unit. This complexity decreases the fuel chargeable to power by 10–15%.

Figure 6.7*D* includes a condensing section on the steam turbine. The condensing section permits increased power-to-heat ratio but at the expense of higher fuel chargeable to power.

6.3.1 Heat-Recovery Steam Generators

Two basic types of heat recovery steam generators are unfired and supplementary fired. The unfired HRSG uses only energy in the gas-turbine exhaust. Hence steam production is a function of the mass flow and exhaust temperature of the gas turbine.

In a supplementary fired HRSG, fuel is burned to increase the exhaust temperature of the gas turbine. The exhaust from a gas turbine typically contains sufficient excess air that can be used for the combustion process. Since the combustion air is significantly preheated, the incremental fuel required is typically less than that required in an ambient-air steam generator to generate the same amount of steam.

A single-pressure level HRSG typically comprises three sections, as illustrated in Figure 6.8. Although this figure illustrates a forced circulation HRSG, natural circulation HRSGs are more commonly used in the United

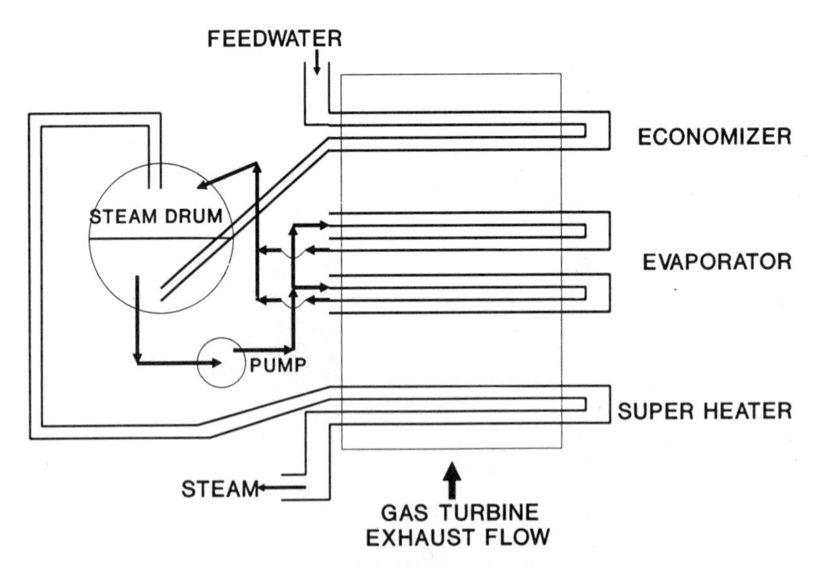

FIGURE 6.8 Basic elements of a single-pressure heat-recovery steam generator (HRSG).

States. The gas-turbine exhaust enters the HRSG in the superheater section and exits through the economizer section. The feedwater enters the HRSG in the economizer, which raises its temperature level almost to that of a saturated liquid at the design pressure of the steam drum.

In the evaporator section, additional energy is absorbed by the fluid and steam is generated. Steam and water separate in the steam drum. The saturated steam from the steam drum is further heated in the superheater. While the water–steam mixture is being heated, the converse is occurring in the gas turbine exhaust path. The gas typically enters the HRSG at 1000°F and leaves at 250–500°F.

Figure 6.9 presents an example of a temperature energy recovery diagram of an unfired single-pressure HRSG providing 850 psia, 800°F steam for use in the steam-turbine portion of a combined cycle. Gas flow from the combustion turbine exhaust enters at 1000°F. Feedwater enters at point *A* at 230°F with an enthalpy of 198 Btu/lb. The final enthalpy of the steam at 825°F is 1396 Btu/lb, for a total enthalpy change of 1198 Btu/lb. The feedwater is heated in the economizer to 525°F with an enthalpy of 510 Btu/lb, (point *B*), giving the fraction of total energy from feedwater heating of (510 − 198)/(1198), or 26%. The saturated steam leaves the evaporator (point *C*) with an enthalpy of 1199; the fraction of energy used to this point is (1199 − 198)/1198, or 84%. The superheater adds an additional 16% of the total energy in increasing the steam temperature from 525 to 800°F. These points are plotted in Figure 6.9.

When the HRSG is designed to generate more steam, more energy is recovered and the gas-turbine exhaust leaves at a lower temperature. Since the spe-

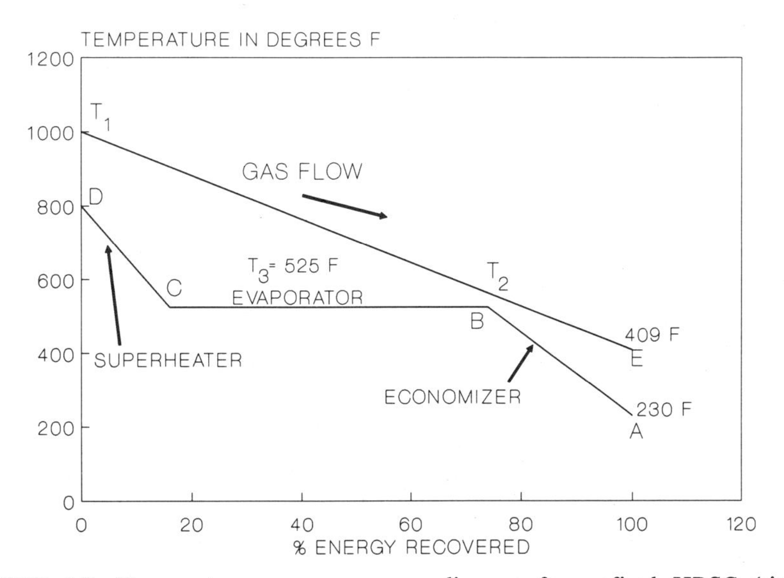

FIGURE 6.9 Temperature energy recovery diagram for unfired HRSG (single-pressure).

cific heat of air and combustion products can be approximated as an "ideal gas," a straight line can be drawn from the gas-turbine conditions of 1000°F at point *D* to the HRSG stack temperature at point *E*. The stack temperature can be decreased (ideally) only to the point where the gas line intersects the steam line at point *B*. If the gas line is below the steam line, then no heat transfer can take place. The intersection of point T_2 and point *B* is the limiting point, or "pinch point," at maximum steam production. Because a finite temperature difference must exist to transfer heat from the gas to the steam, point T_2 is typically 10–30° above point *B*.

The HRSG effectiveness is defined as a measure of the relative gas temperature difference in the superheater and evaporator sections of the cycle as:

$$E = \text{Effectiveness} = \frac{T_1 - T_2}{T_1 - T_3}$$

The maximum steam generation can then be computed by applying an energy-balance equation:

$$m_{\text{STM}} = \frac{m_{\text{GAS}} \cdot C_p \cdot (T_1 - T_3) \cdot E \cdot (1 - L)}{h_{\text{STM1}} - h_{\text{LIQ3}}} \tag{6.2}$$

where m_{STM} = steam flow in lb/hour
m_{GAS} = gas turbine flow in lb/hour
C_p = specific heat of the gas in Btu/lb°F
T_1 = gas turbine exhaust gas temperature to the HRSG
T_3 = saturation temperature of the steam
E = HRSG effectiveness
L = fraction of radiation and unaccounted losses
h_{STM1} = enthalpy of the superheated or saturated steam at the design pressure and temperature
h_{LIQ3} = enthalpy of the saturated liquid steam at the design pressure

Example 6.2. Calculate the maximum steam production for a HRSG having the gas-turbine exhaust temperature of 1005°F, 1,000,000 lb/hour, a Cp = .265, and 800 psia, 800°F steam conditions. Assume 92% HRSG effectiveness and 1.5% radiation loss.

SOLUTION. The enthalpy of the steam is:

h_{STM1} = 1.399 Btu/lb
h_{LIQ3} = 510 Btu/lb
T_3 = 518°F

$$m_{\text{STM}} = \frac{1{,}000{,}000\ (.265)(1005-518)(.92)(.985)}{(1399 - 510)}$$

$$m_{\text{STM}} = 131{,}552 \text{ lb/hour}$$

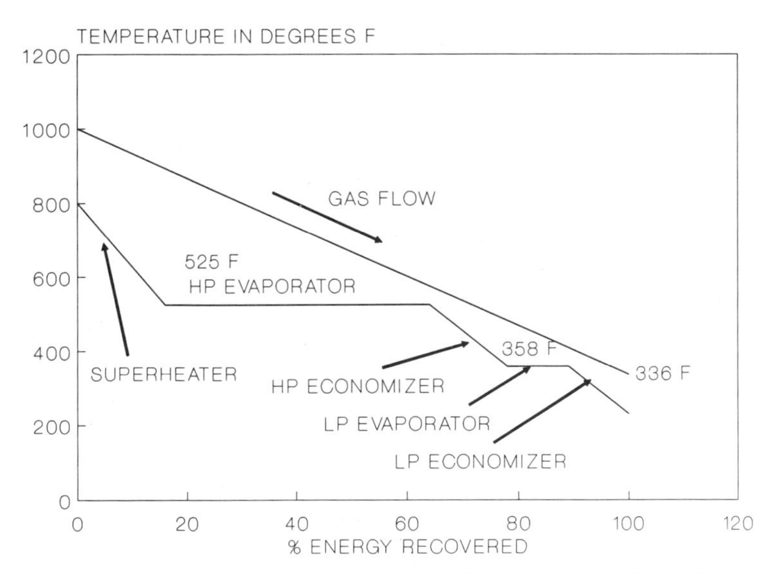

FIGURE 6.10 Temperature energy recovery diagram for unfired HRSG (two-pressure).

Figure 6.10 presents a temperature energy-recovery diagram for a two-pressure level unfired HRSG. A two-pressure level HRSG, shown in Figure 6.7*C* and 6.7*D*, is applicable when the lower-pressure steam may be used for process heat or admittance to a steam turbine. By using two pressure levels, the single pinch point can be replaced by two less limiting pinch points. Consequently, more heat is extracted from the gas as it is further cooled and more steam is produced.

6.3.2 Supplemental Fired HRSG

If more steam is needed, the HRSG can be supplementary fired. Supplemental firing increases the gas temperature prior to entering the superheater section of the HRSG.

Supplementary firing is achieved by placing an auxiliary burner in the exhaust gas duct prior to the superheater section of the HRSG. When the gas temperature is less than about 1700°F, the HRSG is essentially a convective heat exchanger similar in design to an unfired HRSG. For gas temperature above 1700°F (fully fired design), HRSG design is similar to that of a power boiler with radiant and convective heat-transfer surfaces. A fully fired HRSG can typically generate six to seven times the steam available from an unfired HRSG (McMahan, 1987).

The fuel input to raise the gas temperature can be approximated as:

$$\text{Fuel Btu/hour (LHV)} = m_{\text{gas}} \cdot C_p(T_1 - T_{\text{EXH}}) \tag{6.3}$$

where m_{GAS} = gas-turbine exhaust flow in lb/hour
C_p = specific heat of the gas in Btu/lb°F
T_1 = desired final temperature into the HRSG in °F
T_{EXH} = gas-turbine exhaust temperature
LHV = lower heating value of fuel

The lower heating value (LHV) excludes the water condensation heat produced during combustion. The heat of condensation generally is not recovered because the combustion products are exhausted to the stack at temperatures above 212°F. However, U.S. practice is to include the heat of condensation in boiler efficiency calculations, fuel heat contents, and so on. Including the heat of the condensation is referred to as the "higher heating value" (HHV). The ratio of the HHV to LHV for natural gas is typically 1.11, 1.06 for oil, and 1.03 for coal.

Example 6.3. Calculate the amount of fuel required to increase the gas-flow temperature from 1005°F, 1,000,000 lb/hour to 1400°F.

SOLUTION. The C_p of the exhaust gas in the 1005°F–1400°F range is .285 Btu/lb°F. The supplemental fired fuel is:

$$\begin{aligned} \text{Fuel Btu/hour} &= 1{,}000{,}000 \cdot .285 \cdot (1400 - 1005) \\ &= 112.6 \text{ million Btu/hour (LHV)} \end{aligned}$$

The temperature–energy-recovery diagram for this case and that of Example 6.2 is presented in Figure 6.11. For supplementary fired HRSG designs, the limit could become the exhaust stack gas temperature rather than the pinch point. The HRSG effectiveness decreases with supplemental firing because of a higher gas temperature at the pinch point with supplemental firing as can be noted in Figure 16.11. If the fuel contains sulfur, a gas temperature margin above the dew point of sulfuric acid should be maintained to prevent sulfuric acid corrosion in the HRSG economizer.

Example 6.4. Calculate the steam flow to process for a supplementary fired HRSG (Example 6.3) for steam at 800 psia, 800°F. Contrast the fuel saved by supplemental firing to generating the supplemental steam in an ambient boiler of 85% efficiency. Assume an 88% HRSG effectiveness.

SOLUTION. The steam generation with supplemental firing is calculated on the basis of Equation (6.2):

$$\begin{aligned} m_{STM} &= 1{,}000{,}000 \cdot .985 \\ &\quad \cdot [.265 \cdot (1005 - 518) + .285 \cdot (1400 - 1005)] \cdot \frac{.88}{1399 - 510} \\ &= 235{,}596 \text{ lb/hour} \end{aligned}$$

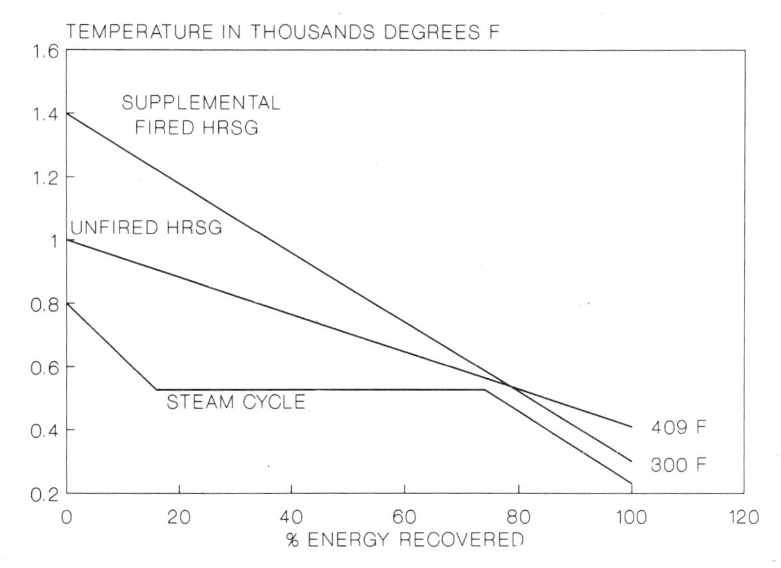

FIGURE 6.11 Temperature energy-recovery diagram for unfired and supplementary-fired HRSG.

Supplemental firing generated 235,596 − 131,552 lb/hour, or 104,044 lb/hour.

Production of the supplementary steam with an ambient air-fired boiler would require heating the feedwater from 230°F (H = 198 Btu/lb) to the final steam conditions.

$$\text{Ambient boiler fuel Btu/hour} = 104{,}044 \text{ lb/hour} \cdot \frac{1399 - 198}{.85}$$

$$= 147.0 \text{ million Btu/hour (HHV)}$$

where HHV = designation for the fuel higher heating value. Thus, supplemental firing of the HRSG saved (147.0 − 112.6 • 1.11) MBtu/hour, or 22.0 MBtu/hour, a 15% fuel savings (a 1.11 conversion factor was used to convert the HRSG fuel to higher heating value).

When the process steam conditions are lower, the present advantage of supplemental firing decreases.

6.3.3 Cycle Configurations

Steam turbines traditionally have been custom manufactured to meet specific facility requirements. Steam-turbine megawatt capability, heat rate, steam extraction points, and other factors are usually specified in the design and manufacturing phase. Gas turbines, however, are traditionally manufactured on a fixed-model design basis. A manufacturer typically has several models with specified power outputs and efficiencies. Gas-turbine cogeneration cycles are usually designed around a specified gas-turbine model.

Several alternative cycles can be constructed for a cogeneration facility using various combinations of gas turbines, HRSGs, and steam turbines. Figure 6.12 presents a performance envelope for various gas-turbine cogeneration cycles using a specified gas-turbine model. The graph plots the megawatt power output versus the net heat to process. The values presented on this graph are for illustration only.

The simplest cycle is point *A*, which is a gas turbine exhausting into an unfired HRSG that supplies steam directly at the process pressure level. For a sample cycle, the fuel chargeable to power (FCP) is 5620 Btu/kWh. Note that the power-to-heat ratio is 180 kW/MBtu, which is approximately two to three times that of a steam-turbine cycle alone. The megawatt output is from the gas turbine only and process heat is from the HRSG. The net heat-to-process can be increased by supplementary firing the HRSG, which is point *B*, and a FCP of 5280.

If increased power generation is desired, higher pressure and temperature HRSG steam can be first expanded through a noncondensing steam turbine and then delivered to the process. Increased megawatt power output is gained at a small expense of net heat-to-process. These points are labeled *C* and *D* for unfired and supplementary fired HRSGs, respectively. Note that these cycles are very kW efficient, with the lowest fuel chargeable to power.

If additional kilowatts are required, then part or all of the HRSG steam may be expanded to a condenser. When all the steam is condensed, the net heat to process is zero and maximum kilowatts are produced. However, this is at the expense of a much higher fuel chargeable to power (points *E* and *F*).

Gas-turbine cycle flexibilities, in conjunction with the availability of several

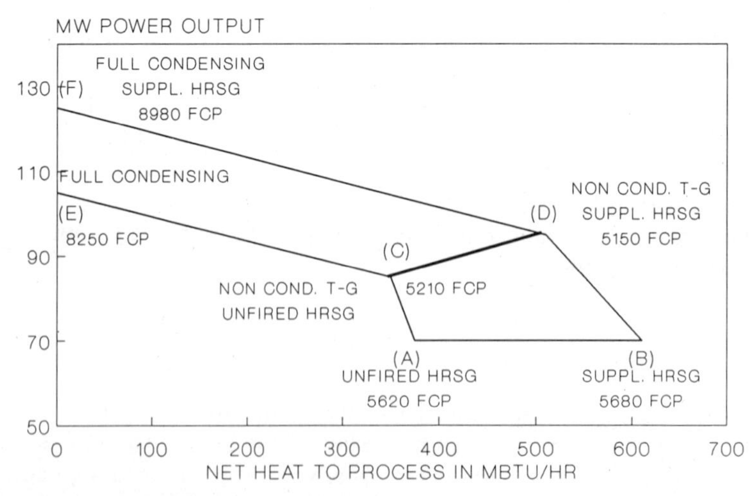

FIGURE 6.12 Performance envelope for various gas-turbine cogeneration cycles.

gas-turbine models, permits a broad range of coverage for industrial process and power needs.

6.3.4 Typical Gas-Turbine Cogeneration System Costs

Table 6.2 illustrates several gas-turbine and HRSG system configurations for different gas-turbine models. The table presents typical values based on 80°F ambient-air temperature at sea level. For specific project evaluation, detailed manufacturer-specific values at site conditions must be used.

Table 6.3 presents typical total installed costs (including 15% plant cost for AFDC and land) of cogeneration systems corresponding to the gas-turbine models of Table 6.2. Table 6.3 is divided into two types of cycles: (1) simple gas-turbine (GT) and HRSG cycle with no steam turbine; and (2) GT, HRSG, combined cycles and steam turbines producing additional kilowatts. In cycle 1, the HRSG produces steam in the 150–300-psig range. In cycle 2, the HRSG produces steam at higher temperatures and pressures, which is expanded to a noncondensing steam turbine.

Many cogeneration cycle applications are designed for industrial plants with

TABLE 6.2 Typical Gas-Turbine Cycle Configurations

Gas-Turbine Model	Model 5	Model 6	Model 7	Model 25	Model 50
Gas turbine only (80°F, sea level)					
Fuel input (MBtu/hour) (HHV)	322	422	890	209	298
Output (kW)	22,400	33,900	73,200	19,400	27,300
Exhaust flow (thousand lb/hour)	926	1,045	2,203	503	888
Exhaust temperature (°F)	938	1,027	1,009	980	819
Thermal efficiency of GT (HHV)	24%	27%	28%	31%	31%
Unfired HRSG	*Steam Production (1000 lb/hour)*				
Pressure (psig/°F)					
160/371	145	191	393	85.1	108
420/655	117	158	323	69.7	83
895/830	100	140	284	60.5	
1315/905	100	132	266		
and 160/371[a]	28	30	70		
Supplementary fired HRSG to 1600°F: fuel input MBtu/hour	202	200	104		227
	Steam Production (1000 lb/hour)				
420/655	288	325	157		277
895/830	269	303	147		258
1315/905	261	295	143		251

[a]Two-pressure HRSG.

TABLE 6.3 Gas-Turbine Cycles[a]
TYPICAL INDUSTRIAL COGENERATION TOTEL — INSTALLED COSTS (MILLIONS 1990 $) (PROCESS HEAT IN 150–300-PSIG RANGE)

System	Model 5	Model 6	Model 7	Model 25	Model 50
Gas-turbine cycle					
GT low-pressure HRSG					
Unfired	22.8	28.2	48.7	23.4	31.0
Supplementary fired to 1600°F	26.2	31.6	54.7	26.0	35.0
GT combined cycle HRSG with Steam-turbine cycle					
Unfired	34.6	42.2	70.7	31.3	34.6
Supplementary fired to 1600°F	44.6	51.5	88.8	37.9	38.6

[a]Includes interest accrued during construction, which represents 15% of plant costs. Site location on U.S. Gulf Coast labor area.

an existing process steam boiler plant. If a retrofit cogeneration plant were to be installed, significant cost savings could be achieved by using existing steam piping and facilities. A savings in the range of 20% is typical.

Operation and maintenance are additional costs. Average yearly maintenance cost is typically estimated at 3% of the total installed cost. The per annum operating labor cost for the Model 6 GT with a HRSG is typically estimated at $1.3M, or $1.5M in a combined-cycle mode. Operating labor cost is relatively independent of gas-turbine model size. Doubling the gas-turbine model size (i.e., from a Model 6 to a Model 7) increases operating labor by only 25%.

6.4 COGENERATION REGULATIONS

The United States Public Utilities Regulatory Policy Act (PURPA) of 1978 significantly changed the cogeneration market. The objective of PURPA is to encourage conservation and effective utilization of energy resources. Cogeneration is one means to achieve that goal. PURPA requires that an electric utility must interconnect with a qualifying PURPA cogeneration facility, purchase the power output at prices up to the utility's avoided cost, and provide nondiscriminatory rates for backup capacity. In addition, the cogenerator is not subject to the state public service commission or Federal Energy Regulatory Commission regulation.

Prior to PURPA, utilities were under no legal obligation to purchase cogenerated power and utility buy-back rates were negotiated. In many instances, the incremental investment to produce additional power at a cogeneration facility was not economically attractive. Therefore, most pre-PURPA cogeneration facilities were designed to generate only enough power to serve the industrial facility's electrical needs.

With the advent of PURPA legislation, a significant amount of power from cogeneration facilities is sold back to the utility power grid. The extent of the power sold is dependent on the utility's buy-back rate.

PURPA legislation established minimum efficiency and fuel use criteria to qualify under the legislation. The PURPA efficiency definition is:

$$\text{eff}_{\text{PURPA}} = \frac{\text{power} + .5 \cdot \text{NHP}}{\text{fuel}} \tag{6.4}$$

where $\text{eff}_{\text{PURPA}}$ = PURPA efficiency
power = power output in MBtu/hour
NHP = net heat to process in MBtu/hour
fuel = fuel heat input in MBtu/hour (LHV)

The lower heating value is used in the PURPA calculations.

Qualification as a PURPA topping cogeneration facility requires satisfaction of the criteria presented in Table 6.4. Steam-turbine cycles burning coal, wood, refuse, or waste products generally meet the criteria. Steam-turbine cycles burning oil and natural gas qualify when using noncondensing cycles, and usually qualify when at least 50% of the steam is delivered to a process at 50 psi and 50% is expanded through a turbine to a condenser. Higher process pressures permit less steam to be condensed to qualify under PURPA (Limaze, 1985).

Gas-turbine combined cycle configurations generally meet PURPA efficiency standards even with low process steam requirements. A cycle producing 850°F steam in a HRSG and expanding the steam in a turbine to a 150-psi process requirement has PURPA efficiencies in the 50–60% range. Many new gas-turbine models can achieve PURPA's required efficiency values with only 5–10% of the steam to process and 90–95% of the steam applied to a condensing steam turbine (Kovacik, 1984).

TABLE 6.4 Criteria for Qualifying as a PURPA Topping Cycle Cogenerator

Criteria	No Oil or Gas Consumed	Oil or Gas Consumed
Net heat to process	5% of total energy output	5% of total energy output
Efficiency	None	45% if NHP <15% of total energy output 42.5% if NHP >15% of total energy output
Utility ownership	≤50%	≤50%

Qualification as a PURPA cogenerator restricts utility ownership to less than 50%. This precludes utilities from having financial and management control of the facility. Utility ownership greater than 50% renders the cogeneration facility as non-PURPA qualifying, and thereby subject to state and FERC regulation as a utility enterprise. Industrial-sized cogeneration facilities must also meet the same new source performance emission standards as utility plants for SO_X, NO_X, and particulates.

The Tax Reform Act of 1986 includes provisions for cogeneration facilities as presented in Table 6.5.

6.5 INDUSTRIAL ECONOMIC ANALYSIS METHODS

The financial objectives of industrial companies are to maximize earnings per share consistent with good business practices. Industrial economic analysis procedure is focused toward this objective. The method most widely used to analyze these financial obligations is the discounted cash flow rate-of-return method.

The discounted cash-flow rate-of-return method (DCFR) begins by calculating revenues, subtracting expenses, and computing earnings for each year over a several-year horizon period or over the life of a project. Each year's earnings are discounted back to the year that the investment is to be made. This process is repeated using several values of discount rates. The discount rate for which the cumulative present worth net revenues equals the initial investment is defined as the DCFR.

Example 6.5. A $1000 investment is made at the beginning of year 1. The investment has a 3-year life with no net salvage. Depreciation rates are given in Table 6.6. The income tax rate is 35%. The investment saves $300 each year in operating expenses. It also improves production output by $200 in year 1 and $100 in years 2 through 3. Calculate the DCFR.

SOLUTION. The DCFR method assumes that investment capital is available. The actual means of financing the investment is not considered in the analysis, only in deciding whether or not to invest.

Three key financial statements are prepared: income tax statement, income statement, and cash report. Table 6.7 presents these data. The cash flow is discounted by 5, 10, and 20%/year, and the cumulative discounted cash flow is computed. Since a discount rate of 10% has a cumulative discounted cash flow that nearly equals the initial investment, the DCFR of this project is 10%.

The DCFR interpretation is that the industrial company would be indifferent to investing $1000 for 3 years in this project versus earning 10%/year in a bank. It also means that if the cost of money for capital investment is more than 10%, the company should not invest, since the return on investment is

TABLE 6.5 Tax Depreciation Rules

Type of Cogeneration System	ADR Guideline Life	Tax Depreciation Life	Tax Depreciation Method
Simple cycle GT and HRSG	20	15	150% declining balance, switch to straight line
Combined cycle and steam-turbine cycles			
MW Output used predominately by owner	22	15	150% declining balance, switch to straight line
MW output predominately sold	28	20	150% declining balance, switch to straight line
Waste reduction and resource recovery (excluding turbine facilities)	10	7	200% declining balance, switch to straight line

TABLE 6.6 Depreciation Rates for Example

Year	Book	Tax
1	.34	.40
2	.33	.35
3	.33	.25

only 10%. All other projects should be evaluated using DCFR and ranked from highest DCFR to lowest. All projects with returns larger than the cost of capital could be implemented if risk of failure is assumed to be zero. Realistically, the project risk is not zero and a DCFR greater than the cost of money is required.

TABLE 6.7 DCFR Calculation Example[a]

	Year 1	Year 2	Year 3	
Income Tax Statement				
Δ Revenue	200	100	100	
Less Δ expenses				
Δ Operating expense	−300	−300	−300	
Δ Tax depreciation	400	350	250	
Δ Taxable income	100	350	150	
Δ Income tax	35	18	52	
Income Statement				
Δ Revenue	200	100	100	
Less Δ expenses				
Δ Operating expense	−300	−300	−300	
Δ Book depreciation	340	330	330	
Δ Income tax	35	18	52	
Δ Deferred income taxes	21	7	− 28	
Δ Net income	104	45	46	
Funds Statement				
Δ Net income	104	45	46	
Δ Book depreciation	340	330	330	
Δ Deferred income tax	21	7	− 28	
Δ Cash flow	465	382	348	
Discounted Rates				Cumulative Discounted Cash Flow
5%	442.8	346.5	300.6	1089.9
10%	422.7	315.7	261.4	999.8
15%	404.3	288.8	228.8	921.9

[a]Changes (Δ) in quantities are shown in this table.

A short-cut cash-flow calculation can be made by examining the entries in Table 6.7. Both book depreciation and deferred taxes are entries that are subtracted in the income statement and added back into the funds report. Therefore, they may be eliminated from the calculation. The DCFR is independent of the book depreciation and deferred taxes.

The DCFR method is the workhorse tool for detailed industrial economic analysis. For screening-type analysis, the investment pay-back method described in Chapter 4 is often used because of its ease in calculation.

6.6 ECONOMICS OF COGENERATION

An industrial facility has several options for providing both steam and electricity. It can:

1. Generate steam only in an ambient-air boiler and purchase power from the utility.
2. Cogenerate steam and supply plant electricity needs.
3. Cogenerate steam, supply plant electricity needs, and sell additional electricity back to the utility.
4. Cogenerate steam, supply partial plant electricity needs and buy additional power from the local utility.

Each cogeneration alternative may be supplied by a steam cycle (Figure 6.3), or gas-turbine combined-cycle system (Figure 6.7), depending on the fuels available and power-to-heat-ratio needs.

The cogeneration facility may be owned by the industrial company, owned by an investor–developer, jointly owned by the industrial company and a third-party investor–developer, or jointly owned with an electric utility. Ownership issues are dependent on the industrial company's financing capability, desire to share the cogeneration business risk, and other business reasons.

The choice of a system is dependent on the project cost, hours per year of plant operation, quantity of steam demand, project electrical and steam performance, fuel cost and availability, electricity rates, and desired DCFR. The following examples explain the influences of these factors.

Example 6.7. A new industrial facility built in 1990 requires 191,000 lb/hour of saturated steam at 150 psig with a process return of 150°F. The facility can purchase coal at 1.5 $/MBtu or natural gas at 3.50 $/MBtu. The industrial facility also requires 40 MW of power. The utility industrial electricity rate is 5.0 ¢/kWh and the utility power buy-back rate is 4.0 ¢/kWh. The industrial facility plans to operate for 8000 hours/year.

The industrial company can evaluate four options:

1. Install a gas-fired ambient-air boiler for the steam needs of the plant and purchase 40 MW of electricity from the utility.

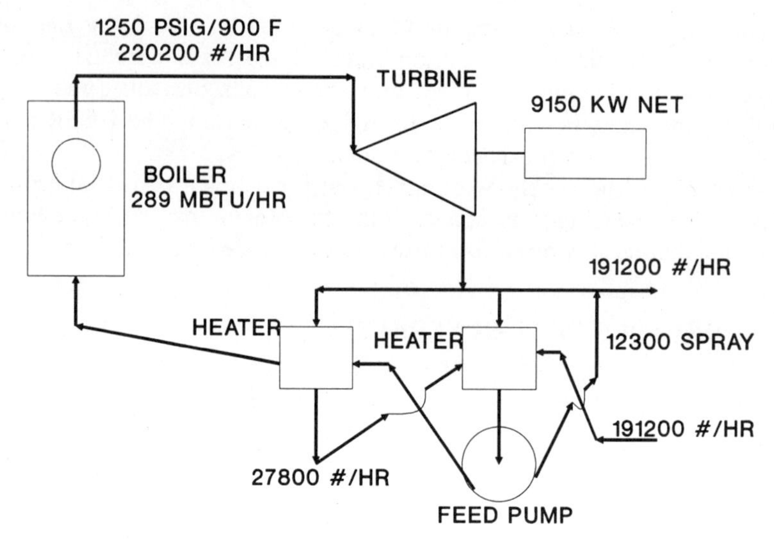

FIGURE 6.13 Coal-fired cogeneration system configuration.

2. Install a coal-fired cogeneration system sized to provide steam needs with any supplementary electricity purchased from the utility (Figure 6.13).
3. Install a Model 6 simple-cycle gas turbine and HRSG to supply process steam (Figure 6.14).
4. Install a combined-cycle Model 7 gas turbine, HRSG, and steam turbine (Figure 6.15).

The analysis will determine which system the industrial company should purchase.

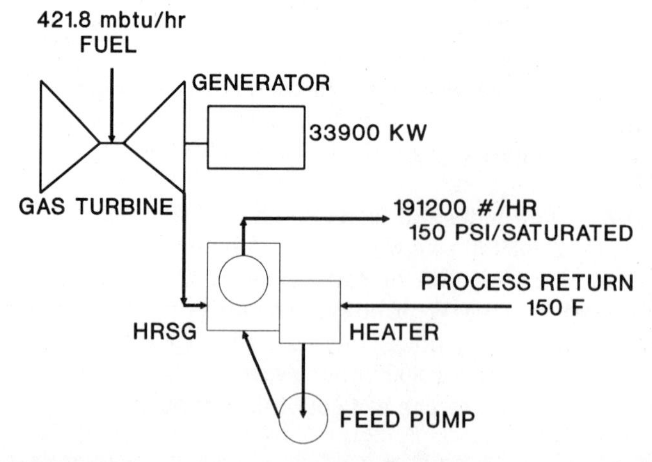

FIGURE 6.14 Model 6 gas turbine–HRSG system configuration.

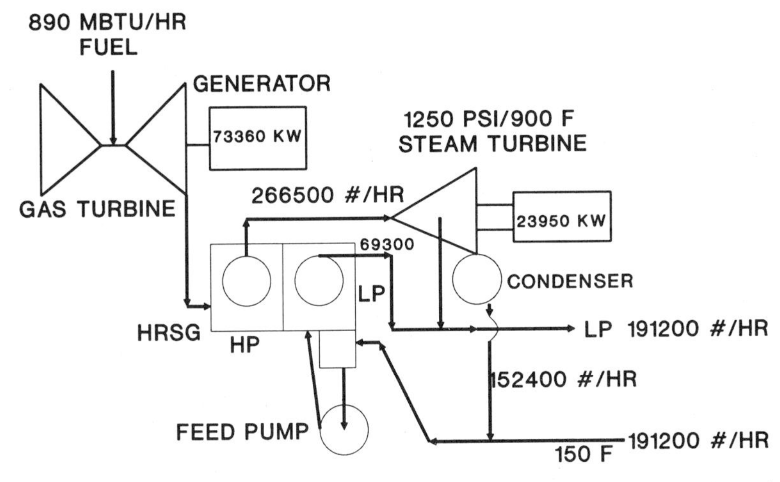

FIGURE 6.15 Model 7 gas turbine–HRSG–steam-turbine system configuration.

SOLUTION. The system performance for the coal-fueled cycle of Figure 6.13 was obtained from data in Table 6.1. Information for the gas-fired cycles of Figures 6.14 and 6.15 were obtained from data in Table 6.2.

The cycle performance results and costs are summarized in Table 6.8. The calculations for 1 year of operating costs are presented at the bottom of Table 6.8.

Considering the gas-fired boiler as a reference case, the cogeneration cycles all have an operating cost advantage relative to the gas-fired boiler. The investment pay-back on the incremental cogeneration investment is presented. The Model 6 has the best incremental investment pay-back of 2.05 years.

A 15-year discounted cash flow return-on-investment calculation was also performed on these projects, based on a 5%/year cost escalation from 1990. The results are shown as the bottom line in Table 6.8. The Model 6 has the best incremental DCFR.

Cogeneration cycles are generally good candidates for investments at new medium- to large-industrial facilities requiring both process heat and electricity.

Example 6.8. An existing facility currently uses an ambient-air gas-fired boiler to produce steam at 150 psig and 191,000 lb/hour. Using the same data as in Example 6.7, compute the benefits of retrofitting a new cogeneration facility. Use an 18% capital cost reduction in cogeneration investment costs to reflect the use of existing facilities.

TABLE 6.8 Summary of Industrial Thermal Cycles for New Cogeneration Plant (M = Million)

	(1) Gas-Fired Boiler	(2) Coal-Fired Cogen System with Scrubber	(3) Model 6 GT-HRSG System	(4) Model 7 GT-HRSG-CC System
Fuel input (MBtu/hour)	245	289	422	890
Fuel cost ($/MBtu)	3.5	1.5	3.5	3.5
Kilowatts produced	0	9,150	33,900	96,310
Kilowatts purchased	40,000	30,850	6,100	−56,310
Utility electric rate (¢/kWh)	5.0	5.0	5.0	4.0
Installed cost (M $)	12.30	36.70	28.20	70.70
Annual operating labor cost (M $/year)	0.80	1.90	1.30	1.90
Annual maintenance cost (M $/year)	0.36	0.96	0.72	1.80
Annual operating hours	8,000	8,000	8,000	8,000
		Economic Results		
Fuel cost (M $/year)	6.86	3.47	11.82	24.92
Utility power purchase (M $/year)	16.00	12.34	2.44	−18.02
O&M cost (M $/year)	1.16	2.86	2.02	3.70
Total annual cost (M $/year)	24.02	18.67	16.28	10.60
Cogeneration annual benefit (M $/year)	Reference	5.35	7.74	13.42
Investment pay-back (years)	Reference	4.56	2.05	4.35
Project DCFR (%/year)	Reference	19	38	19

SOLUTION. Table 6.9 presents the economic results for these cases for the first operating year, 1990. Note that the Model 6 GT-HRSG system provides the best economic results. A new Model 6 cogeneration system provides a 3-year investment pay-back and a 27% year DCFR.

The cogeneration project economic benefits are dependent on economic ground rules. Figures 6.16, 6.17, and 6.18 present the sensitivities to electricity price charged to the industrial company, utility electricity buy-back price, and natural gas fuel price, respectively, while holding the other parameters fixed.

As the electricity price to the industrial company increases (Figure 6.17), the cogeneration investment pay-back decreases because the cogeneration facility produces electricity and saves the utility power purchase costs.

When the utility buy-back electric rate increases (Figure 6.16), the invest-

TABLE 6.9 Summary of Economic Results for Retrofit Cogeneration Plant

	(1) Gas-Fired Boiler	(2) Coal-Fired Cogen System with Scrubber	(3) Model 6 GT-HRSG System	(4) Model 7 GT-HRSG-CC System
Fuel input (MBtu/hour)	245	289	422	890
Fuel cost ($/MBtu)	3.5	1.5	3.5	3.5
Kilowatts produced	0	9,150	33,900	96,310
Kilowatts purchased	40,000	30,850	6,100	−56,310
Utility electric rate (¢/kWh)	5.0	5.0	5.0	4.0
Installed cost (M $)	0.00	30.09	23.12	57.97
Annual operating labor cost (M $/year)	0.80	1.90	1.30	1.90
Annual maintenance cost (M $/year)	0.36	0.96	0.72	1.80
Annual operating hours	8,000	8,000	8,000	8,000
		Economic Results		
Fuel cost (M $/year)	6.86	3.47	11.82	24.92
Utility power purchase (M $/year)	16.00	12.34	2.44	−18.02
O&M cost (M $/year)	1.16	2.86	2.02	3.70
Total annual cost (M $/year)	24.02	18.67	16.28	10.60
Cogeneration annual benefit (M $/year)	Reference	5.35	7.74	13.42
Investment pay-back (years)	Reference	5.62	2.99	4.32
Project DCFR (%/year)	Reference	15	27	20

ment pay-back of Model 7 GT-HRSG-CC decreases because this cycle has excess power which is to be sold to the utility. An increased buy-back price produces an increased revenue stream from the sale of electricity back to the utility, decreasing the investment pay-back. For high electric buy-back rates (typical of an avoided energy and capacity buy-back price as discussed in Chapter 15), cogeneration projects are economical when power-to-heat ratios exceed the industrial company's plant requirement.

When the price of natural gas increases, the investment pay-back of natural gas-fueled cogeneration projects increases, as shown in Figure 6.18. If the price of electricity is constant, while the gas price increases, then the cost saving from electricity production is less. Consequently, the investment pay-back increases. The coal-steam cogeneration cycle investment pay-back decreases, however, relative to a gas-fired boiler.

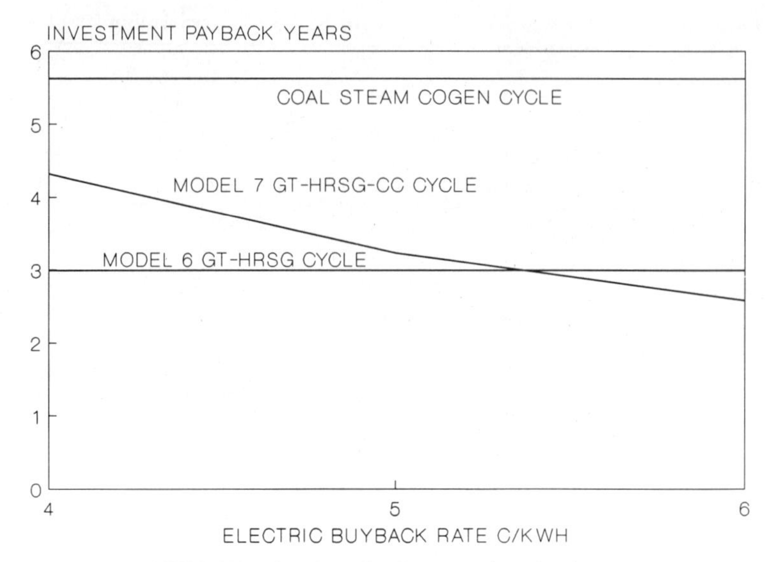

FIGURE 6.16 Pay-back versus buy-back rate.

The three examples shown as Figures 6.16, 6.17, and 6.18 illustrate the sensitivity to *one* economic parameter only. In reality, all of these parameters are coupled because electricity price is dependent on fuel cost and the buy-back rate is dependent on electricity price. The degree of coupling is dependent on the specific utility. If the utility is largely coal-fired, then the electricity price and buy-back rate are largely independent of the natural gas price.

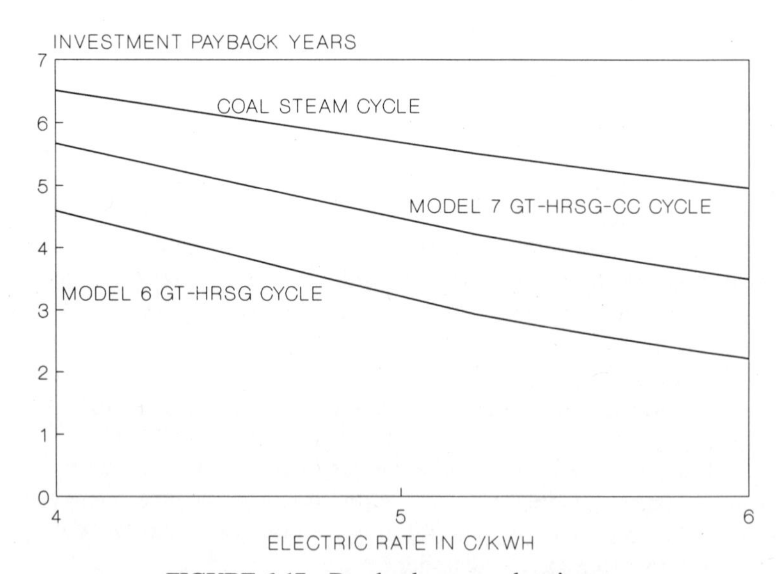

FIGURE 6.17 Pay-back versus electric rate.

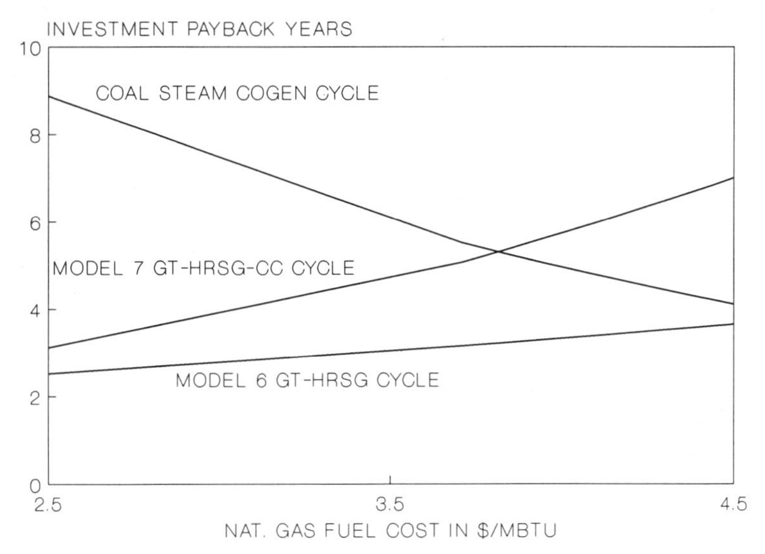

FIGURE 6.18 Pay-back versus fuel cost.

This example had 8000 operating hours/year, which is representative of a plant with continuous "three-shift" operation less (minus) 4 weeks for planned shutdowns and outages. The economic attractiveness of cogeneration decreases as the annual operating hours decrease because the fixed charges on the plant investment is constant. In practice, viable cogeneration projects typically have very high annual operating hours.

The plant size of a cogeneration project influences the overall project economics. The favorable economics of size of the project plant cost favors industrial projects with large (≥ 100,000 lb/hour) steam demands.

In the examples of this section, it was assumed that the cogeneration project was delivered on a "turnkey" basis. The plant investment is made in the beginning of the first year of operation. Plant construction is usually completed over a 2–3 year period, during which project progress payments are made. A more precise DCFR calculation would account for this payment cash flow. This effect of progress payments typically leads to a lower DCFR than indicated by the examples presented here.

6.7 ADDITIONAL COGENERATION ISSUES

The previous example illustrated the principles used in the economic evaluation of cogeneration systems. Practical cogeneration systems must be designed to provide reliable steam and power. The interruption of steam and power due to equipment breakdown cannot only lead to an extended loss of industrial production but, for some processes, can result in damage to plant equipment due to product solidifying or curing in process piping.

Cogeneration plant availability is generally in the 85–95% range. Scheduled plant maintenance shutdowns are typically conducted in 2–4 weeks per year. Forced shutdowns, due to unscheduled plant equipment failures, typically account for one to 3 weeks/year. Since the planned maintenance shutdown on an industrial facility generally coincides with the shutdown period of cogeneration plant, the cogeneration plant planned unavailability does not impact industrial production. Forced shutdowns, however, can significantly impact industrial production and may result in losses of tens to hundreds of thousands of dollars a day.

Provision for backup steam and electricity are made when potential losses are significant. Backup for steam and electricity may be achieved by installing either (1) additional parallel (independent) capacity in the cogeneration system or (2) a backup ambient-air fired boiler for steam production, or use of an existing boiler, and purchasing electricity from the local utility.

One approach to improving cogeneration system reliability is to install two half-size cogeneration systems. The economies of scale in plant size places a penalty on this approach. For example, two Model 6 gas-turbine systems produce 95% of the steam and power of one Model 7 gas-turbine system, yet two Model 6 systems have a 15% higher capital cost.

In retrofit cogeneration applications, the facility already has an existing boiler or steam source that could be operated for steam supply backup. The steam backup source is made available from a cold shutdown condition or, if quick response is needed, the backup source may be operated in a hot mode by firing the boiler at minimum fuel or by bottling up (banking) the boiler. Electricity purchases from the utility are typically the most economical means of providing backup electricity.

One element involved in developing a successful cogeneration project is the electricity buy-back rate as shown earlier in Example 6.8. If the buy-back rate is low, then the cogeneration plant is designed to provide only enough power to meet the industrial plant needs. If the buy-back rate is high, then the cogeneration plant is designed to produce power for sale back to the utility.

Often investment requirements for cogeneration systems are larger than an industrial company can spend, or industrial companies may prefer making investments in industrial process equipment rather than in cogeneration equipment. In this case, cogeneration plants can be developed in conjunction with a utility or third-party developer. The third-party developer can have an equity position in the plant and provide financing. Financing a cogeneration project requires the project to provide good economic benefits (DCFRs in the 20% + range) and contain business risks to acceptable levels. Containing business risks requires three critical elements, assured price and demand availability for:

- The steam product
- The electricity
- The fuel source over the project's financing period, typically 15 years.

If any of these three elements are interrupted, or significantly changed, then project viability is significantly impacted and possibly results in the inability of the principals to meet the financing interest and principle repayments.

One means to achieve financing is to obtain long-term (i.e., 15 years) take or pay contracts for steam demand, industrial electricity demand, utility electricity buy-back rates, and fuel contracts. Firm steam and electricity demand contracts by the industrial company place some constraints on any process changes or business demand changes. If project economics are favorable, then a steam and electricity contract may be developed with broad ranges of acceptable demands.

Long-term firm fuel contracts may be negotiated with suppliers. Most contract prices are determined by a commodity escalation index. This places some risk exposure to the project which is mollified by the fact that alternative steam and electricity methods are also influenced by changes in cost of fuel.

For cogeneration projects that result in the sale of significant amounts of electricity back to the utility, a long-term firm electricity buy-back rate is often needed to assure project economic viability. Some utilities have established long-term electricity buy-back rate schedules, while others do not provide firm long-term electricity buy-back rates.

6.8. UTILITY COGENERATION INCENTIVES

The utility position on cogeneration is divided. Some utilities view cogeneration as an electricity market competitor and thereby discourage its application within their service territory. Cogeneration reduces electricity sales to high-volume industrial customers, requires the utility to buy back power at avoided cost rates and leads to a loss of utility control over the generation sources essential to providing reliable service. Avoided cost buy-back rates provide few benefits to rate payers and tend to raise average utility electric rates.

Other utilities view cogeneration as a complimentary source of electricity. Cogeneration can provide capacity to a utility that can then avoid the construction of capital-intensive power plants. In some states, utilities may establish unregulated subsidiaries to participate in congeneration projects that generate additional earnings.

Each of these positions has merits that depend on the specific economic and utility system environments. The next example illustrates these merits.

Example 6.9. Consider the utility economic viewpoint of the cogeneration project of Example 6.8 for a Model 6 cogeneration plant. Assume that utility marginal cost is 4¢/kWh and that the industrial rate is 5¢/kWh. The utility is evaluating three alternatives:

1. Not to participate in the cogeneration project that is implemented by an industrial company or third-party developer.

2. Assume a 49% equity position in the project. The project is to be 50% financed with bonds at 10%/year, and the project will provide steam at a 25% discount to the industrial company.
3. Discourage the cogeneration project by reducing the industrial company's electric rates so that the project has an unattractive return on investment, which, for this example, is less than 20% return on equity in the first year.

Evaluate the stakeholder benefits (utility, industrial company, and utility rate payer) for these three alternatives.

SOLUTION. The decision to participate in the project should be based on the incremental costs and savings of the cogeneration project. Table 6.10 presents an *incremental* utility income statement for the alternatives of cogeneration and no cogeneration. The income statement is based on an assumed existing utility investment of $18M, of which $9M is equity, to serve this customer along with other normal utility costs. With no cogeneration, the utility achieves a 14% return on equity.

Consider case 1 when the cogeneration facility is to be installed. The results in Table 6.9, suggest that utility electricity sales would decline, but the fixed costs of serving the customer would remain constant. This results in a negative

TABLE 6.10 First-Year Incremental Utility Tax and Income Statement

Tax Statement	No Cogeneration (M $)	With Cogeneration (M $)
Revenues (@ 5¢/kWh)	16.00	2.44
Less		
Incremental cost (@4¢/kWh)	12.80	1.95
Depreciation	0.40	0.40
Interest	0.70	0.70
Taxable income	2.10	−0.61
Income taxes (@ 40%)	0.84	−0.24
Income Statement		
Revenues (@ 5¢/kWh)	16.00	2.44
Less		
Incremental cost (@ 4¢/kWh)	12.80	1.95
Depreciation	0.40	0.40
Taxes	0.84	−0.24
Interest	0.70	0.70
Net income	1.26	−0.37
Net equity investment	9.00	9.00
Return on equity (%/year)	14.00	−4.08

4.08% return on equity (ROE) for this utility in the first year. Therefore, the utility faces a potential loss of earnings if the cogeneration plant facility is constructed.

The utility rate payer is also impacted by the cogeneration project. The loss of energy sales and kilowatt demand charges to the industrial customer initially result in a lower ROE to the utility. In a utility-regulated environment, the electricity revenue requirement would be increased at the next rate proceeding, allowing the utility to earn its regulated ROE. The increased revenue requirement would be allocated to various customer service classes. If the utility achieves a lower ROE in servicing a customer, or class of customers, the net effect can be an electric rate increase for that remaining class of customers.

In alternative 2, the utility could form an unregulated joint venture (in states that permit unregulated utility subsidiaries) with the industrial company to construct the cogeneration facility. This is attractive to the industrial facility because the utility has expertise in constructing and operating power plant facilities. The joint venture would help spread the business risk by reducing the industrial company's overall investment. The utility also benefits from the increased earnings potential.

Assume that a separate joint venture cogeneration project company is formed that sells the steam and power "over the fence" to the industrial company. As an incentive to the industrial company, the industrial pays for the steam at a 25% discount of the fuel and O&M of the existing boiler facility. In practice, the actual steam and/or electricity discount is negotiable between both parties.

The joint venture cogeneration project tax and income statements for the first year are shown in Table 6.11. Three statements are shown: for the joint venture project, for the utility (incremental), and for the industrial company (incremental). Table 6.9 was used to obtain the fuel, power and O&M costs.

From the industrial company, the joint venture receives a steam payment (at 25% discount) and electricity revenues from 40,000 kW of capacity. The joint venture incurs the fuel cost of the Model 6 GT-HRSG system, 6100 kW of purchased capacity, and the O&M costs. The cogeneration project cost is depreciated at 5% in the first year. The interest expense on the 50% debt portion of the $23.12M facility is $1.16M/year. The result is a first-year cogeneration project net income of $2.06M or 17.78%/year return on equity.

The incremental utility income is augmented by the utility's 49% share of the cogeneration joint venture ("other net income") leading to a utility incremental ROE of 4.37%/year.

The industrial company experiences a decrease in fuel costs and O&M resulting in industrial company benefits for a $1.2M/year after tax cost reduction and $1.05M joint venture earnings (other net income) share. The result is an attractive ROE of 38.18%/year for the industrial company.

In the third alternative, the utility would reduce its electricity rates to the industrial so as to render the cogeneration project unattractive. A first-year ROE of less than 20%/year is assumed to be an unattractive return.

TABLE 6.11 First-Year Joint Venture Cogeneration Plant Tax and Income Statements

Tax Statement	Joint Venture Cogeneration Project	Incremental Utility	Incremental Industrial
Revenues			
Steam (@ 75% fuel and O&M)	6.02		
Electricity	16.00	2.44	
Total	22.02	2.44	0.00
Less			
Fuel cost	11.82	1.95	−0.84
Purchased power	2.44	0.00	0.00
O&M	2.02	0.00	−1.16
Depreciation	1.16	0.40	0.00
Interest	1.16	0.70	0.00
Taxable income	3.43	−0.61	2.01
Income taxes (@ 40%)	1.37	−0.24	0.80
Income Statement			
Revenues	22.02	2.44	0.00
Less			
Fuel cost	11.82	1.95	−0.84
Purchased power	2.44	0.00	0.00
O&M	2.02	0.00	−1.16
Depreciation	1.16	0.40	0.00
Taxes	1.37	−0.24	0.80
Interest	1.16	0.70	0.00
Net income	2.06	−0.37	1.20
Other net income	0.00	1.01	1.05
Total net income	2.06	0.64	2.25
Net equity investment	11.56	14.67	5.90
Return on equity (%/year)	17.78	4.37	38.18

The first step in determining the benefits of each alternative is to calculate the net income of the cogeneration project if the industrial company owned the entire project. Data from Table 6.9 are used, and results are presented in Table 6.12 for the 5¢/kWh electric rate case. The industrial-owned cogeneration project has a 28.19%/year return on equity in the first year.

The next step is to decrease the electricity price of purchased power to the point where the cogeneration plant would be economically unattractive (less than a 20% ROE in the example). An electric rate of 4.5¢/kWh leads to a 19.88%/year ROE. This is a 10% utility rate reduction. The corresponding

TABLE 6.12 First-Year Incremental (Industrial Only) Tax and Income Statements Under Two Electric Rates

Tax Statement	(5¢/kWh) Current Electric Rates	(4.5¢/kWh) Lower Electric Rates
Revenues		
Steam (@ 75% fuel and O&M)		
Electricity		
Total	0	0
Less		
Fuel cost	4.96	4.96
Purchased power	−13.56	−11.96
O&M	0.86	0.86
Depreciation	1.16	1.16
Interest	1.16	1.16
Taxable income	5.43	3.83
Income taxes (@ 40%)	2.17	1.53
Income Statement		
Revenues	0.0	0.0
Less		
Fuel cost	4.96	4.96
Purchased power	−13.56	−11.96
O&M	0.86	0.86
Depreciation	1.16	1.16
Taxes	2.17	1.53
Interest	1.16	1.16
Net income	3.26	2.30
Other net income	0.00	0.00
Total net income	3.26	2.30
Net equity investment	11.56	11.56
Return on equity (%/year)	28.19	19.88

utility income statement is presented in Table 6.13. This results in an incremental utility ROE of 3.33%/year.

The results of the original utility situation and three alternatives are summarized in Table 6.14. With no cogeneration project, the utility continues to sell power at existing rates and earns its regulated return on equity. The regulated and incremental return on equity are different because of the difference between average and incremental costs.

If the cogeneration project is constructed with no utility participation, then the utility has a negative 4%/year ROE. If the utility participates in cogenera-

TABLE 6.13 First-Year Utility Incremental Tax and Income Statements with a Lower Electric Rate to Industrial Company

Tax Statement	(4.5 ¢/kWh) Lower Electric Rate
Revenues	14.4
Less	
Incremental cost	12.80
Depreciation	0.40
Interest	0.70
Taxable income	0.50
Income taxes (@ 40%)	0.20
Income Statement	
Revenues	14.4
Less	
Incremental cost	12.80
Depreciation	0.40
Taxes	0.20
Interest	0.70
Net income	0.30
Net equity investment	9.00
Return on equity (%/year)	3.33

tion projects, or reduces its rates to effectively cancel the cogeneration project, then the incremental ROE is positive but is only 3% or 4%/year.

The industrial company, when in a joint venture with the utility, increases its ROE and reduces its investment.

Some utilities have assumed an anticogeneration posture. If successful, the

TABLE 6.14 Summary of Cogeneration Alternatives

	Utility Incremental First-Year ROE	Industrial Incremental First-Year ROE
No-cogeneration project	+14%	—
Cogeneration project		
1. No utility participation	−4%	28%
2. 49% Utility participation	+4%	38%
3. Lower utility rates and cause project to be canceled	+3%	20%

14% ROE can be preserved. If the utility succeeds in only preventing 35% of the cogeneration projects, then their average ROE is approximately 14 •.35 − 4 • .65, or 2.3%/year (assuming all projects to be the same size, cost, etc.). This average ROE is less than the other alternatives.

The preceding example illustrated several utility alternatives. The first-year ROE was used as an economic decision criteria merely because it was easy to calculate. A DCFR calculation is a more appropriate criterion. As indicated by the decision sensitivity to the economic cost factors depicted in Figures 6.16 through 6.18, these conclusions apply for only one set of conditions. Local site conditions and economic factors can influence the conclusion.

In summary, cogeneration is a competitive threat or opportunity to electric utilities as much as interstate trucking is to the railroads. Utilities have several alternatives available and must choose the response most beneficial to the rate payer and utility financial interests.

6.9 U.S. POTENTIAL COGENERATION

Table 6.15 presents a listing of the number of potential industrial cogeneration sites by state for two process steam levels. In the United States in 1987, there are approximately 6382 industrial sites with steam flow requirements greater than 10,000 lb/hour and 2036 with steam requirements greater than 50,000 lb/hour (McMahan, 1987). Because of the economies of size, industrial cogeneration is more applicable to the larger-sized industrial plants. Over the 10-year period from 1976 to 1985, 254 projects (with ≥ 5 MW capacity) had been ordered.

The U.S. Department of Energy prepared a national scoping analysis forecast of U.S. cogeneration potential (*Cogeneration World*, 1982). The analysis identified 3323 potential feasible projects that would have a total net heat to process of 423 billion pounds per hour and could total 60,000 MW of capacity. While this is a forecast of the feasible amount, economic requirements for DCFRs in the range of 20%, containment of business risk, and site-specific factors can limit the actual potential to less than half of this amount. Individual service area forecasts of cogeneration may be made by surveying industrial facility steam and electricity process needs and conducting a scoping project economic analysis of these sites using the techniques illustrated in this chapter.

APPENDIX. THERMODYNAMICS REVIEW

A brief engineering thermodynamics review is provided as a foundation for the steam-turbine and gas-turbine performance calculations.

The First Law of Thermodynamics for a steady-state closed flow process is

TABLE 6.15 U.S. Industrial Cogeneration Forecast Assessment

NUMBER OF INDUSTRIAL SITES BY STATE (DEPARTMENT OF ENERGY FORECAST 1980–2000)

	Steam Use +10 K lb/hour	Steam Use +50 K lb/hour	Cogeneration Orders 1976–1985 5 MW or Larger	Number of Plants Potential	Billions Steam lb/hour Potential	Potential MW Power Capacity	MW/Plant
USA total	6382	2036	254	3323	423.7	60709	18.3
Alabama	133	52	8	98	18.7	1658	16.9
Arizona	26	9	0	24	1.3	110	4.6
Arkansas	107	35	2	39	9.6	1120	28.7
California	448	138	59	382	27.3	8537	22.3
Colorado	34	15	3	17	1.1	235	13.8
Connecticut	76	2	1	47	1.4	370	7.9
Delaware	25	9	1	15	1.9	426	28.4
District of Columbia	1	0	0	0	0.0	0	
Florida	123	36	20	77	13.2	1917	24.9
Georgia	193	60	7	113	18.5	1318	11.7
Idaho	45	17	4	20	1.2	430	21.5
Illinois	361	81	3	181	15.2	2452	13.5
Indiana	225	68	1	61	11.9	1595	26.1
Iowa	102	45	3	51	4.4	451	8.8
Kansas	64	23	2	29	4.9	976	33.7
Kentucky	45	25	1	41	5.9	638	15.6
Louisiana	146	87	17	94	49.4	6202	66.0
Maine	72	26	7	63	8.8	1678	26.6
Maryland	70	18	2	18	2.3	274	15.2

Massachusetts	173	30	6	134	3.2	1168	8.7
Michigan	345	122	10	121	12.8	1345	11.1
Minnesota	112	33	3	42	3.9	456	10.9
Mississippi	82	27	3	51	8.3	1580	31.0
Missouri	116	30	0	53	3.4	506	9.5
Montana	0	0	0	10	1.0	211	21.1
Nebraska	35	15	0	20	0.9	85	4.3
Nevada	4	2	0	2	0.0	2	1.0
New Hampshire	34	7	1	26	1.3	296	11.4
New Jersey	185	45	5	125	9.5	2323	18.6
New Mexico	21	6	0	20	1.2	119	6.0
New York	334	92	6	156	7.6	1304	8.4
North Carolina	269	52	6	121	10.4	1030	8.5
North Dakota	8	5	0	1	0.0	1	1.0
Ohio	437	143	5	156	14.5	2280	14.6
Oklahoma	49	17	2	28	6.2	668	23.9
Oregon	188	71	0	81	5.4	647	8.0
Pennsylvania	389	102	6	214	19.4	4172	19.5
Rhode Island	36	3	0	24	0.6	280	11.7
South Carolina	137	47	5	82	9.7	757	9.2
South Dakota	6	1	0	3	0.0	2	0.7
Tennessee	172	46	4	47	7.2	1694	36.0
Texas	364	184	29	186	68.8	5878	31.6
Utah	29	11	0	7	1.2	145	20.7
Vermont	17	16	1	13	0.3	103	7.9
Virginia	140	47	8	82	11.9	1359	16.6
Washington	147	61	1	51	4.3	813	15.9
West Virginia	50	11	1	19	4.8	361	19.0

a statement of conservation of energy of a working fluid (e.g., steam, water, air).

$$q_{IN} - \omega_{OUT} = \triangle h + \triangle KE + \triangle PE$$

where q_{IN} = heat input to a process in Btu/lb
ω_{OUT} = work produced (output) in Btu/lb
$\triangle h$ = change in enthalpy during the process in Btu/lb; often called the "total heat content of a fluid"
$\triangle KE$ = change in kinetic energy during the process in Btu/lb
$\triangle PE$ = change in potential energy during the process in Btu/lb

The change in kinetic energy and potential energy is usually small in power plant applications.

Enthalpy is a property of a working fluid and is defined as $h = u + p \cdot v \cdot F$,

where u = internal energy of the working fluid in Btu/lb,
p = pressure of the fluid in psia,
v = specific volume, or the inverse of the fluid density, in feet^3/lb, and
F = a .185 conversion factor from foot/lb • ft^2/in^2 to Btu.

Tables of the enthalpy of steam may be found in several reference sources (Keenan and Keyes, 1963). For "ideal gas" working fluids, the enthalpy change may be calculated as:

$$\triangle h = \int_{T_1}^{T_2} C_p(T)dT = \overline{C}_p \cdot (T_2 - T_1)$$

where $C_p(t)$ = specific heat of the gas (at constant pressure) in Btu/lb/°F (C_p is dependent on the fluid temperature)
T_1, T_2 = temperatures before and after the process change in °F
$\overline{C}_p$ = average specific heat over the T_1, T_2 temperature range.

In practice, air and gas-turbine exhaust gases may be approximated as an "ideal gas."

Another key thermodynamic property of a fluid is entropy, defined as

$$\triangle s = \int_{\text{state 1}}^{\text{state 2}} \frac{dq}{T}$$

where $\triangle$s = entropy change in Btu/lb/°R (degrees Rankine)
dq = incremental heat input in proceeding from fluid state 1 to state 2, in Btu/lb/°R
T = absolute temperature in °R

It is shown in elementary thermodynamics textbooks that in an ideal (loss less) process in which no heat is added, the entropy is constant. This is used in calculating the thermodynamic cycles of turbine engines.

Because enthalpy and entropy are key properties of the working fluid, it is useful to graph these two properties, termed a "Mollier diagram." The higher enthalpy region of the Mollier diagram for steam is shown in Figure 6.19. Enthalpy is indicated in the ordinate and entropy in the abscissa. Lines of constant fluid pressure and temperature are also shown on the diagram. As water is heated it undergoes a phase change from liquid (water) to gas (steam). In this phase change, steam may have a range of "qualities." Steam may have no moisture in it and thereby be referred to as "dry saturated vapor," or it may be all moisture and be referred to as "wet saturated liquid." Steam under the saturation line in Figure 16.19 is in this phase-change region. For steam under the saturation line "dome," steam pressure and temperature are perfectly correlated. A given temperature specifies a pressure. Table 6.16 provides steam data in this region.

Example. An "ideal" steam turbine expands steam from temperatures of 1000°F and 1000 psia pressure to a pressure of 54 psia. Compute the steam turbine output per pound of steam.

SOLUTION. The "ideal" turbine expands the steam down a line of constant entropy. From thermodynamic tables of steam (Keenan and Keyes, 1963) or Figure 6.19, we obtain:

State 1 $T = 1000°F$

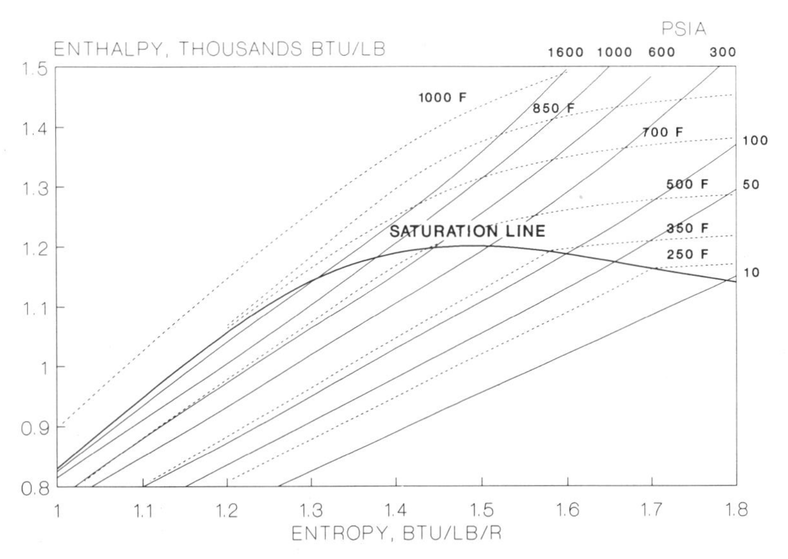

FIGURE 6.19 Mollier diagram for steam in the higher enthaply region.

TABLE 6.16 Saturated Steam: Pressure Table

Pressure (psia)	Temperature (°F)	Specific Volume: Saturated Liquid (ft³/lb)	Enthalpy: Saturated Liquid (Btu/lb)	Enthalpy: Saturated Vapor Btu/lb	Entropy: Saturated Liquid (Btu/lb/°R)	Entropy: Saturated Vapor (Btu/lb/°R)
10	193	0.0166	161	1143	0.284	1.787
50	281	0.0173	250	1174	0.411	1.659
100	327	0.0177	298	1187	0.474	1.603
150	358	0.0181	330	1194	0.514	1.569
200	382	0.0184	355	1198	0.544	1.545
300	417	0.0189	394	1202	0.588	1.510
400	444	0.0193	424	1205	0.621	1.484
600	486	0.0201	472	1203	0.672	1.445
1000	544	0.0216	542	1192	0.742	1.392
1400	587	0.0231	599	1173	0.796	1.345

$$P = 1000 \text{ psia}$$
$$h = 1505 \text{ Btu/lb/°R}$$
$$s = 1.652 \text{ Btu/lb}$$

State 2 is defined by an entropy of 1.652 and a pressure of 54 psia. From thermodynamic tables:

State 2 $s = 1.652$ Btu/lb°R
$P = 54$ psia
$T = 286$°F
$h = 1176$ Btu/lb

From the First Law of Thermodynamics:

$$\Delta\omega_{OUT} = \Delta h = 1505.1 - 1175.6 = 329 \text{ Btu/lb}$$

The units may be converted to kilowatts by dividing by the conversation factor, 3412 Btu/kWh

$$\Delta\omega_{OUT} = 329 \text{ Btu/lb}/3412 = 0.096 \text{ kWh/lb}$$

Example. For the same steam turbine example above, compute the properties of state 2 if the turbine is 85% efficient.

SOLUTION. The turbine will convert only 85% of the available thermal energy to mechanical work. Thus the turbine work output is

$$\Delta\omega_{OUT} = 329. \cdot .85 = 280 \text{ Btu/lb}$$

In electrical units this is:

$$\Delta\omega_{OUT} = 280/3412 = 0.082 \text{ kWh/lb}$$

The final enthalpy of State 2 must then be

$$h_{STATE\ 2} = 1505 - 280 = 1225 \text{ Btu/lb}$$

Given the final pressure of 54 psia and the enthalpy of 1225 Btu/lb, the final temperature and entropy can be found from the thermodynamic tables for a state with 54 psia and $h = 1225$ Btu/lb (or Figure 6.19).

State 2 $P = 54$ psia
$h = 1225$ Btu/lb
$T = 380°F$
$s = 1.71$ Bu/lb/°R

The properties of state 2 are used as the starting point for the next thermodynamic process.

Thermodynamic processes are linked together to form a heat "cycle" in which the thermodynamic fluid is typically heated, expanded in a turbine, used for process heat (or cooled rejecting low-quality residual heat), compressed, and then returned to the beginning of the cycle to be heated again. In this way, heat is added and mechanical energy is produced.

The Rankine cycle is used in steam-turbine cycles. Figure 6.3 illustrates a simplified steam cogeneration cycle. The feedwater enters the boiler (6) and is heated at constant pressure (1). The turbine expands the steam (processes 1 to 2) at constant entropy (if the turbine is "ideal"). The turbine produces work to drive the electrical generator. Most of the steam is delivered to the industrial process and is returned as saturated water. The returned water is mixed with the part of the turbine exhaust steam to partially heat the feedwater prior to returning to the boiler. A heater improves the thermodynamic efficiency. A boiler feed pump increases the pressure of the water to the boiler pressure and the cycle is completed.

REFERENCES

Cogeneration World, "Industrial Cogeneration Potential (1980–2000)," Vol. 1, No. 2 (April 1982), p. 16.

Gentner, R. T., "Cogeneration in the Petrochemical Industry," AICE 1988 Spring National Meeting, March 1988.

Keenan, J. H., and F. G. Keyes, *Thermodynamic Properties of Steam*, Wiley, New York, 1963.

Kovacik, J. M., "Cogeneration Application Considerations," GE Gas Turbine State-of-the-Art Seminar, Saratoga, NY, September 1984.

Limaze, D. R., ed., *Planning Cogeneration Systems*, Fairmont Press, Atlanta, 1985.

McMahan, R. H., *Cogeneration*, Marcel Dekker, New York, 1987.

PROBLEMS

1 An industrial facility is considering the addition of a coal-fired steam cycle cogeneration addition. The facility desires a process steam capability of 200,000 #/hour of steam at 50 psig. The water is returned at 230°F. The boiler is 85% efficient and operates at 1000 psig and 825°F. Assume one open feedwater heater. Calculate the power-to-heat ratio and fuel chargeable to power.

2 Calculate the maximum steam production for a HRSG having a gas inlet temperature of 1050°F, 1,000,000 #/hour flow, and producing steam at 150°F saturated steam. Assume a 92% HRSG effectiveness and a 1.5% radiation loss.

3 Calculate the fuel required and additional steam generated by supplemental firing the HRSG in Problem 2 (above) to 1400°F.

4 An industrial company is evaluating the benefits of a gas-turbine cycle cogeneration plant steam cost improvement alternative. The investment is $28M (in 1990 $) and leads to a $8M/year (in 1990 $) savings in operating cost. The operating cost savings are projected to increase at 6%/year for the next 10 years. Calculate the DCFR over the first 5 years of operation using the tax depreciation rules of Table 6.5.

5 An industrial company is building a new facility and requires 320,000 #/hour of steam at 400 psig. The company is considering a gas-fired packaged boiler and two cogeneration cycles. The company is considering a Model 7 unfired HRSG cogeneration system or a Model 6 HRSG system with supplemental firing to 1600°F (use the data in Table 6.2 for cycle data and assume a 20-psig pressure drop from the HRSG to the process facility). The fuel cost is 3.5 $/MBtu in 1990 and is projected to increase at 5%/year. The industrial company has a 45-MW electricity requirement. Electricity is purchased from the utility at 6¢/kWh and may be sold to the utility at 5.5¢/kWh. Electricity rates are projected to increase at 5%/year. Using the cost data in Section 6.2.4 and 6.4.3, calculate the investment pay-back period for the cogeneration cycles relative to a packaged boiler.

7

ELECTRICITY LOAD-DEMAND FORECASTING

The electric utility industry planning process begins with the electricity load-demand forecast. The demand for electricity initiates actions by utilities to add or retire generation, transmission, or distribution capacity. Because of the long lead time required to license and construct new utility equipment, decisions must be made from 2–10 years in advance of the need for a new utility plant. Table 7.1 illustrates a typical range of decision lead times for several types of new utility equipment.

These long lead times require that the utility planning horizon be at least 10 years. Since utility decisions involve an economic analysis of the operating and investment costs, the utility planning horizon may range from 15 to 30 years into the future. Forecasts with these long lead times are quite a challenge in light of the uncertainties in national, regional, and local economic growth, coupled with uncertainties in electricity usage patterns and conservation trends.

7.1 LOAD FORECAST SEGMENTATION

Forecasting load demand is a difficult procedure and combines art with science. While forecasting tools are available to aid in the process, the key contributions of forecasters are their knowledge of electricity consumers and an understanding of the way they use electricity and other competing energy forms.

A major principle is that consumers make decisions regarding the purchase of electricity. Consumers evaluate electricity in relation to the many facets of their energy needs, including price, availability, reliability, convenience, and

TABLE 7.1 Decision Lead Times

	Lead Time (Years)
Coal fired power plant	6–10
Combustion-turbine power plant	2–3
Transmission line	2–4
Distribution network expansion	1–2

cleanliness. Accurate forecasts require that these and future factors be quantitatively understood.

Typically, the load forecasting task is performed in two steps. First, an economic and demographic forecast is prepared, and then the electricity usage forecast is developed (illustrated in Figure 7.1).

The demand for electricity is dependent on the magnitude and growth of the economy. Robust economic growth creates more jobs, which leads to increased population in a service territory, which, in turn leads to consumers who use more electricity.

The segmentation in Figure 7.1 is helpful in understanding uncertainties in the load forecast. If the economic forecast is in error, then the load forecast will probably be in error. If the electricity usage [per customer or per gross national product dollar (GNP $)] is in error, then the load forecast will likely also be in error. An accurate forecast requires both an accurate economic forecast and an accurate electricity usage forecast.

7.2 OVERVIEW OF LOAD FORECASTING METHODS

The output from a load forecast generally includes a forecast of the annual energy sales (in kilowatt-hours) and the annual peak demand (in kilowatts). Most utilities forecast the annual energy sales first and then use the energy

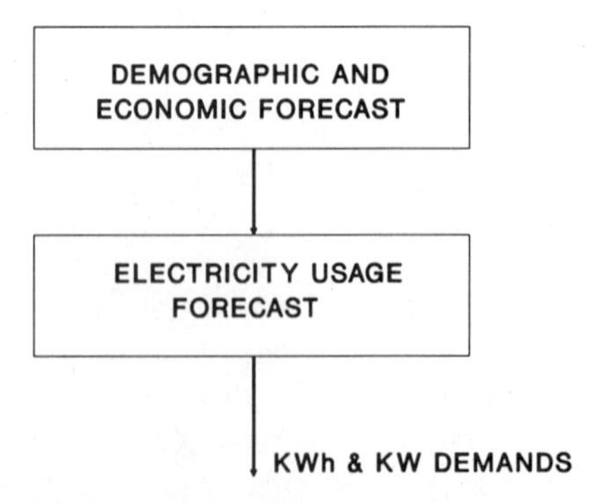

FIGURE 7.1 Load forecasting segments.

sales forecast in determining the annual peak demand forecast. The annual energy sales data is the integration of the hourly loads during the year and is therefore less prone to weather and spurious effects.

The peak demand forecast (in kilowatts) is typically derived by an analysis of the load factor [ratio of average kilowatts to peak kilowatts or kilowatt-hour sales/(8760 • peak kilowatts)]. Once the load factor is forecast, the peak-demand forecast is derived by multiplying the energy forecast times the load factor and dividing by 8760.

$$\text{Peak load} = \frac{\text{energy sales}}{8760 \cdot \text{load factor}} \tag{7.1}$$

There are three widely used methods in load (energy) forecasting: (1) econometric regression analysis, (2) appliance saturation methods, and (3) end-use energy methods. The usefulness of each method depends on data availability, customer segmentations, and degree of detail required.

Econometric regression analysis uses historical annual energy and economic data to determine customer elasticities. An elasticity is a measure of how a customer will change a purchasing pattern in response to a change in price, convenience, reliability, and other factors. Based on customer elasticities, and assuming that these elasticities do not change through time, an energy forecast is made. This method is broadly applied in forecasting kilowatt-hour energy consumption.

The appliance saturation method is an ''engineering''-type methodology. Load research surveys are made to determine the number of customers with a certain appliance (i.e., central air conditioning) and the typical annual energy used by the appliance. Then, on the basis of a forecast of the number of appliances expected in the future, together with a forecast of how the annual energy usage per appliance will change, the energy load forecast is made. This method is commonly used to forecast residential sector energy sales.

The end-use energy method is similar to the appliance saturation method, except that instead of using an appliance as the forecast basis, the basis is an end-use process. For example, this method can be used to forecast the commercial sector. The floor space and kilowatt-hour energy consumption of the principal electric devices per square foot (space heating and cooling, lighting, and auxiliaries) is determined on the basis of a load research survey. Based on a forecast of the floor space, the energy sales forecast is developed.

Each of these methods finds widespread application. When 10–15 years of historical data are available, the econometric regression methods are applicable. When detailed appliance and end-use data are available, then these methods are practical. In some cases, hybrid models are used. For example, a regression model may be used to forecast the percentage of new housing with central air conditioning based on family disposable net income. This would then be used in an appliance saturation model.

This chapter presents econometric load forecasting techniques. Chapter 8 presents the appliance saturation and end-use energy forecasting methods and peak-load forecasting methods. Chapter 8 also discusses load management.

7.3 ECONOMETRIC LOAD FORECASTING

The objective of econometric load forecasting is to forecast electricity sales based on a macroeconomic forecast. The macroeconomic driving variables (also called "drivers") include the gross national product (GNP), industrial production, and electricity price.

7.3.1 Single-Variable Econometric Equations

As an illustration of this procedure, consider an example in which the average residential kilowatt-hour sales per customer is to be forecast. Assume that kilowatt-hour sales/customer is driven entirely by the economic variable, income per household. Theoretically, as income per household increases, people spend more money on electricity-consuming appliances, which leads to more electricity sales.

Table 7.2 presents historical data over a 7-year period (1980–1986). The objective is to develop a forecasting equation:

$$\text{Sales per customer} = A + \beta \cdot \text{income per household} \tag{7.2}$$

where A and β are coefficients to be determined based on the historical trends between sales per customer and income per household.

The forecast will be based on the assumption that the future relationship between sales per customer and income per household will be the same as the historical relationship.

The mathematical equations for the forecasting equation can be derived

TABLE 7.2 Econometric Load Forecasting Example

i	Year	Y Sales per Customer	X Income per Household
1	1980	40	100
2	1981	45	200
3	1982	50	300
4	1983	65	400
5	1984	70	500
6	1985	70	600
7	1986	80	700

more conveniently and interpreted more intuitively if expressed in normalized mathematical format. The normalized mathematical format is:

$$Y_i = A + \beta \cdot (X_i - \overline{X}) + \epsilon_i = \hat{Y} + \epsilon_i \qquad \text{for} \quad i = 1, \ldots, 7 \tag{7.3}$$

where Y_i = actual value of sales per customer (dependent variable)
$\hat{Y}_i$ = predicted (estimated) value of sales per customer
ϵ_i = the error between the actual and the predicted dependent variable
X_i = income per household (independent variable or driver variable)
$\overline{X}$ = mean value of X_i
A, β = constants to be determined

The mean value of X_i is defined as

$$\overline{X_i} = \frac{1}{7} \sum_{i=1}^{7} X_i \tag{7.4}$$

This chapter will use the following superscript notations:

^ denotes the predicted value,
‾ denotes the mean (average value),
No superscript denotes the raw value, and
Lower case denotes a normalized variable that is equal to the raw value less the mean value.

One way of determining the values of A and β is to select these values such that the square of the error between the actual sales per customer and the predicted sales per customer is minimized. This criteria can be expressed as:

$$\min \sum_{i=1}^{7} (Y_i - \hat{Y}_i)^2 \tag{7.5}$$

By substituting Equation (7.2):

$$= \min \sum_{i=1}^{7} [Y_i - A - \beta (X_i - \overline{X})]^2 \tag{7.6}$$

If the derivative of Equation (7.6) with respect to A is calculated and set to zero to achieve a minimum, the result is:

$$\sum_{i=1}^{7} 2(-1) [Y_i - A - \beta (X_i - \overline{X})] = 0 \tag{7.7}$$

Note that it can also be expressed as:

$$\sum_{i=1}^{7} (\epsilon_i) = 0 \tag{7.8}$$

The coefficients A and β are determined such that the sum of the errors is zero.

Since $\sum_{i=1}^{7} X_i = 7 \cdot \overline{X}$, then the β term is eliminated in Equation (7.7), and

$$\boxed{A = \frac{\sum_{i=1}^{7} Y_i}{7}} = 60 \text{ in the example.} \tag{7.9}$$

Thus, the value of the A coefficient is merely the average of the Y_i values. Differentiating Equation (7.6) with respect to β yields:

$$\sum_{i=1}^{7} -(2)(X_i - \overline{X})[Y_i - A - \beta(X_i - \overline{X})] = 0 \tag{7.10}$$

Note that this can also be expressed by substitution of Equation (7.3) as:

$$\sum_{i=1}^{7} (X_i - \overline{X})\,\epsilon_i = 0 \tag{7.11}$$

The coefficients are determined such that the product of the error times the independent variable is zero.

Since $\sum_{i=1}^{7} (X_i - \overline{X}) = 0$, solving for β in Equation (7.10) yields:

$$\boxed{\beta = \frac{\sum_{i=1}^{7} Y_i (X_i - \overline{X})}{\sum_{i=1}^{7} (X_i - \overline{X})^2}} = 0.68 \text{ in the example} \tag{7.12}$$

These calculations can be conveniently done using a worksheet as shown in Table 7.3.

First, the expected value of the driver variable X is calculated, as performed in column 2. The driver variable is then normalized by subtracting the mean value (column 3). The dependent variable is shown in column 4. The average of column 4 is computed, which represents the A coefficient.

The product x_iY_i is computed in column 5, which is needed for the numerator in Equation (7.12). The x_ix_i product is computed in column 6, which is

TABLE 7.3 Sample Calculations[a]

1	2	3	4	5	6	7	8	9
						Predicted	Error	Error Squared
Year	X_i	$X_i - \overline{X} = x_i$	Y_i	$x_i Y_i$	x_i^2	$\hat{Y}_i = A + \beta x_i$	$Y_i - \hat{Y}_i$	$(Y_i - \hat{Y}_i)^2$
1980	100	−300	40	−12,000	90,000	39.64	0.36	0.13
1981	200	−200	45	−9,000	40,000	46.43	−1.43	2.04
1982	300	−100	50	−5,000	10,000	53.21	−3.21	10.33
1983	400	0	65	0	0	60.00	5.00	25.00
1984	500	100	70	7,000	10,000	66.78	3.21	10.33
1985	600	200	70	14,000	40,000	73.57	−3.57	12.75
1986	700	300	80	24,000	90,000	80.36	−0.36	0.13
	ΣX_i		420	$\Sigma x_i Y_i$	Σx_i^2		0.00	$X(Y_i - \hat{Y}_i)^2$
	= 2800			= 19,000	280,000			= 60.71

$$\overline{X}_i = \frac{2800}{7}$$

$$\overline{X} = 400$$

$$A = \frac{420}{7}$$

$$\boxed{A = 60}$$

$$\beta = \frac{\Sigma x_i Y_i}{x_i^2} = \frac{19,000}{280,000}$$

$$\boxed{\beta = .0678}$$

$$\sigma^2 = \frac{\Sigma(\text{error})^2}{N - 2} = \frac{60.71}{5} = 12.14$$

$$\boxed{\sigma = 3.48}$$

[a] Uppercase X_i, indicates raw data; lowercase, x_i, indicates raw data less the mean value, that is, normalized.

needed in the denominator of Equation (7.12). Thus, the β value can be determined.

Once A and β are determined, the predicted dependent values can be calculated as shown in column 7. The error between the predicted values and the actual data is computed by subtracting column 7 from column 4. Finally, the error squared is calculated in column 8. The error squared will be used later in the analysis.

7.3.2 Measurement of Relationship Fit

Figure 7.2 presents a graph of the fitted regression line and contrasts this with the actual data points. The regression line is a reasonably good fit to the data.

It is necessary to quantify how good the fit is between the regression line and the data. One measure that can be calculated is called the "R-squared coefficient."

To examine the goodness of the fit of the regression equation, examine the case in Figure 7.3, which illustrates a regression line approximating some data. In Figure 7.3, several distances can be measured: (1) the distance between the data point and the regression line, (2) the distance between the regression line and the average Y value, and (3) the distance between the data point and the average Y value. From inspection, Equation (7.13) states that the distance between data point i to the average Y value is equal to the sum of the distances between data point i and the regression line and the distance between the regression line and the average Y value.

$$(Y_i - \overline{Y}) = (Y_i - \hat{Y}_i) + (\hat{Y}_i - \overline{Y}) \tag{7.13}$$

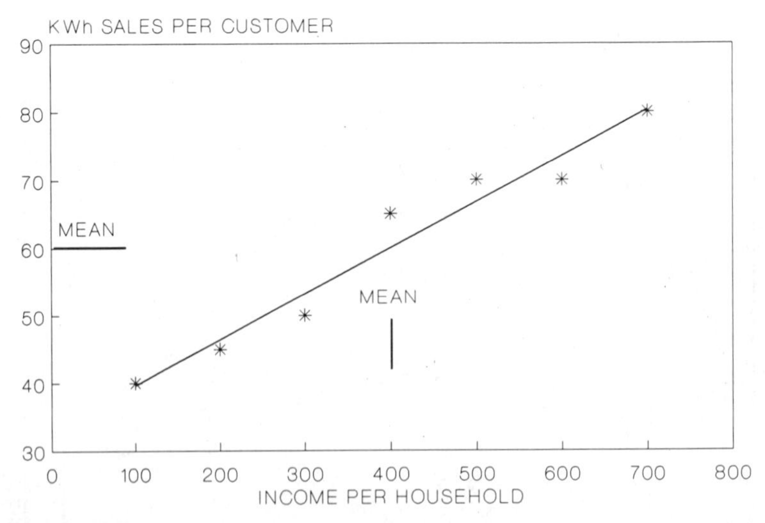

FIGURE 7.2 A sample fitted regression line contrasted with actual data points.

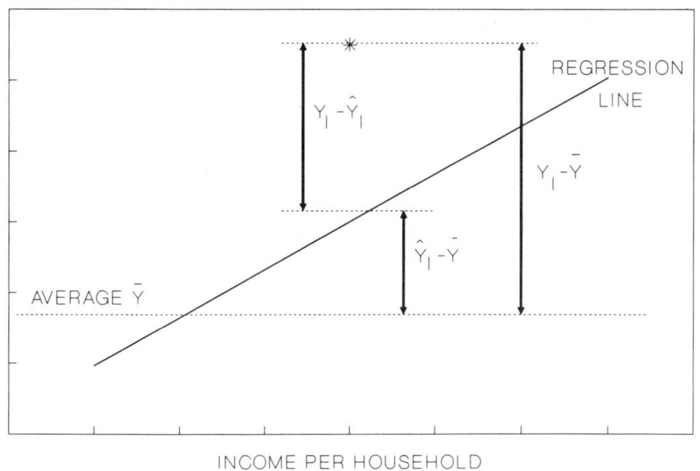

FIGURE 7.3 A fitted regression line approximating sample data.

If Equation (7.13) is squared and summed over all data points, Equation (7.14) results.

$$\sum_{i=1}^{n} (Y_i - \overline{Y})^2 = \sum_{i=1}^{n} (Y_i - \hat{Y}_i)^2 + \sum_{i=1}^{n} (\hat{Y}_i - \overline{Y})^2 + 2 \sum_{i=1}^{n} (Y_i - \hat{Y}_i)(\hat{Y}_i - \overline{Y}) \tag{7.14}$$

Equation (7.14) can be simplified; note that the last term, the cross-products, will be equal to zero.

$$\sum_{i=1}^{n} (Y_i - \hat{Y}_i)(\hat{Y} - \overline{Y}) = \sum_{i=1}^{n} (\text{ERROR}_i)(A + \beta x_i - \overline{Y}) \tag{7.15}$$

However, also note:

$\sum_{i=1}^{n} (\text{ERROR}_i) = 0$ by the construction of the A regression coefficient [Equation (7.8)]

$\sum_{i=1}^{n} (\text{ERROR}_i x_i) = 0$ by the construction of the A and β regression coefficients [Equation (7.11)]

Thus:

$$\Sigma(Y_i - \overline{Y})^2 = \Sigma(Y_i - \hat{Y}_i)^2 + \Sigma(\hat{Y}_i - \overline{Y})^2 \tag{7.16}$$

or

$$\text{Total variation in } Y = \text{residual variation in } Y + \text{explained variation in } Y \tag{7.17}$$

Equation (7.16) states that the total variation of the data points about the average Y value is equal to the residual variation between the data points and the regression line plus the explained variation in Y, which is the distance between the data points and the average Y value. The quantity R^2 is now defined as the proportion of the variation of the data about the mean explained by the regression equation, that is, the explained variation divided by the total variation as presented in Equation (7.18). The range of values for R^2 lies between 0 and 1. An R^2 of 0 indicates that the regression equation has done nothing to explain the total variation in Y, while an R^2 of 1 indicates that the regression equation lies directly on top of all the data points. Thus, it is desirable to obtain R^2 values as close to 1.0 as possible.

$$R^2 = \frac{\text{explained variation}}{\text{total variation}} = \frac{\Sigma(\hat{Y} - \overline{Y})^2}{\Sigma(Y - \overline{Y})^2} \tag{7.18}$$

Take the results from the previous example and compute the R^2 measure of goodness of fit. This calculation is illustrated in Table 7.4. The result is an R^2 coefficient of .959, which indicates that a significant part of the variation in the data has been explained by the regression line.

$$R^2 = \frac{1294.6}{1350} = .9590$$

TABLE 7.4 The R^2 Measure of Goodness of Fit—Based on Example Data

$$R^2 = \frac{\Sigma(\hat{Y} - \overline{Y})^2}{\Sigma(Y - \overline{Y})^2}$$

X_i	Y	$\hat{Y}$	Regression about Mean $(\hat{Y} - \overline{Y})$	$(\hat{Y}_2 - \overline{Y}_2)^2$	Actual about Mean $(Y - \overline{Y})$	$(Y - \overline{Y})^2$
100	40	39.6	−20.4	416.2	−20	400
200	45	46.4	−13.6	184.9	−15	225
300	50	53.2	−6.8	46.2	−10	100
400	65	60.0	0	0	5	25
500	70	66.8	6.8	46.2	10	100
600	70	73.6	13.6	184.9	10	100
700	80	80.4	20.4	416.2	20	400
				1294.6		1350

Values of R^2 for a good fit can vary from .9 to 1.0. Of course, this depends on the quantity of data and its variability. If there are only four pieces of data, fitting a straight line through four pieces of data could yield an R^2 of .90, which could be entirely unacceptable in some cases. In other cases, an R^2 of .7 would be a significant accomplishment if the line was to fit through 100 data points.

7.3.3 Confidence Intervals

The next step is to determine a statistical confidence interval on the regression coefficients, A and β. The mathematical procedure has estimated these coefficients. However, there is statistical uncertainty that A and β are the "true" values of the relationship between income per household and kilowatt-hour electricity sales. The "true" values of A and β could be determined if an infinite amount of data were available (rather than seven data samples) and there were complete understanding of the relationships of all other drivers that influence the kilowatt-hour sales. Thus, the A and β calculated have a confidence interval (or tolerance) as:

$$A_{\text{true}} = A_{\text{calculated}} \pm \text{confidence interval of } A \tag{7.19}$$

$$\beta_{\text{true}} = \beta_{\text{calculated}} \pm \text{confidence interval of } \beta \tag{7.20}$$

Before proceeding to calculate the confidence interval, it is important to review a few statistical preliminaries. First of all, the expected value or mean of X_i is the average value, Equation (7.21).

$$\overline{X} = \frac{1}{N} \sum_{i=1}^{N} X_i \tag{7.21}$$

where N is a *sample* of data. The deviation of X_i from the mean (normalized variable) is defined by Equation (7.22):

$$\text{Deviation}_i = (X_i - \overline{X}) \tag{7.22}$$

The mean-square deviation is an important quantity that measures the average squared deviation, Equation (7.23).

$$\text{Mean-square deviation} = \frac{1}{N} \sum_{i=1}^{N} (X_i - \overline{X})^2 \tag{7.23}$$

The mean-square deviation is a useful measure of the *data sample.* It may be desirable, however, to have an estimate of the mean-square deviation of the *entire population* (keep in mind that the seven data-point example is only a sample from a larger population of reality). This data sample estimate must be inflated slightly to calculate the mean square deviation of the whole population. Advanced texts on statistics demonstrate this (see Guttman, Wilks, and Hunter, 1965).

In the seven data-point sample, only seven independent parameters of the data sample can be calculated. The mean of the data sample is one parameter; only six independent parameters remain. Statisticians refer to this by stating that one "degree of freedom" has been consumed in the calculation of the mean, leaving $N - 1$ degrees of freedom remaining. When the mean square deviation of the entire population is to be estimated based on the data sample, it must be calculated as:

$$\begin{aligned}\text{Population estimate of mean-square deviation} &= \frac{1}{\text{degrees of freedom}} \sum_{i=1}^{N} (X_i - \overline{X})^2 \\ &= \frac{1}{N-1} \sum_{i=1}^{N} (X_i - \overline{X})^2 \qquad (7.24)\end{aligned}$$

The population mean-square deviation is also referred to as the "variance of the population."

$$\text{Variance} = \sigma^2 = \frac{1}{N-1} \sum_{i=1}^{N} (X_i - \overline{X})^2 \qquad (7.25)$$

The population standard deviation is also widely used. The standard deviation is defined as the square root of the variance, that is, σ (sigma).

In the previous analysis, a regression coefficient β has been determined based on the data. However, this value of β is only an estimate of the true regression coefficient. Figure 7.4 presents a graph of the true regression coefficient β versus the probability that the calculated β lies near the true β.

The probability that the true β lies within a specified confidence interval from the calculated β can be determined. This is done by calculating the statistical variance of the true β about the calculated β. Equation (7.26) indicates that the prediction of the dependent variable is:

$$\hat{Y}_i = A + \beta \cdot (X_i - \overline{X}) \qquad (7.26)$$

where A is calculated as the average of the data points and β is calculated as shown in Equation (7.27).

$$\text{Where } A = \overline{Y}_i, \qquad \beta = \frac{\Sigma \hat{Y}_i x_i}{\Sigma (x_i)^2} = \frac{\Sigma y_i x_i}{\Sigma x_i^2} \qquad (7.27)$$

Substitute for the calculation of β, noting that β is equal to the summation of the $y_i x_i$ divided by the summation of the x_i^2. By substituting Equation (7.3)

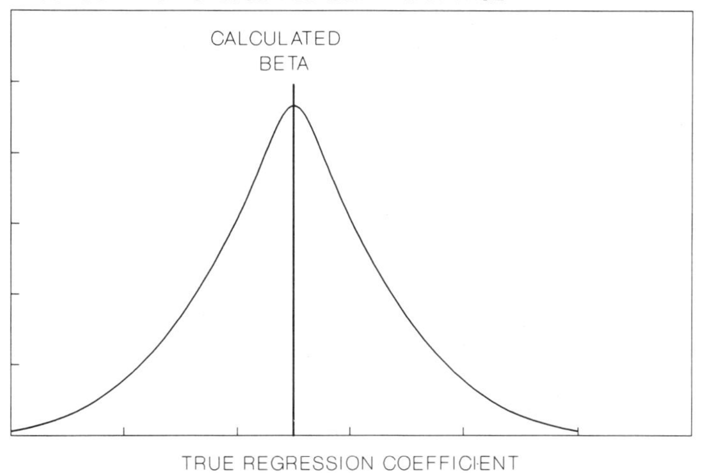

FIGURE 7.4 Expected value of ß based on the true regression coefficient.

into this relationship, an expression can be obtained for β as indicated in Equation (7.28).

$$\beta = \frac{\Sigma y_i x_i}{\Sigma x_i^2} = \frac{\Sigma x_i^2 \beta_{\text{true}} + x_i \epsilon_i}{\Sigma x_i^2}$$

$$= \frac{\Sigma x_i^2 \beta_{\text{true}} + x_i \epsilon_i}{\Sigma x_i^2} \tag{7.28}$$

Simplifying Equation (7.28) yields Equation (7.29).

$$\beta = \beta_{\text{true}} + \frac{\Sigma x_i \epsilon_i}{\Sigma x_i^2} \tag{7.29}$$

Simplifying further yields Equation (7.30), which states that the difference between $\beta - \beta_{\text{true}}$ is equal to the summation of the x_i times the errors divided by the summation of the x_i^2.

$$\beta - \beta_{\text{true}} = \frac{\Sigma x_i \epsilon_i}{\Sigma x_i^2} \tag{7.30}$$

To calculate the variance of $\beta - \beta_{\text{true}}$, square the expression in Equation (7.30) as shown in Equation (7.31).

$$(\beta - \beta_{\text{true}})^2 = \frac{(\Sigma x_i \epsilon_i)^2}{(\Sigma x_i^2)^2} \tag{7.31}$$

An important concept must be discussed at this point. The seven data points in the example are only a sample of data points from a larger total population. Each data point, x_i, may be interpreted as a random variable that just happened to be selected for inclusion in this seven data-point sample. Hence, the expectation of Equation (7.31) for any data-point sample from the population will yield the variance of β.

Taking the expected value of Equation (7.31) yields the variance of β as shown in Equation (7.32).

$$\text{Expected value } (\beta - \beta_{\text{true}})^2 = \text{var}(\beta)$$

$$= \text{expected value } \frac{\{\Sigma x_i \epsilon_i^2\}}{\{\Sigma x_i{}^2\}^2}$$

$$= \frac{\text{expected value } \{x_1^2\, \epsilon_1^2 + x_1 x_2 \epsilon_1 \epsilon_2 + x_2^2\, \epsilon_2^2 + \cdots\}}{(\Sigma x_1^2)^2} \tag{7.32}$$

Note that in Equation (7.32), the summation of x_i^2 in the denominator can be extracted, and an expression results that has terms involving the expected value of a series of products of x_i^2 times the error$_i^2$. If it is assumed that the error terms are uncorrelated with regard to each other, which is a very reasonable assumption, then the expected value of error$_j$ times error$_k$ will be equal to zero [Equation (7.33)].

$$\text{Expected value } \{\epsilon_j\, \epsilon_k\} = 0 \tag{7.33}$$

Equation (7.32) can, therefore, be simplified. The variance of β is then equal to the expected value of x_i^2 times the error$_i^2$ [Equation (7.34)]. If it is assumed that the variance of the error terms, namely, ϵ_i^2 is constant and equal to a value σ^2, then Equation (7.34) can be simplified to Equation (7.35). The variance of β is equal to the variance of the error terms divided by the summation of the x_i^2.

$$\text{var }(\beta) = \frac{\text{expected value } \{x_1^2\, \epsilon_1^2 + x_2^2\, \epsilon_2^2 + x_3^2\, \epsilon_3^2 + \cdots\}}{(\Sigma x_1^2)^2}$$

$$= \frac{\Sigma x_i^2 \text{ expected value } \{\epsilon_i\}^2}{(\Sigma x_i^2)^2} \tag{7.34}$$

$$\text{var}(\beta) = \frac{\sigma^2\, \Sigma x^2}{(\Sigma x^2)^2} \tag{7.35}$$

$$\text{var}(\beta) = \frac{\sigma^2}{\Sigma x_i^2} \tag{7.36}$$

where σ^2 = variance of the error = expected value $\{\epsilon_i^2\}$.

Furthermore, the standard deviation is merely the square root of the variance and is equal to σ divided by the square root of the x_i^2 [Equation (7.37)].

$$\text{Standard deviation } (\beta) = \frac{\sigma}{\sqrt{\Sigma x_i^2}} \tag{7.37}$$

It is now appropriate to develop the concept of confidence limits or confidence intervals. By the central-limit theorem of statistics, the distribution of the mean value of β from the true mean will be a normal curve. This was shown in Figure 7.4. The area under the normal curve is a measure of the probability that the calculated β lies within the range of the true β. Table 7.5 shows some typical values of the area under the normal curve as a function number of standard deviations from the mean. For example, 95% of the probability under the normal curve is within ± 1.96 standard deviations from the mean value. Equation (7.38) provides a measure of the confidence of the true β. This equation states that the confidence interval of the true β is equal to the calculated β plus or minus the standard deviation of β times the number of standard deviations associated with a given level of confidence. The 95% confidence interval of the true value of β is found by adding to the calculated β ± 1.96 times the standard deviation of β.

$$\text{Confidence interval of } \beta_{\text{true}} = \beta_{\text{calculated}} \pm (\text{standard deviation } \beta) \cdot (\text{\# of standard deviations}) \tag{7.38}$$

Note that the normal curve is the correct curve to use to measure confidence intervals if there are an infinite number of degrees of freedom, (i.e., an infinite number of data points). However, since there is not an infinite number of data points, only a finite number, it is more correct to use a variation of the normal curve called "student's t distribution."

The student's t distribution has a shape similar to that of the normal curve, except that it is more spread out and is a two-parameter curve. Table 7.6 presents confidence interval data for a 95% confidence level and 90% confidence level as a function of the number of degrees of freedom.

The student's t distribution depends on the degrees of freedom. The degrees

TABLE 7.5 Typical Values of the Area Under the Normal Curve

(Area) Confidence (%)	Number of Standard Deviations
50	0.68
90	1.65
95	1.96
98	2.33

TABLE 7.6 Student's *t* Distribution

	Number of Standard Deviations	
Degrees of Freedom	95% Confidence	90% Confidence
5	2.571	2.015
10	2.228	1.812
15	2.131	1.753
20	2.086	1.725
25	2.060	1.708
30	2.042	1.697
∞	1.960	1.645

of freedom are equal to the number of data points in the sample minus the number of regression coefficients. If there were 10 degrees of freedom, 2.228 standard deviations would be required to encompass the 95% confidence interval. If, however, there were an infinite number of degrees of freedom, then the result would be that of the normal curve.

Use this theory with the example in Table 7.3 and calculate the confidence interval for the β coefficient. Using the results of Equation (7.37), the standard deviation is calculated as the standard deviation of the error divided by the square root of the summation of the x_i squared. Using the results in Table 7.3, the σ is calculated to be 3.48 and the summation of the x_i squared is calculated to be 280,000. Equation (7.39) presents this calculation of the standard deviation of β as equal to .00658. Equations (7.40) and (7.41) show the β calculations.

$$\text{Standard deviation} = \frac{\sigma}{\sqrt{\Sigma x_i^2}} = \frac{3.48}{\sqrt{280000}} = .00658 \tag{7.39}$$

$$\beta(\text{true}) = .068 \pm 2.571 \cdot .00658 \tag{7.40}$$

$$\boxed{\beta(\text{true}) = .068 \pm .017} \tag{7.41}$$

This result provides a very useful measure of how accurate the value of β is. If the true β had a very large confidence interval, it could be anticipated that the regression analysis would not be particularly good. On the other hand, if the confidence interval value is very small, it could be anticipated that the regression analysis would be very good.

7.3.4 The *t* Statistic

This concept is carried one step further when a parameter called the "*t* statistic" is defined. The *t* statistic is the number of standard deviations that would render uncertainty as to whether the true β was equal to zero, often referred to as the null hypothesis. A value for t is calculated such that the true value of β is equal to zero, as shown in Equation (7.42). The resulting t is equal to 10.38, as shown in Equation (7.43). The t statistic is equal to the regression coefficient divided by the standard deviation.

The t statistic measures the significance of each variable in the regression analysis. A t statistic of 10 would indicate a value of 99.9% confidence that the β calculated is indeed not equal to zero. Expressed in another way, the tolerance on the β is very small. Normally, a t statistic of 2, which would be typical of a 95% confidence interval, can be used as a cutoff point on which to judge whether the regression coefficient is a significant explanatory (driver) variable.

$$\beta = 0 = .068 \pm t \cdot .00658 \tag{7.42}$$

$$t = \frac{.068}{.00658} = 10.38 \tag{7.43}$$

$$\boxed{t_{\text{statistic}} = \frac{\text{regression coefficient}}{\text{standard deviation}}} \tag{7.44}$$

Example. A regression analysis was performed, and the results are shown in Equation (7.45).

$$\begin{aligned} \text{Beer sales} &= 56 + .15 \cdot \text{GNP} \\ t_{\text{statistic}} &= 1.1 \end{aligned} \tag{7.45}$$

The regression suggests that beer sales are equal to a constant plus .15 times the GNP, with a t statistic of 1.1. This is poor regression, because the t statistic is only 1.1, which indicates that there is less than a 90% chance that GNP is the key driver to beer sales.

Suppose that beer sales were calculated using a different regression model where sales were equal to 34 plus .2 times disposable income, and the t statistic was equal to 4.6, as shown in Equation (7.46).

$$\begin{aligned} \text{Beer sales} &= 34 + .20 \cdot \text{disposable income} \\ t_{\text{statistic}} &= 4.6 \end{aligned} \tag{7.46}$$

This is a better regression because it explains the relationship between beer sales as being driven by disposable income. Because it has a t statistic of 4.6, it may be confidently inferred that disposable income influences beer sales.

Previous sections have established methods to calculate regression coefficients and confidence interval criteria. The regression equation can now be used to forecast the kilowatt-hour sales per household given a forecast of the income per household.

7.3.5 Multiple Regression

In many cases, one driver or explanatory variable does not fully explain kilowatt-hour usage, the dependent variable. In this case, a multiple regression is necessary in which several explanatory factors can influence the dependent variable.

The methods used for multiple regression are an extension of those used for single regression. When the equations for multiple regression are written in matrix notation, they appear identical to the case of single regression.

In this section, Y is the dependent variable to be forecast, such as kilowatt-hours; X, W, and Z are independent (explanatory or driver) variables; and A, B, C, and D are regression coefficients.

Equation (7.47) presents the formulation for a multiple regression in which Y may be kilowatt-hours, X may be income per household, W may be electricity price, and Z may be fossil-fuel price.

$$Y = A + \beta(X - \overline{X}) + C(W - \overline{W}) + D \cdot (Z - \overline{Z}) + \epsilon \qquad (7.47)$$

where $A = \overline{Y}$(average) and ϵ = error between the actual and the predicted dependent variable.

These equations can be rewritten in matrix notation, in which y is equal to a vector x_i times a vector B. Equation (7.48) is also shown in subscripted scalar notation: y_i is equal to the summation over all the k regression coefficients of the independent variables, x_i, times the β_k.

$$\begin{aligned} y_i &= [x_i]\,[\beta] + \epsilon_i && \text{(matrix notation)} \\ &= \sum_k x_{i,k}\,\beta_k + \epsilon_i && \text{(scalar notation)} \qquad (7.48) \end{aligned}$$

where $[x_i]$ = vector of K explanatory (driver) variable with element $x_{i,k}$
$[\beta]$ = one-by-K matrix (vector) of coefficients with element β_k
y_i = normalized dependent variable (explained variable)
i = subscript index for data points
k = subscript index for the kth regression coefficient
[ ] = brackets indicating a matrix

The error in the regression equation can be expressed as:

$$\text{Error}_i = y_i - [x_i][\beta] \qquad (7.49)$$

The regression coefficients are chosen to best fit the data. The typical criteria is a least-square criteria to minimize the error squared over all data points i.

Equation (7.50) presents this criterion in both subscripted notation and matrix notation.

$$\min \{\sum_i \text{error}_i^2 = \sum_i [y_i^2 - 2y_i(\sum_k x_{i,k}\beta_k) + (\sum_k x_{i,k}\beta_k)^2]\}$$

$$= \sum_i [y_i^2 - 2y_i[x_i][\beta] + [x_i][\beta]^2]\} \qquad (7.50)$$

To find the β coefficients that minimize the error squared, Equation (7.50) can be differentiated with respect to the β coefficient, β_k, and the result set equal to zero. Equation (7.51) results.

$$\sum_i 2y_i x_{i,k} = \sum_i \frac{d}{d\beta_k}(x_{i,1}\beta_1 + x_{i,2}\beta_2 + \cdots)^2$$

$$= 2\sum_i x_{i,k} \bullet \{x_{i,1}\beta_1 + x_{i,2}\beta_2 + x_{i,3}\beta_3 + \cdots\}$$

$$\text{for} \quad k = 1, \ldots, k \qquad (7.51)$$

Algebraic simplification of Equation (7.51) leads to Equation (7.52).

$$\sum_i y_i x_{i,k} = (\sum_i x_{i,1}x_{i,k})\,\beta_1 + (\sum_i x_{i,2}x_{i,k})\beta_2 + \cdots$$

$$+ (\sum_i x_{i,k}^2)\,\beta_k + \cdots \qquad \text{for} \quad k = 1, \ldots, k \qquad (7.52)$$

Equation (7.52) is a series of algebraic coupled equations that are to be solved for each β_k.

The notation of Equation (7.52) can be simplified by using two definitions, as shown in Equations (7.53) and (7.54).

Let:

$$W_{1k} = \sum_i x_{i,1}\, x_{i,k} = [X^T][X]_{1k}$$

$$= \text{matrix of rank equal to the number of regression coefficients} \qquad (7.53)$$

$$x_k^T Y = \sum_i x_{i,k}\, y_i = [X^T][Y]_k$$

$$= \text{vector of rank equal to the number of regression coefficients}$$

$$[\beta] = \text{vector of regression coefficients} \qquad (7.54)$$

where a superscript T refers to a matrix transpose operation.

Equation (7.53) defines the W product matrices that relate the cross-products between the independent drivers. The W matrix is a correlation matrix between the explanatory variables. The Y^TX matrix of Equation (7.54) is a vector whose rank equals the number of regression coefficients and is a product of the explained variable times the driver variable.

Equation (7.52) can be rewritten in algebraic notation in Equation (7.55). The term X_k^TY is equal to the summation over all the regression coefficients of the W_{kP} matrix times the β_p coefficient.

$$X_k^TY = \sum_P W_{kP}\,\beta_P \qquad \text{for} \qquad k = 1, \ldots, k \tag{7.55}$$

The matrix notation appears in Equation (7.56).

$$[X^TY] = [W]\,[\beta] \tag{7.56}$$

The β vector can be solved by inverting the W matrix, as shown in Equation (7.57), or in algebraic notation in Equation (7.58). The β vector is equal to the inverse of the W matrix times the XY vector.

$$[\beta] = [W]^{-1}\,[X^TY] \tag{7.57}$$

or

$$\beta_P = [W_{kP}]^{-1}\,[X_k^TY] \tag{7.58}$$

where $[W_{kP}]^{-1}$ denotes the matrix inversion of the W_{kP} matrix.

7.3.6 Multiple Regression Confidence Intervals

Having solved for the regression coefficient β, calculate the variance of β. This can be easily done by using Equation (7.58) or (7.57), in which the value of Y_i is substituted as X times the true β plus an ϵ, or error term, as shown in Equation (7.59).

$$[\beta] = [W]^{-1}\,[X^T]\{[X][\beta_{\text{true}}] + \epsilon]\} \tag{7.59}$$

Equation (7.59) can be simplified by noting that W inverse is the inverse of the correlation matrix of the driving variables.

$$[W]^{-1} = \{[X^T][X]\}^{-1} \tag{7.60}$$

Thus, $\beta - \beta_{\text{true}}$ is equal to the inverse W matrix times the vector X times the error, ϵ.

Thus:

$$[\beta - \beta_{\text{true}}] = W^{-1}\,[X^T][\epsilon] \tag{7.61}$$

If the $\beta - \beta_{\text{true}}$ vector is multiplied by the transpose of the $\beta - \beta_{\text{true}}$ vector, a square matrix of $\beta - \beta_{\text{true}}$ results. This square product is shown in Equation (7.62).

$$[\beta - \beta_{\text{true}}]\,[\beta - \beta_{\text{true}}]^T = W^{-1}\,[X^T][\epsilon][\epsilon]^T[X^T]^T[W^{-1}]^T \tag{7.62}$$

The variance of $\beta - \beta_{true}$ is calculated by taking the expected value of Equation (7.62). As in the single regression model case, the expected value is over the entire population of data. By assuming that the error term has an expected constant variance, the $\epsilon\epsilon^T$ term may be extracted from the matrix multiplication in Equation (7.62). The result can be simplified by noting that W^{-1} times X^TX is merely the identity matrix. This results in Equation (7.63). The variance of $\beta - \beta_{true}$ is equal to σ^2 (which is the expected variance of the error) times the inverse of the W matrix. Equation (7.64) presents these results in an algebraic notation.

$$\text{var}[\beta - \beta_{true}] = \sigma^2 W^{-1} \tag{7.63}$$

$$\begin{bmatrix} \text{var}(\beta_1) & \text{cov}(\beta_1\beta_2) & \cdot \\ \cdot & \text{var}(\beta_2) & \cdot \\ \cdot & \cdot & \cdot \\ \cdot & \cdot & \text{var}(\beta_k) \end{bmatrix} = \sigma^2 [W]^{-1} \tag{7.64}$$

The standard deviation of any element in the variance matrix can be calculated by taking the square root. Thus, the standard deviation of β_k is equal to σ (which is the variance of the error), times the square root of the inverse of the W_{kk} matrix [Equation (7.65)].

$$\text{Standard deviation } (\beta_k) = \sigma \sqrt{[W_{kk}]^{-1}} \tag{7.65}$$

In Equation (7.66) the true β_k, is equal to the calculated β_k ± the standard deviation times t, where t corresponds to the number of standard deviations of specified confidence interval of the student's t distribution.

$$\beta_{k\,true} = \beta_k \pm \sigma \sqrt{[W_{kk}]^{-1}} \cdot t \tag{7.66}$$

where t = number of standard deviations corresponding to a specified confidence interval and number of degrees of freedom of the student's t distribution

$[W_{kk}]^{-1}$ = kth diagonal element of the inverse of the correlation matrix $[X^T][X]$

Confidence intervals can now be calculated as in the single regression case. The t test can be calculated for each one of the β_k regression coefficients, as shown in Equation (7.67).

$$t_k = \frac{\beta_k}{\sigma \sqrt{[W_{kk}]^{-1}}} \tag{7.67}$$

where β_k = kth regression coefficient

σ = standard deviation of the error

$[W_{kk}]^{-1}$ = diagonal kth element of the inverse of the correlation matrix

t_k = t test coefficient for the kth regression coefficient.

Again, as with the single regression case, statistically significant regression coefficents generally have t tests greater than 2.

7.3.7 Multiple Regression Example

This section extends the example used in Section 7.3 to include a second explanatory variable, the price of oil. The same data as presented in Table 7.2 but augmented by the second explanatory variable, the price of oil, is presented in Table 7.7. The first step is to compute the average kilowatt-hour sales per customer, income per household, and price of oil, as shown at the bottom of Table 7.6. The next step is to compute all the matrix products defined in Equations (7.53) and (7.54), between the sales per customer, income per household, and price of oil, as shown in Table 7.8.

After computing all the matrix terms, they can be inserted into the set of regression equations of Equation (7.56):

$$[W][\beta] = [X^T Y] \tag{7.68}$$

Inserting the numerical values leads to:

$$\begin{bmatrix} 280{,}000 & -500 \\ -500 & 24 \end{bmatrix} \begin{bmatrix} \beta_X \\ \beta_W \end{bmatrix} = \begin{bmatrix} 19{,}000 \\ -20 \end{bmatrix} \tag{7.69}$$

TABLE 7.7 Econometric Load Forecasting Example with the Addition of Data on the Price of Oil

Year	(Y) kWh Sales per Customer	(X) $ Income per Household	(Z) Competition: Price of Oil
1980	40	100	36
1981	45	200	33
1982	50	300	37
1983	65	400	37
1984	70	500	34
1985	70	600	32
1986	80	700	36
Total	420	2800	245
Average	60	400	35

TABLE 7.8 Cross-Product of Table 7.6 Regression Coefficients

Y_i	Income x_i	Oil Price z_i	$Y_i x_i$	$Y_i z_i$	x_i^2	z_i^2	$x_i z_i$
40	−300	1	−12,000	40	90,000	1	−300
45	−200	−2	−9,000	−90	40,000	4	400
50	−100	2	5,000	100	10,000	4	−200
65	0	2	0	130	0	4	0
70	100	−1	7,000	−70	10,000	1	−100
70	200	−3	14,000	−210	40,000	9	−600
80	300	1	24,000	80	90,000	1	300
$\overline{Y} = 60$			19,000	−20	280,000	24	−500

This is a set of two linear equations for β_X and β_z that can be easily solved. Since the W matrix needs to be inverted to obtain the T test values for statistical significance, the W matrix is inverted first.

A matrix inversion technique called "lower–upper triangular factorization" will be briefly described. Inverting a matrix can be easily done by noting that the inverse of a matrix times the matrix itself equals an identity matrix, as shown in Equation (7.70).

$$A^{-1}A = A\,A^{-1} = I = \begin{bmatrix} 1 & 0 & 0 & 0 & \cdot & 0 \\ 0 & 1 & 0 & 0 & \cdot & 0 \\ 0 & 0 & 1 & 0 & \cdot & 0 \\ \cdot & \cdot & \cdot & \cdot & \cdot & \cdot \\ 0 & 0 & 0 & 0 & \cdot & 1 \end{bmatrix} \tag{7.70}$$

Consider a matrix A with coefficients a_{11}, a_{12}, and so forth, multiplied by another matrix, W, with coefficients, W_{11}, W_{12}, W_{13}, and so on. If the product of this matrix equals the identity matrix, then the W matrix must be the inverse matrix, as shown in Equation (7.71).

$$\begin{bmatrix} a_{11} & a_{12} & a_{13} & \cdot \\ a_{21} & a_{22} & a_{23} & \cdot \\ a_{31} & a_{32} & a_{33} & \cdot \\ \cdot & \cdot & \cdot & \end{bmatrix} \begin{bmatrix} W_{11} & W_{12} & W_{13} & \cdot \\ W_{21} & W_{22} & W_{23} & \cdot \\ W_{31} & W_{32} & W_{33} & \cdot \\ \cdot & \cdot & \cdot & \end{bmatrix} = \begin{bmatrix} 1 & 0 & 0 & \cdot \\ 0 & 1 & 0 & \cdot \\ 0 & 0 & 1 & \cdot \\ \cdot & \cdot & \cdot & \cdot \end{bmatrix} \tag{7.71}$$

The coefficients of the W matrix can be solved in terms of matrix A. One method is a lower–upper factorization procedure to solve for the W coefficients in terms of the A coefficients. In the process of lower factorization, matrix A and the right-hand side could be modified to appear as shown in Equation (7.72).

$$\begin{bmatrix} a' & & a'_s & & \\ 0 & 0 & & & \\ \cdot & \cdot & \cdot & a_{n'-1,n-1} & a_{n'-1,n} \\ 0 & 0 & 0 & & a'_{nn} \end{bmatrix} \begin{bmatrix} W_{11} & W_{12} & W_{13} & \cdot \\ W_{21} & W_{22} & W_{23} & \cdot \\ & & & \\ W_{31} & W_{32} & W_{33} & \cdot \\ \cdot & \cdot & \cdot & \end{bmatrix} = \begin{bmatrix} 0 & & 0 & 0 \\ & & & 0 \\ & & & \\ & \text{Numbers} & & 0 \\ & & & \end{bmatrix} \tag{7.72}$$

The A matrix includes terms in only the upper triangular part of the matrix, while the lower triangular part of the A matrix has zeros. In the process of triangularizing the A matrix, the right-hand side of the equation contains zeros in the upper triangular quadrant but numerical values in the lower triangular section of the matrix.

After concluding the process of upper triangularization of the A matrix, the W coefficients can be solved by using a reverse substitution method.

Now apply this technique to the example using the results from Equation (7.69). The matrix inversion equation is shown in Equation (7.73).

$$\begin{matrix}(1)\\(2)\end{matrix}\begin{bmatrix}280{,}000 & -500\\ -500 & 24\end{bmatrix}\begin{bmatrix}W_{11} & W_{12}\\ W_{21} & W_{22}\end{bmatrix}=\begin{bmatrix}1 & 0\\ 0 & 1\end{bmatrix} \tag{7.73}$$

The first step is to multiply row 1 of Equation (7.73) by 500 divided by 280,000 and add the results of that multiplication to row 2 of Equation (7.73). The resulting matrix is illustrated in Equation (7.74).

$$\begin{matrix}(1)\\(2)\end{matrix}\begin{bmatrix}280{,}000 & -500\\ 0 & 23.10\end{bmatrix}\begin{bmatrix}W_{11} & W_{12}\\ W_{21} & W_{22}\end{bmatrix}=\begin{bmatrix}1 & 0\\ .00178 & 1\end{bmatrix} \tag{7.74}$$

The first row of Equation (7.74) is the same as in Equation (7.73). However, the second row now contains a zero because of the substitution procedure that was applied on Equation (7.73). Furthermore, note that the A matrix is now upper triangular and the right-hand side of the equation is lower triangular. The terms W_{21} and W_{22} can now be solved as shown in Equation (7.75).

$$\begin{aligned}W_{21} &= .00178/23.10 = .000077\\ W_{22} &= 1/23.10 = .0433\end{aligned} \tag{7.75}$$

The reverse substitution procedure can be continued in Equation (7.76).

$$\begin{aligned}280{,}000W_{11} - 500\,(.000077) &= 1; \qquad W_{11} = .0000037\\ 280{,}000W_{12} - 500\,(.0433) &= 0; \qquad W_{12} = .0000773\end{aligned} \tag{7.76}$$

Thus, the inverse elements of the W matrix have been calculated and Equation (7.57) can be applied. Using the results from Equation (7.69), the inverse matrix just calculated, the β values can be calculated.

Thus, $[\beta] = [W^{-1}][X^T Y]$.

$$\begin{bmatrix}\beta_X\\ \beta_Z\end{bmatrix}=\begin{bmatrix}.0000037 & .0000773\\ .000077 & .0433\end{bmatrix}\begin{bmatrix}19{,}000\\ -20\end{bmatrix} \tag{7.77}$$

$$\beta_X = 19{,}000 \cdot .0000037 + (-20) \cdot .0000773 = .0688$$

$$\beta_Z = 19{,}000 \cdot .000077 + (-20) \cdot .0433 = .597$$

This procedure results in the multiple regression equation of $Y = 60 + .0688x + .597z$.

The coefficients β_X and β_W can now be used to calculate the predicted data points. Table 7.9 shows the calculation of the predicted Y. The predicted values appear in column 4 of Table 7.9.

The error-squared term can be calculated and used to obtain confidence intervals. This is calculated by subtracting the predicted value minus the actual data points and squaring the results, as shown in column 5 of Table 7.8. The estimate of the population variance is equal to 52.22 divided by the degrees of freedom or 7 − 3. The variance in the error is equal to 13.1. Consequently, the standard deviation, σ, is equal to 3.61.

$$\sigma^2 = \frac{52.22}{7 - 3} = 13.1$$

(*Note*: 7 data points and 3 degrees of freedom consumed by $\overline{Y}$, β_X, and B_Z)

$$\sigma = \sqrt{13.1} = 3.61$$

The t test for the X explanatory variable, income per household, is equal to the β coefficient divided by σ divided by the inverse of the diagonal element of the W inverse matrix. The t test result (Equation (7.78)) for income per household is 9.89 and for oil price is .795.

$$t_x = \frac{.0687}{3.61\sqrt{.0000037}} = 9.89 \text{ (income)}$$

$$t_w = \frac{.597}{3.61\sqrt{.0433}} = .795 \text{ (oil price)} \tag{7.78}$$

The R^2 is 1 minus 52.2 divided by the total variation 26,550 and is equal to .998, which is an excellent value.

$$R^2 = 1 - \frac{52.2}{26,550} = .998 \tag{7.79}$$

TABLE 7.9 Calculation of the Predicted Y

(1) Y_i	(2) $\hat{x}_i$	(3) z_i	(4) Predicted $\hat{Y}$	(5) $(Y - \hat{Y})^2$	(6) Y^2
40	−300	1	39.96	.00017	1,600
45	−200	−2	45.05	.00436	2,025
50	−100	2	54.31	18.69	2,500
65	0	2	61.19	14.48	4,225
70	100	−1	66.28	13.89	4,900
70	200	−3	71.97	3.80	4,900
80	300	1	81.24	1.45	6,400
				52.22	26,550

In conclusion, the low t test for oil price indicates that it is not a significant driver of load demand in this illustrative example because it has a T value of much less than 2. Unless there are strong theoretical grounds for keeping it in the regression, it should not be included.

7.3.8 Determining Which Are the Key Driving Variables

This section has illustrated the use of multiple regression techniques to determine the coefficients that provide the best fit of the data over the data range. The criterion for determining the regression equation is minimization of the least-squared error. The results of a regression analysis might be as illustrated in Equation (7.80), in which the regression involves four explanatory variables. The t statistics for these are also shown as part of Equation (7.80).

$$Y = 10.6 + 28.1X_1 + 4.0X_2 + 12.7X_3 + 0.84X_4$$

standard deviation	=	2.6	11.4	1.5	14.1	0.76	
t statistic	=	4.1	2.5	2.6	0.9	1.1	(7.80)

The issue is which explanatory variables should be kept. On the basis of the t statistic, begin by eliminating the least significant variable from the equation, that is, variable X_3. Then recalculate the regression equation coefficients and the associated t statistics. If any of the variables are still not statistically significant, eliminate the least significant variable and recalculate the coefficients and their t statistics. (Note that it is necessary to recalculate the coefficients and their t statistics after each iteration because they will change as variables are dropped from the equation.) Proceed in this manner until all of the insignificant variables have been eliminated from the regression equation.

7.4 TYPES OF CUSTOMER CLASSES

In load forecasting it is very important to segment kilowatt-hour usages into homogeneous groups with similar consumption patterns. One of the first segment cuts is typically division into the three broad classes of customers: residential, commercial, and industrial. These classes are typified as:

Residential—single-family, duplex, and two-family, mobile homes, apartment buildings

Commercial—shopping centers, hospitals, schools, office buildings, service-oriented businesses, master-metered apartments

Industrial—large manufacturing, small manufacturing, mining, food processing, etc.

Sometimes it is useful to further segment the customer classes in order to understand and accurately forecast the kilowatt-hour trends of the customers. For example, the kilowatt-hour consumption trends of steel-producing industries are significantly different from the trends that occur in the food industry. If these two industries were lumped together, it would be very difficult to understand and forecast future energy use. By examining these industries individually, forecasts can be made that accurately describe how these industries will respond to external environmental forces such as the price of electricity, the price of oil and the growth of the gross national product (GNP). Similarly, in the residential and commercial sectors, it is sometimes useful to further segment these classes, for example, electrically heated homes versus those without electric heat.

Tables 7.10, 7.11, and 7.12 provide typical driving forces that affect the energy consumption of each of these classes. Electricity price is, for example, a major influence in all customer classes. Sometimes the price of oil or natural gas is a significant driving force. Disposable income or disposable income per household is a key driving variable in the residential class but has little influence on the industrial energy consumption. For the commercial and industrial classes, there are driving variables that will influence a specific class but no other. For these reasons, it is important to segregate the energy consumption load segments into distinct homogeneous classes.

7.5 MODEL STRUCTURES

Until this point, a simplified mathematical hypothesized relationship (i.e., linear) between energy usage and the explanatory variables has been used. However, the form of the hypothesized relationship between energy usage and explanatory variables is extremely important, especially for long-forecast time periods. The form of the hypothesized relationship has an important bearing on the assumptions that are made in the forecast interval during which the regression equations are used.

TABLE 7.10 Residential Class—Likely Driving Forces

Price of electricity
Number of customers
Appliance saturation usage with no competitive fuels
Price of No. 2 oil
Price of natural gas
Disposable income
Appliance price index
Air conditioner saturation
People per household
Cooling and heating degree days

TABLE 7.11 Commercial Class—Likely Driving Forces

Price of electricity
Disposable income per household
Occupied office space
Commercial employment
Number of residential customers
Price of competing fuels; oil and gas
Space heating saturation
Air conditioning saturation
School-age population
Government expenditures
Business fixed investment
Personal consumption expenditures
Cooling and heating degree days

The most widely used formulation is a constant elasticity model form, as shown in Equation (7.81).

$$\mathrm{kWh} = A_0 \cdot (X)^{\beta 1} \cdot (Y)^{\beta 2} \cdot (Z)^{\beta 3} \tag{7.81}$$

where X, Y, and Z are explanatory variables and $\beta 1$, $\beta 2$, and $\beta 3$ are coefficients to be determined by the regression analysis.

The key reason for examining this type of formulation for relating energy usage to explanatory variables is the assumption of constant elasticity. Before defining "elasticity," take the partial derivative of Equation (7.81) with respect to one of the explanatory variables; in this case let it be X. The result is Equation (7.82), and its simplification leads to Equation (7.83).

$$\frac{\partial\ \mathrm{kWh}}{\partial X} = A_0 \cdot (\mathrm{Y})^{\beta 2} \cdot Z^{\beta 3} \cdot \beta 1 (X)^{\beta 1 - 1} \tag{7.82}$$

$$\frac{(\partial\ \mathrm{kWh})/\mathrm{kWh}}{(\partial X)/X} = \beta 1 \tag{7.83}$$

TABLE 7.12 Industrial Class—Likely Driving Forces

Level of industrial output
Price of electricity
Industrial employment
Output per worker
Earnings in manufacturing
Oil and gas prices
Environmental laws and regulations
Cooling and heating degree days

Equation (7.83) is very compact. It states that a percentage change in X, the explanatory variable, will cause a $\beta 1$ percentage change in energy usage. The $\beta 1$ coefficient is called the "elasticity of energy usage," with respect to explanatory variable X. Economists have traditionally used this equation formulation because it assumes a constant elasticity over the forecast period. For example, if X is electricity price, the model formulation assumes that the price elasticity that customers have demonstrated over the historical period during which data are collected will be the same in the future. Customers are assumed to have the same behavioral response to a future increase in the price of electricity as they have had to an increase in the price of electricity in the past.

In its current form of Equation (7.81), the constant elasticity model is not in a linear regresion model form. However, the equation can be transformed into a linear form by taking a natural logarithm of Equation (7.81). The result is Equation (7.84). Note that this equation is now linear in terms of the regression coefficients.

$$\ln(\text{kWh}) = \ln(A_0) + \beta 1 \cdot \ln(X) + \beta 2 \cdot \ln(Y) + \beta 3 \cdot \ln(Z) \qquad (7.84)$$

A variant of the double logarithmic formulation of Equation (7.81) is a partial adjustment model, as illustrated in Equation (7.85).

$$\text{kWh}(t) = A_0 \cdot \text{kWh}(t - 1)\lambda \cdot \text{X}(t)^{\beta 1} \cdot Y(t)^{\beta 2} \cdot Z(t)^{\beta 3} \qquad (7.85)$$

where t is a year and $t - 1$ is the previous year and λ, $\beta 1$, $\beta 2$, and $\beta 3$ are regression coefficients.

In this partial adjustment model, the energy usage at some time period are related to the energy usage at a previous time period times the explanatory variables raised to a power β.

This type of model gives the energy usage response inertial effects, with both a short-term and a long-term elasticity effect. This model responds to a change in electricity price or other driving variables; not with an instantaneous step change in energy usage, but with a ramped response.

This inertial effect sometimes can better simulate the customer's inertial effect as a result of the price change. A customer may receive a price signal and not respond immediately, but may take a year or two years to respond to it by installing more efficient electricity-consuming equipment or changing consumption habits. For example, when electricity prices increased significantly following oil price increases in 1974, it took residential customers several years to respond by retrofitting their houses with increased levels of thermal insulation. This partial-adjustment model attempts to simulate these kinds of transient effects.

To see this effect, simplify Equation (7.85) to a single term, $X(t)$, neglecting $Y(t)$ and $Z(t)$. If the logarithms of both sides are taken, Equation (7.86) results.

$$\ln[\text{kWh}(t)] = \ln(A_0) + \lambda \ln[\text{kWh}(t - 1)] + \beta 1 \cdot \ln X(t) \qquad (7.86)$$

Taking the derivative with respect to X_t, the short-term effect of a change in the explanatory variable X is equal to $\beta 1$, or the short-term elasticity of the explanatory variable X. Now assume a step change of magnitude dX at time zero. Then solve for the change in the kilowatt-hours as a result of this small change in dX at time t_0. The result of this effect at time t_0 is shown in Equation (7.87).

$$\frac{d}{dt}\{\ln[\text{kWh}(t_0)\} = \beta 1 \cdot \frac{d}{dt}\{\ln[\text{X}(t_0)]\} \tag{7.87}$$

Now proceed one more year at $t = t_0 + 1$ using Equation (7.86). The effect of a step change in X at time t_0 is shown in Equation (7.88).

$$\begin{aligned}\frac{d}{dt}\{\ln[\text{kWh}(t_0 + 1)]\} &= \lambda \frac{d}{dt}[\ln \text{kWh}(t_0)] + \beta 1 \frac{d}{dt}\{\ln[X(t_0)]\} \\ &= (\lambda\, \beta 1 + \beta 1) \cdot \frac{d}{dt} \ln[X(t_0)] \\ &= (\lambda + 1)\, \beta 1 \cdot \frac{d}{dt}\{\ln[X(t_0)]\}\end{aligned} \tag{7.88}$$

The effect at time $t = t_0 + 2$ is shown in Equation (7.89).

$$\frac{d}{dt}\{\ln[\text{kWh}(t_0 + 2)]\} = [\lambda \cdot (\lambda + 1) + 1]\, \beta 1 \frac{d}{dt}\{\ln[X(t_0)]\} \tag{7.89}$$

If this were generalized for an infinite number of time steps, the result would be as shown in Equation (7.90). The change in the energy usage at infinity is equal to $1/1 - \lambda$ times $\beta 1$ times the change in the explanatory variable. Thus, the long-term elasticity is equal to $\beta 1$ divided by $1 - \lambda$, as shown in Equation (7.91).

$$\frac{d}{dt}[\ln \text{kWh}(t_0 + \infty)] = \frac{1}{1 - \lambda} \beta 1 \cdot \frac{d}{dt}[\ln X(t_0)] \tag{7.90}$$

$$\frac{\frac{d}{dt}[\ln \text{kWh}(t_0 + \infty)]}{\frac{d}{dt}[\ln X(t_0)]} = \frac{1}{1 - \lambda} \cdot \beta 1 \tag{7.91}$$

Thus, the partial adjustment model has both a short-term and a long-term elasticity, as shown in Table 7.13.

As shown in Table 7.13, the shortcoming of this method is that each elastic-

TABLE 7.13 Short- and Long-Term Elasticities Based on the Partial Adjustment Model

Short-term elasticities =	$\beta 1$;	$\beta 2$;	$\beta 3$
Long-term elasticities =	$\frac{\beta 1}{1 - \lambda}$;	$\frac{\beta 2}{1 - \lambda}$;	$\frac{\beta 3}{1 - \lambda}$

ity is forced to have both a short-term and a long-term elasticity that may not be physically correct in all situations. For example, the energy usage in the industrial sector is immediately affected by the industrial production index, which has a short-term elasticity. The energy usage may also be driven by short-term and a long-term electricity price. Using the partial adjustment model, it forces both industrial production and electricity price to have both short-term and long-term elasticity. This may not be correct, and can cause the model to produce a poor forecast.

An alternative to the partial adjustment model is to insert dependent-variable time lag into the equation. A time lag is shown in Equation (7.92), where energy usage is expressed in terms of explanatory variables. However, the explanatory variables may be lagged (or leaded) in time relative to the kilowatt-hours.

$$\text{kWh}(t) = A_0 X(t)^{\beta 1} \cdot Y(t - 1)^{\beta 2} \cdot Z(t - 2)^{\beta 3} \tag{7.92}$$

where a 1-year lag is on variable Y and a 2-year lag is on variable Z.

The lag gives the model an inertial response time. For example, if Y is the electricity price, a price shock in year t_0 would not influence the energy usage in that year; rather, its force would be experienced a year later. This kind of dynamics is very useful in representing consumer inertial effects, where it may take 1 or 2 years for customers to change their electricity-consuming habits in response to a change in electricity price or some other stimuli.

In summary, model structures are key elements in the forecast equation. The hypothesized structure between explanatory variables and kilowatt-hours has a very strong bearing on the quality of the kilowatt-hour forecast.

7.6 ECONOMETRIC LOAD FORECASTING METHODOLOGY—OVERVIEW

Thus far, regression analysis theory and model structure have been reviewed. Now couple these two factors and summarize the econometric load forecasting procedure. This procedure is summarized in Table 7.14 by five steps.

In step 1, the likely driving variables are defined. In doing so, this becomes the first segmentation of the class of kilowatt-hour-consuming customers. Next, examine these customers to develop a list of the major events that cause

TABLE 7.14 Econometric Load Forecasting Procedure

Step	Activity
1	Define the likely explanatory variables (GNP, disposable income, etc.)
2	Define the functional relationship of explanatory variables to load demand
3	Research time series of these variables
4	Perform multiple regression analysis
5	Test and validate model

them to consume or not consume electricity. A broad list is developed that further analysis can prune down to focus on the true key driving variables. At this stage, the objective is to identify up to 10 key driving variables for each load segment being forecasted.

In step 2, the functional relationship between those driving variables and the energy usage is defined. In most cases of econometric load forecasting, the formulation will be of the double logarithmic form. It may have a partial flow adjustment or lag effect, or it may have a combination of several lag effects. It is important at this stage to understand what key assumptions are being made by the functional relationship between the driving variables and the historical load demand, because these assumptions will be applied during the forecast period.

Step 3 involves research of the data and collecting time series for all the driving variables and the dependent variable, energy usage. Usually, data are collected 20 years back in order to forecast 10 or 15 years into the future. To have confidence in a forecast for the next 15 years, one must have confidence that the last 20 years of data can be explained. There are exceptions to this, because the market for electricity has changed dramatically over the last 10 years. Therefore, the data from 10 to 20 years prior to today may be based on a market that is now completely changed. However, to be sure that the regression equations are sufficiently broad to respond to future forces, they must have successfully responded to forces that have acted on them in the past.

In step 4, the multiple regression analysis is performed. In this case, all the products, cross-products, and matrix inverses are calculated. Computer programs are available to readily perform these calculations. The regression coefficients, t tests, and R^2 tests will be calculated. An examination of these equations, confidence intervals, and t tests will lead to decisions regarding which of the driving variables are kept in the regression and which are not.

There are two general processes to approach the regression analysis when deciding which variables to keep and not to keep. The processes are called "stepwise regression." A bottom-up approach begins with one likely driving variable. The objective is then to find the most significant second driving variable to add. After adding the second most significant variable, one can proceed to identify the best third likely driving variable to add to the regression analysis. Thus, the regression model is built up by starting with a small number of explanatory variables, and then by adding significant explanatory variables one by one until only those that have very poor t tests or in which the R^2 is not being significantly improved by the addition of another driving variable are left out.

The alternative approach, a top-down approach, applies stepwise regression, but in the opposite direction. All the likely driving variables are put in the regression equation. Then the t tests are examined, and the variable with the worst t test is removed from the equation (<2.0). The regression is recalculated with only those driving variables that remain. The number of explanatory variables is scaled down until the remaining explanatory variables all have acceptable t tests. This rule cannot always be applied because in some cases a variable may have a t test of 1.5 and may be a significant factor in the future, although it has not played a significant role in the past. In such a case, it may be kept in the regression, even though it does not have a statistically significant t test.

Finally, in step 5, the model is tested and validated. This is one of the most important aspects of econometric load forecasting. The error term can be plotted by comparing the forecasted results with the actual data over the historical data period. Errors can be examined to determine whether the forecast is adequately following all the kilowatt-hour trends over the last 20 years.

Another test is to examine a subset of the data to forecast historical time periods. For example, if the data set included data from 1965 through 1990, then one test might use data from 1965 through 1985; the regression analysis would be developed on the basis of this limited data set. The kilowatt-hours are then forecast for 1986 through 1990 and contrasted with what actually occurred. This test gives a measure of the stability of the regression coefficients. If the value of the regression coefficients changes significantly from the smaller data set to the full data set, then all of the trends have not been fully captured. If this is the case, then the analysis would begin again at step 1, and steps 1 and 2 would be reexamined to determine whether all the driving variables that are influencing kilowatt-hours have been defined. Perhaps some variables were omitted, or perhaps it was not understood how they related to the dependent variable.

7.7 AN EXAMPLE

For the following example, forecast the total energy usage used from U.S. manufacturing based on actual data from 1963 through 1984.

The key explanatory variables are the Federal Reserve Board index of total industrial production, the price of electricity to the industrial sector expressed in real 1972 cents per kilowatt-hour, and the average price of oil and gas to industrial companies expressed in real 1972 cents per mega–British thermal units (MBtu). Oil and gas are fuels that are competitors of electricity in the manufacturing sector.

To gain insight into electricity use in the manufacturing sector, define an electricity intensity variable as shown in Equation (7.93). The kilowatt-hours of energy usage equal the kilowatt-hours of energy usage per unit of industrial production times the level of industrial production. The industrial production index is a measure of the output from industry. It factors the tons of steel, number of tires made, tons of food produced, and similar variables into a composite index.

$$\text{kWh} = (\text{kWh/industrial production index}) \cdot (\text{industrial production index}) \tag{7.93}$$

By expressing the kilowatt-hours energy usage in terms of kilowatt-hours energy usage per unit of industrial production, some insight is gained into the effects of conservation, fuel substitution, and technology change. A model for projecting the kilowatt-hours per unit of industrial production is shown in Equation (7.94). Assume a double logarithmic model, where the kilowatt-hours per unit of industrial production are related to the industrial production level, the electricity price, and the oil & gas price.

$$(\text{kWh/industrial production index}) = A_0 \cdot (\text{industrial production index})^{\beta 1} \cdot (\text{electricity price})^{\beta 2} \cdot (\text{oil \& gas price})^{\beta 3} \tag{7.94}$$

Table 7.15 presents the historical time-series data for this analysis. Column 1 shows the kilowatt-hours used in the manufacturing sector (expressed in billions), column 2 is the industrial production index, column 3 shows the real electricity price, column 4 shows the real oil & gas price, and the last column shows kilowatt-hours per industrial production.

This data is plotted in Figure 7.5. The left axis shows kilowatt-hours per

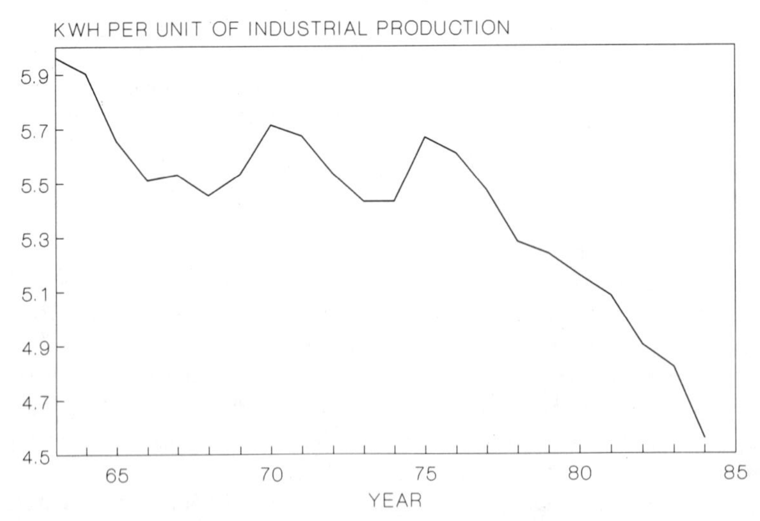

FIGURE 7.5 Electricity intensity in manufacturing data plotted based on kilowatt-hours per industrial production.

TABLE 7.15 Regression Analysis of Manufacturing Electricity Usage

Year	Billions of kWh Used in Manufacturing	Industrial Production Index	Real Electric Price ¢/kWh	Real Oil & Gas Price ¢/MBtu	kWh/Industrial Production Index
1963	459	77	1.01	31.6	5.96
1964	484	82	0.98	31.1	5.90
1965	509	90	0.95	30.7	5.65
1966	540	98	0.92	29.8	5.51
1967	553	100	0.90	29.0	5.53
1968	578	106	0.86	28.3	5.45
1969	614	111	0.83	27.3	5.53
1970	617	108	0.82	28.8	5.71
1971	624	110	0.85	35.0	5.67
1972	664	120	0.87	37.1	5.53
1973	706	130	0.88	42.4	5.43
1974	706	130	1.05	78.8	5.43
1975	663	117	1.19	85.5	5.66
1976	729	130	1.21	87.7	5.60
1977	755	138	1.28	96.4	5.47
1978	771	146	1.32	91.1	5.28
1979	796	152	1.34	109.1	5.23
1980	758	147	1.39	129.8	5.15
1981	762	150	1.48	147.0	5.08
1982	676	138	1.62	141.2	4.89
1983	708	147	1.59	136.0	4.81
1984	747	164	1.59	132.7	4.55

unit of industrial production as a function of year. The real price of electricity is plotted in cents per kilowatt-hour, as shown on the right scale of the graph. In general, industry has been conserving electricity per unit of industrial production. For every increase in industrial production, kilowatt-hours have not increased on a one-to-one basis; instead, there has been a declining trend. After 1975, the slope of this trend becomes increasingly steep as a result of increased conservation from years 1975 through 1984. The real price of electricity has increased significantly during this period as well, thus suggesting that kilowatt-hours per industrial production are sensitive to the price of electricity.

An examination of Table 7.15 data reveals that electricity price and the prices of oil and gas have almost the same trend through time. In statistical terms, these variables are colinear. When two variables have the same type of trend, it typically will mean that only one of them will be statistically significant in the regression equation and the other one may have to be discarded.

The next step is to take the logarithm of the data, compute the mean or average value of the logarithm, and convert the data to a normalized form by subtracting the mean value. Table 7.16 illustrates this procedure.

In column 1 of Table 7.16, the dependent variable Y, or kilowatt-hours per unit of industrial production, is shown expressed as the logarithm less the mean value of the logarithm of Y. Columns 2, 3, and 4 are the explanatory variables—industrial production, electric price, and oil & gas price—expressed in the normalized form of the logarithm minus the mean of the logarithm.

The next step is to compute the products YX and XX^T. This is shown to the right in Table 7.16, where the YX product is computed as the kilowatt-hours times each of the independent variables (the product of 1 times 2, 1 times 3, and 1 times 4); similarly the XX terms, which are column 2 times 2, column 2 times 3, column 2 times 4, column 3 times 3, column 3 times 4, and column 4 times 4. These products have been summed over the 1963 to 1984 data points. The result is a YX and XX^T matrix.

The next step is to calculate the inverse of the XX^T matrix. This is shown in Table 7.17. At the top of Table 7.17, the Y^TX and XX^T matrix terms are reproduced from Table 7.16. First, assemble the XX^T in the matrix. Then triangularize the matrix, perform a reverse substitution to calculate the inverse matrix terms, and verify that the inverse is correct by multiplying times the XX^T matrix to come up with the identity matrix. These steps are shown in the middle and bottom of Table 7.17.

Based on the inverse of the XX^T matrix, the β coefficients can be calculated as shown in Table 7.18. The inverse of the XX^T matrix, which was calculated in Table 7.17, is multiplied by the X^TY product, which was calculated in Table 7.16. Performing the matrix multiplication results in β coefficients equal to $-.373$ for industrial production, $-.472$ for electricity price, and .187 for oil & gas price.

Now consider the interpretation of these β values. The electricity price β of $-.47$ means that for every 1% increase in electricity price, a $-.47$% decrease

TABLE 7.16 Regression Analysis of Manufacturing Electricity Usage

	Logarithm Less Mean Values				Product Weighting Factors in the Regression								
					X^TY			X^TX					
Year	kWh Independent (1)	Independent Product (2)	Electricity (3)	Oil & Gas (4)	kWh Times (1*2)	Independent (1*3)	Variables (1*4)	Independent (2*2)	Variables (2*3)	(2*4)	(3*3)	(3*4)	(4*4)
				Sum	−0.23285	−0.25145	−0.68283	0.921987	0.746289	2.475324	1.187749	3.138878	9.199700
63	0.098394	−0.44263	−0.08383	−0.62284	−0.04355	−0.00824	−0.06128	0.195925	0.37107	0.275693	0.007028	0.052215	0.387938
64	0.088515	−0.37972	−0.11793	−0.63879	−0.03361	−0.01043	−0.05654	0.144187	0.044784	0.242564	0.013909	0.075339	0.408061
65	0.045787	−0.28663	−0.14903	−0.65174	−0.01312	−0.00682	−0.02984	0.082157	0.042716	0.186809	0.022210	0.097129	0.424767
66	0.019751	−0.20147	−0.18111	−0.68149	−0.00397	−0.00357	−0.01346	0.040591	0.036490	0.137302	0.032803	0.123431	0.464436
67	0.023337	−0.18126	−0.20309	−0.70870	−0.00423	−0.00473	−0.01653	0.032858	0.036815	0.128467	0.041248	0.143936	0.502268
68	0.009284	−0.12300	−0.24855	−0.73314	−0.00114	−0.00230	−0.00680	0.015129	0.030573	0.090177	0.061781	0.182229	0.537498
69	0.023614	−0.07690	−0.28406	−0.76911	−0.00181	−0.00670	−0.01816	0.005915	0.021847	0.059152	0.080693	0.218480	0.591542
70	0.055887	−0.10430	−0.29618	−0.71562	−0.00582	−0.01655	−0.03999	0.010880	0.030895	0.074646	0.087727	0.211960	0.512125
71	0.048819	−0.08595	−0.26025	−0.52065	−0.00419	−0.01270	−0.02541	0.007389	0.022371	0.044755	0.067733	0.135503	0.271083
72	0.023939	0.001051	−0.23699	−0.46238	0.000025	−0.00567	−0.01106	0.000001	−0.00024	−0.00048	0.056168	0.109585	0.213802
73	0.005230	0.081094	−0.22557	−0.32885	0.000424	−0.00117	−0.00171	0.006576	−0.01829	−0.02666	0.050881	0.074180	0.108146
74	0.005230	0.081094	−0.04894	0.290908	0.000424	−0.00025	0.001521	0.006576	−0.00396	0.023591	0.002395	−0.01423	0.084627
75	0.047750	−0.02426	0.076216	0.372511	−0.00115	0.003639	0.017787	0.000588	−0.00184	−0.00903	0.005808	0.028391	0.138765
76	0.037288	0.081094	0.092883	0.397917	0.003023	0.003463	0.014837	0.006576	0.007532	0.032268	0.008627	0.036959	0.158338
77	0.012613	0.140813	0.149123	0.492501	0.001776	0.001880	0.006212	0.019828	0.020998	0.069350	0.022237	0.073443	0.242557
78	−0.02276	0.197166	0.179894	0.435953	−0.00448	−0.00409	−0.00992	0.038874	0.035469	0.085955	0.032362	0.078425	0.190055
79	−0.03113	0.237440	0.194932	0.616260	−0.00739	−0.00606	−0.01918	0.056377	0.046284	0.146325	0.037998	0.120129	0.379776
80	−0.04659	0.203992	0.231566	0.789990	−0.00950	−0.01079	−0.03681	0.041612	0.047237	0.161152	0.053623	0.182935	0.624084
81	−0.06153	0.224195	0.294305	0.914428	−0.01379	−0.01811	−0.05627	0.050263	0.065981	0.205010	0.086615	0.269120	0.836178
82	−0.09791	0.140813	0.384689	0.874172	−0.01378	−0.03766	−0.08559	0.019828	0.054169	0.123095	0.147985	0.336284	0.764178
83	−0.11483	0.203992	0.365997	0.836650	−0.02342	−0.04203	−0.09608	0.041612	0.074660	0.170670	0.133953	0.306211	0.699983
84	−0.17065	0.313426	0.365997	0.812086	−0.05348	−0.06245	−0.13858	0.098236	0.114713	0.254529	0.133953	0.297221	0.659484

TABLE 7.17 Regression Procedure Calculation of $[YX]$ and $[XX^T]$

$[YX]$	$[XX^T]$
$\begin{bmatrix} -.23285 \\ -.25145 \\ -.68293 \end{bmatrix}$	$\begin{bmatrix} .92198 & .74628 & 2.475 \\ .74628 & 1.1877 & 3.1388 \\ 2.4753 & 3.1388 & 9.1997 \end{bmatrix}$

Calculation of $[XY^T]$ inverse:

Triangularize the matrix
Reverse substitute to calculate the matrix terms
Verify the inverse

in kilowatt-hours per unit of industrial production results. Similarly, for every 1% increase in oil & gas price, a .18 increase in kilowatt-hours per industrial production results. This is because oil and gas compete with electricity. If the competing prices increase, then more opportunity for electricity exists. These β coefficients can be viewed as elasticity coefficients.

Finally, the industrial production β indicates that the kilowatt-hours per unit of industrial production are inversely proportional to the industrial production level. As industrial production rises, manufacturing companies will add new equipment that reduces the electricity usage per unit of production output. The new additional equipment is more electricity efficient than existing equipment (new industrial processes are also more electricity efficient). Thus, an additional ton of steel is produced more efficiently than the average means of producing steel.

Based on the β coefficients, the calculated kilowatt-hours per unit of industrial production can be computed. Use Equation (7.95) to calculate Y, the logarithm of kilowatt-hours per unit of industrial production.

$$(y) = \beta 1 \cdot \ln(\text{industrial production}) + \beta 2 \cdot \ln(\text{electric price}) + \beta 3 \cdot \ln(\text{oil \& gas price}) \tag{7.95}$$

Next, calculate the error squared in the calculated results and compute the sum of the square of the errors over the 1963–1984 time period (22 data points). The variance of the error, as shown in Equation (7.96), is then calcu-

TABLE 7.18 Regression Procedure

1. Calculate the β coeffients
2. Compute the β coefficients by $\beta = [XX^T]^{-1}\,[X^TY]$

$$\underset{[XX^T]^{-1}}{\begin{bmatrix} 5.77157 & 4.85684 & -3.21005 \\ 4.85684 & 12.64984 & -5.62285 \\ -3.21005 & -5.62285 & 2.89089 \end{bmatrix}} \underset{[X^TY]}{\begin{bmatrix} -0.23285 \\ -0.25145 \\ -0.68293 \end{bmatrix}} = \beta \text{ values}$$

	β values
=	−0.37291 (industrial production)
=	−0.47168 (electricity price)
=	0.18703 (oil & gas price)

lated as the summation of the square of the errors divided by the degrees of freedom. In this case, there are 22 data points minus 4 degrees of freedom, the 3 regression coefficients, plus the average of Y.

This calculation is shown in Table 7.19, with the result that the σ^2 is equal to .000710.

$$\sigma^2 = \frac{\sum_{i=1}^{22} \epsilon_i^2}{(22 - 4)} \tag{7.96}$$

The t test can now be applied using Equation (7.97).

$$T_k = \frac{\beta_k}{\sqrt{\sigma^2 [XX^T_{kk}]^{-1}}} \tag{7.97}$$

TABLE 7.19 Calculation of Error Squared

Year	Actual (y)	Calculated (y)	$(y_{\text{actual}} - y_{\text{calculated}})^2$
1963	.098394	.088110	.000105
1964	.088515	.077753	.000115
1965	.045787	.055282	.000090
1966	.019751	.033095	.000178
1967	.023337	.030839	.000056
1968	.009284	.025983	.000278
1969	.023614	.018815	.000023
1970	.055887	.044754	.000123
1971	.048819	.057430	.000074
1972	.023939	.024911	.000000
1973	.005230	.014647	.000088
1974	.005230	.047257	.001766
1975	.047750	.042773	.000024
1976	.037288	.000373	.001362
1977	.012613	−.03073	.001878
1978	−.02276	−.07683	.002923
1979	−.03113	−.06522	.001162
1980	−.04659	−.03753	.000082
1981	−.06153	−.05139	.000103
1982	−.09791	−.07045	.000753
1983	−.11483	−.09222	.000511
1984	−.17065	−.13762	.001090
			.012795

$$\sigma^2 = \frac{.012795}{18} = .000710$$

TABLE 7.20 ***T*-Test Calculation for β Coefficients**

	$[XX^T]^{-1}$	
5.771578	4.856845	−3.21005
4.856845	12.64984	−5.62285
−3.21005	−5.62285	2.890898

Beta Values		*t* Test	
β1	−0.37291	−5.82190	(industrial production)
β2	−0.47168	−4.97408	(electricity price)
β3	0.187039	4.125926	(oil & gas price)

where σ^2 = error variance and $[XX^T{}_{kk}]^{-1}$ = kth diagonal element of the inverse of the $[XX^T]$ matrix.

These results are shown in Table 7.20. The inverse of the XX^T matrix, the values of β and the *t* test are computed using Equation (7.97). The *t* test for industrial production is 5.82; for electricity price is 4.97; and for oil and gas price is 4.12. The conclusion from this *t*-test analysis is that all the regression coefficients are statistically significant because they have a *t* test greater than 2.

Figure 7.6 shows the graphical comparison of the logarithm of kilowatt-hours per unit of industrial production versus year for the actual and fitted curves. The fitted curve follows many of the trends.

Having verified that the model does provide a reasonable fit to the physical phenomena in the manufacturing sector, this model can be used to forecast

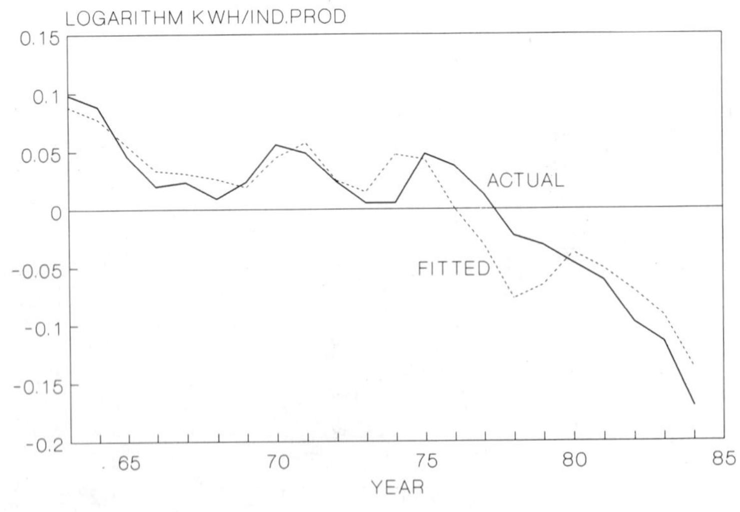

FIGURE 7.6 Comparison of actual versus fitted kilowatt-hours per industrial production.

future kilowatt-hours. The first step is to project the future industrial production, the future price of electricity, and the future price of oil and natural gas.

The forecast of industrial production is typically made by a large-scale econometric model supported by one of the national forecast services, from universities or econometric consultants. The price of electricity can be assessed by a financial analysis of the electric utility or electric utility industry. The price of oil and natural gas can be based on the supply-and-demand perspective of the energy balance situation, and is forecasted by many energy consulting organizations.

Substituting these forecasts into the model, a kilowatt-hour forecast is generated as shown in Figure 7.7. This forecast projects that the logarithm of kilowatt-hours per unit of industrial production will exhibit a declining trend through time, but the slope is not as steep as it has traditionally been through the late 1970s and early 1980s. The principal reason for the declining slope in the 1990s is the significant price increase of oil and gas fuels projected to occur during the forecast.

In Figure 7.8, the inverse of the logarithm was taken and multiplied by the industrial production. The result is the total kilowatt-hours projected to be consumed by the manufacturing sector through the year 2000. This forecast shows that the kilowatt-hours used in manufacturing had a significant dip in 1982, recovered through the 1985–1986 period, and then continue to grow through the 1980s and 1990s as industrial production increases and requires more total kilowatt-hours.

This example was shown for illustration only. It provides a practical example of the methods that can be used in an econometrically based forecast of the manufacturing sector. In actual practice, the manufacturing sector is com-

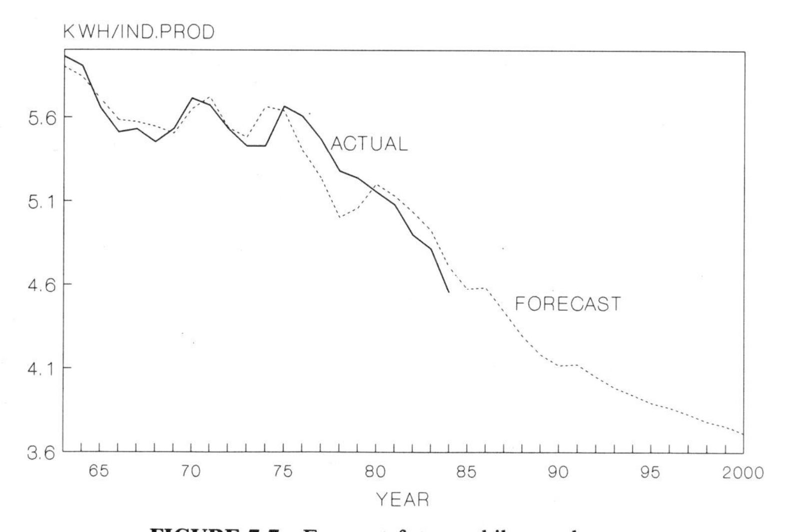

FIGURE 7.7 Forecast future—kilowatt-hours.

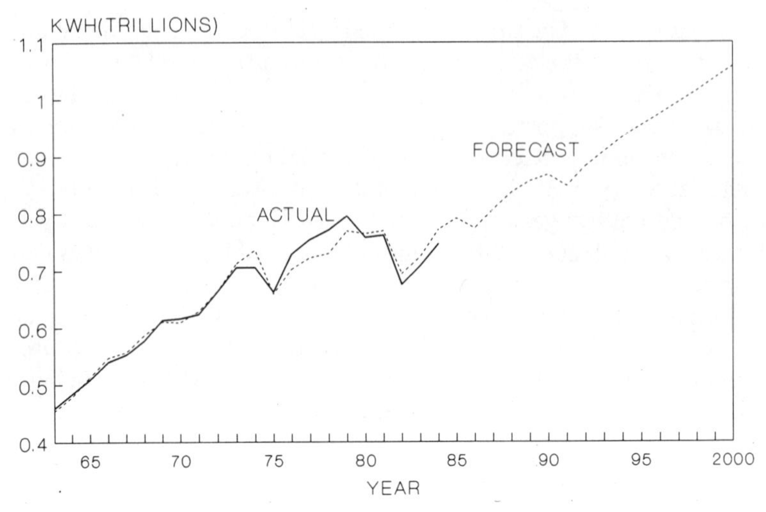

FIGURE 7.8 Comparison of actual versus forecast kilowatt-hours.

prised of many industries, each with a unique electricity consumption pattern. If data are available, it is always beneficial to segment the manufacturing sector into smaller industrial subclassifications.

For example, while the average electricity used in manufacturing per unit of industrial production has been exhibiting a significant decline through time, there are industries within the manufacturing sector that have opposite trends. For example, the food industry manifests a specific trend, as illustrated in Figure 7.9. This industry has been motivated by the increased use of conve-

FIGURE 7.9 Industrial electric intensity for the food industry.

nience and frozen foods, which has led to more electricity use per unit of production.

7.8 LIMITATIONS OF ECONOMETRIC LOAD FORECASTING PROCEDURES

Econometric forecasting methods are extremely powerful forecasting tools. However, some of the limitations that are implicit in their application must be kept in mind. A key limitation is that the future is forecast based on relationships that took place in the past.

Identification of the driving or explanatory variables is key. One problem that always haunts econometric forecasting is the identification of trends between several variables. The issue that arises is whether the relationship between driving variables is really causal or is it casual (or coincidental). For example, a regression analysis could conclude that the GNP is equal to 1000 plus 5.7 times the beer consumption. Statistically, from the relationship between the GNP and beer consumption, someone might conclude that to decrease unemployment and provide a better lifestyle for everyone, all that would be needed is to increase the nation's beer consumption. This is hardly a plain fact. The relationship between beer consumption and GNP is truly casual or coincidental. The same type of problem can impact kilowatt-hour sales forecasting. Therefore, forecasting must be done with care and with sound reasons for supporting the explanatory or driving variables and their impact on kilowatt-hours.

Another factor in identifying the driving variables is the role of new emerging variables. Will new driving variables be important in the future, even though they weren't important in the past? For example, if an explanatory variable has not been important in the past and has not influenced kilowatt-hours, then the regression analysis cannot be expected to reveal any relationship between that driving variable and kilowatt-hours.

This chapter has demonstrated and emphasized some of the mathematical techniques used in performing econometric regression load forecasting. The process is very cumbersome, involving multiplications and matrix inversion. However, statistical packages and econometric modeling programs are available for mainframe and personal computers that relegate all the mathematical manipulations to the computer. Thus, regression analysis and econometric load forecasting model building can be done very quickly and easily because of the automated nature of these statistical modeling tools.

BIBLIOGRAPHY

Billinton, R., *Power System Reliability Evaluation,* Gordon & Breach, Science Publishers, New York, 1970.

Billinton, R., R. J. Ringlee, and A. J. Wood, *Power-System Reliability Calculations,* MIT Press, Cambridge, MA, 1973.

Draper, N. R., and H. Smith, *Applied Regression Analysis,* Wiley, New York, 1966.

Guttman, I., S. S. Wilks, and J. S. Hunter, *Introductory Engineering Statistics,* Wiley, New York, 1965.

Mendenhall, W., *Introduction to Linear Models and the Design and Analysis of Experiments,* Duxburg Press, Belmont, Cal.

Mosteller, F., R. E. K. Rourke, and G. B. Thomas, Jr., *Probability and Statistics,* Addison-Wesley, Reading, MA, 1967.

Nelson, W., *Applied Life Data Analysis,* Wiley, New York, 1982.

Pindyck, R. S., and D. L. Rubinfeld, *Econometric Models and Economic Forecasts,* McGraw-Hill, New York, 1976.

Wonnacott, R. J., and T. H. Wonnacott, *Introductory Statistics,* Wiley, New York, 1969.

PROBLEMS

1 Data have been collected on heat pump saturation over the past 7 years as shown in the table below. Assuming that the saturation level will continue to grow at a linear rate, what would you predict the saturation of heat pumps to be in year 10 and year 20?

Year	Heat Pump Saturation (%)
1	1
2	2
3	4
4	10
5	12
6	12
7	14

2 A generating unit has been experiencing forced outage rates over the past 5 years as shown in table below. Since this generating unit is already 30 years old, it is highly suspected that this unit is now in the wear-out phase of its life. Fit a parabola of the form $Y = B_0 + B_1X + B_2X^2 + \epsilon$ (where X represents the year) to predict what the forced outage rate of the generating unit will be in year 7.

Year	Forced Outage Rate (%)
1	5
2	5
3	6
4	8
5	10

3 The table below shows the amount of money spent on maintenance costs each year by an electric utility, the consumer price index, and also the total amount of installed capacity on the system during each of the years. It is hypothesized that the total maintenance costs are directly related to both the consumer price index and the total installed capacity each year. If this is true, what would you project the maintenance cost to be in the future when the consumer price index is projected to be 175 and there will be 17.5 GW (gigawatts) installed on the system?

Annual Maintenance Costs (M $)	Consumer Price Index	Installed Capacity (GW)
226	95	12.0
236	100	12.0
248	105	12.8
262	111	13.2
278	118	14.0
292	125	14.0
309	132	15.0

4 The objective of this problem is to economically forecast residential load based on the 1970 to 1981 data series. The key explanatory variables are:

- Personal consumption expenditures—a component of the GNP that measures the dollars spent nationally on durable and nondurable products (food, refrigerators, etc.).

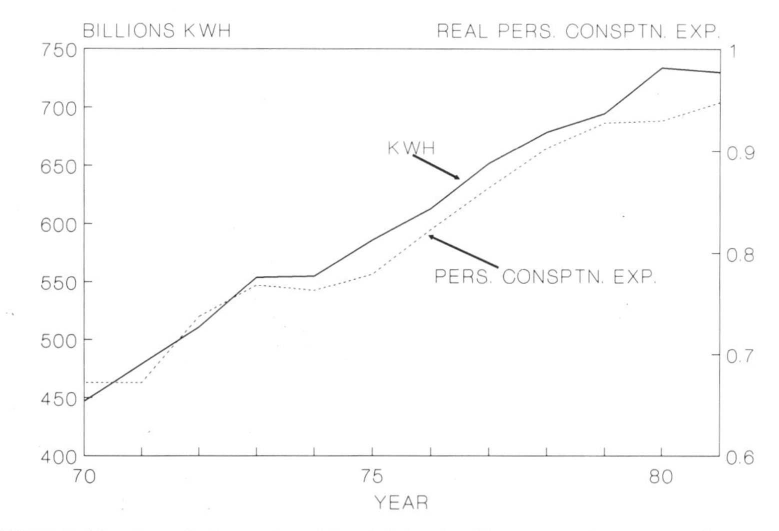

FIGURE 7.10 Correlation of residential load with personal consumption expenditures.

- Real electricity price.
- Real oil & gas price.

Use a double logarithmic model.

$$\text{kWh} = A_0 \,(\text{personal consumption expenditures})^{\beta 1} \cdot (\text{electric price})^{\beta 2} \cdot (\text{oil \& gas price})^{\beta 3}$$

Using 1970–1981 data, forecast the residential kilowatt-hour growth.

Figure 7.10 illustrates the residential load and personal consumption expenditure history. Use the data of Table 7.21.

TABLE 7.21 Regression Analysis of Residential E

Year	kWh Residential	Personal Consumption Expenditures	Real Electric Price ¢/kWh
1970	447	672	0.82
1971	479	672	0.85
1972	511	737	0.87
1973	554	768	0.88
1974	555	763	1.05
1975	586	779	1.19
1976	613	823	1.21
1977	652	864	1.28
1978	679	903	1.32
1979	695	928	1.34
1980	734	930	1.39
1981	730	948	1.48
1982		976	1.62
1983		1006	1.59
1984		1036	1.59
1985		1067	1.60
1986		1099	1.61
1987		1132	1.61
1988		1166	1.62
1989		1201	1.63
1990		1237	1.63
1991		1274	1.64
1992		1312	1.65
1993		1352	1.67
1994		1392	1.68
1995		1434	1.70
1996		1477	1.71
1997		1521	1.73
1998		1567	1.75
1999		1614	1.76
2000		1662	1.78

8

LOAD FORECASTING II

8.1 END-USE ELECTRICITY MODELS

End-use electricity models are physical engineering-based methods often used in forecasting the residential, and sometimes commercial and industrial, service classes. The basic concept of end-use methodology is forecasting the annual number of additions of electricity-consuming devices and the electricity energy consumption per device. When these two quantities are multiplied together, the annual amount of energy sales is obtained.

This end-use methodology is then replicated for all electricity-consuming devices to develop an energy sales forecast for the targeted service class. For the residential sector, the method is referred to as an "appliance saturation method." The method is widely used and provides insightful conclusions regarding how electricity is and will be used for each appliance and sector.

8.1.1 Residential Appliance Saturation Method

8.1.1.1 Method Overview. Figure 8.1 presents a flowchart of the appliance saturation model. Changes in the size and composition of the population lead to changes in the number of households. An increase in the number of households leads to a need for new housing and new construction. New housing results in an increased saturation (number) of air conditioners, washers and dryers, and other electricity-consuming devices.

A house simulation model is sometimes used when considering how electricity is used within a residence. This model can provide an assessment of the heating and cooling loads within the house. The simulation represents the house heating balance such that the sum of the solar heating gain plus space

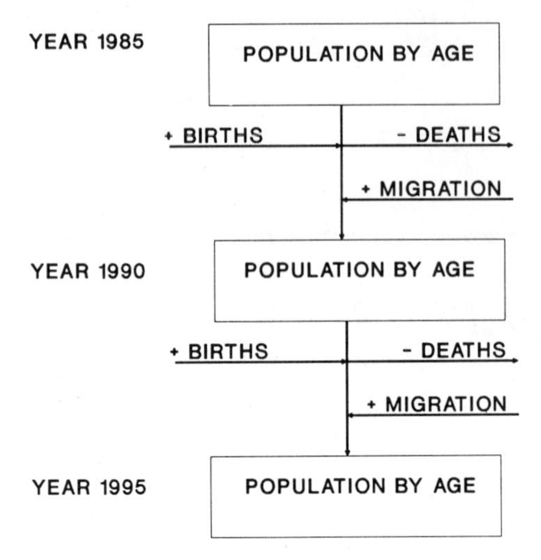

FIGURE 8.1 Major elements considered in projecting population by age group.

heating and cooling energy equals the heat loss through walls, windows, roofs, and air intrusion.

This assessment permits an analysis of electricity consumption in space heating and cooling and how it may be influenced by changes in the future. These may be changes in desired indoor temperature level, changes in building insulation standards, or changes in the weather. Although long-term changes in the weather may not be anticipated, a model that can simulate the weather effects based on historical weather data is very useful for validating the model results with actual history.

The energy usage per appliance is then multiplied by the number of appliances to obtain the total energy usage.

8.1.1.2 Demographics. The first step in the appliance saturation method is to forecast the number of households in the service territory. Normally, the number of households can be computed by forecasting or projecting the population by age groups and the household formation for each age group. The number of households are necessary because appliance saturation is based on households rather than total population. The major elements of this population projection are shown in Figure 8.2.

Table 8.1 presents population calculations based on Figure 8.2. A 1985 population statistic begins the analysis, and survival rates, migration rates, and births are applied to each of the age segments. The result is a demographic forecast for the year 1990.

Households are forecast by applying household formation rates to each age category. Population and household projections are available from state planning agencies, the Federal government, the Bureau of Labor Statistics, and some county organizations.

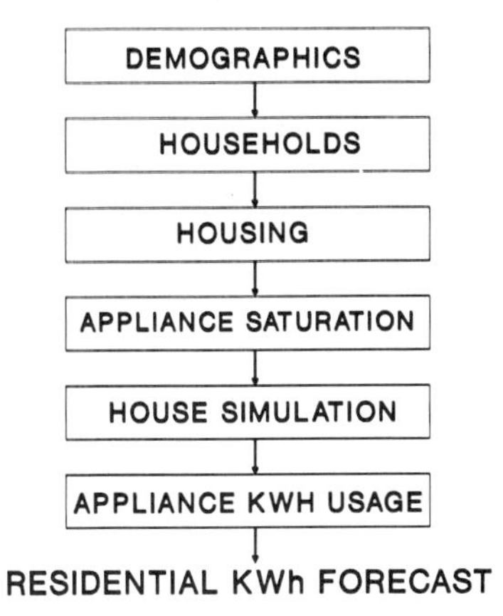

FIGURE 8.2 End-use energy model flowchart.

Table 8.2 presents the U.S. statistics and projections from 1951 through 1991. Historically, population per household or family size in the United States has been shrinking. The shinking population per household trend is based on smaller family sizes, increased divorce rate, and more one-person households as a result of later ages for first marriages.

The population in the United States is growing very slowly at less than 1%/year. Because of the age distribution, however, household growth is 1.25–1.50%/year. Currently, households are growing at a faster growth rate than the actual population.

Households also closely approximate the number of residential customers. Nearly every household will be a residential customer, except for master-metered residential housing. This is somewhat offset by some households with more than one residence, such as summer or vacation residences.

TABLE 8.1 Basic Population Characteristics for a Demographic Forecast

1985					1990	
Age	Population	5-Year Survival Rate	Inward Net Migration Rate	Births	Population	Age
				1800	1800	0–5
0–5	2000	.997	.05	—	2094	6–10
6–10	2400	.998	.10	—	2635	11–15
11–15	2500	.998	.10	—	2745	16–20
16–20	2200	.997	.05	—	2303	21–25
•	•	•	•	•	•	•
•	•	•	•	•	•	•
•	•	•	•	•	•	•

TABLE 8.2 U.S. Population and Household Characteristics Resulting in Customer Base Size

	Population	Households	Population per Household
1951	3020	899	3.36
1961	3428	1048	3.27
1971	3587	1158	3.10
1976	3680	1248	2.95
1981	3791	1348	2.81
Projected 1991	4026	1514	2.66
	Average annual growth rate %/year		
1951–1976	.83	1.38	−.54
1976–1991	.64	1.29	−.74

8.1.1.3 Appliance Saturation Method Procedure. The appliance saturation method can be summarized into several steps, as outlined in Table 8.3.

Step 1 involves forecasting of the future number of households within the service territory. Then, in step 2, the current number of appliances (saturation level) is determined in the service area using census reports, Department of Commerce reports on the state or federal level, and independent market surveys. Utilities have traditionally conducted appliance saturation surveys by including questionnaires in a stratified sample with the monthly billing. This is a convenient and relatively inexpensive means of collecting good data.

In step 3 the potential for new electric devices is forecast. In developing a 15- or 20-year forecast, it is important to envision what new electricity-consuming devices will become available and become widely used. New electricity-consuming devices that may evolve into society over the years include electric vehicles, home computers, and residential automation and control.

In step 4, the future penetration of new appliances is forecast on the basis

TABLE 8.3 Appliance Saturation Method

Step	Activity
1	Forecast number of households in service territory
2	Determine current saturation level of appliances in service area
3	Forecast new electricity-consuming devices
4	Forecast future penetration of appliances (including new sales, conversions, and retirements)
5	Determine electricity usage of existing appliances
6	Forecast future efficiency improvements in electricity usage per appliance
7	Forecast electricity sales
8	Validate the forecast

of an extrapolation of current trends or on an econometric model of the appliance. For example, the historical data of central air conditioning in the service territory can be examined, and a regression analysis of the percentage of new houses with central air conditioning as a function of household income and price of electricity can be performed. Retirements of appliances must also be forecast because they may have a higher (or lower) energy consumption than a new replacement.

In step 5 the kilowatt-hour usage per appliance of the existing stock of appliances is determined. Data sources include the Association of Edison Illuminating Companies (AEIC), Edison Electric Institute (EEI), and the Electric Power Research Institute (EPRI). Utilities, through their load-research programs, can monitor the energy use per appliance within the service territory.

Step 6 involves forecasting of future efficiency improvements in appliances. This can be developed through discussions with appliance manufacturers or through regression analysis based on historical usage over the last 10 or 15 years. The Department of Energy (DOE), as part of its conservation program, had proposed efficiency standards for new appliances in 1980 (*Federal Register,* 1980) and efficiency standards were established in 1987 (*National Appliance Energy Conservation Act,* 1987). On the average, appliances are expected to consume 25% less electricity or natural gas after the standards take effect in the early 1990s. Manufacturers of appliances have initiated designs aimed at achieving these efficiency standards. Energy use data under actual conditions must be used in contrast to manufacturers values that may be based on ideal "test stand" data. Weather conditions and household income levels can also influence appliance energy usage.

Depending on the marketplace economy, energy prices and appliance model choice, future efficiency improvements may or may not occur rapidly. In some cases, load forecasters have developed regression models of appliance efficiency choice: if energy prices are low and household income is high, then larger appliances are purchased and overall efficiency stays the same; if energy prices are high, consumers purchase higher efficiency appliances.

In step 7 the total energy sales are forecast using the data of the previous steps. The forecast energy sales is computed as the product of appliance saturation times kilowatt-hours per appliance times the number of appliances. The total residential energy sales forecast for each year of the forecast period results from the summation of the results for each of the major electricity-consuming appliances.

Step 8 is the forecast validation. This is typically accomplished by validating the forecast over the last five or 10 years of data and ensuring that the forecast method replicates history. Historical validation often requires knowledge about the yearly heating and cooling degree days to replicate the annual variations in heating and air conditioning energy usage. If the historical validation does replicate history, then the forecast achieves some measure of validity based on the driving variables being forecast correctly. If the forecast does not replicate history, the methodology and data must be reexamined.

Example 8.1. Table 8.4 presents data for the sales of refrigerators in the United States from 1960 through 1986 and includes a projection of the number of refrigerators to be sold from 1987 through the year 2000. Assume that refrigerators have an average life of 17 years, which has been the historical trend (see Tippet and Ruffin, "Service Life Expentancy of Household Appliances," 1975).

Table 8.5 presents the refrigerator energy consumption in kilowatt-hours per refrigerator for selected years from 1960 through the year 2000.

Prepare a forecast of refrigerator kilowatt-hour electricity usage.

SOLUTION. First, some background on Table 8.5. Through the 1960s, refrigerator volume in cubic feet was increasing. Frost-free refrigerators were also emerging; today they comprise 85% of the market. Frost-free refrigerators also use about 50% more electricity than the manual-defrost refrigerators. Thus, the average kilowatt-hours per unit of new refrigerators increased dramatically during the 1960s. Through the 1970s, however, energy awareness increased, resulting in a market demand for more efficient refrigerators. Con-

TABLE 8.4 Refrigerator Example (U.S. Data)

Year	Million Refrigerators	Year	Million Refrigerators
1960	2.9	1980	4.9
1961	3.0	1981	4.7
1962	3.2	1982	4.2
1963	3.3	1983	5.1
1964	3.5	1984	6.1
1965	3.6	1985	5.8
1966	3.8	1986	5.5
1967	3.9	1987	5.6
1968	4.0	1988	6.1
1969	4.1	1989	6.4
1970	4.2	1990	6.3
1971	4.3	1991	5.8
1972	4.3	1992	6.0
1973	4.3	1993	6.3
1974	4.3	1994	6.2
1975	4.4	1995	7.6
1976	4.5	1996	7.0
1977	5.2	1997	6.6
1978	5.5	1998	6.3
1979	5.4	1999	5.9
		2000	6.7

TABLE 8.5 Refrigerator Consumption (U.S. Data)

Year	Annual kWh/Unit
1960	1100
1970	1500
1980	1700
1990	1500
2000	1400

sequently, more insulation and improved refrigeration cycles were adopted. Therefore, kilowatt-hour consumption is forecast to have reached a peak in 1980 followed by a decreasing trend through the 1990s to the year 2000.

The results from Tables 8.4 and 8.5 are used in Table 8.6 to calculate the change in kilowatt-hour consumption due to refrigerators in years 1980, 1990, and 2000.

In 1980, 4.9 million refrigerators were sold, each one consuming 1700 kWh/year. That same year, there were 3.3 million refrigerators retired that were purchased in 1963. The average 1963-vintage refrigerator consumed 1190 kilowatt-hours. A net change of 4.4 billion kWh resulted from refrigerators in 1980 (4.9 million • 1700 - 3.3 million • 1190 = +4.4 billion kWh).

In the year 2000, 6.7 million refrigerators are forecast to be sold, each refrigerator consuming 1400 kilowatt-hours. However, 5.1 million are forecast for retirement, each one consuming 1623 kWh per unit. A change of 1.1 million kWh will result for the year 2000. Even though the sales of new refrigerators are projected to be *high* in year 2000, the net change in electricity usage for refrigerators is forecast to be *low* because of the installation of more efficient refrigerators and the retirement of older, less efficient refrigerators. Thus, the load growth contribution from refrigerators is projected to be small in the future. In some areas of the United States, utilities are projecting negative kWh growth in refrigerator usage.

Table 8.7 shows the same results as Table 8.6, but on a yearly basis. This table is typical of a load-forecast model for refrigerators. The forecast model

TABLE 8.6 Refrigerator Example
EXAMPLE CALCULATIONS (U.S. DATA)

	Sales of New Refrigerators (Millions)	kWh/Unit Sold	(17-Year Life) Retirements of Refrigerators (Millions)	kWh/Unit Retired	Annual Net Change in kWh (Billions)
1980	4.9	1700	3.3	1190	4.4
1990	6.3	1500	4.3	1547	2.8
2000	6.7	1400	5.1	1623	1.1

TABLE 8.7 Refrigerator kWh Forecast

Year	Residential Customers	Annual Refrigerator Sales (Millions)	Annual Refrigerator Retirements (Millions)	New Refrigerator Sales (Millions)	New Refrigerator kWh/Unit	Retired Refrigerator kWh/Unit	Change in Refrigerator kWh (Billions)	Total Refrigerator Units (Millions)	Refrigerator Saturation (%)	Average kWh/ Unit	Refrigerator kWh (Billions)
1977	74.1							84.5	114.0		132.9
1978	76.0	5.5	3.0	2.5	1660	1110	5.8	87.0	114.5	1594	138.7
1979	77.3	5.4	3.2	2.2	1680	1116	5.5	89.2	115.4	1617	144.2
1980	79.1	4.9	3.3	1.6	1700	1191	4.4	90.8	114.8	1637	148.6
1981	80.8	4.7	3.5	1.2	1680	1227	3.6	92.0	113.9	1654	152.2
1982	82.5	4.2	3.6	0.6	1660	1270	2.4	92.6	112.2	1670	154.6
1983	85.2	5.1	3.8	1.3	1640	1333	3.3	93.9	110.2	1682	157.9
1984	86.9	6.1	3.9	2.2	1620	1354	4.6	96.1	110.6	1691	162.5
1985	88.5	5.8	4.0	1.8	1600	1420	3.6	97.9	110.6	1697	166.1
1986	90.2	5.5	4.1	1.4	1580	1461	2.7	99.3	110.1	1700	168.8
1987	91.8	5.6	4.2	1.4	1560	1485	2.5	100.7	109.7	1701	171.3
1988	93.5	6.1	4.3	1.8	1540	1533	2.8	102.5	109.6	1699	174.1
1989	95.1	6.4	4.3	2.1	1520	1541	3.1	104.6	110.0	1694	177.2
1990	96.8	6.3	4.3	2.0	1500	1547	2.8	106.6	110.1	1689	180.0
1991	98.3	5.8	4.3	1.5	1490	1591	1.8	108.1	110.0	1682	181.8
1992	99.8	6.0	4.4	1.6	1480	1586	1.9	109.7	109.9	1675	183.7
1993	101.2	6.3	4.5	1.8	1470	1636	1.9	111.5	110.2	1665	185.6
1994	102.7	6.2	5.2	1.0	1460	1625	0.6	112.5	109.5	1655	186.2
1995	104.2	7.6	5.5	2.1	1450	1658	1.9	114.6	110.0	1641	188.1
1996	105.7	7.0	5.4	1.6	1440	1681	1.0	116.2	109.9	1627	189.1
1997	107.2	6.6	4.9	1.7	1430	1702	1.1	117.9	110.0	1613	190.2
1998	108.7	6.3	4.7	1.6	1420	1669	1.1	119.5	109.9	1601	191.3
1999	110.2	5.9	4.2	1.7	1410	1671	1.3	121.2	110.0	1589	192.6
2000	111.7	6.7	5.1	1.6	1400	1624	1.1	122.8	109.9	1577	193.7

calculates the refrigerator saturation, which is the number of refrigerators divided by the number of residential households (customers). The average kilowatt-hours per refrigerator is calculated, and is forecast to decrease through the 1980s and 1990s.

The same type of methodology applied to refrigerators can be applied to other residential appliances. Table 8.8 presents typical U.S. national appliance saturation and consumption values for 1985 and projected for 1995. The annual consumption values are averaged from several sources. The 1995 projections are based on estimates of projected change of use arising from smaller households and projected technology change.

Utility-specific appliance saturation and appliance consumption data must be used in preparing a utility-specific load forecast. Appliance saturation surveys and load-research data is key to this forecast. Many utilities further segment the appliance list to include televisions, fans, circulation motors, lighting, and other appliances. Further segmentation of residential customers is also done for single-family housing, multifamily housing, and mobile homes.

New appliance sales are characterized by several factors, including the penetration rate on new housing, conversions, and replacements.

$$\begin{aligned}\text{New appliance sales} &= \text{penetration rate} \cdot \text{number of new housing starts} \\ &\quad + \text{conversions in existing homes} \\ &\quad + \text{replacements of failures or retirements in existing homes.}\end{aligned} \tag{8.1}$$

A key element of appliance saturation projections is the penetration rates within new construction. The penetration rate is defined as the number of new appliances per new house completion. In many cases, the penetration rate has been stable or slowly changing over the last decade. Examples include refrigerators, ranges, freezers, and dryers.

Appliance penetration of some appliances is influenced by a "pull-through" effect. Centrally air conditioned houses typically have a higher penetration rate of heat pumps because the incremental investment cost of a heat pump (that can both cool in summer and heat in winter) is very small ($200–$500) relative to installation of only central air conditioning. In this way, central air conditioning "pulls through" the heat pump. Figure 8.3 illustrates annual penetration rates for the heat pump/air conditioning example based on U.S. statistics.

Electric water heaters are also "pulled through" by electric heating. If natural gas (or oil) is not being used for space heating, then most builders prefer to not incur the additional expense of a natural gas hookup just for gas hot water heating. Electric dryers and electric ranges are also influenced by the electricity "pull-through" effect. Room air conditioning is influenced by an inverse "pull-through" of central air conditioning. As existing homes convert to central air, room air conditioners are retired. Thus, the central air conditioning appliance sales forecast is a key element of the residential kilowatt-hours per forecast because of the "pull-through" effects.

TABLE 8.8 Typical Appliance Data

	1985 National Appliance Saturation (%)	1985 Annual Consumption (New Appliance) kWh/Appliance	1995 Projected National Appliance Saturation (%)	1995 Projected Annual Consumption (New Appliance) kWh/Appliance
Electric ranges	54	900	57	850
Dryers	46	950	48	900
Refrigerators	111	1600	111	1500
Freezers	36	1120	27	1050
Central air conditioning	33	2250	44	1960
Room air conditioning	45	1130	35	980
Hot-water heaters	41	3970	45	3700
Heat-pump single family	6	5780	15	5000
Resistance heating single family	13	1272	11	1200
Miscellaneous	100	800	100	900

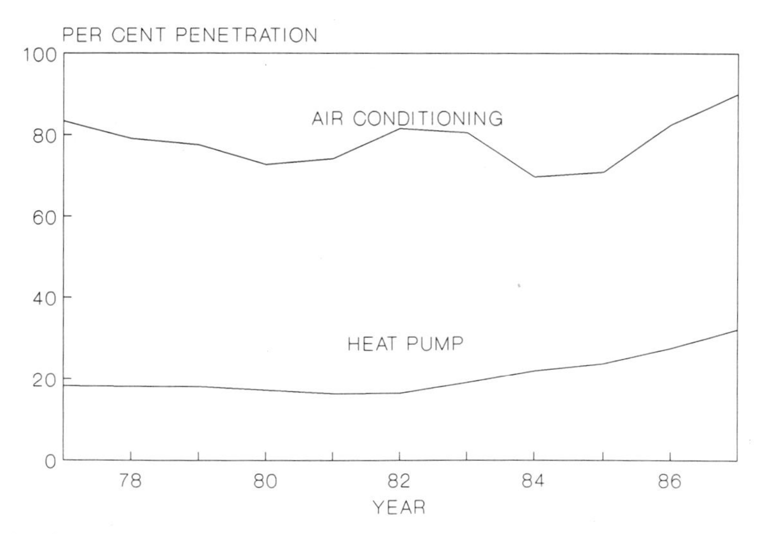

FIGURE 8.3 Annual penetration rates for new construction—"pull-through" effect of central air conditioning on heat pumps (U.S. data).

8.1.1.4 House Space Conditioning Model. Air conditioning and electric heating loads per residence are generally determined from load research studies. Future annual kilowatt-hour energy requirements may change as a result of not only efficiency improvements in heating and cooling systems but also changes in new construction building standards, retrofit construction, indoor temperature settings, and house size.

One method used to evaluate changes in space conditioning kilowatt-hour energy per house requirements is a simplified model of the thermal heat balance of a house. While detailed space conditioning methods are available for steady-state and transient building analysis, a simplified model is often adequate for forecasting purposes (see American Society of Heating, Refrigeration and Air Conditioning Engineers, *Handbook of Fundamentals,* 1976).

House heat loss and gain may be represented by a set of equations that represent heat loss and gain through the ceiling (roof), walls, glass, solar insulation, air infiltration, and internal house heat generation. Heat-loss equations are shown in Equations (8.3) through (8.9). The heat conduction equations (walls, ceilings, doors, glass) are of the form:

$$Q = U \cdot A \cdot \Delta t \tag{8.2}$$

where Q = heat transfer in Btu/foot^2
U = thermal conductivity in Btu/foot^2/°F
A = heat-transfer area in feet^2
Δt = effective temperature difference between the indoor surface and the outdoor surface (the effective outdoor temperature is equal

to the outdoor temperature plus an adjustment for solar insulation and wind speed).

The ceiling Equation, (8.3), calculates heat loss through ceilings into the attic area. The temperature in the attic is higher than the outdoor temperature during the summer as a result of solar insulation on the roof. Table 8.9 illustrates typical values for a house for summer and winter days.

The wall, glass, and door equations are similar to the ceiling equation. The heat loss through the floor to the basement is typically small and is not calculated.

Solar gain through the windows is represented as a product of the glass area times solar intensity flux, *SF,* for a given latitude and time of year times a solar shading factor, *SH,* that accounts for the orientation of the glass relative to the sun, roof line shading, and internal shading with curtains and blinds.

The simplified house thermal energy equations are:

Ceiling	$Q_C = U_C \cdot A_C \cdot \Delta t_C$	(8.3)
Walls	$Q_W = U_W \cdot A_W \cdot \Delta t_W$	(8.4)
Glass	$Q_G = U_G \cdot f_G \cdot A_W \cdot \Delta t_O$	(8.5)
Doors	$Q_D = U_D \cdot f_D \cdot A_D \cdot \Delta t_D$	(8.6)
Solar	$Q_S = A_G \cdot SF \cdot SH$	(8.7)
Air infiltration	$Q_I = .018 \cdot V \cdot ACH \cdot \Delta t_O$	(8.8)
Heat generated by Appliances and People	$Q_H = PPH \cdot 225 + 1200$	(8.9)

where subscript H = internal to house
D = doors

TABLE 8.9 Typical Values Single-Family Conventional Housing

	Summer	Winter
Floor space	1700	1700 feet^2
Fraction of single-level homes (multilevel homes is the complement number)	35%	35%
Ceiling area	1060	1060 feet^2
Wall area	1900	1900 feet^2
Glass percentage	12%	12%
Solar insulation (net *SF* * *SH*)	40	30 Btu/hour/foot^2
Air changes	0.1	1.0 changes/hour
Indoor temperature	77	69°F
Outdoor temperature	97	29°F
Ceiling temperature	113	39°F
Outside wall temperature	100	30°F

C = ceiling
W = walls
G = glass in windows
O = outdoor (temperature)

and Δt = temperature difference
f_X = fraction of wall area in subscript X
SF = solar flux intensity Btu/foot^2
SH = solar shading factor for windows
V = volume of the house, feet^3
ACH = number of air changes per hour
PPH = number of "adult size" people per household
A = heat transfer area in feet^2

Air infiltration is modeled as the product of the number of air changes per hour times the volume of the house times the volumetric heat capacity of air times the indoor/outdoor temperature difference. The air changes per hour in the winter are much larger than in the summer due to higher winds and higher temperature differences. Typical winter air changes are one change per hour. Some houses built with very tight construction and with high insulation values have winter air changes as low as 0.1/hour. This has prompted some health concerns and the suggestion for supplemental ventilation, especially if the home is heated with a fossil fuel (see Elkins and Wensman, "Natural Ventilation of Modern Tightly Constructed Homes," 1971).

The final thermal equation is the heat generated internally to the structure by people and appliances. Typical values of 225 Btu/hour for people and 1200 Btu/hour for appliances are used in Equation (8.9).

The thermal energy transfer from the house must be supplied by space conditioning equipment. The Btu/hour per degree for heating and cooling are:

$$\text{Btu/hour heating} = q_{HH}$$
$$= \frac{Q_C + Q_W + Q_G + Q_D - Q_S - Q_I - Q_H}{\Delta t_0} \tag{8.10}$$

$$\text{Btu/hour cooling} = q_{HC}$$
$$= \frac{Q_C + Q_W + Q_G + Q_D + Q_S + Q_I + Q_H}{\Delta t_0} \tag{8.11}$$

Annual Btu heating and cooling can be computed by using the degree-day method. The American Gas Association (American Gas Association, *House Heating*), determined that gas fuel consumption varies directly as the difference between the mean outdoor temperature and 65°F. Degree days (based on a 65°F reference) can provice an estimate of annual heating usage. The annual electricity usage in kWh/year for heating is:

$$\text{kWh/year heating} = q_{HH} \cdot 24 \cdot HDD/3412/SPF \tag{8.12}$$

where q_{HH} = house Btu/year/°F heating needs
24 = hours per day
HDD = heating degree days per year
3412 = Btu-to-kWh conversion factor
SPF = seasonal performance factor for electric heating

The seasonal performance factor is a measure of the "efficiency" of an electric heating device during an entire heating season. Resistance heating has a 1.0 *SPF,* while a heat pump has an *SPF* of 2.4 in moderate climates. In warm climates, the *SPF* is larger as a result of higher thermodynamic efficiencies in higher outside ambient temperature and less need for supplemental resistance heating (with a *SPF* of 1.0) during cold weather.

The annual electricity used in cooling can be calculated as:

$$\text{kWh/year cooling} = q_{HC} \cdot 24 \cdot CDD \cdot \frac{LHF/3412}{EER/3.412} \tag{8.13}$$

where q_{HC} = house Btu/hour/°F cooling needs
CDD = cooling degree-days per year
LHF = latent heat factor
EER = cooling energy efficiency ratio

Thermal cooling needs are first computed on the basis of the sensible-heat load (temperature). Air conditioners also reduce humidity in the air, which places an additional cooling burden, typically 20–30% of the sensible-heat load.

The cooling energy efficiency ratio is defined as the thousands of Btu of cooling per kilowatt-hour of electricity input. Typically, the *EER* is 7.5 for central air conditioning.

Figure 8.4 presents the sensitivity of heating and cooling electricity requirements to the wall insulation value in a house. The typical *R* value for framed wall using 2 × 4 studs on 16-inch centers with 3-inch insulation and 1/2-inch sheathing is $R = 11$. Increasing the wall *R* value from 5 to 10 decreases the annual electricity usage by approximately 25%. Increasing the wall *R* value from 10 to 15 decreases the annual heat pump and air conditioning energy by approximately 12%. As the wall *R* increases, there is diminishing returns unless other areas of the house are thermally improved.

Figure 8.5 presents the sensitivity of heating and cooling electricity usage to the indoor thermostat setting. Changing the summer thermostat setting has a large percentage change on the kilowatt-hours consumed. The outside to indoor temperature differences are small in the summer (10–20 degrees), and a few degrees of thermostat change will result in a large percentage change on

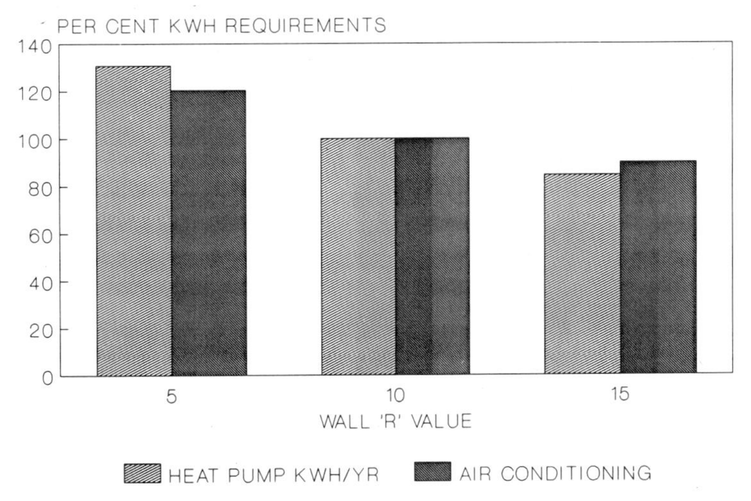

FIGURE 8.4 Annual kilowatt-hour sensitivity to wall insulation.

the outside to indoor temperature difference, which will cause a large percentage variation in the cooling requirements.

In summary, a simplified thermal house simulation model can be useful in projecting electricity changes due to future changes in house floor space, insulation levels, and thermostat settings.

8.1.1.5 Econometric End-Use Method. An alternative method of determining the annual (or monthly) energy use per appliance is by a statistical regression of individual customer usages with the individual customer's appliance inventory and the usage influence variables (Villacis et al., 1988).

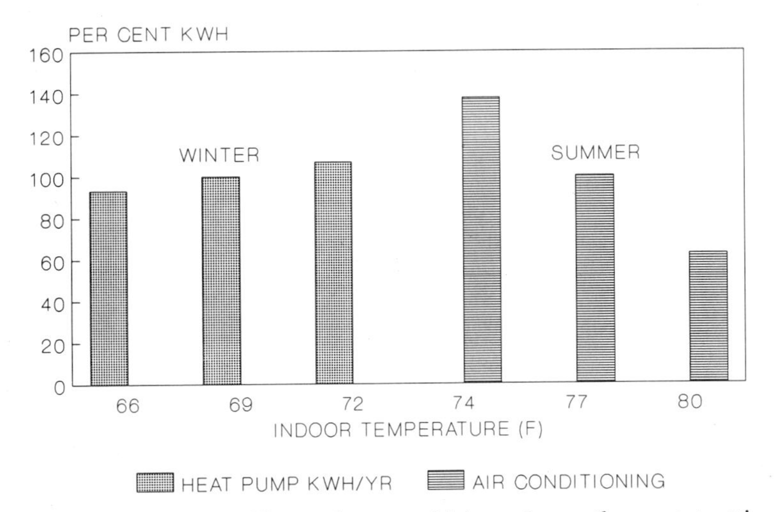

FIGURE 8.5 Annual killowatt-hour sensitivity to house thermostat setting.

The equation for the energy usage by appliance i for one customer, c, is

$$\text{kWh}_{ic} = A_{ic} \cdot [E_{i0} + \sum_{j=1}^{m} E_{ij} \cdot X_{jc}] \tag{8.14}$$

where kWh_{ic} = annual (or monthly) kWh energy usage by customer c

A_{ic} = appliance saturation for appliance i for customer c (1 = ownership, 0 = nonownership)

E_{i0} = constant energy usage for appliance i

E_{ij} = energy usage coefficient as driven by usage influence variable j

X_{jc} = usage influence variable j (including electricity price, real household income, persons per household, floor space, dwelling characteristics, and heating and cooling degree days)

m = total number of appliance types

The total electric energy consumed from all i appliances for customer c is

$$\text{Total kWh}_c = \sum_{i=1}^{I} \text{kWh}_{ic} \tag{8.15}$$

Data on the usage influence variables X_{ic} and appliance saturation A_{ic} are collected for a large sample of the residential customers. Data are also obtained on the annual electricity usage of each customer (typically from utility billing records).

Equation (8.14) substituted into Equation (8.15) yields an equation with a matrix of unknown energy-use coefficients, E_{i0} and E_{ij}. If here are 15 end-use appliances modeled, and 10 influence variables per appliance, then there are 150 unknown energy use coefficients. These 150 coefficients are determined by statistically regressing annual (or monthly) electric energy usage of thousands of customers (Villacis et al., 1988).

The same forecasting steps as outlined in the appliance saturation method may then be used to forecast future electricity sales.

8.1.1.6 Summary. The residential appliance saturation (end-use) method is a very practical method for load forecasting. The method is especially conducive for sensitivity analysis in constructing end-use management programs. An end-use management program may consist of utility conservation promotions to reduce the peak load or utility sales promotions to increase the off-peak load demands.

8.1.2 End-Use Floor-Space Method

The end-use floor-space device method is similar to the appliance saturation method, except that each end-use function (lighting, air conditioning, electric devices, etc.) in an industrial or commercial sector is examined. This section illustrates the application to the commercial sector.

Commercial and industrial definitions distinguish between manufactured products (autos, chemicals) and provided services (insurance, hospitals, stores). Electric utilities typically provide service to the demand classifications of "large light and power" and "small light and power." Since industrial companies typically have large demands, a large fraction of customers falls under a "large light and power" service classification. Many large commercial businesses (office buildings, hospitals, shopping malls) also fall under this same classification. Commercial businesses that grow over the years and have increased electric demand are often reclassified from a "small power" to a "large power" classification. With load forecasts based on service demand classifications, reclassification can present problems. Forecasting based on a "true" economic sector definition of industrial and commercial eliminates this problem but introduces the problem of obtaining historical data series.

Table 8.10 presents an eight-step commercial sector load forecast procedure using an end-use floor-space methodology.

In step 1, classify the building by commercial activity type (stores, office buildings, etc.). In step 2, classify the end-uses in each building type (lighting, space conditioning, auxiliaries, process, etc.).

In step 3, determine the floor area by building type and in step 4, the electricity kilowatt-hour end-use per floor area. These are especially difficult tasks in the commercial sector because of the paucity of data and nonhomogenity of building types. The references at the chapter end contain sources for national and regional survey data. An econometric approach based on the method of Section 8.1.1.5 may also be applied.

Tables 8.11 and 8.12 present U.S. national commercial floor space and typical kWh/ft^2 end-use data, respectively, that are needed for steps 3 and 4 (*Non-Residential Building Energy Consumption Survey-Building Characteristics,* 1981; *Energy Use in Office Buildings,* 1981; *Energy Consumption in Commercial Industries by Census Division,* 1976; *Physical Characteristics, Energy*

TABLE 8.10 Commercial Sector Load-Forecast Procedure

Step	Activity
1	Classify building types
2	Classify end uses
3	Determine building floor area
4	Determine electric end-use penetration rate and annual electricity usage per floor area
5	Forecast future building floor area additions
6	Forecast future electric penetration rate and annual electricity end use per floor area
7	Forecast electricity sales
8	Validate the forecast

TABLE 8.11 Commercial Sector Floor-Space Estimates (U.S. Data) (MILLIONS SQUARE FEET)

	1975	1985	1995
Schools	3802	4579	5938
Government buildings	916	1157	1308
Hotels and motels	2659	3187	3966
Office buildings	2657	5135	9149
Houses of worship	804	1105	1461
Warehouses	2381	4532	6373
Libraries	386	449	560
Dormitories	287	373	527
Automobile service buildings	793	1427	2326
Laboratories	263	427	662
Stores	3662	5655	7591
Health facilities	1516	2177	2993
Amusement	1842	2290	2797
Miscellaneous	1012	1437	1877
Total (in ft^2)	22987	33938	47538

Consumptions and Related Institutional Factors in the Commercial Sector, 1977).

In step 5 a forecast of the future building floor area additions is made. This may be achieved by evaluating building trends, surveying customers, or by a regression analysis driven by a macroeconomic forecast. Figure 8.6 presents an example of a regression analysis of national office building floor-space ad-

TABLE 8.12 Typical Commercial Annual kWh/foot^2 End-Use Data

	Heating	Heating Pull-Through	A/C[a]	Auxiliary	Lighting	Other
Schools	13.0	2.5	0.7	3.1	3.7	1.4
Government buildings	6.7	2.0	2.4	1.4	7.4	1.3
Hotels & motels	16.1	4.7	3.0	2.6	5.5	1.2
Office buildings	7.9	2.5	2.8	1.8	9.8	1.7
Houses of worship	15.0	1.4	7.2	2.7	8.3	0.7
Warehouses	4.7	0.4	24.6	0.7	3.1	0.2
Libraries	13.0	2.3	0.6	3.0	3.6	1.3
Dormitories	16.1	4.7	3.0	2.6	5.5	1.2
Automobile service buildings	6.2	0.4	3.6	0.5	2.3	0.1
Laboratories	7.8	2.6	2.8	2.1	9.8	1.7
Stores	22.4	30.8	5.8	13.9	12.9	2.8
Health facilities	8.9	9.5	5.3	15.2	13.4	5.1
Amusement buildings	8.4	2.6	3.0	1.8	9.8	1.7
Miscellaneous	7.9	2.6	2.8	2.1	9.8	1.7

[a]Air conditioning.

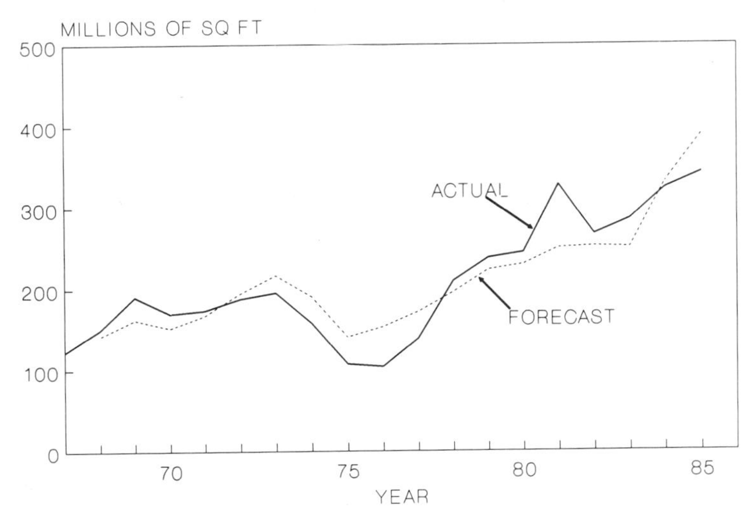

FIGURE 8.6 Office building floor space additions comparison of actual with model forecast.

ditions and contrasts this with actual data. The driving macroeconomic variables of the forecast regression are industrial production index and real commercial construction expenditures. This was developed using historical office addition data from F. W. Dodge construction surveys (Data Resources, Inc., 1985).

In step 6 the forecast of future penetration rates for the end uses and also a projection of the electricity usage for each of the end uses modeled in the commercial model is made.

In Step 7 the total electricity sales are forecast. This involves the product of the end-use saturations times the electricity usage per square foot times the square footage estimates for each of the building types.

Step 8 is the forecast validation. Similar to the residential model, this is also accomplished by validating the forecast over the last 5 or 10 years of data to ensure that the forecast replicates history. If the forecast fails to replicate history then the methodology and data must be reexamined.

Table 8.13 presents a spreadsheet forecast of office building kWh consumption on a national basis. The floor-space additions are presented in column 2.

The electric heating includes a contribution from heating directly (column 3) and a "pull-through" contribution (column 4). The annual electricity sales growth in electric heating is the product of the floor-space additions (column 2) times the electricity usage per square foot (column 3 plus column 4) times the penetration (column 5). Air conditioning annual sales growth is the product of electricity per square foot (column 5) times the floor-space additions (column 2) times the penetration rate (column 7). Auxiliaries, lighting, and other parameters are assumed to be included with each building; therefore, the kWh/ft^2 parameter is multiplied only by floor-space additions.

TABLE 8.13 Office Building kWh Forecast

	(1)	(2)	(3)	(4)	(5)	(6)
Year	Floor Space (Feet^2)	Floor Space Additions (Feet^2)	Annual Heating (kWh/Feet^2)	Annual Heating Pull-Through (kWh/Feet^2)	Penetration of Electric Heating (%)	Annual Air Conditioning (kWh/Feet^2)
1971	2,010,000	173,341	9.90	2.90	56	3.50
1972	2,197,632	187,632	9.90	2.90	56	3.50
1973	2,392,702	195,069	9.90	2.90	56	3.50
1974	2,550,388	157,686	9.90	2.90	66	3.50
1975	2,657,095	106,708	9.90	2.90	66	3.50
1976	2,760,717	103,621	9.90	2.90	66	3.50
1977	2,898,837	138,120	9.90	2.90	66	3.50
1978	3,108,151	209,314	9.56	2.84	66	3.38
1979	3,345,731	237,580	9.24	2.78	66	3.27
1980	3,590,184	244,452	8.95	2.73	64	3.17
1981	3,917,904	327,720	8.69	2.68	64	3.07
1982	4,184,674	266,771	8.45	2.64	64	2.99
1983	4,470,196	285,521	8.24	2.60	64	2.91
1984	4,793,440	323,244	8.04	2.56	64	2.84
1985	5,135,492	342,052	7.85	2.53	64	2.78
1986	5,516,708	381,216	7.69	2.49	64	2.72
1987	5,875,729	359,020	7.54	2.47	64	2.66
1988	6,243,685	367,956	7.40	2.44	64	2.62
1989	6,623,011	379,326	7.27	2.42	64	2.57
1990	7,013,465	390,454	7.15	2.40	64	2.53
1991	7,410,833	397,368	7.05	2.38	64	2.49
1992	7,824,616	413,783	6.95	2.36	64	2.46
1993	8,253,077	428,461	6.86	2.34	64	2.43
1994	8,694,612	441,535	6.78	2.33	64	2.40
1995	9,149,072	454,460	6.71	2.32	64	2.37
1996	9,616,577	467,505	6.64	2.30	64	2.35
1997	10,098,617	482,040	6.58	2.29	64	2.33
1998	10,594,227	495,611	6.53	2.28	64	2.31
1999	11,101,939	507,712	6.48	2.27	64	2.29
2000	11,609,650	507,711	6.43	2.26	64	2.27

The retirement of older buildings is based on a 1% retirement rate of buildings. The retired buildings are assumed to have half the kWh/ft^2 electric intensity of more modern buildings. The total new annual kilowatt-hour requirements is the sum of the annual growth in usage less the retired building electricity usage.

While new building floor space has conservation effects from more efficient end uses in new buildings, the overall effect of conservation on *existing* buildings also requires inclusion. The existing building conservation factor (column

(7)	(8)	(9)	(10)	(11)	(12)
Penetration Rate of A/C Including Conversions (%)	Annual Auxiliaries (kWh/Feet^2)	Annual Lighting (kWh/Feet^2)	Other (kWh/Feet^2)	Resulting Composite Annual kWh/Feet^2	Annual Increase in kWh
98	2.10	10.90	1.90	25.44	4.41
98	2.10	10.90	1.90	25.44	4.77
98	2.10	10.90	1.90	25.44	4.96
101	2.10	10.90	1.90	26.83	4.23
101	2.10	10.90	1.90	26.83	2.86
101	2.10	10.90	1.90	26.83	2.78
101	2.10	10.90	1.90	26.83	3.71
101	2.05	10.71	1.86	26.17	5.48
101	2.01	10.54	1.82	25.56	6.07
110	1.97	10.38	1.79	25.06	6.13
110	1.94	10.24	1.76	24.55	8.04
110	1.91	10.10	1.73	24.08	6.42
110	1.88	9.98	1.70	23.66	6.75
110	1.85	9.87	1.68	23.27	7.52
110	1.83	9.77	1.65	22.91	7.84
110	1.81	9.68	1.63	22.59	8.61
110	1.79	9.60	1.62	22.30	8.00
110	1.77	9.52	1.60	22.03	8.10
110	1.75	9.45	1.58	21.78	8.26
110	1.74	9.39	1.57	21.56	8.42
110	1.72	9.33	1.56	21.35	8.48
110	1.71	9.28	1.55	21.16	8.76
110	1.70	9.23	1.54	20.99	8.99
110	1.69	9.18	1.53	20.84	9.20
110	1.68	9.14	1.52	20.69	9.40
110	1.67	9.11	1.51	20.56	9.61
110	1.66	9.07	1.50	20.44	9.85
110	1.65	9.04	1.50	20.33	10.08
110	1.65	9.02	1.49	20.24	10.27
110	1.64	8.99	1.48	20.14	10.23

(*continued*)

15) accounts for changes, including indoor temperature settings, retrofit energy management systems, and lighting changes. Historically, marked changes have occurred in these areas during the early and mid-1970s and early 1980s. The existing building conservation factor (column 15) is multiplied by the total kilowatt-hour requirements excluding existing building conservation (column 14) to yield the total annual electricity requirements (column 16).

Repeating this process and accumulating the results for the other commercial sector segments leads to a total commercial sector load forecast. Figure

TABLE 8.13 (Continued)

	(13)	(14)	(15)	(16)
Year	kWh Loss Due to Building Retirements	Total kWh Requirements Excluding Existing Building Conservation (Billion kWh)	Existing Building Conservation Factor	Total Annual kWh Requirements Including Conservation (Billion kWh)
1971	0.24	52.6	1.00	52.63
1972	0.26	57.1	1.00	57.14
1973	0.29	61.8	1.00	61.82
1974	0.31	65.7	0.95	62.46
1975	0.33	68.2	0.95	64.86
1976	0.34	70.7	0.95	67.18
1977	0.35	74.0	0.95	70.37
1978	0.37	79.1	0.95	75.22
1979	0.40	84.8	0.95	80.61
1980	0.42	90.5	0.95	86.03
1981	0.45	98.1	0.92	90.29
1982	0.49	104.0	0.92	95.75
1983	0.52	110.3	0.92	101.4
1984	0.55	117.2	0.92	107.9
1985	0.59	124.5	0.92	114.5
1986	0.62	132.5	0.92	121.9
1987	0.66	139.8	0.92	128.6
1988	0.70	147.2	0.92	135.4
1989	0.74	154.8	0.92	142.4
1990	0.77	162.4	0.92	149.4
1991	0.81	170.1	0.92	156.5
1992	0.85	178.0	0.92	163.7
1993	0.89	186.1	0.92	171.2
1994	0.93	194.3	0.92	178.8
1995	0.97	202.8	0.92	186.5
1996	1.01	211.4	0.92	194.5
1997	1.06	220.2	0.92	202.6
1998	1.10	229.2	0.92	210.8
1999	1.15	238.3	0.92	219.2
2000	1.19	247.3	0.92	227.5

8.7 contrasts the U.S. national forecast model value with the actual data for the historical years 1971 through 1985. Agreement between the forecast model and actual data is good.

8.1.3 Summary

End-use forecasting models are a widely used method in the residential sector and occasionally used in the commercial sector when load-research data are available. This method provides an accountable means of adding up the com-

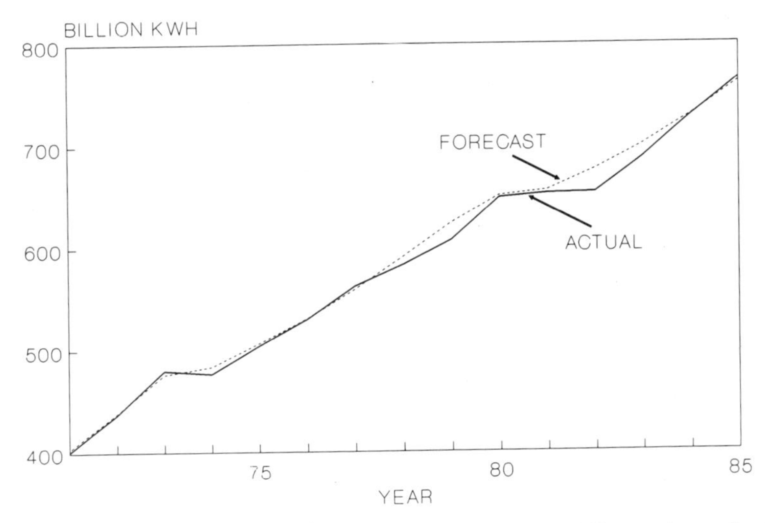

FIGURE 8.7 Comparison of forecast and actual commercial kilowatt-hours for 1971–1985 (U.S. data).

ponents of load demand. It also provides a means to establish the peak hourly demand, which will be discussed in a subsequent section.

8.2 PROJECTING PEAK LOADS

Peak-load demands are generally forecast subsequent to an electricity energy sales forecast. Peak-load data are difficult to forecast directly from macroeconomic variables. Historical peak-load data series have considerable variation from year to year due to weather sensitivity. Even after adjusting weather effects to normal weather (weather effects are often too complex to be completely normalized), peak-load series still have much variability, which makes analysis procedures difficult. Annual energy data have much less variability because 8760 hours are included, which provides an averaging and smoothing effect. In addition, coincident peak-load data by customer segments are not as readily available as annual energy data.

Peak loads are generally forecast by multiplying a load factor forecast by the total annual kWh energy sales forecast or by multiplying individual load factors by each component of the energy sales forecast. Because peak loads are strongly influenced by weather, one of the first tasks is to weather-correct the load data.

8.2.1 Weather-Correcting Peak Load

Figure 8.8 illustrates the trend of peak load to weather variation. The graph of the peak-load trend versus a weather index is almost linearly proportional to the weather index. The actual data points scatter about this trend line.

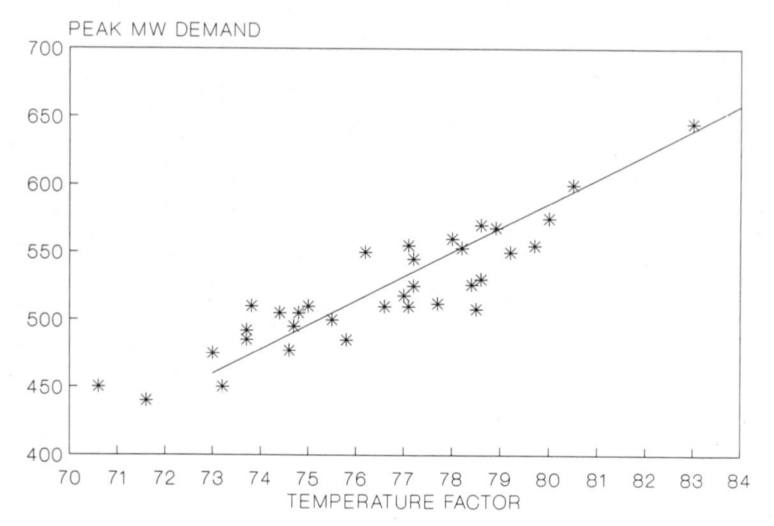

FIGURE 8.8 Weather index—trend of peak-load to weather variation.

Factors that influence the peak load are temperature (dry bulb), wet bulb temperature (to account for humidity), wind speed, solar intensity, weather conditions over the prior 2 days (thermal buildup effect), time of day, and time of year. The coefficients are determined and a weather index is developed using multiple regression analysis on many hours of peak-load data. The weather index is typically expressed in degrees Fahrenheit. One example is:

$$\text{Average temperature} = \frac{1}{2}\,(\text{dry-bulb temperature} + \text{wet-bulb temperature}) \tag{8.16}$$

$$\begin{aligned}\text{Weather index} = {} & .60 \cdot \text{temperature today} + .30 \cdot \text{temperature yesterday} \\ & + .10 \cdot \text{temperature 2 days ago}\end{aligned} \tag{8.17}$$

Once the weather index is defined, a weather standard (or weather normal) is computed as the expected average weather index based on the past 20 or 30 years. The weather standard serves as the benchmark to which the actual peak-load data are normalized to standard weather conditions.

A regression analysis of the peak-load data versus the weather index will give the load/weather sensitivity factor, typically expressed in MW/°F (megawatts per degree Fahrenheit).

Example 8.2. Calculate this year's load forecast error. The projected peak load is 10,000 MW for normal weather. The normal weather index is 79°F and the load/weather sensitivity is 100 MW/°F. A peak load of 10,200 MW occurred on July 17, with a weather index of 83°F.

SOLUTION. The weather was 83° − 79° = 4°F hotter during the peak day. Thus, the load demand adjusted to normal weather is

10,200	actual peak @ 83°F
−400	weather-sensitive contribution (4° • 100 MW/°F)
9,800	adjusted actual peak @ 79°F

The forecast of 10,000 MW is 200 MW in error.

This procedure is useful for reconciliation and correction of peak-load data.

Annual energy sales data also need to be weather-adjusted. These data are normally adjusted by using monthly or annual heating and cooling degree days rather than a weighted temperature humidity index at the time of peak. Heating and cooling degree days have a good correlation to the monthly or annual electricity sales.

8.2.2 Peak-Load Forecasting Methods

The peak-load demands are of key importance for utility capacity planning. Peak-load demands are the result of coincident electric demand consumption by all end-users. The general procedure is to forecast annual electricity sales and then calculate peak demand using a load factor or load-shape synthesis methodology. This procedure is illustrated in Figure 8.9.

The simplest peak-load forecasting method uses the system load factor. The system load factor is defined as:

$$\text{Load factor} = \frac{\text{Average-load demand}}{\text{peak-load demand}}$$

$$= \frac{\text{annual kWh energy}}{\text{peak-load demand} \cdot 8760 \text{ hours/year}} \tag{8.18}$$

Figure 8.10 illustrates the U.S. annual load factor over the last 30 years. The load factor increased during the 1950s as utilities switched from a heavy

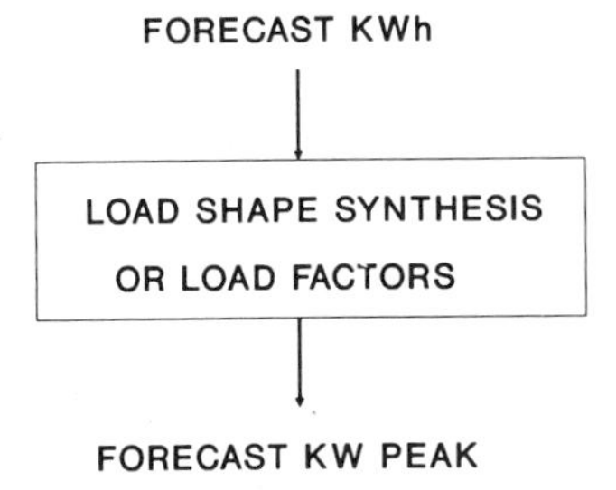

FIGURE 8.9 Peak-load demand forecast procedure.

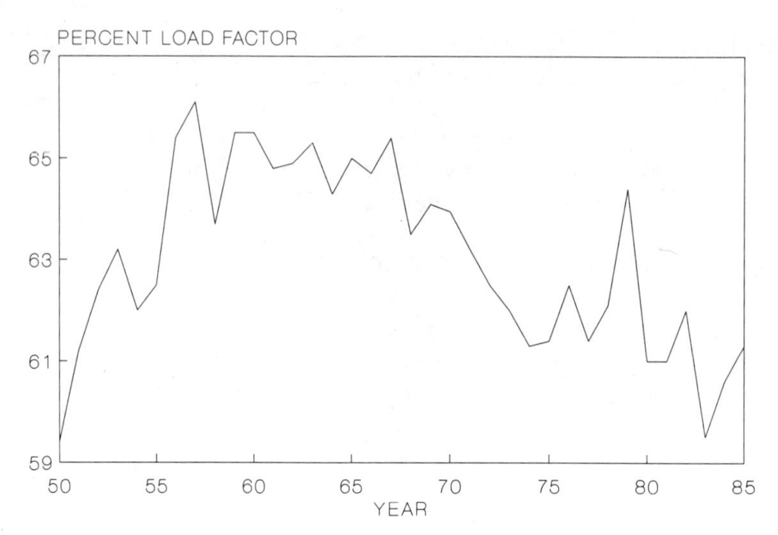

FIGURE 8.10 Annual U.S. load factor from 1950 to 1985.

winter peak to a biseasonal peak load (approximately equal winter and summer peaks). During the 1960s, the load factor decreased as the summer peak became more dominant because of the significant penetration of air conditioning.

Several years of utility-specific weather-normalized load-factor data are used, and a forecast trend is made. The peak load (normal weather) is then forecast as:

$$\text{Peak-load demand} = \frac{\text{annual kWh energy}}{\text{8760 hours/year} \cdot \text{load factor projection}} \tag{8.19}$$

8.2.2.1 End-Use Load Factor Method. The system load factor method can be expanded to provide a more accurate long-term peak-load forecast by using data on end-use load factors. In this method, the peak load contribution for each end-use device is calculated by using the end-use device kWh and end-use device load factor.

Table 8.14 presents typical end-use load factors for normalized weather (see *Report of Member Electric Systems of the New York Power Pool,* 1983). The definition of load factor used in this table is:

$$\text{Load factor(summer)} = \frac{\text{end-use annual kWh consumption}}{8760 \cdot \text{kW demand at time of system summer peak demand}} \tag{8.20}$$

$$\text{Load factor(winter)} = \frac{\text{end-use annual kWh consumption}}{8760 \cdot \text{kW demand at time of system winter peak demand}} \tag{8.21}$$

TABLE 8.14 Typical End-Use Load Factors

	Summer	Winter
	Residential	
Conventional refrigerator	65	65
Frost-free refrigerator	55	55
Freezer	60	80
Window A/C	8	—
Central A/C	12	—
Electric dryer	65	65
Electric range	30	35
Water heater	75	60
Resistance heating	—	17
Heat pump	—	10
Total	45	50
	Commercial	
Space heating	—	15
Air conditioning	20	—
Lighting	55	45
Total	50	55
	Industrial	
	80	85

Both load factor definitions [Equations (8.20) and (8.21)], use *annual* end-use consumption values in the numerator but use diversified end-use demand coincident at the time of the seasonal system peak load in the denominator.

The total system coincident demand can be computed by summing up the kilowatt-hour contributions from each end-use device using these load factor definitions:

Summer peak load =

$$\sum_{\text{end uses, } i} \frac{(\text{annual kWh consumption})_i/(\text{summer load factor})_i}{8760} \tag{8.22}$$

Winter peak load =

$$\sum_{\text{end uses, } i} \frac{(\text{annual kWh consumption})_i/(\text{winter load factor})_i}{8760} \tag{8.23}$$

With these end-use definitions, it is possible to have end-use devices with load factors greater than unity if the peak demand of the device occurs at a time other than system peak.

Example 8.3. A summer peaking end-use device consumes 8760 kWh/year and has a daily summer load profile as shown in Figure 8.11. The power system achieves its peak load during the 14–18 time period. Calculate the load factor of the end-use device.

SOLUTION. During the system peak-load period, the end-use device has a 0.5-kW demand. The summer end-use device load factor (at the time of the summer coincident peak load) is:

$$\text{Load factor(summer)} = \frac{8760}{8760(.5)} = 2.0$$

The end-use load factor data of Table 8.14 is typical only. The load factors have regional variation, especially the space conditioning items. Utility-specific load research data must be used.

Example 8.4. Use the end-use load factors of Table 8.14 and the kWh end-use forecast of Table 8.15 to project the summer and winter peak load in year 1990 and year 2000.

SOLUTION. Table 8.16 presents the peak-load forecast. The system load factor is projected to increase from year 1990 to 2000 from 56.5% to 57.1%. The system load factor changes as a result of different end-use components of the load growing at different rates.

The end-use load factor method is a very useful and practical method to forecast the peak-load demands. It can be further expanded to forecast sea-

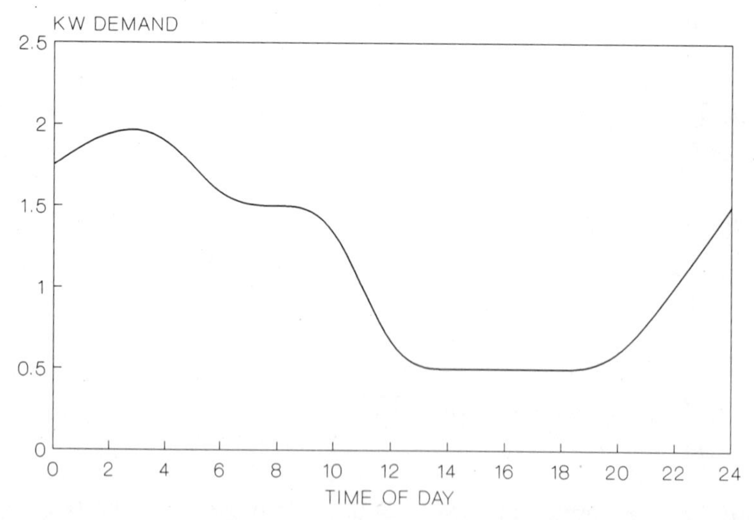

FIGURE 8.11 Daily summer hourly load profile for hypothetical end-use device.

TABLE 8.15 kWh Forecast

Load Forecast	Annual kWh 1990	Annual kWh 2000
Residential		
Conventional refrigerator	500	400
Frostless refrigerator	1,300	1,600
Freezer	400	350
Window A/C	150	140
Central A/C	1,100	1,200
Dryer	600	500
Range	450	710
Water heater	1,600	1,900
Resistance heating	310	300
Heat pump	1,000	1,250
Miscellaneous	290	550
Commercial	7,000	9,300
Industrial	15,000	18,000
Total	30,300	37,000

sonal peak loads, summer, winter, spring, and autumn. Knowledge of the seasonal peak loads is useful for maintenance scheduling and planning, power plant fuel consumption forecasting, and power system reliability forecasting.

An additional generalization is an hourly load-synthesis method. The hourly utility load profile can be synthesized from a bottom-up accumulation of individual end-use kilowatt contributions for each hour similar to the end-use load factor method. Detailed hourly load-research data are required. The result is an hourly forecast of the peak load. This permits an analysis of changes in the daily and hourly load trends in the future.

8.3 LOAD DEMAND MANAGEMENT

Utilities desire to reduce their peak load demands to reduce the need for future capacity additions and to reduce the high fuel costs of serving these peak demands. Utilities also want to increase off-peak valley-hour load demands to improve the utilization of facilities and thereby reduce the overall cost of electricity.

Utilities have several methods to manage end-use demand. Utilities may promote conservation programs or off-peak usage programs through advertisement, low-cost financing, and supplemental incentive payments.

Alternatively, the hourly load demand profiles of the end-user may be modified by the utility providing incentives to the end-use customer. If the utility "incentives" (cost savings, service reliability, etc.) to the customer are greater

TABLE 8.16 Peak-Load Forecast

	Load Factors		Annual kWh		Peak Loads 1990		Peak Loads 2000	
Load Forecast	Summer	Winter	1990	2000	Summer	Winter	Summer	Winter
Residential								
Conventional refrigerator	65	65	500	400	87.8	87.8	70.2	70.2
Frostless refrigerator	55	55	1,300	1,600	269.8	269.8	332.0	332.0
Freezer	60	80	400	350	76.1	57.0	66.5	49.9
Window A/C	8	—	150	140	214.0	0.0	199.7	0.0
Central A/C	12	—	1,100	1,200	1,046.4	0.0	1,141.6	0.0
Dryer	65	65	600	500	105.3	105.3	87.8	87.8
Range	30	35	450	710	171.2	146.7	270.1	231.5
Water heater	75	60	1,600	1,900	243.5	304.4	289.1	361.4
Resistance heating	—	17	310	300	0.0	208.1	0.0	201.4
Heat pump	—	10	1,000	1,250	0.0	1,141.6	0.0	1,426.9
Miscellaneous	70	70	290	550	47.2	47.2	89.6	89.6
Commercial	50	55	7,000	9,300	1,598.1	1,452.8	2,123.2	1,930.2
Industrial	80	85	15,000	18,000	2,140.4	2,014.5	2,568.4	2,417.4
Total			30,300	37,000	6,000.4	5,835.8	7,239.1	7,199.1
System Average Load Factor					56.5%		57.1%	

than the customer "costs" (cost, reliability, inconvenience, etc.) of the modification, then customers will change their electricity usage pattern and there will be a net societal benefit. This type of active utility program is referred to as load demand management (or load management).

Load management may be segmented into two approaches, direct load control, and indirect load control ("Impacts of Several Major Load Management Projects", 1982). Direct load control is achieved by the utility directly disconnecting, reconnecting, or modifying the operation of end-use electric devices.

Indirect load control is achieved via an electric rate structure to encourage the desired load change. Seasonal and time-of-day electricity rate tariffs using two or four seasonal rates per year and two or three rate periods per day in which rates are two to four times more expensive during the peak seasonal and/or daily periods than during the off peak periods can induce a time shift of electricity consumption.

Load management implementation requires additional equipment installed on the customer site for controlling demand in direct load control or a more detailed metering system in indirect load control. In the residential sector, the installed equipment cost ranges from $100 to $300 per customer in 1990 dollars.

8.3.1 Direct Load Control

Examples of direct control application include:

- Cycling central air conditioners 45 minutes on and 15 minutes off each hour.
- Controlling water heaters to be off for several hours.
- Controlling electric space heating loads to be off for several hours.
- Controlling demand limiters, which limit maximum demands.

8.3.1.1 Central Air Conditioner Control. Many utilities are summer peaking with a large contribution due to central air conditioning loads. Controlling the operation of central air conditioners is one means of reducing this peak. The control is typically achieved by a utility signal (radio or power line transmission) that turns off the compressor and compressor fan motor. The circulation fan is permitted uninterrupted operation. The utility signal cycles the air conditioner off typically 7.5 minutes per half hour (25% control) (Brazil and Grimmeth, 1978; Hunt, 1978).

The controlled air conditioners are segmented into groups so that several groups are off while the remainder are on. A few minutes later, the utility signals one "off" group to return to the "on" condition and signals another "on" group to the "off" condition. This group arrangement permits the total utility load changes to be uniform across each hour.

The load control period typically ranges from 4 to 10 hours per day, depending on the length of the utility peak load. During this period, the house indoor temperature may rise several degrees higher than if the air conditioner were not controlled. Figure 8.12 presents an example of an Arkansas house under 25% A/C control during a hot day for a 4-hour control period. This utility experience has typically found 80–90% customer acceptance for 20–30% air conditioner control for a 4- to 5-hour control period.

Air conditioner (A/C) control can yield a load reduction of 0.5–1.0 kW per residence during a hot day. This is dependent on weather conditions, control period, and the percent control times. Figure 8.13 presents a typical A/C load profile with and without load control for a northern utility on a hot day. During the control period (1–5 P.M.), load control provides a 1–1.2-kw load reduction. Following the load-control period, the air conditioner is permitted to catch up and reduce the house temperature back to the desired value. This postcontrol period is referred to as the "energy-pay-back period." During the energy-pay-back period, the load demand of controlled air conditioners is higher than that of the uncontrolled case.

The ratio of the electricity kilowatt-hour energy difference (uncontrolled minus controlled) during the energy payback period divided by the kilowatt-hour energy difference (uncontrolled minus controlled) during the control period is defined as the energy pay-back fraction. Conceptually, if all the house thermal energy during the control period is retained in the house interior, then the energy pay-back fraction would approach unity because the air conditioner would be required to remove the interior thermal energy built up during the pay-back period. However, the energy pay-back period is typically less than unity as the result of several factors, including thermal energy buildup partially

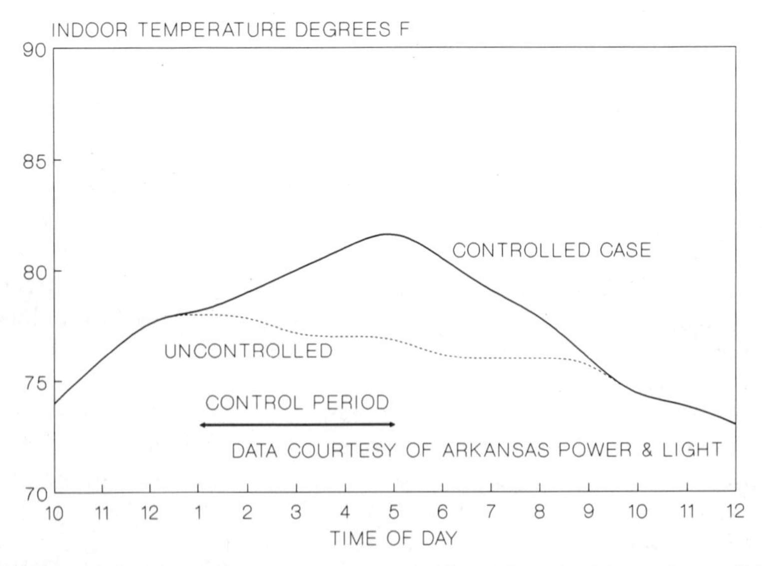

FIGURE 8.12 Typical indoor temperature profile with and without air conditioning control (25% A/C control).

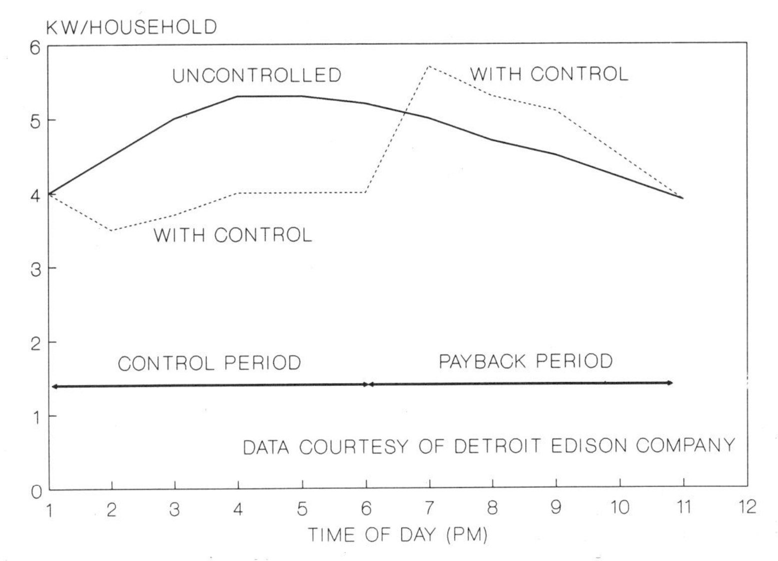

FIGURE 8.13 Air conditioning demand per household with and without 30% control.

becoming dissipated into the house structure and improvement of air conditioner efficiency during the energy-pay-back period because of a lower ambient temperature associated with it being later in the day. Reported values of the energy-pay-back fraction are lower in the northern utilities (Detroit Edison 35%, American Electric Power 50%) and higher in the southern utilities (Arkansas P&L and Mississippi P&L report almost 100%). Figure 8.14 illustrates a southern utility example.

The kilowatt load reduction is proportional to outdoor temperature (or temperature-weather index) as illustrated in Figure 8.15. For very mild days, there is little kW reduction. While the kilowatt reduction increases with outdoor temperature, the number of days per year decreases with increasing outdoor temperature. In actuality, the kilowatt reduction over the entire summer A/C season may average only 0.5 kW/unit.

However, since the system peak load is proportional to outdoor temperature, the A/C reduction is well correlated to system peak load. Those days when the system peak load is high (and load relief is highly desired), the A/C control reduction is also high.

The availability of load-research data on air conditioners with and without control is key to specific utility analysis. The utility has several economic incentives for load management. One economic incentive to reduce load demand during the peak-load period is to save on future generation, transmission, and distribution capacity addition requirements. The economic worth of this peaking capacity saving is in the order of the capital cost of a new gas turbine, or 250 $/kW of A/C control reduction (see Chapters 14 and 15).

Another economic incentive is to reduce system production cost by shifting

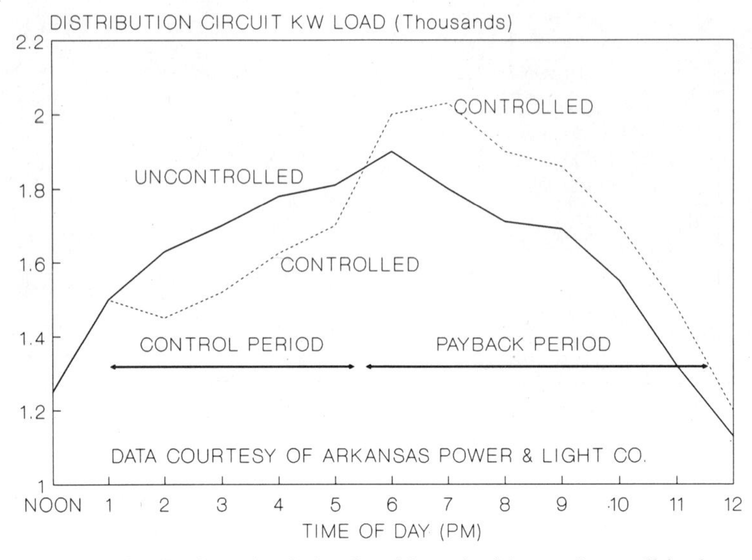

FIGURE 8.14 Distribution circuit loads with and without air conditioning control (A/C control at 35%).

load demand from the high-cost peak periods to the lower-cost off-peak periods. The annual economic worth per air conditioner may be roughly estimated by assuming a 1-kW reduction over a 5-hour control period for 70 days/year. Assume also a 10 $/MWh energy cost saving from generating energy during the off-peak period rather than on-peak. This leads to an estimate of

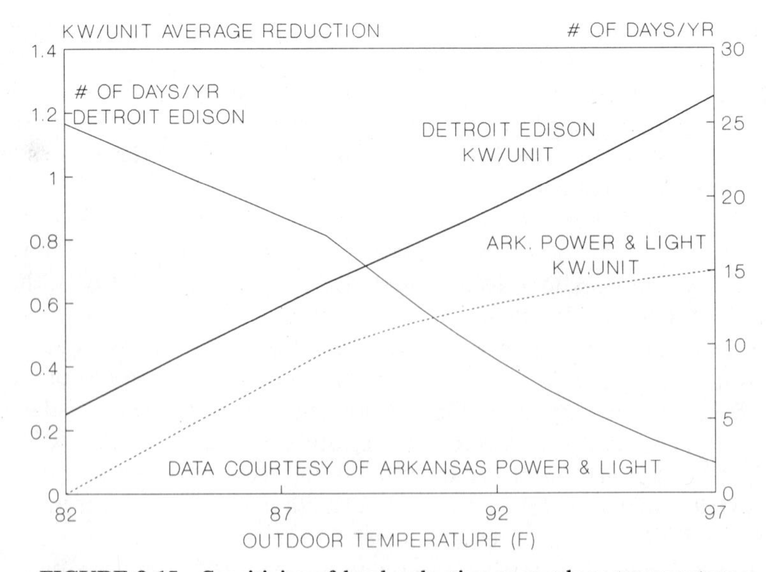

FIGURE 8.15 Sensitivity of load reduction to outdoor temperature.

the annual saving as 0.001 MW • 5 hours/day • 70 days/year • 10 $/MWh = 3.50 $/year. The equivalent capitalized saving using an annual levelized fixed charge rate of 20% is $17.50/kW of A/C control reduction.

Both capacity and production cost savings may be added together. Deductions from the gross savings of load control are O&M (operation and maintenance) expense, customer participation payments, and investment charges on load control equipment. The following example illustrates this.

Example 8.5. Compute the economic merits to a utility for controlling 40,000 air conditioners with a 25% cycling mode. The utility is summer peaking and has a temperature-sensitive load. The hourly power system peak loads and number of summer week days are presented in Table 8.17.

The central air conditioner load demand with and without load control is presented in Table 8.18 for 4 temperature-index days. The control period is 5 hours, 1–6 P.M. The pay-back period is 5 hours, with an energy payback of 80%.

The levelized generation system incremental production cost is presented in Figure 8.16. This characteristic is typical of a power system with 1200 MW of coal unit capacity and 800 MW of oil- and gas-fired steam and gas-turbine capacity. When the load demand is 1700 MW, each *additional* megawatt of generation output costs the power system 55 $/MWh. Similarly, each megawatt decrease of generation output saves 55 $/MWh. One contribution of load management is in reducing peak load demands, thereby saving on high production costs.

The utility plans on installing 40,000 air conditioner controllers.

TABLE 8.17 Power System Summer Weekday Load Demands

Number of days/year	25	25	15	5
Temperature index	97°F	92°F	87°F	82°F
Hour		*MW Load Demands*		
12:00 Noon	1500	1200	1200	1155
1	1700	1355	1280	1205
2	1725	1425	1275	1170
3	1755	1470	1450	1200
4	1700	1430	1340	1115
5	1675	1435	1315	1105
6	1550	1340	1190	980
7	1425	1230	1125	915
8	1330	1150	1060	850
9	1265	1070	980	800
10	1180	1045	925	775
11	1135	1030	895	730
12	1050	975	825	675

TABLE 8.18 A/C kW Demand With and Without Load Control

Number of Days/Year	25		25		15		5	
Temperature Index	97°F Normal	97°F Controlled	92°F Normal	92°F Controlled	87°F Normal	87°F Controlled	82°F Normal	82°F Controlled
Hour	kW Demand per Air Conditioner							
12:00 Noon	3.0	3.0	1.0	1.0	1.0	1.0	0.7	0.7
1	4.0	2.8	1.7	0.7	1.2	0.6	0.7	0.5
2	4.5	3.3	2.5	1.5	1.5	0.9	0.8	0.6
3	4.7	3.5	2.8	1.8	2.0	1.4	1.0	0.8
4	5.0	3.8	3.2	2.2	2.6	2.0	1.1	0.9
5	5.0	3.8	3.4	2.4	2.6	2.0	1.2	1.0
6	5.0	6.2	3.6	4.6	2.6	3.2	1.2	1.4
7	4.5	5.5	3.2	4.1	2.5	3.0	1.1	1.2
8	4.2	5.1	3.0	3.7	2.4	2.8	1.0	1.1
9	4.1	4.9	2.8	3.5	2.2	2.6	1.0	1.1
10	3.7	4.4	2.8	3.4	2.0	2.3	1.0	1.1
11	3.4	3.4	2.7	2.7	1.8	1.8	0.7	0.7
12	3.0	3.0	2.5	2.5	1.5	1.5	0.5	0.5

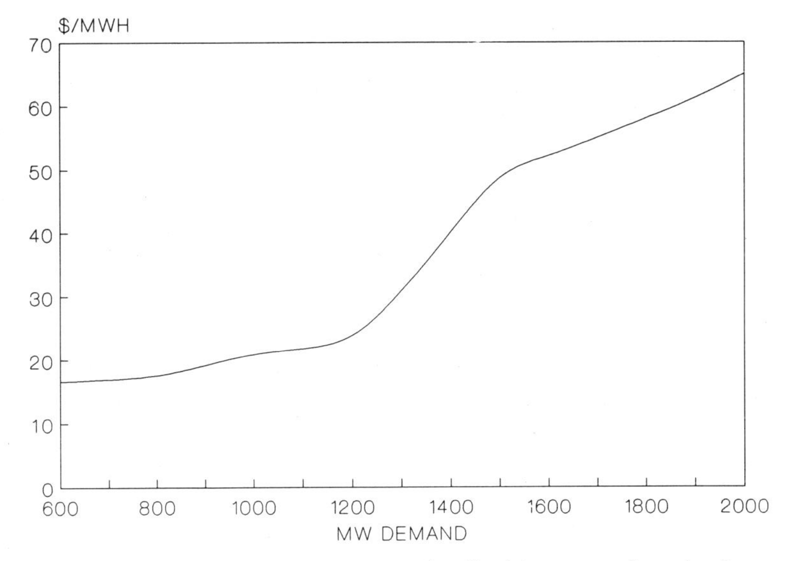

FIGURE 8.16 Sample generation system levelized incremental production cost.

SOLUTION. The system load demand with the load control is presented in Table 8.19. This is calculated by subtracting the air conditioner kilowatts per unit controlled demand from the uncontrolled demand and multiplying by 40,000 controllers. Figure 8.17 graphically contrasts the system load profile with and without load control for two outdoor temperature indexes.

TABLE 8.19 Power System Summary Weekday Load Demands (in MW) with Load Control

Number of Days/Year Temperature Index Hour	25 97°F	25 92°F	15 87°F	5 82°F
		MW Load Demands		
12:00 Noon	1500	1200	1200	1155
1	1652	1315	1256	1197
2	1677	1385	1251	1162
3	1707	1430	1326	1192
4	1652	1390	1316	1107
5	1627	1395	1291	1097
6	1600	1382	1215	988
7	1468	1266	1147	922
8	1368	1182	1079	856
9	1299	1098	997	806
10	1210	1070	940	780
11	1135	1030	895	730
12	1050	975	825	675

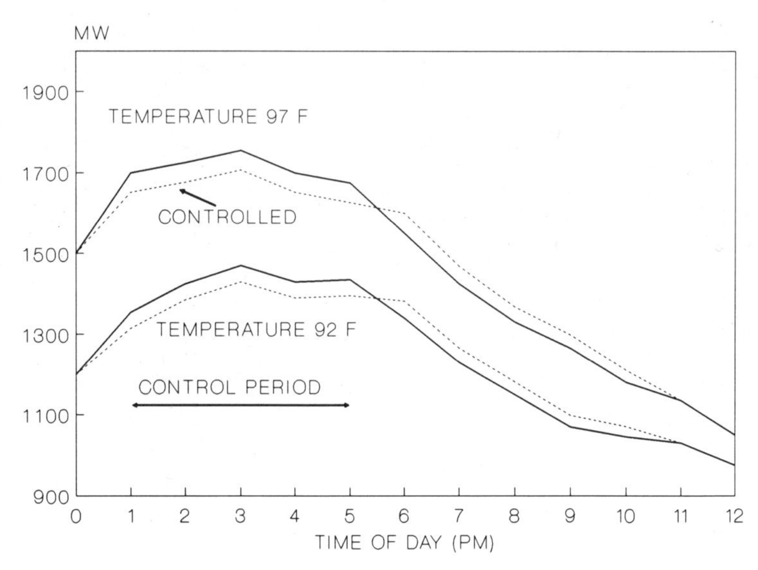

FIGURE 8.17 Sample load profile with and without air conditioner control.

The economic benefits analysis consists of five components: (1) the power system production cost (fuel) savings; plus (2) power system generation, transmission, and distribution capacity savings less; (3) the revenue lost due to lower energy sales; less (4) the load-management O&M expenses; less (5) the customer incentive participation costs.

PRODUCTION COST SAVINGS. The levelized annual system production cost savings is calculated in Table 8.20. This is computed by multiplying the system megawatt difference between the controlled and uncontrolled cases (Table 8.17 less Table 8.19) times the average incremental production cost \$/MWh (Figure 8.17). For example, the hourly savings in hour 3 for the 97°F day is calculated as (1755–1707) • 56.2 \$/MWh = 2697 \$/hour.

During the control period, savings are positive, while during the energy pay-back period, savings are negative. The net production cost (fuel) savings are proportional to the *difference* in the incremental production cost during the control period and the energy-pay-back period. The production cost savings has a contribution due to less energy generation in the load control case because the energy-pay-back fraction (in this example) is 80%. The annual saving accumulates to \$261,842.

POWER SYSTEM GENERATION, TRANSMISSION, AND DISTRIBUTION CAPACITY SAVINGS. The load control permits a 48-MW load reduction during the system's peak day when the temperature is 97°F, and a smaller reduction for lower temperatures. Since lowering the peak load reduces the need for future capacity additions, load control receives a capacity credit for deferring future capacity.

TABLE 8.20 Production Cost Saving from Load Control

Number of Days/Year	25	25	15	5
Temperature Index	97°F	92°F	87°F	82°F
Hour		Hourly Savings ($)		
12:00 Noon	0	0	0	0
1	2,642	1,457	791	234
2	2,669	1,680	766	217
3	2,697	1,840	874	234
4	2,642	1,680	874	195
5	2,590	1,680	817	195
6	−2,567	−1,530	−711	−175
7	−1,814	−1,080	−546	−138
8	−1,331	− 825	−432	−113
9	−1,072	− 640	−350	− 97
10	− 853	− 559	−296	− 85
11	0	0	0	0
12	0	0	0	0
Total $/Day	5,604	3,703	1,787	468
Total $ by day type	140,109	92,587	26,804	2,342
Annual total $	261,842			

Generation capacity is typically planned on the basis of a systemwide reliability measure, such as LOLP (see Chapters 10 and 11). Because the load control produces 48 MW for only 25 summer weekdays, and less during other weather conditions, a power system generation reliability evaluation computes an effective load management capacity credit of 45.6 MW. In this example, generation capacity is given a levelized 50 $/kW/year credit. Thus, the generation capacity saving is 45,600 kW • 50 $/kW or $2,279,000/year.

Transmission capacity is also conceptually saved as a result of the 48-MW peak-load reduction. The credit for transmission depends on the location of the generation capacity that is deferred. A generating plant located remote to the load center leads to a large transmission capacity credit, whereas a plant local to the load center merits only a small credit. In this example, it is assumed to be zero.

The distribution system load typically peaks several hours later than the system peak load. The distribution system peak typically occurs during the energy-pay-back period, and load control may increase the distribution peak load. Thus, a small negative distribution capacity credit may be applicable because of load control (Davis, Krupa, and Diedzic, 1982). In this example, it is assumed to be zero.

REVENUE LOST. In this example, the energy payback fraction is 80%. The load-control case has less delivered energy, and this leads to less revenue collected from the customers. The lost energy may be calculated by subtracting the daily energy for the uncontrolled case (Table 8.17) from the controlled case (Table 8.19), multiplying the difference by the appropriate number of days per year and accumulating the result for the year. These calculations result in an annual energy loss of 2,424 MWh/year.

The levelized residential electric rate is assumed to be 6 ¢/kWh or 60 $/MWh. Multiplying the electric rate times the energy lost yields a revenue lost cost of $145,470.

LOAD MANAGEMENT O&M EXPENSE. The primary cause of load management O&M is repair of the residential receivers. Detroit Edison reports an annual failure rate of 4.5%. Each failure requires dismantling equipment, shop repair (90% of the time) or replacement, and reinstallation.

This example assumes a failure rate of 5% and a levelized installed repair cost of $50/failure. Thus the O&M expense is estimated to be (.05) • 50 $/Unit • 40,000 units, or $100,000/year.

CUSTOMER INCENTIVE PAYMENTS. The A/C load control impacts customer lifestyle and comfort. Indoor temperature typically increases by 2–3°F. Most utilities have found that the customer is willing to accept this in conjunction with an annual participation payment. Table 8.21 lists three typical annual incentive plans.

This example uses a $25/year customer incentive payment. The total annual cost for 40,000 customers is $1,000,000/year.

TOTAL BENEFITS. The total benefits are summarized in Table 8.22. The benefits are itemized in terms of both annual saving and the equivalent capitalized value per load-controlled air conditioner. The capitalized value is obtained by dividing the levelized annual benefits by the levelized annual fixed-charge rate (20%/year in this example) and by the 40,000 controlled air conditioners.

The net capitalized worth is $162 per load control. If the total cost of the load control including the receiver, plus installation, plus the allocated cost of the transmitter is less than $162, then load control is an economic activity for the utility to promote. If the total cost of load control is greater than $162, then load control is not an economic activity to the utility and conventional capacity is more economical.

Note that the component with the largest saving in this example is the capac-

TABLE 8.21 Air Conditioner Annual Customer Incentive Payment

Utility Example	
Arkansas P&L	$30
Commonwealth Edison	$76
Detroit Edison	$39

TABLE 8.22 Economic Benefits Summary of Load Control of A/C

	Levelized Annual Benefits ($/Year)	Equivalent Capitalized Benefits ($/Controlled A/C)
Fuel savings	261,842	32.7
Capacity	2,279,842	285.0
Less revenue lost	−145,470	−18.2
Less load management	−100,000	−12.5
Less customer credit	−1,000,000	−125.0
Total	1,296,214	162.0

ity savings. This example assumed that the utility has an imminent need for additional capacity that load management could potentially alleviate. If the utility has a large reserve margin, then the capacity savings credit would be zero, which significantly diminishes the economic benefits of load control implementation in the near term.

SENSITIVITY CASES. If the load control system were installed in a cooler climate, then the benefits would be smaller. Assume that the summer weather had a distribution as shown in Table 8.23. These data are extrapolated to determine the economic benefits shown in Table 8.24. The net capitalized worth is $104.4 per load control.

If the load control were located in the original climate and utility had *both* a winter and a summer peak, then the capacity saving credit would be diminished because the load control contributes to reducing only half of the utility reliability risk (refer to Chapters 10 and 11), that of the summer only. Some utilities with nearly equal summer and winter peak-load demands have significant generation maintenance activities during the autumn and spring periods, thus rendering these utilities susceptible to reliability risks throughout the year. In this case, the summer A/C load control receives only approximately 25% of the summer peak capacity credit. Table 8.25 presents the economic summary for this case. In this case the total economic benefits are negative.

The fuel savings benefits are proportional to the average incremental cost during the control period and the energy-pay-back period. This is utility-specific and also influences the economic results.

In this example, the load control period was from 1 to 6 PM, and the load control penetration was 40,000 controllers. Figure 8.17 presented the load pro-

TABLE 8.23 Weather Data for Cooler Climate Utility

Number of days/year	5	15	25	25
Temperature index	97°F	92°F	87°F	82°F

TABLE 8.24 Economic Benefits Summary of Load Control of A/C in Cooler Climate

	Levelized Annual Benefits ($/Year)	Equivalent Capitalized Benefits ($/Controlled A/C)
Fuel savings	130,388	16.3
Capacity savings	1,896,515	237.1
Less revenue lost	−91,758	−11.5
Less load management O&M	−100,000	−12.5
Less customer credit	−1,000,000	−125.0
Total	8,359,145	104.4

TABLE 8.25 Economic Benefits Summary of Load Control of A/C in Utility with Uniform Reliability Risk Throughout the Year

	Equivalent Capitalized Benefits ($/Controlled A/C)
Fuel savings	32.7
Capacity savings	71.3
Less revenue lost	−18.2
Less load management O&M	−12.5
Less customer credit	−125.0
Total	−51.7

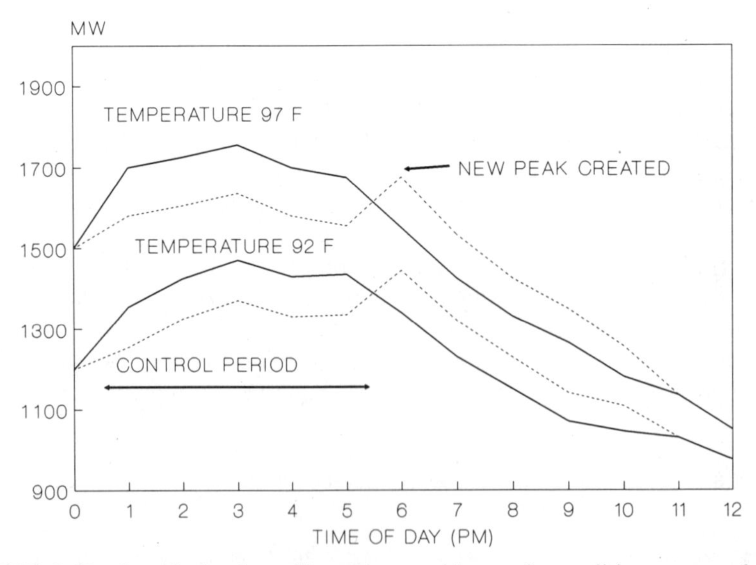

FIGURE 8.18 Sample load profile with an without air conditioner control—high A/C control penetration.

file for the previous example. If a 100,000 penetration is used, then the load control profile is as shown in Figure 8.18. Note that the controlled load during the load-control period (1–6 PM) is reduced, but the postcontrol (energy-payback) period leads to a higher peak than during the control period. This case would lead to a very low capacity credit in the economic analysis.

In this high-penetration case, load control reduced the loads during the load-control period to the point that the loads are higher during the noncontrol periods. One approach is to extend the load-control period beyond the 1–6-PM time period to a 1 to 9-PM time period. This has its drawbacks, because it would lead to higher indoor temperatures and correspondingly increased customer dissatisfaction with the intrusion on lifestyle. Consequently, customers may demand larger incentive payments for participation in the load management program.

Alternatively, the 100,000 load controllers could be grouped into 5-hour control periods with one group of 20,000 beginning control at 1–6 PM and another group of 20,000 beginning at 2–7 PM, another group of 20,000 beginning at 3–8 PM and another group of 40,000 beginning at 4–9 PM. While this group strategy may reduce a new peak forming after 6 PM, at the 3-PM peak load only 60,000 load controllers are active. Thus, the effective megawatt peak reduction constitutes only 6/8 of the total penetration. There is a "diminishing returns" benefit as the load-control penetration becomes large enough to create new peak loads outside of the control period.

8.3.1.2 Central Air Conditioner Control Summary. This example illustrated several economic benefits analysis cases of air conditioning load management. The example used simple "models" of the A/C device and power system production cost model. While these simple methods provide insight into the issues, more accurate and detailed methods are generally required. These methods are presented in Chapters 10 through 15.

The benefits from load management of central air conditioners are utility-specific, as was illustrated in the example. Accurate load-research data are essential for utility-specific analysis.

8.3.2 Electric Water Heater Control Strategy

Electric water heaters may also be load-controlled by turning them off during the control period and then on at the conclusion of the control period. Water heaters, however, have a high installed rating of 4–5 kW. Thus, turning a water heater from a controlled "off" condition to an "on" condition can result in a 4-kW demand, which can lead to an abrupt secondary peak being created. By proper coordination, however, this may be mitigated.

Modern electric water heaters typically have two heater coils, a top coil that heats the top 20% of the water to permit a quick recovery of a small volume of water and a lower element that heats the remainder of the tank. Only one coil can be operative at a time. This two-coil design produces water temperature stratification, with the hotter water on top near the outlet.

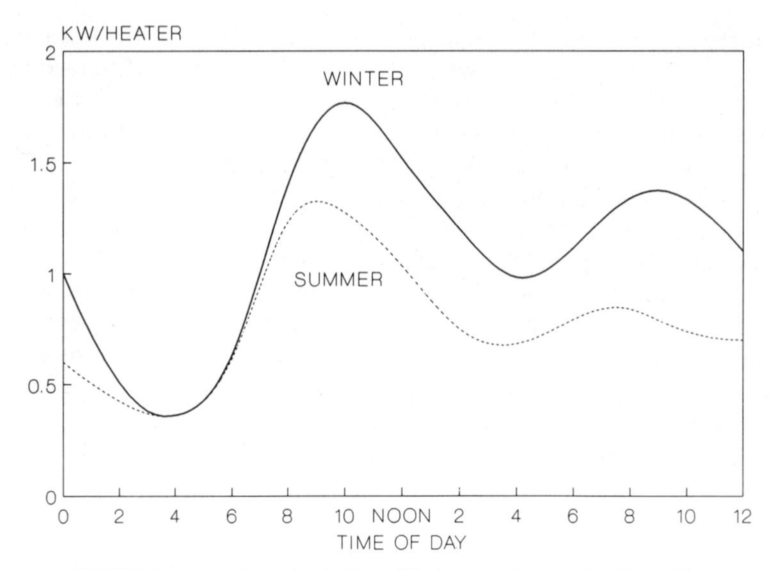

FIGURE 8.19 A typical diversified water-heater load profile.

Figure 8.19 presents a typical diversified water heater load profile. Electricity usage is larger in the winter due to lower inlet water temperatures and lifestyle changes.

Water heater load control may be applied throughout the year to reduce peak loads and for fuel savings. Water heater load control is especially useful for utilities that have nearly equal summer and winter peak-load demands.

Figure 8.20 illustrates a conceptual load control action. Water heaters are

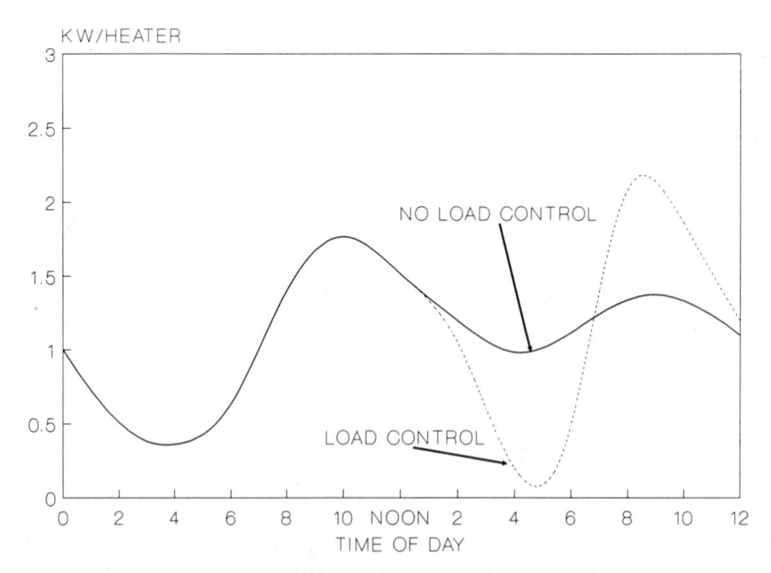

FIGURE 8.20 Conceptual load control action during water-heater control period.

controlled from 2 to 6 PM. Following the control period, the demand is increased for several hours. The water heater energy that was curtailed during the control period is largely paid back in 2 hours following a 4–5-hour control period. When the water temperature is reduced during the control period, people and some programmed appliances mix less cold water with more reduced-temperature hot water to obtain the correct desired end-use water temperature. In this case, if the tank is not exhausted, the energy pay-back following the control period is 100%. Devices such as dishwashers and washing machines set to "hot" do not regulate the hot and cold water. In these cases, the energy-pay-back following the control period is less than 100%. Because the water temperature is slightly less during the control period, the thermal heat losses (approximately 400 watts) are slightly less, resulting in a pay-back of less than 100%.

Typical water heater energy pay-backs range from 70 to 100% in the summer and 60 to 90% in the winter. Power systems typically peak in the summer near 3 PM and in the winter near 5 PM (or later). The water heater control period may be from 1 to 6 PM in the summer and 3 to 7 PM in the winter (or 1 to 10 PM if daily bimodal peaking).

Figure 8.21 presents a typical water heater response following the control period for a water heater in which the utility's energy payback is 65% (Lee and Wikkins, 1982). The parameters of the curve are based on the kilowatt-hour curtailed during the control period. For example, if the normal water heater diversified demand is 1 kW during a 3-hour control period, then the 3-kW curve presents the demand following the control period.

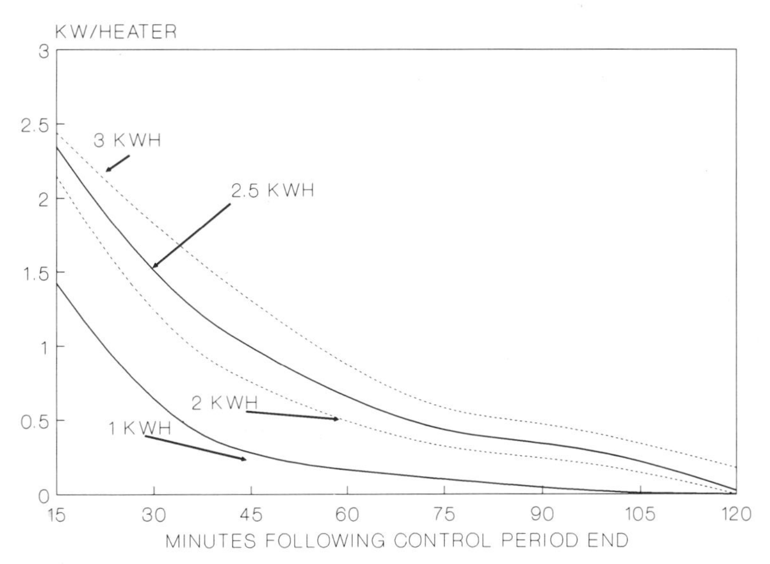

FIGURE 8.21 Water-heater response following control as a function of kilowatt-hour reduction during the control period.

The economic benefits of water heater control are comprised of the same cost components as air conditioner control. However, the capacity savings for water heaters is larger than A/C control for utilities with nearly equal summer and winter peak loads because water heater control can be applied year-round. The production cost savings from water heater control has two influences: it increases as a result of year-round utilization but decreases because of a smaller replacement energy cost difference between on-peak and off-peak production costs since the energy-pay-back period is typically only 2–3 hours.

Example 8.6. Compute the economic benefits of water heater load control on a power system with the characteristics of Example 8.5 and with a winter peak load that is 92% of the summer peak load.

The power system loads are presented in Table 8.26, in which the summer loads are characterized by four weekday types, the winter by three weekday types, and the spring and fall by two weekday types.

Water heater load control characteristics are given in Table 8.27 for the three seasons. The energy pay-backs are 97% in summer, 95% in spring and fall, and 90% in winter. The load control period is 2–6 PM in summer, 3–7 PM in spring and fall, and 5–9 PM in winter.

SOLUTION. Table 8.28 presents system load demand with water heater control for a 40,000 penetration. The peak day in each season with and without load control is presented in Figure 8.22.

Capacity savings, production cost savings, revenue loss, O&M expense, and

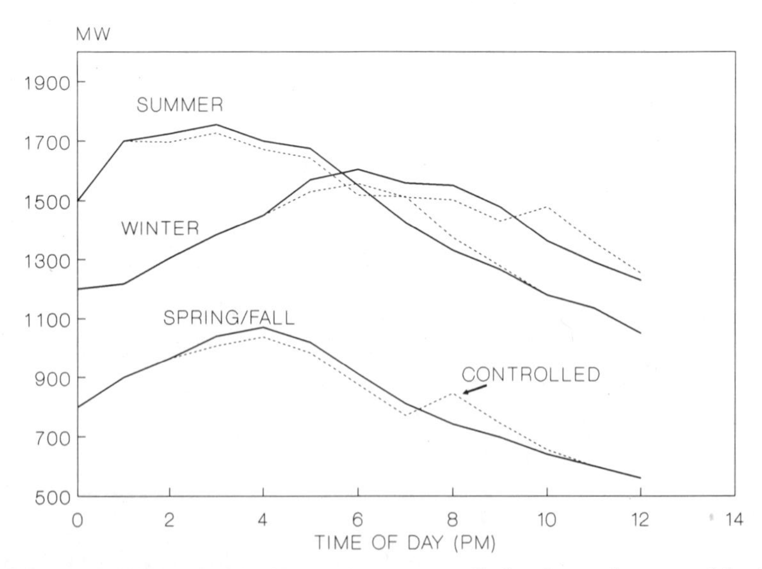

FIGURE 8.22 Load control of water heaters—peak day for each seasonal load profile.

TABLE 8.26 System Load Profiles (No Load Control)

	Summer				Winter			Spring/Fall	
Number of Days/Year	25	25	15	5	25	25	20	70	70
Hour					MW Demand				
0	1500	1200	1200	1155	1200	1075	975	800	700
1	1700	1355	1280	1170	1218	1093	993	900	800
2	1725	1425	1275	1205	1305	1180	1080	964	864
3	1755	1470	1350	1200	1383	1258	1158	1040	940
4	1700	1430	1340	1115	1450	1325	1225	1070	970
5	1675	1435	1315	1105	1570	1445	1345	1020	920
6	1550	1340	1190	980	1605	1480	1402	912	812
7	1425	1230	1125	915	1558	1433	1333	812	712
8	1330	1150	1060	850	1550	1425	1325	742	642
9	1265	1070	980	800	1478	1353	1253	698	598
10	1180	1045	925	775	1363	1238	1138	640	540
11	1135	1030	895	730	1290	1165	1065	600	500
12	1050	975	825	675	1230	1105	1005	560	460

TABLE 8.27 Water Heater Load Demand

	kW Diversified Demand/Water Heater					
	Summer		Winter		Spring/Fall	
Hour	Normal	Controlled	Normal	Controlled	Normal	Controlled
0	1.0	1.0	1.5	1.5	1.0	1.0
1	0.8	0.8	1.3	1.3	0.9	0.9
2	0.7	0	1.2	1.2	0.8	0.8
3	0.7	0	1.1	1.1	0.8	0
4	0.7	0	1.0	1.0	0.8	0
5	0.8	0	1.0	0	0.9	0
6	0.8	0	1.2	0	1.0	0
7	0.9	3.1	1.2	0	1.0	0
8	0.9	2.0	1.2	0	1.0	3.6
9	1.0	1.3	1.2	0	1.0	2.2
10	1.0	1.0	1.3	4.2	1.1	1.5
11	0.9	0.9	1.2	2.9	1.0	1.0
12	0.8	0.8	1.1	1.7	0.8	0.8

cutomer payments are calculated as was done for the air conditioner example. The results are presented in Table 8.29. The water heater capitalized benefits are \$116.9 per controlled heater. If load-control equipment costs exceed \$116.9, then load control is less attractive than a generation capacity addition alternative.

8.3.3 Indirect Load Control

Indirect load control (management) focuses on providing customers with price signals as incentives to change their electricity usage patterns. The price signals may be electric rate tariffs dependent on the season, the maximum monthly demand, the energy consumption, and the time-of-day (or seasonal) demand and/or energy consumption. The customer response (and price signal) is measured at the customer's electric meter.

Implementation of indirect load control requires a well-designed rate tariff structure as well as customer metering capable of measuring the tariff billing components. For large commercial and industrial customers, the cost of metering is a small contribution to the overall cost of electricity. For residential and small commercial customers, the cost of a new sophisticated meter can be a significant cost consideration when evaluating the net benefits of indirect load control.

Among the many factors that cause electric utility costs of operations to vary are season, weather, and time of day. The philosophy of indirect load control is that customers should be made aware of these costs through the electric rate tariff structure so as to achieve the highest overall societal eco-

TABLE 8.28 System Load Profile with Water Heater Control

	Summer				Winter			Spring/Fall	
Number of Days/Year	25	25	15	5	25	25	20	70	70
Hour	MW Demand								
0	1500	1200	1200	1155	1200	1075	975	800	700
1	1700	1355	1280	1170	1218	1093	993	900	800
2	1697	1397	1247	1177	1305	1180	1080	964	864
3	1727	1442	1322	1172	1383	1258	1158	1008	908
4	1672	1402	1312	1087	1450	1325	1225	1038	938
5	1643	1403	1283	1073	1530	1405	1305	984	884
6	1518	1308	1158	948	1557	1432	1354	876	776
7	1513	1318	1213	1003	1510	1385	1285	772	672
8	1374	1194	1104	894	1502	1377	1277	846	746
9	1277	1082	992	812	1430	1305	1205	746	646
10	1180	1045	925	775	1479	1354	1254	656	556
11	1135	1030	895	730	1358	1233	1133	600	500
12	1050	975	825	675	1254	1129	1029	560	460

TABLE 8.29 Water Heater Control—Economic Benefits Summary

	Equivalent Capitalized Benefits ($/Controlled Water Heater)
Fuel savings	62.7
Capacity savings	214.8
Less revenue lost	−23.1
Less load management O&M	−12.5
Less customer credit	−125.0
Total	116.9

nomic efficiency. Presumably, electricity is like any other commodity or service in which a consumer performs an economic analysis to assess the quantity of electricity consumption as a function of price. If electricity price is high, the consumer may reduce the electricity quantity used and substitute something else (conservation equipment, other energy types, etc.). The customer achieves greater overall economic efficiency by adjusting to the high price than would be achieved by not adjusting and paying a higher electricity bill. Similarly, if electricity price is low, the consumer may increase electricity consumption to achieve greater economic efficiency.

Any industry, whether electric utility or manufacturing, has two types of costs: fully allocated costs, based on actual current cost of service; and marginal costs, based on the cost of one additional incremental unit of output. Which cost should be used to send price signals to consumers so that they perform the correct economic decision? A number of economists and electric rate analysts favor marginal price signals (refer to Chapter 5).

Many electric utilities with significant seasonal cost variations have implemented seasonal electric rate tariffs. Because of large variations in hourly load demands, electric utility costs also have a time-of-day cost dependence. Increased societal economic efficiency may be achieved by time-of-day or peak-load pricing of electricity.

Figure 8.23 illustrates the residential hourly load change in an Orange and Rockland utility experiment (Cuccaro and Schuh, 1984). The experiment consisted of two groups, one group consisting of 150 volunteers for the peak-load pricing rate tariff and another group consisting of a 100 customers in a control group with the same characteristics as the first group. The peak-load pricing rate was applicable during the summer only during days when the load was 90% of the summer peak, approximately 15 days/year. The peak-load price was 30¢/kWh higher than the average during the peak-load period (approximately 300% higher), and the off-peak price was approximately 7% less than average. Figure 8.23 illustrates a demand reduction of 0.5 kW during the peak day's peak-pricing period. Average demand reductions for all summer peak-load hours averaged 0.28 kW per customer, or approximately 28%.

Let the peak period price elasticity be defined as the change in peak period

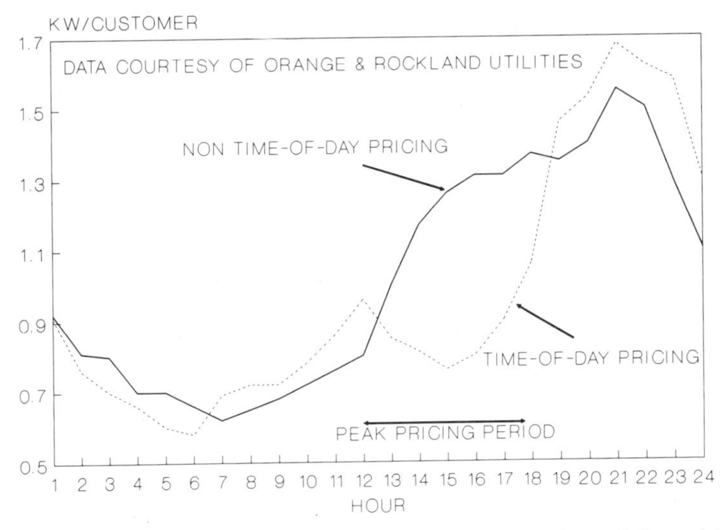

FIGURE 8.23 Residential load demands—with and without time-of-day pricing (1983 peak day).

kW divided by the change in price relative to the conditions of a constant time-of-day price. From this experiment, the peak-period price elasticity is:

$$\text{Peak period price elasticity} = \frac{\ln(1.0 - .28)}{\ln(3.00)} = -.30$$

For a 1% change in peak-period electricity price relative to the average daily price, the peak-period average demand is reduced by 0.3%.

Other rate experiments on residential customers indicate mixed results. A Jersey Central Power and Light study (see Galgon and Carter, "Residential Customer Response to an Optimal Time of Use Rate," 1978–1979) found that an average 6.5% on-peak-energy reduction was achieved by a peak to average price change of 75%. This leads to a peak-period price elasticity estimate of −.12. A Northeast Utilities Service Company study of mandatory peak-load pricing for residential customers (see Meewsen and Brown, "Residential Time of Day Class Load Study," 1984–1985) reports no evidence to indicate peak-load pricing customers shifted energy to the off-peak period.

These load research experiments were based on short-term (1–3 years) response. It is possible that long-term response characteristics may be greater.

The average annual residential consumer's electricity cost savings from peak-load pricing rate tariffs is shown for several experiments:

	Annual Consumer Cost Savings per Year
Orange and Rockland	$32
Central Maine Power	$51
Jersey Central P&L	$25

The residential consumer's incentive to participate in rescheduling the electricity time-of-use profile is dependent on the overall economic savings opportunity and degree of lifestyle intrusion changes required. The enthusiasm and extent of customer participation may be influenced by potential savings as a fraction of household disposable net income.

Peak-load pricing can lead to decreases in overall electricity consumption (kWh energy). Orange and Rockland found a 1.5% decrease in residential annual energy. Energy loss can occur because of different price elasticities during the day.

The economics of residential peak-load pricing can be evaluated using the same cost components as in direct load control. The following example illustrates this.

Example 8.7. A summer peaking electric utility is considering the addition of time-of-day (TOD) meters for 40,000 residential consumers. The peak-load pricing period would apply from June through August, six hours per day. Residential consumers have the following characteristics:

Energy consumed during summer peak hours per customer	800 kWh/year
Energy consumed during summer off-peak hours/customer	2,200 kWh/year
Energy consumed during winter/spring/fall	7,000 kWh/year
Total energy consumed per year per customer	10,000 kWh/year
Estimated peak-load price elasticity	−.30
Estimated off-peak price elasticity	−.10

The proposed TOD rate structure is designed so that the annual customer bill is the same as the existing tariff if the customer does not alter the usage profiles.

The existing and proposed TOD rate tariff is:

	Existing Tariff	Proposed TOD Tariff
Summer peak rate	8¢/kWh	16.00 ¢/kWh
Summer off-peak rate	8¢/kWh	5.09 ¢/kWh
Winter/spring/fall	5¢/kWh	5.00 ¢/kWh

Marginal costs of operation for the utility are:

	Marginal Costs of Operation
Summer peak period	22.5 ¢/kWh
Summer off-peak period	4.0 ¢/kWh
Winter/fall/spring	3.0 ¢/kWh

The failure rate of the TOD meter is estimated at 0.5% per year, and the repair cost is $100.

Evaluate the cost benefit of this application.

SOLUTION. The change in peak kilowatt-hour consumption is:

$$\frac{\text{kWh TOD}}{\text{kWh non-TOD}} = \left(\frac{\text{price TOD}}{\text{price non-TOD}}\right)^{-.30} = \left(\frac{16}{8}\right)^{-.30} = .812$$

The change in off-peak kWh consumption is:

$$\frac{\text{kWh TOD}}{\text{kWh non-TOD}} = \left(\frac{\text{price TOD}}{\text{price non-TOD}}\right)^{-.10} = \left(\frac{5.09}{8}\right)^{-.10} = 1.046$$

The energy consumption and savings are calculated in Table 8.30. The overall utility economic benefits are evaluated as:

	Annual $/Year/Customer	Capitalized Equivalent $/Customer
Utility operation cost saving	30.	187.5
Less loss of revenue	−19.	−118.8
Less O&M expense	− 0.5	− 3.1
Total	$10.5	$ 65.6

The overall economic value for the utility is $65.6 per customer (in equivalent capitalized costs). If the TOD meter can be purchased and installed for less than $65, then the residential peak-period pricing program is economical and should be pursued.

Large commercial and industrial electricity consumers are also responsive to peak-load pricing. Figure 8.24 presents a load profile of cement industry customers in response to TOD rates in Southern California Edison (see Gargan, et al., "Time-of-Use for Very Large Power Commercial and Industrial Customers," 1978–1979). The load study of mandatory time-of-use customer rates contrasted 1977 relative-demand-use prior to time-of-use rates with 1978 relative demand during time-of-use rates. In the cement industry example, peak demand was reduced by an average of 3.5%. The peak-price rate per kilowatt-hour is 62% higher than the average price per kilowatt-hour. The peak-period price elasticity is calculated for the cement industry as −.05. Other industries in that research study were less responsive to peak-period pricing. The average peak-period price elasticity for all large customers was −.02.

This lower peak-period price elasticity in large industrial and commercial customers may be attributable to several factors. In this sector, average load

TABLE 8.30 Energy Consumption and Savings

	Existing Tarriff			New TOD Tariff			Savings		
	Rate (¢)	Energy	Revenue ($/Year)	Rate (¢)	Energy	Revenue ($/Year)	Energy (kWh)	Utility Cost Operation (¢/kWh)	Utility Cost Saving ($)
Summer peak	8.0	800	64	16.0	650	104	150	22.5	34
Summer off-peak	8.0	2,200	176	5.09	2,301	117	−101	4.0	−4
Winter	5.0	7,000	350	5.0	7,000	350	0		0
		10,000	$590		9,951	571	49		30

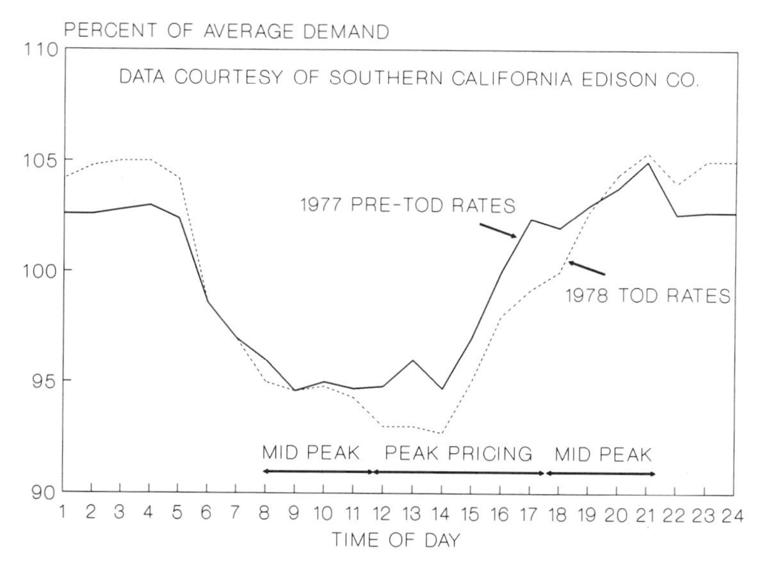

FIGURE 8.24 Large commercial and industrial loads—response to time-of-day rates.

factor is typically already high, and there is less opportunity to improve the usage profile. Changing usage patterns to off-peak requires changing the work periods of industrial workers and commercial businesses. There may be some economic factors (shift premiums) and inertia that restrain change. Large industrial and commercial customers are already metered with a maximum monthly demand (with or without a "ratchet" clause). This is in contrast to most residential and small commercial customers who are metered only on energy.

Large commercial and industrial elasticity gauges the incremental response to TOD demand relative to maximum-demand metering, while residential elasticity gauges the response to time-of-day demand relative to average energy metering. Peak-period price elasticity may also be larger in the long-term than was assessed during short-term experiments.

In summary, the economics of indirect load control are both utility- and customer-specific.

8.4 SUMMARY

This chapter presented two end-use kilowatt-hour load forecasting methodologies: the appliance saturation method, applied to the residential sector, and the floor-space method, applied to the commercial sector. The kilowatt-hour forecast is translated into a peak load (kilowatt) forecast using a load factor methodology. The load factor methodology can also be used to forecast seasonal peak-load values.

Load demand can be partially managed by providing pricing incentives to customers that change electricity consumption profiles. Direct load management provides the utility with direct control of the operation of specific end-use devices and compensates customers with participation payments. For indirect control, the utility provides TOD pricing signals as inducements for customers to change usage profiles. The economics of load management are dependent on the characteristics of the utility power system (fuel mix, reserve margin, etc.), end-use load characteristics, and the ability to modify end-use load profiles.

REFERENCES

American Gas Association, *House Heating,* Industrial Gas Series, 3rd ed., 1968.

American Society of Heating, Refrigeration, and Air Conditioning Engineers (345 East 47th Street, New York, 10017), *Handbook of Fundamentals,* 1976.

Brazil, G. W., and S. H. Grimmeth, "Load Management Through the Radio Control of Air Conditioners," EEI Electrical Systems and Equipment Committee Meeting (1978, Akron, OH).

Cuccaro J. E., and J. C. Schuh, "Residential Peak-Activated Class Load Study," Report of the Load Research Committee (1984–1985) Association of Edison Illuminating Companies.

Data Resources, Inc., "The DRI Energy Review," Lexington, MA.

Davis, M. W., T. J. Krupa, and M. J. Diedzic, "The Economics of Direct Control of Residential Loads on the Design and Operation of the Distribution System, Part I, II, III," IEEE-PES 82SM439-8, 82SM440-6, 82SM441-4 (1982 Summer Meeting, San Francisco).

Elkins, R. H., and C. E. Wensman, "Natural Ventilation of Modern, Tightly Constructed Homes," American Gas Association—Institute of Gas Technology Conference on Natural Gas Research and Technology, Paper 8, Session V (1971).

Energy Information Systems, *Non-Residential Building Energy Consumption Survey: Building Characteristics,* Department of Energy, DOE/EIA-246 (1981).

Federal Register, Vol. 45, No. 127, Monday, June 30, 1980.

General Electric Company, *Energy Use in Office Buildings,* Department of Energy Contract DOE/CS/20189 (1981).

Galgon, R. J., and E. F. Carter, "Residential Customer Response to an Optimal Time of Use Rate," Report of the Load Research Committee (1978–1979), Assoc. Edison Illuminating Companies.

Gargan, L., W. J. Powell, J. Hung, and R. Daniels, "Time-of-Use for Very Large Power Commercial and Industrial Customers," Report of Load Research Committee (1978–1979), Assoc. Edison Illuminating Companies.

Hunt, T., "Mississippi Power and Light Company's Load Management Test Project," EEI Electrical Systems Equipment Committee Meeting (1978, Akron, OH).

"Impacts of Several Major Load Management Projects," Load Management Working Group Report, IEEE-PES 82 WM 134-5 (1982 Winter Meeting, New York City).

Jack Faucett Associates, *Energy Consumption in Commercial Industries by Census Division—1974,* (Chevy Chase, MD, JACKAU-76-143-2 (1976).

Lee, S. H., and C. L. Wikkins, "A Practical Approach to Appliance Load Control Analysis: A Water Heater Case Study," IEEE-PES 82SM308-5 (Summer Meeting, 1982, San Francisco).

Meewsen, D. S., and R. H. Brown, "Residential Time of Day Class Load Study," Report of the Load Research Committee (1984–1985), Assoc. Edison Illuminating Companies.

New York Power Pool (Guilderland, NY), *Report of Member Electric Systems of the New York Power Pool (Pursuant to Section 5-112),* 1983, Vol. 1.

Office of Energy Conservation and Environment, *Physical Characteristics, Energy Consumptions, and Related Institutional Factors in the Commercial Sector,* Department of Energy, FEA/D-7/040 (1977).

Tippet, K. S., and M. D. Ruffin, "Service Life Expectancy of Household Appliances," Family Economics Review Highlights, Summer 1975, U.S. Department of Agriculture.

U.S. Code, *National Appliance Energy Conservation Act,* U.S. Code, Vol. 3, Sec. 5, 1987.

Villacis, E., M. Norman, and K. Gainer, "AEP System's Residential End-Use Forecasting Model—An Econometric Approach," Proceedings of the American Power Conference, 1988.

PROBLEMS

1 For an electric utility with 500,000 residential customers in 1990 and a projected growth in residential customers of 2%/year through 2000, calculate the growth in residential energy consumption from 1990 to 2000 assuming the energy usage patterns and appliance saturations as shown in the table below.

Appliance	1990 Saturation (%)	1990—Average Annual Energy Consumption (kWh/Appliance)	2000 Saturation (%)	2000—Average Annual Energy Consumption (kWh/Appliance)
Electric ranges	55	875	60	850
Dryers	47	925	55	875
Refrigerators	111	1550	111	1450
Freezers	35	1100	40	1000
Central A/C	35	2100	45	1900
Room A/C	40	1100	30	950
Hot water heater	40	4000	45	3500
Electric heat	20	2500	25	2200
Miscellaneous	100	800	100	1000

2 a If you currently use 3000 kWh/year to electrically heat your home at a cost of 8¢/kWh, how much money will you save over the next 5 years by increasing your home wall insulation level from R-10 to R-15? Use the data on Figure 8.4 and assume that your electric rates will increase at 5%/year.

b Now, assume that you also have air conditioning that uses 2000 kWh/year. How much can you afford to spend in increasing the wall insulation from R-10 to R-15, assuming that you would like to recover your initial investment in four years?

3 Assume that your electric utility has a load/weather sensitivity factor of 50 MW/°F. You and your boss have both made load forecasts for each of the past 10 days as shown below. Whose forecast, on the average, is closest to what actually happened, both with and without accounting for weather deviations?

Day	Your Prediction	Boss's Prediction	Actual Load	Expected Temperature	Actual Temperature
1	12,000	12,050	12,000	80	78
2	12,000	12,010	12,400	80	90
3	12,045	12,050	12,240	81	85
4	12,095	12,100	12,245	82	85
5	12,090	12,100	12,100	82	80
6	12,140	12,130	12,180	83	83
7	12,190	12,190	12,155	84	83
8	12,185	12,190	12,000	84	80
9	12,235	12,250	11,700	85	75
10	12,300	12,310	12,100	84	80

4 Assume that an electric utility has projected energy usage for the various end uses by their customers as shown in the table below. Assuming the same seasonal load factors for each of the end uses, as given in Table 8.14, calculate the average annual growth rate in both energy and peak load from 1990 to 2000, and also calculate the load factor in 1990 and 2000.

Load Forecast (kWh)	1900	2000
Residential		
Conventional refrigerator	500	400
Frostless refrigerator	1,300	1,600
Freezer	400	350
Window A/C	300	250
Central A/C	2,000	1,500
Dryer	600	500
Range	600	500

Load Forecast (kWh)	1900	2000
Water heater	1,600	1,400
Resistance heating	310	300
Heat pump	1,000	1,250
Miscellaneous	290	550
Commercial	7,000	9,500
Industrial	20,000	26,000
Total	35,900	44,100

5 A utility has done a survey of 1000 customers with air conditioning in which they asked their residential customers whether or not they would participate in a load management program under several different customer incentive payment programs. Their responses are shown below.

	Response	
Payment	Will Participate	Won't Participate
$25	300	700
$30	400	600
$40	500	500

If this utility has 100,000 residential customers to whom they plan to offer this load management program, what customer incentive payment should they offer in order to maximize the utility's annual savings from the program. (Assume the same data as given in Example 8.5 and Table 8.22. Also assume that the fuel savings, capacity savings, revenue lost, and load management O&M are directly proportional to the number of customers.)

9

POWER PLANT RELIABILITY CHARACTERISTICS

The objective of an electric utility is to provide electrical service to its customers in both an economical and reliable manner. Although it is not possible to provide service that is 100% reliable because of random equipment failures, service reliability can approach 100% if equipment is well maintained and investments are made in redundant equipment and backup systems. System redundancy is costly, of course, and there is a tradeoff between economical service and reliable service. This chapter introduces the foundations of power system reliability evaluations. Subsequent chapters will examine generation system reliability measures in more detail.

9.1 BASIC RELIABILITY CONCEPTS

"Probability" is a word widely used in our society, encompassing everything from the meteorologist's daily precipitation forecast to the power system engineer's estimate of service failure probability.

Probability is measured on a scale from 0 to 1. The probability of an event occurring can be determined from an experiment in which it is observed how frequently the event occurs within the population of the experiment. This frequency interpretation of probability is defined as:

$$\begin{aligned} P &= \text{probability of an event} \\ &= \text{limit as \# experiments} \rightarrow \infty \left\{ \frac{\text{event occurrences}}{\text{\# experiments}} \right\} \end{aligned} \tag{9.1}$$

The probability of a number 4 appearing on a six-sided dice, for example, would be determined by rolling the dice thousands of times and noting the frequency with which the number 4 appears. An engineer may be tempted to shortcut the process and reason that the probably of a number 4 is 1 in six, which is true if the dice is fair. However, only by conducting thousands of experiments can this be assured.

This set of conducted experiments is referred to as the "statistical event population." In practice, the population can be segmented into various categories, which leads to the concept of conditional probabilities.

For example, a statistical population may consist of a large sample of steam power plants in the United States. The failure statistics of this sample power plant population is given in Table 9.1.

The failure probability of the statistical population is 85/685 = .124 from Equation (9.1). However, the conditional failure probability of coal-fired power plants is 60/360 = .167. In other words, the probability of a failure for a power plant that burns coal is .167. In mathematical notation, this is written:

$$P(\text{failure}|\text{coal-fired}) = .167$$

Practical systems involve many components that interact with each other. To compute the probability of success or failure for the system, methods are needed to combine component probabilities into a system probability.

There are several rules for combining probabilities (Mosteller and Rourke, 1967). These are based on two important definitions:

- *Independent Events.* Two events are independent if the occurrence of one event does not affect the occurrence of the second event. For example, the two events of rolling a 6 with one die and drawing an ace from a deck of cards are independent events.
- *Mutually Exclusive Events.* If one event precludes the occurrence of a second event, then they are mutually exclusive. For example, the two events of drawing an ace of spades from a deck of cards and drawing an ace of spades from the same deck of cards without replacing the first card are mutually exclusively events.

TABLE 9.1 Statistical Population Example: Steam Power Plants in the United States

	Failures	Nonfailures	Total
Oil-fired	10	100	110
Gas-Fired	15	200	215
Coal-Fired	60	300	360
Total	85	600	685

The rules for combining probabilities are:

- *Simultaneous Occurrence.* The simultaneous occurrence of two or more *independent events* is the product of their probabilities:

$$P(A \text{ and } B \text{ and } C) = P(A) \cdot P(B) \cdot P(C) \tag{9.2}$$

- *Union of Occurrence of Mutually Exclusive Events.* If two or more events are mutually exclusive, then the probability of them occurring is:

$$P(A \text{ or } B \text{ or } C) = P(A) + P(B) + P(C) \tag{9.3}$$

Because it is assumed that these occurrences are mutually exclusive, either only A occurs, only B occurs, or only C occurs. Mutual exclusion prohibits both A and B occurring together, or A and C, or B and C.

At first, Equation (9.3) may seem of little value because of its mutually exclusive requirements. One way of evaluating complicated systems, however, is to use a "probability state enumeration" method in which each exclusive reliability condition or "state" of a system is evaluated. Since the system can only be in one condition or "state" at any one time, the reliability of the system can be determined by using Equation (9.3).

Consider, for example, deriving the probability equation corollary to Equation (9.3) under the condition that events A, B, and C are *not* mutually exclusive, but are independent. To do this, state enumeration of the probabilities is performed in which all the possibilities are examined, as in Table 9.2. The probability of each state is computed using Equation (9.2), where the probability of event i occurring is $P(i)$ and thus the probability that event i does not occur is $1 - P(i)$.

The probability of either A or B or C occurring can be computed by adding

TABLE 9.2 State Enumeration of *A*, *B*, and *C*[a]

State	Event A	Event B	Event C	Probability of the State
1	+	+	+	$P(A)P(B)P(C)$
2	+	+	−	$P(A)P(B)[1 - P(C)]$
3	+	−	+	$P(A)[1 - P(B)]\,P(C)$
4	−	+	+	$[1 - P(A)]P(B)P(C)$
5	+	−	−	$P(A)[1 - P(B)][1 - P(C)]$
6	−	+	−	$[1 - P(A)]P(B)[1 - P(C)]$
7	−	−	+	$[1 - P(A)][1 - P(B)]P(C)$
8	−	−	−	$[1 - P(A)][1 - P(B)][1 - P(C)]$
				sum = 1.0

[a]Plus (+) = occurs; minus (−) = does not occur.

up the probabilities for states in which either A, B, or C occurs singularly or simultaneously, 1 through 7, using Equation (9.3). Equation (9.3) can be applied because all the individual states in Table 9.2 are mutually exclusive.

The algebra can be simplified by noting that the summation of the probability of all the states is unity. Thus, the addition of states 1 through 7 is equal to unity minus state 8.

$$
\begin{aligned}
P(A \text{ or } B \text{ or } C) &= 1 - [1 - P(A)][1 - P(B)][1 - P(C)] \\
&= 1 - \{1 - P(A) - P(B) + P(A)P(B) - P(C) \\
&\quad + P(A)P(C) + P(B)P(C) - P(A)P(B)P(C)\} \\
&= P(A) + P(B) + P(C) - P(A)P(B) - P(A)P(C) \\
&\qquad - P(B)P(C) + P(A)P(B)P(C)
\end{aligned}
$$

- *Conditional Probabilities.* If the occurrence of an event A is conditionally dependent on one or more mutually exclusive events, then the probability of event A is:

$$
\begin{aligned}
P(A) &= P(A|B) \cdot P(B) \\
&\quad + P(A|C) \cdot P(C) \\
&\quad + P(A|D) \cdot P(D)
\end{aligned}
\qquad (9.4)
$$

where $P(A|*)$ is the conditional probability for event A given that event $*$ has occurred.

From the data in Table 9.1, the conditional probabilities of failure of power plants is:

$$
\begin{aligned}
P(\text{failure}|\text{ oil-fired}) &= 10/110 = .0909 \\
P(\text{failure}|\text{ gas-fired}) &= 15/215 = .0677 \\
P(\text{failure}|\text{coal-fired}) &= 60/360 = .1667
\end{aligned}
$$

Furthermore, the probability of a power plant being fired on these fuels is (from Table 9.1):

$$
\begin{aligned}
P(\text{oil-fired}) &= 110/685 = .1606 \\
P(\text{gas-fired}) &= 215/685 = .3139 \\
P(\text{coal-fired}) &= 360/685 = .5255
\end{aligned}
$$

Then, the probability of a failure at any power plant in the total plant population is by Equation (9.4):

$$
\begin{aligned}
P(\text{failure}) &= P(\text{failure}|\text{oil-fired}) \cdot P(\text{oil-fired}) \\
&\quad + P(\text{failure}|\text{gas-fired}) \cdot P(\text{gas-fired}) \\
&\quad + P(\text{failure}|\text{coal-fired}) \cdot P(\text{coal-fired}) \\
&= .0901 \cdot .1606 + .0677 \cdot .3139 + .1667 \cdot .5255 \\
&= .1241
\end{aligned}
$$

9.2 RELIABILITY EVALUATIONS OF PRIMAL SYSTEMS

Many systems can be represented as networks in which components are connected in series, parallel, or a combination of these formats. It is useful to first evaluate the reliability of series and parallel systems before analyzing more complex systems.

A series system, as illustrated in Figure 9.1, may be comprised of N components. For the system to be a success, each component must be a success. The probability of a system success can be calculated using Equation (9.2) as the product of the success probabilities of each component.

$$P(\text{system success}) = P(A = \text{success}) \cdot P(B = \text{success}) \cdots \cdot P(N = \text{success})$$

For a five-component system with each component having a 75% success probability, the system success probability is $.75 \cdot .75 \cdot .75 \cdot .75 \cdot .75 = .237$.

A parallel system is illustrated in Figure 9.2. The only way the system may fail is by having every one of the components fail simultaneously. Because the probability of a failure is one minus the success probability, the system success probability can be found by using Equation (9.2).

$$P(\text{system fails}) = P(A \text{ fails}) \cdot P(B \text{ fails}) \cdots \cdot P(N \text{ fails})$$

or

$$P(\text{system success}) = 1 - P(\text{system fails}) = 1 - [1 - P(A = \text{success})] \cdot [1 - P(B = \text{success})] \cdot \cdots [1 - P(N = \text{success})]$$

For a five-component system with each component having a 75% success probability, the system success probability is $1 - [(1 - 75) \cdot (1 - .75) \cdot (1 - .75) \cdot (1 - .75) \cdot (1 - .75) = .999$.

9.3 RELIABILITY OF COMPLEX SYSTEMS—STATE ENUMERATION

A system reliability analysis diagram of a generating unit would show a system containing both series and parallel components. Some of the parallel components might also operate at partial rather than full capacity. For example, many generating units have two 60% capability boiler fuel-pump turbines in

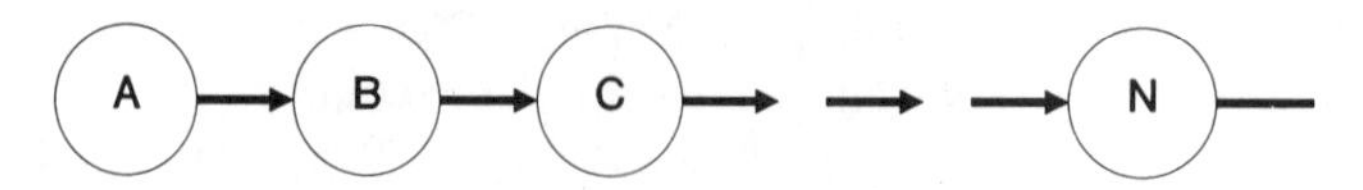

FIGURE 9.1 Series system.

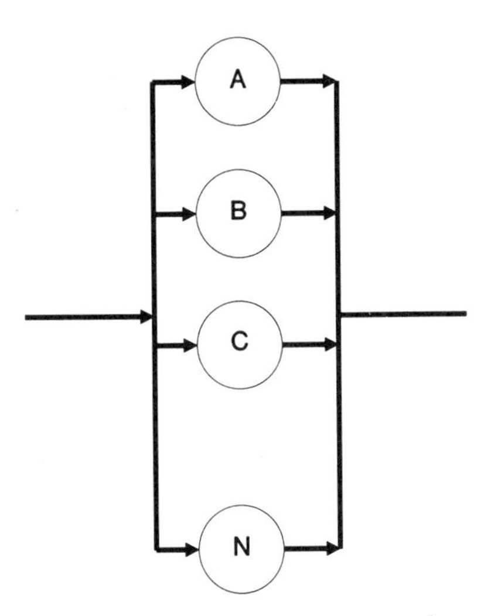

FIGURE 9.2 Parallel system.

parallel so that if one boiler feed-pump turbine fails, the generating unit could still produce 60% output.

The reliability measure of complex systems needs to evaluate system reliability at full capacity as well as at other system conditions when partial or derated capacity is delivered to the power system. "Equivalent" reliability accounts for the contributions of derated capacity output by weighting reliability by the output contribution relative to full capacity.

In the reliability of complex system, the state enumeration method is often extremely useful. Each possible mutually exclusive state of the system is identified. Associated with each state is the capacity result. The probability of each state can be computed by using Equation (9.2), (9.3), or (9.4).

For the case of the two 60% capacity boiler feed pumps, assume each pump has 95% reliability. The equivalent system reliability can be computed using a state enumeration, as in Table 9.3. The results from the state enumeration can

TABLE 9.3 State Enumeration

State #	Pump A	Pump B	Total Pump Capability (%)	System Capability (%)	Probability of State	Capability-Weighted Probability
1	Works	Works	120	100	.95 • .95 = .9025	.9025 • 1.0 = .9025
2	Failed	Works	60	60	.05 • .95 = .0475	.0475 • .6 = .0285
3	Works	Failed	60	60	.95 • .05 = .0475	.0475 • .6 = .0285
4	Failed	Failed	0	0	.05 • .05 = .0025	.0025 • .0 = 0
						Equivalent reliability = .9595

TABLE 9.4 Capacity State Table

Capacity State % Available	% Unavailable	Probability
0	100	.0025
60	40	.0950
100	0	.9025

be grouped by capacity states to describe the partial or derated states of the boiler feed-pump system. The capacity state table (Table 9.4) will be utilized in the power system reliability calculations of Chapter 10.

In power systems, the unreliability of a component is generally measured as the fraction of the time that it is not available for service. This is called the "forced outage rate." Since power system components are generally repaired if they fail, time unavailability measurement is appropriate. The following example illustrates the outage state enumeration method on a combined-cycle power plant.

Outage State Enumeration Example. In this example, the equivalent forced outage rate of a combined-cycle generating unit is evaluated. The 320-MW combined-cycle unit consists of two 100-MW simple-cycle gas turbines whose exhaust into a heat recovery steam generator provides steam to drive a 120-MW steam-turbine generator. The hot gas turbine exhaust gases (typically 900°–1000°F) are input into a heat-recovery steam generator; the steam is expanded through a steam turbine that drives a 120-MW generator. This system is illustrated in Figure 9.3.

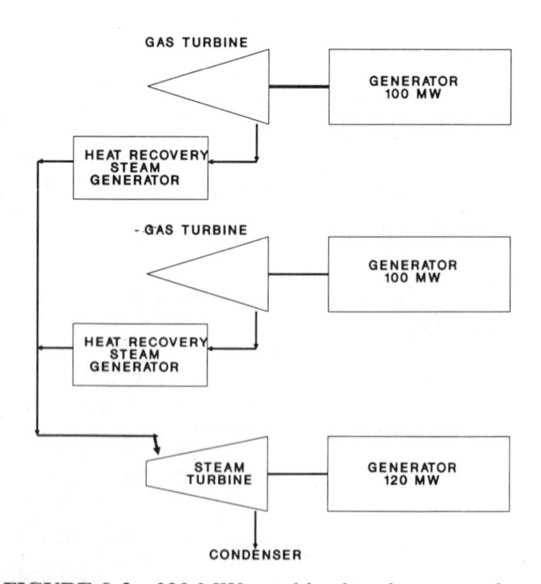

FIGURE 9.3 320-MW combined-cycle power plant.

In this case, several observations are made:

1. The loss of a gas turbine generator results in a 160-MW loss—100 MW from the gas-turbine generator and 60 MW from the steam-turbine generator.
2. The loss of a single heat-recovery steam generator leads to a 60-MW loss.
3. The loss of a steam turbine generator results in a 120-MW loss.

The forced outage rates of the components are:

- Gas-turbine generator: 6%.
- Heat-recovery steam generator: 3%.
- Steam-turbine generator: 4%.

Table 9.5 illustrates the outage state enumeration of this example unit. It is assumed that the reliability of each gas turbine is the same (similarly, for each of the heat-recovery steam generators). Thus, some of the outage states have a permutation of 2, indicating that an identical state exists if the label of the duplicate gas turbine or steam generator is interchanged. Because there are five distinct components, there are 2^5 or 32 outage states.

After identifying the state probabilities, the states with identical megawatt power (MW) out of service can be grouped together, as shown in the capacity state table (Table 9.6). The equivalent unreliability, or equivalent forced outage rate, of the system can then be evaluated as the MW-weighted probability of failure divided by the MW of full capacity. The result is that the combined-cycle unit has a 8.4% equivalent forced outage rate.

Some subsystems consist of many components arranged in series. For example, a turbine rotor has many buckets (or blades), and if any one fails, the turbine rotor subsystem fails. When the N components are identical, the probability of subsystem success is given by Equation (9.2) as:

$$P(\text{subsystem success}) = [1 - P(\text{FAIL})]^N$$

and the subsystem failure probability is:

$$P(\text{subsystem failure}) = 1 - [1 - P(\text{FAIL})]^N$$

However, it is often useful to know the probability of one component (bucket or blade) failing, two components failing, and so on.

A state enumeration of these combinations could be performed, but if N is large, the computation is formidable. Fortunately, however, when the components are identical, the binomial coefficient can be used to determine the number of similar outage states.

TABLE 9.5 Outage State Enumeration of 320-MW Combined-Cycle Generating Unit

State	State Permutations	States (1 = success, 0 = fail)					MW Out of Service	Probabilities	Total
		GT	HRSG	GT	HRSG	ST-G			
1	1	1	1	1	1	1	0	.94 .97 .94 .97 .96	.798124
2	1	1	1	1	1	0	120	.94 .97 .94 .97 .04	.033255
3	2	1	1	1	0	1	60	.94 .97 .94 .03 .96	.049368
4	2	1	1	0	1	1	160	.94 .97 .06 .97 .96	.101888
5	2	0	0	1	1	1	160	.06 .03 .94 .97 .96	.003151
6	1	0	1	0	1	1	320	.06 .97 .06 .97 .96	.003251
7	1	1	0	1	0	1	120	.94 .03 .94 .03 .96	.000763
8	2	0	1	1	0	1	220	.06 .97 .94 .03 .96	.003151
9	2	1	1	1	0	0	120	.94 .97 .94 .03 .04	.002057
10	2	1	1	0	1	0	220	.94 .97 .06 .97 .04	.004245
11	1	1	0	1	0	0	120	.94 .03 .94 .03 .04	.000031
12	1	0	1	0	1	0	320	.06 .97 .06 .97 .04	.000135
13	2	0	0	1	1	0	220	.06 .03 .94 .97 .04	.000131
14	2	0	1	1	0	0	220	.06 .97 .94 .03 .04	.000131
15	2	0	0	1	0	1	220	.06 .03 .94 .03 .96	.000097
16	2	0	0	0	1	1	320	.06 .03 .06 .97 .96	.000201
17	2	0	0	1	0	0	220	.06 .03 .94 .03 .04	.000004
18	2	0	0	0	1	0	320	.06 .03 .06 .97 .04	.000008
19	1	0	0	0	0	1	320	.06 .03 .06 .03 .96	.000003
20	1	0	0	0	0	0	320	.06 .03 .06 .03 .04	.000000
	32								1.000000

TABLE 9.6 Capacity State Table

MW Out of Service	Probability	MW-Weighted Probability
0	0.798124	0
60	0.049368	2.96208
120	0.036106	4.33272
160	0.105039	16.80624
220	0.007759	1.70698
320	0.003598	1.15136
	1.000000	26.95938

Equivalent unreliability = 26.96/320 = 0.084248

The binomial coefficient (BC) is:

$$BC(n,N) = \frac{N!}{n!(N - n)!}$$

where N = total number of components and n = number of identical states.

For example, for $N = 4$ and $n = 1$:

$$BC(1,4) = \frac{4 \cdot 3 \cdot 2 \cdot 1}{1 \cdot 3 \cdot 2 \cdot 1} = 4$$

Thus, with four components, there are four combinations of the state that would have one component out of service (failure) and three components in service (success).

The binomial coefficient can be easily derived. The number of distinctive ways in which N distinguishable components can be arranged in different states is $N!$. Suppose that a particular state with N components is examined. The first condition, such as one component failed (or success), can be occupied by any one of N components. The second condition, such as one component failed (or success), could be achieved by any one of the $N - 1$ remaining components. Finally, the Nth condition could be achieved only by the last object. Hence, all the available conditions can be achieved by $N(N - 1)(N - 2) \dots 1 = N!$ possible ways.

This enumeration considers each component as distinguishable. However, if n components are indistinguishable, such as n failures, then there are $n!$ permutations of these components among themselves. Similarly, there are $(N - n)!$ permutations of the opposite case, such as a success. By dividing the total number of distinguishable arrangements of components by the number $n!(N - n)!$ of indistinguishable arrangements, one obtains the total number of distinct states.

9.4 EQUIPMENT RELIABILITY TRENDS

Mechanical and electronic components exhibit failures dependent upon equipment age or duty. Figure 9.4 illustrates the typical "*bathtub*" reliability curve. The failure rate begins at a higher rate when the equipment is new. Early failures ("bugs") or problems occur, and the equipment is repaired. The early debugging period constitutes phase 1. After debugging, the equipment enters the mature operating period, phase 2, when equipment performance is very reliable. Toward the end of the mature operating period, the wearout phase, phase 3, is entered. The failure rate increases as a result of accumulated operating stresses. The equipment is operated into this wearout period usually until its unreliability or other factors (such as high operation and maintenance cost, inefficiency, and/or technological obsolescence) require its retirement or major refurbishment.

Generating units also exhibit the bathtub reliability trend, as illustrated in Figure 9.5 for units 50–200 MW in size. The forced outage rate exhibits a debugging period for the first several years. Then, the generating unit operates reliably until its age approaches 30 years. After 30 years, the generating unit enters a wearout phase. Generating units are traditionally retired after 40–45 operating years or are refurbished during or prior to entering the wearout phase. Not every component in a generating unit follows the same reliability curve. For example, the turbine generator and the total generating unit experience different "bathtub" characteristics.

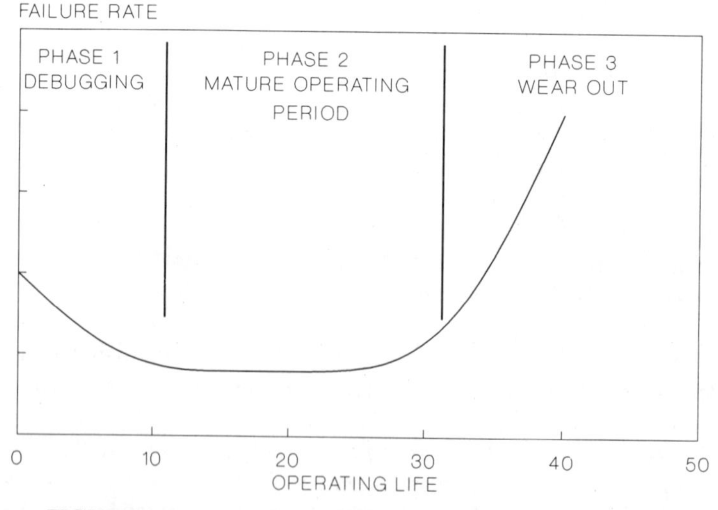

FIGURE 9.4 Typical reliability aging curve (bathtub curve).

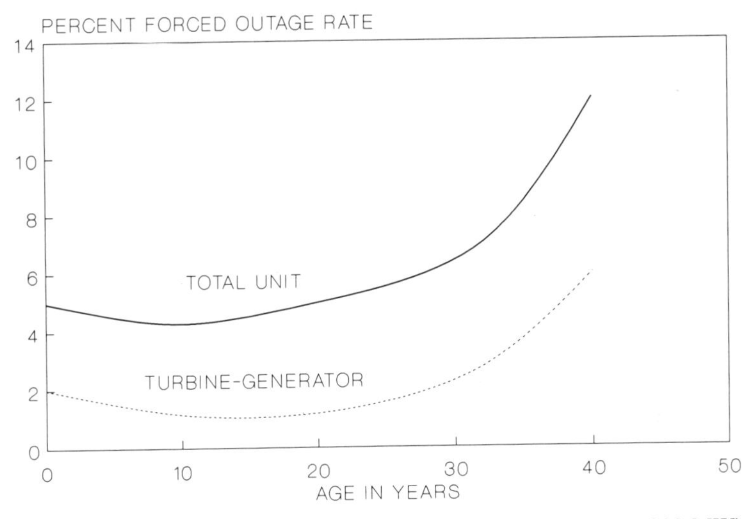

FIGURE 9.5 Steam plant forced outage rate trend (units 50–200 MW).

9.5 FAILURE-RATE MODELS

It is sometimes necessary to forecast the failure probability of a power plant component. This is especially true as a power plant nears the wearout phase. One approach to forecasting failure probability is to describe the component reliability performance in terms of a failure-rate model.

A distinction must be made between a "nonrepaired" component and a "repaired" component. In a power system, components are usually "repaired." From a statistical viewpoint, it is very important to determine how they are repaired. A "nonrepaired" component is one in which the failed component is not replaced in *like* kind. The component may be replaced by a new technology component that doesn't have the same failure mode. A "repaired" component is one that is replaced (or repaired) in like kind but has its aging (bathtub) curve "clock" reset to age 0.

9.6 NONREPAIRED FAILURE-RATE MODELS

One of the simpler probability models is called an "exponential" model. The model assumes that the failure rate is independent of time or plant age. This model is described as:

$$\frac{d\,p(t)/dt}{p(t)} = -\frac{1}{\alpha} \tag{9.5}$$

where $p(t)$ = probability of a component surviving to age t.

α = constant (estimated from historical failure data)

$\frac{d\,p(t)}{dt}$ = the time derivative at which the surviving probability changes with age (associated with failures)

Equation (9.5) states that the failure rate (failures at age t divided by the remaining survivors at age t) is equal to a constant, $1/\alpha$.

Equation (9.5) can be integrated to yield:

$$p(t) = \exp[-\frac{t}{\alpha}] \tag{9.6}$$

where the probability of a survivor is unity at age $t = 0$ and decreases with age.

The cumulative probability of failures $F(t)$ is one minus the number of survivors:

$$F(t) = 1 - \exp(-\frac{t}{\alpha}) \tag{9.7}$$

The time derivative of the failures is the rate per unit time of failures:

$$f(t) = \frac{d\,F(t)}{dt} = \frac{1}{\alpha}\exp(\frac{t}{\alpha}) \tag{9.8}$$

The mean time to failure can be computed by multiplying the age times the probability of failure [Equation (9.8)] and integrating over all ages:

$$\text{Mean time to failure} = \int_0^\infty t\,(\frac{1}{\alpha})\exp(\frac{-t}{\alpha})\,dt$$

$$= -t\exp\frac{-t}{\alpha}\Big|_0^\infty - \int_0^\infty \exp(\frac{-t}{\alpha})\,dt$$

$$= \alpha \tag{9.9}$$

The mean-time-to-failure is the reciprocal of the failure rate in the case of an " exponential" model.

The hazard function is often used in component reliability analysis. This is defined as the failure rate or time derivative of the failures divided by the number of survivors:

$$h(t) = \frac{d\,F(t)/dt}{1 - F(t)} \tag{9.10}$$

The hazard function measures the rate at which failures are occurring within the population of survivors at a specific age.

For the "exponential" failure model, the hazard function is:

$$h(t) = \frac{1/\alpha \exp(t/\alpha)}{\exp(t/\alpha)} = \frac{1}{\alpha} \tag{9.11}$$

Thus, the hazard function, or failure rate, for the "exponential model" is a constant.

The cumulative hazard function is defined as:

$$H(t) = \int_{-\infty}^{t} h(t)dt \tag{9.12}$$

Substituting the definition of the hazard function, Equation (9.10), into Equation (9.12) yields:

$$H(t) = \int_{-\infty}^{t} [\frac{[d\,F(t)/dt]dt}{1 - f(t)}] = -\ln\,[1 - F(t)] \tag{9.13}$$

This can be rearranged by taking the antilogarithm of Equation (9.13) as:

$$F(t) = 1 - \exp[-\,H(t)] \tag{9.14}$$

A knowledge of the cumulative hazard function can be used to obtain the cumulative failure probability.

Example. The failure rate, hazard function, of a power plant component is estimated to be .05 per year. What is the probability that the component will fail over the next 10 years?

Using Equation (9.11), we obtain $h(t) = .05$. Thus, $H(t) = .05 \cdot t$, where t = age in years. From Equation (9.14), the cumulative failure probability is:

$$F(t) = 1 - \exp[-.05 \cdot t]$$

The cumulative failure rate after 10 years is:

$$F(10) = 1 - \exp[-.05 \cdot 10] = .393$$

The exponential model with constant hazard rate, or failure rate, is often too simple to describe the reliability consequences of mechanical or electrical stresses acting on a component. Other models finding application include the Weibull, normal, and log-normal model.

The Weibull model is widely used to model the failures of components. It has widespread usage because it models a broad range of failure trends. The Weibull model assumes that the hazard (failure rate) is:

$$h(t) = \left(\frac{\beta}{\alpha}\right)\left[\left(\frac{t}{\alpha}\right)^{\beta - 1}\right] \tag{9.15}$$

The Weibull failure rate is proportional to age. The β parameter is a shape parameter that provides the age dependence and the α parameter is an age scale parameter. Figure 9.6 illustrates the hazard function for several values of β.

If β is unity, then the failure rate is constant with age, as the exponential model. If β is greater than unity, the failure rate increases with age. If β is less than unity, then the failure rate decreases with age.

The cumulative Weibull hazard function is:

$$H(t) = \left(\frac{t}{\alpha}\right)^{\beta} \tag{9.16}$$

The cumulative failure probability for the Weibull hazard function can be computed using Equation (9.14) as:

$$F(t) = 1 - \exp\left[-\left(\frac{t}{\alpha}\right)^{\beta}\right] \tag{9.17}$$

The Weibull distribution is widely used for reliability models because of its flexibility. For the special case of $\beta = 1$, the Weibull takes the form of exponential distribution. For β in the range of 3–4, the Weibull is close to that of

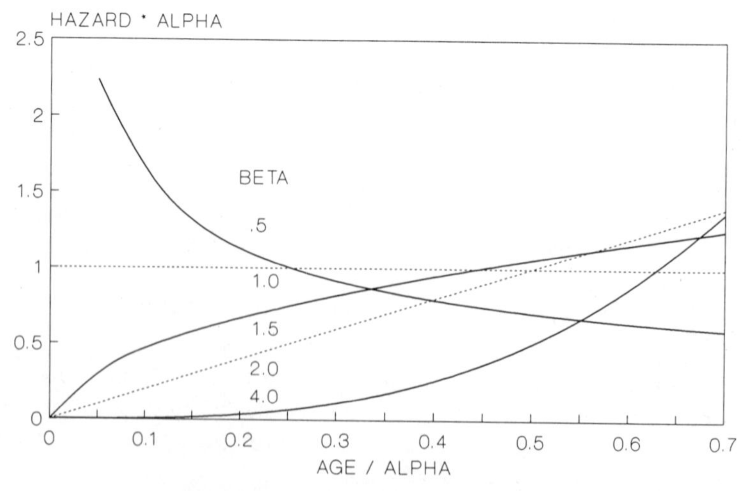

FIGURE 9.6 Weibull hazard function.

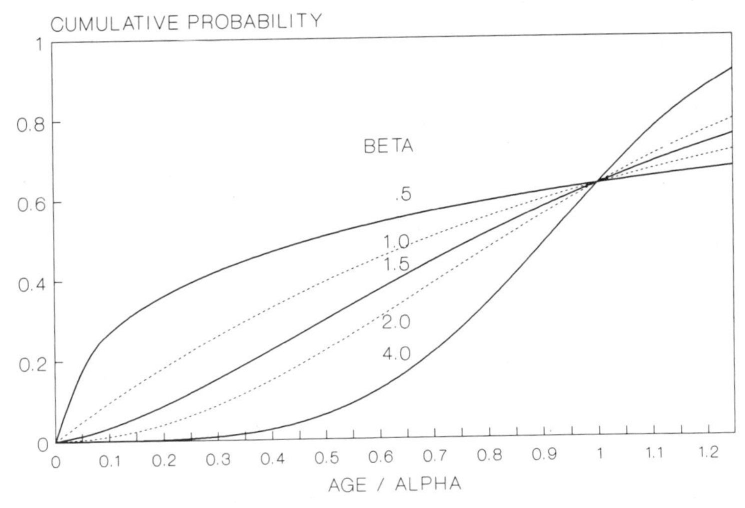

FIGURE 9.7 Weibull cumulative probability.

the normal distribution. Figure 9.7 illustrates the cumulative failure probability for several values of β.

Other hazard distributions are discussed by Wayne Nelson (see *Applied Life Data Analysis*, 1985). For reliability models, however, Nelson suggests trying the Weibull distribution first.

Example. A power plant component has been found to follow a Weibull distribution with $\alpha = 500$ and $\beta = 1.5$. The power plant has 60 of these components, any one of which would cause the power plant to experience a forced outage (failure) of 20 days. What is the cumulative probability of a forced outage over the next 10 years that is attributable to this system of components?

SOLUTION. Using the Weibull failure distribution:

$$F(10) = 1 - \exp\left[-\left(\frac{10}{500}\right)^{1.5}\right]$$

$$= .002824$$

There are 60 components in series, and since any one failure would cause the system to fail, the probability of not failing (component success) is:

$$P(\text{component success}) = 1 - .002824 = .997176$$

The probability of a system success is computed by using Equation (9.5)

$$P(\text{system success}) = (.997176)^{60} = .843935$$

$$P(\text{system failure}) = (1 - .843935) = .156$$

Thus, the probability of a forced outage over the next 10 years is 0.156.

A power plant typically has many systems consisting of many components in series. Any one failure can lead to a system failure. Such system components include: turbine blades, boiler tubes, condenser tubes, and generator stator bars.

9.7 CONDITIONAL FAILURE PROBABILITIES (NONREPAIRABLE COMPONENT)

Often the reliability model is known over a product life span. At some point, it may be necessary to calculate the failure probability given that the component has survived to age t_0.

To derive this equation, note that the hazard function is defined as the failures per unit time divided by the survivors. Define the following terms as:

$F(t)$ = cumulative failure probability to age t
$F(t|t_0)$ = conditional cumulative failure probability to age t, given that no failures have occurred through age t_0
$h(t)$ = hazard rate at age t
$H(t)$ = cumulative hazard function through age t
$H(t|t_0)$ = conditional cumulative hazard function through age t, given that no failures occurred through age t_0

Then, the conditional cumulative hazard function is defined using Equation (9.10) and defining the conditional cumulative failure probability as:

$$H(t|t_0) = \int_{t_0}^{t} h(t)dt = \int_{t_0}^{t} \frac{[d\, F(t|t_0)/dt]}{1 - F(t|t_0)} \tag{9.18}$$

Integration leads to:

$$H(t|t_0) = -\ln\left[\frac{1 - F(t_0|t_0)}{1 - F(t|t_0)}\right] \tag{9.19}$$

$$= -\ln[1 - F(t|t_0)] \tag{9.20}$$

since $F(t_0|t_0) = 0$. Also note that:

$$H(t|t_0) = \int_{-\infty}^{t} dt\, h(t) - \int_{-\infty}^{t_0} dt\, h(t) = \ln\left[\frac{1 - F(t)}{1 - F(t_0)}\right] \tag{9.21}$$

Thus:

$$- \ln \left[\frac{1 - F(t)}{1 - F(t_0)} \right] = - \ln \left[1 - F(t|t_0)\right] \tag{9.22}$$

and

$$1 - F(t|t_0) = \frac{1 - F(t)}{1 - F(t_0)} \tag{9.23}$$

or

$$F(t|t_0) = \frac{F(t) - F(t_0)}{1 - F(t_0)} \tag{9.24}$$

Equation (9.24) states that to calculate the failure probability given that no failure has occurred through age t_0, the cumulative nonconditional failure probability at age t_0 is subtracted from the nonconditional failure probability at age t, and the result is divided by the probable survivors to time t_0.

Example. A generator rotating field is 15 years old and has not failed. Statistics on generators of similar manufacture show that the failure hazard rate is characterized by a Weibull distribution with an $\alpha = 40$ years and a $\beta = 2$. Calculate the cumulative probability of failure for this generator rotating field over the next 10 years (age 16–25 years).

SOLUTION. Use Equation (9.17) for calculating the Weibull cumulative probability:

$$F(15) = .131$$

$$F(25) = .323$$

The conditional failure probability from Equation (9.24) is:

$$F(25/\text{no failure at } 15) = \frac{.323 - .131}{1 - .131} = .221$$

Thus, there is a 22% failure probability over the next 10 years.

9.8 DETERMINING RELIABILITY MODELS FROM FAILURE DATA

The reliability models presented were based on a specified hazard rate characterization. This section presents an easy method for determining the parameters of a reliability model.

From Equation (9.16), the cumulative hazard function for the Weibull distribution is:

$$H(t) = (\frac{t}{\alpha})^{\beta} \tag{9.25}$$

Taking logarithms of this yields:

$$\log H(t) = \beta \log (t) - \beta \log (\alpha) \tag{9.26}$$

or

$$\log (t) = \frac{1}{\beta} \log H(t) + \log (\alpha) \tag{9.27}$$

Equation (9.27) is a straight line on graph paper that is double logarithmic in $\log(t)$ and $\log H(t)$.

The α parameter can be determined as the age when $H(t) = 1.0$. The β parameter is the inverse logarithmic slope of the line.

The procedure for determining α and β can best be illustrated with an example.

"Repairable Case" Example. What is the probability of component failure on generating unit 9 over the next 10 years? The component failure history of a population of nine power plants is shown in Table 9.7. The data are presented on the failures and nonfailures of the component.

SOLUTION. The data in Table 9.7 show when the component began service and when either the component had a failure or the observation period ended. In this case, the data were collected through 1987. The number of exposure years is calculated as the ending year minus the starting year.

In the case of a component failure, the component would be repaired with a component similar to the original component, and the generating unit would be placed back in service. For example, data case 1 extends from 1959 to 1970, when a failure occurred. The component was repaired (case 1A) in 1970 and did not have a failure through 1987.

The first step in the analysis is to assume that the failure probability is dependent on exposure years. The second step is to rearrange the data monotonically in order of increasing exposure years as shown in Table 9.8.

Next, calculate the failure rate or hazard rate. The failure rate, hazard, is defined as the number of failures divided by the number of survivors during a time interval of one year. The first failure is case 3 and occurs at an exposure of 5 years. The failure hazard is calculated as one failure per year divided by 11 component cases that survived to an exposure of 5 years or more.

TABLE 9.7 Component Failure History of Population of Nine Power Plants

Generating Unit Data Case	Starting Year	Ending Year	Exposure Years	Component Failure (1 = Yes); (0 = No)
1	1959	1970	11	1
1A	1970	1987	17	0
2	1962	1987	25	0
3	1964	1969	5	1
3A	1969	1979	10	1
3B	1979	1987	8	0
4	1968	1987	19	0
5	1970	1985	15	1
5A	1985	1987	2	0
6	1975	1987	12	0
7	1978	1987	9	0
8	1978	1986	8	1
8A	1986	1987	1	0
9	1983	1987	4	0

Since the graphical technique requires the cumulative hazard, the cumulative value is computed in the last column of Table 9.8.

The six failure data points are plotted in Figure 9.8. The α parameter is calculated as the year when the cumulative hazard is 100%. Substitution for $H(t) = 100\%$, or $H(t) = 1.0$ per unit, then by Equation (9.27), $\log(t) = \log(\alpha)$, or α is the year when $H(t) = 1$. In this case, $\alpha = 16$ years. The β parameter is the inverse logarithmic slope of the line.

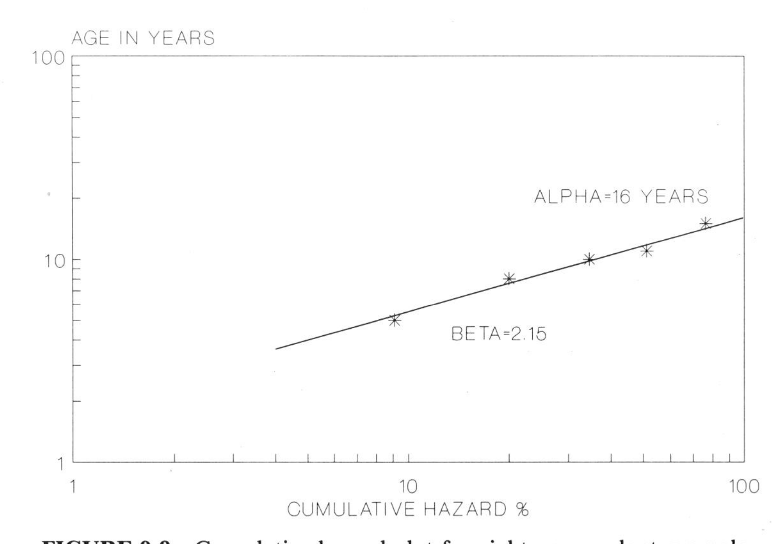

FIGURE 9.8 Cumulative hazard plot for eight-power-plant example.

TABLE 9.8 Monotonically Ordered by Increasing Exposure Year

Generating Unit Data Case	Starting Year	Ending Year	Exposure Years	Failure (1 = Yes); (0 = No)	Number of Cases with an Exposure = Case	Failure Hazard (%)	Cumulative Hazard (%)
8A	1986	1987	1	0	14	0	0
5	1985	1987	2	0	13	0	0
9	1983	1987	4	0	12	0	0
3	1964	1969	5	1	11	9.090909	9.09091
3B	1979	1987	8	0	10	0	9.09091
8	1978	1986	8	1	9	11.11111	20.20202
7	1978	1987	9	0	8	0	20.20202
3A	1969	1979	10	1	7	14.28571	34.48773
1	1959	1970	11	1	6	16.66666	51.15440
6	1975	1987	12	0	5	0	51.15440
5	1970	1985	15	1	4	25	76.15440
1A	1970	1987	17	0	3	0	76.15440
4	1968	1987	19	0	2	0	76.15440
2	1962	1987	25	0	1	0	76.15440

Taking the inverse slope from 10 to 100% hazard yields:

$$\beta = \frac{\log 100 - \log 10}{\log 16 - \log 5.5} = \frac{2 - 1}{1.204 - 0.740} = 2.15$$

The β of 2.15 indicates that the failure rate increases significantly with age (see Figure 9.6).

Consider now the specific generating unit number 9 for which the failure probability over the next 10 years is to be determined. The component has had four years of experience. From the graph, the cumulative hazard at four and 14 years is:

$$H(4) = 5\% = .05$$
$$H(14) = 75\% = .75$$

The cumulative failure probability from Equation (9.13) is:

$$\mathrm{F}(4) = 1 - \exp(-.05) = .049$$
$$\mathrm{F}(14) = 1 - \exp(-.75) = .528$$

Using Equation (9.24), the conditional failure probability for unit 9 is:

$$F(14|4) = \frac{.528 - .049}{1 - .049} = .503 = 50.3\%$$

Thus, there is a 50% probability of a component failure on unit 9 over the next 10 years.

The key assumption here is that the failures are proportional to chronological age. From engineering considerations, failures would theoretically be proportional to plant usage. A plant used for only 3 months per year to meet summer peak loads would be expected to have less "aging" than a two-shift plant operated for 12 months per year.

The key exposure parameter is the accumulated plant duty rather than chronological age. A better measure of plant duty is the product of plant chronological age times the plant capacity factor. While this measure does not fully account for the effects of plant cycling on plant duty, it does account for the influence of plant usage on failures.

The last example of this section presents generator stator bar insulation failure data.

Example. Compute the generator stator bar girth-crack insulation probability of unit 20 during the 1985–1995 time period. Unit 20, installed in 1957, is expected to operate at 75% capacity factor over the next 10 years.

SOLUTION. Each of these generators has 100 stator bars in it, and an insulation failure of any one would lead to a forced outage. The bars were originally insulated with asphalt. Today, mica-based insulation is used.

The failure statistics are shown graphically on Figure 9.9. The circled numbers show the number of bars replaced when a failure occurred.

After the early 1960s, any failed bars were replaced with mica-based insulation that totally eliminates the possibility of failure.

Table 9.9 presents the data depicted in Figure 9.9 in tabular form. Each generator has 100 bars represented. These 100 bars are segmented into subcases. In case 9, for example, there was one bar that operated from 1950 to 1969 (19 exposure years) and failed in 1969. Four other bars operated over the same exposure period and were replaced with mica-based insulation in 1969.

```
                                 YEAR

1944 46 48 50 52 54 56 58 60 62 64 66 68 70 72 76 80 84  Case
        ----------------------------------------------     1
          ---------------
                         A ---------------------------     2
          --------------
                        A-----------------------------     3
        -----------7-----------------3----------------     4
          ---------------------------1----------------     5
             ----------------------------------            6
          --------------------------------------------     7
        ----------------------------------------------     8
          ----------------------------5----                9
             ----------------------------------           10
          --------------------------------------------    11
             -----------------------------------------    12
          --------------------------------------------    13
                --------------------------------------    14
                ---------------9----------4-----M---      15
              ----------------------------------------    16
                --------------------------------------    17
                    ----------------------------------    18
                      --------------------------------    19
Huntley 67 HP         ---------------3----------------    20
                       ---------1---------------------    21
                       -------------------------------    22
                       -------------------------------    23
                         -----------------------------    24

CODE:

O  Service Date              M   Rewound with New Technology
A  Asphalt Rewind                Material
E  Early Failure Generator   B   Generator Out of Service
                             X   Rewound Cause Other than Girth
                                 Cracks
```

FIGURE 9.9 Girth cracks in 168-inch core-length asphalt-insulated bars.

The five replacement bars with mica-based insulation operated from 1969 to 1971, when the generator was removed from service. There were 95 bars that operated from 1950 to 1971 that were not modified.

The analysis procedure is the same as in the previous example, except that the capacity-factor-weighted age of each generating unit is used, and the failure probability of one (1) stator bar is to be computed. Once the failure probability of one bar is computed, the failure probability of the entire generator can be computed using the series probability rule [Equation (9.5)].

Table 9.10 presents the failure data sorted by capacity factor weighted age. The failure hazard and cumulative hazard are computed in the last columns of the table.

Figure 9.10 presents the girth-crack failure data plotted on Weibull hazard paper. The Weibull parameters for one stator bar are α = 560 years and:

$$\beta = \left(\frac{\log 100 - \log 10}{\log 560 - \log 115} \right) = 1.45$$

Unit 20 has an age of 28 years and a capacity-factor-weighted age of 20.39 years. At a 75% future capacity factor, the future capacity-factor-weighted age in 10 years is 27.89 years.

The cumulative hazard per bar at these two ages is:

$$H(20.39) - 0.7\% = .007$$
$$H(27.89) = 1.1\% = .011$$

The cumulative failure probability per generator bar is:

$$F(20.39) = 1 - \exp(-.007) = .007$$
$$F(27.89) = 1 - \exp(-.001) = .011$$

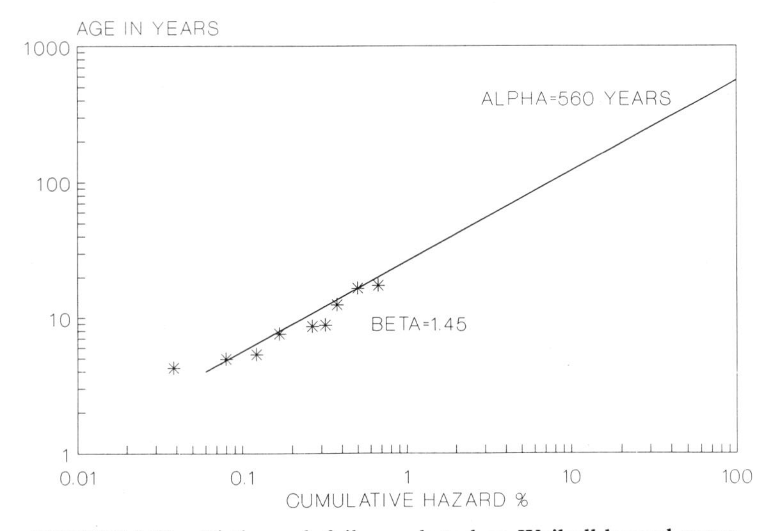

FIGURE 9.10 Girth-crack failures plotted on Weibull hazard paper.

TABLE 9.9 Girth-Crack Data from Example

Data Case	(2) Subcase Description	(3) Number of Bars	(4) Starting Year	(5) Ending Year	(6) Exposure Years	(7) Failure Event	(8) Mica-Based Bar (1 = Yes)	(9) Comments
1	1A	100	48	85	37		0	
2	2A	1	49	60	11	1	0	
	2B	99	49	60	11		0	
	2C	100	60	85	25		0	New asphalt rewind
3	3A	100	49	59	10		0	
	3B	100	59	85	26		0	Asphalt rewind cause other than girth crack
4	4A	1	48	56	8	1	0	
	4B	6	48	56	8		0	
	4C	7	56	85	29		0	Replaced asphalt winding
	4D	1	48	70	22	1	0	
	4E	2	48	70	22		0	
	4F	3	70	85	15		1	
	4E	90	48	85	37		0	
5	5A	1	49	70	21	1	0	
	5B	1	70	85	15		1	
	5C	99	49	85	36		0	
6	6A	100	51	76	25		0	
7	7A	100	50	85	35		0	
8	8A	100	49	85	36		0	
9	9A	1	50	69	19	1	0	
	9B	4	50	69	19		0	
	9C	5	69	71	2		1	
	9D	95	50	71	21		0	
10	10A	100	51	76	25		0	

11	11A	100	50	85	35		0
12	12A	100	51	85	34		0
13	13A	100	50	85	35		0
14	14A	100	53	85	32		0
15	15A	1	53	64	11	1	0
	15B	8	53	64	11		0
	15C	9	64	80	16		1
	15D	1	53	72	19	1	0
	15E	3	53	72	19		0
	15F	4	72	80	8		1
	15G	1	53	80	27	1	0
	15H	86	53	80	27		0
	15I	100	80	85	5		1
16	16A	100	52	85	33		0
17	17A	100	53	85	32		0
18	18A	100	56	85	29		0
19	19A	100	57	85	28		0
20	20A	1	57	67	10	1	0
	20B	2	57	67	10		0
	20C	3	67	85	18		1
	20D	97	57	85	28		0
21	21A	1	58	63	5	1	0
	21B	1	63	85	22		1
	21C	99	58	85	27		0
22	22A	100	58	85	27		0
23	23A	100	58	85	27		0
24	24A	100	59	85	26		0

TABLE 9.10 Data Sorted by Capacity-Factor-Weighted Age

Data Case	(2) Subcase Description	(3) Number of Bars	(4) Starting Year	(5) Ending Year	(6) Capacity Factor Exposure (Years)
	15F	4	72	80	1.03
	9C	5	69	71	1.40
	15I	100	80	85	2.90
	15C	9	64	80	3.73
21	21A	1	58	63	4.25
12	12A	100	51	85	4.37
11	11A	100	50	85	4.62
15	15A	1	53	64	4.95
	15B	8	53	64	4.95
4	4A	1	48	56	5.35
	4B	6	48	56	5.35
14	14A	100	53	85	5.62
23	23A	100	58	85	6.42
	5B	1	70	85	7.01
	4F	3	70	85	7.48
	15D	1	53	72	7.57
	15E	3	53	72	7.57
3	3A	100	49	59	8.00
20	20A	1	57	67	8.15
	20B	2	57	67	8.15
	15G	1	53	80	8.59
	15H	86	53	80	8.59
	2B	99	49	60	8.80
2	2A	1	49	60	8.80
24	24A	100	59	85	11.84
	20C	3	67	85	12.27
9	9A	1	50	69	12.39
6	6A	100	51	76	12.98
19	19A	100	57	85	13.13
	21B	1	63	85	13.27
10	10A	100	51	76	13.56
16	16A	100	52	85	13.61
7	7A	100	50	85	14.80
	9B	4	50	69	14.92
17	17A	100	53	85	15.00
13	13A	100	50	85	15.25
1	1A	100	48	85	15.99
	9D	95	50	71	16.32
	2C	100	60	85	16.43
5	5A	1	49	70	16.43

(7) Failure Event	(8) Mica-Based Replace (1 = Yes)	(9) Cumulative Bars	(10) Number of Asphalt Bars	(11) Failure Hazard	(12) Cumulative Hazard
	1	0	2607	0	0
	1	0	2607	0	0
	1	0	2607	0	0
	1	0	2607	0	0
1	0	1	2606	0.000383	0.00383
	0	101	2506	0	0.00383
	0	201	2406	0	0.00383
1	0	202	2405	0.000415	0.000799
	0	210	2397	0	0.000799
1	0	211	2396	0.000417	0.001216
	0	217	2390	0	0.001216
	0	317	2290	0	0.001216
	0	417	2190	0	0.001216
	1	417	2190	0	0.001216
	1	417	2190	0	0.001216
1	0	418	2189	0.000456	0.001673
	0	421	2186	0	0.001673
	0	521	2086	0	1.001673
1	0	522	2085	0.000479	0.002153
	0	524	2083	0	0.002153
1	0	525	2082	0.000480	0.002633
	0	611	1996	0	0.002633
	0	710	1897	0	0.002633
1	0	711	1896	0.000527	0.003161
	0	811	1796	0	0.003161
	1	811	1796	0	0.003161
1	0	812	1795	0.000557	0.003718
	0	912	1695	0	0.003718
	0	1012	1595	0	0.003718
	1	1012	1595	0	0.003718
	0	1112	1495	0	0.003718
	0	1212	1395	0	0.003718
	0	1312	1295	0	0.003718
	0	1316	1291	0	0.003718
	0	1416	1191	0	0.003718
	0	1516	1091	0	0.003718
	0	1616	991	0	0.003718
	0	1711	896	0	0.003718
	0	1811	796	0	0.003718
1	0	1812	795	0.001257	0.004976

(continued)

TABLE 9.10 (Continued)

Data Case	(2) Subcase Description	(3) Number of Bars	(4) Starting Year	(5) Ending Year	(6) Capacity Factor Exposure (Years)
22	22A	100	58	85	17.07
	3B	100	59	85	17.23
	4E	2	48	70	17.27
	4D	1	48	70	17.27
18	18A	100	56	85	17.45
	21C	99	58	85	17.47
8	8A	100	49	85	18.26
	4C	7	56	85	18.37
	20D	97	57	85	20.39
	5C	99	49	85	23.33
	4E	90	48	85	24.73

The conditional failure probability for the next 10 years is:

$$F(27.89|20.39) = \frac{.011 - .007}{1 - .007} = .004$$

The data of Figure 9.9 or Table 9.9 show that in unit 20 three bars were replaced with mica-based insulation which are not subject to failure, leaving 97 bars that are subject to failure. The probability of success of the entire generator (all 97 bars) is:

$$P(\text{success}) = (1 - .004)^{97} = .678$$

The probability of failure of the generator over the next 10 years is:

$$P(\text{failure}) = (1 - .678) = .322 = 32.2\%$$

The 32.2% probability is the probability of any failure, including one, two, three, or more failures over the 10-year period. For example, one stator bar could fail in year 3 and another in year 7. The binomial coefficient can be used to compute the probability of one and only one failure, two and only two failures, and so forth.

The probability of one and only one failure (out of 97 stator bars) is:

$$97 \cdot (1 - .004)^{96} \cdot .004 = .264 = 26.4\%$$

(7) Failure Event	(8) Mica-Based Replace (1 = Yes)	(9) Cumulative Bars	(10) Number of Asphalt Bars	(11) Failure Hazard	(12) Cumulative Hazard
	0	1912	695	0	0.004976
	0	2012	595	0	0.004976
	0	2014	593	0	0.004976
1	0	2015	592	0.001689	0.006665
	0	2115	492	0	0.006665
	0	2214	393	0	0.006665
	0	2314	293	0	0.006665
	0	2321	286	0	0.006665
	0	2418	189	0	0.006665
	0	2517	90	0	0.006665
	0	2607	0	0	0.000000

The probability of two and only two failures is:

$$\frac{97 \cdot 96}{2} \cdot (1 - .004)^{95} \cdot (.004)^2 = .051 = 5.1\%$$

The probability of three or more failures can be calculated as the total failure probability less the probability for one and two failures.

$$P(3 \text{ or more failures}) = 32.2 - 26.4 - 5.1 = .7\%$$

9.9 REPAIRABLE FAILURE-RATE MODELS

Repairable failure-rate models give an added dimension to the problem. The repaired component can fail again. This is typically the case in "like kind" power plant repairs.

As in the non-repairable case, the hazard function for the individual component can be defined as:

$$h(t) = \frac{\text{probable time rate of failures}}{\text{probability of surviving}} \quad (9.28)$$

$$h(t) = \frac{d\,F(t)/dt}{1 - F(t)} \quad (9.29)$$

The other hazard formulas [Equations (9.9), (9.13), and (9.24)] apply also. The cumulative probability of a failure is:

$$F(t) = 1 - \exp[-H(t)] \tag{9.30}$$

The probability of a first failure during the age interval t to $t + dt$ is:

$$\text{prob(first failure)} = \frac{d\,F_1(t)}{dt} \cdot dt \tag{9.31}$$

where $F_1(t)$ is the cumulative failure probability for the first failure.

After the component is repaired, it has a new cumulative failure probability, $F_2(t)$. The probability of a second failure, in the age interval t to $t + dt$, given that the first failure occurred in the interval t_0 to $t_0 + dt_0$ is given by:

$$\text{prob}\left(\begin{matrix}\text{second failure at } t \text{ with}\\ \text{first failure at } t_0\end{matrix}\right) = \frac{d\,F_1(t_0)}{dt_0}\,dt_0\,\frac{d\,F_2(t - t_0)}{dt}\,dt \tag{9.32}$$

The probability of a second failure in the interval t to $t+dt$, regardless of when the first failure occurred, is:

$$\text{prob}\left(\begin{matrix}\text{second failure:}\\ t \text{ to } t + dt\end{matrix}\right) = \int_0^t dt_1\,\frac{d\,F_1(t_1)}{dt_1}\,\frac{d\,F_2(t - t_1)}{dt}\,dt \tag{9.33}$$

The integral is called a "convolution" integral.

The cumulative probability of a second failure is the integral of the convolution integral, or:

$$\text{Cumulative prob}\left(\begin{matrix}\text{second failure}\\ \text{at } t_0\end{matrix}\right)$$

$$= \int_0^{t_0} dt_2 \int_0^{t_0} dt_1\,\frac{d\,F_1(t_1)}{dt_1}\,\frac{d\,F_2(t_2 - t_1)}{dt_2} \tag{9.34}$$

Examine this equation for the special case of an exponential distribution in which the repaired component is identical to the original failed component. Thus:

$$F_1(t) = F_2(t) = [1 - \exp(-\frac{t}{\alpha})] \tag{9.35}$$

then

Cumulative Prob (second failure)

$$= \int_0^{t_0} dt_2 \int_0^{t_2} dt_1 \frac{1}{\alpha^2} \exp\left(-\frac{t_1}{\alpha}\right) \cdot \exp\left(-\frac{t_2 - t_1}{\alpha}\right)$$

$$= 1 - \exp\left(-\frac{t}{\alpha}\right) - \frac{t}{\alpha} \exp\left(-\frac{t}{\alpha}\right) \tag{9.36}$$

Similarly, the cumulative probability of a third failure is:

Cumulative Prob (third failure)

$$= 1 - \left[\exp\left(-\frac{t}{\alpha}\right) + \frac{t}{\alpha} \exp\left(-\frac{t}{\alpha}\right) + \frac{t^2}{2} \exp\left(-\frac{t}{\alpha}\right)\right]$$

$$= 1 - \left[\exp\left(-\frac{t}{\alpha}\right)\right] \cdot \left[1 + \frac{t}{\alpha} + \frac{t^2}{2}\right] \tag{9.37}$$

Figure 9.11 illustrates the cumulative failure probability for the first, second, third, and fourth failures. The time derivative of the cumulative failure probability time, or failure probability density, is shown in Figure 9.12. Note that at age = 0, the failure density of the first failure is $1/\alpha$. Figure 9.12 has its ordinate scaled by α.

The probability of a first failure decreases with increasing age in an expo-

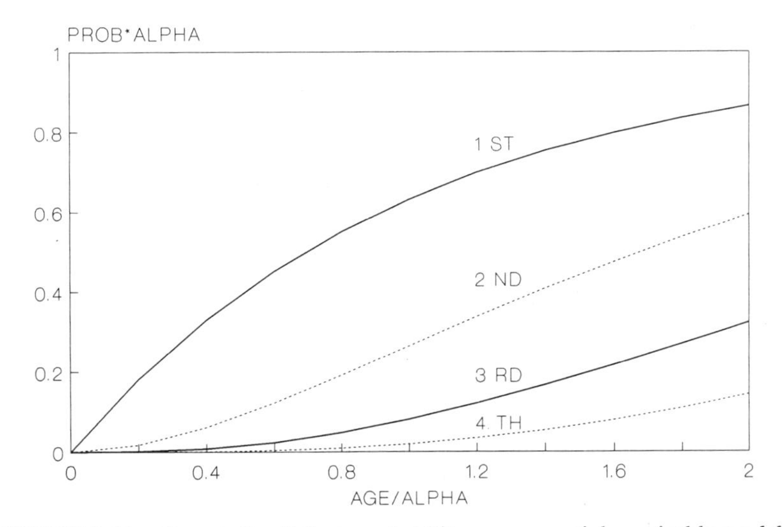

FIGURE 9.11 Cumulative failure probability—exponential repairable model.

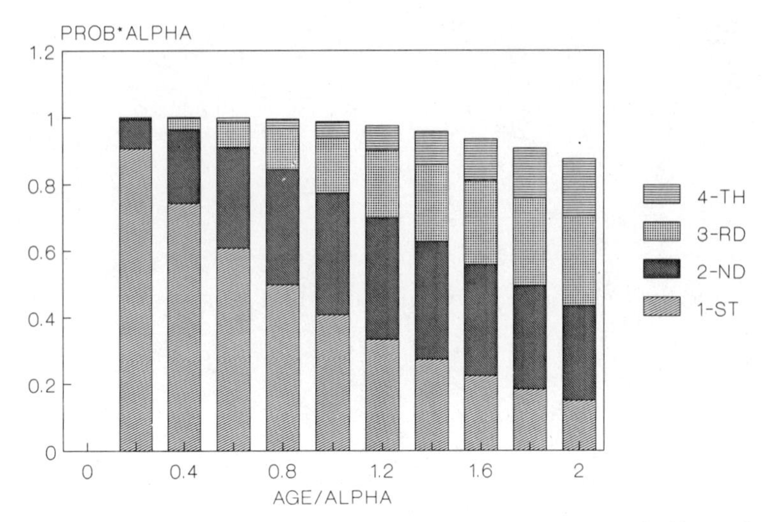

FIGURE 9.12 Failure probability density—exponential repairable model.

nential model, as shown in Figure 9.12. However, since any first failures are repaired, the probability of a second failure first increases and then decreases at higher ages. It is interesting to note that the failure probability per unit time (failure density) is constant because the failure hazard rate is constant and all failures are assumed to be repaired in like kind. The graph of Figure 9.12 would show this if the fifth, sixth, and further failures were also plotted.

Example. A component will be repaired in like kind when it fails. It will require 2.5 days to repair. The component can be represented as an exponential failure model with $\alpha = 5$ years. Compute the probable number of weeks per year that the component will be unavailable for service due to a forced maintenance outage.

SOLUTION. Since the component will be repaired in like kind, then the graph of Figure 9.12 applies. The annual probability of failure is 1/5. Thus, the annual expected days of forced maintenance are $1/5 \cdot 2.5 = 1/2$ day/year.

If the repaired component has a lower failure rate, then the probability density, failure probability per year, exhibits a transition period where it changes from the characteristics of the original component to the characteristics of the repaired component. Figure 9.13 illustrates the exponential model in which the repaired component (for second, third, fourth, etc.) has an α that is twice that of the original component or half the failure hazard rate.

Example. Using the data of Table 9.7, compute the cumulative number of component failures over the next 10 and 15 years for generating unit 9. As-

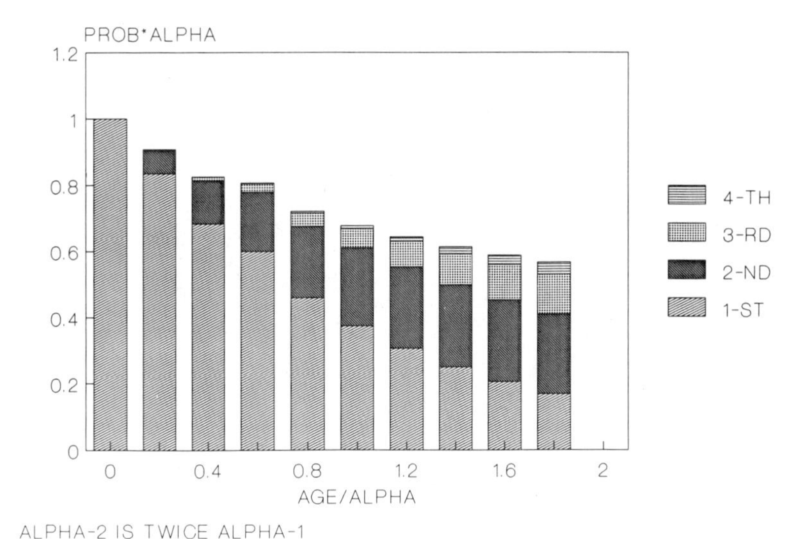

FIGURE 9.13 Failure probability density—exponential repairable model.

sume the original component is repaired when it fails, and that the repaired component has an α that is twice that of the original component.

SOLUTION. The failure data on the original component revealed that a Weibull distribution with $\alpha = 16$ years; $\beta = 2.15$ describes the component reliability performance. The repaired component is assumed to have an $\alpha = 32$ years and $\beta = 2.15$.

Because the original component on generating unit 9 had 4 years of age without any failures, the conditional probability describes its cumulative failure probability:

$$F_1(t|4) = \frac{[1 - \exp[(t + 4/16)^{2.15}] - .049]}{(1 - .049)} \tag{9.38}$$

where .049 is the unconditional Weibull cumulative probability at age equal to four years [using Equation (9.24)].

$$F_2(t) = 1 - \exp[(\frac{-t}{32})^{2.15}] = F_3(t) = F_4(t), \cdots \tag{9.39}$$

The convolution integral for the Weibull distribution cannot be analytically integrated as the exponential distribution. Therefore, the convolution integral was numerically computed.

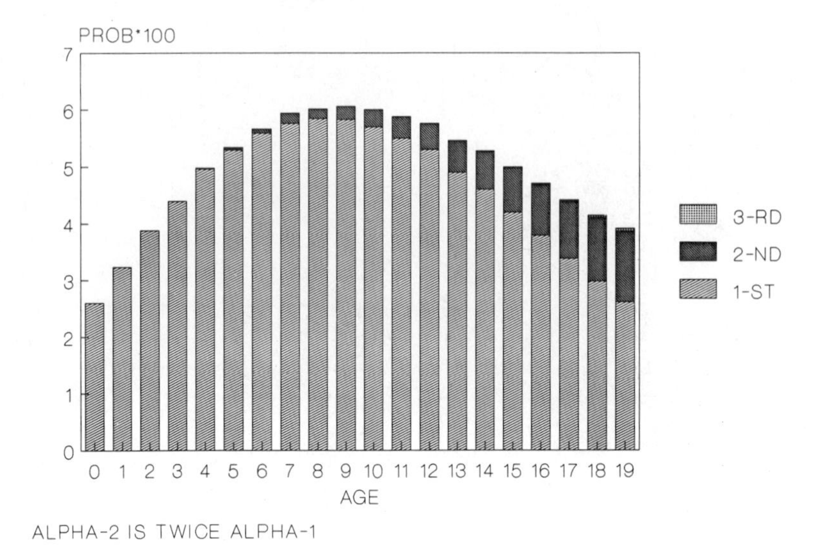

FIGURE 9.14 Failure probability density—Weibull repairable model.

Figure 9.14 illustrates the annual results. The cumulative expected number of failures over the next 10 and 15 years is shown in Table 9.11.

The repairable component exhibits a failure probability density that may exhibit an initial cyclical trend with age, as Figure 9.14 illustrates. However, it can be shown (see Bozovisky, 1961, p. 58) that after several repairs, the failure rate of a repairable component will behave exponentially regardless of the underlying failure distribution. Mathematically, the convolution integration essentially broadens ("smooths out") the underlying failure distribution. Consequently, after three or four mean failure periods, the failure rate appears uniform.

It can be shown that the failure rate of a repairable component after several mean failure periods approaches 1/mean time to failure.

In summary, repairable reliability models are generally applicable to power plant components. Failure data can be used to determine the reliability model of the original component. If the repaired component has the same reliability characteristics, then the probability of second, third, fourth, and additional

TABLE 9.11 Cumulative Expected Failures

	10 Years	15 Years
Original component (first) failure	.504	.748
Repaired component (second) failure	.010	.038
Repaired component (third) failure	0	.003
Total	.514	.786

failures can be computed using the convolution integration. Often, the repaired component has different reliability characteristics than the original component. Design calculations (stress analysis, erosion calculations, etc.) and experimental testing are typically used to estimate the improved reliability characteristics.

9.10 PROJECTING FAILURE PROBABILITIES

The previous section illustrated how statistical reliability models of power plant components can be constructed from the statistics of the component failures and nonfailures. The reliability model can then be applied to project future failure probabilities.

Several factors should be considered in this analysis:

1. The components in the data sample should be homogeneous—of the same size, construction, design, and operating environments. If repaired components have different reliability characteristics, then the statistics of the original component and those of the repaired component should be analyzed separately.
2. The data sample should be as complete as possible and should include data on failures as well as nonfailures. Often, failures are reported more accurately and nonfailures are sometimes not reported at all. Since the failure rate is to be determined and defined as the number of failures divided by the number of survivors, data on the survivors (successes) are just as important as data on failures.
3. The method illustrated a graphical solution for the Weibull reliability model. Regression analysis (Chapter 7) could be used to fit the best (least square) straight line through the data and, in addition, provide statistical parameters such as confidence intervals.
4. The method illustrated a single parameter reliability model in which unit age was the independent explanatory variable. It is possible to include two or three explanatory variables in the reliability model if there is a sufficient number of failure data points to obtain good confidence levels. A multiple regression analysis procedure could be used.
5. The failure data should cover a range of plant ages or other explanatory variables that are within the range of the forecast parameter. A reliability model forecast should not be made on the basis of an extrapolation outside the range of failure data experience.
6. Components may have several modes of failure. The key to projecting reliability lies in understanding the physical failure mechanism of each mode. A reliability model can then be built for each failure mode. Some failure modes may not have appeared to date.
7. A generating unit can be represented by a network of components (or

subsystems). There is always a danger in omitting one or more components or system failure interaction among the components. The degraded condition of one component (e.g., feedwater demineralizer) can lead to an accelerated failure of another component, such as boiler tube leaks and failures.

8. Reliability models are only one approach to forecasting future failure probabilities. Employing engineering design methods (stress mechanics, etc.) is another excellent approach.

Statistical reliability models provide a very useful method for projecting future failure probabilities when used and interpreted within their range of applicability.

9.11 CONSEQUENCES OF COMPONENT FAILURE

A component failure can lead to a power plant subsystem or system failure or degradation. Power system equipment is normally repaired on-site, or the component is repaired off-site at a manufacturer's factory or service shop.

When the component is being repaired, the utility incurs both direct and indirect costs (Marsh, 1980). The direct costs are those associated with the repair parts and the maintenance labor cost to install the repaired parts.

Indirect costs, however, are usually much larger than direct costs. Purchased power to replace the power that the unit would have produced is an indirect cost. The cost of replacement power is typically much higher than the generating cost of the failed unit. A failure that requires a long repair period (4–6 weeks) may cost the utility $200 thousand in direct costs and over $5 million in indirect costs. From a power system viewpoint, the more important factor is the product of the probability of failure times the number of days to repair the failure.

The percent of time that the generating unit is out of service because of a component failure is called the "component forced outage rate." The expected number of hours that the unit will be out of service due to the component failure is [forced outage hours (FOH) per year]:

$$\text{FOH(component)} = \text{(probability of failure during the year)} \cdot \text{(mean time (hours) to repair)} \tag{9.40}$$

In practice, it is often useful to replace the yearly variations of failure probability by the asymptomatic value of 1/mean time to failure. Then the asymptotic forced outage hours can be calculated as the mean time to repair divided by the mean time to failure.

The forced outage rate is defined as the fraction of time the generating unit (or component) is unavailable for service. Assume that the generating unit provides SH service hours per year. Because the unit is on forced outage for

FOH hours per year, the unit was required (or demanded) to be in service for SH + FOH hours per year. Thus, the fraction of required time the unit was on forced outage is:

$$\text{Forced outage rate} = \text{FOR} = \frac{\text{FOH}}{\text{SH} + \text{FOH}} \tag{9.41}$$

9.12 GENERATING UNIT RELIABILITY AND AVAILABILITY MEASURES

A generating unit has several operating conditions. It can be in service and operating, be available for service and not operating (reserve shutdown), or be unavailable for service (on outage).

According to the North American Electric Reliability Council (NERC), there are several contributors to outages:

Forced outage — "The occurrence of a component failure or other condition that requires removal of the unit from service immediately or up to and including the next weekend."

Maintenance outage — "The removal of a unit from service to perform work on specific components which could have been postponed past the next weekend. This is work done to prevent a potential forced outage and that could not be postponed from season to season."

Planned outage — "The removal of a unit from service for inspection and/or general overhaul of one or more major equipment groups. This is work which is usually scheduled well in advance (e.g., annual boiler overhaul, five-year turbine overhaul)."

In addition to the complete outage of a generating unit, outages can also occur which result in the derating of the unit.

Forced derated outage — "The occurrence of a component failure or other condition which requires that the load on the unit be reduced 2% or more immediately or up to and including the next weekend."

Planned derated outage — "The occurrence of a component failure or other condition which requires that the load on the unit be reduced 2% or more, but where this reduction could be postponed past the next weekend."

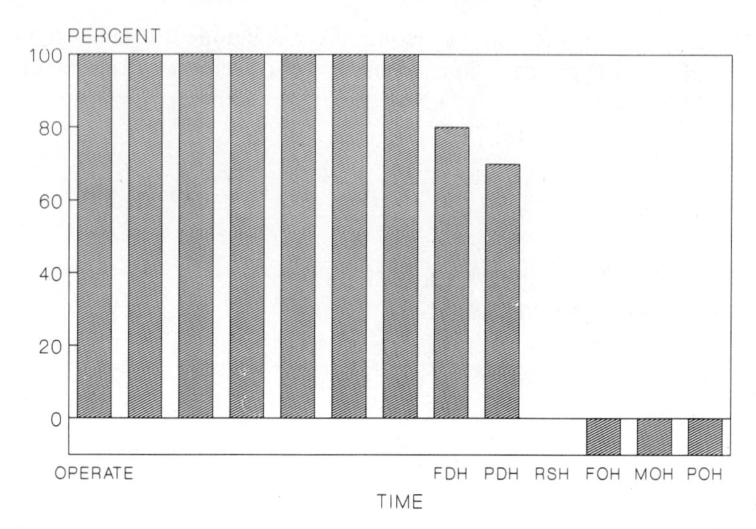

FIGURE 9.15 Generating unit availability—based on NERC definitions.

Figure 9.15 illustrates the definitions of NERC symbols and terms.

AH = available hours
FDH = clock hours accumulated during a forced derating outage
FOH = forced outage hours
MOH = maintenance outage hours
OPER = total operating time at 100% availability (not a NERC definition)
PDH = clock hours accumulated during a planned derating outage
PH = period hours (usually 8760 hours, i.e., 1 year)
POH = planned outage hours
RSH = total economy outage hours (reserve shutdown hours)
SH = service hours

From these definitions of times, the following equations of available and service hours are:

$$\begin{aligned} \text{SH} &= \text{PH} - \text{POH} - \text{MOH} - \text{FOR} - \text{RSH} \\ &= \text{OPER} + \text{FDH} + \text{PDH} \end{aligned} \tag{9.42}$$

$$\text{AH} = \text{SH} + \text{RSH} \tag{9.43}$$

The definition of outage rates (%) are more useful than outage hours. In terms of the above symbols, the definitions that are generally used are:

$$\text{FOR} = \text{forced outage rate} = \frac{\text{FOH}}{\text{FOH} + \text{SH}} \cdot 100\% \tag{9.44}$$

The forced outage rate is the fraction of the time the unit is not available for service when required. It is a measure of the random unavailability of a generating unit. The denominator is the hours that the unit was required to be operating. Thus, the forced outage rate measures the ratio of the hours the unit was unavailable due to a forced outage divided by the exposure time when the unit was supposed to be running.

The planned outage factor (or rate) is the fraction of the period (generally a year) that the generating unit is on planned outage. It is a measure of the unavailability of a generating unit due to annual planned maintenance.

$$\text{POF} = \text{planned outage factor} = \frac{\text{POH}}{\text{PH}} \cdot 100\% \tag{9.45}$$

The scheduled outage factor (or rate) is often used to define the sum of maintenance and planned outages. The scheduled outage factor (or rate) is:

$$\text{SOF} = \frac{(\text{POH} + \text{MOH})}{\text{PH}} \cdot 100\% \tag{9.46}$$

The preceding definitions apply to the full outages of generating units. Since partial derating outages can contribute to the loss of reliability or availability of a generating unit, provision is made to derive the equivalent forced and planned outage rates.

One procedure for incorporating derating outages is to weight the partial outage hours by the percent magnitude of the reduction. The result is the equivalent outage hours of full outage time due to derating outages. Table 9.12 presents data for 400–599 MW coal units from 1975 to 1984 (refer to NERC, 1985).

TABLE 9.12 Coal Units 400–599 MW (1975–1984) Based on NERC Derating Outage Data

% Available	% Unavailable	Unplanned Derating Hours	Planned Derating Hours
0–20	80–100	10	1
20–30	70–80	11	1
30–40	60–70	35	4
40–50	50–60	73	14
50–60	40–50	161	22
60–70	30–40	178	20
70–80	20–30	307	24
80–90	10–20	587	71
90–100	0–10	756	60
Total Hours		3118	217
Equivalent derating outage hours		434(EUDH)	55(EPDH)

The full outage data for coal units 400 to 599 from 1975 to 1984 are:

$$
\begin{aligned}
\text{PH} &= 8574\\
\text{SH} &= 6323\\
\text{FOH} &= 738\\
\text{MOH} &= 288\\
\text{POH} &= 944
\end{aligned}
$$

The NERC definition of equivalent forced outage rate ("Report on Equipment Availability for the Ten-Year Period—1976–1985," p. 97) is:

$$\text{EFOR} = \frac{\text{FOH} + \text{EUDH}}{\text{FOH} + \text{SH}} \tag{9.47}$$

where EUDH is the percent megawatt weighted equivalent unplanned (forced) derating outage hours.

Using data from Table 9.13, this leads to:

$$\text{EFOR} = \frac{738 + 434}{738 + 6323} = 16.7\% \tag{9.48}$$

The "full" forced outage rate is:

$$\text{FOR} = \frac{738}{738 + 6323} = 10.5\%$$

Derating hours significantly contribute to the total equivalent forced outage rate.

The equivalent planned outage rate can be defined as:

$$\text{EPOR} = \frac{\text{POH} + \text{EPDH}}{\text{PH}} \tag{9.49}$$

9.13 RANDOM-OUTAGE RATE AND SCHEDULED-OUTAGE RATE

In generation system reliability analysis, generating unit outages must be segregated into a random outage rate and a scheduled outage rate. The random outage rate impact on system reliability can be analyzed by means of probabilistic mathematics, while the scheduled outage rate impact can be analyzed by applying deterministic mathematics. These two aspects will be discussed in the next chapter. Clearly, forced outages are included in the random outage category, while planned outages are included in the scheduled category.

In addition to forced outage rate and planned outage rate definitions, the category of maintenance outages needs to be assigned. Maintenance outages fall into a gray area. The outage could have been postponed beyond the next weekend, but not postponed to the next planned outage. There are differing opinions as to whether maintenance outages should be included with the random outage rate or the scheduled outage rate. Although both practices and combinations are used in the industry, the convention of including it with the forced outage rate is more widely used.

The definitions of the random outage and scheduled outage rates are:

$$\text{Random outage rate} = \frac{\text{FOH} + \text{MOH} \cdot \text{fraction}}{\text{FOH} + \text{MOH} \cdot \text{fraction} + \text{service hours}} \quad (9.50)$$

$$\text{Scheduled outage rate} = \frac{\text{POH} + \text{MOH} \cdot (1 - \text{fraction})}{\text{PH}} \quad (9.51)$$

where fraction = fraction of the maintenance outage hours that are included in the random outage rate definition (0 to 1.0).

While the correct outage rate terminology for generation system reliability analysis is "random" and "scheduled" outage rates, the utility industry uses random and forced outage rates interchangeably as well as scheduled and planned. In the remainder of this book these terms will also be used interchangeably, but with the full understanding that maintenance outages have been properly included in the numerical values of random and scheduled outage rates.

9.14 UNITS USED IN PEAKING SERVICE

The previous outage rate definitions are most applicable to generating units with high capacity factors. For units used in peaking service, an adjustment must be made. Consider the case of gas turbines. The forced outage rate of gas turbines is reported by NERC ("Equipment Availability Report for 1984") as being 54%. However, to directly use this number in a power system reliability analysis would be inaccurate. The reasons for this are discussed in this section.

Figure 9.16 illustrates an example of the operating experience of a gas turbine. As is the case with many gas turbines, it is called on to deliver power only a few hours per day.

Over the weekend, the unit is on reserve shutdown. On Monday, it operates for 4 hours and then returns to reserve shutdown. On Tuesday, the unit fails to start and is characterized as being on forced outage. When on forced outage, all hours from Tuesday through Wednesday, when repairs are completed, are being counted (approximately 24 hours) as forced outage hours. On Wednesday, the unit returns to service and operates for 4 hours. On Thursday

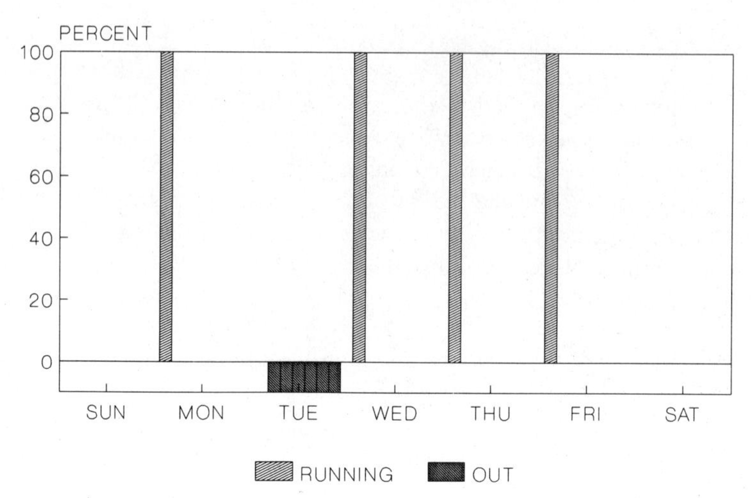

FIGURE 9.16 Gas-turbine schedule with a one-day outage.

and Friday, the unit serves load for 4 hours. The tabulation of the unit's operating experience in this example is:

Weekly Data	Hours
Service hours (SH) (4 days)	216
Reserve shutdown (RSH)	128
Forced outage (FOH)	24
Total (PH) period of hours	168

The NERC definition of forced outage rate for this unit for this week would be:

$$\text{FOR} = \frac{\text{FOH}}{\text{FOH} + \text{SH}} = \frac{24}{24 + 16} = 60\% \tag{9.52}$$

However, does this forced outage rate number mean that the probability that the unit is unavailable when called on to deliver capacity is 60% for this week? Actually, the unit was called on five times and failed only once. In a reliability analysis, the more appropriate value for representing the random unavailability of the gas turbine would be:

$$\text{(FOR) forced outage rate} = 1/5 = 20\% \tag{9.53}$$

The problem with using Equation (9.52) directly is that when the gas turbine is on forced outage, most of the elapsed hours that are counted as forced outage hours would have been reserve shutdown hours, had the unit been available. Consequently, one method of developing the appropriate forced outage rate for gas turbines is to use the following formula:

$$\text{FOR} = \frac{\text{FOH} \cdot (D/24)}{\text{FOH} \cdot (D/24) + \text{SH}} \tag{9.54}$$

where FOR = forced outage rate
FOH = forced outage hours
SH = service hours
D = duty cycle hours per day (i.e., 4 hours in the previous example).

The multiplication of the FOH by the D/24 results in the accounting of forced outage hours only when the unit would have been expected to deliver power. In the preceding example, the FOR would be evaluated as:

$$\text{FOR} = \frac{24 \cdot 4/24}{24 \cdot 4/24 + 16} = 20\% \tag{9.55}$$

The first step in computing the forced outage rate of a peaking unit is to evaluate its duty cycle hours per day. This may be computed as its service hours divided by the number of starts.

The definition of the forced outage rate of Equation (9.53) can be rearranged to read as:

$$\text{FOR} = \frac{\text{FOH}}{\text{FOH} + 24 \cdot (\text{\# of starts})} \tag{9.56}$$

Another way to estimate the gas turbine forced outage rate is to compute the number of unsuccessful starts divided by the number of attempted starts. This provides an estimate because almost all of the forced outages on gas turbines are associated with failures to start. Using 1984 NERC data ("Equipment Availability Report for 1984"), the gas turbine forced outage rate is estimated to be FOR = 10.8%.

In conclusion, for units used in peaking service, an adjustment to the forced outage data is recommended when calculating random outage rate for generation reliability planning. The article by Albrecht et al. ("Gas Turbines Require Different Outage Criteria," 1970) further amplifies the discussion on forced outage data.

9.15 AVAILABILITY AND EQUIVALENT AVAILABILITY

The availability of a generating unit is the percentage of time the unit is available to serve load.

$$\text{Availability} = \frac{\text{Available hours}}{\text{period hours}} = \frac{\text{PH} - \text{POH} - \text{MOH} - \text{FOH}}{\text{PH}} \tag{9.57}$$

The equivalent availability of a generating unit is the fraction of time the unit is able to serve an equivalent full load. Equivalent availability accounts for equivalent derating outages.

$$\text{Equivalent availability} = \frac{\text{PH} - \text{POH} - \text{MOH} - \text{FOH} - \text{EUDH} - \text{EPDH}}{\text{PH}} \tag{9.58}$$

In the definitions of availability and equivalent availability, reserve shutdown hours are counted as time that the unit is available. It is often useful to compute the maximum availability in the case of no reserve shutdown hours (RSH), or RSH = 0.

Maximum equivalent availability can be defined as:

$$\begin{aligned} \text{Maximum Equivalent Availability} &= \left.\frac{\text{PH} - \text{POH} - \text{MOH} - \text{FOH} - \text{EUDH} - \text{EPDH}}{\text{PH}}\right|_{\text{RSH}=0} \\ &= \left.\frac{\text{SH} - \text{EUDH} - \text{EPDH}}{\text{PH}}\right|_{\text{RSH}=0} \\ &= \frac{\text{SH} - \text{EUDH} - \text{EPDH}}{\text{SH} + \text{EUDH} + \text{FOH} + \text{MOH} \cdot \text{FRAC} - \text{EPDH} - \text{EUDH}} \\ &\quad \cdot \frac{\text{PH} - \text{POH} - \text{MOH} \cdot (1 - \text{FRAC}) - \text{EPDH}}{\text{PH}} \end{aligned} \tag{9.59}$$

where FRAC is the fraction of maintenance outage hours classified as contributing to the random outage rate.

Note that the identity SH + FOH + MOH + POH = PH was used.

Let the equivalent random outage rate be defined as:

$$\text{EROR} = \frac{\text{FOH} + \text{MOH} \cdot \text{FRAC} + \text{EUDH}}{\text{SH} + \text{FOH} + \text{MOH} \cdot \text{FRAC} - \text{EPDH}} \tag{9.60}$$

Note that in this definition of equivalent random outage rate, equivalent planned derating hours are subtracted from the denominator. This is because the service hours include operation at full capability plus operating during derated conditions. Consequently, the denominator represents the equivalent full-load exposure time that the unit may experience a random outage.

Let the equivalent scheduled outage rate be defined as:

$$\text{ESOR} = \frac{\text{POH} + \text{MOH} \cdot (1 - \text{FRAC}) + \text{EPDH}}{\text{PH}} \tag{9.61}$$

Then, by substituting Equations (9.60) and (9.61) into (9.59), the maximum equivalent availability is:

$$\text{Maximum equivalent availability} = (1 - \text{EROR}) \cdot (1 - \text{ESOR}) \tag{9.62}$$

If derating outages are not considered, maximum availability can be shown to be equal to:

$$\text{Maximum availability} = (1 - \text{ROR}) \cdot (1 - \text{SOR}) \tag{9.63}$$

As an example of these definitions, consider the 1984 performance of oil-fired steam units of 200–299 MW.

Period hours	8784
Service hours	6331
Reserve shutdown hours	1106
Available hours	7438
Forced outage hours	457
Planned outage hours	729
Maintenance outage hours	159
Unit availability	$\frac{7438}{8784} = 84.7\%$
Forced outage rate	$\frac{457}{457 + 6331} = 6.7\%$
Planned and maintenance outage rate	$\frac{729 + 159}{8784} = 10.1\%$
Maximum availability	$(1 - .067) \cdot (1 - .101) = 83.9\%$

Maximum availability is the highest availability the unit could operate at if it were always operated except for outages. The operating availability is 84.7%, higher than the maximum availability because the operating availability assumes that the unit is 100% reliable during reserve shutdown periods.

Maximum availability and the maximum equivalent availability are useful parameters for comparison of the operating performance of generating units. The maximum equivalent availability characterizes the maximum output generation from a unit.

9.16 TYPICAL OUTAGE DATA

Typical outage rate data of a 300 to 400 MW coal-fired fossil-steam unit are presented in Table 9.13.

The data presented in Table 9.13 are typical of a mature generating unit. When new generating units are first placed into service, however, they exhibit an immaturity effect for the first several years. During this time outage rates can be significantly higher than mature values. Typically, it is during this period that many of the "bugs" are worked out of the generating unit. Also, as units approach 25 years of age, they enter a wear out phase (Stoll, 1985) in which the reliability and availability decrease.

The immaturity and aging effects are important to consider in generating system reliability analysis.

9.17 POWER PLANT RELIABILITY CHARACTERISTICS—CHAPTER SUMMARY

This chapter has presented the foundations of reliability mathematics that will be used extensively in the remaining chapters. It has illustrated how to evaluate the reliability of complex generating units using the outage state enumeration method. Reliability models of components that find application in projecting future component and plant reliability have been discussed. Finally, electric utility reliability, and availability definitions of outage rate data were presented.

TABLE 9.13 Typical Outage Data for a 300–400-MW Coal-Fired Fossil-Steam Unit (1984 NERC Data)

Full forced outage rate (%)	9%
Equivalent forced outage rate (%)	13.2%
Average time to repair the unit: forced outage (days)	1.8 days
Average number of full forced outages per year (1 year)	14 per year
Full forced + maintenance outage rate	12%
Average time to repair the unit: maintenance outage (days)	2.3 days
Average number of maintenance outages per year (1 year)	4.3 per year
Planned outage rate (%)	9.6%
Average number of planned outages per year	1.6 per year

REFERENCES

Albrecht, P. F., W. D. March, and F. H. Kindl, "Gas Turbines Require Different Outage Criteria," *Electrical World*, April 27, 1970.

Billinton, R., *Power System Reliability Evaluation*, Gordon & Breach, Science Publishers, New York, 1970.

Bozovsky, I., *Reliability Theory and Practice*, Prentice-Hall, Englewood Cliffs, NJ, 1961.

Marsh, W. D., "Economics of Electric Utility Power Generation," Oxford University Press, New York, 1980.

Mosteller, F., R. E. K. Rourke, and G. B. Thomas, Jr., *Probability and Statistics*, Addison-Wesley, Reading, MA, 1967.

Nelson, W., *Applied Life Data Analysis*, Wiley, New York, 1982.

NERC, "1985 Generating Availability Data System Reports," North American Electric Reliability Council, Research Park (Terhune Road, Princeton, NJ 08540-3573).

Stoll, H. G., "The Economics of Power Plant Life Extension," 12th Reliability, Availability and Maintainability (INTER-RAM) Conference, Baltimore (1985).

PROBLEMS

1 A power system has four power plants. Two are oil-fired, one is gas-fired, and the other is coal-fired. What is the probability that all four power plants will be operating, assuming the reliability data given in Table 9.1?

2 A power plant has three feedwater pumps, each with a failure rate hazard function of .03 per year. If each of the pumps are assumed to be independent of each other, what is the probability that at the end of 20 years all three pumps will have been replaced? (Assume an exponential failure model.)

3 A power plant has 20 components in series that must all be in operation for the power plant to operate. It has been found that these components all follow a Weibull distribution pattern with $\alpha = 1000$ years and $\beta = 0.5$. When any one of these components fail, the power plant will be on forced outage for 6 weeks to repair the failure, with a repair cost of $70,000. When this power plant is not running, an additional $20,000 per day operation expense is incurred by operating more expensive power plants. What is the expected increased cost to be incurred over the next 20 years as a result of the failure of this power plant due to these components?

4 Assuming the same failure probability data for Unit 20 as given in the example contained in Table 9.9, and assuming that a failure will result in an outage cost of $750,000, how much outage cost would you expect over the next 20 years due to stator bar girth cracks if Unit 20 is projected to operate at 40% capacity factor? How much more outage costs result if the unit is operated at 70% capacity factor over the same time period? Assume that its previous operating capacity factor is the same in both cases and as in the example data in Table 9.9.

5 A gas turbine has the following outage statistics over the past year:

Service hours	800
Forced outage hours	400
Reserve shutdown hours	7560

Assuming duty cycles of 24, 12, 8, 6, 4, and 2 hours of operation per day, calculate the various forced outage rates.

10

GENERATION SYSTEM RELIABILITY

This chapter presents methods for quantifying generation system reliability and determining the quantity of capacity required to achieve a specified reliability level.

Generation system reliability analysis quantifies the electricity service reliability provided from the generation system. The quantity of generation capacity (MW) required to achieve a desired quality of electricity service is of key interest. Generation capacity adequacy is conveniently measured in terms of the generation reserve margin. Generation reserve margin is defined as:

$$\text{Generation reserve margin (\%)} = \left(\frac{\text{Generation capacity in service at time of peak load (MW)} - \text{peak load}}{\text{peak load (MW)}}\right) \cdot 100$$

To illustrate this equation, take, for example, a power system with a peak load of 10,000 MW and an installed capacity of 12,000 MW at time of peak. That system would have a generation reserve margin of $(12{,}000/10{,}000 - 1.0) \cdot 100 = 20\%$.

Generation capacity margin, another measure of generation reserve, is also used. Although similar to reserve margin, the denominator for capacity margin is different. Capacity margin is defined as:

$$\text{Generation capacity margin (\%)} = \left(\frac{\text{capacity in service} - \text{peak load}}{\text{capacity in service}}\right) \cdot 100$$

For the same example, the capacity margin is (12,000 − 10,000/12,000) • 100 = 16.7%.

Why are generation reserves required by the utility power system? Several factors contributing to the need for reserve generation include generation outages, load uncertainty, less-than-perfect interconnection reliability to neighboring power systems, uncertainty in the installation dates for new generation, and potential transmission line outages within the power system (Albrecht et al., 1981).

10.1 HISTORICAL MEASURES OF POWER SYSTEM RELIABILITY

When planning generation, the question naturally arises as to how much generation capacity is required to serve the load demand. Three historical methods used to determine capacity are presented in this section.

10.1.1 Percent Reserve Evaluation

The earliest method and most easily computed criterion for evaluation of generation system adequacy is the percent generation reserve margin approach. This method is sensitive to only two factors at one point in time.

Percent reserve evaluation computes the generation capacity exceeding annual peak load. It is calculated by comparing the total installed generating capacity at peak with the peak load. The criterion is based on past experience requiring reserve margins in the range of 15–25% to meet demand. Satisfactorily meeting load demand meant that the frequency and magnitude of emergency power purchases from neighboring power systems was reasonable and/or the number of curtailments was small.

There are, however, disadvantages to the percent reserves approach. It is insensitive to forced outage rates and unit size considerations, as well as to differing load characteristics of power systems. Although this approach is a useful step in the analysis of generation reserve problems, it does not provide a complete answer to how much generation capacity is required to adequately serve load demands.

10.1.2 Loss-of-the-Largest-Generating-Unit Method

Loss-of-the-largest-generating-unit method provides a degree of sophistication over the percent reserve margin method by reflecting the effect of unit size on reserve requirements. With the loss-of-the-largest-unit method, required reserve margin is calculated by adding the size of the largest unit divided by the peak load plus a constant reserve value.

For example, if reserve requirements are 15% plus the largest unit, and the largest unit is 500 MW in a power system with a 5000-MW peak load, then the reserve requirement is 15% + 500/5000•(100%), or 25%. This approach begins to explicitly recognize the impact of a single outage, that is, loss of the largest generating unit. Probabilistic measures are necessary to extend this method to include multiple simultaneous outages.

Loss-of-the-largest-unit method, although simple, has a distinct advantage over the generation reserve margin method. As larger units are added to a system, the percent reserves for a system are implicitly increased by this method, as is proper.

10.1.3 Loss-of-Load-Probability Method

In 1947, Calabresse suggested a probabilistic approach to determining required reserves (Calabresse, 1947). This technique examines the probabilities of simultaneous outages of generating units that, together with a model of daily peak-hour loads, determines the number of days per year of expected capacity shortages (AIEE Committee, 1961).

The resulting measure, termed "loss-of-load-probability" (LOLP) index, provides a consistent and sensitive measure of generation system reliability. The term "loss of load probability" is misleading in two respects. First, this index is not a probability, but is an expected value of the number of days per year of capacity deficiency. Second, it is not a loss of load, but rather a deficiency of installed available capacity. Despite the misnomers, LOLP is the most widely accepted approach for determining reserves in the utility industry today.

Loss of load probability consists of two segments in which the generating unit unavailability is characterized by random outage rate (forced) and scheduled outage rate. The effect of random outages is evaluated probabilistically, while that of scheduled outages is evaluated deterministically (Endrenyi, 1978).

10.1.4 Comparison

These three methods—percent reserve, loss of largest unit, and loss of load probability—are used to evaluate how much additional capacity is needed to support additional load demand growth. Figure 10.1 presents a graph of the increase in system load made possible by a generation addition of various sizes. This is referred to as the "load-carrying capability" of a generating unit. If the unit had perfectly reliable capacity, then a 600-MW-size unit could support 600 MW in load increase.

In the percent reserve method, where the reserve margin requirement is 25%, a 600-MW unit could support a 600/1.25 or 480-MW load increment.

In the loss-of-largest-unit method, a 600-MW unit allows a load growth increment of only 400 MW when the largest unit in the power system is 400 MW.

In the LOLP method, the load-carrying capability of the unit is based on the forced outage rate of the generating unit and planned outage rate and the characteristics of the other units in the power system. As the unit size in-

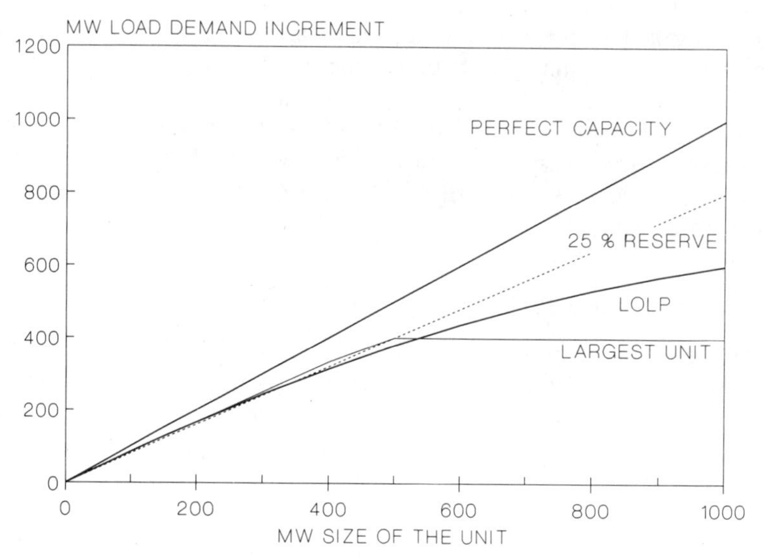

FIGURE 10.1 Comparison of methods to determine the load-carrying capability of a new unit.

creases, the LOLP method follows the percent reserve curve. For larger sizes, however, it follows the loss-of-largest-generating-unit method.

In summary, the (probabilistic) LOLP method, or variant thereof, is the most widely accepted approach in the utility industry for evaluating generation capacity requirements.

10.2 LOSS-OF-LOAD-PROBABILITY (LOLP) METHODOLOGY (RANDOM OUTAGES)

The LOLP method employs the fundamental data of unit capacity, random outage rate, and load demand. Consider this three-unit example. The generation system is comprised of three units:

	Capacity	FOR	Innage Rate
Unit A	50	.05	.95
Unit B	100	.07	.93
Unit C	200	.10	.90
System	350		

For this three-unit system, there are eight combinations of generating units on outage and in service. Table 10.1, an outage state enumeration table, enumerates all of these states and the probability of each. For example, the probability that no generating unit is on outage (all units in service) is the product

TABLE 10.1 Outage State Enumeration

On Outage	MW on Outage	In Service	Probability	
None	0	A, B, C	.95 • .93 • .90 =	.79515
A	50	B, C	.05 • .93 • .90 =	.04185
B	100	A, C	.95 • .07 • .90 =	.05985
C	200	A, B	.95 • .93 • .10 =	.08835
A, B	150	C	.05 • .07 • .90 =	.00315
A, C	250	B	.05 • .93 • .10 =	.00465
B, C	300	A	.95 • .07 • .10 =	.00665
A, B, C	350	None	.05 • .07 • .10 =	.00035
				1.00000

of the innage rates for units A, B, and C. Similarly, another state may be that unit A is an outage and B and C are in service. The probability of this state is the product of the forced outage rate of A times the innage rates of B and C. Table 10.1 presents this state enumeration. Note that the sum of the state probabilities is unity. Table 10.1 can be ordered in monotonically increasing order of megawatts of on outage as presented in Table 10.2.

Consider evaluating the probability of not being able to supply a 220-MW load demand. If 220 MW or less of capacity is in service, a 220-MW load cannot be served. Since the capacity of the three-unit system is 350 MW, the load could not be supplied if (350 − 220) or 130 MW of capacity *or more* is on outage. According to data from Table 10.2, the probability of 130 MW *or more* on outage is:

$$.00315 + .08835 + .00465 + .00665 + .00035 = .10315$$

Hence, the probability of not meeting load demand is .10315.

Since computation of the probability of not meeting load demand requires an evaluation of MW or *more* on outage, Table 10.2 can be written as a cumu-

TABLE 10.2 Units on Outage Based on Monotonically Increasing Order

MW on Outage	MW in Service	Probability
0	350	.79515
50	300	.04185
100	250	.05985
150	200	.00315
200	150	.08835
250	100	.00465
300	50	.00665
350	0	.00035

TABLE 10.3 Cumulative Outage Probability

X MW or More on Outage	Probability of X MW or More on Outage
0	1.00
50	.20485
100	.16300
150	.10315
200	.10000
250	.01165
300	.00700
350	.00035

lative outage table as presented in Table 10.3. The entry of 0 MW or more on outage is the summation of Table 10.2 from 0 to 350 MW. The entry of 50 MW or more is the summation of Table 10.2 from 50 to 350 MW.

Thus, to evaluate the probability of *not* meeting a 220-MW load, the required capacity on outage would be:

$$350 - 220 = 130 \text{ MW or more on outage}$$

From Table 10.3, the probability of not meeting load is .10315.

10.2.1 Weekly Index

It may be desired to evaluate the expected number of times the power system would not be able to supply system load demand for a week. Take, for example, a week with daily peak loads as indicated in columns 1 and 2 of Table 10.4.

Using the results from Table 10.3, the probability of not meeting load each

TABLE 10.4 Weekly Probability Index Based on Cumulative Outage Data and Daily Peak Load

Day	Load	Capacity or More on Outage	Probability of Not Meeting Load
Sunday	140	210	.01165
Monday	280	70	.16300
Tuesday	240	110	.10315
Wednesday	220	130	.10315
Thursday	260	90	.16300
Friday	290	60	.16300
Saturday	130	220	.01165
			.71860

day can be computed as presented in Table 10.4. Summing the probability for each day yields a weekly number of 0.7186 days/week. This is interpreted as an expected value. Namely, on the average, the power system is unable to serve the load demand 0.7186 times (almost once) every week.

Although the weekly index was used as an example, the utility industry generally uses an annual index (LOLP), which is the summation of daily probabilities, or often termed "daily risks," over the entire year. Typical values of LOLP lie in the 0.1 days/year to 1.0 day/year range depending on the required reliability of service (AIEE Committee, 1961).

10.3 LOLP CONVOLUTION ALGORITHM

While the procedure described in the previous section is conceptually correct, the actual computational algorithm requires enumeration of 2^N states, where N is the number of generating units in the power system. If there are 30 units in the system, this would mean an enumeration of more than a billion capacity states. This is computationally impractical. Rather than an enumeration process, a convolution process can be developed.

The objective is to construct a table of the probability of X MW on outage. This can be derived from the conditional probability Equation (9.4). Assume that the table has already been constructed for several generating units and that a new table that includes one more unit must be completed. Let:

$P^{old}(X)$ = probability of X MW on outage for the old (existing) table
$P^{new}(X)$ = probability of X MW on outage for the new table
C = MW capacity of the unit that is to be added (the reliability characteristic of the unit is two-state, either on outage or in service)
FOR = equivalent random (forced) outage rate of the unit to be added.

Consider some value of X MW on outage. In order to have X MW on outage, the original system could have X MW on outage *and* the additional unit could have 0 MW on outage. Or, the additional unit could have C MW on outage and the original system could have $(X - C)$ MW on outage. Table 10.5 illustrates both these scenarios.

Since cases 1 and 2 are mutually exclusive, the probability of these two cases

TABLE 10.5 System Outage Probability with an Additional Unit

	Original System		Unit to be Added	
Case	MW on Outage	System Probability	MW on Outage	Probability
1	X	$P^{old}(X)$	O	(1 − FOR)
2	$X - C$	$P^{old}(X - C)$	C	FOR

is the sum of the probabilities of cases 1 and 2. Thus, the probability of X MW on outage for the old (existing) set of generating units *and* the additional unit is:

$$P^{new}(X) = P^{old}(X) \cdot (1 - \text{FOR}) + P^{old}(X - C) \cdot \text{FOR} \quad \text{if} \quad X \geq C \tag{10.1}$$

Suppose, however, that $X < C$. Then Equation (10.1) needs to be modified to:

$$P^{new}(X) = P^{old}(X) \cdot (1 - \text{FOR}) \quad \text{if} \quad X < C \tag{10.2}$$

since the probability of negative megawattage on outage is zero.

However, for a cumulative probability table:

$$\text{CPROB}\ (X) = \int_x^{\infty} P(z)dz$$

Thus, Equation (10.1) can be integrated to yield:

$$\text{CPROB}^{new}(X) = \text{CPROB}^{old}(X)\ (1 - \text{FOR}) + \text{CPROB}^{old}(X - C) \cdot \text{FOR} \quad \text{if} \quad X \geq C \tag{10.3}$$

If $X < C$, then the integration must proceed in two parts:

$$\int_x^{C} P(z)dz + \int_C^{\infty} P(z)dz$$

Performing this analysis yields:

$$\text{CPROB}^{new}(X) = \text{CPROB}^{old}(X)\ (1 - \text{FOR}) + \int_C^{\infty} P^{old}(X - C) \cdot \text{FOR} \tag{10.4}$$

or

$$\text{CPROB}^{new}(X) = \text{CPROB}^{old}(X)\ (1 - \text{FOR}) + \text{CPROB}(0) \cdot \text{FOR}$$

Equations (10.4) and (10.3) are often combined into a single equation along with the *convention* that CPROB(neg.numbers) = CPROB(0) = 1.0. Thus, the recursive convolution formula for the cumulative probability of X MW or more on outage is:

$$\text{CPROB}^{new}(X) = (1 - \text{FOR}) \cdot \text{CPROB}^{old}(X) + \text{FOR} \cdot \text{CPROB}^{old}(X - C) \tag{10.5}$$

and

$$\text{CPROB (neg.numbers)} = \text{CPROB}(0) = 1.0$$

The case for which Equation (10.5) was derived was based on a two-state characterization of a generating unit:

State	MW in Service	Probability
1	C	1 − FOR
2	0	FOR

The two-state characterization typically uses the equivalent random outage rate. In some cases, when high accuracy is demanded, a multistate characterization of a generating unit is used as illustrated in Table 10.6.

The outage state can be described as:

C_N = MW out of service for capacity (derating) state N
FOR_N = random outage rate for capacity (derating) state N

Then Equation (10.5) can be generalized to:

$$\text{CPROB}^{\text{new}}(X) = \sum_{i=1}^{N} \text{FOR}_i \cdot \text{CPROB}^{\text{old}}(X - C_i);$$
$$\text{CPROB}^{\text{old}}(\text{neg}) = \text{CPROB}^{\text{old}}(0) = 1.0 \tag{10.6}$$

Thus, the multistate probability equations are not conceptually different from the two state equations. However, the computational time does increase proportionally to the number of generating unit probability states used.

Typical normalized values for a fossil-steam unit are presented in Table 10.7.

TABLE 10.6 Multistate Generating Unit Characterization

State	MW in Service	MW Out of Service	Probability
1	$C - C_1$	C_1	FOR_1
2	$C - C_2$	C_2	FOR_2
3	$C - C_3$	C_3	FOR_3
•			
•			
•			
N	$C - C_N$	C_N	FOR_N
			1.0000

TABLE 10.7 Typical Normalized Multistate Steam Unit Data

State	MW in Service (%)	MW Out of Service (%)	Probability (%)
1	0	100	9
2	60	40	3
3	75	25	2
4	85	15	7
5	95	5	10
6	100	0	69
			100

The computational procedure for solving either Equation (10.5) or (10.6) begins by establishing a probability megawatt step size. Step size is the megawatt increment of the probability table. In the previous example, the step size was 50 MW. Typically, step size is calculated as the capacity of a power system divided by 1000 (or 5000, in very detailed cases). Thus, there are 1000 entries in the outage probability table. For example, a 5000-MW-capacity power system would have 5-MW step sizes. A 92-MW generating unit would be represented as a 90-MW unit with a 2-MW truncation error, and a 109-MW unit would be represented as a 110-MW unit with a −1-MW truncation error. Most computer algorithms ignore the truncation error because the sum approaches zero for a system with many units.

The procedure begins by initializing the cumulative outage table for no generating units. The cumulative outage table is initialized to unity for 0 MW or more on outage and 0 for all other megawatt values or more on outage. Equation (10.5) or (10.6) is then applied repetitively for each generating unit.

As an example of the application of the recursive convolution formula for the cumulative outage table, consider the system with three generating units described earlier in Section 10.3.

Step 1. Initialize the Cumulative Outage Table to unity, as shown in Table 10.8.

TABLE 10.8 Initial Cumulative Outage Table

MW or More on Outage	Probability of MW or More on Outage
0	1.0
50	0
100	0
150	0
200	0
250	0
300	0
350	0

Step 2. Sequentially convolve generating units A, B, and C using Equation (10.5).

The three-unit system outlined in Table 10.9 illustrates the construction concepts.

A more typical result is illustrated in Figure 10.2. Note that the outage table plots as a nearly straight line on semilog paper. The cumulative outage probability tends to be convex downward.

For electric utilities, the LOLP index is typically on the order of 0.1–1.0 days/year, which is generally equivalent to 15–25% capacity reserve. Since there are approximately 100 peak-load days per year, the daily probability is approximately 0.001–0.01 days/day.

It is important to note that in this probability range, .001–.01, the cumulative outage table is approximately linear on semilog paper. With a good approximation, the logarithm of the cumulative outage table can be represented as:

$$\ln(\text{probability of} \geq X \text{ MW on outage}) = \ln(a_0) - \frac{X}{M} \tag{10.7}$$

or

$$\text{Probability } (\geq X \text{ MW on outage}) = a_0 \exp\left(\frac{-X}{M}\right) \tag{10.8}$$

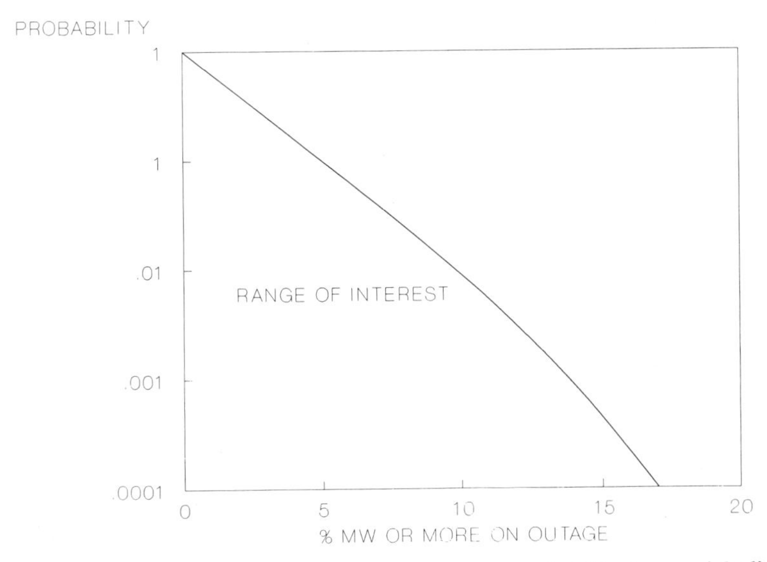

FIGURE 10.2 Typical trend of the cumulative outage probability is a straight line on semi-log paper.

TABLE 10.9 Probability of MW or More on Outage

MW or More on Outage	Initialize Table	Unit A Convolved (50 MW 5% FOR)	Unit B Convolved (100 MW 7% FOR)	Final Results Unit C Convolved (200 MW 10% FOR)
0	1.0	1 • .95 + 1 • .05 = 1.0	1 • .93 + 1 • .07 = 1.0	1 • .9 + 1 • .1 = 1.0
50	0	0 • .95 + 1 • .05 = .05	.05 • .93 + 1 • .07 = .1165	.1165 • .9 + 1 • .1 = .20485
100	0	0 • .95 + 0 • .05 = 0	0 • .93 + 1 • .07 = .07	.07 • .9 + 1 • .1 = .16300
150	0	0	0 • .93 + .05 • .07 = .0035	.0035 • .9 + 1 • .1 = .10315
200	0	0	0 • .93 + .05 • 0 = 0	0 • .9 + 1 • .1 = .1000
250	0	0	0	0 • .9 + .1165 • .1 = .01165
300	0	0	0	0 • .9 + .07 • .1 = .0070
350	0	0	0	0 • .9 + .0035 • .1 = .00035
400	0	0	0	0 • .9 + .000 • .1 = 0

where a_0 is the y intercept value of 0 MW on outage for an exponential approximation and M is the logarithmic slope of the cumulative outage table. This result is helpful in developing approximate methods of generating reliability analysis.

10.4 MAINTENANCE SCHEDULING (LOLP METHOD—SCHEDULED OUTAGES)

In Section 10.3, the random failure characteristics of a generating unit system were evaluated and the convolution method for computing cumulative outage probability tables was presented. The convolution process is performed for all units that are available for service during a given week. If a generating unit is not available for service because of a scheduled outage during that week, it is not included in the convolution process.

Since generating units must be maintained and inspected, the generation planner must schedule planned outages during the year. Several factors entering into this scheduling analysis include: seasonal load-demand profile, amount of maintenance to be done on all generating units, size of the units, elapsed time from last maintenance, and availability of maintenance crews.

In operations planning (with time horizons of up to 2 years), the maintenance scheduling process is very detailed. Operations planning includes such factors as the availability of maintenance crews, elapsed time from last maintenance, amount of maintenance, and the number of units on maintenance at the same time at each generating plant (Skrotzki, 1954).

The short-term maintenance process is continually updated. Should a generating unit experience a long forced outage, a decision may be made to perform some or all of its planned maintenance during the same time as the forced outage. If this decision is made, then the annual maintenance schedule for the power system can be reshuffled to further improve system reliability as well as decrease system production costs.

In longer-range power system planning (2–30 years), maintenance scheduling is usually analyzed with less detail than in operations planning. The principal considerations are:

- Seasonal load-demand profile.
- Planned outage requirements for each generating unit.
- Elapsed time since the last planned outage for each unit.
- Generating unit size.

Planned outage requirements of power plants typically have a cyclical pattern. Major outages of a steam plant are conducted every 4–6 years, during which time extensive inspections and repairs are performed. Major outages may require 5–10 weeks. During interim years, plant inspections and repairs may require only 3–5 weeks to complete.

A traditional long-range planning technique is to schedule the generation maintenance outages to levelize the load plus capacity on outage over a period of a year, as illustrated in Figure 10.3.

A logical maintenance procedure schedules generating units for maintenance so that available percent generation capacity reserve is the same for all weeks. This procedure attempts to have the lowest annual LOLP. Intuitively it works because the weekly (or daily) LOLP is exponentially proportional to the weekly reserve. If the weekly reserve margin is not uniform, then a better schedule could be developed to rearrange maintenance off of the week with the lowest reserve (highest weekly LOLP) and move it to a week with the highest reserve (lowest weekly LOLP). Because of the exponential relationship of LOLP to reserve level, a smaller *annual* LOLP would result. Table 10.10 illustrates two maintenance schedules for a 5-week period. Note that the schedule with a uniform reserve has the lower LOLP.

The maintenance scheduling procedure that seeks to levelize the weekly generation capacity reserves can be expressed as:

$$\text{Procedure} = \text{levelized weekly capacity reserves for each week in the year}$$

$$\left(\frac{\text{installed capacity} - \text{weekly capacity on maintenance} - \text{weekly peak load}}{\text{installed capacity}}\right) \tag{10.9}$$

Note that the denominator is a constant for every week in the year. Thus, if the numerator is to be a constant for every week in the year, as well, then:

$$\text{Constant} = \text{weekly peak load} + \text{weekly capacity on maintenance} \tag{10.10}$$

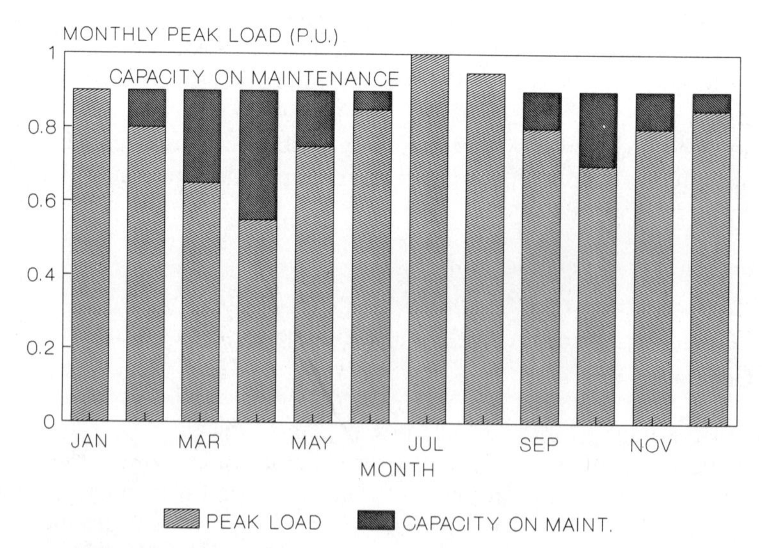

FIGURE 10.3 Maintenance scheduling to levelize reserves.

TABLE 10.10 Maintenance Scheduling Example

Week	Schedule 1 Weekly Reserves	Schedule 1 Weekly LOLP	Schedule 2 Weekly Reserves	Schedule 2 Weekly LOLP
1	20%	.02	15%	.01
2	10%	.005	15%	.01
3	15%	.01	15%	.01
4	15%	.01	15%	.01
5	15%	.01	15%	.01
		.055		.05

Hence, maintenance should be scheduled such that weekly peak load plus weekly capacity on maintenance is a constant. This principle is the basis for a maintenance scheduling process.

A constraint on maintenance scheduling is the minimum and maximum time available between planned outages on generating units. Typically, utilities prefer annual outages with 10–14-month intervals.

While Figure 10.3 illustrates a perfectly levelized maintenance schedule, in practical applications, perfectly levelized load plus capacity on outage is seldom achievable because of discrete generating unit sizes. Therefore, maintenance scheduling must be adapted to account for these effects. The most widely used algorithm for scheduling maintenance consists of four steps:

1. Arrange generating units by size, with the largest first and the smallest last.
2. Schedule the largest generating unit for maintenance during periods of lowest load, consistent with the minimum and maximum time constraints.
3. Adjust weekly peak load by the generating unit capacity on maintenance.
4. Repeat steps 2 through 4 until all generating units are scheduled.

The following example illustrates these steps.

Example (Scheduled Maintenance Algorithm). The power system is comprised of the generating units shown in Table 10.11.

The estimated monthly peak loads are:

Month	J	F	M	A	M	J
Peak Load (MW)	3800	3700	3300	3300	3800	4300

Month	J	A	S	O	N	D
Peak Load (MW)	4300	4400	4100	3600	3600	4000

TABLE 10.11 Maintenance Scheduling Data

Unit	MW Size	Months Since Last Outage (as of January)	Outage Constraints Minimum Months	Outage Constraints Maximum Months	Months of Maintenance
1	1000	9	9	15	2
2	800	6	10	14	1
3	300	10	10	14	1
4	500	0	12	14	1
5	800	11	10	14	2
6	1000	4	9	15	2
7	500	3	10	14	1
8	100	6	10	14	1
9	200	9	10	14	1
10	100	6	10	14	1

Determine the maintenance schedule for these generating units.

SOLUTION. The objective of the maintenance schedule is to levelize the load plus capacity on maintenance over the year.

Rank the order of the generating units by size and compute the maintenance period outage constraint. The results of this are shown in Table 10.12. The calculation procedure is demonstrated in Table 10.13.

Figure 10.4 graphically illustrates how the maintenance schedule fits into the valleys of the monthly peak load. While this example illustrates a maintenance scheduling procedure based upon a monthly model of load and capacity

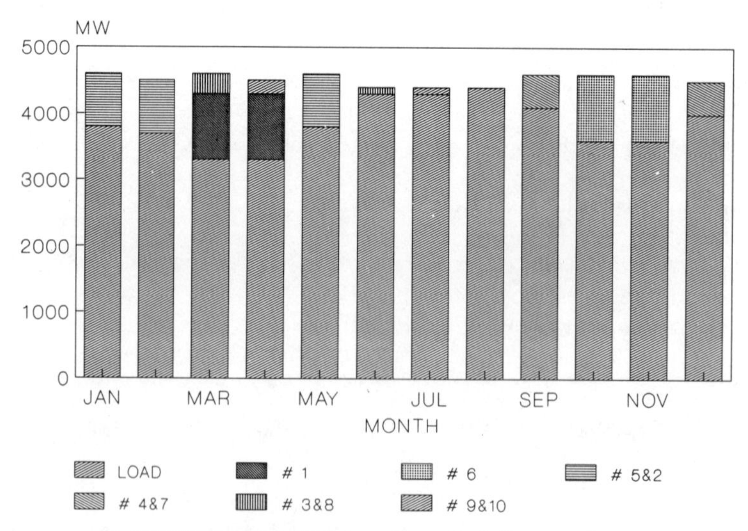

FIGURE 10.4 Maintenance scheduling results.

TABLE 10.12 Scheduling Order of Units with Earliest and Latest Possible Month

			As of January	Outage Constraints			Available Maintenance Period	
Rank	Unit	MW Size	Months Since Last Outage	Minimum Months	Maximum Months	Months of Maintenance	Earliest Month	Latest Month
1	1	1000	9	9	15	2	Jan	Jun
2	6	1000	4	9	15	2	Jun	Nov
3	2	800	6	10	14	1	May	Aug
4	5	800	11	10	14	2	Jan	Mar
5	4	500	0	10	14	1	Nov	Feb
6	7	500	3	10	14	1	Aug	Nov
7	3	300	10	10	14	1	Jan	Apr
8	9	200	9	10	14	1	Feb	May
9	8	100	6	10	14	1	May	Aug
10	10	100	6	10	14	1	May	Aug

TABLE 10.13 Maintenance Scheduling Procedure[a]

Rank No.	MW Size	Unit No.	Duration of Maintenance	Constraint Months	Month of Maintenance
		0			Loads
					Comment:
1	1000	1	2	Jan–Jun	Mar–April
					Comment:
2	1000	6	2	Jun–Nov	Oct–Nov
3	800	5	2	Jan–Mar	Jan–Feb
4	800	2	1	May–Aug	May
5	500	4	1	Nov–Feb	Dec
6	500	7	1	Aug–Nov	Sep
7	300	3	1	Jan–Apr	Mar
8	200	9	1	Feb–May	Apr
9	100	8	1	May–Aug	Jun
10	100	10	1	May–Aug	Jul
Total					

[a]Set up a table of the monthly load plus capacity on maintenance.

Answer:	Unit No.	Maintenance Period
	1	Mar–Apr
	2	May
	3	Mar
	4	Dec
	5	Jan–Feb
	6	Oct–Nov
	7	Sep
	8	Jun
	9	Apr
	10	Jul

outages, most power system reliability analysis models use a weekly outage time and a weekly load model. Hence, maintenance scheduling is usually performed on a weekly basis.

10.5 ANNUAL LOLP CALCULATIONS

Step 1. The first step in computing the annual LOLP is to compute the annual maintenance schedule of generation equipment (Section 10.4).

Step 2. The capacity outage table can then be built based on the recursive convolution algorithm (Equation 10.5) using only the capacity available for service during a particular week.

<table>
<tr><th>J</th><th>F</th><th>M</th><th>A</th><th>M</th><th>J</th><th>J</th><th>A</th><th>S</th><th>O</th><th>N</th><th>D</th><th>J</th></tr>
<tr><td>3800</td><td>3700</td><td>3300</td><td>3300</td><td>3800</td><td>4300</td><td>4300</td><td>4400</td><td>4100</td><td>3600</td><td>3600</td><td>4000</td><td>3800</td></tr>
<tr><td colspan="13">Take first unit, put in largest off-peak valley, test if within outage constraint, and accumulate load and cap on outage</td></tr>
<tr><td>3800</td><td>3700</td><td>4300</td><td>4300</td><td>3800</td><td>4300</td><td>4300</td><td>4400</td><td>4100</td><td>3600</td><td>3600</td><td>4000</td><td>3800</td></tr>
<tr><td colspan="13">Take second unit, put in largest off-peak valley, test if within outage constrating, etc.</td></tr>
<tr><td>3800</td><td>3700</td><td>4300</td><td>4300</td><td>3800</td><td>4300</td><td>4300</td><td>4400</td><td>4100</td><td>4600</td><td>4600</td><td>4000</td><td>3800</td></tr>
<tr><td>4600</td><td>4500</td><td></td><td></td><td></td><td></td><td></td><td></td><td></td><td></td><td></td><td></td><td></td></tr>
<tr><td></td><td></td><td></td><td></td><td>4600</td><td></td><td></td><td></td><td></td><td></td><td></td><td></td><td></td></tr>
<tr><td></td><td></td><td></td><td></td><td></td><td></td><td></td><td></td><td></td><td></td><td></td><td>4500</td><td></td></tr>
<tr><td></td><td></td><td></td><td></td><td></td><td></td><td></td><td></td><td>4600</td><td></td><td></td><td></td><td></td></tr>
<tr><td></td><td></td><td>4600</td><td></td><td></td><td></td><td></td><td></td><td></td><td></td><td></td><td></td><td></td></tr>
<tr><td></td><td></td><td></td><td>4500</td><td></td><td></td><td></td><td></td><td></td><td></td><td></td><td></td><td></td></tr>
<tr><td></td><td></td><td></td><td></td><td></td><td>4400</td><td></td><td></td><td></td><td></td><td></td><td></td><td></td></tr>
<tr><td></td><td></td><td></td><td></td><td></td><td></td><td>4400</td><td></td><td></td><td></td><td></td><td></td><td></td></tr>
<tr><td>4600</td><td>4500</td><td>4600</td><td>4500</td><td>4600</td><td>4400</td><td>4400</td><td>4400</td><td>4600</td><td>4600</td><td>4600</td><td>4500</td><td>3800</td></tr>
</table>

Step 3. After the capacity outage table has been developed, daily outage probabilities can be computed (as in Section 10.2) and accumulated in a weekly index.

Step 4. The process can be repeated for each week in the year (repeat steps 2 and 3).

This algorithm is easily performed by digital computer.

The most time-consuming process (for both manual evaluations and computer processing) is construction of the capacity outage table. Since capacity in service can change weekly as a result of maintenance and generating unit additions during the year, the capacity outage table needs to be reconstructed each week. Fortunately, there are shortcuts for this process.

The table is first constructed using *all* of the units in the power system via Equation (10.5) (recursive convolution). The resulting table (referred to as the "master table") is then stored.

Referring to Equation (10.5), the convolution algorithm is rewritten as:

$$\begin{aligned} \text{CPROB}^{\text{new}}(Z) &= \text{CPROB}^{\text{old}}(Z) \cdot (1 - \text{FOR}_\text{B}) \\ &\quad + \text{CPROB}^{\text{old}}(Z - \text{capacity}_\text{B}) \cdot \text{FOR}_\text{B}; \\ \text{CPROB}(Z) &= 1.0 \quad \text{for} \quad Z \leq 0 \end{aligned} \tag{10.11}$$

where the old/new designation refers to the cumulative table before/after unit B was convolved.

Consider now *subtracting* a unit from the new (or master table), and solving for the old table before unit B was installed.

$$\mathrm{CPROB}^{\mathrm{old}}(Z) = \frac{\mathrm{CPROB}^{\mathrm{master}}(Z)}{(1 - \mathrm{FOR_B})} - \mathrm{CPROB}^{\mathrm{old}}(Z - \mathrm{Capacity_B}) \cdot \frac{\mathrm{FOR_B}}{(1 - \mathrm{FOR_B})};$$

$$\mathrm{CPROB}^{\mathrm{old}}(Z) = 1.0 \quad \text{for} \quad Z < 0 \qquad (10.12)$$

The following example illustrates this calculation using the three-unit example of Section 10.3. The removal of unit C (200 MW; 10% FOR) is presented in Table 10.14.

The procedure of subtracting units on maintenance from the master outage table is a quick means of developing a new outage capacity table. Some caution must be exercised for this procedure, however. Note that the algorithm, Equation (10.8), involves a subtraction of two tables. This subtraction can lead to computational instability from numerical truncation because the subtraction of table values of the same order of magnitude may lead to loss of accuracy and, in some cases, negative probability values. This computation process is particularly unstable when the forced outage rate is greater than 50%. A similar equation for partial outage states has more computational stability problems. Double precisioning the calculations will help, but will not solve this problem for all applications.

10.5.1 Short-Term Load Forecast Uncertainty

Load uncertainty is one contributor to the need for reserves. Load demand for electricity is weather-dependent (Figure 10.5). During the summer, load

TABLE 10.14 Removing a Unit From the "Master" Table

MW or More on Outage	Master Table	Subtraction of Unit C 200 MW, 10% FOR	
0	1.0	1.0	= 1.0/.9 − 1.0 • .1/.9
50	.20485	.1165	= .20485/.9 − 1.0 • .1/.9
100	.16300	.07	= .16300/.9 − 1.0 • .1/.9
150	.10315	.0035	= .10315/.9 − 1.0 • .1/.9
200	.10000	0	= .1000/.9 − 1.0 • .1/.9
250	.01165	0	= .01165/.9 − .1165 • .1/.9
300	.0070	0	= .0070/.9 − .07 • .1/.9
350	.00035	0	= .00035/.9 − .0035 • .1/.9
400	0		

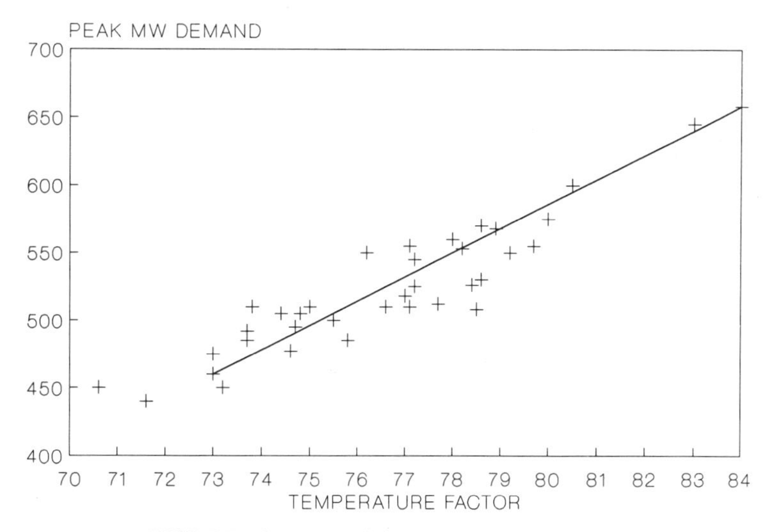

FIGURE 10.5 Load–temperature correlation.

demand increases with high temperatures. Similarly, during the winter, load demand increases with lower temperatures.

The peak temperature, however, is not constant each year, but tends to follow a normal distribution around the average temperature. Figure 10.6 illustrates 30 years of peak summer temperature history for a northeastern city, where the average peak temperature is 86.7°F. According to Figure 10.6, there is a 33% probability that the peak temperature will be 2°F higher than the

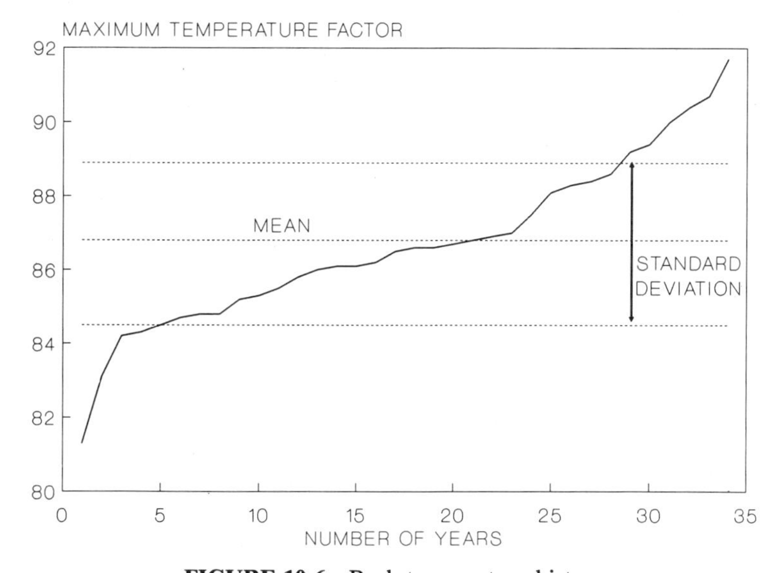

FIGURE 10.6 Peak temperature history.

average. From Figure 10.5, the slope of peak load versus temperature is 18.5 MW/°F. Hence, there is a 33% probability that the load will be 37 MW higher (a 5% higher peak load) than the load for the average summer. Since utilities are expected to provide electric service during a broad range of weather conditions, there must be reserve generation to cover weather-related load uncertainty.

The LOLP discussion has been based on the assumption that hourly demand was specified deterministically. Inclusion of load forecasting uncertainty is easily integrated into the computational procedure, as illustrated in Figure 10.7. At each demand point in the uncertainty distribution, the LOLP is calculated. The equivalent daily LOLP is then determined by weighting the LOLP result at each demand point by the probability distribution value (Garver, 1970).

This procedure involves minor additional computational effort to explicitly represent short-term load forecast uncertainty (Gupta and Yamada, 1972).

10.6 GENERATION ADDITIONS PLANNING

Using the LOLP technique, generation system planners can evaluate generation system reliability and determine how much capacity is required to obtain a specified level of LOLP. As demand grows over time, generation additions are timed such that the LOLP does not exceed the design criterion.

Figure 10.8 plots annual LOLP versus annual peak load for a specific generation system (readily available from LOLP computer program results). As in-

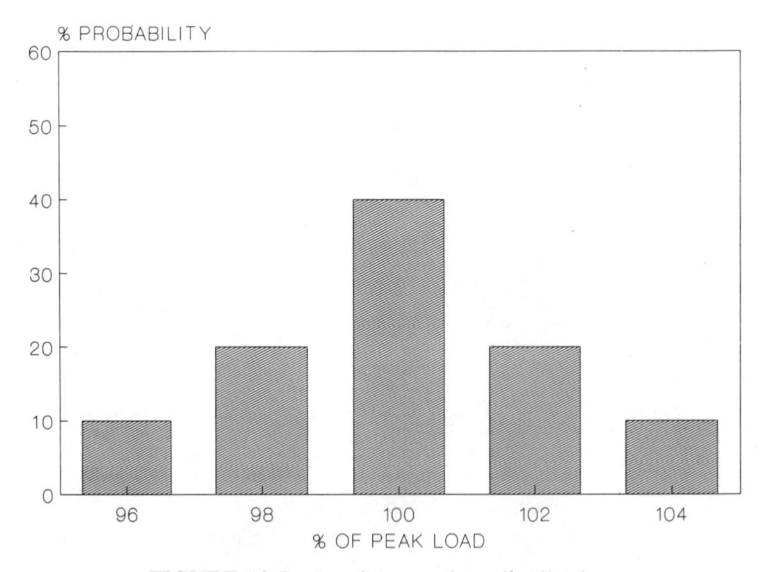

FIGURE 10.7 Load uncertainty distribution.

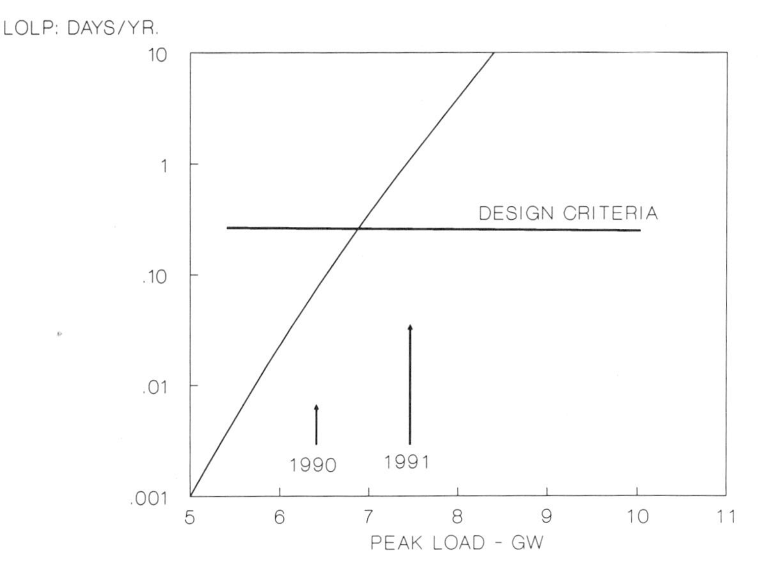

FIGURE 10.8 Influence of peak load on LOLP.

dicated by the graph as an almost straight line on semilog paper, LOLP varies exponentially with load changes. The design criterion in this example case is 0.1 days/year. With the 1990 peak load as indicated on the graph, the generation system is able to meet that load at a level of reliability better than 0.1 days/year. Therefore, no additional capacity is required.

In 1991, annual load growth is predicted to increase the peak demand to an extent to where the generation system cannot maintain the desired 0.1 days/year LOLP. In anticipation, a unit addition would be scheduled for a 1991 installation. What happens to the LOLP versus peak load curve?

With the new unit addition installed, the curve shifts to the right as in Figure 10.9. In 1991, the LOLP decreases from 0.8 days/year to 0.01 days/year with the unit addition. This is well below the desired 0.1 days/year criterion established by the utility system planner.

What was the unit addition worth? Although the first addition decreased the LOLP from 0.8 to 0.01 days/year, it is not consistent to describe the unit worth as 0.79 days/year, because this is the value only for the 1991 peak load. If the peak should vary, then this days/year measure (difference) would change, and the measure would become meaningless. It is more reasonable to measure the worth of a generating unit in megawatts.

The effective load-carrying concept solves this problem by focusing on the horizontal distance (MW) between the two LOLP versus peak-load curves (Garver, 1966). This difference, measured along the horizontal axis, is in megawatts. Since the system is designed to 0.1, the megawatt difference is measured on that axis to be consistent, as shown in Figure 10.10.

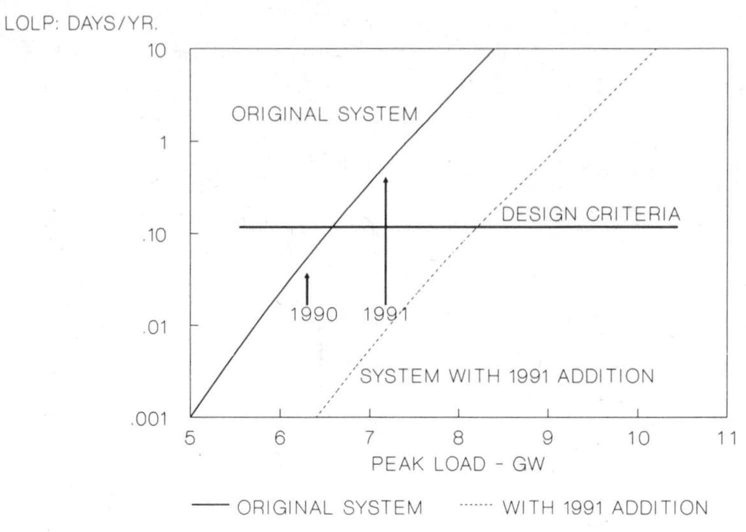

FIGURE 10.9 Influence of peak load on LOLP with new unit addition.

10.7 APPROXIMATE CALCULATION OF EFFECTIVE CAPACITY AND APPLICATIONS

The concept of effective load-carrying capacity of a generating unit is defined as the amount of additional peak load that a generating unit permits a power system to carry at the same LOLP index. While an LOLP computer analysis can evaluate the effective load-carrying capability of a generating unit, an approximate formula is very useful.

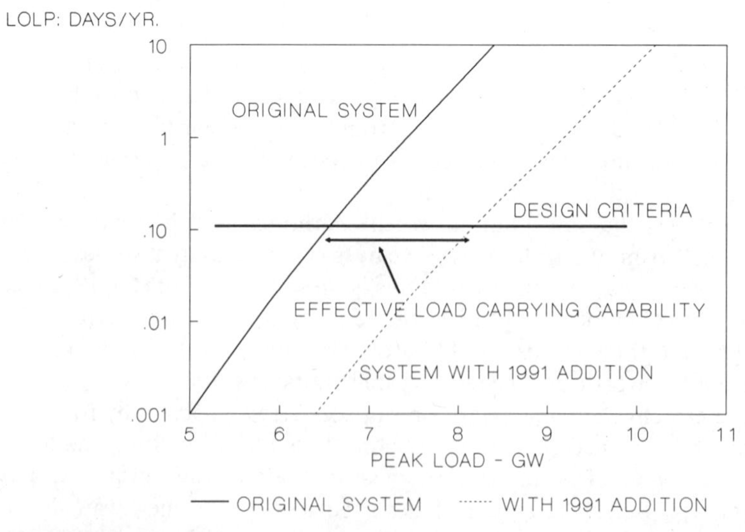

FIGURE 10.10 Effective load-carrying capability of a generating unit.

The principle assumption for deriving an approximate estimate of the effective capacity of a generating unit is that the cumulative outage table has an exponential character in the vicinity of the target LOLP value. Figure 10.11 illustrates this assumption. The assumption of an exponential character can be represented as $\text{CPROB}(Z) = a_0 \exp(-Z/M)$, where M is the characteristic inverse logarithmic slope of the cumulative probability outage table and Z is the megawatt(s) or more on outage. The following discussion presents a derivation of the approximation formula of effective generating unit capability. For a more rigorous derivation, see Garver (1966).

The power system has a total capability SYSCAP, and could serve a load demand, LOAD, prior to an addition of capacity. For this load demand, then, the capacity on outage that would lead to a shortage of capacity is SYSCAP $-$ LOAD $=$ reserve_1. The cumulative probability of this is CPROB (reserve_1) (point A in Figure 10.11).

Consider adding a generating unit of capacity, C, and forced outage rate, FOR. By the recursive convolution formula, the new outage table after the unit is added is:

$$\text{CPROB}^{\text{new}}(\text{reserve}_1) = (1 - \text{FOR}) \cdot \text{CPROB}^{\text{old}}(\text{reserve}_1) + \text{FOR} \cdot \text{CPROB}^{\text{old}}(\text{reserve}_1 - C)$$

or

$$\text{CPROB}^{\text{new}}(\text{reserve}_1) = (1 - \text{FOR}) \cdot \text{CPROB}^{\text{old}}(\text{reserve}_1) + \text{FOR} \cdot \exp\left(\frac{C}{M}\right) \cdot \text{CPROB}^{\text{old}}(\text{reserve}_1) \tag{10.13}$$

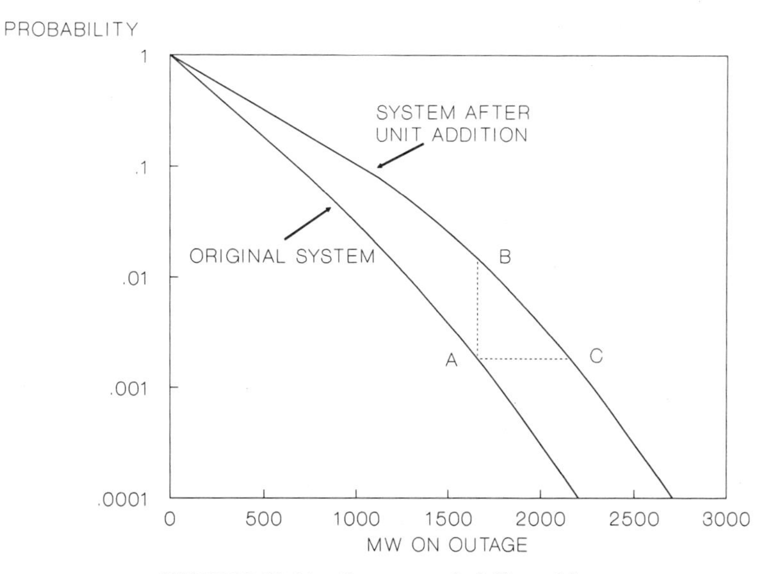

FIGURE 10.11 Outage probability table.

This corresponds to point B in Figure 10.11.

However, we are not interested in the cumulative outage table at reserve_1, but rather the reserve value (reserve_2) at point C, which has the same probability as the original system of point A.

The exponential approximation is again used for calculating:

$$\text{CPROB}^{\text{new}}(\text{reserve}_1) = \text{CPROB}^{\text{new}}(\text{reserve}_2) \cdot \exp\left(\frac{(\text{reserve}_2 - \text{reserve}_1)}{M}\right)$$

Letting $\text{CPROB}^{\text{new}}(\text{reserve}_2) = \text{CPROB}^{\text{old}}(\text{reserve}_1)$ and substituting in Equation (10.11) yields:

$$\exp\left(\frac{(\text{reserve}_2 - \text{reserve}_1)}{M}\right) = (1 - \text{FOR}) + \text{FOR} \exp\left(\frac{C}{M}\right) \quad (10.14)$$

Taking logarithms yields:

$$(\text{Reserve}_2 - \text{Reserve}_1) = M \left\{\ell\text{n}\left[(1 - \text{FOR}) + \text{FOR} \cdot \exp\left(\frac{C}{M}\right)\right]\right\} \quad (10.15)$$

but

$$\begin{aligned} \text{Reserve}_2 &= \text{SYSCAP} + \text{C} - \text{LOAD} - \text{LCC}; \\ \text{Reserve}_1 &= \text{SYSCAP} - \text{LOAD} \end{aligned} \quad (10.16)$$

where LCC is the incremental load that the generating unit permits the system to carry, that is, the load-carrying capability.

Solving for this incremental load, LCC, yields:

$$\text{LCC} = C - M\,\ell\text{n}\left[(1 - \text{FOR}) + \text{FOR} \cdot \exp\left(\frac{C}{M}\right)\right] \quad (10.17)$$

This equation is useful because the effective capability of a generating unit can be analytically evaluated. Valuable insight can be gained by this. For example, Figure 10.12 illustrates the effect of unit size and forced outage. The larger-size 1000-MW generating unit has a much faster decline in the effective load-carrying capability than do two smaller 500-MW units. Thus, from a system reliability perspective, it is better to have two smaller units than one large unit that is twice the size. Subsequent chapters will discuss other factors that influence unit size decisions.

The approximate calculation of the effective load-carrying capability is also useful in maintenance scheduling (Garver, 1972; Stremel, 1981). In Section 10.4, the maintenance schedule was developed from levelizing (over a 1-year period) the sum of the load plus capacity on maintenance (levelized reserves

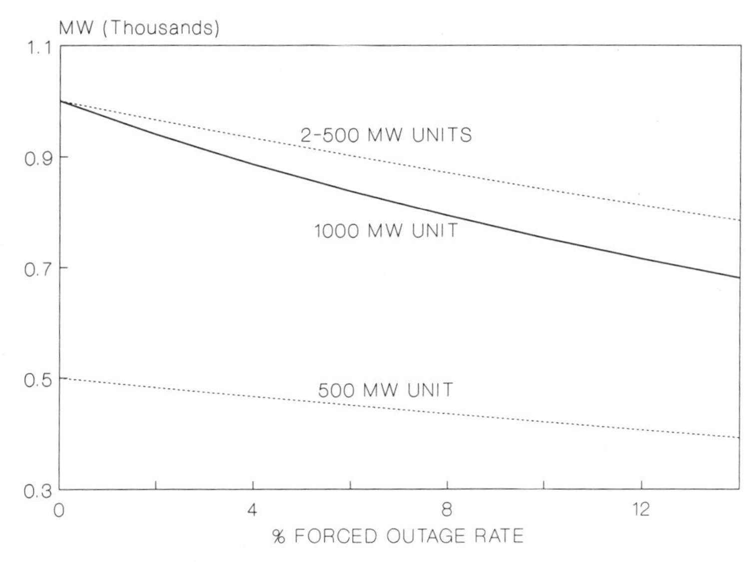

FIGURE 10.12 Effective load-carrying capability as a function of unit size and FOR.

approach). With the approximate formula for effective load-carrying capability, maintenance scheduling can be performed to minimize the risk of capacity deficiencies (LOLP) over the time period of 1 year.

A second potential consideration in maintenance scheduling is the concept of equivalent peak load. Equivalent peak load is a single load that, if encountered for each day during the week, would produce the same LOLP risk contribution as the seven actual daily peak loads.

The equivalent peak load can be derived by assuming the risk proportional to the daily peak load. Then letting:

N days = number of days per maintenance interval (typically 7)
Load (day) = daily peak load
$\overline{\text{Load}}$ = effective weekly peak load
M = M slope

$$N \text{ days} \cdot \exp\left(\frac{\overline{\text{Load}}}{M}\right) = \sum_{\text{day}=1}^{N \text{ days}} \left\{\exp\left[\frac{\text{load(day)}}{M}\right]\right\} \tag{10.18}$$

or

$$\text{Equivalent load} = \overline{\text{Load}} = M \cdot \ln\left\{\frac{1}{N \text{ days}} \sum_{\text{day}=1}^{N \text{ days}} \exp\left[\frac{\text{load(day)}}{M}\right]\right\} \tag{10.19}$$

The equivalent load is then used in the maintenance scheduling algorithm.

The following example illustrates this method using the same data as the example in Section 10.4.

Example (Scheduling Maintenance to Levelized Risk). The power system is comprised of the following generating units:

TABLE 10.15 Generating Unit Data

Unit	MW Size	Months of Maintenance	Forced Outage Rate
1	1000	2	.15
2	800	1	.10
3	300	1	.05
4	500	1	.07
5	800	2	.10
6	1000	2	.15
7	500	1	.07
8	100	1	.05
9	200	1	.06
10	100	1	.05
Total	5300		

The equivalent monthly peak loads are:

Month	J	F	M	A	M	J
Peak Load (MW)	3800	3700	3300	3300	3800	4300

Month	J	A	S	O	N	D
Peak Load (MW)	4300	4400	4100	3600	3600	4000

Determine the maintenance schedule for these generating units. The M slope for this system is 500 MW. Assume in this example that there are *no* maintenance period outage constraints.

SOLUTION. The objective of the maintenance schedule is to levelize the load plus effective capacity on maintenance over the year.

Rank the order of the generating units by size and calculate the effective capability using Equation (10.17), as shown in Table 10.16.

The calculation process is shown in Table 10.17.

Figure 10.13 presents the maintenance schedule resulting from using a levelized LOLP risk method. Figure 10.14 is a comparison of the levelized LOLP method and the levelized reserve method (both cases without outage period constraints). This comparison reveals that the levelized LOLP method conducts more maintenance in the off-season valley months and less during the peak season periods of June through August.

TABLE 10.16 Maintenance Scheduling Priority

Rank	Unit No.	MW Size	Months of Maintenance	FOR	Effective Capability
1	1	1000	2	.15	664
2	6	1000	2	.15	664
3	5	800	2	.10	633
4	2	800	1	.10	633
5	4	500	1	.07	443
6	7	500	1	.07	443
7	3	300	1	.05	280
8	9	200	1	.06	185
9	8	100	1	.05	94
10	10	100	1	.05	94

The levelized reserve method assumes that a 1000-MW unit on outage is equivalent to a peak-load increase of 1000 MW. However, the levelized LOLP method calculates that a 1000-MW unit is equivalent to a load increase of 664 MW. Thus, the levelized LOLP method places more capacity on maintenance during the off season valley months because of the effective capacity calculation.

The benefit of the levelized LOLP risk method is that it results in a maintenance schedule that yields a lower annual LOLP value than the levelized reserve method. Consequently, the levelized LOLP risk method of maintenance scheduling is more widely used within the utility industry.

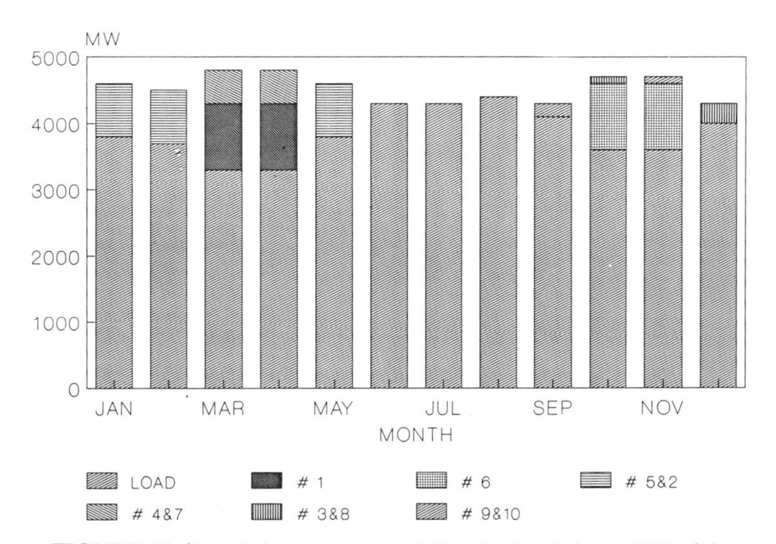

FIGURE 10.13 Maintenance scheduling by levelizing LOLP risk.

TABLE 10.17 Maintenance Calculational Procedure[a]

Rank No. Loads	Unit No.	MW Effective Capacity	Duration of Maintenance	Month of Maintenance
1	1	664	2	Mar–Apr
2	6	664	2	Oct–Nov
3	5	663	2	Jan–Feb
4	2	663	1	May
5	4	443	1	Mar
6	7	443	1	Apr
7	3	280	1	Dec
8	9	185	1	Sep
9	8	94	1	Oct
10	10	94	1	Nov

[a]Set up a table of the monthly load plus effective capacity on maintenance.

10.8 LOSS-OF-ENERGY PROBABILITY

The LOLP method is the most widely used measure of generation system reliability (NERC, 1985). Generally, this method evaluates 365 daily peak loads during the year. (Weekends generally have load demands much less than weekdays and consequently have small contributions to the annual LOLP.) Using 365 peak loads gives a measure that is expressible in terms of days/year expected value of capacity deficiency.

Some utilities prefer to compute the LOLP for every hour of the year

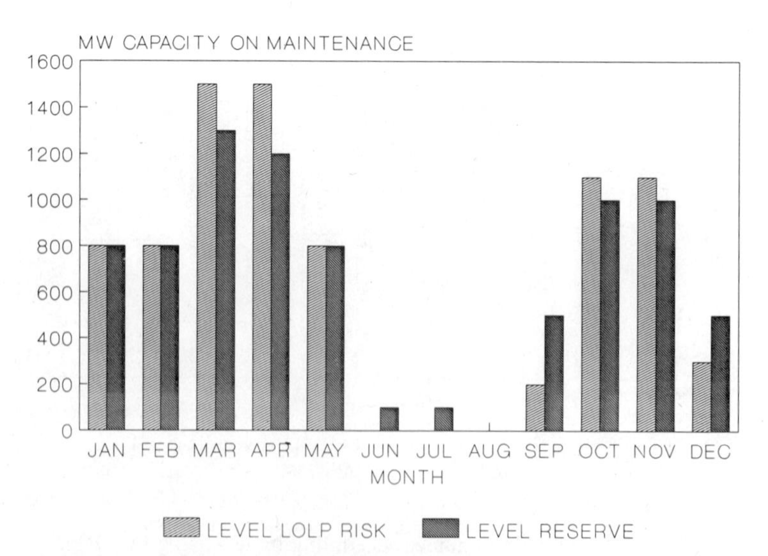

FIGURE 10.14 Maintenance scheduling comparison of LOLP risk versus reserve.

J	F	M	A	M	J	J	A	S	O	N	D
3800	3700	3300	3300	3800	4300	4300	4400	4100	3600	3600	4000
3800	3700	3964	3964	3800	4300	4300	4400	4100	3600	3600	4000
3800	3700	3964	3964	3800	4300	4300	4400	4100	4264	4264	4000
4433	4333	3964	3964	3800	4300	4300	4400	4100	4264	4264	4000
4433	4333	3964	3964	4433	4300	4300	4400	4100	4264	4264	4000
4433	4333	4407	3964	4433	4300	4300	4400	4100	4264	4264	4000
4433	4333	4407	4407	4433	4300	4300	4400	4100	4264	4264	4000
4433	4333	4407	4407	4433	4300	4300	4400	4100	4264	4264	4280
4433	4333	4407	4407	4433	4300	4300	4400	4285	4264	4264	4280
4433	4333	4407	4407	4433	4300	4300	4400	4285	4358	4264	4280
4433	4333	4407	4407	4433	4300	4300	4400	4285	4358	4358	4280

(8760), rather than for 365 daily peak loads. The resulting measure is generally expressed in hours/year expected value of capacity deficiency. The 8760 hourly LOLP method permits the reliability index to be sensitive to changes in daily load profiles due to evolving load profile changes or load management control. The hourly measure also permits the reliability index to be sensitive to alternative designs of energy-limited generating units, such as pumped-storage hydro, and generating units where hourly output is not constant, such as solar and wind-generating units (Albrecht et al., 1981).

Although the LOLP index yields a measure of the expected number of capacity deficiencies per year, it doesn't provide information as to the extent of the capacity deficiency. For example, was the expected deficiency 100 or 1000 MWh? To provide a measure of the extent of the capacity deficiency, an alternative measure to LOLP is the loss-of-energy probability, LOEP. The LOEP index is generally measured in terms of the expected load energy unable to be served divided by system load requirements.

The LOEP method can be illustrated using the example of Section 10.3 and a load of 170 MW. Using the *exact* outage table (*not* the cumulative outage table) of Table 10.1, the unserved energy computation is presented in Table 10.18. Columns 1 and 2 are from Table 10.1. Column 3 is the megawatts reserves available for service; 350 MW capacity less the capacity on outage (column 1). The unserved load, column 4, is the amount that the load exceeds the available capacity (170-MW load less column 3). The expected unserved load is the product of the probability of the outage state, column 2, times the unserved load, column 4. The total expected unserved load is the summation over all the outage states. For this one load,

$$\text{LOEP} = \frac{2.95}{170} = .01735$$

TABLE 10.18 Calculation of LOEP for Load of 170 MW ("Exact" Outage Table)

(1) X MW on Outage	(2) Probability of Capacity on Outage	(3) MW Reserves	(4) Unserved Load MW	(5) Expected Unserved Load (MWh)
0	.79515	350		
50	.04185	300		
100	.05985	250		
150	.00315	200		
200	.08835	150	20	1.767
250	.00465	100	70	0.3255
300	.00665	50	120	0.798
350	.00035	0	170	
400	0	0		0.0595
				2.95

The other 8760-hour loads are evaluated similarly. The hourly expected unserved energy is accumulated for the year and divided by the annual energy.

For large power systems, it can be shown that, to a good approximation, the unserved energy is equal to the M-slope times the hourly LOLP.

Assume that the cumulative outage table can be represented as:

$$\text{CPROB}(Z) = a_0 \exp\left(-\frac{(Z - \text{capacity} + \text{load})}{M}\right) \tag{10.20}$$

where Z = MW capacity or more on outage
capacity = system total capacity
load = system load demand
a_0 = constant
M = characteristic system logarithmic slope M

The hourly LOLP is then:

$$\text{LOLP} = \text{CPROB}\,(Z = \text{capacity} - \text{load}) = a_0 \tag{10.21}$$

The exact outage table is the negative differential of the cumulative table.

$$P(Z) = \frac{a_0}{M} \exp\left(-\frac{(Z - \text{capacity} + \text{load})}{M}\right) \tag{10.22}$$

The expected unserved load energy is the integral of the exact outage probability times the unserved load. This can be expressed as:

Expected unserved energy =

$$\int_{Z\,=\,\text{capacity}\,-\,\text{load}}^{\infty} dz \cdot \left[\frac{a_0}{M} \exp\left[-\frac{(Z - \text{capacity} + \text{load})}{M} \right] \cdot (Z - \text{capacity} + \text{load}) \right] \quad (10.23)$$

This last integral can be rewritten by substituting

$$t = \frac{(Z - \text{capacity} + \text{load})}{M}$$

$$\text{Expected unserved energy} = a_0 M \int_0^{\infty} dt \cdot \exp(-t) \cdot t \quad (10.24)$$

$$= a_0 M = M \cdot \text{hourly LOLP}$$

While the previous example illustrates expected unserved energy for only one load, the same approximation results from a more detailed analysis considering the 8760 loads during the year.

The average load not served can be estimated by:

$$\text{Expected load not served} = \frac{\text{unserved energy (MWh)}}{\text{hourly LOLP (hour)}} \quad (10.25)$$

Using Equation (10.22), the expected load not served is estimated by:

$$\text{Expected load not served} \approx M$$

Thus, the average load not served during the year is equal to the M slope characteristic of the generation system.

10.8.1 Frequency and Duration Method

The frequency-and-duration method is an extension of the LOLP method. It computes the expected length of time of a capacity deficiency (i.e., 1.5 days) and the expected frequently (number of times per year) of a capacity deficiency. The hourly loss of load probability and the frequency and duration are related by:

$$\text{HLOLP} = \text{duration} \cdot \text{frequency} \quad (10.26)$$

To calculate the frequency and duration index, additional generating unit data on the mean-time-to-failure and mean-time-to-repair is required. This data is used to calculate expected repair rate and failure rate.

$$\text{Repair rate per day} = \frac{1}{\text{mean time to repair}}$$

$$\text{Failure rate per day} = \frac{1}{\text{mean time to failure}} \tag{10.27}$$

Generating units can be modeled as two-state devices, failed or operating, with transition rates to describe the rates at which failed units are repaired and operating units fail. The state transition model and appropriate load-demand model are convolved to compute the probability and frequency of capacity shortages. This method is discussed in detail in Billinton and Allan (1984).

Although the frequency-and-duration method provides an additional generation reliability statistic, it has not found widespread utility industry application.

10.9 SUMMARY

This chapter introduced the principles of generation system reliability. Each generating unit is characterized as having capacity, a planned (scheduled/maintenance) outage rate, and a random outage rate (forced outage rate). Deterministic maintenance-planning methods are used to account for planned outage rate characteristics. Probabilistic methods are used to account for random outage rate characteristics. The resulting measure of generation system reliability is the LOLP (or LOLE). It is defined as the expected number of days per year without sufficient capacity to meet the load demand.

REFERENCES

AIEE Committee Report, "Applications of Probability Methods to Generating Capacity Problems," AIEE Transactions, Pt. III, *Power Apparatus and Systems*, Vol. 79, February 1961, pp. 1165–1177.

Albrecht, P. F., L. L. Garver, G. A. Jordan, A. D. Patton, and P. R. Van Horne, "Reliability Indices for Power Systems," *EPRI EL-1773,* Project 1353-1, Final Report, March 1981.

Billinton, R., and R. N. Allan, *Reliability Evaluation of Power Systems*, Plenum Press, New York, 1984.

Calabresse, G., "Generating Reserve Capacity Determined by the Probability Method," *AIEE Transactions*, Vol. 66, 1947, pp. 1439–1450.

Endrenyi, J., *Reliability Models in Electric Power Systems*, Wiley, New York, 1978, Chapter 6.

Garver, L. L., "Effective Load Carrying Capability of Generating Units," *IEEE Transactions on Power Apparatus and Systems*, Vol. PAS-85, August 1966, pp. 910–919.

Garver, L. L., "Reserve Planning Using Outage Probabilities and Load Uncertainties," *IEEE Transactions on Power Apparatus and Systems*, Vol. PAS-89, April 1970, pp. 514–521.

Garver, L. L., "Adjusting Maintenance Schedules to Levelized Risk," *IEEE Transactions on Power Apparatus and Systems*, Vol. 91, September–October 1972, pp. 2057–1963.

Gupta, P. C., and K. Yamada, *"Adaptive Short-Term Forecasting of Hourly Loads Using Weather Information," IEEE Transactions on Power Apparatus and Systems*, Vol. PAS-91, 1972, pp. 2085–2094.

Skrotzki, B. G. A. (ed.), *Electric System Operation*, McGraw-Hill, New York, 1954, pp. 36–41.

NERC, *Reliability Concepts in Bulk Power Electric Systems*, North American Electric Reliability Council, Reliability Criteria Subcommittee, February 1985, p. 8.

Stremel, J. P., "Maintenance Scheduling for Generation System Planning," *IEEE Transactions on Power Apparatus and Systems*, Vol. PAS-100, March 1981, pp. 1410–1419.

PROBLEMS

1 An electric utility had the following load demand and capacity in 1989.

Year	Peak Load (MW)	Generating Capacity (MW)	Generation Reserve Margin(%)
1989	5000	6000	20

The load demand is projected to increase by 4.88% each year. The utility desires to maintain a generation reserve margin no less than *15%* and wants to add 600-MW units. Since the utility wants to add only 600-MW units, in how many years will the utility need to add capacity so that the generation reserve margin is no less than 15%? Plan only to the year 2000.

2 The cumulative outage probability table for a system with three generating units

	Rating	Forced Outage Rate
Unit A	100	.01
Unit B	150	.02
Unit C	200	.03

is given in the following table:

MW or More on Outage	Probability of $\geq X$ MW on Outage
0	1.000
50	.058906
100	.058906
150	.049400
200	.030194
250	.001088
300	.000894
350	.000600
400	.000006
450	.000006
500	0

a Compute a new outage table for the addition of a 150-MW unit with a 10% forced outage rate.

b For a load of 425 MW, what is the probability that there is insufficient capacity to supply this load demand.

3 Figure 10.15 presents results from several LOLP studies and plots the LOLP versus system peak load demand. A new generating unit has the following characteristics:

MW size	200 MW
Forced outage rate	10%

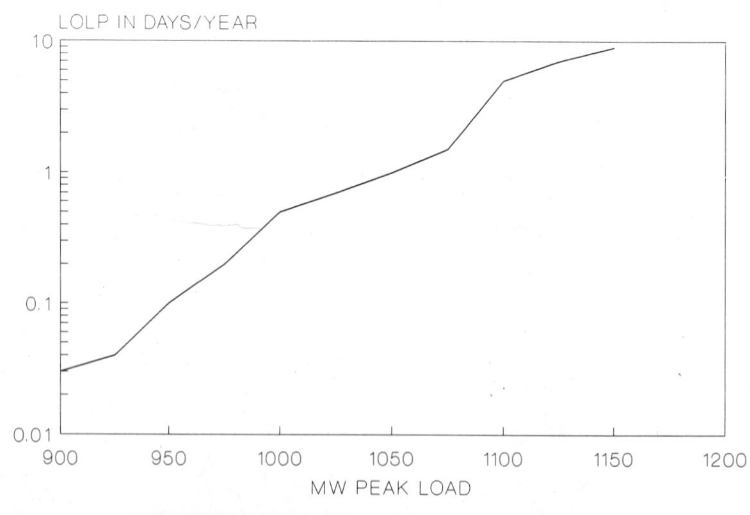

FIGURE 10.15 LOLP versus system peak load.

The utility desires an LOLP of 0.5 days per year.

a From Figure 10.15, estimate the *M* slope.

b Estimate the effective load-carrying capability of the new generating unit.

4 Use the attached cumulative outage table and a total capacity of 1245 MW. Assume no planned maintenance. Compute the annual LOLP, assuming that the load shape is:

Days per year	Load – MW
10	1000
20	950
60	900
170	850
105 (weekend)	600
365 days/year	

Cumulative Outage Table

Entry	MW Capability	Probability
1	0.	.10000000E 01
2	10.0	.23089914E 00
3	20.0	.23089914E 00
4	30.0	.19966581E 00
5	40.0	.19966581E 00
6	50.0	.19927025E 00
7	60.0	.15218244E 00
8	70.0	.15216054E 00
9	80.0	.15026560E 00
10	90.0	.13457266E 00
11	100.0	.13454844E 00
12	110.0	.10155818E 00
13	120.0	.70166211E-01
14	130.0	.68844283E-01
15	140.0	.66608478E-01
16	150.0	.66591806E-01
17	160.0	.17520338E-01
18	170.0	.15598319E-01
19	180.0	.13607199E-01
20	190.0	.12868856E-01
21	200.0	.12203024E-01
22	210.0	.89327460E-02
23	220.0	.75861027E-02

(continued)

Entry	MW Capability	Probability
24	230.0	.71337563E-02
25	240.0	.60794676E-02
26	250.0	.60255665E-02
27	260.0	.39835972E-02
28	270.0	.19811676E-02
29	280.0	.18797913E-02
30	290.0	.17323501E-02
31	300.0	.17035731E-02
32	310.0	.61387288E-03
33	320.0	.48041825E-03
34	330.0	.42280331E-03
35	340.0	.37658491E-03
36	350.0	.33464912E-03
37	360.0	.25167150E-03
38	370.0	.16834439E-03
39	380.0	.14460853E-03
40	390.0	.11924176E-03
41	400.0	.11566989E-03
42	410.0	.72816920E-04
43	420.0	.28637519E-04
44	430.0	.25573323E-04
45	440.0	.22132621E-04
46	450.0	.20389648E-04
47	460.0	.10911952E-04
48	470.0	.75434750E-05
49	480.0	.63280232E-05
50	490.0	.53433258E-05

5 The power system is comprised of the following generating units:

			Outage Constraints		
Unit	MW Size	Months Since the Last Outage (as of January)	Minimum Months	Maximum Months	Months of Maintenance
1	1000	6	9	15	2
2	800	6	10	14	1
3	300	2	10	14	1
4	500	6	10	14	1
5	800	10	10	14	2
6	1000	8	9	15	2

Unit	MW Size	Months Since the Last Outage (as of January)	Outage Constraints Minimum Months	Maximum Months	Months of Maintenance
7	500	6	10	14	1
8	100	6	10	14	1
9	200	11	10	14	1
10	100	6	10	14	1

The *M* slope for this system is 500 MW. The estimated peak monthly loads are:

Month	J	F	M	A	M	J
Peak Load (MW)	3800	3700	3300	3300	3800	4300

Month	J	A	S	O	N	D
Peak Load (MW)	4300	4400	4100	3600	3600	4000

Determine the maintenance schedule for these generating units using the levelized LOLP method.

6 A power system has an installed *capacity of 3000 MW* in year *1990*. The utility is strongly summer peaking, and the winter loads are less than 70% of the summer load. The utility can do all of the planned maintenance during the winter. Hence, the winter loads can be neglected. The summer load is projected to be (assume the summer peak load occurs 100 days/year):

Year	Peak Load (MW)	Days/Year
1990	2050	100
1991	2150	100
1992	2250	100
1993	2350	100
1994	2450	100
1995	2550	100
1996	2650	100
1997	2750	100

The cumulative outage probability is presented in the following table:

MW More on Outage	Probability (Original System)
0	1.0
100	.8
200	.6
300	.4
400	.23
500	.09
600	.05
700	.03
800	.01
900	.008
1000	.004
1100	.002
1200	.001

a What is the annual LOLP in each year?

b The utility is planning to add 300-MW units with 10% forced outage rates. The utility has a LOLP criterion of 2 days/year. in what years does the utility need to add 300-MW units?

c What is the LOLP after addition of the 300-MW units?

d What is the effective load-carrying capacity of the 300-MW units at 2 days/year LOLP?

(You can determine your answer by using semilog graph paper.)

11

GENERATION SYSTEM RELIABILITY—II

This chapter expands on generation system reliability analysis and includes topics of optimum reliability levels, emergency operations procedure criteria, multiarea reliability evaluation with transmission constraints, Monte Carlo reliability evaluations, and Gaussian approximations.

11.1 ELECTRICITY INTERRUPTION IMPACTS

This section examines the consumer costs from electricity service interruptions. With large-scale electrification in industrial and commercial establishments, the cost of an interruption is significant. Consumer-interruption cost data can be used as one factor in power system planning analysis for generation, transmission, and distribution equipment addition decisions.

Economic loss due to power interruption can be assessed through microeconomic analysis of the consumer or a utility survey of the consumers. A survey of consumers can be phased to evaluate three value interpretations: (1) what is the total direct consumer cost of an outage, (2) what would the consumer be willing to pay for "backup" service that would eliminate any outage, and (3) what payment would the consumer accept for service that had a higher outage frequency. Conceptually, an economically motivated consumer would respond identically to each of these survey value interpretations. However, surveys reveal that consumers, especially residential consumers, place a higher value on reliable service (interpretation 2) than the direct cost associated with the outage (interpretation 1). These consumers also desire a larger payment for poorer reliability (interpretation 3) than the actual incurred cost of the interruption (interpretation 1).

Interruption cost has a dollar per interruption component as well as a cost component that varies with interruption duration. The interruption duration cost dependence is not necessarily a linear function of duration. Since utility outage duration is typically 4 hours in length, the 4-hour outage cost is a convenient value for reliability analysis. The cost is then quoted on a cost per unserved energy, $/kWh.

Economic loss to the industrial class of customers is in the form of production losses (wages and salaries), inventory losses, and repair of damaged equipment. A summary of surveys on the cost of short-term industrial interruptions is presented in Table 11.1 (The National Electric Reliability Study, 1981).

According to these surveys, the cost of interruption has a very broad distribution, with an average of 5.7 $/kWh interrupted.

Commercial-class short-term interruption costs have a range as broad as that of the industrial class. Office buildings typically have a high cost as computers, typewriters, and office machinery stop during an interruption, thus curtailing higher-paid office work. Table 11.2 presents a survey of four studies (The National Electric Reliability Study, 1981).

Residential-class interruption costs have a much lower short-term interruption cost, ranging from an Ontario Hydro study reporting 0.2$/kWh to an average outage cost value of 4 $/kWh (Stephenson and Walters, 1987).

The composite consumer costs of a service interruption can vary over a broad range. The 1977 New York City blackout involving a 5750-MW loss for 25 hours had an economic impact of $345 million (in 1977$) (National Electric Reliability Study, 1981). Expressed in 1987 dollars and assuming a 70% daily load factor, the interruption cost is 6.8$/kWh.

Note that the reported cost to the consumer of an interruption is approximately 100 times the average price of electricity (6$/kWh interruption cost vs. 6¢/kWh electricity price). These data can be applied to a broad range of benefit–cost analyses in generation, transmission and distribution system design.

TABLE 11.1 Industrial Service Interruption Costs

Source	Study Year	Value 1987 ($/kWh)
Modern manufacturing	1969	3.0
Gannon (IEEE)	1971	17.1
Telson	1973	3.5
IEEE	1974	8.2
Kaufman	1975	2.0
Telson (New York)	1975	2.8
Environmental Analysis	1975	2.2
Meyers/SRI	1976	2.8
Ontario Hydro	1976	7.4
Yabroff/SRI	1980	13.2
Average		5.7

TABLE 11.2 Commercial Interruption Costs

Source	Study Year	Value 1987 ($/kWh)
Gannon (IEEE)	1975	18.7
Meyers (SRI)	1976	1.6
Ontario Hydro	1979	2.0
Yabroff (SRI)	1980	1.0
Average		5.8

11.2 OPTIMUM RELIABILITY LEVEL

The previous chapter has shown how to calculate the annual LOLP. However, to what value of LOLP should the system be designed?

One design approach is to base the design target on a historical reveiw. The annual LOLP for the last 10 years is calculated using actual load demands, forced outage rates, and maintenance schedules. Next, power system operations personnel and/or utility customers are interviewed to determine which, if any, of these historical years they believed generation system reliability to be inadequate. Contrasting the years of inadequacy with the calculated LOLP values provides a rough guide to the desired LOLP index.

Another approach is an analytical one. The total cost of electricity for several different levels of the LOLP index is calculated. The LOLP value that provides the lowest total cost is selected. This procedure consists of four steps:

1. Determine utility cost to improve reliability.
2. Determine cost savings to electricity-consuming customers for improved reliability.
3. Compute total cost as the sum of steps 1 and 2.
4. Find the minimum cost by repeating steps 1, 2, and 3 for alternate reliability levels.

The following example illustrates the calculation. First, a series of LOLP cases would be performed on the power system for a given peak load. Various amounts of peaking-type capacity would be added to adjust the LOLP. Peaking-type capacity, such as combustion turbines or load management, has low capital cost, and typically operates up to 1500 hours/year on a utility load curve. Because of low capital cost, the most economic way to regulate the annual LOLP values is by adding or subtracting peaking capacity MW (see Chapter 15).

Assume that in the base case the annual LOLP is 10 days/year. If the M slope is 217 MW, then approximately 500 MW of additional combustion turbines (10 combustion turbines of 50 MW), or other peaking-type capacity or demand reductions, lead to an LOLP of one day per year. The unserved energy at 10 days LOLP is 13,020 MWh [assuming a 6-hour effective peak-load dura-

TABLE 11.3 Example 11.1: Utility Cost to Improve Reliability (*M* SLOPE 217 MW AND 6-HOUR PEAK DURATION)

Days/Year LOLP	Unserved Energy (MWh)	Additional GT Capacity	Additional Annual Capital Cost Charges @ 250 $/kW M/year[a]	Additional Annual GT Fuel Cost @ 50 $/MWh M/year	Additional Total Cost $/year M/year
10.0	13,020	Base case (0)		Base case (0)	Base case (0)
3.26	4,114	250	12.5	.445	12.945
1.0	1,302	500	25	.585	25.585
0.32	411	750	37.5	.630	38.130
0.1	130	1,000	50	.644	50.644

[a]M = millions.

tion and using Equation (10.22)]. The results are shown in Table 11.3 for five LOLP values. For example, the unserved energy at a LOLP of 10 days/year is

$$10 \text{ days/year} \cdot 6 \text{ hours/day} \cdot 217 \text{ MW slope} = 13{,}020 \text{ MWh}$$

Peaking capacity typically costs approximately 250 $/kW (1987 $). The annual owning cost based on an annual levelized fixed charge rate of 20% is 50 $/kW/year. Annual capital cost charges are then $25M/year for a 500 MW addition.

The gas turbines would operate to serve the energy that was previously unserved. Thus, the fuel cost (at 50$/MWh levelized operating cost) is computed (for a 1.0 LOLP) as

$$(13{,}020 - 1302) \text{ MWh} \cdot 50 \text{ \$/MWh} = \$.585\text{M/year}$$

The incremental cost of supply is the sum of the capital and fuel cost of the additional combustion turbines.

Electricity consumers benefit from improved LOLP because they have less unserved energy. During periods of unserved demand, industrial and commercial production ceases, causing an economic loss. Table 11.4 presents calculations of the interruption cost to consumers for various LOLP values based on a levelized 5$/kWh interruption cost.

The total cost is the cost to the utility plus cost to the consumer. These values are added together in Table 11.5, and, as components, are presented in Figure 11.1. Note that there is an optimal cost justified value of LOLP near one day per year. Consumer cost decreases as reliability increases (LOLP decreases) because the consumer incurs less outage-related costs. Utility cost increases linearly (on a semilog scale) with increasing LOLP. The total cost to

TABLE 11.4 Consumer Cost

LOLP (Days/Year)	Unserved Energy (MWh)	Cost to Consumers @ 5$/kWh (M/year)
10	13,020	$65.10
3.16	4,114	20.57
1.0	1,302	6.51
0.32	411	2.06
0.1	130	0.65

TABLE 11.5 Total Electricity Cost to the Consumer

	Millions of $ per Year		
LOLP	Utility Cost	Consumer Cost	Total Cost
10	Base (0)	65.10	65.10
3.16	12.945	20.57	33.52
1.0	25.585	6.51	32.10
0.32	38.130	2.06	40.19
0.1	50.644	0.65	51.29

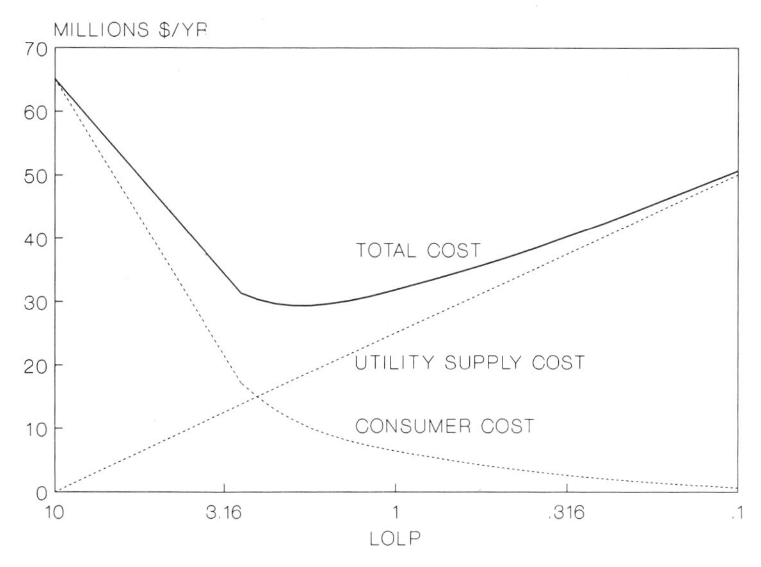

FIGURE 11.1 Total LOLP cost to the consumer is the sum of the utility cost and consumer cost due to outage.

the consumer is the sum of the utility cost (which the consumer pays for through the electricity bill) and the consumer cost due to outages.

These results can be generalized. The total consumer cost of electricity can be approximated for reserve margin analysis as:

$$\text{Total consumer cost (LOLP)} = \text{HLOEP} \cdot \text{outcost} + \text{annual capacity charges} \cdot \text{reserve margin (LOLP)} + \text{other factors (LOLP)} + \text{constant} \tag{11.1}$$

where HLOEP = annual hourly loss of energy probability
outcost = consumer cost of unserved energy in \$/MWh
Annual capacity charges = annual cost of providing peaking capacity in \$/KW/year
Reserve margin (LOLP) = reserve margin required as a function of LOLP
Other factors (LOLP) = other factors (usually small) as a function of LOLP, including peaking capacity fuel and O&M costs
Constant = a constant that accounts for all other non-reserve margin (or LOLP) related costs

The optimum reserve margin can be found by differentiating the total consumer cost as a function of the LOLP index and setting the derivative to zero. This yields:

$$0 = \frac{\partial\ \text{HLOEP}}{\partial\ \text{LOLP}} \cdot \text{outcost} + \text{annual capacity charges} \cdot \frac{\partial\ \text{reserve margin}}{\partial\ \text{LOLP}} \tag{11.2}$$

If we assume that the LOLP characteristic can be well approximated by an exponential curve, LOLP = exp(−reserve margin/M), and that the hourly loss of energy probability is equal to the daily LOLP times the M slope times the equivalent peak hours of load demand, then this yields:

$$M \cdot \text{equivalent peak hours} \cdot \text{outcost} - \text{Annual capacity charges} \cdot \frac{M}{\text{LOLP}_{opt}} = 0 \tag{11.3}$$

where M = characteristic inverse logarithmic M slope
Equivalent peak hours = equivalent daily peak-load duration in hours
LOLP_{OPT} = optimum LOLP value

Rearranging yields:

$$\text{LOLP}_{\text{opt}} = \frac{\text{annual capacity charges}}{\text{equivalent peak hours} \cdot \text{outcost}} \tag{11.4}$$

For the example just examined, this would result in:

$$\text{LOLP}_{\text{OPT}} = \frac{250\ \$\text{kW/year} \cdot (.20)}{6\ \text{hours} \cdot 5\ \$/\text{kWh}} = 1.7\ \text{days/year}$$

This example used the M slope equation along with the unserved energy equation for calculating numerical values. More precise values could be obtained using detailed LOLP computer programs. However, the trends would be similar.

11.3 SUPPLEMENTAL MEASURES OF RELIABILITY

Although the most common measurement index of system reliability has been the LOLP, there has been a need to obtain a physical interpretation of this index to the expected emergency operating procedures of a utility, such as the number of voltage reductions or emergency purchases from neighboring utilities (Moisan and Kenney, 1976).

The LOLP measure simply evaluates the expected number of occurrences when the operating reserve is equal to or less than zero. Operating reserves are defined as: Operating reserves = installed capacity − capacity on outage − load demand. In practice, however, utilities would not let operating reserves drop to zero without initiating emergency actions to prevent the curtailment of electric service. Table 11.6 presents a typical list of emergency actions.

TABLE 11.6 Typical Emergency Operating Procedures

Step	Procedure	Effect of Procedure
1	Emergency purchases	Increase capacity
2	10-minute reserve to zero	Allow operating reserve to decrease to largest generating unit
3	5% voltage reduction	Load relief ≈ 3% peak load
4	Industrial interruptible load curtailment	Load relief ≈ 4% peak load
5	General public appeals	Load relief ≈ 1.5% peak load
6	Operating reserve to zero	Allow operation reserve to decrease to zero
7	8% voltage reduction	Load relief ≈ 1% peak load
8	Customer disconnection	Rotation of disconnection among customers

The methodology for calculating the expected number of emergency actions per year is essentially the same procedure as that used for LOLP computations. Figure 11.2 pesents a graph of the cumulative outage table as a function of "MW or more on outage." Megawatt(s) or more on outage is expressed in terms of % or *less* operating reserves. The conventional LOLP calculation determines the LOLP at zero or less operating reserves.

In Figure 11.2, the first emergency operating procedure (EOP) is emergency purchases when utility operating reserves decrease to 10% or less. Assume that a perfectly reliable 3% emergency purchase can be contracted. From Figure 11.2, this has the probability of .01 of occurring (point 1*a*). After executing the emergency purchase, the utility has increased its operating reserve by 3% (point 1*b*) to 13%. The new cumulative outage curve (after the purchase) is exactly parallel to the old curve, except that it is displayed by 3% to the left. This follows by noting that from Equation (10.5), the new outage probability table is the same as the old probability table when the unit being convolved has a zero outage rate (perfectly reliable).

The next action point is (from Table 11.6) when the operating reserves decrease to the largest generating unit. In this example, suppose that the largest generating unit represents 5% of the operating reserves. If the operating reserve were to decrease to 5% (point 2 in Figure 11.2), the utility would need to take additional action. From Table 11.6, the next action is a 5% voltage reduction, which provides a 3% load reduction, and, consequently, a 3% increase in operating reserves (point 3). The probability of the voltage reduction is .001 days/day.

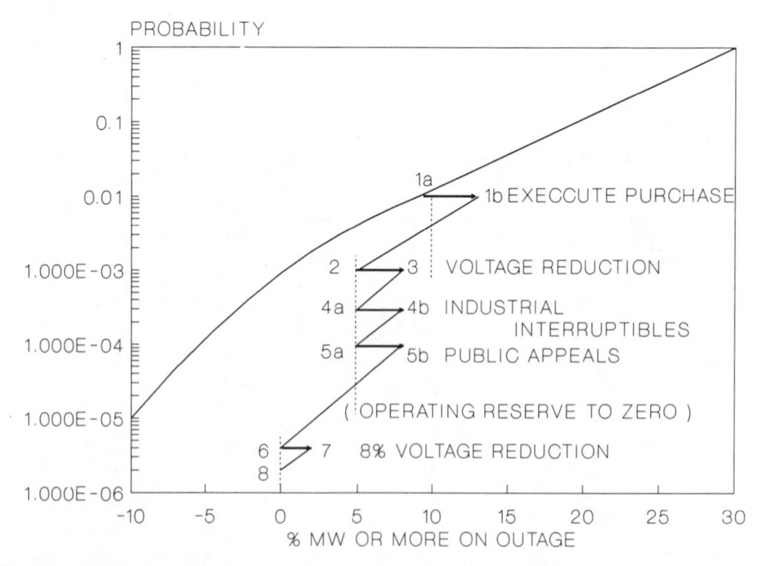

FIGURE 11.2 The cumulative outage table is used to calculate the frequency of emergency operating procedures.

The emergency operating procedure is repeated as above for steps 4 and 5. In step 6, the operating reserve is allowed to reduce to zero. After this step, the next step is an 8% voltage reduction.

Following the voltage reduction, rotation of customer disconnections would begin. This can be executed in several strata: 10, 20, and 30% of the customers.

This procedure illustrated the computation for 1 day (or 1 hour). This procedure can be repeated for the other 365 days (or 8760 hours) and accumulated into an annual value for an expected number of emergency actions per year. A typical result is illustrated in Figure 11.3, where the expected number of actions is contrasted with the LOLP index. Some utilities have established their LOLP index based on the results of emergency operating procedures. For example, if the criterion is established as .1 days/year customer appeals for electricity conservation, this would correspond to a traditional .2 days/year LOLP index using the Figure 11.3 data.

The results of Figure 11.3 are illustrative only. The actual amount of load relief derived from each emergency action varies for each utility and depends on the composition of the load. Field tests have illustrated that loads on some utility feeders can increase following a voltage reduction. Load relief from public appeals can be variable as well, and may diminish as increased frequency of use renders customers callous to requests. Utility-specific load and customer research is required to establish these values.

Example 11.2. Table 11.7 presents the cumulative outage data of a 1245 MW utility whose largest unit is 50 MW. Compute the probability of having to

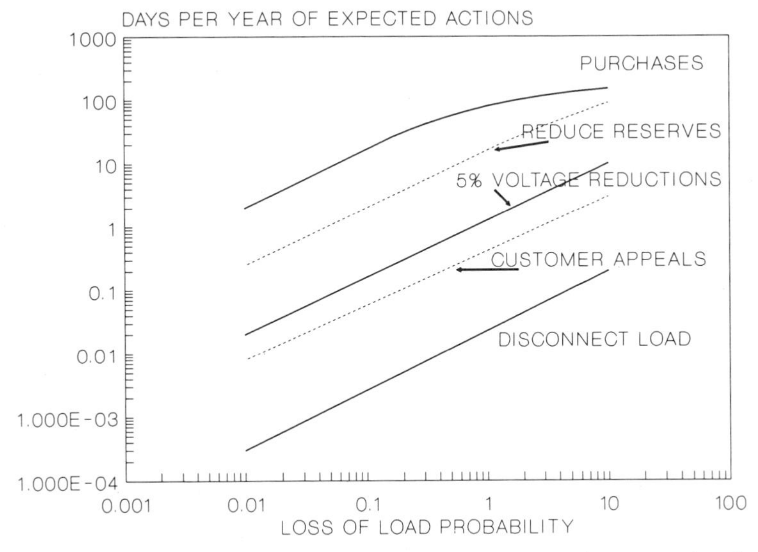

FIGURE 11.3 Annual frequency of emergency procedures is proportional to the LOLP.

TABLE 11.7 Cumulative Outage Table

Entry	MW Capability	Probability
1	0.	.10000000E 01
2	10.0	.23089914E 00
3	20.0	.23089914E 00
4	30.0	.19966581E 00
5	40.0	.19966581E 00
6	50.0	.19927025E 00
7	60.0	.15218244E 00
8	70.0	.15216054E 00
9	80.0	.15026560E 00
10	90.0	.13457266E 00
11	100.0	.13454844E 00
12	110.0	.10155818E 00
13	120.0	.70166211E-01
14	130.0	.68844283E-01
15	140.0	.66608478E-01
16	150.0	.66591806E-01
17	160.0	.17520338E-01
18	170.0	.15598319E-01
19	180.0	.13607199E-01
20	190.0	.12868856E-01
21	200.0	.12203024E-01
22	210.0	.89327460E-02
23	220.0	.75861027E-02
24	230.0	.71337563E-02
25	240.0	.60794676E-02
26	250.0	.60255665E-02
27	260.0	.39835972E-02
28	270.0	.19811676E-02
29	280.0	.18797913E-02
30	290.0	.17323501E-02
31	300.0	.17035731E-02
32	310.0	.61387288E-03
33	320.0	.48041825E-03
34	330.0	.42280331E-03
35	340.0	.37658491E-03
36	350.0	.33464912E-03
37	360.0	.25167150E-03
38	370.0	.16834439E-03
39	380.0	.14460853E-03
40	390.0	.11924176E-03
41	400.0	.11566989E-03
42	410.0	.72816920E-04
43	420.0	.28637519E-04
44	430.0	.25573323E-04
45	440.0	.22132621E-04
46	450.0	.20389648E-04

TABLE 11.7 Cumulative Outage Table

Entry	MW Capability	Probability
47	460.0	.10911952E-04
48	470.0	.75434750E-05
49	480.0	.63280232E-05
50	490.0	.53433258E-05

execute each of the "emergency operating procedures" of Table 11.8 when the peak load is 1000 MW and when the operating reserve is equal to that of the largest unit.

SOLUTION:

Step 1

Load at first procedure = 1000 MW

Reserve requirement = (largest unit) MW

Compute capacity on outage to initiate a purchase = (1245 − 1000) − 50 = 195

Frequency of purchase = probability (195) = .012203

Step 2

Capacity after purchase = 1245 + 100 = 1345

What capacity on outage would cause the next action?

Capacity on outage = (1345 − 1000) − 50 = 295

Frequency of voltage-reduction action = probability (295) = .001703

Step 3

Capacity after voltage reduction = 1345

Load after Voltage Reduction = 1000 − 30 = 970

TABLE 11.8 Emergency Operating Procedures for the Example

Step	Activity
0	Maintain reserve of at least the largest unit (50 MW)
1	Purchase capacity via tielines (100 MW)
2	5% voltage reduction: benefit 30 MW
3	Curtail large interruptible loads: benefit 40 MW
4	Reduce operating reserve to zero (50 MW)
5	Disconnect 20% of residential load: benefit 60 MW

Capacity on outage to cause the next action = (1345 − 970) − 50 = 325
Frequency of large interruptible loads = probability (325) = .0004228

Step 4

Capacity after large interruptible loads disconnection = 1345
Load after disconnection = 970 − 40 = 930
Capacity on outage to cause next action = (1345 − 930) − 50 = 365
Frequency of operating reserves to zero = probability (365) = .000168

Step 5

Capacity = 1345
Load = 930
Operating reserve = 0
Capacity on outage to cause next action = (1345 − 930) = 415
Frequency of 20% residential disconnect = probability (415) = .0000286

11.4 MULTIAREA RELIABILITY CALCULATION

In the previous sections of this chapter, the reliability of an electric utility or power pool was analyzed without explicit consideration of interconnections with neighboring utilities and power pools. Interconnections can play a significant role in reducing the reserve requirements of utilities and power pools.

Power system reliability is generally improved by interconnection with other power systems. When other power systems have surplus capacity, the surplus capacity can be used by transmitting power over the interconnections to the utility in need.

Interconnections work well because of the diversity of outages and load demands. One utility probably will not have a large amount of capacity on forced outage during the same week as a neighboring utility. A utility or power pool may experience a specific peak load demand several hours earlier or later than another utility.

When interconnected, each utility can maintain the same generation system reliability as on a non-interconnected basis, but with a lower reserve margin (Adamson et al., 1977).

Multiarea reliability analysis is typically based on a "transportation model" of the interconnection. In this case, the interconnection is specified as having a fixed megawatt transfer limit. This is an approximation of a true interconnection, which may consist of several transmission lines connecting the utility networks. The actual transfer limit is dependent on several factors, including transmission network characteristics and configuration, system load demands, and power output of the generating units on line.

The multiarea calculation using the "transportation model" proceeds by first calculating the cumulative outage probability table of each utility or area. The outage probability tables of each area are then convolved to compute the outage probability table of one utility interconnected to another.

11.4.1 Two-Area Equations

The notation that will be used is:

$\text{CPROB}_{\text{A}}(X)$ = cumulative outage probability in system A of having X MW or more on outage

$\text{CPROB}_{\text{B}}(X)$ = cumulative outage probability in system B of having X MW or more on outage

LOLP_{A} = probability that system A will be unable to serve the load demand

$P_{\text{A}}(\text{X})$ = exact outage probability of system A in having X MW on outage

$P_{\text{B}}(X)$ = exact outage probability of system B in having X MW on outage

R_{A} = reserve of system A (capacity on line − load) to cause load curtailment or emergency action

R_{B} = reserve of system B to cause curtailment of interconnection power transfer to system B

TIEMAX = maximum transfer limit between area A and B

The loss of load proability of system A (or an emergency action) by system A itself is:

$$\text{LOLP} = \int_{R_{\text{A}}}^{\infty} P_{\text{A}}(X)dx = \text{CPROB}_{\text{A}}(R_{\text{A}}) \tag{11.5}$$

Consider the case when system A is deficient by T MW. System B can help system A except when system B has more than (R_{B} − T) MW on outage. Thus the probability that system B cannot help is:

Probability that system B cannot provide A with T MW =

$$\int_{R_{\text{B}}-T}^{\infty} dx\, P_{\text{B}}(X) = \text{CPROB}_{\text{B}}\,(R_{\text{B}} - T) \tag{11.6}$$

Therefore, the probability that system A interconnected to system B will be deficient by T MW is the probability that system A is deficient by T MW times the probability that system B *cannot* provide T MW. Mathematically this is:

$$\begin{aligned}\text{Probability that interconnected A is deficient by } T \text{ MW} &= P_{\text{A}}(T) \int_{R_{\text{B}}-T}^{\infty} dx\, P_{\text{B}}(x) \\ &= P_{\text{A}}\,(T)\, \text{CPROB}_{\text{B}}(R_{\text{B}} - T)\end{aligned} \tag{11.7}$$

If the value of T MW is larger than TIEMAX, then the probability that system B *cannot* provide T MW is unity.

If Equation (11.7) is integrated over all T MW from zero up to the tieline value of TIEMAX, the result is the expected probability of interconnected system A not serving the load demand, or LOLP_A.

$$\text{LOLP}_A = \left[\int_{R_A}^{R_A + \text{TIEMAX}} dx\, P_A(X) \int_{R_B - X}^{\infty} P_B(y)dy \right] + \left[\int_{R_A + \text{TIEMAX}}^{\infty} dx\, P_A(X) \right] \tag{11.8}$$

$$\text{LOLP}_A = \int_{R_A}^{R_A + \text{TIEMAX}} dx\, P_A(X)\, \text{CPROB}_B(R_B - X) + \int_{R_A + \text{TIEMAX}}^{\infty} dx\, P_A(X) \tag{11.9}$$

This equation can be simplified by adding and subtracting:

$$\int_{R_A}^{R_A + \text{TIEMAX}} dx\, P_A(X) \tag{11.10}$$

which leads to:

$$\begin{aligned}\text{LOLP}_A &= \int_{R_A}^{\infty} P_A(X)dx - \int_{R_A}^{R_A + \text{TIEMAX}} dx\, P_A(X) \cdot [1 - \text{CPROB}_B(R_B - X)] \\ &= \text{CPROB}_A(R_A) - \int_{R_A}^{R_A + \text{TIEMAX}} dx\, P_A(X) \cdot [1 - \text{CPROB}_B(R_B - X)]\end{aligned} \tag{11.11}$$

Thus, the last equation for the LOLP_A has an interesting interpretation. The interconnected system A probability is equal to the LOLP of system A noninterconnected *less* the help it receives from system B. Since $\text{CPROB}(R_B - X)$ is the probability that system B *cannot* provide X MW, then the term $[1 - \text{CPROB}(R_B - X)]$ is the probability that system B *can* provide X MW. The term $P_A(X)$ is how much the probability of system A would decrease if it were interconnected to a perfectly reliable source of power. The product accounts for the reliability of the source of power from system B.

11.4.2 Two-Area LOLP Calculation Example

Two small power systems are interconnected by a 120-MW tie, as illustrated in Figure 11.4. Normally, the two power systems have different types and sizes

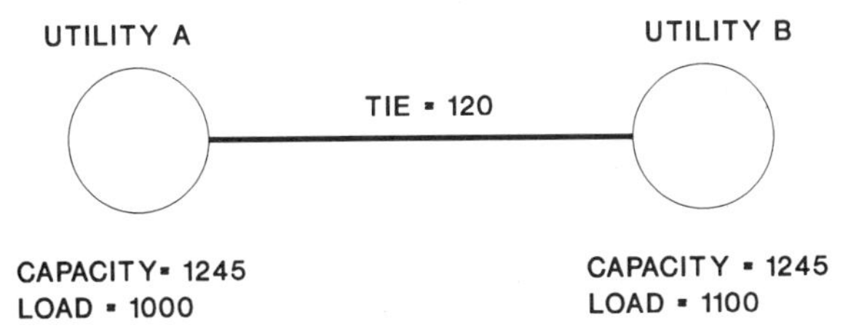

FIGURE 11.4 Two-area LOLP calculation example, two small power systems interconnected by a 120-MW tie.

of capacity and their outage probability tables are different. For this example, however, assume that systems A and B are identical except that they have different loads. The cumulative outage probability table for both utilities is the same and is presented in Table 11.9.

The two utilities will sell reserve power to each other up to the limit of their operating reserve requirements of 60 MW. If utility A were to have 100 MW of reserves, utility A would sell 100 − 60 = 40 MW to utility B, if utility B requires the power. Similarly, the two utilities will purchase power from each when their operating reserves fall to 60 MW or less.

Problem. In part a, determine the daily probability during the peak hour of having insufficient capacity to meet the load (LOLP) of system A as interconnected to system B and, similarly, in part b, determine the daily LOLP of system B as interconnected to system A.

TABLE 11.9 Cumulative Outage Probability Data for the Utilities in Figure 11.4

MW or More on Outage	Probability of X MW or More on Outage
0	1.00000
30	.19966
60	.15218
90	.13457
120	.07016
150	.06659
180	.01360
210	.00893
240	.00608
270	.00198
300	.00170
330	.00042
360	.00025
390	.00012
420	.00003
450	.00002

11.4.2.1 Solution: Part A—Utility A

1. First determine the LOLP of utility A (with no interconnection). A deficiency of capacity in A occurs when there is 1245 − 1000 = 245 MW or more on outage. From Table 11.9:

$$\text{LOLP}_{\text{A}} \text{ (no interconnections)} = .00198$$

2. Next, consider the LOLP of utility A *if* utility B were capable of supplying *perfectly reliable* capacity to utility A. If utility B supplied 30 MW of perfectly reliable capacity, then a deficiency of capacity in A occurs when

$$1245 + 30 - 1000 = 275 \text{ MW or more is on outage}$$

The daily probability of insufficient capacity to meet the load demand is then (from Table 11.9) = .00170 days/day. Table 11.10 (column 2) presents the results of the daily probability calculations of insufficient capacity to meet the load demand for several values of perfectly reliable capacity assistance from utility B. Column 3 of Table 11.10 presents the incremental probability change for 30-MW increments of *perfectly reliable* tie capacity. This is found by subtracting adjacent rows in column 2.

3. However, utility B is not perfectly reliable. Hence, it is necessary to calculate the probability that the utility B has sufficient operating reserves to sell power to utility A. The daily probability of insufficient generation in utility B (for no interconnections from utility A) is:

$$\text{Daily probability (1245 − 1100 MW or more on outage)}$$
$$= \text{(145 MW or more on outage)} = .06659$$

TABLE 11.10 Daily Probability of Capacity Efficiency vs. Tieline Capacity

(1) Perfectly Reliable Tieline Capacity from B (MW or More)	(2) Daily Probability (Days/Day) of Capacity Deficiencies in Utility A	(3) Daily Incremental Probability for 30 MW Increments of Additional Tie
0	.00198	—
30	.00170	+ .00028
60	.00042	+ .00128
90	.00025	+ .00017
120	.00012	+ .00013

However, utility B must maintain a 60-MW operating reserve. The daily probability of utility B failing to have its operating reserve requirement of 60 MW is:

$$\text{Daily probability } (1245 - 1100 - 60 \text{ MW or more on outage}) =$$
$$\text{daily probability } (85) = .13457$$

If utility B were to meet its operating reserve requirement *and* sell 30 MW to utility A, then utility B must have *less* than:

$$1245 - 1100 - 60 - 30 = 85 - 30 = 55 \text{ MW on outage}$$

The probability that utility B has 55 MW or more on outage is (from Table 11.9) = .15218. Hence, the probability that utility B has *less* than 55 MW on outage (which means B can supply 30 MW or more of capacity) is the compliment of this probability or:

$$1 - .15218 = .84782$$

Table 11.11 contains the results of the daily probability of B providing power to A for several values of interchange.

4. The daily probability of utility A having insufficient capacity to meet the load ($LOLP_A$) as interconnected with utility B can be computed by applying the results of Tables 11.10 and 11.11.

Column 3 of Table 11.10 is the incremental probability change of utility A for 30-MW increments of perfectly reliable capacity. This is multiplied by the probability that utility B can supply these increments (column 3 of Table 11.11).

The daily probability of A having insufficient capacity to meet the load

TABLE 11.11 Daily Probability of Providing Power

(1) MW Transfer of Power to Utility A	(2) Probability of Not Being Able to Provide MW of Power to A	(3) Daily Probability Providing MW Power to A
0		
30	.15218	.84782
60	.19966	.80034
90	1.00000	0

demand for a 30-MW tie is $LOLP_A$ (30 MW tie) = .00198 − .00028 * .84782 = .00174. For the full 120 MW tie, the $LOLP_A$ (120 MW tie) is:

$$\begin{aligned} LOLP_A = .00198 &- .00028 * .84782 \\ &- .00128 * .80034 \\ &- .00017 * 0.0 \\ &- .00013 * 0.0 \end{aligned}$$

$$LOLP_A = .000718 \text{ days/day}$$

Utility A sees the value of the tie to utility B as being worth (.00198 − .000718) = .00126 days/day. From Table 11.9, this tie benefit would correspond to approximately 60 MW of effective load-carrying capacity. This was estimated by proceeding down Table 11.9 from 270 MW on outage and noting the additional MW on outage until the probability is .000718. Physically, this means that utility A would be indifferent (for reliability purposes) with regard to the 120 MW tie to utility B or 60 MW of reliable generating capacity.

Another question that can be asked is: for this hour, what is the expected MW that would flow over this tie? Utility A would have reached its operating reserve requirement (60 MW) at 185 MW or more on outage. This would occur with a probability of .00893 days/day. Thus, utility A needs to purchase power from utility B with a probability of .00893 days/day. Utility B has a probability of .84782 of supplying 30 MW or more over the tie. Table 11.12 summarizes this result for several values of the interchange power.

To evaluate the megawattage of purchase by A, the megawatts or more cumulative probability table in Column (3), Table 11.12, needs to be converted into an exact table of megawatt purchased. This is done in column 2 of Table 11.13 by subtracting adjacent entries of the cumulative probability table of column 3 in Table 11.12. The expected megawatt power purchased can then be evaluated as the sum of the triple products of megawatt power purchased, probability of A purchasing power, and probability of B selling power, column 4 of Table 11.13.

TABLE 11.12 Summary of Interchange Power Transaction Probabilities

(1) Capacity or More on Outage	(2) Required MW or More Purchased from B	(3) Probability of A Purchasing MW or More Power (Table 11.8)	(4) Probability of B Selling MW or More Power to A (Table 11.10)
185	30	.00893	.84782
215	60	.00608	.80034
245	90	.00198	0
275	120	.00170	0
305	150	.00042	0

TABLE 11.13 Daily Probability of Tie Use

(1) Purchase MW Power	(2) Probability of A Purchasing the Exact MW Power	(3) Probability of B Selling MW or More Power	(4) Expected MW Power Purchased by A	(5) Probability of TIE Use
30	.00285	.84782	.072489 = .84782 • .00285*30	.002416
60	.0041	.800034	.196884 = .80034 • .0041*60	.003281
90	.00028	0		
120	.00128	0		
			.269373	.005697

The daily probability of using the tie per day can be calculated using Table 11.13. It is the product of columns 2 and 3. This probability is .005697 days/day.

Dividing the expected MW days/day by the daily probability of using the tie, days/day, yields a measure of the expected megawatt tie flow. This yields a value of (.269373/.005697) = 47.3 MW as the expected tie flow during the peak hour of the day.

11.4.2.2 Solution: Part B—Determine the LOLP of Utility B

1. A deficiency in utility B occurs when

$$1245 - 1100 = 145 \text{ MW or more on outage}$$

2. Table 11.14 presents the incremental probability change for utility B for the capacity.
3. The daily probability of utility A failing to serve its load, 1000 MW, and having a 60 MW of operating reserve is:

$$\text{Daily Probability } (1245 - 1000 - 60) = \text{daily probability } (185) = .00893$$

Table 11.5 presents the probability of utility A being able to meet its operating reserve requirement and sell power to utility B.
The resulting daily probability ($LOLP_B$) of B not being capable of serving the load demand when interconnected with A is:

$$\begin{aligned}\text{Daily probability} &= .06659 - .05299 \cdot .98640 - .00467 \cdot .93341 \\ &\quad - .00285 \cdot .92984 - .00410 \cdot .86543 = .003763\end{aligned}$$

Hence, utility B sees the value of the tie to utility A as being worth (.06659 − .003763) = .062827 days/day. This corresponds to approximately 100 MW of effective load-carrying capacity from Table 11.9.

TABLE 11.14 Incremental Probability Change for Utility B

Tieline Capacity from A (MW)	Daily Probability of Capacity Deficiencies in Utility B	Daily Incremental Probability Changes for 30-MW Increments of Additional Tie
0	.06659	—
30	.01360	.05299
60	.00893	.00467
90	.00608	.00285
120	.00198	.00410

TABLE 11.15 Probability of A Selling Power to B

MW Transfer or More of Power to Utility B	Probability of A Not Being Able to Provide MW or More of Power	Probability of A Providing MW Power or More
0		
30	.01360	.98640
60	.06659	.93341
90	.07016	.92984
120	.13457	.86543

11.4.2.3 Interconnection Benefits. Comparison of the 100-MW benefit that utility B achieves with the 60-MW benefit that utility A achieves leads to the conclusion that not all utilities in an interconnected system benefit equally. The benefit is proportional to the reserves of the *sending* utility.

To further amplify on the general characteristics of interconnection benefits, consider the same power system but with the two utilities having equal load demand. Consider varying the load demand to generate a range of conditions. Table 11.16 presents the results. Of special note is the equivalent megawatt benefit from the tie. Figure 11.5 illustrates this trend.

The tie megawatt benefit is small when the reserve margin is low because the probability of either system having surplus capacity for tie assistance is small. The intersection of zero megawatt tie benefit at 6% capacity reserve margin results from the 60-MW operating reserve requirements of the sending utility. As the reserve margin becomes larger, the probability for tie assistance increases; hence, the megawatt benefit increases. The megawatt tie benefit for high megawatt reserve margins is limited by the capacity of the interchange, 120 MW in this example.

The expected load supplied by the tie is also of importance (Figure 11.6).

TABLE 11.16 Interconnection Benefits Based on Varying Load Conditions

Load − MW (area A = area B)	1160	1100	1000	940
Gross capacity reserve margin (%)	7.3	13.2	24.5	32.4
Single-area LOLP	.13457	.06659	.00198	.00042
Interconnected LOLP	.13457	.017926	.000238	.000025
LOLP benefit from tie	0	.04866	.00174	.000395
MW benefit from tie	0	30	90	110
Expected MW tie transferred	0	1.809	.4703	.1141
Expected load supplied by the tie(%)	0	.16%	.05%	.01%

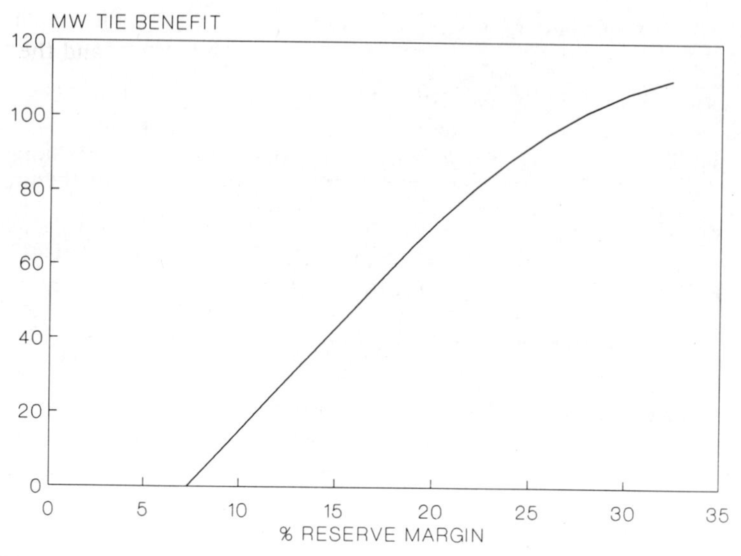

FIGURE 11.5 Effective megawatt tie benefit as a function of reserve margin.

At *very* low reserve margins, the interconnection flow is limited by the lack of surplus capacity in the sending utility. At low reserves (10%–20%) the tie is used extensively. At higher reserve margins (30%), the tie has a larger megawatt worth (Figure 11.5) but the tie flow is less because the receiving utility has no need to call on it.

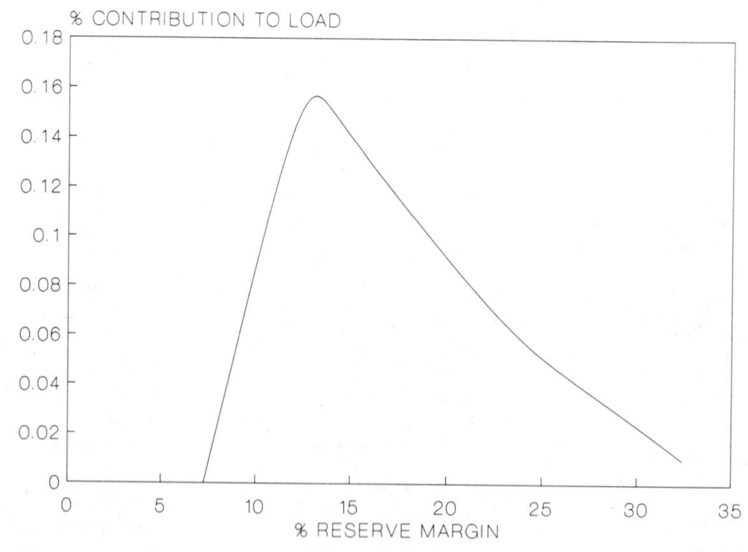

FIGURE 11.6 Contribution of tie to serving load (expressed as a percent of peak load).

While the preceding discussion illustrated two-area reliability calculations for one hour, the process may be repeated for an entire year, and the results accumulated into an annual LOLP value.

In summary, the two-area reliability calculation, using a transportation model of the interconnection, can be readily performed by "convolving" the cumulative outage probability tables of each area. The value of the interconnection can be measured in terms of the effective megawatt capacity of the tie. Benefits from the interconnection may be different for each utility and are proportional to the reserve margin of the *sending* utility.

11.4.3 Pooled Maintenance

Another form of diversity that aids utilities is the potential for load diversity; that is, load peaking in one utility at a different time (or with less intensity) than the load peaks in another utility. Daily load diversity can be a contributing benefit to interconnected utilities where one may peak a few hours earlier than the adjacent utility.

While daily load diversity has a contributing benefit, another potential is seasonal diversity. Utilities in the north with peak loads largely occurring in the winter can find benefits by interconnecting with southern utilities which peak in the summer. An example of this is the seasonal diversity between the U.S. Pacific Northwest and the Pacific Southwest. While seasonal diversity may first appear as a potential opportunity for reliability enhancement from an inspection of the seasonal load demands, it is important to include considerations of the annual generating unit maintenance requirements of each system. These considerations can diminish the reliability enhancement potential. As an example, consider two utilities (with infinite electrical transmission capability between them) having a capacity of 12 GW each and seasonal load demands as presented in Table 11.17. While each utility has a 10-GW peak load itself (taken separately together, 20 GW), the interconnected system has a peak of only 18 GW. It appears that a 2-GW diversity benefit would result from interconnection.

However, consider that the system average planned outage rate is 8.33% (one month of scheduled maintenance or 12 GW-months of maintenance). The resulting loads plus capacity on maintenance are presented in Table 11.18. Because of annual maintenance requirements, the diversity benefit has shrunk from 2 to 1 GW.

TABLE 11.17 GW Monthly Peak-Load Demand

Month	J	F	M	A	M	J	J	A	S	O	N	D
Northern utility	10	9	8	7	7	7	8	8	7	8	9	10
Southern utility	6	6	6	7	6	9	10	10	8	7	7	7
Interconnected load	16	15	14	14	13	16	18	18	15	15	16	17

TABLE 11.18 GW Monthly Peak Load[a] Plus Capacity on Maintenance[b]

Month		J	F	M	A	M	J	J	A	S	O	N	D
Northern utility	L	10	9	8	7	7	7	8	8	7	8	9	10
	M	0	0	1	2	2	2	1	1	2	1	0	0
	L+M	10	9	9	9	9	9	9	9	9	9	9	10
Southern utility	L	6	6	6	7	6	9	10	10	8	7	7	7
	M	2	2	2	1	2	0	0	0	0	1	1	1
	L+M	8	8	8	8	8	9	10	10	8	8	8	8
Interconnected L & M		18	17	17	17	17	18	19	19	17	17	17	18

[a]L = Load.
[b]M = maintenance.

The example presented in Table 11.18 illustrates that the maintenance requirements dilute the potential diversity benefit. However, the maintenance scheduling calculation in Table 11.18 was based on each utility scheduling its maintenance to levelize monthly reserves. Thus, each utility was performing maintenance during the other utility's peak load, which contributed to the dilution of diversity benefit (note, e.g., the month of August). If maintenance were performed on a pooled basis to levelize the interconnected reserves, diversity could be improved. Table 11.19 presents a pooled maintenance schedule.

From inspection, it appears that 2 GW of diversity has been achieved by pooled maintenance scheduling. However, the interconnected peak load plus capacity on maintenance is 18 GW for 7 months (Table 11.19) versus 19 GW for 2 months in Table 11.18. Since the annual LOLP is an expected value over the whole year, the LOLP for pooled maintenance does not squeeze the extra full gigawatt of load diversity.

TABLE 11.19 Pooled Maintenance—GW Monthly Peak Load Plus Capacity on Maintenance

Month		J	F	M	A	M	J	J	A	S	O	N	D
Northern utility	L	10	9	8	7	7	7	8	8	7	8	9	10
	M	0	0	1	2	2	2	0	0	2	2	1	0
	L+M	10	9	9	9	9	9	8	8	9	10	10	10
Southern utility	L	6	6	6	7	6	9	10	10	8	7	7	7
	M	1	2	2	2	3	0	0	0	0	0	1	1
	L+M	7	8	8	9	9	9	10	10	8	7	8	8
Interconnected L & M		17	17	17	18	18	18	18	18	17	17	18	18

To evaluate this would require a detailed LOLP computer simulation. For illustration, assume that the monthly outage probability is proportional to:

$$\text{Monthly outage probability} \approx \exp \frac{(\text{Peak load} + \text{capacity on maintenance} - 18)}{M} \qquad (11.12)$$

where $M = .500$ GW (typical of a large system) and 18 is a scale factor, rendering the relative monthly outage probability unity for an 18 GW load plus capacity on maintenance.

The *relative* monthly outage probabilities and annual sums are presented in Table 11.20 for separate utility maintenance and pooled maintenance practice.

To translate the difference in annual outage probability (18.725 vs. 7.675) to an equivalent peak load, substitute into Equation (11.12) the following:

$$\frac{18.725}{7.675} = \exp(\Delta\text{load}/500) \qquad \text{and} \qquad \Delta\text{load} = 0.446 \text{ GW} \quad (11.13)$$

Thus, while the pooled maintenance procedure appeared to reduce the load by 1 GW, compared to the utility basis, the reliability effect was only 0.446 GW.

In summary, seasonal load diversity can lead to improved system reliability. The benefits of diversity can be enhanced by conducting a pooled planned maintenance procedure (El-Sheikhi and Billinton, 1983).

11.5 MULTIAREA RELIABILITY ANALYSIS

Thus far this section two-area reliability analysis has been presented. However, the same methodology used in the two-area case can be extended to three-area (delta or wye connection—Figure 11.7) and several areas.

TABLE 11.20 Relative Outage Probability

Month												
Maintenance Practice												
J	F	M	A	M	J	J	A	S	O	N	D	Annual Sum
Utility Basis												
1.0	.135	.135	.135	.135	1.0	7.39	7.39	.135	.135	.135	1.0	18.725
Pooled Basis												
.135	.135	.135	1.0	1.0	1.0	1.0	1.0	.135	.135	1.0	1.0	7.675

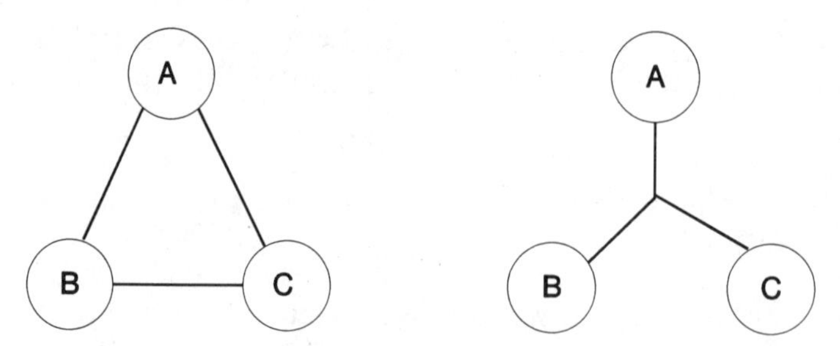

FIGURE 11.7 Three-area utility system interconnection.

When examining more than two areas, it is important to recognize that each utility, power pool, or area may have priorities to whom they provide interchange assistance. For example, consider utility A in Figure 11.7. Utility A may receive assistance from utilities B and C. However, utilities B and C may have priority to interchange assistance with each other before providing assistance to utility A. To calculate the probability that A can receive assistance from B or C, the evaluation must first calculate what is the probability of receiving assistance from B or C, given that B or C is not already providing assistance to each other. An additional set of data for multi area reliability calculations is now required: the interchange assistance priority.

While the methodology is available, the computer execution costs begin to escalate rapidly with an increased number of regions. As an alternative to the application of multiarea reliability analysis, an iterative two-area methodology is useful (Adamson et al., 1977) and provides a very good approximation. As an example, consider the four-area example of Figure 11.8.

Consider utility B first. The iterative process proceeds by first assuming megawatt interconnection worths of ties to utilities A and D. A two-area calculation is then performed for utilities B and C, while representing the ties to A and D as fixed megawatt values. This evaluates the megawatt interconnection worth of utility C. The megawatt interconnection worth of C is then used in conducting a two-area calculation of utilities B and D. A two-area calculation is then made for B and A. This concludes the first iteration. A second iteration

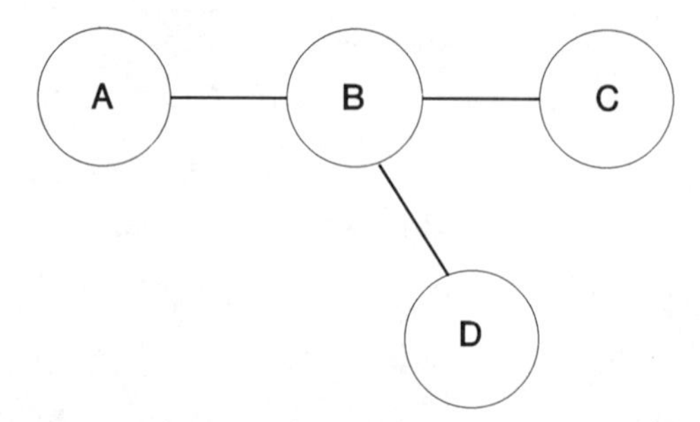

FIGURE 11.8 A four-area utility interconnection.

can then be performed. In general, this process converges well in one iteration; a second iteration may not be needed if the original estimates were reasonably accurate.

In summary, multiarea reliability calculations provide additional insight into reliability credits for interconnections with neighboring utilities.

11.6 MONTE CARLO RELIABILITY EVALUATIONS

The Monte Carlo technique is based on a simulation technique in which histories are developed corresponding to a sampling of the random variables. The Monte Carlo technique is rarely used for single-area generation reliability evaluations because it requires more computer execution time than the analytic LOLP recursive convolution algorithm of Chapter 10. However, the Monte Carlo computation time is largely independent of the number of load areas being evaluated. The Monte Carlo method is suited for a large number of areas with transmission constraints. In addition, this technique can easily combine transmission reliability considerations as well as generation analysis. Consequently, the Monte Carlo technique can be used to evaluate combined generation–transmission system reliability.

With the Monte Carlo technique (a simulation technique), many "histories" are developed corresponding to a sampling of the random variables. For a reliability analysis example, suppose that there were five generating units as shown in Table 11.21. A random number from 0 to 1 is chosen for unit 1. If the random number is less than .10, (10% forced outage rate), then the unit is on forced outage. Otherwise the unit is in service. Four more random numbers are drawn corresponding to units 2 through 5. The results of this case are shown as history number one in Table 11.21.

Table 11.21 presents six histories. The megawatts out of service are computed. Suppose the load demand is 650 MW. If 450 MW (1100 − 650) or

TABLE 11.21 Example Generating Unit Data

Unit	Capacity	Forced Outage Rate	History Number					
			1	2	3	4	5	6
1	100	10%	In	In	Out	In	In	In
2	200	10%	In	In	In	Out	In	In
3	100	5%	Out	In	In	In	In	In
4	500	15%	In	In	In	Out	In	Out
5	200	10%	Out	In	In	In	In	Out
	1100							
Total MW out of service			300	0	100	700	0	700
Contribution to Daily LOLP			0	0	0	1	0	1

more is on outage, the system will not be able to serve the load demand. In this case, history 4 and history 6 would not permit the full load to be served. Therefore, the daily probability of not having sufficient capacity in service to serve load demand is estimated by:

$$\text{Daily LOLP} = \frac{2 \text{ histories}}{6 \text{ histories}} = .333$$

Since only six histories were evaluated in this example, the daily LOLP value is a very crude estimate. What is the accuracy of the daily LOLP estimate? To resolve the question of accuracy, a few statistical sampling theory preliminaries are needed. Many more histories would be required to estimate the Daily LOLP within a suitable accuracy.

Assume that the contribution to a daily LOLP is a random variable. For these five units, there are 2^5 or 32 states, or a population of 32. The *sample* population is six (for six histories).

The daily LOLP is defined as:

$$\text{Daily LOLP} = \Sigma \text{ contribution to daily LOLP per population} \tag{11.14}$$

The sample estimate of the mean uses the sample population as:

$$\text{Daily LOLP}_{\text{SAMPLE}} = \frac{0 + 0 + 0 + 1 + 0 + 1}{6} = \frac{2}{6} = .333$$

The sample variance (square of the sample standard deviation) is:

$$\sigma^2_{\text{SAMPLE}} = \frac{(0 - .333)^2 + (0 - .333)^2 + (0 - .333)^2 + (1 - .333)^2 + (0 - .333)^2 + (1. - 333)^2}{(6 - 1)}$$

$$= .267 \tag{11.15}$$

The daily LOLP was estimated from the data sample of six histories. It is necessary, however, to estimate the standard deviation of the mean daily LOLP.

The mean is defined as:

$$\overline{X} = \frac{X_1 + X_2 + \cdots + X_N}{N} = \frac{X_1}{N} + \frac{X_2}{N} + \frac{X_3}{N} + \cdots + \frac{X_N}{N} \tag{11.16}$$

where X_j = jth sample data point
$\overline{X}$ = mean of the sample
N = number of data points in sample

To estimate how much variation there might be in the sample mean, Equation (11.16), consider each sample data point, X_i, as a random variable. The variance of the sample mean is defined as the expected value of $(X - \overline{X})^2$ over the entire population. Thus, the variance of the mean is:

$$\text{Variance } (\overline{X}) \text{ of the mean } = \frac{1}{N^2} \text{ variance } (X_1)$$

$$+ \frac{1}{N^2} \text{ variance } (X_2) + \ldots + \frac{1}{N^2} \text{ variance } (X_N) \tag{11.17}$$

Since each data point is from the same population, each point has the same variance. Thus:

$$\text{Variance } (\overline{X}) \text{ of the mean } = N \left[\frac{\text{variance } (X)}{N^2} \right] = \frac{\text{variance } (X)}{N}$$

The variance of the mean is then:

$$\text{Variance } (\overline{X}) = \frac{.267}{6} = .044$$

The standard deviation is $\sqrt{.044} = .211$

To find the 90% confidence interval for the LOLP, use the student's t distribution with five (6 − 1) degrees of freedom (refer to Chapter 7). This gives:

$$\begin{aligned} \text{Daily LOLP} &= .333 \pm (.211) \cdot 1.48 \\ &= .333 \pm .312 \end{aligned} \tag{11.18}$$

where 1.48 is the t-distribution value at 90% confidence interval with five degrees of freedom.

Thus, using only six histories, the daily LOLP has a very broad confidence band.

Consider now a more practical case of a large power system in which the daily LOLP is in the .01 to .001 range. Equation (11.14) reduces to:

$$\text{Daily LOLP} = \frac{k}{N} \simeq .001 \tag{11.19}$$

where k = number of histories contributing to LOLP and N = total number of histories. The sample standard deviation in this case [Equation (11.15)] would contain many terms of either $(0 - .001)^2$, which is near zero, or $(1 - .001)^2$, which is near unity. Equation (11.15) can then be approximated by:

$$\sigma^2_{\text{SAMPLE}} = \frac{k(1)^2}{N} = \text{LOLP} \tag{11.20}$$

Thus the variance of the mean is:

$$\text{Variance } (\overline{X}) \text{ of the mean } = \frac{\text{LOLP}}{N} \tag{11.21}$$

and the standard deviation of the mean is the square root of the variance of the mean.

Theoretically, the LOLP statistics follow a binomial probability distribution. Usually the LOLP is much less than unity. For a large number of histories, the normal distribution closely approximates the binomial distribution. Hence, confidence intervals can be developed based upon the normal distribution statistics.

Thus, the 90% confidence level of the LOLP is:

$$\text{LOLP}_{\text{true}} = \text{LOLP} \pm 1.28 \frac{\sqrt{\text{LOLP}}}{\sqrt{N}} \tag{11.22}$$

where 1.28 is the *t*-distribution value at 90% confidence about the standard deviation of the mean.

If the LOLP tolerance band is to be no larger than 10%, then:

$$\text{LOLP} + 1.28 \frac{\sqrt{\text{LOLP}}}{\sqrt{N}} = \text{LOLP} * 1.10$$

$$1.28 \frac{\sqrt{\text{LOLP}}}{\sqrt{N}} = .10 * \text{LOLP} \tag{11.23}$$

or

$$N = \left(\frac{1.28}{.10}\right)^2 \frac{1}{(\text{LOLP})} = \frac{163.8}{\text{LOLP}} \tag{11.24}$$

A daily LOLP of .001 to .01 yields required values of N from 16,384 to 163,840 histories for 10% tolerance band.

The same confidence level equations can be approximately applied to the annual LOLP calculation. Thus, Equation (11.24) also applies to annual LOLP values. To estimate the annual LOLP to a 10% tolerance band requires approximately 163–1638 annual histories (365 daily calculations/year) if the annual LOLP is 1 day/year to .1 day/year, respectively. If a 1% tolerance band is desired approximately 200 times the cases are needed (Equation 11.23).

In practice, as each Monte Carlo history is compiled, a running average

LOLP value is obtained. Histories continue to be performed until the statistics of the running average settle down to the required accuracy.

A large number of histories is required to accurately simulate the LOLP of a power system. Fortunately, there are methods to reduce the number of required histories.

The key concept of reducing the amount of required histories is to insert an approximate solution into the Monte Carlo technique that assists in the sampling of the random variables. The approximate solution is usually obtained from a theoretical or analytical result.

For illustration, suppose the Monte Carlo problem objective is to calculate the average (mean) of $f(x)$, in which $f(x)$ is a difficult function to integrate.

$$\bar{f} = \theta = \int_0^1 f(x)dx \tag{11.25}$$

The "crude" Monte Carlo method as previously discussed would sample random numbers on the X axis from 0 to 1. However, a large number of "histories" are required to calculate $\bar{f}$.

Consider the "importance sampling" technique. The integral of Equation (11.25) is rewritten as:

$$\theta = \int_0^1 f(x)dx = \int_0^1 \frac{f(x)}{g(x)} g(x)dx = \int_0^1 \frac{f(x)}{g(x)} d\,G(x) \tag{11.26}$$

where

$$G(x) = \int_0^x g(y)dy \tag{11.27}$$

and restrict $G(x)$ such that

$$G(1) = 1 \tag{11.28}$$

Restricting $G(x)$ to lie between 0 and 1 means that $G(x)$ can be considered a cumulative probability distribution function.

The objective is to choose a $g(x)$ such that the sampled points are biased toward those sample points that are of most importance. Choosing a $g(x)$ such that $f(x)/g(x)$ has less variation (is almost constant) will achieve the objective. Selection of $g(x)$ can be based on an approximate or theoretical solution to a problem.

Applying the Monte Carlo technique to Equation (11.26) requires two steps. First, a random number is generated for $G(x)$; $G(x)$ is then inverted to obtain the x value that then is used to compute $f(x)/g(x)$.

Consider an example in which $f(x) = \exp(x)$. Assume that $\exp(x)$ is a complicated function to which Monte Carlo is applied. If $g(x) = \frac{2}{3}(1 + x)$. Then $G(x) = \frac{2}{3}[x + (x^2/2)]$. The inverse of $G(x)$ is $x = \sqrt{3G + 1} - 1$.

Suppose that five random numbers are chosen as in Table 11.22. Table 11.22 illustrates the calculations for importance weighting and "crude" Monte Carlo.

Note that the exact answer for the example is 1.718. The importance weighting result achieved a 2% error, whereas the crude Monte Carlo achieved a 6% error. The importance weighting technique is able to significantly reduce error in the estimate.

Another Monte Carlo approach is called the "control variate method." The average or integral of Equation (11.24) is replaced by:

$$\bar{f} = \theta = \int_0^1 f(x)dx = \int_0^1 \phi(x)dx + \int_0^1 [f(x) - \phi(x)]dx \qquad (11.29)$$

In this case, $\phi(x)$ is an approximate solution to the problem. The first integral is known from the approximate solution. The second integral should have less variation than the original problem.

In the case of $f(x) = \exp(x)$, let $\phi(x) = 1 + x$. The calculations are shown in Table 11.23.

$$\text{Integral of } \phi(x) = \int_0^1 1 + x = x + \left.\frac{x^2}{2}\right|_0^1 = 1.5$$

The control variate estimate is 1.5 + .25 = 1.75, which is more accurate than the crude Monte Carlo result from Table 11.22.

The control variate technique is useful for sensitivity cases. In this case, $\phi(x)$ may be a reference case. The Monte Carlo simulation is then applied to the difference between the sensitivity case and the reference case.

Consider the application of the importance weighting method to the power system calculation of weekly LOLP. The cumulative probability of failing to meet load demand is dependent on the load demand, "LOAD," and the random outage rates of the generating units, R. The integral over R is many integrals over each r value of each generating unit. The weekly LOLP is the double integral (or summation) of failure probability over weekly load demands and random outages.

$$\text{Weekly LOLP} = \int_{\substack{0\\ \text{week}}}^{T} dt \int dR \text{ failure}[\text{load}(t), R] \qquad (11.30)$$

Suppose that the failure probability is roughly proportional to g(Load). Then:

$$\text{Weekly LOLP} = \int_{\substack{0\\ \text{week}}}^{T} dt \int dr \frac{\text{failure}[\text{load}(t), R]}{g[\text{Load}(t)]} \cdot g[\text{Load}(t)] \qquad (11.31)$$

TABLE 11.22 Contrast of Importance Weighting With Crude Monte Carlo

Importance Weighting					Crude Monte Carlo Random #	
Random # $G(x)$	x	$f(x)$	$g(x)$	$f(x)/g(x)$	x	$f(x)$
.12	.166	1.18	.776	1.52	.12	1.13
.65	.718	2.05	1.14	1.80	.65	1.92
.87	.90	2.45	1.26	1.94	.87	2.38
.73	.786	2.19	1.19	1.84	.73	2.08
.46	.543	1.72	1.03	1.67	.46	1.58
Average				1.75		1.818

TABLE 11.23 The Control Variate Approach Calculation

Random #	$\phi(x)$	$f(x)$	$f(x) - \phi(x)$
.12	1.12	1.13	.01
.65	1.65	1.92	.27
.87	1.87	2.38	.51
.73	1.73	2.08	.35
.46	1.46	1.58	.12
Average			.25

Consider replacing:

$$\int_0^t dt \int_0^R dR\, g[\text{load}(t)] = G[\text{load}(t),R] \tag{11.32}$$

with $G(\mathrm{T},1) = 1$.

Thus:

$$\text{Weekly LOLP} = \int_{\substack{0 \\ \text{week}}}^T dt \int_R dG[\text{load}(t))] \frac{\text{failure}[\text{load}(t),R]}{g[\text{load}(t)]} \tag{11.33}$$

Random numbers for this calculation can be drawn from the distribution $G[\text{load}(t),R]$. This biases the evaluation to choose failures that significantly contribute to the weekly LOLP.

Consider a simplified example in which:

$$\text{Load}(t) = t \qquad \text{for} \qquad t = 0 \text{ to } 1 \qquad (\mathrm{T} = 1)$$

Suppose that the hypothetical failure probability is given by:

$$\text{Failure(load},R) = \exp[\text{load}(t)] * \exp(-\frac{R}{5})$$

where R is a random variable. Let the importance weight be based on the assumption that the LOLP follows an exponential rule. Let

$$g[\text{load}(t)] = a_o \exp[\text{load}(t)] = a_o \exp(t) \tag{11.34}$$

but:

$$\int_0^1 dt \int_0^1 dR\, g[\text{load}(t)] = a_0[\exp(1) - 1] = 1 \tag{11.35}$$

Thus $a_o = .582$. Hence:

$$g(t) = .582 \exp(t)$$

$$G(t) = .582[\exp(t) - 1]$$

Inverting, this gives:

$$t = \ln[G(t)/.582) + 1]$$

Consider the application of the importance-weighted Monte Carlo technique and evaluating it for five random numbers as shown in Table 11.24.

The exact answer is 1.56. The importance weighting has an error of 1.3%, whereas the "crude" Monte Carlo has an error of 4.7%.

In practical simulations, importance weight may be estimated as

$$g(x) = \exp\left(\frac{\text{load}}{M}\right)$$

where M is the characteristic M-slope of the generation system. This weight biases the Monte Carlo technique toward high-load demands that contribute to an annual LOLP. In this way, fewer Monte Carlo histories are required.

In summary, the Monte Carlo technique has not been used in single-area generation reliability evaluations because the more analytic recursive convolution techniques are much easier to apply and provide exact answers. However, Monte Carlo techniques are being studied for use with multiarea reliability evaluations in which the transmission system is explicitly represented.

11.7 GAUSSIAN APPROXIMATIONS

The recursive convolution technique developed in Chapter 10 is a procedure well adapted to digital computer implementation. The outage table can be rap-

TABLE 11.24 Application of the Importance-Weighted Monte Carlo Technique

Importance Weighting						"Crude" Monte Carlo		
Random # $G(t)$	t	Random # R	$g(t)$	Failure	Failure $g(t)$	t	R	Failure
.12	.187	.65	.700	1.059	1.512	.12	.65	.990
.65	.750	.72	1.23	1.833	1.490	.65	.72	1.66
.87	.914	.07	1.45	2.460	1.70	.87	.07	2.35
.73	.813	.45	1.31	2.061	1.57	.73	.45	1.90
.43	.553	.95	1.014	1.438	1.42	.43	.95	1.27
					1.54			1.634

idly constructed by computer to provide accurate results. While computerized models are the most widely used reliability analysis tools, there is an incentive to develop approximate methods for use in analytic studies. Among these methods are the Gaussian and higher-order approximations to the cumulative outage probability table. Applications for these methods in production simulation evaluations may be found in Chapter 13.

From theoretical considerations, the Gaussian method should be a rough approximation to the real outage probability density function.

The Gaussian approximation can be obtained by applying the method of probability moments. With this method, the first few moments of Gaussian approximation are set equal to the first few moments of the "true" outage probability table. This procedure will be illustrated shortly, following a discussion of probability moments.

Consider a two-state representation of a generating unit, either available with capacity C or unavailable with zero capacity with a probability equal to the forced outage rate, FOR.

The probability *density* function of the megawatts on outage (the probability that the unit has capacity between X and $X + dX$) is:

$$p(x)dx = (1 - \text{FOR}) * \delta(x) + \text{FOR} * \delta(X - C) \tag{11.36}$$

where δ is the Dirac delta function (unit impulse functon) such that:

$$\int_{-\infty}^{\infty} \delta(x)\, F(x) = F(0) \tag{11.37}$$

This probability density function states that the probability of finding X MW of the generating unit on outage is zero, except when $X = 0$ (unit is available) and at $X = C$ (capacity of the unit). This equation could be generalized to the multistate representation of a generating unit as well.

If Equation (11.36) is a true probability density function, its integral over all X MW on outage must be unity. Using the definition of the delta function, the integral:

$$\int_{-\infty}^{\infty} p(x)\, dx =$$

$$\int_{-\infty}^{\infty} [(1 - \text{FOR})\delta(x) + \text{FOR}\ \delta\ (X - C)]\, dx = 1 - \text{FOR} + \text{FOR} = 1 \tag{11.38}$$

Thus, it is, indeed, unity.

Consider evaluating the mean or expected megawatts on outage:

$$\int_{-\infty}^{\infty} xp(x)dx = \int_{-\infty}^{\infty} x[(1 - \text{FOR})\delta(x) + \text{FOR}\ \delta(X - C)]dx = C \cdot \text{FOR} = \overline{X} \tag{11.39}$$

The mean or expected megawatts on outage is the capacity of the unit times the forced outage rate.

Consider evaluating the variance about the mean of the megawatts on outage:

$$<X - \overline{X}>^2 = \int_{-\infty}^{\infty} (X - C \cdot \text{FOR})^2 \cdot p(x)$$

$$<X - \overline{X}>^2 = \int_{-\infty}^{\infty} (X - C \cdot \text{FOR})^2 \cdot [(1 - \text{FOR})\delta(x) + \text{FOR} \cdot \delta(X - C)]dx$$

$$<X - \overline{X}>^2 = +(C \cdot \text{FOR}^2) \cdot (1 - \text{FOR}) + C^2(1 - \text{FOR})^2 \cdot \text{FOR}$$

$$<X - \overline{X}>^2 = (1 - \text{FOR})\,(\text{FOR})C^2 = \sigma^2 \tag{11.40}$$

Similarly, it can be shown that the third and fourth moments about the mean are:

$$<X - \overline{X}>^3 = \int_{-\infty}^{\infty} (X - C \cdot \text{FOR})^3 \cdot p(x) = C^3\,(1 - \text{FOR})(\text{FOR})(1 - 2\text{FOR}) \tag{11.41}$$

$$<X - \overline{X}>^4 = \int_{-\infty}^{\infty} (X - C \cdot \text{FOR})^4 \cdot p(x) = C^4(1 - \text{FOR})(\text{FOR})[1 - 3\,(\text{FOR})(1 - \text{FOR})] \tag{11.42}$$

These relations were shown to be true for one generation unit. If there are many generation units, the same relationships hold except that a summation is used. Hence,

$$\overline{X} = \sum_{N=1}^{\text{total}} C_N * \text{FOR}_N; \qquad \sigma^2 = \sum_{N=1}^{\text{total}} C_N^2\,(1 - \text{FOR}_N)\text{FOR}_N \tag{11.43}$$

$$<X - \overline{X}>^3 = \sum_{N=1}^{\text{total}} C_N^3\,(1 - \text{FOR}_N)(\text{FOR}_N)(1 - 2\text{FOR}_N) \tag{11.44}$$

$$<X - \overline{X}>^4 = \sum_{N=1}^{\text{total}} C_N^4\,(1 - \text{FOR}_N)(\text{FOR}_N)[1 - 3(\text{FOR}_N)(1 - \text{FOR}_N)] \tag{11.45}$$

where the subscript N denotes the characteristic parameters of the Nth generating unit.

The Gaussian approximation to the probability density function is:

$$p(x) \approx \frac{1}{\sqrt{2\pi}}\,\frac{1}{\sigma} \exp\,[-(x - u)^2/\sigma^2] \tag{11.46}$$

where u is the mean of the Gaussian approximation and σ is the standard deviation about the mean of the Gaussian approximation. Through the use of integration formulas, it can be shown that the moments of the Gaussian distribution are:

$$\int_{-\infty}^{\infty} dx \frac{1}{\sqrt{2\pi}} \frac{1}{\sigma} \exp\,[-(x - u)^2/2\sigma^2] = 1.0 \qquad (11.47)$$

$$\int_{-\infty}^{\infty} dx(x - u) \frac{1}{\sqrt{2\pi}} \frac{1}{\sigma} \exp\,[-(x - u)^2/2\sigma^2] = 0 \qquad (11.48)$$

$$\int_{-\infty}^{\infty} dx(x - u)^2 \frac{1}{\sqrt{2\pi}} \frac{1}{\sigma} \exp\,[-(x - u)^2/2\sigma^2] = \sigma^2 \qquad (11.49)$$

$$\int_{-\infty}^{\infty} dx(x - u)^3 \frac{1}{\sqrt{2\pi}} \frac{1}{\sigma} \exp\,[-(x - u)^2/2\sigma^2] = 0 \qquad (11.50)$$

$$\int_{-\infty}^{\infty} dx(x - u)^4 \frac{1}{\sqrt{2\pi}} \frac{1}{\sigma} \exp\,[-(x - u)^2/2\sigma^2] = 3\sigma^4 \qquad (11.51)$$

Using the first two moments, Equation (11.43) and equating moments to Equations (11.48) and (11.49) results in the Gaussian approximation.

As an example, consider the following 3405-MW system comprised of 32 generating units. The moments are evaluated in Table 11.25. The result is:

$$\text{Mean} = 208.6 \text{ MW}$$

$$\sigma^2 = \text{variance} = 232 \text{ MW}$$

The Gaussian approximation is then:

$$p(x) = \frac{1}{\sqrt{2\pi}} \frac{1}{232} \exp\,[-\frac{1}{2} \frac{(x - 208.6)^2}{(232)^2}] \qquad (11.52)$$

The cumulative outage distribution is the integral of the density function:

$$C\ \text{Table}(X) = \int_{x}^{\infty} p(x)dx$$

$$= \text{erfc}\,(\frac{1}{\sqrt{2}} \frac{(x - 208.6)}{232}) \qquad (11.53)$$

where X is the megawatts on outage and ERFC ( ) is the complimentary error function (tabulated in many reference books, i.e., CRC Standard Mathematical Tables).

TABLE 11.25 Generating Unit Data

Unit Size	Number	Forced Outage Rate	(C*FOR)	$(1 - \text{FOR})\text{FOR}(C)^2$
12	5	0.02	1.20	14.11
20	4	0.10	8.00	144.00
50	6	0.01	3.00	148.50
76	4	0.02	6.08	452.83
100	3	0.04	12.00	1152.00
155	4	0.04	24.80	3690.24
197	3	0.05	29.55	5530.28
350	1	0.08	28.00	9016.00
400	2	0.12	96.00	33792.00
			208.63	53939.90

Cap system = 3,405 MW
Mean = 208.6 MW
σ = 232.2 MW = $\sqrt{53939}$

The Gaussian approximation is compared with the exact cumulative outage table (developed by means of the recursive convolution method) in Figure 11.9. The Gaussian approximation is a reasonable approximation in the range of small X MW on outage, but decreases too rapidly with large X MW on outage. For power system reliability analysis, daily probabilities are in the .001–.01 range. The Gaussian approximation alone is somewhat too crude for power systems reliability analysis.

An extension of this method can, however, be developed to improve on this

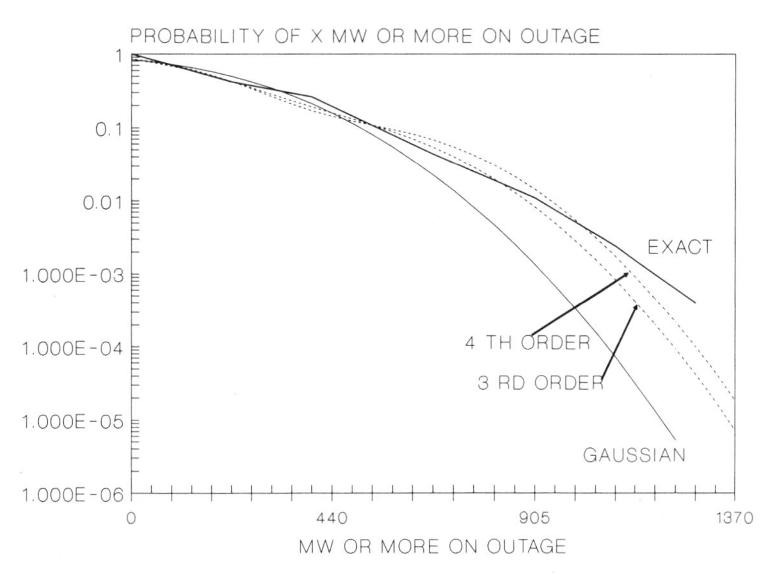

FIGURE 11.9 Comparison of the Gaussian approximations with exact solution.

approximation. Since the Gaussian gives reasonable accuracy for small X MW on outage but not for large X MW on outage, consider adding a few terms that can modify the Gaussian for large X MW on outage. For example:

$$p(x) = \frac{1}{\sqrt{2\pi}} \frac{1}{\sigma} \exp - [\frac{1}{2} \frac{(x-u)^2}{\sigma^2}]$$

$$[b_0 + b_1 X + b_2 X^2 + b_3 X^3 + b_4 X^4 + \ldots] \tag{11.54}$$

where b_0, b_1, b_3, b_4 are coefficients to be determined. Since the Gaussian drops off too fast with large X MW on outage, the effect of the X^n terms will improve this.

While the series X^n is a good basis for expansion, consider a slightly modified expansion that will be more convenient. Differentiate the Gaussian probability density function:

$$\frac{d}{dx} [(\frac{1}{\sqrt{2\pi}} \frac{1}{\sigma} \exp [- \frac{1}{2} \frac{(x-u)^2}{\sigma^2}]$$

$$= \frac{1}{\sqrt{2\pi}} \frac{1}{\sigma} \exp [- \frac{1}{2} \frac{(x-u)^2}{\sigma^2}] [- \frac{(x-u)}{\sigma^2}] \tag{11.55}$$

By differentiating the Gaussian, the results are a Gaussian multiplied by the first power of X. Similarly

$$\frac{d^2}{dx^2} \frac{1}{\sqrt{2\pi}} \frac{1}{\sigma} \exp [- \frac{1}{2} \frac{(x-u)^2}{\sigma^2}]$$

$$= \frac{1}{\sqrt{2\pi}} \frac{1}{\sigma} \exp [- \frac{1}{2} \frac{(x-u)^2}{\sigma^2}] [- \frac{1}{\sigma^2} + \frac{(x-u)^2}{\sigma^4}] \tag{11.56}$$

$$\frac{d^3}{dx^3} \frac{1}{\sqrt{2\pi}} \frac{1}{\sigma} \exp [- \frac{1}{2} \frac{(x-u)^2}{\sigma^2}]$$

$$= \frac{1}{\sqrt{2\pi}} \frac{1}{\sigma} \exp [- \frac{1}{2} \frac{(x-u)^2}{\sigma^2}] [2 \cdot \frac{(x-u)}{\sigma^4} - \frac{(x-u)^3}{\sigma^6}] \tag{11.57}$$

$$\frac{d^4}{dx^4} \frac{1}{\sqrt{2\pi}} \frac{1}{\sigma} \exp [- \frac{1}{2} \frac{(x-u)^2}{\sigma^2}]$$

$$= \frac{1}{\sqrt{2\pi}} \frac{1}{\sigma} \exp [- \frac{1}{2} \frac{(x-u)^2}{\sigma^2}] [\frac{2}{\sigma^4} - 3 \frac{(x-u)^2}{\sigma^6} + \frac{(x-u)^4}{\sigma^8} \tag{11.58}$$

The resulting expansion is then represented as:

$$p(x) = [a_0 + a_1 \frac{d}{dx} + a_2 \frac{d^2}{dx^2} + a_3 \frac{d^3}{dx^3} + a_4 \frac{d^4}{dx^4}]$$

$$\cdot [\frac{1}{\sqrt{2\pi}} \frac{1}{\sigma} \exp [- \frac{1}{2} \frac{(x - u)^2}{\sigma^2}] \tag{11.59}$$

Moments can now be taken of this expression. In view of the derivative term, d/dx, many of the contributions are zero. The moments can be calculated by successive integration by parts.

Taking the zeroth moment, $\int_{-\infty}^{\infty} dx$, produces:

$$\int_{-\infty}^{\infty} p(x)dx = a_0 \tag{11.60}$$

Taking the first moment, $\int_{-\infty}^{\infty} (x - u)dx$, produces:

$$\int_{-\infty}^{\infty} p(x)(x - u)dx = + a_1 \tag{11.61}$$

Similarly, taking the second moment, $\int_{-\infty}^{\infty} (x - u)^2dx$, produces:

$$\int_{-\infty}^{\infty} p(x)(x - u)^2dx = a_0\sigma^2 + 2 a_2 \tag{11.62}$$

The third moment produces:

$$\int_{-\infty}^{\infty} p(x)(x - u)^3dx = -3 a_1\sigma^2 - a_3 \cdot 6 \tag{11.63}$$

The fourth moment produces:

$$\int_{-\infty}^{\infty} p(x)(x - u)^4dx = 3\sigma^4 a_0 + 12a_2\sigma^2 + a_4 \cdot 24 \tag{11.64}$$

Equating the moments of the expansion [Equations (11.60) through (11.64)] to be equal to the moments of probability density [Equations (11.43) through (11.45)] yields:

$$a_o = 1 \tag{11.65}$$

$$a_1 = 0 \tag{11.66}$$

$$a_0\sigma^2 + 2a_2 = \sigma^2; \qquad \text{hence} \qquad a_2 = 0 \tag{11.67}$$

$$\langle X - \overline{X} \rangle^3 = -3a_1\sigma^2 - 6a_3;$$

$$\text{hence} \qquad a_3 = \frac{-\langle X - \overline{X} \rangle^3}{6} \tag{11.68}$$

$$\langle X - \overline{X} \rangle^4 = 3\sigma^4 a_0 + 12a_2\sigma^2 + a_4 \cdot 24;$$

$$\text{hence} \qquad a_4 = \frac{\langle X - \overline{X} \rangle^4}{24} - \frac{\sigma^4}{2} \tag{11.69}$$

As an illustration of this method, consider the previous example. Computing the third and fourth moments of the distribution from Table 11.25 and substituting into Equations (11.65) through (11.69) yields:

$$a_3 = -.1940 \cdot \sigma^3$$
$$a_4 = +.0657 \cdot \sigma^4$$

where $\sigma = 232.2$ MW and $u = 208.6$ MW.

The cumulative outage probability table is the integral, $\int_x^\infty p(x)dx$, of the probability density. Hence:

$$C\ \text{Table}(X) = \text{erfc}[\frac{1}{\sqrt{2}}(\frac{x - 208.6}{232.2})]$$

$$+ \{.194\ \sigma\ [-1 + (\frac{x - u}{\sigma})^2] + .0657\ \sigma\ [2\ (\frac{x - u}{\sigma}) - (\frac{x - u}{\sigma})^3]\}$$

$$\cdot \{\frac{1}{\sqrt{2\pi\sigma}} \exp\ [-\frac{1}{2}\frac{(x - u)^2}{\sigma^2}]\}$$

The results of this expansion (labeled "fourth order") are contrasted with the exact cumulative outage table in Figure 11.9. This fourth-order expansion does result in improvements over the Gaussian. For large X MW on outage, however, this approximation drops off too fast. Yet, within the interest of daily probabilities in range of .001, this method is useful for analytic approximate analysis. For further reading, see Stremel (1981).

11.8 SUMMARY

This chapter has presented techniques for assessment of optimum reliability levels based on customer cost of outages, multiarea realiability calculation with transmission constraints, and approximate reliability calculational procedures. For further reading, see Billinton and Allan (1984), and Endrenyi (1978).

REFERENCES

Adamson, A. M., A. L. Desell, L. L. Garver, and G. A. Jordan, "Generation Reserve Value of Interconnections," *IEEE Transactions on Power Apparatus and Systems,* Vol. PAS-96, March–April 1977, pp. 337–346.

Billinton, R., and R. N. Allan, *Reliability Evaluation of Power Systems,* Plenum Press, New York, 1984.

El-Sheikhi, F. A., and R. Billinton, "Generation Unit Maintenance Scheduling for Single and Interconnected Systems," *IEEE Transactions on Power Apparatus Systems,* Vol. PAS-103, 1983, pp. 1038–1044.

Endrenyi, J., *Reliability Models in Electric Power Systems,* Wiley, New York, 1978, Chapter 6.

Moisan, R. W. and J. F. Kenney, "Expected Need for Emergency Operating Procedures," *IEEE Paper,* A 76 319-4, July 1976.

Stephenson, W., and W. Walters, "Measure of Residential Outage Costs for Reliability Planning," 14th International RAM Conference for Electric Power Industry, 1987, p. 115.

Stremel, J. P., "Sensitivity Study of the Cumulant Method of Calculating Generation System Reliability," *IEEE Transactions on Power Apparatus Systems,* Vol. PAS-100, 1981, pp. 711-713.

The National Electric Reliability Study—Final Report, U.S. Department of Energy DOE/EP0004, 1981.

PROBLEMS

1 Given the cost of interruptions as

Industrial	5.7 $/kWh
Commercial	5.8 $/kWh
Residential	6.8 $/kWh

Select the lowest total cost plan.

		Energy Not Served (MWh)		
Plan	Annual Capital Cost ($$10^6$)	Industrial	Commercial	Residential
A	0	1200	900	600
B	10.0	500	300	200
C	15.0	300	100	100

2 Estimate the optimum LOLP assuming annual capacity charges of 300 $/kW/year, an equivalent peak hour duration of 6 hours, and outage costs of 5, 10, and 15 $/kWh.

3 A utility has a peak load of 950 MW. The utility emergency operating procedure is: (1) maintain a reserve of 100 MW, (2) purchase capacity yielding 50-MW benefit, (3) 5% voltage reduction yielding 40 MW, and (4) disconnect load. Using the cumulative outage Table 11.7, compute the expected probability of initiating the EOP's.

4 Utilities A and B are interconnected by a 90-MW transmission line. Assume that both utilities have 1245-MW capacity. Utility A has a load demand of 1000 MW and utility B has a load demand of 900 MW. The utilities must maintain 100 MW of operating reserve. The utilities will provide assistance to the other utility if its operating reserves exceed 100 MW. Using the data of Table 11.9, compute the isolated and interconnected LOLP of utilities A and B.

5 Utility A is connected to utility B by a 90-MW transmission line and also to Utility C by a 60-MW transmission line. Assume the cumulative outage table for all three utilities is the same and as given in Table 11.9. Utilities B and C will sell power to utility A if they have an operating reserve greater than 100 MW. The capacity and loads of the utilities are:

	Utility A	Utility B	Utility C
Capacity	1245	1245	1245
Load	1000	1100	1050
Reserves	245	145	195

a Calculate the noninterconnected LOLP for this day for each utility.

b Calculate the interconnected LOLP for this day for utility A. (*Hint:* Utility A can receive power from B *or* C.)

12

PRODUCTION SIMULATION

Production simulation consists of modeling future generation system operation to determine the generation operating expenses (production costs) that will be incurred for a future time period. The expenses being evaluated are fuel costs, generation operation and maintenance (O&M) charges, and generation startup costs. The ability to accurately evaluate production expenses is of considerable importance to utility planners because these expenses constitute a significant percentage (40–60%) of the total annual costs of electricity.

12.1 APPLICATION OF PRODUCTION SIMULATION

Production simulation studies find extensive application by the utility industry for generation planning. The major applications include:

1. *System Planning Studies.* These studies are conducted for a 10–20 year study horizon period to evaluate annual operating expenses for a specific plan (or several alternative plans) of generating unit additions. The procedure here is to add annual opeating expenses with those of the capital charges for new generation and transmission; and, by application of present-worth mathematics, determine the total present worth costs of a specific plan (or several alternative plans).
2. *Fuel Budgeting.* These studies are similar to long-range system planning studies except that the horizon periods are 1–5 years. These simulations are used for fuel purchasing decisions, 1–5 year corporate projections, and filing reports to regulatory agencies. Near-term studies are generally performed with greater simulation detail than longer-range studies.

3. *Purchase and Sales Analysis.* Production simulation studies are used to evaluate potential benefits associated with contractual purchases or sales of electricity to another utility.
4. *Cost-of-Service Studies.* Production simulation is very useful in establishing the cost of electricity service for rate analysis. Especially important are the costs of electric service as a function of time of day (TOD) and season.
5. *Load Management Studies.* Studies can be performed to evaluate operating demand profiles brought about by the impact of load management control devices or electricity pricing structures.
6. *System Operating Policy.* Simulation studies can be conducted to evaluate the economic impacts of system operating policies such as overnight shutdown of generating units, extended maintenance policies, environmental impact, and unit operating flexibilities.
7. *Transmission Planning Studies.* Detailed multiarea production simulation studies can be used to evaluate the economic benefit of transmission expansion plans or to determine the impact of modifying the operations security constraints.

12.2 LOAD-DEMAND REPRESENTATIONS

The representation of load demand in a production simulation model generally requires more detail than in power system reliability analysis. In reliability analysis, most of the LOLP contributions are from peak loads during the year since risk (LOLP) is exponentially proportional to load. In production simulation, however, production expenses are contributed by all of the loads during the year. Consequently, for production simulation, emphasis is placed on developing accurate hourly load-model representation.

While accurate hourly load-model representation is important, the amount of computer time (and cost) is almost directly proportional to the number of distinct hourly loads being evaluated. Therefore, there is an incentive to compromise load representation from 8760 hourly loads per year to something less. The degree of compromise is related to the application as well. For detailed budgeting and electric rate analysis, 8760 hourly loads per year may be required, while generation expansion studies may need explicit representation of only 576 hourly loads, or two typical days per month.

A typical load-demand profile is illustrated in Figure 12.1. This profile may be condensed into four day types: Saturday, Sunday, peak weekday, and average weekday. Time chronology of the load demand is important for the more detailed simulation application where recognition is required of the operation and shutdown constraints on generating units, hourly load diversity between interconnected utility companies, reservoir limitations of energy storage units (such as pumped-storage hydro), and TOD characteristics of new generating concepts (such as solar power plants).

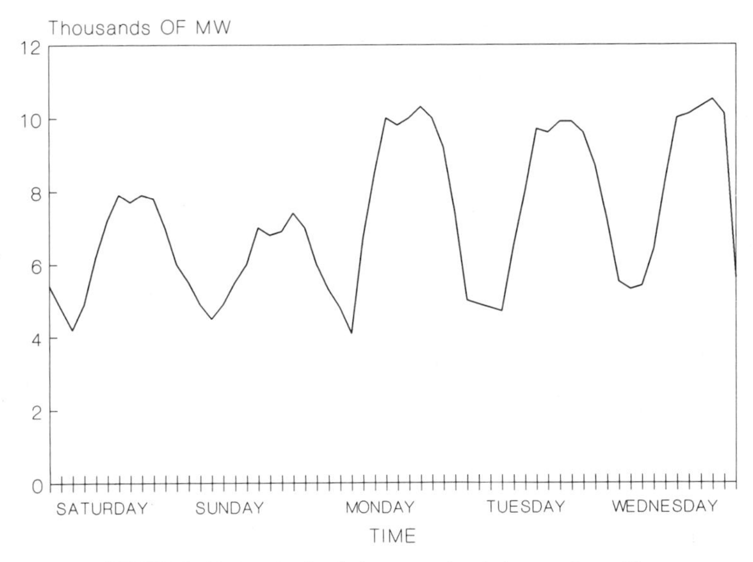

FIGURE 12.1 Typical day-type load-demand profile.

In less detailed simulations, it is convenient to remove the time chronology from the load representation and reorder the load demands for the day in a daily load-duration curve as illustrated in Figure 12.2. Since load demands for Saturdays and Sundays are generally less than those of a given weekday, one widely used computer program uses a load representation that includes one 24-hour weekday load-duration curve and one 24-hour weekend load-duration curve.

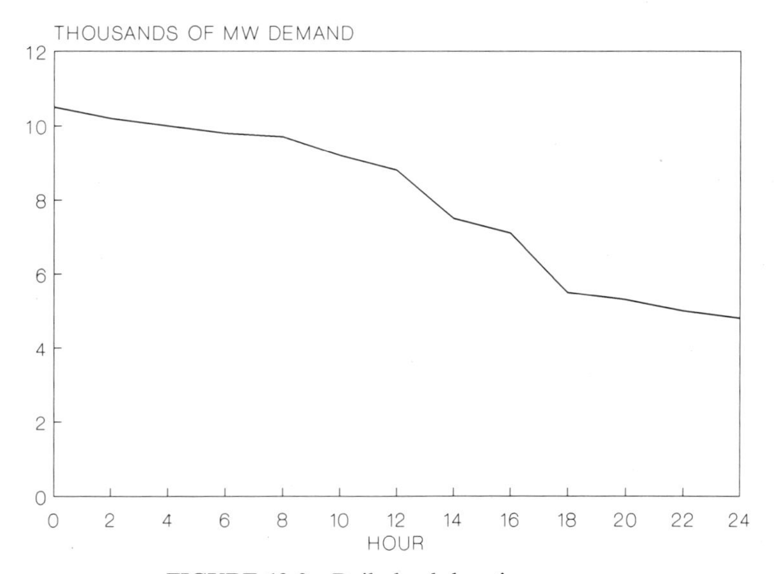

FIGURE 12.2 Daily load-duration curve.

In summary, the choice of the load-model representation has a strong influence on the computer execution costs of performing production simulations. Tradeoffs of simulation fidelity with computer execution costs are typically made by utility planning personnel depending on the application.

12.3 GENERATING UNIT CHARACTERISTICS

The principal performance characteristic of thermal generating units is the required fuel input versus power output. A typical curve is illustrated in Figure 12.3. A generating unit has a minimum stable output, typically 10–30%, for oil and natural gas-fired fossil-steam units and 20–50% for coal-fired fossil-steam units. The ripples in this curve arise from losses associated with initially opening each steam admission valve. The incremental heat rate, which is the instantaneous slope of Figure 12.3 (Δ fuel input/Δ power output), is illustrated in Figure 12.4, where the valve points appear as sharp spikes in the curve. Since the incremental heat rate is large near the valve points, optimum economy is achieved by avoiding operation near these regions. Generally, for both on-line dispatch techniques and in simulation models, the heat rate characteristics are approximated by a series of straight lines or polynomials (trend line in Figures 12.3 and 12.4).

The incremental heat-rate curve should be distinguished from the average heat-rate curve shown in Figure 12.5. The average heat-rate curve is the fuel input divided by the power output. The average heat-rate curve is used more by power plant operators and engineers, while the incremental heat-rate curve is used more by power system dispatch and planning engineers.

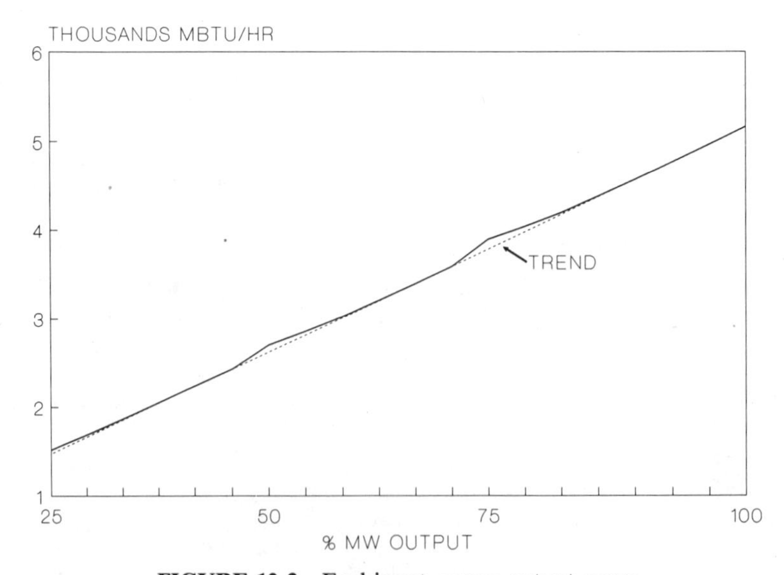

FIGURE 12.3 Fuel input–power output curve.

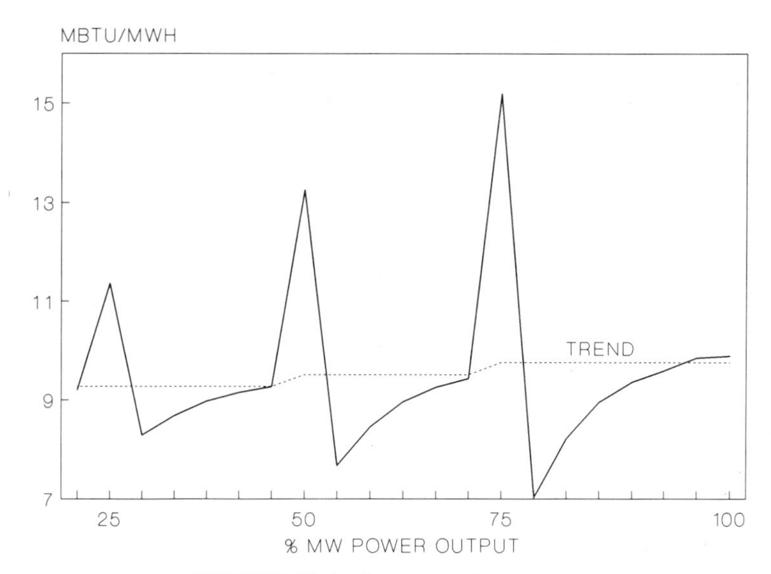

FIGURE 12.4 Incremental heat rate.

The heat-rate of the steam-turbine cycle depends on the cooling water temperature. For small variations, this effect can be represented as a percentage correction over the entire operating range of the plant. For larger variations in cooling-water temperature, such as those between seasons, heat-rate performance can be reevaluated each season on the basis of the mean seasonal cooling-water temperature.

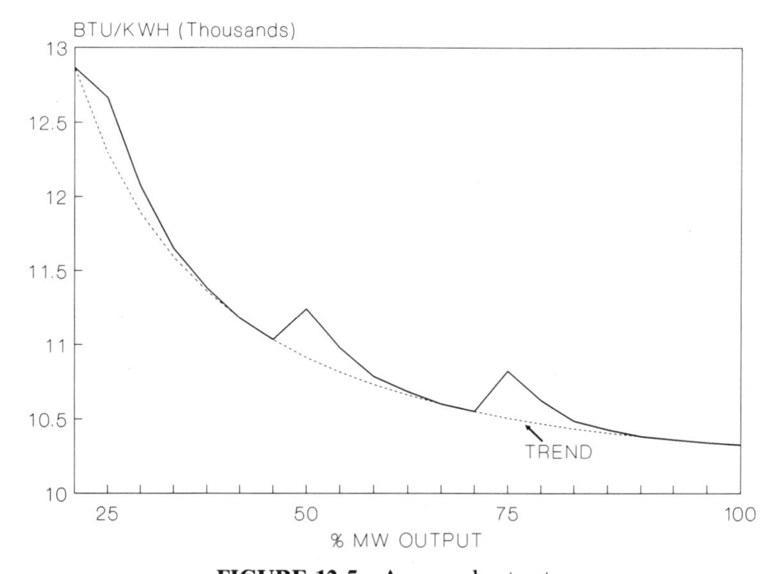

FIGURE 12.5 Average heat rate.

Heat-rate performance is also influenced by the time between maintenance periods. Steam leakages, boiler crud deposition, erosion losses, and other heat losses may accumulate between maintenance intervals, resulting in gradual degradation of net plant heat rate.

Hydro units have similar input/output characteristics as thermal units, as is illustrated in Figure 12.6. There is an absence of ripples in this characteristic, as was noted in the thermal unit input/output characteristic, because all the water-admission valves (wicket gates) are opened as a unit rather than one after another. The effective plant head, which is a parameter in Figure 12.6, is the difference between the forebay elevation (reservoir) and the tail race elevation.

12.4 GENERATION UNIT COMMITMENT

Utilities have daily and weekend load variations that may vary by more than 200% from peak-hour load demand through early morning load valley hours. If all the generating equipment that is on-line for the peak hour would remain on-line for the entire day, then many of these units would be operating at their minimum power limits during the early morning valley hours. Rather than run many of these units at minimum power, it may be more economical to shut these units down overnight. Consequently, economic decisions must be made as to the selection of units to be shut down, the hour of the day they are to be shut down and the hour of the following day that they are to be started up again. In addition to the economic considerations for shut down, other

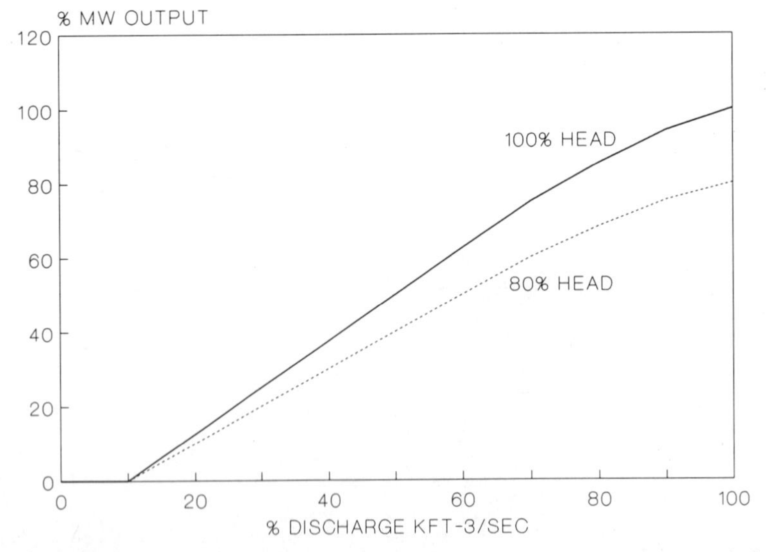

FIGURE 12.6 Hydro-electric unit water input-power output characteristic.

considerations must be reviewed that relate to utility operation policies, physical operating constraints of the units, and utility system reliability. All these considerations enter into the analysis of unit commitment.

12.4.1 Economic Considerations of Unit Commitment

A fundamental principle in developing a preliminary commitment is that the most economic operation tends to result when the fewest number of units are on-line. This can be illustrated by considering the operating cost per megawatt-hour versus the electrical power output characteristic of thermal units, as illustrated in Figure 12.7.

The average operating cost per megawatt-hour is the product of the cost of fuel times the average heat rate plus variable O&M costs. Power is more expensive to generate per kilowatt-hour when the unit operates at low-power output than when it operates at higher output.

A power system with many generating units could operate all of them to serve every load demand. Since the sum of the power output from all of the generators must equal the load, many of the generating units could be operating at low-power output, which results in an expensive operating cost, as noted in Figure 12.7. Alternatively, the power system could start up only enough generating units to meet the load. In this case, all of the generating units are operating at the lower cost region of Figure 12.7. In this case, the total average operating costs are lower. Thus, the minimum operating cost commitment policy is to commit the minimum amount of capacity for service.

Having illustrated that commitment with a minimum number of units tends

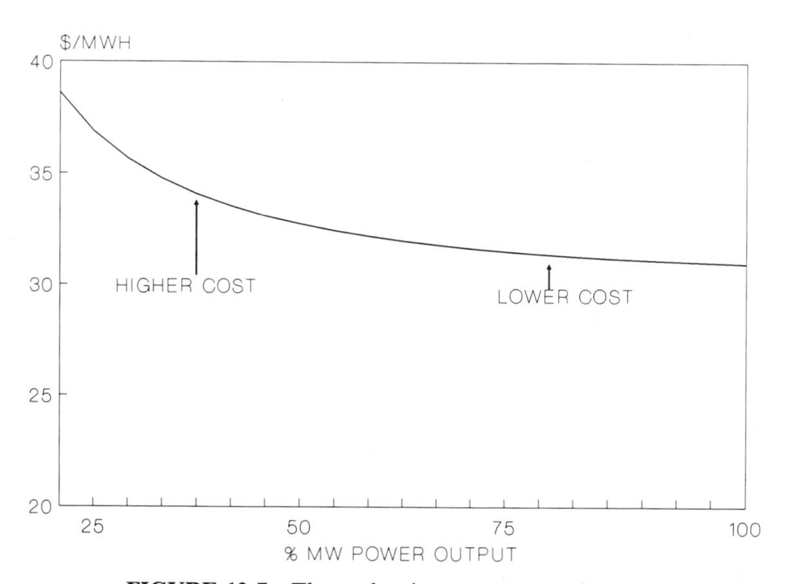

FIGURE 12.7 Thermal unit average operating cost.

to be most economical, we must now decide which units are the best ones to commit for each hour. As a preliminary step, a commitment priority list is developed that ranks the units in order from most economical to least economical. One criterion is the unit's full-load hourly fuel cost per megawatt. For a given hour the priority list is reviewed, in order from lowest to highest $/MWh, committing enough units to serve the load. Refinements to this preliminary commitment will make it more economical.

An economic refinement that may be made to this preliminary commitment is to decrease the total unit startup costs, the costs associated with bringing units from down to operating conditions. The startup procedure may require startup times in the magnitude of 10 minutes (for a combustion turbine) to several hours (for a steam unit), depending on the number of hours it has been on shutdown. Startup cost is associated with supplying energy to bring the unit to operating temperature and pressure. Figure 12.8 illustrates a typical unit startup cost as a function of the number of hours off-line. Since startup cost may be an important consideration in operational planning, one refinement in unit commitment planning is to examine the economic tradeoffs of extending the on-line hours of units to avoid incurring startup costs. While this and other economic refinements are important in on-line operation, these refinements are usually neglected during planning because they influence the total production cost by only .5% or less.

12.4.2 Reliability Considerations of Unit Commitment

In addition to economics, it is important to ensure that there will be sufficient generation on-line to meet load demand during fault as well as normal situa-

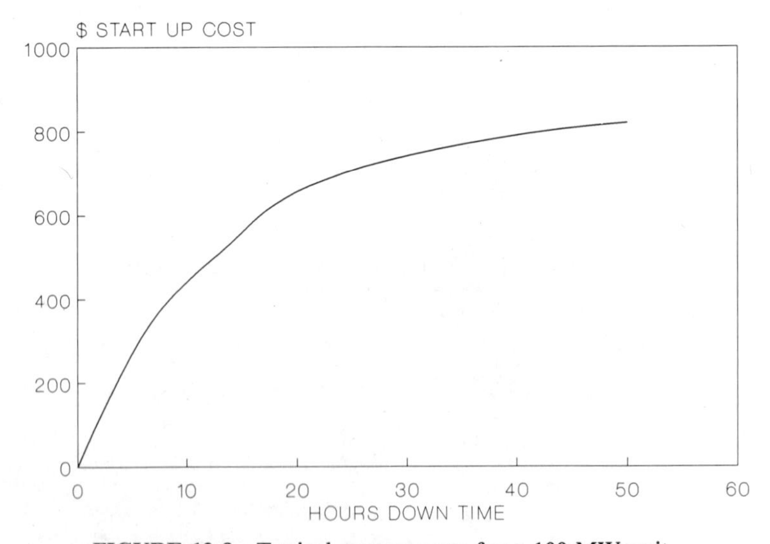

FIGURE 12.8 Typical startup costs for a 100-MW unit.

tions, such as generating unit forced outages, transmission line outages and interconnection emergency demands.

One of the first tasks in real-time unit commitment is to have accurate models for projecting future load demand. Models are generally required for three time spans: (1) several hours in advance, (2) 1–2 days in advance, and (3) 1 week in advance. Several-hour projection is useful for trimming an established unit commitment schedule; 1- or 2-day projection aids in establishing unit startup and shutdown schedules; and, the load projected one week in advance is useful in scheduling any hydro unit on the system. These load forecasts are weather-sensitive and are computed for several distinct geographical areas served by the utility.

With a load forecast, hydro unit (run-of-river, pondage, and PSH) hourly generating schedules are planned. For pondage and pumped storage units, this may involve some iteration between unit commitment and unit dispatch procedures to achieve the most economical schedule. This procedure is discussed further in Section 12.6, "Hydroelectric Generation." The net effect of hydro scheduling is to reduce the peak-load demand on thermal generating units.

With the load forecast and hydro generation schedules established, the thermal unit commitment can be planned to ensure system reliability. A fundamental measure of system reliability is generating capacity reserve, the additional amount of megawatts of capacity required beyond the load demand. The utility must be capable of providing this capacity within a given time interval, generally 10 minutes. Capacity reserve enables the system to serve the load demand even in the case of generating unit forced outages, load forecasting errors, and transmission-line outages or faults. The amount of reserve needed by a utility system or pool is calculated in several ways, and usually results in an on-line reserve requirement of 3–8% of load demand.

For example, one utility power pool requires an on-line reserve equal to the largest generating unit plus 100 MW for immediate response under governor control. Another utility power pool divides its reserve requirements into two parts: (1) a primary reserve with a 10-minute response time and (2) a secondary reserve with a 30 minute response time.

Reserve requirements can be met with generating capacity of two types: (1) on-line spinning reserve and (2) quick-start reserve. Spinning reserve is the amount of additional capacity that results from operating generating units at less than full output. That is, a 1000-MW unit that is operating at 700-MW output provides the system with 300 MW of spinning reserve credit. Quick-start reserve is the amount of capacity that can be started up and deliver power within the required response time. Gas turbines generally fall into this category as they can usually be started up, synchronized, and brought to full output within 10 minutes.

Although both of these reserve types appear equivalent in satisfying reserve requirements, there is some differentiation. Two reasons for differentiation are: (1) the quick-start reserve unit is a less reliable reserve unit since it may fail to start, and (2) electrical transients associated with a fault will be smaller if the unit is connected to the power system. This happens because a connected unit under governor control responds to the fault immediately, while quick-start units may require several minutes before they can aid in damping out fault transients. One power pool requires 1300 MW of 10-minute response time spinning reserve, at least 700 MW or that equal to the largest unit on line, (whichever is larger) must be on-line while the remaining 600 MW of reserve may be carried on quick-start units.

In computing reserve capability, consideration is given to the rate at which units can change their power output. Generating unit response rates (also referred to as "ramp rates") for increasing or decreasing electrical energy output depend on the type of generating unit, the magnitude of power excursion and the rate at which power change is taking place. Generally, the larger the power excursion, the slower the response rate. Thus for spinning reserves of 10 minutes or shorter, unit response time merits consideration.

Spinning reserve is generally shared among all units in the power system. Consider the case of a system with a load of 1000 MW that is to be met with 11 generators, 100 MW each with a response rate of 2%/minute for extended excursions. If 10 of the units are operated at full capacity, the load would be met. Only one unit would remain to provide the spinning reserve for which it could supply only 20% per 10 minutes, or 20 MW of reserve. If, however, all 11 units generated 90.9 MW in meeting the 1000-MW load, then the spinning reserve would be 2%/minute on all 11 units (up to their maximum power output), which would equal 100 MW of reserve capability in 10 minutes. Spinning reserve is most reliable when shared among all units in the power system.

Even if the response rate on one unit were adequate to meet spinning reserve requirements, system reliability considerations dictate having all units share reserve requirements so as not to lose all reserve capability should any one unit experience a forced or transmission outage. In planning simulations, generating units generally are given a maximum nameplate megawatt rating as well as a continuous megawatt rating that is less than nameplate and determined as the largest power the unit may deliver to share reserve requirement among all the committed units.

Area protection is another consideration in unit commitment. With load demand and generating sites spread out over a large service region, adequate generation capability must be ensured within each area to meet the area load demand. The area protection requirement is generally a consideration during nighttime valley load periods. Based on an analysis of area load demand and interarea transmission capability and reliability, an area protection requirement may be formulated as requiring a minimum amount of area generating capacity committed at all times (or a prescribed amount of area generation capacity reserves).

12.4.3 Operating Constraint Considerations in Unit Commitment

In addition to economic and system reliability considerations in unit commitment, consideration should be given to the physical operating constraints of

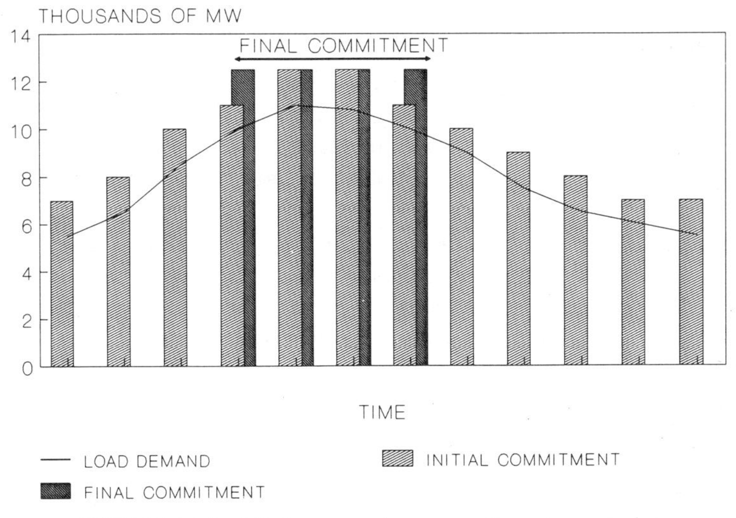

FIGURE 12.9 Minimum uptime commitment constraint.

the generating units. Types of operating constraints include: (1) unit minimum uptime rule, which permits a generating unit to be started up only if it will run for a minimum number of continuous hours (Figure 12.9), and (2) unit minimum down-time rule, which permits a unit to be shut down only if it will remain shut down for a minimum number of continuous hours (Figure 12.10). Both of these rules are largely a matter of judgment and are intended to provide time for temperature equalization in the generating unit, particularly the turbine, thereby controlling material stresses arising from thermal gradients within technical limits. These two rules influence system operation by increas-

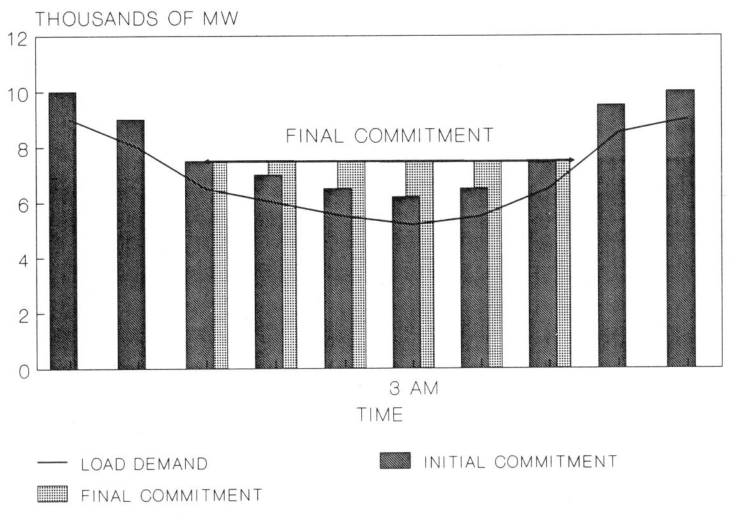

FIGURE 12.10 Minimum down-time commitment constraint.

ing the number of committed units. The minimum uptime rule increases commitment adjacent to peak-load periods and the minimum downtime rule increases commitment during valley-load periods.

Utility operating policy may add other operating constraints to unit commitment. These may include labor rules and crew scheduling policies for units scheduled for operation, and rules limiting the number of units started up each hour.

All system economic, reliability, and operating considerations are involved in unit commitment. Happ and colleagues (1971), illustrate a practical technique for combining these factors to achieve a minimal cost unit commitment. Fortunately, this involved procedure is not required to perform long-term power system simulations. While this section has discussed many of the practical factors in unit commitment, the next section presents a balanced technique for modeling unit commitment in planning analysis.

12.4.4 Simulating Unit Commitment

In modeling unit commitment for long-term planning simulations, the practical considerations are (1) selecting units in the commitment to minimize operating costs, (2) observing area protection rules, (3) maintaining reliable spinning reserve, and (4) observing unit minimum downtime rules. A four-step approach is used:

1. Units available for operation during the week are ranked in a priority list according to full-load operating cost in $/MWh.
2. The area protection rule is modeled as a minimum number of must-run units per power station. Thus, the area protection rule serves to rearrange the economc priority list of step 1 by placing the must-run units at the top of the list.
3. By proceeding according to the priority list, a minimum number of units are committed each hour of the week to meet both (1) the load, using the continuous operating ratings of the units, and (2) the load plus spinning reserve requirement, using the maximum ratings of the units.
4. The hourly commitment determined per step 3 is reviewed for unit minimum downtime violations. If violations exist, the unit with the violation is required to be committed for those hours between which the violation occurred.

Example. Compute the commitment for the following load and generating unit data given in Tables 12.1 and 12.2, respectively.

The spinning reserve requirement is 10% of the current load level. Area protection rules require that one unit be on-line at all times at the City and River plants.

TABLE 12.1 Load Data

Hour	Load (MW)
10 AM- 2 PM	1790
2 PM- 6 PM	1500
6 PM-10 PM	1200
10 PM- 2 AM	1100
2 AM- 6 AM	900
6 AM-10 AM	1500
10 AM- 2 PM	1790

SOLUTION

1. Develop a priority list based on \$/MWh.
2. Modify priority list to account for area protection rules. Place must-run units at the top of the piority list.
3. Commit minimum number of units for each load such that:

$$\sum_{\substack{\text{Committed} \\ \text{Units}}} \text{maximum rating} \geq \text{load} + \text{spinning reserve}$$

$$\sum_{\substack{\text{Committed} \\ \text{Units}}} \text{continuous rating} \geq \text{load}$$

4. Review commitment for minimum downtime violations.

The commitment order for the units developed in steps 1 and 2 is shown in Table 12.3. Note that the unit termed "City 1" was advanced in the priority order to observe the area protection rules.

TABLE 12.2 Unit Data

Plant	Unit	Continuous Rating (MW)	Maximum Rating (MW)	\$/MWh at Full Load	Minimum Down Time (Hours)
City	City 1	400	440	9.0	50
	City 2	200	200	30.0	2
	City 4	100	110	10.0	6
River	River 1	200	200	32.0	2
	River 2	400	430	6.0	50
	River 3	200	220	6.5	10
	River 4	200	220	9.5	10
Shore	Shore 2	200	200	31.0	2
	Shore 3	100	100	31.0	2

TABLE 12.3 Commitment Priority List

Priority	Plant Name	Unit Name	Continuous Rating (MW)	Maximum Rating (MW)	$/MWh	Down Time (Hours)
1	River	River 2	400	430	6.0	50
2	City	City 1	400	440	9.0	50
3	River	River 3	200	220	6.5	10
4	River	River 4	200	220	9.5	10
5	City	City 4	100	110	10.0	6
6	City	City 2	200	200	30.0	2
7	Shore	Shore 2	200	200	31.0	2
8	Shore	Shore 3	100	100	31.0	2
9	River	River 1	200	200	32.0	2

Table 12.4 shows the results of the preliminary commitment obtained by applying the two rules stated in step 3.

Table 12.5 compares the unit minimum downtimes to the actual downtimes determined from the preliminary commitment.

Comparison of column 4 with column 3 in Table 12.5 reveals that commitment unit number 4, River 4, was actually shut down for 4 hours, yet the minimum required shut down was 10 hours. Since this is a minimum downtime violation, River 4 must be kept on-line for the 2–6 AM time period.

The *final* commitment that does not violate minimum downtime is presented in Table 12.6.

TABLE 12.4 Preliminary Commitment Calculation

1	2	3	4	5	6
Hour	Load	Commitment Number Based on Continuous Ratings	Load + Spinning Reserve (10%)	Commitment Number Based on Maximum Ratings	Required Commitment Number Maximum of Columns 3 and 5
10 AM- 2 PM	1790	8	1969	9	9
2 PM- 6 PM	1500	6	1650	7	7
6 PM-10 PM	1200	4	1320	5	5
10 PM- 2 AM	1100	4	1210	4	4
2 AM- 6 AM	900	3	990	3	3
6 AM-10 AM	1500	6	1650	7	7
10 AM- 2 PM	1790	8	1969	9	9

TABLE 12.5 Unit Downtime Calculation

(1) Commitment Number	(2) Unit	(3) Minimum Downtime (Hours)	(4) Downtime per Commitment of Step 3: Column 6 (Hours)
1	River 2	50	Must run
2	City 1	50	Must run
3	River 3	10	0
4	River 4	10	4
5	City 4	6	8
6	City 2	2	12
7	Shore 2	2	12
8	Shore 3	2	20
9	River 1	2	20

12.5 THERMAL UNIT DISPATCH

While unit commitment planning establishes which units will be on-line for each hour, unit dispatch planning determines the share of the load demand that each unit delivers. The purpose of unit dispatch is to minimize the total system operating cost for delivering energy to meet the minute-by-minute system load demand. This minimum cost objective can be achieved by regulating the generation from each unit to take advantage of its unique operating cost characteristics.

12.5.1 Transmission Losses

On the average, transmission losses may result in energy losses of 1–2% of the total electricity generation. When the distance between the generation site and the load-demand centers is large, losses may also be large. Therefore, attention

TABLE 12.6 Final Commitment Number

Hour	Required Commitment Number
10 AM– 2 PM	9
2 PM– 6 PM	7
6 PM–10 PM	5
10 PM– 2 AM	4
2 AM– 6 AM	4
6 AM–10 AM	7
10 AM– 2 PM	9

should be given to transmission losses in scheduling generation. Alternatives exist between meeting load by operating less costly units that are sited more distant from the load center or operating more costly units closer to the load center. In scheduling distant units, the net cost improvement of generating units should equal or exceed transmission loss inefficiency in delivering this energy.

Transmission losses can be evaluated by using a loss formula based on a bus-impedance matrx approach (Wood and Wollenberg, 1984). With this approach, the real and reactive load power demand for each bus is specified along with the generator's real power and voltage magnitude. This procedure is accurate, but computation requirements are large because of the need for an alternating-current (AC) load-flow solution. If transmission losses are to be evaluated every hour or less, this technique can become exceedingly time consuming. For dispatch computers, this is acceptable; for planning studies, a simpler method is used.

Alternatively, transmission losses can be evaluated using a loss equation of the form (Kirchmayer, 1958):

$$P_{\text{loss}} = \sum_m \sum_n P_m B_{mn} P_n + \sum_n B_{n0} P_n + B_{00}$$

where P_n = power output from source n

B_{mn} = mutual loss coefficient between source m and source n (the principal contributing term in the loss formula)

B_{n0} = self-loss coefficient of source n

B_{00} = constant loss regarded as representing the intercept for the imaginary condition of zero system power.

The assumptions on which this formula is derived are:

1. The actual source bus voltages are determined both at peak-load and valley-load conditions, and the bus voltage between these two extremes varies linearly with the total system load.
2. Reactive power generation is determined to meet the specified bus voltage magnitudes during peak-load and valley-load conditions.
3. The substation load demand varies linearly with total load demand from the peak substation load to valley substation load demands.
4. Power factor at each load varies linearly with the total load demand from its value at the peak and its value at the valley.

The B coefficients are generally recomputed annually or whenever significant additional generation or transmission capability is added to the system.

This formula is in a convenient form and permits a rapid evaluation of the transmission losses for a given set of generator (or source) powers.

12.5.2 Economy Dispatch

The objective of unit dispatch is to operate the system to meet load demand in the most economical manner. This criterion is usually interpreted as minimizing system fuel costs, yet consideration can be given to include operation and maintenance costs that vary with the power output of the generating unit.

Let: P_n = power delivered by generating unit n (MW)
$F_n(P_n)$ = required fuel input (MBtu)/hour into unit n for an output electrical power, P_n (MW)
Load = total system load demand (MW)
λ = incremental power cost (\$/MWh), a LaGrange Multiplier
COSINC_n = incremental cost of fuel for unit n (\$/MBtu)
$\text{O\&M}_n$ = incremental operation and maintenance costs for unit n (\$/MWh).
C_0 = fixed costs of operation (\$).

The constraint that system load must be met by power output from the committed generators less transmission losses is expressed by the following equation:

$$\text{Load} = \sum_n P_n - \{ \sum_m \sum_n P_m B_{mn} P_n + \sum_n B_{n0} P_n + B_{00} \} \tag{12.1}$$

The total system operating cost per hour is the summation over all committed units of the product of the fuel cost times the fuel input, plus variable O&M times the megawatt output.

$$\text{Cost (\$)} = \sum_n [F_n(P_n) \cdot \text{COSINC}_n + \text{O\&M}_n \cdot P_n] + C_0 \tag{12.2}$$

If this cost is to be minimized relative to operation of each generating unit's power output, then the first derivative of the system cost with respect to each unit's power output must be zero. Since there is a constraint condition that the load demand must be met, this derivative of system cost must take into account that a decrease in one generator's output is at the expense of an increase in another generator's output. This constraint may be mathematically treated by adjoining the load constraint condition to the cost criteria via a LaGrange multiplier, λ (Kirchmayer, 1959; Hildebrand, 1962).

$$\text{Cost} = \sum_n F_n(P_n)\, \text{COSINC}_n + \text{O\&M}_n \cdot P_n + \lambda \cdot \{\text{LOAD} - \sum P_n + [\sum_m \sum_n P_m B_{mn} P_n + \sum_n B_{n0} P_n + B_{00}]\} \tag{12.3}$$

The derivative of the cost with respect to each unit's generation gives the dispatch condition.

$$0 = \frac{\partial F_n(P_n)}{\partial P_n} \text{COSINC}_n + \text{O\&M}_n - \lambda \left[1 - \sum_m (B_{nm} + B_{mn}) P_m - B_{n0}\right] \tag{12.4}$$

or

$$\left[\frac{\partial F_n(P_n)}{\partial P_n} \cdot \text{COSINC}_n + \text{O\&M}_n\right] \cdot \left[\frac{1}{1 - \sum_m (B_{nm} + B_{mn}) P_m - B_{n0}}\right] = \lambda$$

The term:

$$\sum_m (B_{nm} + B_{mn}) P_m + B_{n0}$$

is the incremental transmission loss for unit n and is denoted by:

$$\frac{\partial P_{\text{loss}}}{\partial P_n}$$

The first term:

$$\frac{\partial F_n(P_n)}{\partial P_n} \cdot \text{COSINC}_n + \text{O\&M}_n$$

is the incremental cost of the generating unit for delivering more power. This equation may be rewritten as:

$$\left[\frac{\partial F_n(P_n)}{\partial P_n} \cdot \text{COSINC} + \text{O\&M}_n\right] \cdot 1/\left[1 - \frac{\partial P_{\text{loss}}}{\partial P_n}\right] = \lambda \tag{12.5}$$

This states that maximum system economy results when (for every committed unit) the incremental generating cost plus the incremental cost of transmission losses are equal (except when a unit operates at minimum or maximum power output).

The solution of the dispatch equation [Equation (12.5)] generally is performed in dispatch control centers for on-line unit dispatch applications. In generation planning simulation, an approximation is usually made to reduce computer simulation costs. This approximation consists of representing the incremental transmission loss term as a constant penalty factor:

$$\frac{1.0}{1.0 - \sum_{m} (B_{nm} + B_{mn}) P_n + B_{n0}} = L_n \tag{12.6}$$

where L_n may be in the range of $1.0 \leq L_n \leq 1.10$.

12.5.3 Fuel Limitations and Hydro Coordination

The development of the dispatch equation in the previous section assumed that there were no fuel limitations that would restrict any unit from delivering dispatched power. For example, a fuel-supply gas contract may have a monthly maximum usage limitation of $1.0 \cdot 10^6$ MCF. If dispatched according to the dispatch equation [Equation (12.6)], gas consumption of $1.5 \cdot 10^6$ MCF may result. This is clearly a violation of the gas contract. To avoid this violation the unit will have to be restrained and used only during peak load, since there is insufficient energy in the gas contract to permit unconstrained operation over the month.

Other fuel-energy limitations exist for coal-, oil-, and nuclear-fired units. Multifueled units, such as gas/oil dual-fired units, with fuel consumption constraints and price differentials between the fuel type are another kind of fuel limitation (Stoll, 1974).

Pondage hydro units also have a fuel limitation, since the monthly or seasonal water inflow must last for the entire month or season. Operation must be adjusted so the unit can deliver energy during the peak load periods throughout that time period. Additionally, minimum flow restrictions must also be met.

Pumped-storage hydro units are fuel limited as well, since they have a finite tank capacity and an economic pumping limit in filling the tank at night. Coordination of hydro units with thermal unit generation will be presented in Section 12.6.

12.5.4 Interchange

Essentially all electric utilities in the U.S. and Canada are interconnected with neighboring utilities. The advantages for interconnection are improved system reliability and system economy. Power system reliability is improved by interconnection because neighboring utilities are capable of supplying capacity reserves. System economy is improved since a neighboring utility may be capable of generating additional power at lower cost or they may purchase power, thereby allowing fuller, more efficient loading of power plants. Hence, energy transactions can be executed that will be of mutual benefit.

There are three major classifications of energy transactions:

1. *Emergency Energy.* Emergency energy is delivered to a neighboring system when the neighboring system has no available generating capability to supply the energy demand. Generally, this energy is billed as the cost of the sender plus an adder (usually 10%).
2. *Economy Energy.* Economy energy is generated by one utility system and substituted for more expensive energy in another system. This energy is supplied from the seller's operating reserve with the purchaser sustaining the cost of transmission. This interchange is transacted with the seller's option, or purchaser's option, with cancellation at any time. In most two-party transactions, this energy is billed on a split-savings basis. Savings are computed as the buyer's decremental cost of generation and transmission less one-half of the savings, and the seller receives incremental cost plus one-half of the savings. These transactions may also be priced at the sellers cost plus either a fixed percent or a fixed $/MWh.
3. *Assured Economy Energy.* Assured economy energy is the same as economy energy except that the seller makes provision for its sale by increasing the generation on line. The energy is dedicated to the purchaser and is not derived from the seller's operating reserve. Nonetheless, should a generation or transmission contingency arise on the seller's system that requires this energy, the interchange contract may be canceled.

Interchange can be a significant contribution to a power system's generation and, consequently, may merit explicit representation in power system simulation studies. "Emergency energy" and "assured economy energy" may be modeled as specified megawatt sales or as purchases on an hourly basis. As such, they serve as a territorial load adder or subtractor. Economy energy is more difficult to model, since knowledge of the interchange utility's incremental cost on an hourly basis is required, as well as transmission capability constraints. If the transmission capacity between parties is large, transmission constraints may be neglected and all the interchange parties may be modeled as one effective company. If the transmission capacity is limited, a multiarea production simulation model would be required (see Chapter 13).

12.5.5 Environmental Dispatch

While the objective of unit dispatch is minimizing production costs in satisfying the load demand, other constraints enter that are essentially noneconomic. Environmental constraints dictate a modification of the dispatch from purely economic grounds. Environmental constraints are basically of two forms: (1) a limitation on environmental pollutants in the fuel on a pollutant per MBtu basis and (2) a limitation on ambient-air quality on the basis of pollutant per cubic meter of air at ground level. The first constraint can be met by fuel specifications. The second constraint, however, depends on the assembly of power plants in each region, the atmospheric dispersion and dilution of pollutants, fuel types, and the generating plant source rate of pollutants per hour. Since the generating plant source rate of pollutants is proportional to power output, the environmental constraint may place limitations on unit dispatch within given areas.

12.5.6 Simulating Economic Unit Dispatch in Production Simulations

In power system studies covering one month through several years, less modeling detail is required than for on-line dispatching. For on-line dispatching, hourly results are required, whereas in power system studies generally monthly results are reported. The monthly results are the integration of more than 700 hourly results over the month, and errors tend to average out. This averaging permits some relaxing of modeling detail.

The thermal characteristics of the generating units were presented in Section 12.3. Figure 12.3 (trend) illustrates a straight-line approximation between valve points of this characteristic. While sectionally quadratic approximations and higher order polynomials have been used, the sectional linear characteristic results in a computationally expedient and accurate model.

The dispatch equation [Equation 12.5], with a constant transmission penalty factor, becomes:

$$[\frac{\partial F_n(P_n)}{\partial P_n} \cdot \text{COSINC} + \text{O\&M}_n] \cdot L_n = \lambda \tag{12.7}$$

Following the model in Figure 12.3, the incremental heat rate model of the generating unit is shown in Figure 12.4 (trend). As can be noted, the incremental heat rate is a series of horizontal line segments between value points. Furthermore, the incremental energy cost (multiplying the incremental heat rate times the fuel cost plus the O&M cost times the penalty factor) is also a series of horizontal linc segments between valve points.

This stepped incremental cost characteristic of units permits construction of an efficient five-step dispatching algorithm:

1. Determine which units are on-line by performing a unit commitment for the specified hour.
2. Add up all the power from the on-line units when they are operating at minimum power points, since all on-line units must run at least at their minimum power points before they can load up additional value loading sections.
3. For all on-line units, rank the valve loading sections in order of smallest incremental cost to largest incremental cost. Plot this incremental cost against power system generation, taking into account the minimum power generation from step 2. This plot is illustrated in Figure 12.11.

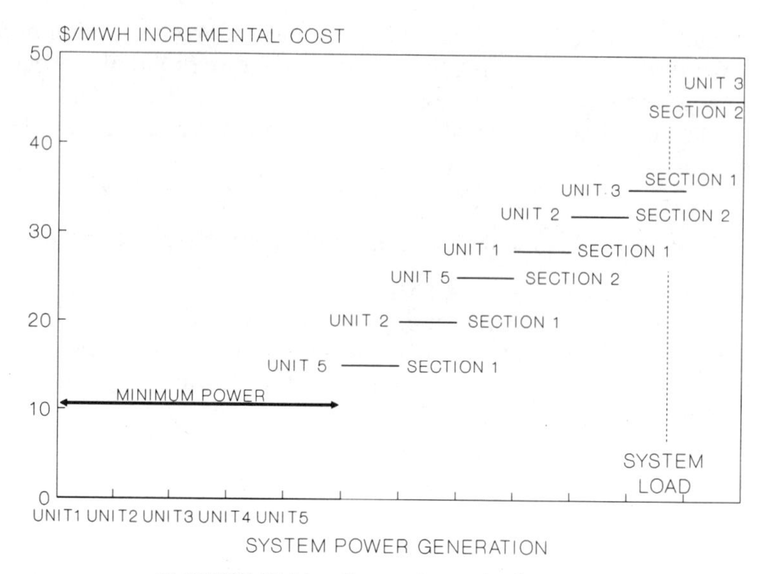

FIGURE 12.11 Computing unit dispatch.

4. Mark on the graph of Figure 12.11 the required system energy generation, which is equal to the system load demand.
5. Any unit with a power section to the left of the load-demand line will be loaded. Each unit's dispatch can be computed by addng up the power on its loading sections to the left of the load, together with its minimum power-point loading.

Example. The hourly load demand is 525 MW. Assume that transmission losses are negligible. The two committed generating units are Shore 1 and City 1. The data is shown in Table 12.7. Compute the fuel costs for the specified hour. (Neglect O&M costs in this example.)

TABLE 12.7 Unit Power Section Data

Unit	Fuel Cost ($/MBtu)	Maximum Power (MW)	Minimum Power (MW)	Fuel at Minimum Power (Mbtu/Hour)	Loading Section Number	Power of Section Number (MW)	$\frac{\partial F}{\partial P}$ (MBtu/MWh)
Shore 1	.50	400	200	2300			
					1	50	10.0
					2	100	11.0
					3	50	12.0
City 1	.55	200	100	1200			
					1	50	9.5
					2	50	11.0

SOLUTION

Step 1. Minimum power = 200 + 100 = 300 MW.

Steps 2 and 3. Rank loading section in order of incremental cost, \$/MWh. Dispatch the sections until the cumulative dispatch equals the load demand.

Unit	Section Number	Incremental Cost (\$/MWh)	Section Capacity (MW)	Section Dispatch (MW)	Cumulative Power (MW)
Minimums					300
Shore 1	1	5.0	50	50	350
City 1	1	5.225	50	50	400
Shore 1	2	5.5	100	100	500
Shore 1	3	6.0	50	25	525
City 1	2	6.05	50	0	525

Step 4. Operating cost:

$$\text{Cost of minimums} = 2300 \cdot .50\ \$/\text{MBtu} + 1200 \cdot .55 = 1810\ \$/\text{hour}$$

$$\text{Cost of sections} =$$
$$5.0 \cdot 50 + 5.225 \cdot 50 + 5.5 \cdot 100 + 6.0 \cdot 25 = 1211.25\ \$/\text{hour}$$

$$\text{Total fuel cost} = 1810 + 1211.25 = 3021.25\ \$/\text{hour}$$

12.6 HYDROELECTRIC GENERATION

The previous section presented the dispatch of thermal generating units without fuel limitations. Since hydroelectric generation has zero fuel cost, it should be dispatched first. Hydroelectric generation is usually energy limited, however, and cannot be dispatched in the same way as a thermal unit.

There are three general types of hydroelectric generation:

1. Run-of-river hydro is hydroelectric generation that has little or no water storage capacity. Electricity generation is dictated entirely on the hour-by-hour river flow conditions.
2. Pondage hydro has some capability for impounding water (such as a dam). Hydroelectric generation can be scheduled by the utility to achieve maximum economy consistent with reservoir limitations and any hourly minimum and/or maximum generation constraints arising from irrigation, navigation, or environmental considerations.

3. Pumped storage hydro has no principal natural water inflow. Rather, an upper reservoir is filled by pumping water from a lower reservoir generally at night or during low load periods. The water is then used to generate hydroelectric power during the day (or during high load-demand periods) and is exhausted to the lower reservoir.

12.6.1 Run-of-River Hydro Generation

Run-of-river hydro is the easiest form of hydro generation to simulate. It is specified on an hour-by-hour basis and is a direct modifier of load. No spinning reserve credit is involved because all available energy is used hourly.

12.6.2 Pondage Hydro

Pondage hydro is usually simulated on a monthly basis with the following characteristics:

- Energy to be generated during the month in megawatt-hours
- Maximum megawatts to be generated.
- Minimum megawatts to be generated.
- Spinning reserve credit allowed.

Subject to the previous constraints, an operating schedule may be developed that will, over the course of the month being studied, use pondage hydro to reduce the peaks of the load cycles. Figure 12.12 illustrates this action for a

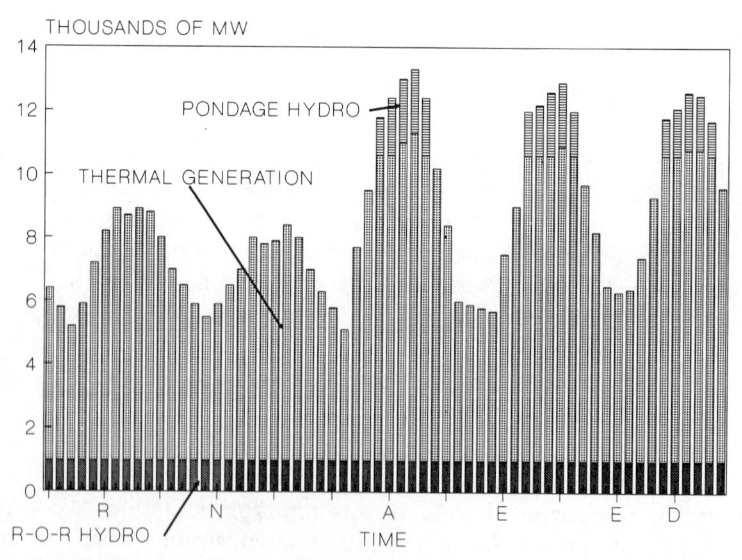

FIGURE 12.12 Scheduling pondage hydro to reduce peaks of the load cycle.

single pondage plant having limited storage, a maximum rating, and a zero minimum rating.

With pondage hydro, it is possible to take spinning reserve credit during hours when the plant is not generating at its maximum rating. A 100-MW pondage plant might be scheduled to produce only 60 MW during some hours of the day or weekend, therefore adding a spinning reserve credit of 40 MW.

Generally the individual pondage hydro plants can be assumed to be independent of each other and can, therefore, be scheduled sequentially. On some systems, however, multiple plants on the same river may have a significant impact on each other. This is particularly true if the reservoirs are small. In this case special logic must be employed to simultaneously schedule the plants.

12.6.3 Pumped Storage Hydro

For development of an economic operating schdule for a pumped storage hydro plant the following information is necessary:

- The hour-by-hour sequence of loads and spinning reserve requirements for the week.
- The array of thermal units available for service during the week and the priority in which these units would be committed to service.
- The spinning reserve credit that could be taken by the pumped storage plant, assuming that no pumped storage hydro generation were called for during the hour in question.

To develop a schedule, the following steps are taken:

Step 1. Develop an hour-by-hour thermal commitment schedule that would result if the pumped storage hydro (PSH) plant were to be operated at zero generation and credit taken for its contribution to spinning reserve. After reviewing this commitment schedule, the maximum and minimum thermal commitment that was made during the week can be determined.

Step 2. Calculate a thermal cost function relating hourly thermal costs ($/hour) to thermal power output (MW). The manner in which this cost curve is built up is shown in Table 12.8. This example has seven units, three of which must be committed on-line. The units are ordered in commitment priority from lowest to highest $/MWh cost. Because of the "must run" requirements, the thermal cost function is developed with these units at minimum load. The operating sections from minimum to maximum power are prioritized along with the other generating units.

In this example, assume that the PSH plant is 400 MW with a 1200-MWh reservoir (3-hour reservoir), and that it has a 66.7% overall cycle efficiency. Without loss of generality, it can be assumed that all of the losses occur in the pumping mode.

TABLE 12.8 Pumped Storage Schedule Development of the Thermal Cost Function

Commitment Priority	Unit Name	Minimum MW Power	Maximum MW Power	$/MWh	Commitment Policy	Cumulative MW Power
1	Lake 3	100		11	Must-run	100
2	City 4	100		18	Must-run	200
3	Shore 2	100		4	Must-run	300
4	Lake 3		200	10		500
5	City 4		200	17		700
6	Lake 1		300	20		1000
7	Lake 2		300	22		1300
8	City 3		300	35		1600
9	City 1		300	37		1900
10	Shore 2		200	40		2100

When the PSH plant generates power, it replaces the need for other generating units to produce power. The value of PSH generation ($/MWh) is the operating cost of the displaced generating units. The top portion of Figure 12.13 illustrates the value of PSH generation. The "generate cost" is based on the thermal cost function developed from the data in Table 12.8.

When the PSH plant is in the pumping mode, it consumes power, the cost of which is equal to the $/MWh cost of the incremental generating unit that is used to pump the PSH plant. However, because the losses are assumed to be accounted for in the pumping mode, the cost of pumping is equal to the cost of the incremental generating unit divided by the 66.7% efficiency of the PSH cycle. The upper curve on Figure 12.13 presents the pump cost, which is the thermal cost function curve divided by .667.

Whenever the "generation cost" value is higher than the pump cost (Figure 12.13), it is economical to execute a pump-generate cycle.

The lower graph in Figure 12.13 presents a daily load cycle rotated 90° from its typical orientation. The loads are plotted on the abscissa and the hours are plotted on the ordinate.

Consider using the PSH to generate during the peak-load hour (hour 16). For a first iteration, use the plant to reduce the thermal peak load in hour 16 to the level of hours 14 and 18. In this case, the thermal load is reduced from 1900 to 1700 MW by using the PSH plant to generate 200 MW. The energy generated during hour 16 must be replaced by pumping during the valley-load periods of hours 4 and 6. Since the plant is 66.7% efficient, 300 MWh of pumping is required to replace the 200 MWh of PSH generation. The translation of the final load points to the upper graph, "pumped storage cost function," indicates that the generation cost value is higher than the pump cost. Thus, this PSH iteration is cost effective.

The next step is to try additional generating and pump cycles. Figure 12.14 presents the final PSH schedule including the reservoir storage level. The

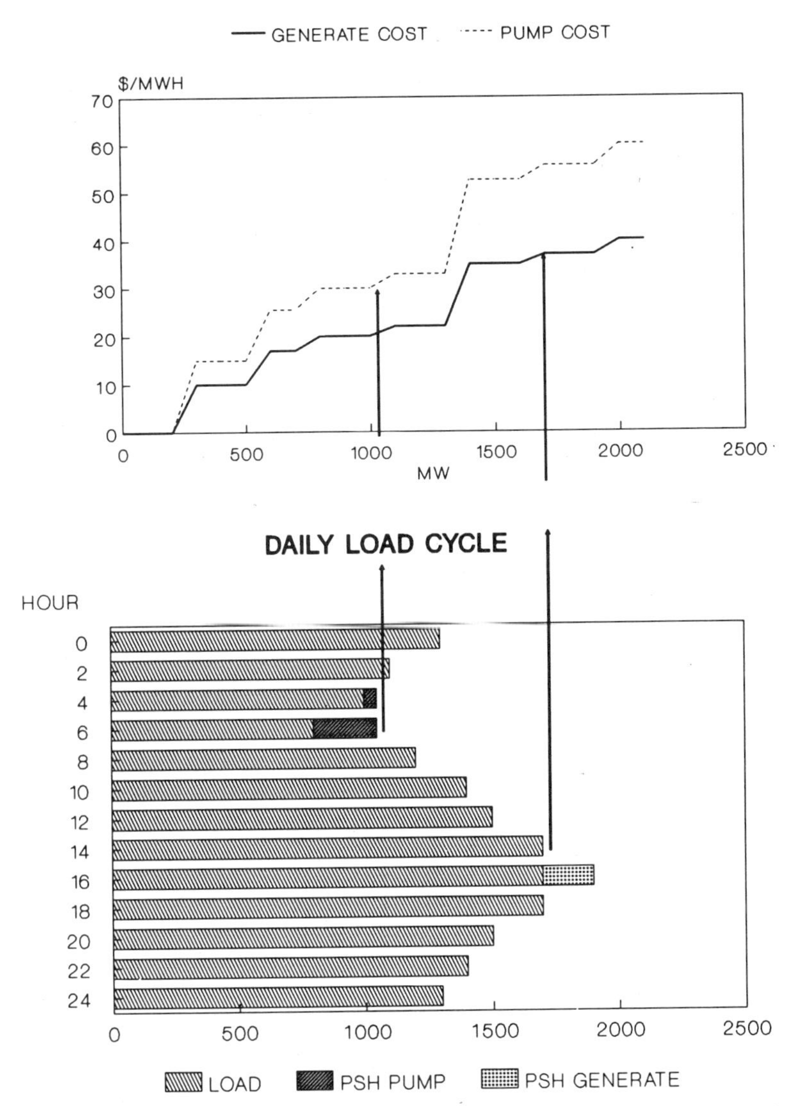

FIGURE 12.13 Scheduling of pumped storage hydro generation using a load curve and cost function.

PSH generation–pump cycle is at the limit of economics. Any additional PSH scheduling would only lead to increased costs.

The PSH schedule can be limited not only by economics, but by generate and pump capacity of the plant and PSH reservoir size as well. The reservoir schedule is shown on the right-hand graph in Figure 12.14. The reservoir is

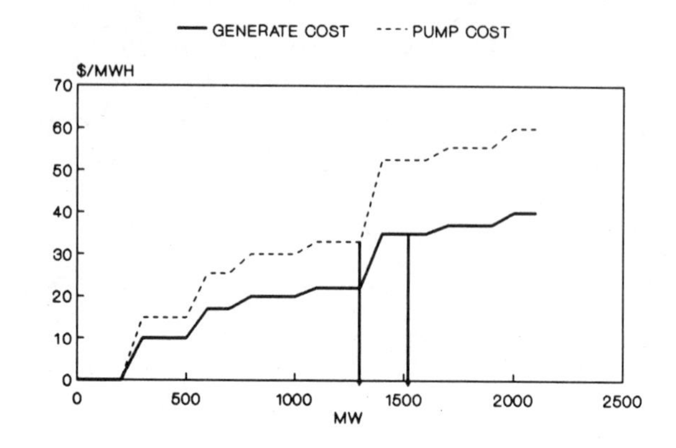

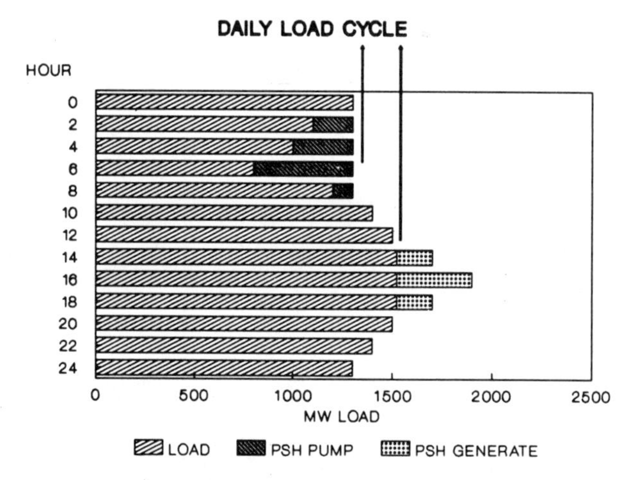

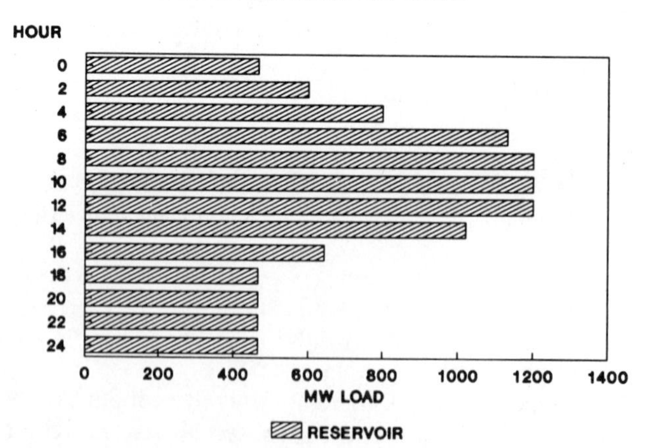

FIGURE 12.14 A pumped storage hydro schedule recognizing reservoir levels.

assumed to be filled by the morning of the day, hour 8 (8:00 AM). It gets depleted during the day as the PSH plant generates power, and is filled at night, hours 2 through 8. The reservoir can constrain the PSH schedule if it runs "dry" (zero energy) or is overfilled. The reservoir condition must be checked at each iteration. If a constraint occurs, then the PSH schedule must be modified to prevent violating the reservoir constraint.

While the previous example illustrates a daily PSH cycle, in practice the PSH cycle is developed on a weekly basis. The simulation method is similar to the example. Generally, with a weekly PSH schedule, weekend time is used to pump the reservoir to a full condition by Monday morning. Pumping is also done during weekday nights, but not enough to fully replenish the reservoir.

A weekly cycle generally relies heavily on refilling the reservoir on the weekend, causing the power system to begin successive weekdays with less and less storage. This declining storage condition can be of concern to the power

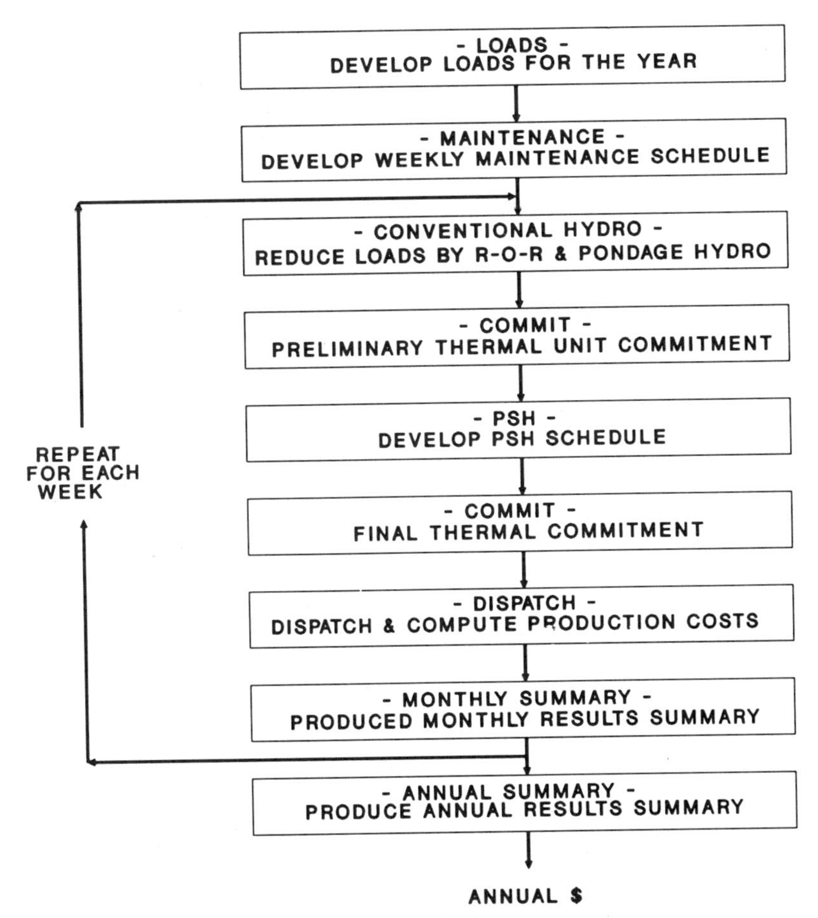

FIGURE 12.15 Annual generation production simulation procedure.

system operator because, in the event of generation outages or higher than anticipated loads, the PSH plant might have insufficient storage to permit full and extended operation. For this reason, some PSH plants are operated such that the reservoir must be refilled each night. This mandatory refill allows the hydro unit to enter each weekday with a full reservoir.

Following the development of the PSH schedule, a thermal commitment is recalculated since fewer units are needed as a result of the PSH generation.

12.7 SUMMARY

This chapter established the principles and techniques used in production simulation. Figure 12.15 presents a flowchart of this procedure. The next chapter expands on the commitment and dispatch section to include treatment of random forced outages and multiarea calculations.

REFERENCES

Happ, H. H., R. C. Johnson, and W. J. Wright, "Large-Scale Hydro-Thermal Unit Commitment Method and Results," *IEEE Transactions on Power Apparatus Systems,* Vol. PAS-90, 1971, pp. 1373–1384.

Hildebrand, F. B., *Advanced Calculus for Applications,* Prentice-Hall, Englewood Cliffs, NJ, 1962.

Kirchmayer, L. K., *Economic Operation of Power Systems,* Wiley, New York, 1958.

Kirchmayer, L. K., *Economic Control of Interconnected Systems,* Wiley, New York, 1959.

Stoll, H. G., "Operational System Planning with Fuel Energy Contract Constraints," IEEE 1974 Winter Power Meeting, C74 020-4.

Wood, A. J., and B. F. Wollenberg, *Power Generation Operation and Control,* Wiley, New York, 1984.

PROBLEMS

1 Using the unit data in Table 12.2, determine the commitment for the following loads:

Hours	Load (MW)
10 AM– 2 PM	1500
2 PM– 6 PM	1750
6 PM–10 PM	1810
10 PM– 2 AM	1350
2 AM– 6 AM	850
6 AM–10 AM	1250
10 AM– 2 PM	1500

Assume that area protection requires that at least one unit from the Shore plant must be on at all times and that the spinning reserve is 50 MW plus 7% of the load.

2 Using the generating unit data in Table 12.7, calculate the cost of serving a 532-MW load. Assume that 10% incremental losses occur from the Shore Plant and that the City Plant has 5% losses. How much would the dispatch cost if there were no losses?

3 Using the loads in Table 12.1 and thermal cost function in Table 12.8, develop a "pumped storage hydro" schedule. Assume that the PSH plant has a 500-MW pump–generate rating, is 70% efficient, and has a 5000-MWh storage capacity that is half-full at the beginning of the day and should end at the same level. Discuss how the action would change if the reservoir was only 3000 MWh and started half-full.

13

PRODUCTION SIMULATION II

This chapter expands on the principles of production simulation presented in the previous chapter; specifically (1) random forced outages and (2) multiarea production simulation, which accounts for energy transactions among utilities with transmission constraints.

13.1 PROBABILISTIC PRODUCTION SIMULATION

Since power plants experience unexpected forced outages, unavailability must be included in production simulation if accurate results are to be obtained. The previous chapter assumed that the power plants were in service except when on scheduled maintenance.

Four methods of accounting for forced outages will be discussed:

1. Derating, when the maximum power plant rating is decreased by the forced outage rate (FOR):

$$\text{Capacity} = \text{rating} \cdot (1 - \text{FOR})$$

2. Maintenance extension, when the scheduled outage rate is increased by the forced outage rate (FOR):

$$\text{Scheduled outage rate} \leftarrow \text{scheduled outage rate} + \text{FOR}$$

3. Monte Carlo, when numerous production simulation (histories) are calculated during each simulated hour using random numbers drawn for each unit in proportion to its FOR.

4. Probabilistic simulation, when extending the probability mathematics developed for generation reliability analysis (Chapter 10) to include an energy forecast. The results include (a) expected energy served by each generator, (b) expected production cost of each unit and the system, and (c) energy unserved.

Derating and maintenance extensions show peaking units running too little and interconnections used too little. The result is that production costs are lower by 5–10% when compared to historical records (Garver, 1978, p. 563).

The Monte Carlo method yields good results, but with a heavy burden of calculation time. The Monte Carlo method is effective in large multiarea simulations.

Probabilistic simulation, developed in the late 1960s, handles generating unit forced outages in production simulation in an efficient method. Probabilistic simulation is similar to the recursive convolution method of generation reliability studies.

13.1.1 Probabilistic Simulation Using Probability Mathematics

To illustrate the probabilistic simulation methodology, consider the three generating-unit power system example in Table 13.1 designed to serve a 200-MW load for one hour. The "TIE" simulates outside help from neighboring utilities via electrical interconnection to avoid unserved energy.

Production cost for the power system will be computed for three methods; two deterministic and one probabilistic.

Deterministic Solution. No inclusion of forced outages:
Table 13.2 presents the solution for this case.

Deterministic Solution. Derating units by forced outage rate:
Table 13.3 presents the solution results for this case.

Probabilistic Solution. Outage state enumeration results:
Table 13.4 presents the solution results for this case.

TABLE 13.1 Generation Data for Three-Unit Example of Probabilistic Production Simulation

	Capacity (MW)			
Unit	Minimum Power	Maximum Power	Forced Outage Rate	($/MWh)
A	0	100	0.10	15
B	0	100	0.20	25
C	0	100	0.25	40
TIE	0	1000	0.0	80

TABLE 13.2 Three-Unit Example of Production Cost Solved Without Forced Outages

Unit	Power Capacity (MW)	Dispatch (MWh)	$/MWh	$
A	100	100	15	1500
B	100	100	25	2500
C	100	0	40	0
TIE	1000	0	80	0
Total		200		$4000

TABLE 13.3 Three-Unit Example of Production Cost Solved by Derating Units by Forced Outage Rate

Unit	Derated Power Capacity (MW)	Dispatch (MWh)	$/MWh	$
A	90	90	15	1350
B	80	80	25	2000
C	75	30	40	1200
TIE	1000	0	80	0
Total		200		4550

TABLE 13.4 Three-Unit Example of Production Cost Solved by Complete Enumeration

Units in Service	Probability of the State	Dispatch of State (MWh)					$
		A	B	C	TIE	Total	
A,B,C,TIE	.540	100	100	0	0	200	4000
B,C,TIE	.060	0	100	100	0	200	6500
A,C,TIE	.135	100	0	100	0	200	5500
A,B,TIE	.180	100	100	0	0	200	4000
C,TIE	.015	0	0	100	100	200	12000
B,TIE	.020	0	100	0	100	200	10500
A,TIE	.045	100	0	0	100	200	9500
TIE	.005	0	0	0	200	200	16000
Expected dispatch		90	80	21	9	200	4910

Production costs increase from $4000/MWh for the deterministic, no FOR, case, to $4910/MWh for the probabilistic case, a 22% increase. Note that in the deterministic solution (Table 13.2), only units A and B are dispatched. In the probabilistic simulation (Table 13.4), all the power plants are operating: unit A = 90 MWh, unit B = 80 MWh, unit C = 21 MWh, and TIE = 9 MWh. The outage state enumeration procedure becomes very computer-time-consuming for a large power system. The Monte Carlo and probabilistic simulation methods are more useful. The probabilistic simulation method will be discussed first.

13.1.2 Probabilistic Convolution Method

The probabilistic simulation method uses a convolution algorithm to compute the probabilistic dispatch in a computer-efficient manner. Two important assumptions are made in the calculation:

1. No power flow limits exist between busses (i.e., there is adequate transmission capacity).
2. Scheduling (commitment and dispatch) order is unaffected by the unit outage pattern.

These two simplifying assumptions permit a manageable solution.

Two methods of viewing probabilistic simulation are helpful. Method 1 uses tables and starts from the load-duration curve alone. Method 2 uses graphs and starts from the dispatch, assuming that there are no forced outages. Both methods will convolve one generating section at a time and yield the same results.

The early developers of the probabilistic convolution method utilized an annual load duration curve (Booth, 1972). The need to account for scheduled maintenance shortened the time modeled to 52 weekly periods (Sager and Wood, 1973). Because the unit commitment changes during the week, it was necessary to further divide the week into constant commitment intervals within the week (typically 20–168 per week). In the examples that follow a single hourly load will be used to simplify the explanation. Load uncertainty and studying several hourly loads together (of the same commitment) use the same procedures that follow.

13.1.2.1 Three-Unit Example, Tabular Visualization. A 200-MW load demand for one hour is divided into increments of constant demand and generation (for this example there are 50-MW increments). If there were no generating units, all the load would be unserved or served by the TIE (interconnection from another utility). Thus, the convolution method begins with a table of unserved demand for this hour, as shown in Table 13.5.

The method then proceeds to sequentially calculate the expected loading of

TABLE 13.5 Example of 200-MWh Unserved Demand Modeled by 50-MW Increments

MW Increments Unserved	Expected Hours
1–50	1.0
51–100	1.0
101–150	1.0
151–200	1.0

each generating unit and its impact on the unserved demand. The probabilistic dispatch is completed when all the energy generation of each unit has been calculated. The results that will be calculated are shown in Table 13.6. The procedure to arrive at Table 13.6 will be described below.

Consider the addition of unit A, with a capacity of 100 MW and a forced outage rate of 10%. The load demand is first segmented into two states: one load state represents the 90% of the time when unit A is available (innage), and the other load state represents the 10% of the time when Unit A is on forced outage, as shown in Table 13.7.

For the segment when unit A is available (90% of the time), the first two 50-MW increments of demand are served. The remaining 100 MW, the third and fourth increments of demand, is still not served. Thus, the remaining unserved demand after unit A's dispatch becomes 100 MW modeled by the 1–50 MW and 51–100 MW steps, shown in Table 13.8.

For the segment when unit A is on outage (10% of the time), the unserved demand remains unchanged (Table 13.8). The total remaining unserved demand is then the sum of the two segments, the "convolution" of unit A's impact, shown in the last column of Table 13.8.

TABLE 13.6 Three-Unit Example of Production Cost Solved by Probabilistic Convolution

Unit	MW Increment	MWh Generation	$/MWh	Cost ($)	Reference
A	1–50	.9 • 50	15	675	Table 13.8
A	51–100	.9 • 50	15	675	Table 13.8
B	1–50	.8 • 50	25	1000	Table 13.10
B	51–100	.8 • 50	25	1000	Table 13.10
C	1–50	.21 • 50	40	420	Table 13.11
C	51–100	.21 • 50	40	420	Table 13.11
TIE	5–50	.085 • 50	80	340	Table 13.11
TIE	51–100	.085 • 50	80	340	Table 13.11
TIE	101–150	.005 • 50	80	20	Table 13.11
TIE	151–200	.005 • 50	80	20	Table 13.11
Expected Dispatch		200 MWh		$4910	

TABLE 13.7 Segmentation of 200 MWh Unserved Demand Before Unit A Dispatched

MW Unserved	Expected Hours During Innage of Unit A (90% of Time)	Expected Hours During Outage of Unit A (10% of Time)
1-50	.9	.1
51-100	.9	.1
101-150	.9	.1
151-200	.9	.1

The energy generation from unit A is recorded in Table 13.6. The total remaining unserved demand of Table 13.8 is applied as the net demand for the remaining generating units. This completes the convolution of unit A's dispatch into the unserved energy table.

Unit B and the remaining demand will now be convolved. Unit B has a 100-MW capacity and a 20% forced outage rate. The total remaining unserved demand (from Table 13.8) is segmented into a 80% innage period and a 20% outage period, as shown in Table 13.9.

When Unit B is available (80% of the time), it generates 100 MW and reduces the remaining unserved demand by 100 MW. Table 13.10 presents unit B's convolution result. Unit B generation is recorded in Table 13.6 and the unserved demand after convolving B is applied to unit C's convolution.

Finally, unit C (100 MW, forced outage rate = 25%) is convolved as shown in Table 13.11. The total remaining unserved demand of Table 13.11 must be served by the TIE. Energy generation is reported in Table 13.6, and the cost agrees with the outage state enumeration method outlined in Table 13.4.

The example may be generalized into a computational algorithm.

TABLE 13.8 Unit A Convolved with 200 MW Unserved Demand

MW	Expected Hours During Innage (90%)	Expected Hours During Outage (10%)	Probability of Total Remaining Unserved Demand
		Served	
1-50	.9 unit A generation (recorded in Table 13.6)		
51-100	.9 unit A generation (recorded in Table 13.6)		
		Unserved	
1-50	.9	.1	1
51-100	.9	.1	1
101-150		.1	.1
151-200		.1	.1

TABLE 13.9 Segmentation of Remaining Unserved Demand after Unit A and Before Unit B Dispatched

MW Unserved	Expected Hours During Innage of Unit B (80% Time)	Expected Hours During Outage of Unit B (20% Time)
1–50	.8 = 1 • .8	.2 = 1 • .2
51–100	.8 = 1 • .8	.2 = 1 • .2
101–150	.08 = .1 • .8	.02 = .1 • .2
151–200	.08 = .1 • .8	.02 = .1 • .2

TABLE 13.10 Unit B Convolved with Remaining Unserved Demand

MW	Expected Hours During Innage (80%)	Expected Hours During Outage (20%)	Probability of Total Remaining Unserved Demand
	Served		
1–50	.8 unit B generation (recorded in Table 13.6)		
51–100	.8 unit B generation (recorded in Table 13.6)		
	Unserved		
1–50	.08	.2	.28
51–100	.08	.2	.28
101–150		.02	.02
151–200		.02	.02

TABLE 13.11 Unit C Convolved with Remaining Unserved Demand

MW	Expected Hours During Innage (75%)	Expected Hours During Outage (25%)	Probability of Total Remaining Unserved Demand
	Served		
1–50	.21 = .28 • .75 unit C generation (recorded in Table 13.6)		
51–100	.21 = .28 • .75 unit C generation (recorded in Table 13.6)		
	Unserved		
1–50	.015 = .02 • .75	.07 = .28 • .25	.085
51–100	.015 = .02 • .75	.07 = .28 • .25	.085
101–150		.005 = .02 • .25	.005
151–200		.005 = .02 • .25	.005

Let:

$PROB^{new}$ (MW) = probable unserved hours at a system demand of MW after the convolution of the next generating unit

$PROB^{old}$ (MW) = probable unserved hours at a system demand of MW before the convolution of the next generating unit, $PROB^{old}$ (negative MW) = 0

$FOR_{unit\ j}$ = forced outage rate of unit J

The tabular method leads to a recursive formula, in which unit J is the next unit in the dispatch order to be convolved.

$$PROB^{new}(MW) = PROB^{old}(MW + Capacity_{unit\ J}) \cdot (1 - FOR_{unit\ J}) + PROB^{old}(MW) \cdot FOR_{unit\ J} \tag{13.1}$$

For example, convolving unit C for the second step of unserved demand table (Table 13.11) would begin with the results from the "total remaining unserved demand" column of Table 13.10 as:

$$PROB^{new}(51 \text{ to } 100) = .02\,(1 - .25) + .28\,(.25) = .015 + .070 = .085$$

When MW + Capacity is beyond the load demand, 200 + 100, for example, then $PROB^{old}$ (MW + capacity) = 0. For example:

$$PROB^{new}(151 \text{ to } 200) = 0 \cdot (1 - .25) + 0.02 \cdot 0.25$$

Also, when X MW is negative but MW + Capacity is positive, − 50 + 100 for example, then the "Served" load is computed. For example, the probability of served load on the 50-MW unit segment of unit C is:

PROB of served load on unit C

$$= PROB^{new}(-50) = PROB^{old}(-50 + 100) \cdot (1 - FOR_{unit\ J})$$

$$= .28 \cdot (1 - .25) = .21$$

The expected power outout from unit J is calculated as the summation over the demand increments corresponding to the dispatch location of unit 3.

$$\text{Expected power output (unit J)} = \sum_{\substack{MW \\ unit\ J}} PROB^{old}(MW + capacity_{unit\ J}) \cdot (1 - FOR_{unit\ J}) \cdot MW \tag{13.2}$$

For example, for unit C the expected output using Table 13.11 is (where 1 − FOR = 1 − .25 = 0.75):

$$\text{Expected power (unit C)} = \text{PROB}^{\text{old}}\,(-\,50 + 100) \cdot 0.75 \cdot 50$$
$$+\ \text{PROB}^{\text{old}}\,(-\,100 + 100) \cdot 0.75 \cdot 50 = 0.28 \cdot 0.75 \cdot 50$$
$$+\ 0.28 \cdot 0.75 \cdot 50 = 10.5 + 10.5 = 21\ \text{MW}$$

13.1.2.2 Three-Unit Example, Graphical Approach. Figure 13.1 begins the graphical visualization of the probabilistic convolution algorithm using a one-hour 200-MW load served by the generating units listed in Table 13.1 and assuming that all units are available (no forced outages). In this case, units A and B serve the load demand for this hour. Time is plotted on the abscissa in Figure 13.1. The one-hour period is represented as 100% running time. In the deterministic production simulation, only units A and B are dispatched for the entire hour; unit C is not needed.

Probabilistic production simulation requires a new interpretation of the graph in Figure 13.1. The abscissa represents the *probability that a generating unit will be called on to operate during the hour.* It may also be interpreted as the probable or expected amount of operating time.

Consider the event that unit A will have a forced outage and will not be available during the hour (unit A has a 10% forced outage rate). When unit A is on forced outage, unit C will be called on to serve the load. Note that unit B cannot serve any additional load because it will be operating for the entire hour, regardless of whether unit A is on forced outage. Since unit A is on forced outage 10% of the time, unit C must be called on to operate for 10% of the time. This situation is illustrated in Figure 13.2.

Now consider the event in which unit B has an outage (unit B has a 20% forced outage rate). Referring to Figure 13.3, when unit B outages, Unit C

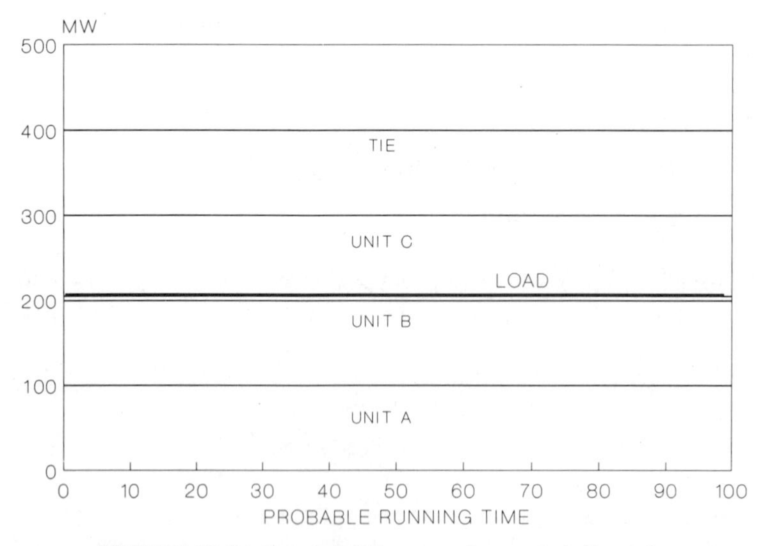

FIGURE 13.1 Deterministic case of expected dispatch.

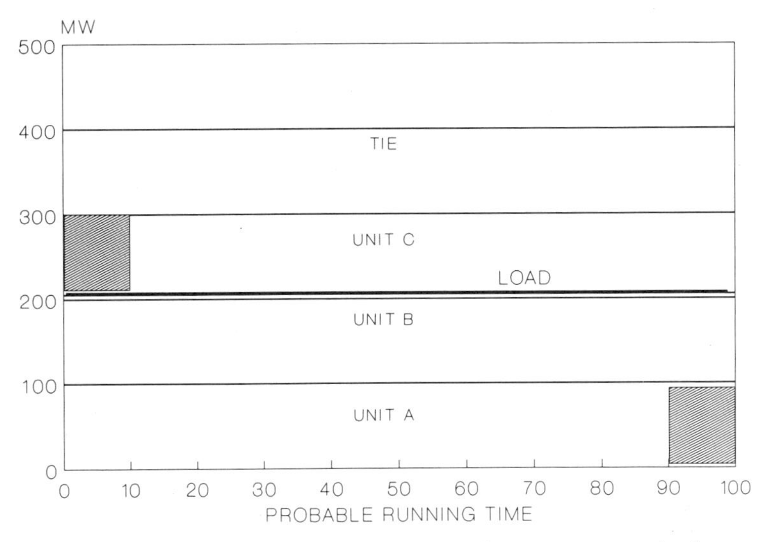

FIGURE 13.2 Expected dispatch with unit A outage convolved.

will be called on to serve the outage energy of unit B. However, unit C cannot always serve load from the outage of unit B since unit C may already be operating as a result of the outage of unit A. In the event that unit C cannot serve all of unit B's outage energy, the TIE will be called on. Therefore, either unit C or the TIE could serve load in the event of an outage of unit B.

The next step is to calculate the probability that unit C will be called on to generate as a result of unit B's outage. Since unit C has a probable operating

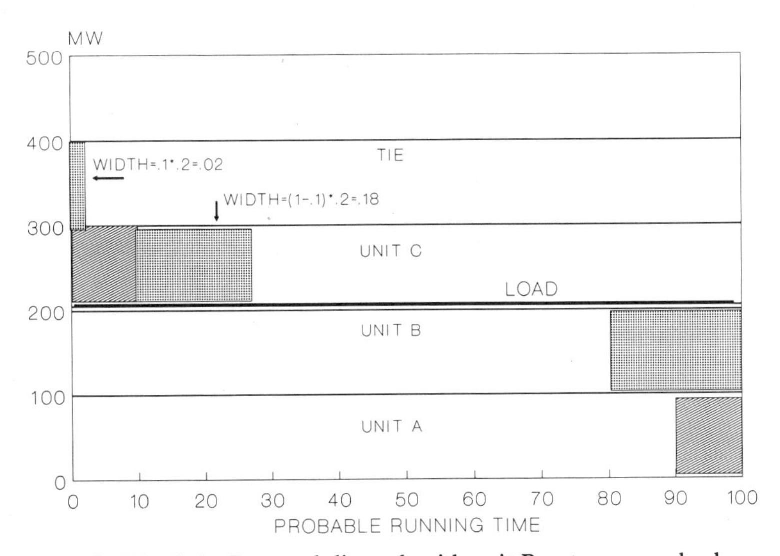

FIGURE 13.3 Expected dispatch with unit B outage convolved.

time of 10%, unit C has a 90% probability of being able to serve load if called on as the result of an outage of unit B. Since unit B has a forced outage rate of 20%, unit C is expected to serve 90% of the outage energy, or .90 • .20 = .18. Since unit C can be called on only 90% of the time to deliver load in the event of a unit B outage, the TIE must be called on the balance of the time, or .10 = (1.0 − .9). Thus, the outage energy of unit B that is served by the TIE is .1 • .2 = .02.

Just as units A and B are susceptible to forced outages, unit C can also have forced outages (unit C has a 25% forced outage rate). Unit C is called on to serve load 28% of the time, as shown in Figure 13.4. Since unit C has a forced outage rate of 25%, the TIE must be called on to serve load 7% of the time (.28 × .25 = .07). The TIE is already operating at 100 MW for 2% of the time, so the probability that the TIE can serve load due to outages on C is (.28 − .02)/.28 = .9286. Therefore, the TIE operates .07 • .9286 = .065 of the time to serve outages on C. Similarly, the next 100 MW of the TIE must serve .07 • (2/28), or .005 of the time due to outages on C. The final result is the "energy summary" in Table 13.6.

The technique illustrates how the effective load demand is convolved up the load curve. The result is that all of the generating units have a probability of being called on to serve load.

13.1.3 Multipower Section Representations on Probabilistic Simulations

The prior presentation dealt with power plants that were represented by one power section, from zero minimum to full maximum power. However, most generating units have a minimum loading; for steam units, this minimum is

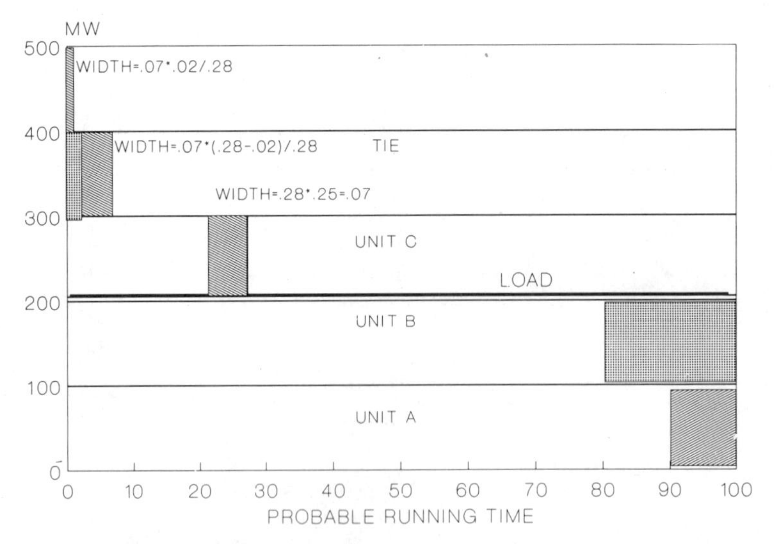

FIGURE 13.4 Expected dispatch with unit C outage convolved.

from 15% to 50% of the maximum rating. It may also be desirable to represent the heat rate characteristic by several power sections. Where power plants are represented by more than one power section, a modification to the recursive convolution algorithm is needed.

As an illustration, consider the two-unit power system in Table 13.12:

TABLE 13.12 Generation Data for Two-Units-with-Minimum-Sections Example of Probabilistic Production Simulation

Unit	Minimum Output (MW)	Maximum Output (MW)	Fuel at Minimum Output (MBtu/Hour)	Incremental Heat Rate from Minimum to Maximum Power (MBtu/MWh)	Forced Outage Rate	Fuel Cost ($/MBtu)
A	50	100	300	10.0	.10	200
B	50	150	300	10.0	.20	300
TIE	0	1000	—	10.0	0	600

The expected production dispatch for a demand of 200 MWh in Table 13.5 will be computed by using the dispatch priority outlined in Table 13.13.

First A_{min} and then B_{min} are convolved with the remaining energy demand (Table 13.14).

The convolution of $A_{section}$ could follow A_{min} and B_{min}, except that $A_{section}$ and A_{min} belong to the same generating unit. Thus A_{min} and $A_{section}$ will outage simultaneously. This fact must be accounted for in the simulation process.

The multistate character of unit A is recognized by first deconvolving A_{min} and then convoluting the combined rating of A_{min} *and* A_{sect}.

Deconvolve A_{min} by recalling Equation (13.1), repeated here:

$$\text{PROB}^{new}(\text{MW}) = \text{PROB}^{old}(\text{MW} + C) \cdot (1 - \text{FOR}) + \text{PROB}^{old}(\text{MW}) \cdot (\text{FOR}) \quad (13.3)$$

where C = capacity of the unit being deconvolved and FOR = the unit's forced outage rate. Solving for $\text{PROB}^{old}(\text{MW})$ yields:

$$\text{PROB}^{old}(\text{MW}) = \frac{\text{PROB}^{new}(\text{MW}) - (1 - \text{FOR}) \cdot \text{PROB}^{old}(\text{MW} + C)}{\text{FOR}} \quad (13.4)$$

TABLE 13.13 Priority List for Two Units-with-Minimum-Sections Example

Unit/Section	MW
A_{min}	50
B_{min}	50
$A_{section}$	50
$B_{section}$	100

TABLE 13.14 Units A_{min} and B_{min} Convolved with 200-MWh Unserved Demand

MW	Unit	Initial Hours	A_{min} Convolved (10% FOR)	B_{min} Convolved (20% FOR)
		Served		
1–50	A_{min}		0.9	
			—	
1–50	B_{min}			0.8
				—
		Unserved		
1–50		1.0	1.0	1.0
51–100		1.0	1.0	1.0
101–150		1.0	1.0	.28
151–200		1.0	0.1	.02

The calculations are shown in Table 13.15. Use Equation (13.4) to deconvolve A_{min} for the unserved demand with unit A removed. The "$PROB^{old}$" of Table 13.15 is completed by starting at the "151–200" entry and using $PROB^{old}(MW + C) = 0.0$ when $MW + C$ is larger than the demand, 200 MW in this example. The sequence of calculations proceeds up the table as indicated by the sequence numbers in Table 13.15.

The convolution process for the entire unit ($A_{min} + A_{section}$) is now performed as illustrated in Table 13.16.

Deconvoluting B_{min} and convoluting B_{min} and $B_{section}$ is performed in Table 13.17.

TABLE 13.15 Unserved Demand While A_{min} Is Being Deconvolved

MW Unserved	A_{min} and B_{min} Convolved $PROB^{new}$	B_{min} Convolved A_{min} Deconvolved Capacity = 50 MW FOR = .10 $PROB^{old}$	Computation Sequence
0–50	1.0	1.0	(4)
51–100	1.0	1.0	(3)
101–150	.28	1.0	(2)
151–200	.02	0.2	(1)

Computation Sequence

1. $(.02 - 0 \cdot 0.9)/0.1 = 0.2$
2. $(.28 - 0.2 \cdot 0.9)/0.1 = 1.0$
3. $(1.0 - 1.0 \cdot 0.9)/9.1 = 1.0$
4. $(1.0 - 1.0 \cdot 0.9)/9.1 = 1.0$

TABLE 13.16 Unit A_{min} Plus $A_{section}$ Convolved with Remaining Unserved Demand[a]

MW	Expected Hours During Innage (90%)	Expected Hours During Outage (10%)	Probability of Total Remaining Unserved Demand
	Served		
1–50	A_{min} is not calculated; Table 13.14 has the correct value		
50–100	$A_{section}$ 1.0 • 0.9		
	Unserved		
1–50	1.0 • 0.9	1.0 • 0.1	1.00
51–100	0.2 • 0.9	1.0 • 0.1	.20
101–150		1.0 • 0.1	.10
151–200		0.2 • 0.1	.02

[a]*Notes:* (1) Use the A_{min} answer from the previous Table 13.14, where it was correctly computed; (2) transfer $A_{section}$ answer to the table of served energy (Table 13.18).

TABLE 13.17 Unit B_{min} Deconvolved, Then Unit B_{min} + $B_{section}$ Convolved with Remaining Unserved Demand

MW	A_{min} + $A_{section}$ and B_{min} Convolved	Deconvolve B_{min} C = 50 FOR = .2	Computation Sequence	Convolve B_{min} + $B_{section}$ C = 500 FOR = .2	Computation Sequence
		Served			
1–50	B_{min} is not calculated; Table 13.14 has the correct value				
51–100				0.80	10
101–151				0.08	9
		Unserved			
1–50	1.00	1.0	4	0.28	8
51–100	.28	1.0	3	0.20	7
101–150	.10	0.1	2	0.02	6
151–200	.02	0.1	1	0.02	5

Computation Sequence

1. (.02 − 0.0 • 0.8)/.2 = 0.1, Equation (13.4)
2. (.10 − 0.1 • 0.8)/.2 = 0.1, Equation (13.4)
3. (.28 − 0.1 • 0.8)/.2 = 1.0, Equation (13.4)
4. (1.00 − 1.0 • 0.8)/.2 = 1.0, Equation (13.4)
5. (0 • 0.8 + 0.1 • 0.2) = 0.02, Equation (13.1)
6. (0 • 0.8 + 0.1 • 0.2) = 0.02, Equation (13.1)
7. (0.1 • 0.8 + 1.0 • 0.2) = 0.20, Equation (13.1)
8. (0.1 • 0.8 + 1.0 • 0.2) = 0.28, Equation (13.1)
9. (0.1 • 0.8 + 0 • 0.2) = 0.08, Equation (13.1)
10. (1.0 • 0.8 + 0 • 0.2) = 0.80, Equation (13.1)

The two-units-with-minimum-sections example is summarized in Table 13.18. Note that the total generation equals the energy demand, 200 MWh.

In summary, recognition of the simultaneous outage of multiple power sections requires a deconvolution step and a subsequent convolution step.

13.1.4 Synthetic Outage Distribution Approximation

The recursive convolution method is computationally time-consuming. Typically, a recursive convolution method would involve 5–10 times the necessary computer time compared to a deterministic dispatch calculation. Consequently, a faster approximate method is often used for large (≥ 50 units) systems.

Computer speedup is accomplished by approximating the cumulative outage distribution by a synthetic distribution such as the binomial distribution. The basis for this method is the special case in which all the generating units have the same size and forced outage rate. If this is the case, then the probability that n out of a total of N units are on forced outage is merely the binomial probability distribution.

$$\text{Prob}(n \text{ on outage}) = \frac{N!}{n!\,(N-n)!}(\text{FOR})^n(1-\text{FOR})^{N-n} \qquad (13.5)$$

In this case, N units are being called on to serve the load demand.

Suppose that there are four units of 100 MW with a 10% forced outage rate (all called "unit A") serving a 400-MW load. The binomial distribution probability of zero to four units on outage is shown in Table 13.19.

The cumulative outage probability for this example is given in Table 13.20.

If 100–400 MW are on outage, then units above these first four units must be called on to serve the 400-MW demand. Figure 13.5 illustrates this loading.

TABLE 13.18 Two-Units-with-Minimum-Sections Example of Production Cost Solved by Probabilistic Convolution

Unit	MW Increment	MWh Generation	Reference
A_{min}	1–50	.9 • 50	Table 13.14
$A_{section}$	51–100	.9 • 50	Table 13.16
B_{min}	1–50	.8 • 50	Table 13.14
$B_{section}$	51–100	.8 • 50	Table 13.17
$B_{section}$	101–150	.08 • 50	Table 13.17
TIE	1–50	.28 • 50	Table 13.17
TIE	51–100	.20 • 50	Table 13.17
TIE	101–150	.02 • 50	Table 13.17
TIE	151–200	.02 • 50	Table 13.17
Expected dispatch		200 MWh	

TABLE 13.19 Binominal Probability Distribution with Four Identical 100-MW Units

Units on Outage	MW on Outage	Probability
0	0	.6561
1	100	.2916
2	200	.0486
3	300	.0036
4	400	.0001
		1.0000

TABLE 13.20 Cumulative Outage Probability with Four Identical Units

Units or More on Outage	MW or More on Outage	Cumulative Probability
0	0	1.00
1	100	.3439
2	200	.0523
3	300	.0037
4	400	.0001

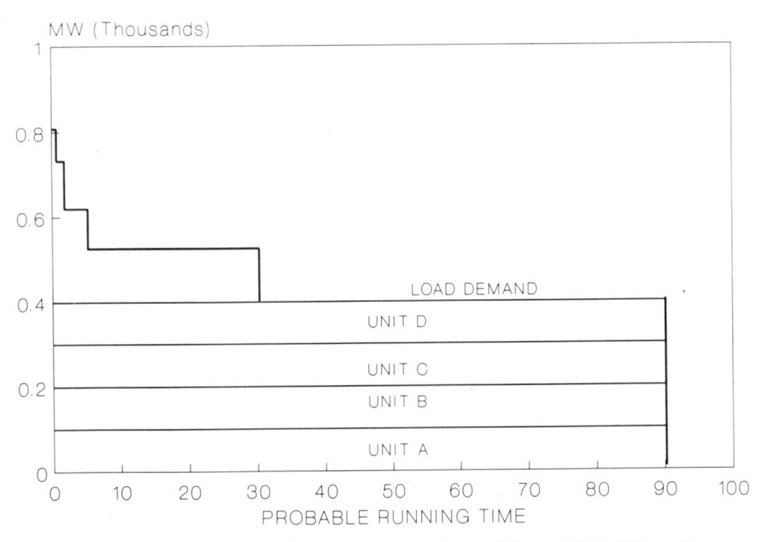

FIGURE 13.5 Binomial approximation of four 100-MW units.

Thus, for identical units dispatched to serve the original load, the binomial distribution can be used to calculate the outage energy distribution due to base-load unit outages.

Energy generation by units above the load line is dispatched to serve the outage energy from base-load units. Because these generating units also have forced outage rates, they should be convolved as well. However, a suitable approximation is to derate these units by the forced outage rate.

Although binomial distribution is correct if the initially loaded units are identical, it is a very good approximation of nonidentical units results if average megawatt size and average outage rates are used.

Consider the example in Table 13.21 of a utility with four generating units and an emergency energy cost (TIE cost) of \$50/MWh.

Consider the case of a 250-MW load. The average megawatt size of the initially dispatched units serving a 250-MW load is:

$$\frac{100 + 100 + 50}{3} = 83.33 \text{ MW}$$

The average forced outage energy of the initially dispatched units is:

$$(100 \cdot .1 + 100 \cdot .15 + 50 \cdot .10) = 30 \text{ MWh}$$

The average forced outage rate is then:

$$\frac{30 \text{ MWh}}{250 \text{ MWh}} = .12$$

The binomial probability distribution for three units is shown in Table 13.22.

The cumulative probability distribution is shown in Table 13.23.

The outage probability distribution can then be regarded as placing additional loads on units above the initially dispatched units on the power system. The loads and probabilities are shown in Table 13.24.

Power plants can be arranged in order of dispatch. In units above the deterministic load, the forced outage rate is included by derating the unit's capacity, Table 13.25.

TABLE 13.21 Four-Nonidentical-Units Example for Binomial Distribution

Unit	MW Capacity	Forced Outage Rate	Operating Cost (\$/MWh)
A	100	.10	20
B	100	.15	25
C	150	.10	30
D	50	.20	35
TIE	1000	0	50

TABLE 13.22 Three-Unit Example of Binominal Probability Unit Size = 83.33 MW, 0.12 FOR

Units on Outage	MW on Outage	Probability	Outage Energy (MWh)
0	0.00	.681472	
1	83.33	.278784	23.231
2	166.66	.038016	6.337
3	249.99	.001728	.432
		1.000000	30 MWh

TABLE 13.23 Three-Unit Example of Cumulative Probability Unit Size = 83.33 MW, 0.12 FOR

Units or More on Outage	MW or More on Outage	Cumulative Probability
0	0	1.0
1	83.33	.318528
2	166.66	.039744
3	249.99	.001728

TABLE 13.24 Additional Loads Above Initial Dispatch for Three-Unit Example

Load	Probability of Being Called on to Serve Load
250	1.0
333.33	.318528
416.66	.039744
499.99	.001728

TABLE 13.25 Cumulative Derated Capacity for Four-Unit Example

Unit	Capacity	Derated Capacity	Cumulative Derated Capacity
A	100	100	100
B	100	100	200
C	50	50	250
	Deterministic Load		
C	100	90	340
D	50	40	380
Emergency TIE	1000	1000	1380

The system loading diagram outlined in Table 13.26 is computed as follows. For the 333.33-MW load of Table 13.24 with a .318528 probability, the segment of unit C above the deterministic load would operate for (333.33 − 250) = 83.33 MW. For the 416.66-MW load, unit C would operate for an additional (340 − 333.33), or 6.67 MW for .039744 hour. Unit D would operate for 40 MW, and the emergency TIE would operate for (416.66 − 380), or 36.66 MW. For the 499.99 MW load, the emergency TIE operates for an additional (499.99 − 416.66), or 83.33 MW.

The total cost of the binomial dispatch is $6215.45. The cost of the dispatch using the exact recursive convolution is $6262, or .7% more than the approximate, binomial dispatch.

For more typical power systems with 50–100 generating units, the binomial dispatch typically achieves an accuracy in the range of .25–.50% and a computational speed advantage of a factor of 5, compared to the exact recursive convolution method. The binomial method is a useful technique for incorporating the effects of random forced outages.

While the binomial approximation is very useful, binomial distribution for a large number of units approaches the Poisson distribution. The Poisson distribution coefficients are easier to calculate than the binomial distribution coefficients. Thus, the Poisson dispatch technique is also used (Marsh et al., 1974). Higher-order Gaussian distribution methods (see Chapter 11) are also used.

13.2 MULTIAREA PRODUCTION SIMULATION

Utilities in the United States are interconnected with other utilities, in some cases, through a formalized power pool. Interconnection not only offers reliability benefits but operating cost benefits as well. Since neighboring utilities may be able to generate power at some periods of the day at a lower cost than another given utility, there is incentive for utilities to purchase power from others. There can be mutual benefits to each party. If the power may be pur-

TABLE 13.26 Four-Unit Summary Based on the Binomial Approximation

Unit	MWh	Availability	Expected MWh	Expected Cost ($)
A	100	.90	90	1800
B	100	.85	85	2125
C	50	.90	45	1350
C	83.33 • 318 + 6.67 • .040	1.0	26.76	802.8
D	40 • .040	1.0	1.6	56
TIE	36.66 • .04 + 83.3 • .002		1.633	81.65
			250	6215.45

chased at less than the purchaser's generating cost and sold at a higher than seller's generating cost, then there are mutual incentives for both purchaser and seller to conduct the business transaction.

If a utility conducts a large number of power purchases and sales or desires to analyze such an opportunity, then a multiarea production simulation is required. This procedure simulates the operation of each utility, calculates when economy interchanges may occur, and calculates the interchange purchase and sale pricing. When there are transmission limitations on the interchange of power, the multiarea production simulation must recognize these and operate the multiarea power system to assure that there are no constraint violations.

In Chapter 12 and in this chapter thus far, electric utility simulations were conducted according to the assumption that there was adequate transmission capacity among all the generating units in the utility (or power pool) power system. This section introduces two transmission system models used to represent transmission limitations: transportation models and linear electric transmission models. The linear electrical transmission model will be discussed first, followed by details of how the model is created. Security constrained dispatching is modeled in Section 13.2.6. Beginning in Section 13.2.10, the nonelectrical transportation model is presented and illustrated for a two-company case. Interchange billing is discussed in Section 13.2.12. In Section 13.2.13 the same two companies are modeled by a three-bus transmission network and the costs differences noted. Finally, the solution method using linear programming is discussed.

13.2.1 Transmission Constraint Models

The transmission network interconnecting utilities may be comprised of many transmission lines of many voltage levels. A theoretically correct procedure would be to model all the interconnecting lines and internal lines of each utility and solve the multiarea production simulation using an AC power-flow solution that recognizes thermal, voltage, and stability constraints in conjunction with commitment and dispatch principles. However, an AC power-flow technique is too computer-resource-intensive to use with a simulation of 8760 hours/year. Therefore, the exact AC power-flow equations are generally reduced in detail to a DC (linear) power-flow model or reduced further to a "transportation model." To examine these transmission models in more detail, the discussion begins with the AC load-flow formulation, which is discussed in greater detail in Chapter 16.

13.2.2 AC Power Flow

Let: P_K = active power into bus K
Q_k = reactive power into bus k
E_k = complex voltage of bus k
Y_{ki} = complex admittance between bus k and bus i

Y_{kk} = complex admittance between bus k and ground
I_{ki} = complex current from bus k to bus i

The AC power-flow equation is:

$$P_k + jQ_k = \sum_i E_k I_{ki}^*$$

$$= \sum_{\substack{i \\ i \neq k}} E_k\{[E_k - E_i] \cdot Y_{ki}\}^* + E_k^2 \cdot Y_{kk} \tag{13.6}$$

$$P_k + jQ_k = \sum_{\substack{i \\ i \neq k}} |E_k|\ell^{j\theta k}\{[|E_k|\ell^{j\theta k} - |E_i|\ell^{j\theta i}] \cdot [g_{ki} + jb_{ki}]\}^*$$

$$-j|E_k|^2 B_{kk} \tag{13.7}$$

where b_{ki} = bus capacitance, g_{ki} = bus conductance, and * = complex conjugate.

The active power terms can then be written as:

$$P_k = \sum_i [E_k^2\, g_{ki} - g_{ki} E_k E_i \cos(\theta_k - \theta_i) - b_{ki}|E_k|\,|E_i| \sin(\theta_k + \theta_i)] \tag{13.8}$$

13.2.3 DC (Linear) Power Flow

The DC power-flow (also referred to as "linear power flow") equations can be written by assuming that the voltages are all near one per unit,

$$|E_k| = 1 \tag{13.9}$$

and that the resistance of the line is small compared to the reactance. This reduces to:

$$Y_{ki} = g_{ki} + jb_{ki} = \frac{r_{kj}}{r_{ki}^2 + x_{ki}^2} + \frac{-jx_{ki}}{r_{ki}^2 + x_{ki}^2} \tag{13.10}$$

where r = per unit line resistance and x = per unit line reactance (lowercase x). With these assumptions,

$$g_{ki} \sim 0 \quad \text{and} \quad b_{ki} = \frac{1}{x_{ki}} \tag{13.11}$$

Furthermore, assume that the angles across the lines are small. Thus:

$$\sin(\theta_k - \theta_i) = \theta_k - \theta_i \quad \text{and} \quad \cos(\theta_k - \theta_i) \sim 1 \tag{13.12}$$

These assumptions reduce Equation (13.8) to the DC power-flow Equation (13.13):

$$P_k = \sum_i \frac{1}{x_{ki}} (\theta_k - \theta_i) = \sum_i B'_{ki} \theta_i \tag{13.13}$$

in which:

$$B'_{kk} = \sum_i \frac{1}{x_{ki}} \quad \text{and} \quad B'_{ki} = \frac{-1}{x_{ki}}$$

These equations form an $N \times N$ matrix, of which only $N - 1$ is independent. The coefficients of Equation (13.13) satisfy:

$$B'_{kk} = - \sum_{k \neq i} B'_{ki} \tag{13.14}$$

Consider adding up all the rows of the B matrix. Consider the Pth column after adding all the rows from Equation (13.14).

$$P\text{th column sum of rows} = B'_{1p} + B'_{2p} + B'_{pp} + \cdots B'_{np} \tag{13.15}$$

$$\begin{aligned} P\text{th column sum of rows} &= \sum_{i \neq p} B'_{ip} + B'_{pp} \\ &= 0 \end{aligned} \tag{13.16}$$

Only $N - 1$ row equations are linearly independent because the sum of the rows of the B matrix is zero, which is a statement of the conservation of power into a mode. The angle at one bus (one equation) can be specified arbitrarily (usually 0). This one bus is referred to as the "slack," "swing," or "reference" bus. In this text, "slack" will refer to the one bus with its angle specified.

To solve these equations, set the slack (or reference) bus angle to zero, $\theta_{\text{slack}} = 0$. Equation (13.13) can now be written in matrix formulation as:

$$P_k = \sum_i \mathrm{B}'_{ki} \; \theta_i \tag{13.17}$$

Delete the slack bus equation and rewrite (13.17):

$$P_k = \sum_i \mathrm{B}_{ki} \, \theta_i \tag{13.18}$$

for all k not equal to the slack bus. Equation (13.18) can be solved for the angles in terms of the bus powers (13.19):

$$\theta_k = \sum_i \overline{X}_{ki} P_i \qquad (13.19)$$

where $\overline{X}_{ki}$ (uppercase X_{ki}) is the inverse of the B_{ki} matrix.

If the generation power is known, then the angle of each bus can be calculated. From the bus angles, the transmission flows on the transmission lines can be calculated.

13.2.4 Limitations of Power Flow

Transmission line constraints can be characterized by three considerations:

1. Thermal limitations due to high currents that may cause the conductor temperature to rise and thereby change metallurgical properties or even stretch or break (usually applicable to short lines).
2. Voltage stability limitations in which heavy line loading leads to low line voltages that ultimately cannot support the power flow ("voltage collapse") (usually applicable to 100–200-mi lines)
3. Angular stability limitations in which the angular displacement approaches the steady state, dynamic, or transient stability limit (usually applicable to long lines).

On electrical systems, typically only a small fraction (0–10%) are limiting. Recognizing which lines are limiting is a major assignment of electrical utility transmission planners, and there are no easy methods. Repeated use of ac power flow and stability programs is required. With a list of lines likely to be limiting (thermal, voltage or stability), a power-flow constraint can be expressed as:

$$\text{Minimum rating} \leq \text{power flow} \leq \text{maximum rating} \qquad (13.20)$$

where:

$$\text{Power on line } k,i = F_{ki} = (\theta_k - \theta_i) \cdot \frac{1}{x_{ki}} \qquad (13.21)$$

(Note: $\theta_{\text{slack}} = 0$.)

A transmission description of the multiarea production simulation includes those transmission lines on the interface between utilities *and* lines internal to the utility that may be limiting. Figure 13.6 illustrates a case in which three transmission lines make up the interface from utility A to utility B. Lines 4 and 5 and lines 6 and 7 are internal to each respective utility but are known to be potential limitations to the power transfers. These limitations require shifting power generation from the most economic schedule to a schedule as

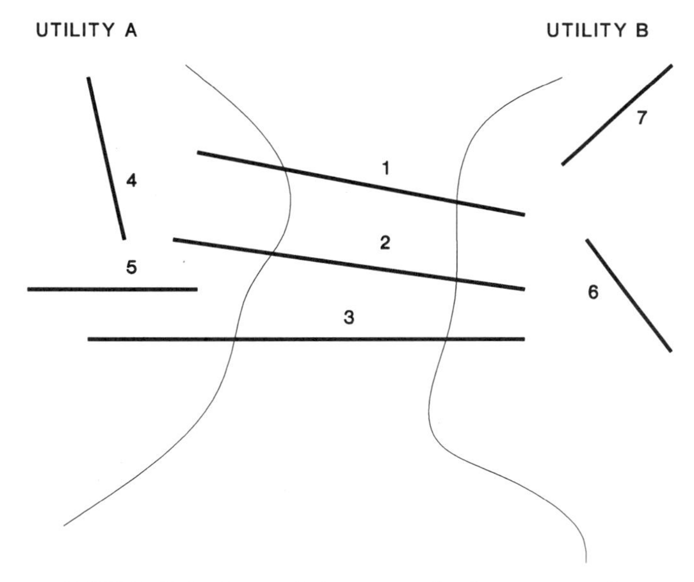

FIGURE 13.6 Transmission constrained power transfer.

near economic as can be allowed by power flows on internal and interface lines.

13.2.5 Generation Shift Factors

Generation shift factors (GSF) help to resolve the question: "How will the flow on a certain line change when an amount of generation is shifted from a certain generator to the reference bus?" The "reference" bus in a GSF calculation is usually *not* the same as the "slack" bus of the AC power flow. The certain line will be denoted as *km* and the certain generator as *i*.

The flow on a line can be calculated by using Equations (13.21) and (13.19).

$$\sum_i (\overline{X}_{ki} P_i - \overline{X}_{mi} P_i) \frac{1}{x_{km}} = \text{power on line } k,m$$

or

$$\sum_i \frac{(\overline{X}_{ki} - \overline{X}_{mi})}{x_{km}} P_i = \text{power on line } k,m \tag{13.22}$$

The generation shift factor is defined as:

$$\boxed{\text{Generation shift factor} = a_{km,i} = \frac{\overline{X}_{ki} - \overline{X}_{mi}}{x_{km}}} \tag{13.23}$$

where km is the line and i is the generator. (*Note:* $X_{ki} = 0$, or $X_{mi} = 0$ when k or m = reference bus.)

The procedure begins with a solved AC power flow (base). The DC power flow is used as a linear variation from the exact AC power flow base. Then the base flow plus the linear variations must be between the line rating limits:

$$\min_{k,m} \leq P_{\text{base}} + \Sigma\, a_{km,i} \cdot \Delta P_i \leq \max_{k,m} \tag{13.24}$$

Because the generating unit power output at each bus with a generating unit is controllable, Equation (13.24) can calculate which unit's power must be changed (ΔP_i) to assure that the transmission-line flows are within the limits and how much that change should be.

If the bus power is changed in Equation (13.24) to satisfy the transmission constraint, bus power of the reference bus must also be changed. Because of the conservation of power requirements, generation equals load. However, by changing two generators to the reference bus in the same calculation, power can be shifted between any two buses.

13.2.6 Security Dispatch

In power system operations, it is typical to operate the power system such that no transmission-line constraints would be violated if any transmission line were to suddenly outage ("trip"). Being prepared for any reasonable tripping is referred to as a "security constraint." This consideration can be evaluated by using line outage distribution factors to calculate how the power flow on one line would be redistributed onto other lines in the network in the event of line outage.

13.2.7 Line Outage Factors

Consider the case of a transmission line under an outage condition as shown in Figure 13.7. Part *a* shows the power normally flowing on the line. Part *b* shows the power flow when the line is on outage, or "tripped." The effect of a line outage can also be represented as a power injection into both ends of the line while leaving the line in service, part *c*. The effect of a line outage could be calculated by recomputing the $\overline{X}$ matrix or by using the power injection approach. The power injection approach is much easier to calculate given that there is a solution available for the line in service. Consider calculating the flows on the remaining lines using part *c*.

The flow on the line from k to m is:

$$F_{km} = (\theta_k - \theta_m) \cdot 1/x_{km} = \sum_i \frac{\overline{X}_{ki}P_i - \overline{X}_{mi}P_i}{x_{km}}$$

$$= \sum_i \frac{1}{x_{km}} (\overline{X}_{ki} - \overline{X}_{mi})P_i \tag{13.25}$$

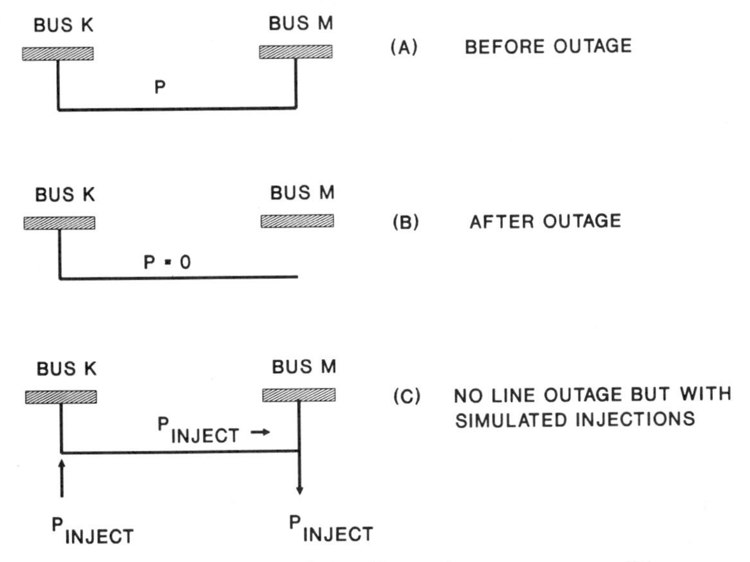

FIGURE 13.7 Transmission line under an outage condition.

Suppose that a change is made to the power injection on bus k and bus m. These injections, ΔP, should be such that

$$\Delta P_k = -\Delta P_m$$
$$\Delta P_i = 0, \qquad i \neq k \neq m \tag{13.26}$$

The incremental power flow is, from Equation(13.25):

$$\Delta F_{km} = \frac{(\overline{X}_{kk} - \overline{X}_{km} - \overline{X}_{mk} + \overline{X}_{mm})\,\Delta P_k}{x_{km}}$$
$$= (\overline{X}_{kk} - 2\overline{X}_{km} + \overline{X}_{mm})\frac{1}{x_{km}}(\Delta P_k) \tag{13.27}$$

The incremental power flow plus the original flow should equal the power injection in order to prevent flow through the line modes.

$$\Delta P_k = F_{km} + \Delta F_{km} \tag{13.28}$$

$$\Delta P_k = F_{km} + (\overline{X}_{kk} - 2\overline{X}_{km} + \overline{X}_{mm})\frac{1}{x_{km}}\Delta P_k \tag{13.29}$$

$$\boxed{\Delta P_k = [1 - \frac{1}{x_{km}}(\overline{X}_{kk} - 2\overline{X}_{km} + \overline{X}_{mm})]^{-1} \cdot F_{km}} \tag{13.30}$$

When the line about to be on outage is connected to the reference bus:

$$F_{km} = \frac{\theta_k}{x_{km}} = \Sigma \frac{\overline{X}_{ki}}{x_{km}} P_i \qquad \text{(bus } m = \text{reference bus)} \tag{13.31}$$

Suppose that the injection on bus k is increased by ΔP_k. Since the conservation of power automatically applies to the reference bus, $\Delta P_{ref} - \Delta P_k$. The incremental power flow is, from Equation (13.31):

$$\Delta F_{km} = \frac{\overline{X}_{kk}}{x_{km}} \Delta P_k \qquad (m = \text{reference bus}) \tag{13.32}$$

The incremental power flow plus the original flow should equal the power injection in order to have no flow through the modes.

$$\Delta P_k = F_{kn} + \Delta F_{kn} \qquad (m = \text{reference bus}) \tag{13.33}$$

$$\Delta P_k = F_{kn} + \frac{\overline{X}_{kk}}{x_{km}} \Delta P_k \qquad (m = \text{reference bus}) \tag{13.34}$$

$$\boxed{\Delta P_k = (1 - \frac{\overline{X}_{kk}}{x_{km}})^{-1} \bullet F_{km}} \qquad (m = \text{reference bus}) \tag{13.35}$$

Consider how this line outage and power injection will influence power flows on other transmission lines. The change in flow on line o,p due to the power injections ΔP_k and ΔP_m is from Equation (13.25):

$$\Delta F_{op} = \frac{1}{x_{op}} [\overline{X}_{ok} - \overline{X}_{om} - \overline{X}_{pk} + \overline{X}_{pm}] \Delta P_k \tag{13.36}$$

Substituting Equation (13.36) into Equation (13.30) yields

$$\Delta F_{op} = d_{op,km} \bullet F_{km} \tag{13.37}$$

where $d_{op,km}$ is a line outage distribution factor describing how the power flow in a line, km, would be distributed over other transmission lines, op:

$$\boxed{d_{op,km} = \frac{(x_{op})^{-1} [\overline{X}_{ok} - \overline{X}_{om} - \overline{X}_{pk} + \overline{X}_{pm}]}{1 - (x_{km})^{-1} [\overline{X}_{kk} - 2\overline{X}_{km} + \overline{X}_{mm}]}}$$

$$\text{for} \qquad o,p,k,m \neq \text{reference bus} \tag{13.38}$$

Consider the case when the line outage occurs on a line that is connected to the reference bus. The flow on other lines can be calculated as

$$\Delta F_{op} = \frac{1}{x_{op}} [\overline{X}_{ok} - \overline{X}_{pk}] \, \Delta P_k \tag{13.39}$$

Substituting Equation (13.39) into Equation (13.35) and using the definition in Equation (13.37) yields the line outage distribution factor:

$$\boxed{d_{op,km} = \frac{(x_{op})^{-1} [\overline{X}_{ok} - \overline{X}_{pk}]}{[1 - \overline{X}_{kk}/x_{km}]}} \quad \text{when } m = \text{reference bus} \tag{13.40}$$

Consider the final case of the line outage distribution factor for a line, op, connected to the reference bus, either o or p. From Equation (13.22B), the change in flow is:

$$\Delta F_{op} = \frac{1}{x_{op}} [\overline{X}_{ok} - \overline{X}_{om}] \, \Delta P_k \tag{13.41}$$

This yields the line outage distribution factor of:

$$\boxed{d_{op,km} = \frac{(x_{op})^{-1} [\overline{X}_{ok} - \overline{X}_{om}]}{1 - (x_{km})^{-1} [\overline{X}_{kk} - 2\overline{X}_{km} + \overline{X}_{mm}]}}$$

$$\text{when } p = \text{reference bus} \tag{13.42}$$

Consider the case when both lines are connected to the reference bus. From Equation (13.22B), the change in flow is:

$$\Delta F_{op} = \frac{1}{x_{op}} [\overline{X}_{ok}] \, \Delta P_k \qquad p = \text{reference bus} \tag{13.43}$$

This yields:

$$\boxed{d_{op,km} = \frac{(x_{op})^{-1} [\overline{X}_{ok}]}{[1 - \overline{X}_{kk}/x_{km}]}} \quad \begin{array}{l} p = \text{reference bus} \\ m = \text{reference bus} \end{array} \tag{13.44}$$

These formulas can be reduced to a single formula by expanding the definition of the $\overline{X}$ matrix by a row and column of zeros to represent the reference bus.

$$\overline{X} \leftarrow \begin{bmatrix} 0 & 0 & 0 & 0 & 0 & 0 & 0 & 0 \\ 0 & & & & & & & \\ 0 & & & & \overline{X} & & & \\ 0 & & & & & & & \\ 0 & & & & & & & \end{bmatrix}$$

The formula is then reduced to Equation (13.38). Thus, the total power flow on line o,p when line k,m is on outage is:

$$F_{op}(km = \text{out}) = \sum_i \left[\frac{(\overline{X}_{oi} - \overline{X}_{pi})}{x_{op}} + \frac{d_{op,km} \cdot (\overline{X}_{ki} - \overline{X}_{mi})}{x_{km}} \right] \cdot P_i \qquad (13.45)$$

or

$$F_{op}(km = \text{out}) = \Sigma\, (a_{op,i} + d_{op,km,i}) \cdot P_i$$

Hence, a series of transmission-line flow constraints can be established as in Table 13.27.

13.2.8 Distribution Factor Example

The two components of the distribution factors: (1) generation shift factors (GSF) and (2) line outage factors, will be illustrated in the following example.

Figure 13.8 illustrates a 200-mile transmission network comprised of 345-kV lines. The impedance of each line is .10 on a 100-MVA (megavolt-ampere) base.

A *reference case* AC power flow has

822-MW net generation–load at bus 1 at an angle 0 (reference bus)
−200-MW net generation–load at bus 2 at an angle −14.0°
−600-MW net generation–load at bus 3 at an angle −16.4°
22-MW losses
248-MW on line 1–2
287-MW on line 1–3;1
287-MW on line 1–3;2
42-MW on line 2–3

The DC load flow will be used to calculate the change in power flow from the reference case AC power flow for other hours. The net generation–load at the buses for this hour is shown in Table 13.28.

TABLE 13.27 Transmission-Line Flow Constraints

Equation	Constraint	Criteria
(13.24) $\lvert F_{op}\rvert \leq$	$PMAX_{op}$	Normal all lines in
(13.46) $\lvert F_{op}(km = \text{out})\rvert \leq$	$PMAXCONTGENCY_{op}$ for km = set of lines to be checked for outage criteria	Contingency criteria for system security

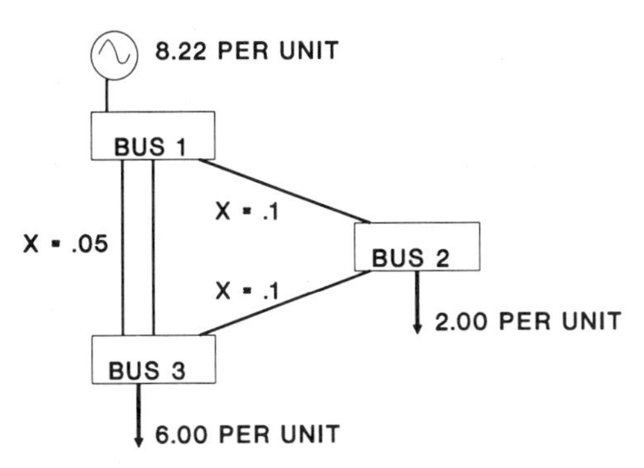

FIGURE 13.8 Three-bus electrical transmission network example.

13.2.9 Example Linear (DC) Delta Power Flow Calculation

Solve the following for the change in line flow from the AC power flow reference case: (1) determine the GSF for bus 2, and (2) determine the line outage distribution factor for outaging lines 2–3 and 1–2.

The DC power flow, Equation (13.13):

$$\Delta P_k = \Sigma \frac{1}{x_{k1}} (\Delta\theta_k - \Delta\theta_i) \tag{13.46}$$

is set up as the matrix:

$$\begin{bmatrix} 2 \\ -1 \\ -1 \end{bmatrix} = \begin{bmatrix} \frac{1}{.1} + \frac{1}{.05} & -\frac{1}{.1} & -\frac{1}{.05} \\ -\frac{1}{.1} & \frac{1}{.1} + \frac{1}{.1} & -\frac{1}{.1} \\ -\frac{1}{.05} & -\frac{1}{.1} & \frac{1}{.1} + \frac{1}{.05} \end{bmatrix} \begin{bmatrix} \theta_1 \\ \theta_2 \\ \theta_3 \end{bmatrix} \tag{13.47}$$

This equation can be rewritten as:

$$\Delta P_k = B_{ki}\, \Delta\theta_i \tag{13.48}$$

TABLE 13.28 Net Changes in Generation for the Three-Bus Network

Bus	AC Power-Flow Reference	Bus Generation During This Hour	Change in Bus Generation
1	8.2 pu[a]	10.2 pu	+2.0 pu
2	−2.0 pu	−3.0 pu	−1.0 pu
3	−6.0 pu	−7.0 pu	−1.0 pu

[a]Per unit.

This reduces to:

$$\begin{bmatrix} 2 \\ -1 \\ -1 \end{bmatrix} = \begin{bmatrix} 30 & -10 & -20 \\ -10 & 20 & -10 \\ -20 & -10 & 30 \end{bmatrix} \begin{bmatrix} \theta_1 \\ \theta_2 \\ \theta_2 \end{bmatrix} \tag{13.49}$$

The first bus equation will be chosen as the reference and θ_1 set to zero. Equation (13.49) becomes:

$$\begin{bmatrix} P_2 \\ P_3 \end{bmatrix} = \begin{bmatrix} 20 & -10 \\ -10 & 30 \end{bmatrix} \begin{bmatrix} \theta_2 \\ \theta_3 \end{bmatrix} \tag{13.50}$$

By linear elimination of variables, this small 2 • 2 matrix can be inverted:

$$P_2 + 2P_3 = 50\ \theta_3 \tag{13.51}$$

Rearranging yields:

$$\theta_3 = \frac{1}{50} P_2 + \frac{2}{50} P_3 \tag{13.52}$$

Reverse substitution into Equation (13.52) yields:

$$P_2 = 20\ \theta_2 - \frac{10}{50} P_2 - \frac{20}{50} P_3 \tag{13.53}$$

Rearranging yields:

$$\theta_2 = \frac{3}{50} P_2 + \frac{1}{50} P_3$$

Thus, the change in angle is given by:

$$\Delta\theta_k = \Sigma\, \overline{X}_{ki}\, \Delta P_i$$

and

$$\overline{X} = \begin{bmatrix} \frac{3}{50} & \frac{1}{50} \\ \frac{1}{50} & \frac{2}{50} \end{bmatrix}$$

Therefore:

$$\begin{bmatrix} \Delta\theta_2 \\ \\ \Delta\theta_3 \end{bmatrix} = \begin{bmatrix} \frac{3}{50} & \frac{1}{50} \\ \\ \frac{1}{50} & \frac{2}{50} \end{bmatrix} \begin{bmatrix} -1 \\ \\ -1 \end{bmatrix} = \begin{bmatrix} \frac{-4}{50} \\ \\ \frac{-3}{50} \end{bmatrix} \rightarrow \begin{bmatrix} -4.6^\circ \\ \\ -3.4^\circ \end{bmatrix}$$

The delta line flows (Figure 13.9) from Equation (13.21) are:

$$\text{Delta flow line 1–2} = \frac{0 - (-.08)}{.1} = +.8 \text{ pu}$$

$$\text{Delta flow line 2–3} = \frac{-.08 + .06}{.1} = -.2 \text{ pu}$$

$$\text{Delta flow line (1–3)} = \frac{0 + .06}{.05} = 1.2 \text{ pu}$$

The GSF for bus $i = 2$ from Equation (13.23) are:

$$a_{km,i} = \frac{\overline{X}_{ki} - \overline{X}_{mi}}{x_{km}}$$

$$a_{21,2} = \frac{\overline{X}_{22} - \overline{X}_{21}}{x_{21}} = \frac{\overline{X}_{22} - 0}{x_{21}} = \frac{3/50}{.1} = \frac{3}{5}$$

$$a_{23,2} = \frac{\overline{X}_{22} - \overline{X}_{32}}{x_{23}} = \frac{(3/50) - (1/50)}{.1} = \frac{2}{5}$$

$$a_{31,2} = \frac{\overline{X}_{32}}{x_{31}} = \frac{1/50}{.05} = \frac{2}{5}$$

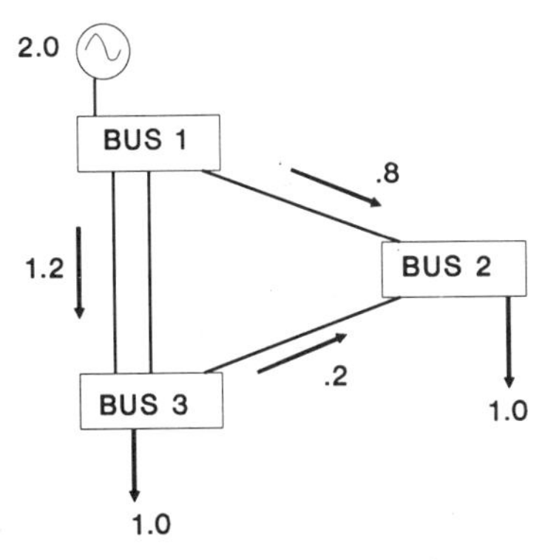

FIGURE 13.9 Change in flow due to changing generation and load.

The line outage distribution factors for outaging line 2–3 using Equation (13.42) are:

$$d_{21,23} = \frac{(1/x_{21})[\overline{X}_{22} - \overline{X}_{23}]}{1 - (1/x_{23})\,[\overline{X}_{22} - 2\overline{X}_{23} + \overline{X}_{33}]}$$

$$= \frac{1/.1\ [3/50 - 1/50]}{1 - 1/.1\ [3/50 - 2/50 + 2/50]} = \frac{2/5}{2/5} = -1$$

$$d_{31,23} = \frac{(1/x_{31})\ [\overline{X}_{32} - \overline{X}_{33}]}{1 - (1/x_{23})\ [\overline{X}_{22} - 2\overline{X}_{23} + \overline{X}_{33}]}$$

$$= \frac{(1/.05)\ [1/50 - 2/50]}{2/5} = \frac{-2/5}{2/5} = -1$$

The line outage distribution factor for outaging line 2–1 using Equation (13.44) is:

$$d_{31,21} = \frac{(1/x_{31})\,\overline{X}_{32}}{1 - \overline{X}_{22}/x_{21}} = \frac{(1/.05)(1/50)}{1 - 3/50/.1} = \frac{2/5}{2/5} = 1$$

using Equation (13.40):

$$d_{23,21} = \frac{(1/x_{23}\ [\overline{X}_{22} - \overline{X}_{32}]}{1 - \overline{X}_{22}/x_{21}} = \frac{(1/.1)\ [(3/50) - (1/50)]}{2/5} = 1$$

The total flow on line 3–1 for all lines in service is:

− 5.74 pu	base case flow
− 1.20 pu	delta case flow
− 6.94 pu	total

The total flow on line 2–3 is .042 pu (base case flow) plus −.2 pu (delta case flow) = −.158 pu.

The flow on line 3–1 when line 2–3 is out of service is:

− 5.74 pu	base case flow
− 1.20 pu	delta case flow
(−1) • (−.158)	line outage factor times line flow
−6.882 pu	total

The flow on line 3–1 when line 2–1 is out of service is:

− 5.74 pu	base case flow
−1.20 pu	delta case flow
(1) • (−3.28)	line outage factor time line flow
10.22 pu total	

In summary, the DC power-flow method provides a linear set of equations that can be used to represent the power flow, constraint conditions, and security constraints on all or selected transmission lines.

13.2.10 Transportation Constraint Models

Sections 13.2.3 through 13.2.9 discussed the DC load-flow network constraint as an approximation to the AC load-flow equations. A further approximation is the transportation network model. In this model, power flow on a line is assumed to be limited only by the characteristic of the line regardless of the loadings of other lines in the power system. This is analogous to power being transported in trucks on a highway. The electrical constraint equation is that the sum of the net power into a bus must be zero:

$$P_k + \sum_i F_{ki} = 0 \tag{13.54}$$

where P_i = power injected into bus k and F_{ki} = power flow on transmission lines connected to bus k.

The transmission-line constraints are then expressed as:

$$|F_{ki}| \leq P_{\text{max},ki} \tag{13.55}$$

With this method, the entire interface between utilities is usually represented as a single "line" rather than individual lines that constitute the interconnection. The interconnection constraint is represented as:

$$|F_{\text{utility } k, \text{ Utility } i}| \leq P_{\text{max}} \tag{13.56}$$

The advantage of this transportation model is the ease with which it can be analyzed. Its disadvantage is that electrical networks cannot be correctly modeled in this manner.

13.2.11 Transportation Model Constraint Example

In this section, a two-area example is presented to illustrate the solution techniques necessary to solve the multiarea problem. This example utilizes the transportation model of the transmission network constraints.

In performing two-area production simulation with the transportation model of the interconnection, the conservation of power condition accounts for the transmission power inflow/outflow as:

$$\text{Load}_{\text{area A}} = \text{generation}_{\text{area A}} - \text{transmission}_{\text{A}\rightarrow\text{B}} \tag{13.57}$$

$$\text{Load}_{\text{area B}} = \text{generation}_{\text{area B}} + \text{transmission}_{\text{A}\rightarrow\text{B}} \tag{13.58}$$

For multiarea simulations, the same types of equation as (13.57) and (13.58) apply, but in a more general form where transmission flows from all areas are included.

To illustrate the methods of multiarea production simulation, a two-area example is presented. For simplicity, assume that the minimum power output of the power plants is zero. The power plants in Table 13.29 are committed for service in this hour.

The load and interconnection capability are shown in Figure 13.10. The production costs for the power pool and utilities A and B will be computed for this one hour.

First, order units in rank of $/MWh (see Table 13.30).

The principles of two-area dispatch are:

- The generation of the entire pool is ordered in an incremental $/MWh loading order, as shown in Table 13.30.
- Generation priority is given to serving a utility's own loads first.
- Excess generation from one utility can be sold over the interconnection up to the interconnection capacity.
- The price of the purchased/sold energy is negotiable, but generally priced on a split-savings basis.

TABLE 13.29 Utility A and B Generators

Generating Units	Rating	$/MWh
	Utility A	
A1	100	5.0
A2	200	10.0
A3	100	25.0
A4	200	30.0
	Utility B	
B1	200	6.0
B2	300	11.0
B3	300	12.0

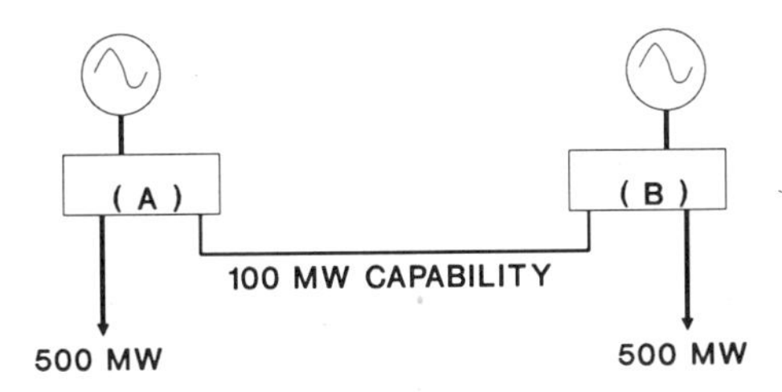

FIGURE 13.10 Two-utility interface example.

To facilitate dispatching, a tabular array is useful, where the unit loading (output) and transmission loading are explicitly monitored for constraint violations. The dispatch (Table 13.31), is created by proceeding down the pool priority list (Table 13.30), adding generation until:

$$\text{Generation} - \text{load} = 0 \tag{13.59}$$

while:

$$|\text{Transfer MW}| \leq 100 \tag{13.60}$$

Total power pool production cost can be computed by multiplying the unit loading by the $/MWh cost, $10,700/hour in this example. The cost to utility A is $5000/hour and to utility B, $5700/hour.

13.2.12 Interchange Billing

Interchange of power between companies requires interchange billing, generally computed on a split-savings basis. To arrive at the proper billing rate, the power system operation must be evaluated for no interchange between utilities. The no-interchange dispatch of company A is shown in Table 13.32. The incremental cost for utility A generating an additional 100 MW of power rather

TABLE 13.30 Two-Company Generation Ranked in $/MWh Order

Order	Unit Name	Utility	Rating	Cumulative MW	$/MWh
1	A1	A	100	100	5.0
2	B1	B	200	300	6.0
3	A2	A	200	500	10.0
4	B2	B	300	800	11.0
5	B3	B	300	1100	12.0
6	A3	A	100	1200	25.0
7	A4	A	200	1400	30.0

TABLE 13.31 Two-Utility Dispatch Using Table 13.30 Priority List

Utility	Rank #	Unit Capacity	Unit Loading	\$/Mwh[a]	Transfer Flow $T_{A \to B}$	Cumulative Generation Area A	Cumulative Generation Area B	Net Generation–Load Area A	Net Generation–Load Area B	Net Generation–Load Pool
	0				0	0	0	−500	−500	−1000
A	1	100	100	5.0	0	100	0	−400	−500	−900
B	2	200	200	6.0	0	100	200	−400	−300	−700
A	3	200	200	10.0	0	300	200	−200	−300	−500
B	4	300	300	11.0	0	300	500	−200	0	−200
B	5	300	100	12.0	−100	300	600	−100	100	−100
A	6	100	100	25.0	−100	400	600	−100	100	0
A	7	200	0	30.0	—	—	—	—	—	—

[a]Total production cost = 10,700 \$/hour.

TABLE 13.32 No-Interchange Dispatch of Utility A

Priority	Unit	Rating	Loading	\$/MWh[a]	Generation–Load
0					−500
1	A1	100	100	5.0	−400
2	A2	200	200	10.0	−200
3	A3	100	100	25.0	−100
4	A4	200	100	30.0	0

[a]Total production cost = 8000 \$/hour.

than purchasing it is the principal component of interchange billing. If the load had been 400 MW, then unit A4 would not have operated with an incremental savings of \$30/MWh.

The no-interchange dispatch of utility B is shown in Table 13.33. The incremental cost of generating 100 MW additional from unit B3 is \$12.0/MWh.

Billing rate for energy is generally based on the split incremental savings method (one-half of the seller's incremental cost plus one-half of the buyer's incremental cost per megawatt-hour).

$$\text{Billing rate} = 1/2\ (12.0 + 30) = \$21/\text{MWh}$$

Thus, the purchases and sales are:

$$\text{Utility A purchase} = \$21/\text{MWh} \cdot 100\ \text{MW} = \$2100/\text{hour}$$

$$\text{Utility B sales} = \$21/\text{MWh} \cdot 100\ \text{MW} = \$2100/\text{hour}$$

Production cost for each utility can be calculated by using the pooled dispatch of Table 13.31 and the interchange billing cost above.

$$\begin{aligned} \text{Utility A} &= \text{generation cost} + \text{purchase cost} \\ &= 5000 + 2100 \\ &= \$7100/\text{hour} \end{aligned}$$

$$\begin{aligned} \text{Utility B} &= \text{generation cost} - \text{sales revenue} \\ &= 5700 - 2100 \\ &= \$3600/\text{hour} \end{aligned}$$

TABLE 13.33 No-Interchange Dispatch of Utility B

Priority	Unit	Rating	Loading	\$/MWh[a]	Generation–Load
0					−500
1	B1	200	200	6.0	−300
2	B2	300	300	11.0	0
3	B3	300		12.0	

[a]Total production cost = 4500 \$/hour.

TABLE 13.34 Comparison (in $/Hour) of Two-Utility Dispatch Options

	No Interconnection	With Interconnection	Benefit of Interconnection
Utility A	8,000	7,100	900
Utility B	4,500	3,600	900
Total pool	12,500	10,700	1,800

Table 13.34 summarizes the total production cost comparison. Note that both the sending and receiving utilities benefit from the interconnection.

When more than two utilities are involved, the interchange billing still typically uses split incremental savings. The billing formula is utility and pool specific. A typical price formula, however, is based on the pooled average incremental selling price of all of the selling utilities, and the purchasing price formula is based on the pooled average incremental price of all the purchasing utilities.

If

$CPURS_j$ = decremental replacement energy cost of purchasing electricity to utility company j in $/MWh

$CSELL_i$ = incremental energy cost of producing electricity to utility company j in $/MWh

$ENERGY_i$ = energy interchange for utility i in MWh

The pool average purchasing cost is

$$\text{POOL PURS.} = \sum_{\substack{\text{purchasing} \\ \text{utilities} \\ j}} CPURS_j \cdot ENERGY_j / \sum_{\substack{\text{purchasing} \\ \text{utilities} \\ j}} ENERGY_j \tag{13.61}$$

The pooled average selling cost is:

$$\text{POOL SELL} = \sum_{\substack{\text{selling} \\ \text{utilities} \\ j}} CSELL_i \cdot ENERGY_i / \sum_{\substack{\text{selling} \\ \text{utilities} \\ j}} ENERGY_i \tag{13.62}$$

The billing incremental rate ($/MWh) to purchasing utility j, is the split savings between the utilities' costs and the pooled average.

$$\text{Purchasing cost}_j = 1/2\,(CPURS_j + \text{POOL SELL}) \tag{13.63}$$

The selling cost credit ($/MWh) to selling utility j, is the split savings between the utilities' incremental costs and the pooled average.

$$\text{Selling credit cost}_j = 1/2\,(CSELL_j + \text{POOL PURS}) \tag{13.64}$$

TABLE 13.35 Five-Utility Example Incremental Costs and Interchange Flows[a]

Utility	$/MWh Incremental Cost	MW Net Purchase (−) Net Sales (+)
A	12	+100
B	16	+150
C	25	−100
D	30	− 50
E	35	−100

[a]The average pooled purchase costs are $30/MWh; the average pooled selling costs are $14.4/MWh.

An example of five utilities is given in Tables 13.35 and 13.36. Three utilities are buying a total of 250 MW. The total billing nets to zero with all five utilities benefiting.

13.2.13 DC Load-Flow Constraint Example

Consider evaluating the two-utility example of Section 13.2.11 using the simplified electrical network model in Figure 13.11. This example will assume that the ac power-flow reference case is zero. The network is comprised of 230-kV lines 300 miles long. Bus 1 belongs to utility B, and bus 2 and 3 lie in Utility A. A 100-MVA base is used (see Chapter 16).

The simplified network of Figure 13.11 has only three lines, and all will be evaluated as possible constraints. In a larger and more realistic network, not all of lines need to be evaluated as possible constraints. However, all lines must be used to calculate the shift factors.

Equation (13.65) presents the DC load-flow formulation to begin the shift factor calculations. The inverse matrix is presented in Equation (13.66).

TABLE 13.36 Five-Utility Example of Billing

Utility	Billing Cost ($/MWh)	MW Purchase (−) Sale (+)	Billing Total $
A	21	100	2100 sale
B	23	150	3450 sale
C	19.7	−100	−1970 purchase
D	22.2	− 50	−1110 purchase
E	24.7	−100	−2470 purchase
Pool net		0	0

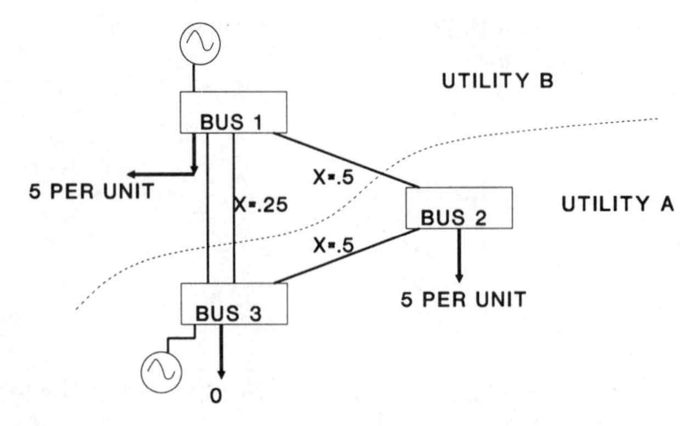

FIGURE 13.11 Two-utility, three-bus example (100-MVA base).

$$\begin{bmatrix} P_B - 5 \\ -5 \\ P_a \end{bmatrix} = \begin{bmatrix} 6 & -2 & -4 \\ -2 & -4 & -2 \\ -4 & -2 & 6 \end{bmatrix} \begin{bmatrix} \theta_1 \\ \theta_2 \\ \theta_3 \end{bmatrix} \tag{13.65}$$

$$\overline{X} = \begin{bmatrix} 0 & 0 & 0 \\ 0 & .3 & .1 \\ 0 & .1 & .2 \end{bmatrix} \tag{13.66}$$

The flow on the lines is calculated as:

$$F_{21} = \frac{1}{.5} \cdot [.3 \cdot (-5) + .1\ P_a] = -3.0 + .2P_a \tag{13.67}$$

$$F_{23} = \frac{1}{.5} [.2 \cdot (-5) + (-.1)\ P_a] = -2.0 - .2P_a \tag{13.68}$$

$$F_{31} = \frac{1}{.25} [.1 \cdot (-5) + -.2\ P_a] = -2.0 - .8P_a \tag{13.69}$$

The constraints on the line flows are given as:

$$|F_{21}| \leq 2.0 \text{ pu} = 200 \text{ MW} \tag{13.70}$$

$$|F_{23}| \leq 3.5 \text{ pu} = 350 \text{ MW} \tag{13.71}$$

$$|F_{31}| \leq 4.0 \text{ pu} = 400 \text{ MW} \tag{13.72}$$

Assume that there are no line outage contingency constraints.

The dispatch solution begins by proceeding down the dispatch priority list of Table 13.30 until the cumulative dispatch equals the 1000-MW load. In this case:

$$\text{Generation}_A = 300 \text{ MW}; \qquad \text{Generation}_B = 700 \text{ MW}$$

Inserting this dispatch into Equations (13.67) through (13.69) yields the transmission flows:

$$F_{21} = -.24 \text{ pu} = -240 \text{ MW}$$

$$F_{23} = -.26 \text{ pu} = -260 \text{ MW}$$

$$F_{31} = .04 \text{ pu} = 40 \text{ MW}$$

Note that the constraint on line F_{21} of 200 MW is violated. In this two-utility, three-bus case, the violation will be corrected by inspection. Larger problems require optimization methods such as linear programming, discussed in the next section. The constraint can be satisfied if Equation (13.67) is made into an equality constraint. Solving Equation (13.67) as an equality yields:

$$-2.0 = -3.0 + .2\, P_a$$

$$P_a = 500$$

The transmission flows are then:

$$F_{21} = -.20 = -200 \text{ MW}$$

$$F_{23} = -.30 = -300 \text{ MW}$$

$$F_{31} = .20 = 200 \text{ MW}$$

With 500-MW generation at bus 3, utility A, the flow constraints are satisfied. Because of the transmission constraint, utility A must generate 200 MW more than is economical. The production dispatch is then calculated by proceeding back up the dispatch priority table (Table 13.30) and removing the most expensive 200 MW of generation from utility B as shown in Table 13.37.

The total system dispatch cost is $12,500 compared to $10,700 in Table 13.31. No power can be transferred from utility B to utility A. Delivering 500 MW from bus 3 to bus 2, both of utility A, will cause 300 MW to flow directly on line 3 to 2 and 200 MW to loop flow from 3 to 1 to 2. This loop flow completely loads line 1 to 2 and effectively blocks any power delivery from utility B to utility A. Thus the transfer limit assumption of 100 MW shown in Figure 13.10 is in error, at least for this load combination.

13.2.14 Solution Methods for Multiarea Production Simulation

Sections 13.2.11 and 13.2.13 presented simplified illustrations of multiarea transmission constraints for production simulation. While the two-utility ex-

TABLE 13.37 Two-Utility, Three-Bus Example Dispatch and Production Cost

Utility	Rank #	Unit Capacity	Unit Loading	$/MWh	Cost ($)
A	1	100	100	5	500
B	2	200	200	6	1,200
A	3	200	200	10	2,000
B	4	300	300	11	3,300
B	5	300	0	12	0
A	6	100	100	25	2,500
A	7	200	100	30	3,000
Totals			1000		12,500

ample can be solved easily, the multiarea problem (with three or more power plants) is more difficult to solve because of the complexity of transmission flow constraints. These constraints are of the form:

$$|a_{11}\, P_1 + a_{12}\, P_2 + \cdots + | \leq P_{1:\max}$$

$$\vdots$$

$$|a_{n1}\, P_1 + a_{n2}\, P_2 + \cdots + | \leq P_{N:\max}$$

where $P_1, P_2, \ldots, P_N$ is the power output at each generating units bus.

In the multiarea case there are many P bus values such as $P_1, P_2, \ldots P_N - 1$ (e.g., 500) that can be used to satisfy the constraint. During the solution process transmission constraints may be violated, and arbitrarily choosing a P_k value can lead to constraint violations on *other* transmission lines.

A formalized solution method is required and linear programming is a general solution technique well suited to these problems. The theory is explained in Hillier and Lieberman (1974), and the FORTRAN code is detailed in Land and Powell (1973).

The problem can be formulated as:

$$\text{Minimize} \sum_{i=1}^{N} C_i\, \Delta P_i \tag{13.73}$$

where C_1 = cost of power from generating unit i in $/MWh and ΔP_i = change in power output from generating unit i in megawatts from the AC power-flow reference case. Nonlinear generation characteristics such as an increasing $/MWh operating cost characteristic can be represented using piecewise-linear representations.

The utility constraint that the load demand must be served is:

$$\sum_{i=1}^{N} P_i + \Delta P_i = \text{load} \tag{13.74}$$

The transmission constraints can be represented as:

$$F_{op} + \sum_{i=1}^{N} (a_{op,i} + d_{op,kn}a_{kn,i})\,\Delta P_i \le \text{PMAX}_{op} \qquad \text{for } op = 1\text{–}n$$
$$\text{for } kn = 1\text{–}k \tag{13.75}$$

where $a_{op,i}$ = generation shift for line op for generating unit i
$d_{op,kn}$ = line outage distribution factor for line op when line kn is outaged
ΔP_i = change in power from the reference case for generator i
PMAX_{op} = maximum transmission flow constraint on line op
F_{op} = reference AC power flow

The constraints of Equation (13.75) are written for the key transmission lines that are to be monitored for constraint violations. When the constraint does not include a contingency, the $d_{op,kn}$ term is omitted.

Large power pool studies (up to 4000 buses) monitor up to 300 key transmission lines, including contingency constraints. The selection of which lines to monitor with which lines tripped has been studied by many authors, including McClelland (1978), who produced the computer program named "Contingency Analysis Procedure" (CAP); Ejebe and Wollenberg (1979); and Irisarri et al. (1979). These authors address the problem of analyzing the order of 4000 buses, 8000 lines and thus approximately $(8000)^2$ single contingencies and approximately $(8000)^3$ double contingencies to identify the 50–300 most critical transmission line combinations to monitor.

In summary, the linear programming method algorithm permits a rapid solution to multiarea production simulation problems. For further reading about linear programming and the secure dispatch problem, see Wood and Wollenberg (1984, Chapter 11), Chan and Yip (1979), and Stott et al. (1979).

While the discussion on multiarea production simulation focused on generation dispatch, it must be kept in mind that *both* the commitment process and the dispatch process must be simulated by use of multiarea techniques.

13.3 MONTE CARLO METHOD

Monte Carlo techniques are able to model both transmission flow limits and random forced outages together in production simulation. The recursive convolution method cannot include the distribution factor model of flow limits. The Monte Carlo method provides the additional capability to include other realistic factors such as:

- Partial forced events that both reduce the output and change all the incremental heat-rate data.
- Ramp rates—the limitation in moving the generator output from a previous hour's operating point.
- Demand side load management.

See Patton and Singh (1984) for a 1984 application of Monte Carlo methods to load management.

The application of Monte Carlo techniques to production simulation is similar to that used in reliability techniques (see Chapter 11, Section 11.6). A random number is chosen for each generating unit that is not on scheduled maintenance. If the random number is less than the unit forced outage rate, then the generating unit is on forced outage. If the random number is larger, then the unit is available and in service. After completing this procedure for all of the units, the multiarea production simulation (both commitment and dispatch) is performed as discussed in Section 13.2. This generates one "history."

In practice, random numbers are usually generated on a weekly basis. Units on outage and units in service remain unchanged for the entire week. This simulates reality since generating units typically have an average forced outage duration of 4–5 days. Developing a Monte Carlo schedule on a weekly basis also reduces computer execution time because priority and dispatch-order lists need to be generated only once a week.

How many "histories," then, are required to have the Monte Carlo technique provide the required production cost accuracy? To provide insight into the "how many histories" issue, consider an example. Suppose that the standard deviation of production cost due to random forced outages is .03 (per unit). The standard deviation of the mean production cost estimate, following the logic of Equations (11.6) through (11.21), is then:

$$\text{Standard deviation of the mean} = \frac{.03}{\sqrt{N}} \tag{13.76}$$

where N = number of histories. The confidence level of the production cost is then:

$$\text{Production cost} = 1.00 \pm t_N \frac{.03}{\sqrt{N}} \text{ (pu)} \tag{13.77}$$

where t_N is the Student's t-test value at the desired confidence level. For a .5% accuracy,

$$1.005 = 1.00 + t_N \frac{.03}{\sqrt{N}} \tag{13.78}$$

$$\sqrt{N} = \frac{t_N\,.03}{.005} \qquad (13.79)$$

$$N = 36(t_N)^2 \qquad (13.80)$$

For a 95% confidence band ($t_N = 1.96$), the histories required is:

$$N = 36 \cdot (1.96)^2 = 138 \qquad (13.81)$$

Assuming that the production cost is approximately the same for each week and 52 weeks per year are simulated, then 2.66 (138/52) annual Monte Carlo simulations are required. However, weekly production costs are not the same for each week. Therefore, a larger number of histories must be computed.

Relatively good accuracies in total production cost can be achieved with Monte Carlo simulations with only a few annual histories. However, individual generating unit results are more susceptible to a wider variance than the total annual cost (Wang, 1988).

13.4 FURTHER READING

The work on probabilistic production simulation is summarized in Booth (1972) and Sager and Wood (1973) and extended to fitting probability distributions to the load-duration curve (Rau et al., 1980; Stremel et al., 1980). The transportation model of transfer limits was developed by Adamson et al. (1978). Distribution factors and their application to transfer limits are discussed by Landgren and Anderson (1976) and McClelland (1978). They are applied to production costing by Desell et al. (1984) and Garver et al. (1984). The factors considered in production simulation are summarized by Garver (1977, Chapter 4).

REFERENCES

Adamson, A. M., and A. L. Desell, J. F. Kenney, and L. L. Garver, "Inclusion of Inter-Area Transmission and Production Costing Simulation," *IEEE Transactions on Power Apparatus and Systems,* Vol. PAS-97, 1978, pp. 1481–1488.

Booth, R. R., "Power System Simulation Model Based on Probability Analysis," *IEEE Transactions on Power Apparatus and Systems,* Vol. PAS-91, January–February 1972, pp. 62–69.

Chan, S. M., and E. Yip, "A Solution of the Transmission Limited Dispatch Problem by Sparse Linear Programming," *IEEE Transactions on Power Apparatus and Systems,* Vol. PAS-98, May/June 1979, pp. 1044–1053.

Desell, A. L., E. C. McClelland, K. Tammar, and P. R. Van Horne, "Transmission Constrained Production Cost Analysis in Power System Planning," *IEEE Transac-*

tions on Power Apparatus and Systems, Vol. PAS-103, August 1984, pp. 1291–2198.

Ejebe, G. C., and B. F. Wollenberg, "Automatic Contingency Selection," *IEEE Transactions on Power Apparatus and Systems,* Vol. PAS-98, No. 1, January–February 1979.

Garver, L. L., "Factors Affecting the Planning of Interconnected Electric Utility Bulk Power Systems," Part II, Chapter IV, *Factors Influencing Electric Utility Expansion,* U.S. Department of Energy, Conf-77-869, Vol. II, 1977.

Garver, L. L., "The Electric Utilities," Chapter II-6 of *Handbook of Operations Research, Models and Applications,* Vol. II, Joseph J. Moder and Salah E. Elmaghraby, eds., Van Nostrand Reinhold, New York, 1978.

Garver, L. L., G. A. Jordan, J. L. McDermott, and R. M. Sigley, "The Modeling of Transmission Limits in Production Simulation," *Proceedings of the American Power Conference,* Vol. 46, 1984, pp. 408–414.

Hillier, F. S., G. J. Lieberman, *Operations Research,* 2nd ed., Holden Day, San Francisco, 1974.

Irisarri, G., A. M. Sasson, and D. Levner, "Automatic Contingency Selection for On-Line Security Analysis—Real-Time Tests," *IEEE Transactions on Power Apparatus and Systems,* Vol. PAS-98, No. 5, September–October 1979.

Land, A., and S. Powell, *FORTRAN Codes for Mathematical Programming,* Wiley, New York, 1973.

Landgren, G. L., and C. W. Anderson, "Maximum Transmission Grid Loading Using Linear Programming," *IEEE Tutorial Course,* 76CH1107-2-PWR, 1976, pp. 30–38.

Marsh, W. D., R. W. Moisan, and R. C. Murrell, "Perspectives on the Design and Application of Generating Planning Programs," *Proceedings of the Nuclear Utilities Planning Methods Symposium,* Chattanooga, TN, January 1974.

McClelland, E. C., "Computer Code for a Generalized and User Oriented Power Transmission System Analysis Procedure," *Canadian Communications and Power Conference,* IEEE Press, 1978, pp. 504–507.

Patton, A. D., and C. Singh, "Evaluation of Load Management Effects Using the OPCON Generation Reliability," *IEEE Transactions on PAS,* Vol. PAS-103, November 1984, pp. 3230–3238.

Rau, N. W., P. Toy, and K. F. Shenk, "Expected Energy Production Costs by the Method of Moments," *IEEE Transactions on Power Apparatus and Systems,* Vol. PAS-99, September-October 1980, pp. 1908–1917.

Sager, M. A., and A. J. Wood, "Power System Production Cost Calculations—Sample Studies Recognizing Forced Outages," *IEEE Transactions on Power Apparatus and Systems,* Vol. PAS-92, January–February 1973, pp. 154–158.

Stott, B., J. L. Marinho, and O. Alsac, "Review of Linear Programming Applied to Power System Rescheduling," *Proceedings 1979 PICA Conference,* IEEE Document No. 790CH1381-3-PWR, pp. 142–154.

Stremel, J. P., R. T. Jenkins, R. A. Babb, and W. D. Bayless, "Production Costing Using the Cumulant Method of Representing the Equivalent Load Curve," *IEEE Transactions on Power Apparatus and Systems,* Vol. PAS-98, September–October 1980, pp. 147–1956.

Wang, L., "Approximate Confidence Bounds on Monte Carlo Simulation Results for Energy Production," *IEEE Transactions on PAS,* Paper 88WM215-6, presented at IEEE/PES 1988 Winter Meeting, New York City.

Wood, A. J., and B. F. Wollenberg, *Power Generation Operation and Control,* Wiley, New York, 1984.

PROBLEMS

1 Given the power system data:

Unit	Capacity	Forced Outage Rate	$/MWH
A	100	.1	15
B	150	.2	25
C	200	.25	40
TIE	1000	0	80

Calculate the production cost for a 200-MW demand.

a Give the deterministic solution, no outages.

b Give the deterministic solution with unit deratings.

c Give the probabilistic solution using state enumeration.

d Give the probabilistic solution using the convolution algorithm.

2 Calculate the probabilistic production cost using the convolution algorithm (Section 13.1.3) for a 225-MW load:

Unit	Minimum Output (MW)	Maximum Output (MW)	Fuel at Minimum Power (MBtu/ Hour)	Incremental Heat Rate from Minimum to Maximum Power (MBtu/MWh)	Forced Outage Rate	Fuel Cost (¢/MBtu)
A	50	150	300	10	.1	200
B	50	100	300	10	.2	300
TIE	0	1000		10	0	600

3. Use the binomial distribution method to calculate the probabilistic production cost for a 300-MW load.

Unit	MW Capacity	Forced Outage Rate	Operating Cost ($/MWh)
A	100	.15	20
B	100	.15	25
C	150	.20	30
D	50	.10	35
TIE	1000	.10	50

4. In the distribution factor example of Section 13.2.9, change the bus generating values during this hour from Table 13.28 values to problem 4 values:

Bus	Table 13.28	Problem 4
1	10.2	11.2
2	−3.0	−3.0
3	−7.0	−8.0

a What per unit flow will occur on line 3–1 with all lines in service?

b What per unit flow will occur on line 3–1 with line 2–3 tripped?

c What per unit flow will occur on line 3–1 with line 2–1 tripped?

5. In the distribution factor example of Section 13.2.9, compute the GSF with bus 3 (not 1) as the reference bus.

a Invert the 2 • 2 matrix to obtain $\overline{X}$.

b Calculate the three delta line flows.

c Calculate the shift factors $a_{21,2}$, $a_{23,2}$, and $a_{31,2}$.

d Calculate the line outage factors $d_{21,23}$, $d_{31,23}$, $d_{31,21}$, and $d_{23,21}$.

e Calculate the total flow on line 3–1, all lines in; 2–3 all lines in; line 3–1 with 2–3 tripped, line 3–1 with 2–1 tripped.

6. In the transportation model example of Section 13.2.11, the load in area B is reduced from 500 to 400 MW, in Figure 13.10.

a Compute the production cost as in Tables 13.30 and 13.31.

b Compute the split savings as in Table 13.34.

7. A five-utility pool has net sales and costs of:

Utility	MW Net Sale	$/MWh Incremental Costs
A	150	12
B	100	16
C	− 50	25

Utility	MW Net Sale	$/MWh Incremental Costs
D	−100	30
E	−100	35

Calculate the billing totals based on a pooled average split savings method [Equation (13.63)].

14

GENERATION PLANNING

Reliability analysis, production simulation, and investment cost analysis were studied in previous chapters. In this chapter, these three analyses are integrated to address the issue of generation planning. The issues in generation planning are when and how much generation equipment needs to be installed, as well as what kind of equipment it should be (i.e., coal, gas turbine, combined-cycle). Total costs and financial implications of a generation plan are also important issues.

This chapter first provides an overview of generation planning issues. subsequent sections discuss alternative generation planning methodologies that range from simple levelized bus-bar approaches through detailed mathematical optimization.

14.1 GENERATION PLANNING ISSUES—OVERVIEW

The issue of when and how much generation capacity is needed is addressed by a reliability analysis, as illustrated in Figure 14.1. In this chart, the loss-of-load probability (LOLP) in days per year is plotted against the peak load in megawatts. The target LOLP design criteria of .1 days/year is shown as a horizontal bar. In 1990, the original system, experiencing a peak load of 6100 MV, will be able to meet the reliability target of less than .1 days/year. In 1991, however, as the load grows the LOLP increases from .03 days/year up to .3 days/year, which is greater than the target criteria. Consequently, another generating unit needs to be installed. In 1991, 800 MW of capacity is added to the system, which is sufficient until the load grows again. By the next year, the load has grown, the LOLP level again increases up to .3 days/year,

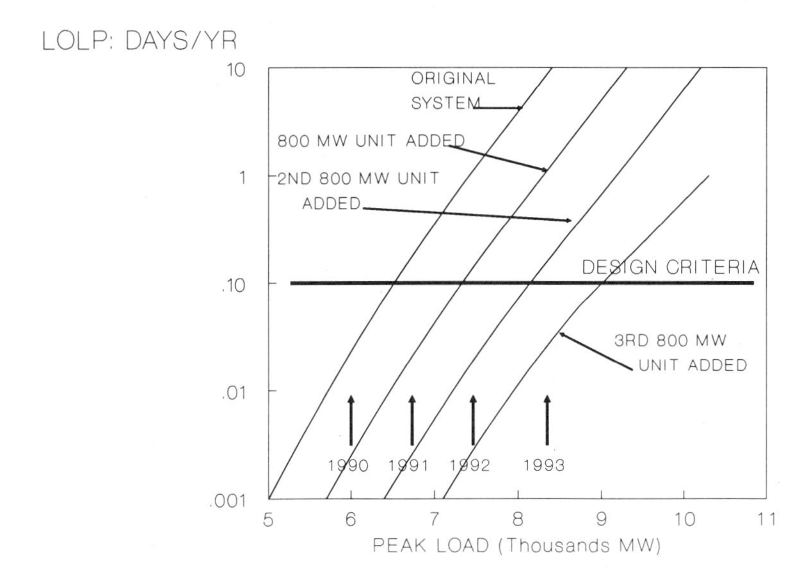

FIGURE 14.1 Capacity planning using loss-of-load probability.

and a second unit needs to be added. This procedure is repeated each year through to the horizon year.

The issue of what kind of generating equipment (i.e., nuclear, coal, gas turbine or hydroelectric) would be the most economical addition is addressed by combining a production cost analysis with an investment cost analysis. Figure 14.2 illustrates an example in which the cost of a 500-MW coal unit addition is compared to a 450-MW combined-cycle unit addition. In this example, the coal-unit option requires a greater capacity addition than the combined-

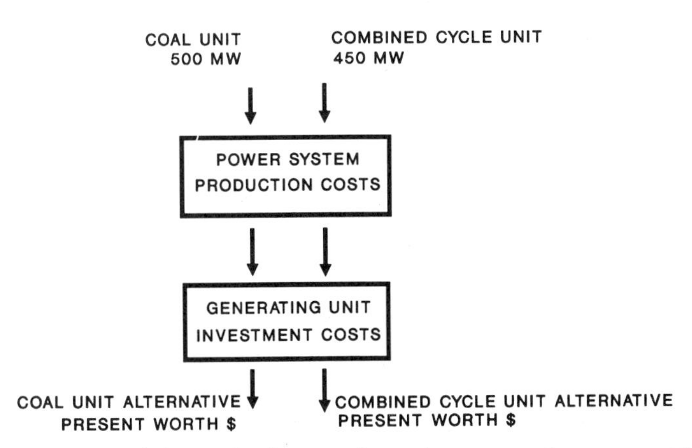

FIGURE 14.2 Determining optimal generation unit type—coal versus combined-cycle unit example.

cycle option because the coal unit has a higher forced outage rate (typically 15%), whereas a combined-cycle has a forced outage rate of about 7%.

Having identified how much capacity of each type needs to be added, the issue of which is the least costly can be identified. A production simulation of the power system through the horizon year is performed, first with the 500-MW coal unit, and then with the 450-MW combined-cycle unit. An investment cost analysis is done for both options. The system production costs and investment cost are added together to obtain a total cost for the coal-unit alternative and a total cost for the combined-cycle alternative. The cumulative present worth of the alternatives would be compared, and the addition that had the lowest cost would be selected. This procedure is repeated for each year of the study.

Having determined an optimal or least-cost plan, another issue to be addressed is financial implications. In Figure 14.3, the external financing of plan A and plan B are both shown. Suppose plan A is the most economical alternative on the basis of having the least cumulative present worth of revenue requirements and plan B a second alternative. However, plan B may have lower external financing requirements because plan A is a highly capital-intensive plan. Where plan A may use coal power plants or large hydroelectric facilities, plan B may use gas turbine or combined-cycle capacity additions that are less costly on a first-cost basis.

While plan A may by more cost-effective, if the utility has difficulty in finding and raising large sums of capital, then plan A may not be feasible, and the utility may have to resort to plan B, which it can finance and implement.

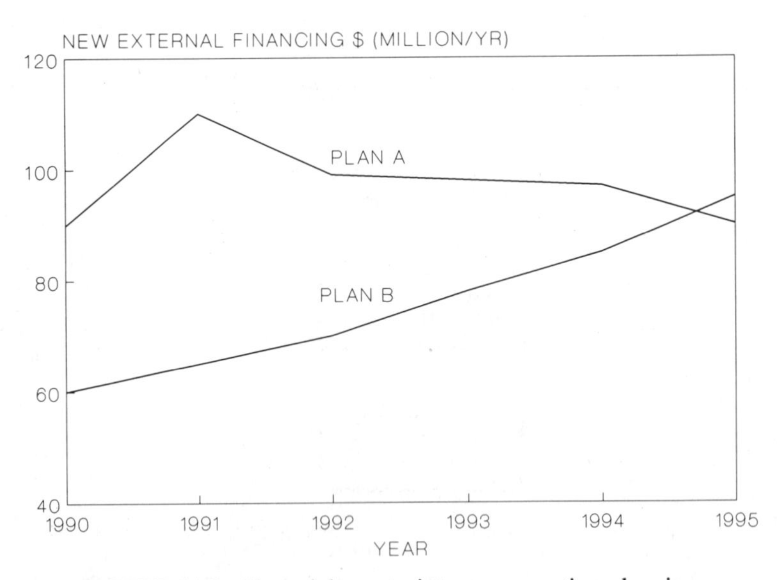

FIGURE 14.3 Financial constraints on generation planning.

A financial simulation of the utility for several generation alternatives may be helpful in identifying these potential constraints.

Business risk is another issue that influences generation planning. Generation additions are capital intensive, consume large quantities of fuel, and have 40 years or more of service life. A business decision regarding each generation addition must account for future fuel cost inflation and the future regulatory climate. These and other business environment issues (such as unknown future load demand) may be highly uncertain. There is a risk that a business decision made today may prove to be a costly error because the future business environment evolves differently than what had been forecasted.

For example, in Figure 14.4, plan A has a lower cost than plan B, when fuel cost is assumed to inflate at the reference forecast value (i.e., 5%/year). However, high inflation impacts plan A to a greater extent. Thus, with regard to fuel price escalation, plan A has more business risk than plan B. Depending on the degree of uncertainty in the fuel price, a utility may wish to choose plan B to avoid potential exposure to much higher costs. A decision-tree or risk-analysis theory is very helpful in evaluating generation business decisions in an uncertain environment.

Typical generation planning parameters of power plants built in the United States that meet U.S. environmental and regulatory requirements are presented in Table 14.1. These parameters have a broad range (±20%) and vary depending on the specific design, technology, and regional location of the generating unit. The several examples in this chapter use values with this range

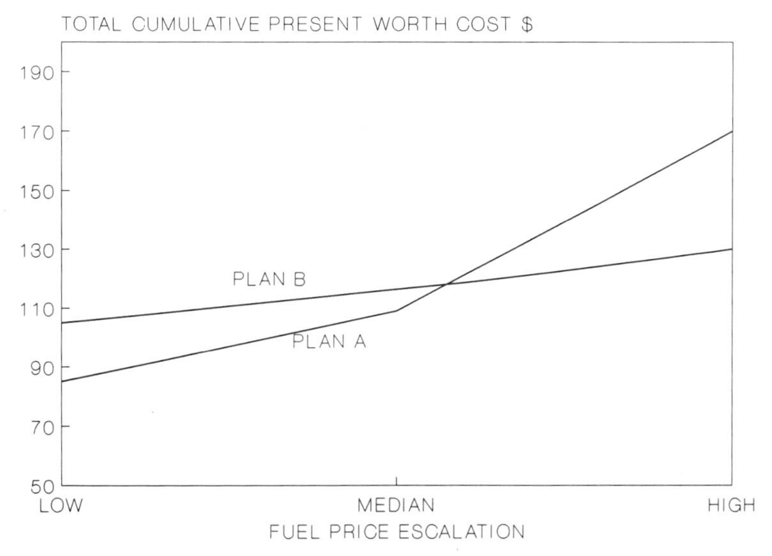

FIGURE 14.4 Risk exposure in generation planning based on fuel price escalation.

TABLE 14.1 Typical Generating Unit Parameters
(ALL COSTS IN 1990 $)

Generation Type	Typical MW Size	Capitalized Plant Cost ($/kW)	Construction Lead Time (Years)	Heat Rate (Btu/kWh)	Fuel Cost ($/MBtu)	Fuel Type	Equivalent Forced Outage Rate (%)	Equivalent Scheduled Outage Rate (%)	O&M Fixed ($/kW/Year)	Cost Variable ($/MWh)
Nuclear	1200	2400	10	10,400	1.25	Uranium	20	15	25	8
Pulverized coal steam	500	1400	6	9,900	2.25	Coal	12	12	20	5
Atmospheric fluidized bed	400	1400	6	9,800	2.25	Coal	14	12	17	6
Gas turbine	100	350	2	11,200	4.00	Natural gas	7	7	1	5
Combined-cycle	300	600	4	7,800	4.00	Natural gas	8	8	9	3
Coal-gasification combined-cycle	300	1500	6	9,500	2.25	Coal	12	10	25	4
Pumped storage hydro	300	1200	6	—	—		5	5	5	2
Conventional hydro	300	1700	6	—	—		3	4	5	2

for illustration. The *Technical Assessment Guide* reference (Tag, 1986) provides typical data ranges for numerous plant technology types.

14.1.1 Conventional Pulverized-Coal Steam-Generating Unit

Figure 14.5 illustrates a typical functional design of a conventional pulverized-coal steam-generating unit incorporating a "wet" flue-gas cleanup system. Coal is crushed, pulverized, and fed into the steam generator (boiler) to produce steam. The steam is admitted to the steam turbine at 1000°F (2400-psi subcritical design or 3500-psi supercritical design) and expanded. The steam expansion produces work which drives the turbine and generator to produce electricity. The expanded steam is condensed to water in the condenser and returned to the steam generator (boiler) to complete the cycle. The condenser is cooled by circulating water from cooling towers.

Flue-gas from the combustion of the coal in the steam generator is passed through an electrostatic precipitator (or a baghouse collector) to remove particulates. The flue-gas then passes through a flue-gas desulfurization unit, or scrubber, to remove sulfur dioxide (SO_2) products from the coal-combustion flue-gas. Desulfurization is typically achieved by mixing the flue-gas with limestone or lime, which reacts with sulfur and is collected as a solid or liquid slurry. After scrubbing, the flue-gas is exhausted through a tall stack.

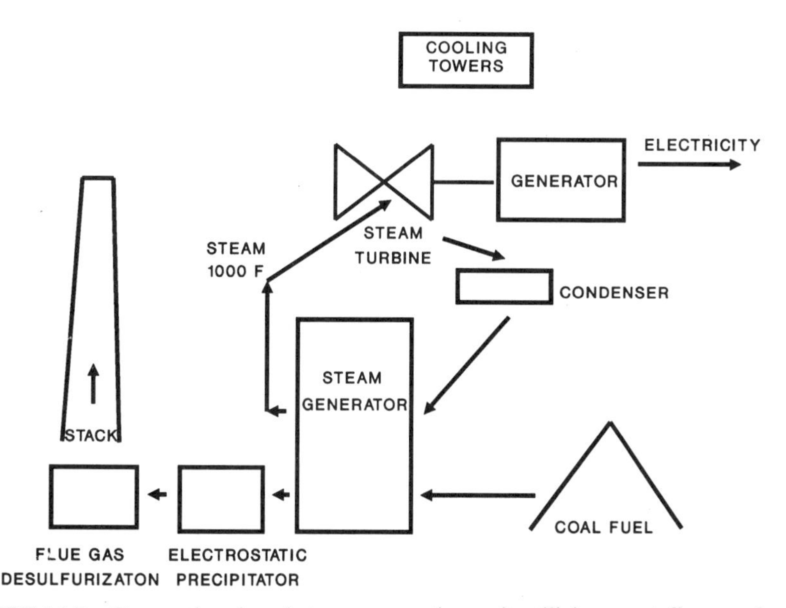

FIGURE 14.5 Conventional coal-steam generating unit utilizing a wet flue-gas desulfurization system.

14.1.2 Atmospheric Fluidized-Bed Coal Steam-Generating Unit

An atmospheric fluidized-bed coal steam-generating unit is a more recent technologic addition to the utility power system (Figure 14.6). The key difference between this and the pulverized-coal steam unit of Figure 14.5, is the design of the steam generator. The fluidized-bed steam generator is able to combust coal in the presence of limestone, thereby removing SO_2 in the combustion process rather than with a postcombustion process such as a scrubber.

14.1.3 Simple-Cycle Combustion-Turbine Generating Unit

A simple-cycle combustion-turbine generating unit (Figure 14.7) is also referred to as a "gas turbine." A compressor compresses air to an 11:1 compression ratio (more for some gas-turbine designs) and mixes this with fuel oil or natural gas in a burner stage. The hot gases ($\geq 2000°F$) are expanded in a power turbine that drives the compressor and electrical generator and exhausted in the range of 1000°F to a stack. Gas turbines are manufactured in sizes up to 150 MW.

There are two families of gas turbine designs, "heavy duty" and "aircraft derivative." The heavy duty type utilizes a single shaft for the compressor and power turbine. The aircraft derivative operates with compression ratios of 20–30 and utilizes a two-shaft design in which one power turbine drives the compressor and another turbine shaft drives the generator. The aircraft derivative can be approximately 10% more efficient than the heavy-duty one but costs approximately 25% more.

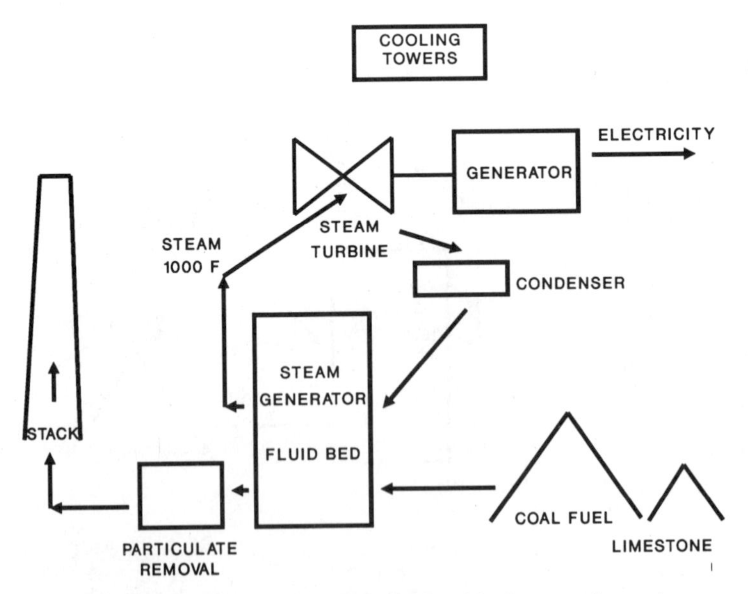

FIGURE 14.6 Atmospheric fluidized-bed generating unit.

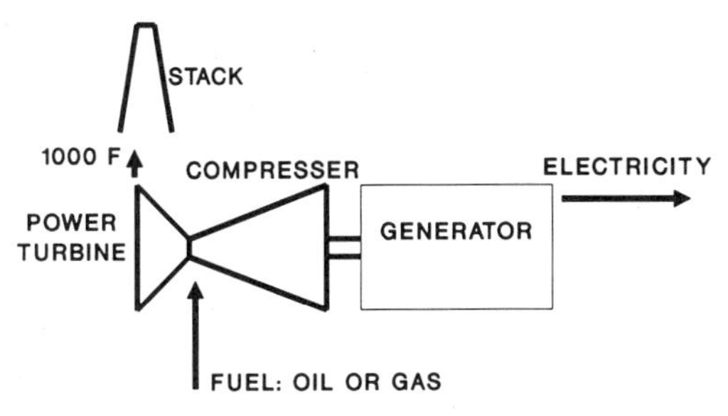

FIGURE 14.7 Simple-cycle combustion turbine generating unit.

14.1.4 Combined-Cycle Generating Units

The combined-cycle generating unit is a hybrid of the gas turbine and steam-turbine cycles. The heat contained in the 1000°F exhaust of a combustion turbine is fed through a heat-recovery steam generator (boiler), as shown in Figure 14.8. The heat-recovery steam generator (HRSG) produces steam in the range of 850°F, which is expanded in a steam-turbine cycle.

The combined-cycle typically obtains two-thirds of its megawatt output from the gas turbine and one-third from the steam-turbine cycle. The combined-cycle can be built in modular stages, a gas turbine unit can be installed first, followed by a combined-cycle add-on several years later. The combined-cycle add-on increases plant output by approximately 50% of the gas turbine rating (one-third steam-turbine/two-thirds gas-turbine contribution) and improves the heat rate from the 11,000-Btu/kWh range (of a gas turbine) to the 7800-Btu/kWh range.

14.1.5 Coal-Gasification Combined-Cycle Generating Unit

The combined-cycle generating unit can be further modified if a coal fuel is desired (rather than an oil or natural gas fuel) by the addition of a coal-gasifier

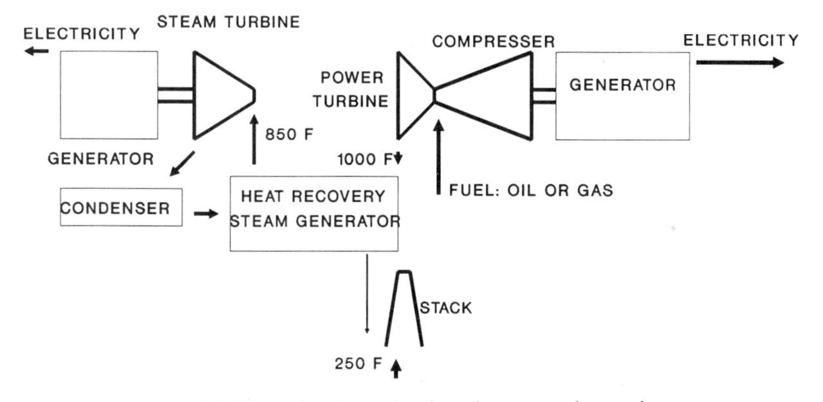

FIGURE 14.8 Combined-cycle generating unit.

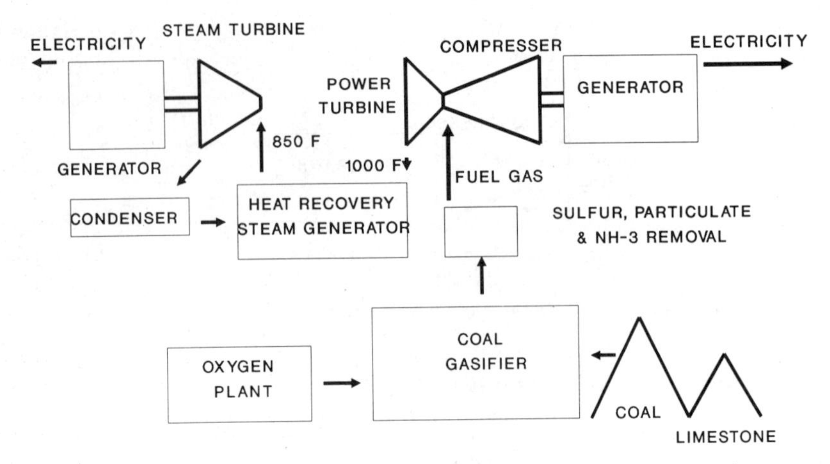

FIGURE 14.9 Coal-gasification combined-cycle generating unit.

unit (Figure 14.9). The coal gasifier is a recent technologic addition to the utility power industry. It partially combusts coal and limestone in an oxygen-rich environment to produce a low or medium Btu-content coal-gas fuel. The coal-gas fuel has sulfur and particulates removed from the gas prior to complete combustion in the gas turbine. The coal gasifier can be built as a modular add-on to the combined-cycle unit. Gasifier add-ons typically reduce the megawatt output by 5%, and the heat rate increases by 1500 Btu/kWh or more as a result of the auxiliary power and steam requirements of the coal gasifier.

It is important to note that technologies using coal as a fuel are more capital-cost-intensive than are oil- or natural gas-fueled technologies. However, the fuel cost of coal is less expensive than that of oil or natural gas. The utility industry uses the diversity of capital cost and fuel cost to economically serve a broad range of customer load demands.

14.2 GENERATION PLANNING METHODOLOGY

Generation planning and analysis can be performed by using a spectrum of methodologies that range from very simple analysis, such as levelized bus-bar cost, to much more detailed analysis involving reliability, production cost, investment cost, and financial analysis. These methods, in level of detail, are categorized into four types:

- Levelized bus-bar cost.
- Screening curves analysis.
- Manual evaluation of power system reliability, production cost, and investment cost.
- Automated power system reliability, production cost, and investment cost analysis.

The levelized bus-bar cost method analyses generating unit decisions on a unit basis only, not recognizing how the units may be operated in a power system. This method is useful for preliminary economic comparisons of the costs of generating unit alternatives. This method is also the most elementary and thereby the easiest to understand and communicate.

While levelized bus-bar and screening curve analyses are useful for initial feasibility studies, detailed generation planning analyses must use methods based on the tools of power system reliability, power system production cost, and investment cost.

14.2.1 Levelizing

Generation planning is typically conducted over a 10–30-year time period. The treatment of inflation is very important since decisions can be marked influenced by its effects.

One simplified method of accounting for inflation when using the levelized bus-bar cost method is to use levelized cost values. The cost levelizing process converts a yearly escalating series of costs into a single, constant, present-worth equivalent value, as illustrated in Figure 14.10. In this example, a fuel-cost value in $/MBtu begins at $2.0/MBtu and escalates from 1990 at 5% per year. By the year 2010, the fuel cost is $5.31/MBtu.

Often the inflation rate is assumed to be constant over a 10-, 20- or 30-year period. Consequently, the analysis can be simplified significantly, and the levelized fuel cost is easily computed.

For the example, the levelized fuel cost of 2.845 is the present-worth average of the escalating fuel costs over the 1990–2010 time period. This concept was

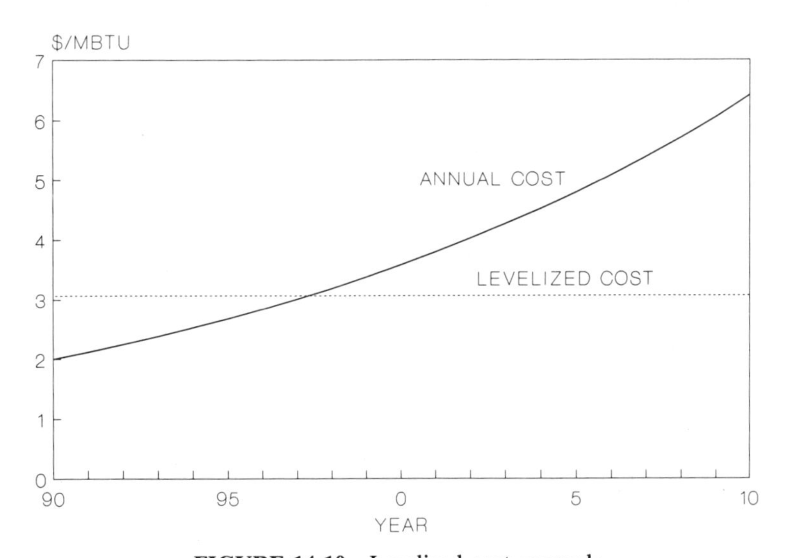

FIGURE 14.10 Levelized cost example.

discussed in Chapter 3 and is calculated by using the uniform levelizing value. Using Equation (3.25) and a present-worth discount rate of 10%; we obtain:

$$U = \text{fuel cost} \cdot \frac{1 - [(1 + a)/(1 + i)]^N}{(i - a)} \cdot \text{CRF} \tag{14.1}$$

where a = inflation rate
CRF = capital recovery factor
i = present worth rate
N = number of years of levelization
U = levelized fuel cost

$$U = 2.0\ \$/\text{MBtu} \cdot \frac{[1 - (1.05/1.10)^{20}]}{(.10 - .05)} \cdot \frac{(.1)(1.10)^{20}}{[(1.10)^{20} - 1]}$$

$$= 2.0\ \$/\text{MBtu} \cdot 1.423 = 2.845\ \$/\text{MBtu}$$

The levelizing factor of 1.423 is the per-unit multiplier that translates the starting-year value of the fuel cost to the levelized value. Figure 14.11 illustrates the levelizing factor as a function of the inflation rate and present-worth rate (PWR) in the case of a 10-year levelization period.

Another consideration that is involved in generation planning is the levelized annual fixed-charge rate. The annual fixed-charge rate is a number which, when multiplied by initial capital cost of the equipment, yields the levelized annual cost of owning a power plant, substation, or transmission lines. The fixed-charge rate includes depreciation, return on investment, local taxes (or contribution in lieu of taxes), and federal taxes. The levelized annual fixed-charge rate is similar to the levelized fuel cost and provides a uniform annual payment over the life of the plant. The present worth of the uniform payments is equal to the present worth of the actual payments.

The levelized fixed-charge rate using today's money cost is typically 20%/year for investor-owned utilities, 15%/year for publicly owned utilities, and 13%/year for municipally or government-owned utilities. Chapter 4 presents a detailed discussion of the calculations of the levelized fixed-charge rate.

14.2.2 Levelized Bus-Bar Analysis

In the levelized bus-bar analysis, the cost per kilowatt-hour or cost per year of operating a generating unit plant is computed. This permits a direct economic comparison of one generating unit versus another.

The levelized bus-bar method may be illustrated through an example. This example utilizes cost data typical of a higher-construction-cost region of the United States (northeast, north central, middle Atlantic regions). Table 14.2 contains data on a 400-MW coal-fired power plant. The heat rate, fuel cost, plant cost, and levelized fixed-charge rate are given. The objective is to compute the annual levelized owning cost in dollars per year, based on plant operation at 70% capacity factor and on a 20-year economic evaluation.

The first step is to compute the levelized fuel cost. In this example, the fuel cost is two dollars, and, from Equation (14.1), the levelizing factor is 1.537. The product of these is the levelized fuel cost of 3.074 $/MBtu. A similar calculation is performed to levelized the O&M cost.

The analysis involves four steps, as illustrated in Table 14.3. The first step is to compute the annual levelized production cost. This is computed by multiplying the megawatt size of the unit times 8760 hours/year times 70% capacity factor. This product gives the annual megawatt-hours generation. This is multiplied by the heat rate of 9.5 MBtu/MWh times the levelized fuel cost in dollars per MBtu. The result is an annual levelized production cost of 71.63 million dollars per year ($71.63 M/year).

TABLE 14.2 Sample Data for a Unit Addition
400-MW COAL-FIRED GENERATING UNIT (1990 DOLLARS)

Heat rate = 9500 Btu/kWh
Fuel cost = 2.0 $/MBtu
Plant cost = 1500 $/kW
O&M Cost
Fixed 20 $/kW/year
Variable 5 $/MWh
Levelized fixed-charge rate = 20%/year
Present-worth rate = 10%/year
Fuel price escalation = 6%/year
Capacity factor = 70%

TABLE 14.3 Coal Plant Addition

Solution

1. Computation of annual levelized fuel costs
   $fuel = 400 MW • 8760 hours/year • (.70 capacity factor)
   • 9.500 MBtu/MWh • (2.0 $/MBtu)
   • (1.537 levelized factor)
   = $71.63M$/year
2. Computation of annual levelized O&M costs
   Variable = 400 MW • 8760 hours/year • .70 capacity factor
   • $5/MWh • 1.537 = $18.86M $/year
   Fixed = 400,000 kW • 20 $/kW/year • 1.537 = $12.28M/year
   O&M total = 18.86 + 12.28 = $31.14M/year
3. Computation of annual levelized investment costs
   $ invested = (400,000 kW) • 1500 $/kW • (.20 FCR)
   = $120M/year
4. Total costs (@ 70% capacity factor) = fuel + O&M + investment
   = 71.63 + 31.14 + 120.0
   = $222.8M/year
   Total costs (@40 capacity factor) = $184.0M/year

The next step is to compute the levelized fixed and variable O&M cost. The result is an annual levelized O&M cost of $31.14 M/year.

The third step is to compute the annual levelized investment cost. This is computed by multiplying the rating of the unit (400,000 kW) times $1500/kW plant cost for a coal unit times a 20% fixed-charge rate (FCR). This yields 120 million dollars per year as the annual levelized investment cost.

Finally, the total costs are computed by adding the investment, fuel, and O&M costs together. The total owning cost for this alternative is $222.8 M/year.

It is sometimes useful to convert this levelized annual cost into dollars per megawatt-hour. This is done by dividing the owning cost by the annual megawatt-hour generation of 2,450,000 MWh. The resulting levelized annual generation cost is 90.9 dollars per megawatt-hour (or 9.09¢/kWh).

The example assumed that the unit would operate at 70% capacity factor. For use later in this section, it is necessary to compute the annual cost if this unit were to operate at 40% capacity factor. This is accomplished by noting that the fuel and variable O&M costs are linearly proportional to the capacity factor, as presented in Table 14.3. Thus, the fuel cost can be ratioed by 40% divided by 70% times $71,629,000/year. At 40% capacity factor, the fuel cost is $40.930M/year and the variable O&M cost is $10.778M/year. The remaining costs, fixed O&M and investment cost, were not dependent on operating hours and do not have to be recomputed for the 40% capacity factor case. The total costs at 40% capacity factor are the sum of fuel cost, variable O&M cost, fixed O&M cost, and the investment cost, or $184.0M/year.

TABLE 14.4 Sample Data for a Combined-Cycle Unit Addition

400-MW COMBINED-CYCLE UNIT (1990 DOLLARS)
Heat rate = 8500 Btu/kWh
Fuel cost = 5.0 $/MBtu
Plant cost = 700 $/kW
O&M cost
Fixed 9 $/kW/year
Variable 3 $/MWh
Levelized fixed-charge rate = 20%/year
Present-worth rate = 10%/year
Fuel price escalation = 6%/year
Capacity factor = 70%

Now examine an alternative generating unit, a 400-MW combined-cycle unit, based on the data presented in Table 14.4. In Table 14.5, the levelized annual owning cost calculations of this generation addition alternative are presented. At 70% capacity factor, the total cost is $233.1M/year; at 40% capacity factor, the total cost is $159.6M/year.

Similarly, data for a 400-MW gas-turbine plant alternative are shown in Table 14.6. This generating plan has a higher fuel cost and lower plant cost than does the coal-fired steam plant examined in Table 14.3. The annual owning costs are computed in Table 14.7. At 70% capacity factor, the total cost is $296.3M/year; at 40% capacity factor, the total cost is $181.6M/year.

The results for the three alternatives are summarized in Table 14.8. At 70% capacity factor, the coal power plant alternative is the least expensive.

The levelized bus-bar method in this example illustrates that the coal unit has the least present-worth cost over the 20 years based on a 70% capacity factor. It has the least cumulative present-worth cost because *levelized* costs were used. It is interesting to examine the actual (nonlevelized) annual costs

TABLE 14.5 Combined-Cycle Evaluation

1. Annual fuel cost = 400 • 8760 • .70 • 8.5 • (5.0 $/MBtu) • (1.537 levelizing factor)
   = $160.22M/year
2. Annual O&M cost
   Variable = 400 • 8760 • .7 • 3 $/MWh • 1.537 = 11,309,860 $/year
   Fixed = (400,000 kW) • 9 $/kW/year • 1.537 = 5,533,200 $/year
   Total = $16.84M/year
3. Annual investment cost = 400,000 • 700 • .20
   = $56M/year
4. Total cost (@ 70% capacity factor) = $233.1M/year
   Total cost (@ 40% capacity factor) = $159.6M/year

TABLE 14.6 Sample Data for a Gas-Turbine Unit Addition

400 MW OF GAS-TURBINE CAPACITY (1990 DOLLARS)
Heat rate = 11,000 Btu/kWh
Fuel cost = 6.0 $/MBtu
Plant cost = 350 $/kW
O&M cost
Fixed 1 $/kW/year
Variable 5 $/kW/year
Levelized fixed-charge rate = 20%/year
Present-worth rate = 10%/year
Fuel price escalation = 6%/year
Capacity factor = 70%

TABLE 14.7 Gas-Turbine Example

1. Annual fuel cost = 400 • 8760 • .7 • 11.0 • 6.0 • 1.537 = $248.8M/year
2. Annual O&M Cost
   Variable = 400 • 8760 • .7 • 5 $/MWh • 1.537 = $18.84M/year
   Fixed = 400,000 kW • 1 $/kW/year • 1.537 = $0.31M/year
   Total = $19.46M/year
3. Annual investment cost = $28M/year
4. Total cost (@ 70% capacity factor) = $296.3M/year
   Total cost (@ 40% capacity factor) = $181.6M/year

TABLE 14.8 Levelized Annual Owning Cost ($/year; 70% Capacity Factor)

Alternative	Levelized Owning Costs in Million $/Year
Coal	222.8
Combined-cycle	233.0
Gas-turbine	296.3

for these alternatives. The fuel and O&M costs increase with time as a result of inflation, while the fixed charges decrease slightly with time. Figure 14.12 illustrates the yearly costs over the first 20 years of operation. Note that the coal unit, which has the least cumulative present-worth cost, does not have the lowest cost in the first year. The costs of the coal unit alternative are composed of high investment charges that do not escalate with time and relatively low fuel and O&M charges. Consequently, a capital-intensive alternative such as the coal unit will have a yearly cost that does not escalate rapidly with annual fuel and O&M cost escalation. A gas-turbine unit is, on the other hand, a low-capital but high-fuel cost-intensive alternative, has a yearly cost that is strongly influenced by the annual fuel and O&M cost escalation.

In summary, the levelized bus-bar method permits a direct unit economic comparison at a constant capacity factor. This method does not account for the fact that alternative generation types may not be dispatched at the same capacity factor because of operating cost differences. Consequently, a more detailed methodology is required. In spite of this, the levelized bus-bar method is still useful for screening-type analysis of alternatives that would operate with similar capacity factors. The method is also useful because of its simplicity, and it can be easily presented to the general public or other interested parties.

14.3 SCREENING CURVE ANALYSIS

A key assumption of the levelized bus-bar method is that the capacity factor of each generating unit is one constant value. The "screening curve analysis

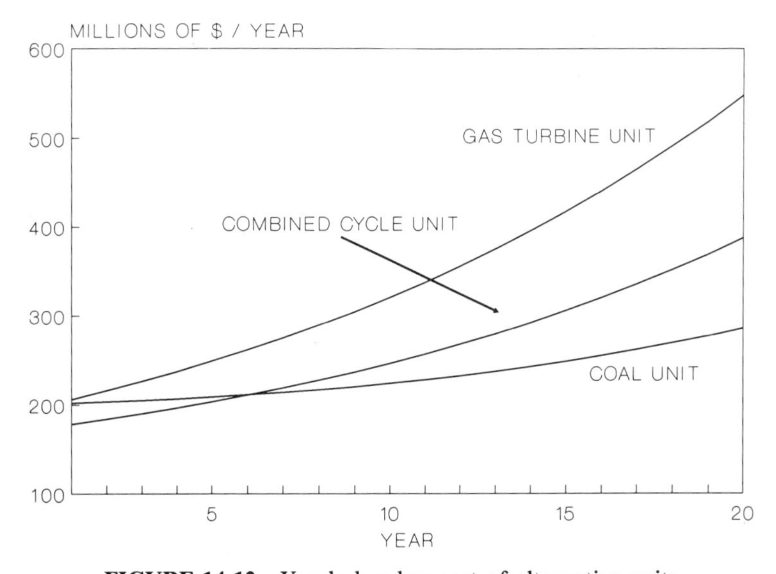

FIGURE 14.12 Yearly bus-bar cost of alternative units.

method'' is an extension of the levelized bus-bar method to remove this restriction.

The data presented in the previous examples can be plotted on a graph. The *y* axis is the total levelized annual owning cost in dollars per year and the *x* axis is the plant capacity factor, as illustrated in Figure 14.13. The two levelized bus-bar cost points that were evaluated at 40 and 70% capacity factor for each alternative are plotted on the graph. Since the owning costs are linearly proportional to the capacity factor, the graphs are straight lines.

This graph illustrates that the economic merit of each generating unit is dependent on the capacity factor of unit operation. The example shows that the coal steam plant has a very high plant cost, yet is also very fuel-efficient. Consequently, at capacity factors above 60%, the coal unit has the lowest owning cost. The gas turbine, on the other hand, has a very low capital cost but a high operating cost. If a power plant is needed to run at a capacity factor of less than 25%, the gas turbine is the most economical choice. Similarly, if the capacity factor for a future plant will be between 25 and 60%, then the combined-cycle is the most economical plant.

The screening curve is very useful in understanding the relative economic merits of alternative generation types. However, one must determine the capacity factor at which a new generation type will operate.

The capacity factor for the future power plant is determined by performing a production simulation analysis in which each generation plan alternative is operated in a power system. The parameters of the power system determine the economic loading and the economic capacity factor for the operation of each generating unit.

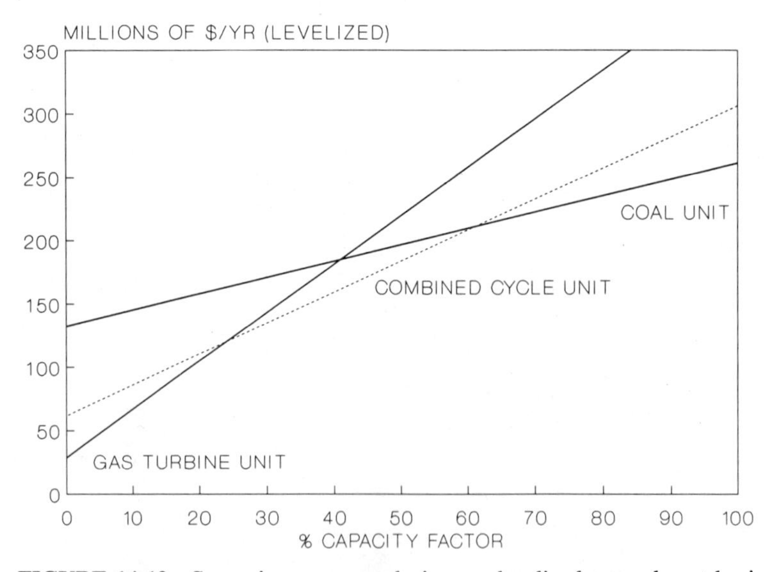

FIGURE 14.13 Screening curve analysis on a levelized annual cost basis.

One approximate method of assessing an optimal generation mix uses an annual load-duration curve in conjunction with a screening curve, as illustrated in Figure 14.14. The screening curve is plotted on the top of the graph, which shows the dollars per year versus capacity factor. On the bottom half of the graph, a load-duration curve plots the megawatt load versus the percent of the year. By projecting the intercepts of the screening curve onto the load-

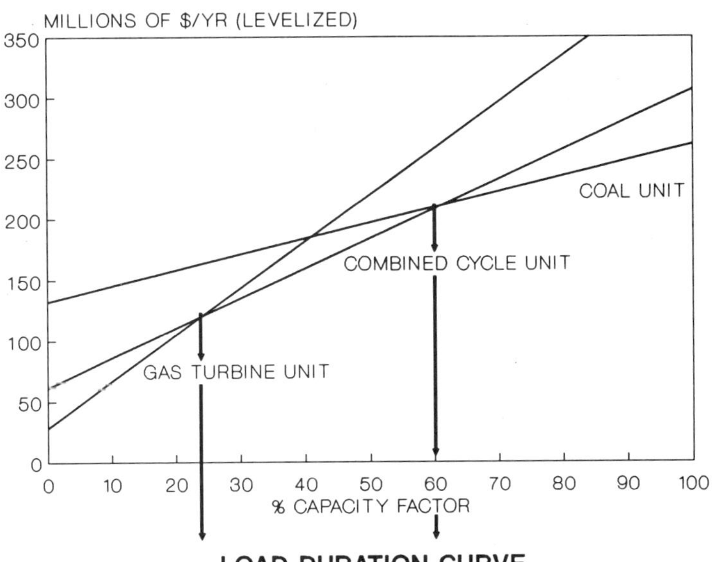

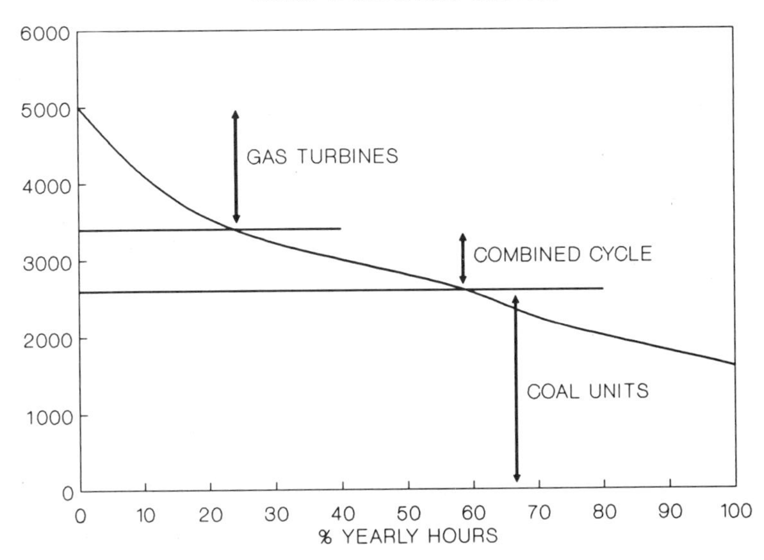

FIGURE 14.14 Screening curve and load-duration curve analysis.

duration curve, the optimal megawatt amount of each type of capacity can be evaluated. This analysis shows that coal plants should represent 2600 MW of generation capacity; combined-cycle power plants, 800 MW (3400 MW − 2600 MW) of capacity; and gas turbines, 1600 MW (5000 MW − 3400 MW) of capacity.

This graph illustrates the generation mix concepts. In a power system, there is an opportunity for several types of generation plants to serve the load demand since load demand is not uniform throughout the year but has weekly, daily, and hourly variations. High-load demands occur only during the hot summer months or the very cold winter months for only a few hours per year. These peak periods demand capacity to serve them. They should be satisfied by peaking-type generation such as gas turbines, which have low capital cost but high operating cost. For this example, the 2600-MW load demands that persist nearly all year (termed "base load") should be met by coal-fired capacity, with a low operating cost.

Suppose that the 5000-MW load-duration curve represents the year 2000. Furthermore, suppose that the existing system requires 5000 MW of capacity in year 2000 and has 3500 MW of existing capacity in year 1990, consisting of 2000 MW of coal, 500 MW of combined-cycle, and 1000 MW of gas-turbine. Furthermore, suppose that these existing units have the same fuel, O&M, and heat rate as the respective new unit candidates. Figure 14.15 shows how the load-duration curve/optimal-mix curve helps to determine how much of each new unit type should be added to the power system from 1990 to the year 2000. For this example, the mix would be approximately 600 MW of coal, 300 MW of combined-cycle, and 600 MW of gas-turbine. Thus, the optimal mix

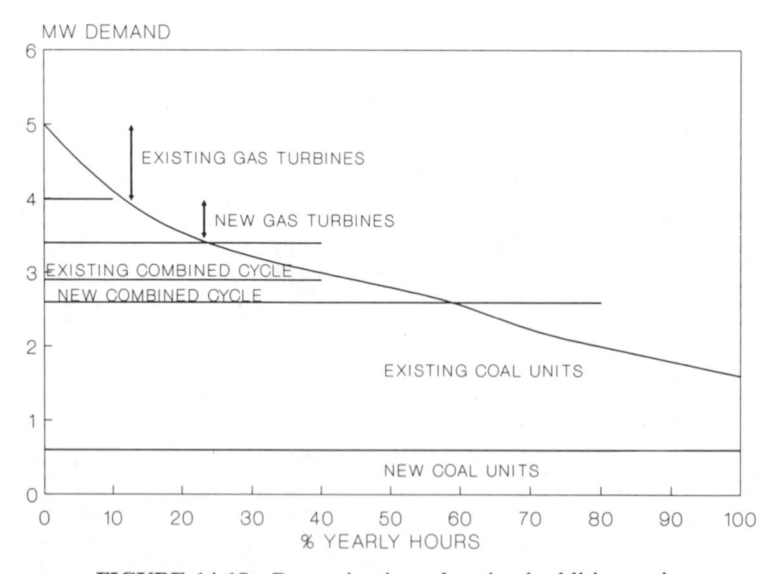

FIGURE 14.15 Determination of optimal additions mix.

of future generation additions is dependent not only on the economic characteristics of the new generating-unit alternatives but also on the existing composition of the generating units.

While this simplified analysis is very useful in understanding the concept of an optimal mix, it neglects some very important factors. The operating characteristics of the existing plants in the power system may be different from those of the new-unit candidates. A detailed economic analysis must consider the operating characteristics of power plants already in the system.

In addition, the capacity factors for both existing and new plants may change in the future. The analysis conducted thus far assumes that the capacity factors will remain fixed over the lifetime of the plant. This is not an unreasonable assumption, but capacity factors of units do change through time as new and more efficient equipment is added and old equipment is retired.

The forced and scheduled outage rates of power plants also need to be explicitly considered. Outages necessitate a requirement for capacity in excess of the 5000-MW peak load. Typical planned reserve margins are in the range of 15–25%.

These limitations can be removed by conducting a more detailed generation analysis, which will be presented in Section 14.5.

14.4 HORIZON-YEAR GENERATION ADDITIONS ANALYSIS

The final approximate generation-addition analysis method to be discussed is the horizon-year method, which is a refinement of the load-duration method. This is a practical method for understanding and interpreting optimal economic generation addition plans. This method may be used on an annual year-by-year basis, or on a horizon-year basis.

The horizon-year method is an approximate procedure when a quick analysis of the total number of additions over a period of time is required. A horizon year and appropriate levelizing factors are selected.

The horizon-year generation analysis procedure begins with the load-duration curve of the horizon year—year 2000, for example.

A list of alternate generation additions is made. A reliability analysis or reserve margin criteria establishes the amount of capacity needed in the horizon year, 2000. The existing capacity of each alternate generation capacity addition is then placed on the load-duration curve, and the levelized annual production cost is computed.

The power system production cost (fuel and O&M) and investment cost of each alternative are added together to compute a total levelized annual cost. The generation addition alternative with the lowest cost is the one selected.

It is important to note that this evaluation is a *system* analysis. Each generation-addition candidate type operates in the power system at a different capacity factor because each candidate type will have differing fuel costs, heat rate, outage rates, and so on. Therefore, power system operation cost is influenced by the type of capacity. Also note that only the investment cost of each future unit alternative needs to be considered. The investment cost of the existing power system is a "sunk" cost. Since it is the same for every alternative, sunk costs need not be evaluated. The decision cost criteria is the sum of the power system production cost and the investment cost of the future units.

Example. A utility desires to install either 588-MW coal units or 555-MW combined-cycle units in the year 1990. The existing system is comprised of coal-steam, oil-steam, and gas-turbines (GT). Determine whether coal or combined-cycle units are the most economical additions to the power system in the single year, 1990.

Data

- Existing capacity

Existing Type	Nameplate Capacity	Available Capacity (MW) Derated by Forced Outage Rate	Levelized Operating Cost $/MWh (1990 $)
Coal-steam	3600	3000	20.00
Oil-steam	4700	4000	50.00
GT	3888	3500	60.00

- The capital costs of the new generation plant in this low-plant-cost region example are:

Type	$/kW (1990 $)
Coal	1000
Combined-cycle	500
GT	250

- The annual levelized fixed-charge rate is 20%/year.
- For simplicity, assume that the generating units can be aggregated by type, and that the incremental operating cost characteristic is a constant from zero to full output. The megawatt capacity rating for evaluating power system production costs is the capacity rating derated by the forced outage rate (FOR) of the units, specifically, MW • (1-FOR). Assume zero scheduled outage rate.
- The load-duration curve is approximated as a simplified step curve as shown in Figure 14.16.

A: Load-Carrying Capability

- The *M*-slope of the power system is 800 MW. Reliability studies have shown that 1000 MW of *effective* load carrying capacity must be added to the power system in 1990. The forced outage rate and size data are:

Type of Addition	Forced Outage Rate	Size (MW)
• Coal	.15	588
• Combined-cycle	.10	555
• Gas-turbine	.10	55.5

Using Garver's effective load-carrying capability formula, calculate the effective load carrying capability of each unit addition type:

$$C_{\text{eff}} = C - M \cdot \ln[(1 - R) + R \cdot e^{(C/M)}]$$

where C = capacity, R = forced outage rate, and M = M-slope.

SOLUTION

Solution Part A. The effective capabilities of the coal, combined-cycle, and gas-turbine additions are:

$$\text{Coal } C_{\text{eff}} = 588 - 800 \cdot \ln\,[.85 + .15e^{588/800}] = 467$$

$$\text{Combined-cycle } C_{\text{eff}} = 555 - 800 \cdot \ln\,[.90 - .1e^{555/800}] = 478$$

$$\text{Gas-turbine } C_{\text{eff}} = 55.5 - 800 \cdot \ln\,[.9 + .1e^{55.5/800}] = 50$$

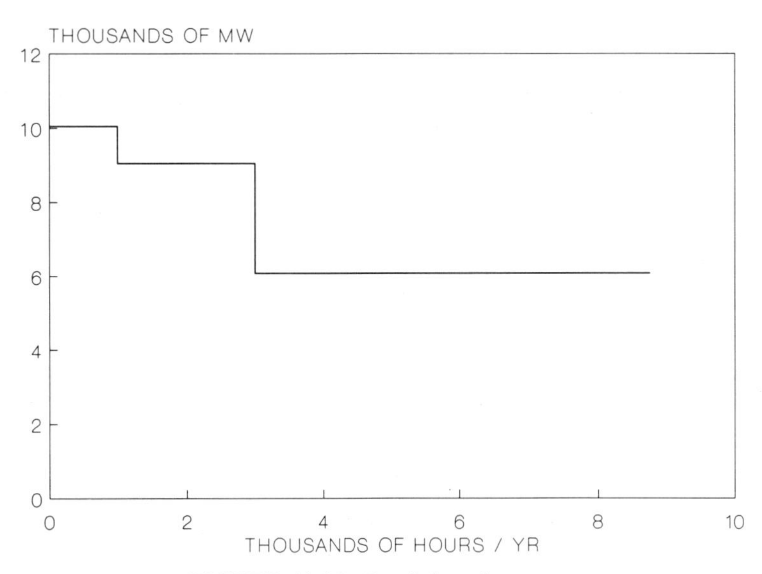

FIGURE 14.16 Load-duration curve.

B: How Much Capacity

1. The utility desires to add only one or two coal units and satisfy the remaining capacity requirements by adding gas-turbine units. How many additional gas turbines are required (integer number) to provide 1000 MW of effective capacity additions to the system?
2. If the utility adds one or two combined-cycle units, how many gas turbines are required to provide 1000 MW of effective capacity additions to the power system?

Solution Part B. The effective capacity required less that of two coal units is $1000 - 2 \times 467 = 66$ MW. Therefore, two gas turbines are also needed.

The required amount of effective capacity less that of two combined-cycle units is:

$$1000 - 478 \cdot 2 = 44 \text{ MW}$$

Thus, one gas turbine is also needed.

C: What Kind of Capacity. Using the results of Part A, determined which alternative [(coal plus GTs) or (combined-cycle plus GTs)] is more economical, based on the simplified annual levelized load duration curve analysis. For production simulation purposes, use the capacity derated by the forced outage rate.

	Existing Capacity (Derated By FOR) (MW)	Potential New Capacity Additions		Levelived Operating Cost ($/MWh)	Capital Costs Of New Plant ($/kW)	Levelized Annual Fixed-Charge Rate
		Nameplate Rating	Derated by FOR			
Coal-steam	3000	588	500	20	1000	.20
Oil-steam	4000	NA	—	50	NA	NA
GT	3500	55.5	50	60	250	.20
Combined-cycle	0	555	500	40	500	.20

Solution C: Evaluation of Coal and GT Alternative. The first step is to place capacity on the load-duration curve (Figure 14.17) in order of increasing $/MWh operating cost, lowest cost first. Coal capacity is the least expensive and, therefore, is the first type of capacity to be loaded, followed by combined-cycle, oil-steam, and gas turbines. The area of each capacity type shown in Figure 14.17 is its annual megawatt-hour generation. The annual (levelized) production cost can be computed by multiplying the annual energy generated times the levelized operating cost, as shown in Table 14.9.

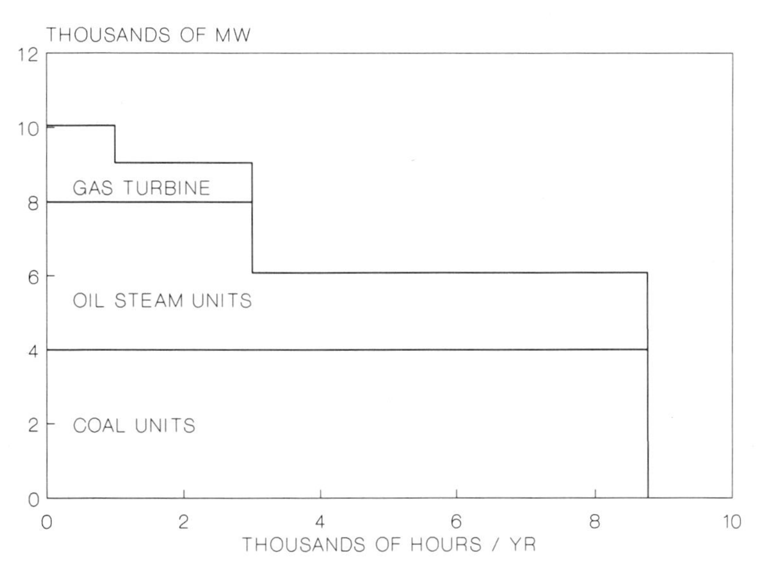

FIGURE 14.17 Production cost based on load-duration curve for the coal alternative.

TABLE 14.9 Calculation of Coal Alternative Decision Cost

COAL ALTERNATIVE

	Coal	Oil-Steam	GT	Combined-Cycle
Capacity (derated by FOR)	3000 + 1000	4000	3500	0

LEVELIZED ANNUAL PRODUCTION COSTS

	Energy MWh	\$/MWh	Million \$/Year
Coal	8760 • 4000	20	700.8
Oil-steam	2000 • 8760 + 2000 • 3000	50	1176.0
GT	1000 • 3000 + 1000 • 1000	60	240.0
			\$2116.8/year

LEVELIZED ANNUAL INVESTMENT COST

[2 • 588000 kW • 1000 \$/kW
+ 2 • 55500 kW • 250 \$/kW] • .20 = \$240.7M/year

Total levelized annual costs = \$2357M/year

Investment cost is calculated by multiplying the capital cost of the coal (and GT) plant times the respective megawatt nameplate rating of capacity times the levelized fixed-charge rate.

The total levelized decision cost is the sum of the levelized annual operating and investment cost of $2357M/year as shown in Table 14.9.

Solution C: Evaluation of Combined-Cycle and GT Alternative. The same calculations can be performed for the combined-cycle alternative as illustrated in Figure 14.18 and Table 14.10. The total levelized annual decision cost is $2406M/year.

Solution C: Summary. Because the total levelized annual decision cost of the coal alternative is less than the combined-cycle alternative, the coal unit alternative should be selected.

In summary, this method is a useful and practical means of analyzing and displaying generation addition plans. However, it is only a screening method, and does not account for the detailed operating characteristics of generating units in meeting the load demand. The next several sections present more detailed analysis methods that are widely used in generation planning. The more simplified analyses are very useful for displaying and understanding the results from more detailed analysis procedures.

TABLE 14.10 Calculation of Combined-Cycle Alternative Decision Costs

COMBINED-CYCLE ALTERNATIVE

	Coal	Oil-Steam	GT	Combined Cycle
Capacity (derated by FOR)	3000	4000	3500	1000

LEVELIZED ANNUAL PRODUCTION COSTS

	Energy	$/MWh	Million $/Year
Coal	8760 • 3000	20	525.6
Combined-cycle	8760 • 1000	40	350.4
Oil-steam	8760 • 2000 + 2000 • 3000	50	1176.0
GT	1000 • 3000 + 1000 • 1000	60 60	240.0
			= $2292.0/year

LEVELIZED ANNUAL INVESTMENT COSTS

[2 • 555000 • 500 $/kW
+ 1 • 55500 • 250 $/kW] • .2 = $113.8 M/year

Total levelized annual costs = $2406 M/year

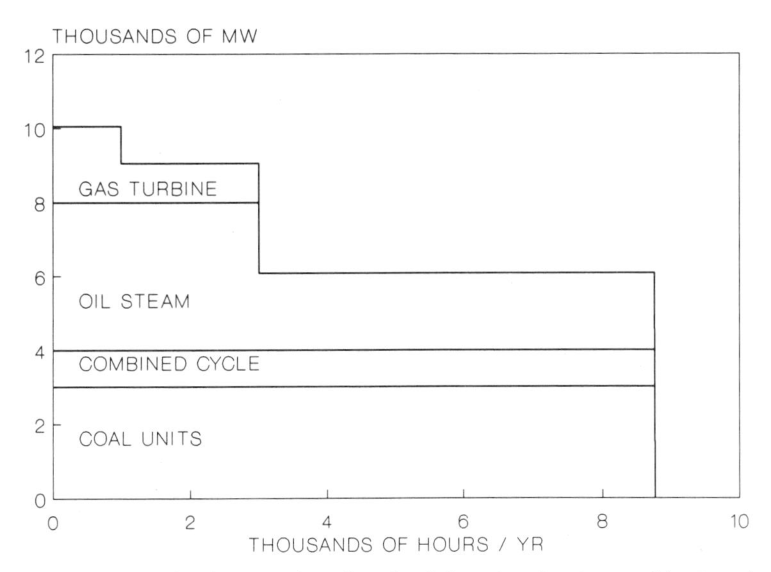

FIGURE 14.18 Production cost based on load duration for the combined-cycle alternative.

14.5 MANUAL GENERATION PLANNING

The previous sections illustrated screening-type techniques used in generation planning. These techniques are useful not only for preliminary studies but also in visualizing the results from more detailed methods. For detailed planning studies, a widely used procedure is manual generation planning, discussed in this section. It combines the disciplines of reliability (LOLP evaluations), system production simulation, and investment costing.

Figure 14.19 illustrates this approach. First, a proposed candidate set of additions is prescribed for each year (a 400-MW coal unit installed in 1990, a 400-MW coal unit installed in 1994, etc.). This proposed schedule of unit additions is analyzed using a power system reliability evaluation (LOLP) each year over a 20-year period from 1990 to 2009. The reliability evaluation provides a LOLP calculation in days per year for each year from 1990 to the year 2009. If the LOLP is less than the desired goal (such as one day per year), then this proposed set of additions meets the reliability target, and the subsequent steps are then followed. If the LOLP from these proposed additions is *not* adequate and exceeds the LOLP target in any study year, then this proposed set of additions would need to be modified to meet the reliability criteria. For example, if the LOLP is inadequate in 1998, then the 400-MW coal unit that would have been installed in 1999 may have to be advanced one year and installed in 1998, or the plan may need to be modified by increasing the megawatt size of several of these generating unit additions, or several gas-turbine units could be added to improve system reliability. This preliminary plan may have been generated using some of the screening techniques discussed previously.

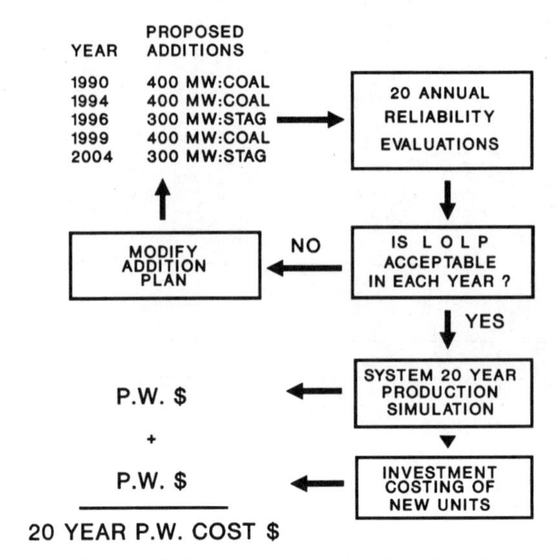

FIGURE 14.19 Manual generation planning.

After the proposed addition plan is modified to have an acceptable LOLP, then production simulation and investment costing procedures can be performed. A system production simulation is performed each year from 1990 to 2009 for this generation plan. The fuel costs and operation costs are printed and summarized each year and the present worth of these costs are calculated in 1990 year dollars. The investment costs of unit additions to be installed are also computed to a 1990-year value. The sum of these two costs, the production cost and the investment cost, are accumulated and yield the 20-year cumulative present-worth cost for the proposed addition plan.

The resulting 20-year cumulative present-worth cost is the cost of merely *one* generation-addition alternative. Other plans need to be examined as well, as is illustrated in Table 14.11. The total cost for plan 1 might be $4000M and for plan 2, $3890M. Similarly, alternative plans 3 and 4 may have higher costs.

TABLE 14.11 Manual Generating Planning

Plan 1		Plan 2		Plan 3	Plan 4
Year	Proposed Additions	Year	Proposed Additions		
1990	400-MW coal	1990	400-MW combined-cycle	—	—
1994	400-MW coal	1994	400-MW combined-cycle	—	—
1996	400-MW combined-cycle	1996	400-MW coal	—	—
1999	400-MW coal	1999	400-MW coal	—	—
2004	300-MW combined-cycle	2004	300-MW combined-cycle	—	—
Cumulative present-worth cost					
$Plan 1 = $4000M		$ Plan 2 = $3890M			

Plan 1, plan 2, plan 3, and so on are plans wherein the capacity addition types are modified, or the timing of power plant additions is modified, or the sizes of units are modified. Whatever the source of the variation, the objective is to find an alternative plan having the lowest cumulative present-worth cost. In this case, plan 2 has the lowest cost and would be chosen.

As shown in this example, manual generation planning involves very detailed analysis. It requires a great deal of manual intervention because there are many alternatives to examine. The next section will discuss automating this enumeration procedure using optimization techniques. While automated generation planning procedures are useful, the manual generation planning technique is still very widely used for examining specific generation planning issues, including unit size, reliability sensitivities, and investment tradeoffs.

14.6 AUTOMATED GENERATION PLANNING

The previous section discussed detailed manual generation planning. In the middle 1960s, this optimization process was automated. There are several approaches to implementing an optimization procedure. One approach is an exact mathematical optimization procedure called "dynamic programming." Other approaches include linear programming, mixed integer–linear programming, and several approximation techniques. This chapter reviews two of these and illustrates the benefits and tradeoffs of each.

Figure 14.20 shows the optimization of generation additions from an overview perspective. Suppose there are three generation alternatives available each year. Let S denote a steam-coal unit option, CC a combined-cycle option, and GT a gas-turbine option. For simplicity, assume that each unit is of the same megawatt size with the same reliability, and that one unit must be added every year to maintain the reliability criteria. Because there are three alternative generation types each year, a 20-year optimization study would generate a tree of more than three billion alternative generation plans, $(3)^{20}$. This huge tree and the three billion alternative plans are much more than present-day computers can reasonably compute. Fortunately, however, there are exact and approximate solutions for resolution of this problem.

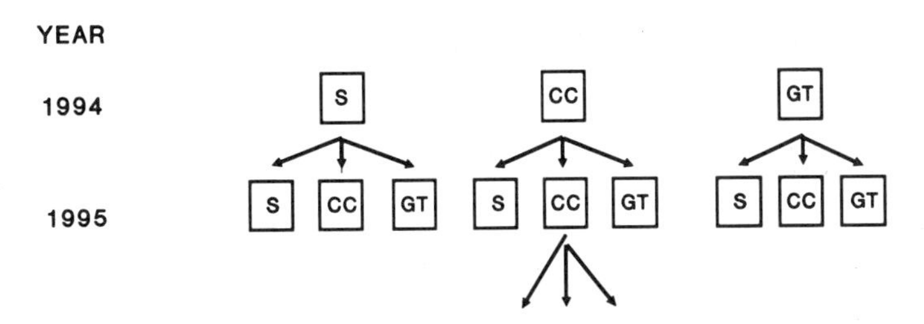

FIGURE 14.20 Generation additions optimization overview.

14.6.1 Dynamic Programming Concept

Before approximation techniques are discussed, it would be helpful to examine dynamic programming, an exact mathematical optimization procedure, and understand how this procedure can be used to solve the generation planning problem. Dynamic programming is an exact optimization technique that can be applied to many types of problems. Examination of a traditional illustrative problem, called the "stagecoach problem," will be helpful in gaining insight to the application of dynamic programming for generation planning.

In the stagecoach problem, a person needs to travel between Washington, D.C., and Los Angeles in the 1850s. The cost of stagecoach tickets from one city to the next is labeled in Figure 14.21. For example, a ticket from Washington, D.C. to St. Louis is $13. The objective is to find the least-cost way to travel between Washington, D.C. and Los Angeles.

The dynamic programming solution technique begins by starting from the destination city of Los Angeles. The objective is to compute the total cost of optimally traveling from any city that connects to Los Angeles. Therefore, it is necessary to find the cost to Los Angeles from Pierre, Denver, or El Paso. The cumulative cost to the destination from these cities is presented in Table 14.12, based on the data from Figure 14.21.

Now take a step (stage) toward the origination city of Washington. Consider the locations of the previous stage, stage 2, where the cities could be Chicago, St. Louis, or Atlanta. From Chicago, one could then proceed to

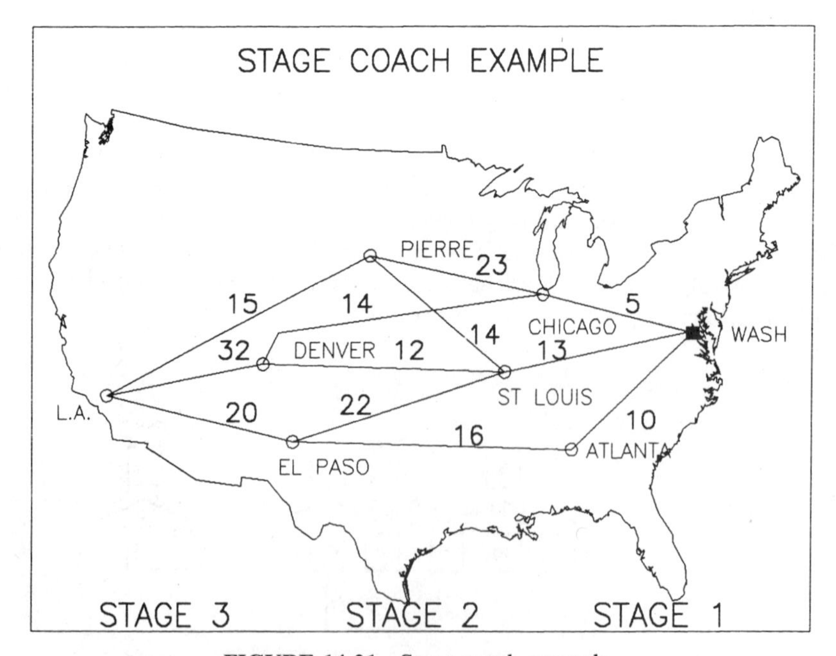

FIGURE 14.21 Stage coach example.

TABLE 14.12 Stagecoach Problem—Stage 3

Current City	Next City	Cumulative Cost to Los Angeles	Optimal Route
Pierre	Los Angeles	15	Los Angeles
Denver	Los Angeles	32	Los Angeles
El Paso	Los Angeles	20	Los Angeles

Pierre or Denver. The cumulative cost to the destination from Chicago through Pierre is $23 plus the cumulative least cost to go from Pierre to Los Angeles, for a total cost of $38. This calculation is illustrated in Table 14.13. Similarly, to go from Chicago through Denver to the destination costs $46. Thus, from Chicago, the least expensive ticket to Los Angeles is $38, and the route is through Pierre. Similar calculations are presented in Table 14.13 for the other cities.

From Washington, the origination city, the ticket cost is computed to Chicago, and added to the cumulative least cost from Chicago to Los Angeles. The ticket from Washington to Chicago costs five dollars, and the least expensive way to travel from Chicago to the destination according to the results in Table 14.11 is $38. Hence, the cumulative is $43. The cost from Washington through St. Louis to Los Angeles is $42, and the cost from Washington through Atlanta to the destination is $46.

On the basis of this information, backtracking will determine the optimal route. Examining the results leads to the conclusion that the least-cost way to travel from Washington to Los Angeles is to travel first from Washington to St. Louis, as shown in Table 14.14, column 5. Table 14.13 indicates that the least expensive route from St. Louis is through Pierre. Once in Pierre, Table 14.12 indicates that the next destination city is Los Angeles. The total cost of this trip is $42, as indicated in Table 14.14, column 4.

Dynamic programming builds a least-cost cumulative table proceeding backward from the destination. At each stage, the least expensive route from that stage to the destination is computed. The process is continued until the origination point is reached. In this example, after the cost from Washington to Los Angeles is computed, the journey can then be recreated by proceeding back through the cumulative least-cost tables to determine the best route through the cities.

The dynamic programming solution technique can be expressed mathematically. Let:

i = the current stage or "state" (city in the prior example)
n = the next stage or "state" (city in the prior example)
C_n = the least cumulative cost for proceeding from state n to the final destination
t_{in} = the route (ticket) cost from stage i to stage n
r_i = optimal route from stage i to the *next* stage

TABLE 14.13 Stagecoach Problem—Stage 2

(1) Current City	(2) Next City	(3) Cumulative Cost to Los Angeles	(4) Cumulative Least Cost to Los Angeles	(5) Optimal Route
Chicago	Pierre	15 + 23 = 38	38	Pierre
	Denver	32 + 14 = 46		
St. Louis	Pierre	15 + 14 = 29	29	Pierre
	Denver	32 + 12 = 44		
	El Paso	20 + 22 = 42		
Atlanta	El Paso	20 + 16 = 36	36	El Paso

The solution technique is a recursive algorithm that proceeds backward from the destination to the source. The dynamic programming algorithm is:

$$C_i = \text{minimum } [C_n + t_{in}] \text{ of all connected routes from } i \text{ to } n \qquad (14.2)$$

The stagecoach example illustrated here is similar to the generation planning problem. In the stagecoach problem, a stage or ticket from one city to the next is analogous to a year in a generation planning problem. Similarly, the stagecoach ticket cost is analogous to the annual production and investment cost of a generation plan. Dynamic programming provides an exact solution to the problem. Since many enumerations of the alternative paths are required, computer execution time is very large. The most time-consuming aspect in generation planning is the computation of the annual production cost for each alternative for each of the paths.

14.6.2 Dynamic Programming Applied to Generation Planning

The first step in applying the dynamic programming method is to define the cost-objective criteria. Usually, a plan is selected that has the least cumulative

TABLE 14.14 Stagecoach Problem—Stage 1

(1) Current City	(2) Next City	(3) Cumulative Cost to Los Angeles	(4) Cumulative Least Cost to Los Angeles	(5) Optimal Route
Washington	Chicago	38 + 5 = 43		St. Louis
	St. Louis	29 + 13 = 42	42	
	Atlanta	36 + 10 = 46		

present-worth cost over the study-period horizon, typically 20–30 years into the future. A study-period horizon that is short leads to a plan that produces short-range results but may position the utility for poorer performance over the midrange and longer term. A very long study-period horizon leads to forsaking of the near term to achieve long-term results. When consideration is given to the uncertainties of economic parameters far out into the future, the selection of a very long study period is further questioned. Consequently, a typical utility study-period horizon is a midrange selection of 15–25 years. This provides a balance of long- and short-range economic goals.

The second step in applying the dynamic programming method is the calculation of the "stage ticket cost." One component is investment charges. In any forecast year (e.g., 1995), the power system may require a capacity addition. The capacity addition could be a coal unit, a combined-cycle unit, a gas-turbine unit, a load-management unit, or a cogeneration unit. Once this capacity addition is committed for service, it becomes a fixed cost. Therefore, the cost of a capacity addition added during the year is equal to the present worth of the fixed charges on that investment from the year the unit is installed until the year of the study-period horizon. Mathematically this can be expressed as:

$$\text{Invest charges}_I = \sum_{y\,=\,\text{installation year}}^{\text{horizon year}} \text{PW}_y \cdot \text{FCR}_y \cdot \text{Plant Cost \$/kW} \cdot \text{kW} \qquad (14.3)$$

where PW_y = present-worth discount factor to the installation year from year y

FCR_y = annual fixed-charge rate in year y

Plant cost $/kW = plant capital cost in $/kW expressed in the installation year $

kW = kW rating of the plant

This can be expressed as:

$$\text{Invest charges} = \text{plant cost \$/kW} \cdot \text{kW} \cdot \overline{\text{FCR}}_N \cdot \text{USF}_N \qquad (14.4)$$

where N = number of years from the installation year to the horizon year

USF_N = uniform series present worth factor for N years

$\overline{\text{FCR}}_N$ = levelized annual fixed-charged rate over N years

The other component of the "stage ticket cost" is the annual production cost of operating the power system during the forecast year. A production simulation that includes capacity additions to the power system is conducted to evaluate the annual fuel and O&M costs.

The total "stage ticket cost" for proceeding on a route with a specific capacity addition is the present worth of the two costs. The costs are typically present-worth discounted to the first year of the study. Thus:

$$\text{Total annual expansion cost in YR} =$$
$$PW_{YR} \cdot [\text{invest charges (capacity addition)} + \text{annual fuel and O\&M cost}]$$

where PW_{YR} = present-worth discount factor from the forecast year to the first year of the study.

The total expansion cost for each alternative (coal unit addition, gas-turbine unit addition, combined-cycle unit addition, load management addition, etc.) is calculated similarly.

For example, in 1995, the power system can proceed via several "routes" to the year 1996 by adding any one of several types of capacity alternatives as illustrated in Figure 14.22.

The process outlined in Figure 14.22 is repeated for each forecast study year through to the horizon year. What results is a large tree, the first two years of which are shown in Figure 14.23.

This "tree" becomes very large very quickly when many capacity addition options are considered. However, several issues add perspective to the solution.

In some cases, only the results for the short-term period (the first 10 years) are of interest but a 20-year study-period horizon is desired. In this case, an extrapolation procedure may be used to approximate the effect of the second 10-year period. The second 10-year period is simulated with zero load growth

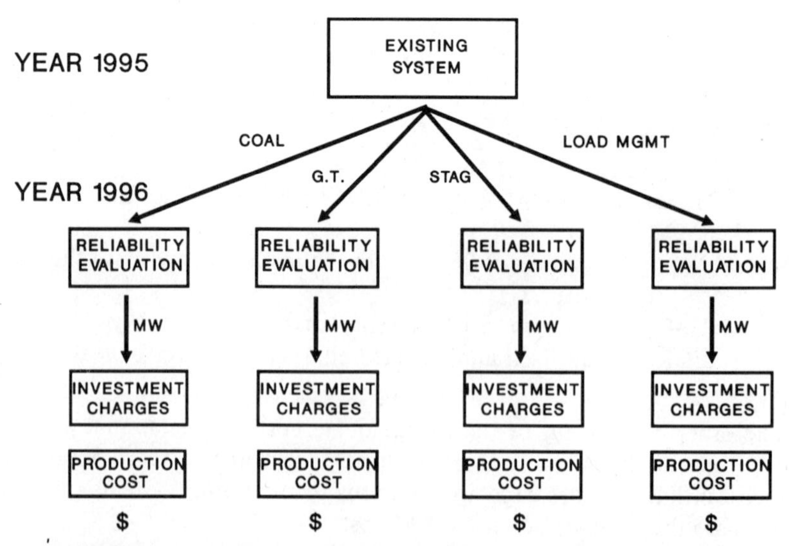

FIGURE 14.22 Capacity addition alternatives in the planning evaluation.

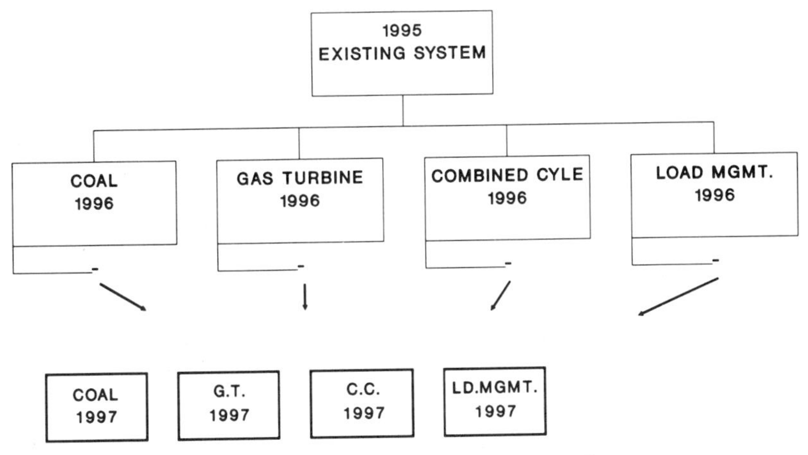

FIGURE 14.23 Forecast-tree example.

and no capacity retirements following the first 10-year period. In this case, no future capacity additions are required during the second 10-year period. Therefore, production cost simulations may be approximated by escalating the annual production costs of the results of the last year of the first 10-year period. In this way, a 10-year study can be performed that closely approximates the first 10-year results of a longer 20-year study.

Another approximation that can be applied is a "search tunnel." When data from similar cases are available, the width of the tree can be pruned so that the optimization is performed along a narrow range of possible "routes" during the study. For example, routes that have more than 35% additions of gas turbines might be pruned from the tree as being *a priori* uneconomical. Establishing a "search tunnel" can significantly decrease the computational time.

Annual production simulations of the power system for each possible alternative route through the study years can be computed and stored in a "library" for inclusion in the dynamic programming solution phase. Creating this library is the most computer time consuming phase. The next phase is to compute the investment charges for the library. The final phase is the dynamic programming phase in which the library is interrogated during the optimal solution process.

The following example illustrates this process.

Example. A power system at the beginning of 1993 has the following generation capacity characteristics:

	Capacity (MW)
Hydroelectric	41
Nuclear	330

	Capacity (MW)
Coal steam	2765
Natural gas steam	569
Gas turbines	171
Pumped storage hydro	268
Power purchase	100
	4244

The utility is considering adding either a coal-fired steam unit and/or natural gas-fired combined-cycle units for future generation capacity. These capacity choices have the following characteristics (1993$):

	Fuel Type	Heat Rate (Btu/kWh)	1993$ O$M Cost Fixed ($/kW/Year)	Variable ($/MWh)	$/kW Plant Capital Cost 1993 $	1985 $
Coal-steam	Coal	9800	24	6	1800	1200
Combined-cycle	Gas	7800	10	3	800	550

Utility load growth is such that *one* 100-MW capacity addition is required *each year* from 1993 through year 2003. Thus, by 1997, five 100-MW unit additions are required.

The fuel cost forecast is shown in Figure 14.24. After having declined during the early 1980s, the fuel cost is projected to increase during the 1990s and beyond. The price of natural gas is forecast to increase at 11%/year, while coal is projected to increase at 6%/year. Plant cost and O&M costs are projected to increase at 5%/year.

The objective is to compute the optimal generation mix of additions for this utility over the 1993–2002 time period by using an extrapolation procedure to give a 20-year study evaluation horizon.

Table 14.15 presents the present-worth production cost library generated by performing production cost simulations of the power system. Parameters for the library are year and number of combined-cycle units in this example. For example, in 1997, the present-worth annual production cost for the power system with three combined-cycle unit additions (and two coal-unit additions, since a total of five units were added over the 5 years) is $223.1011M/year. The production cost table has been extrapolated from year 2003 through 2012. These cost have been present-worth discounted to year 1985.

Table 14.16 presents the investment charge calculation by capacity installation year. For example, the investment charge calculation for a 1997 coal unit addition can be explained as follows. Column 1 presents the escalation index

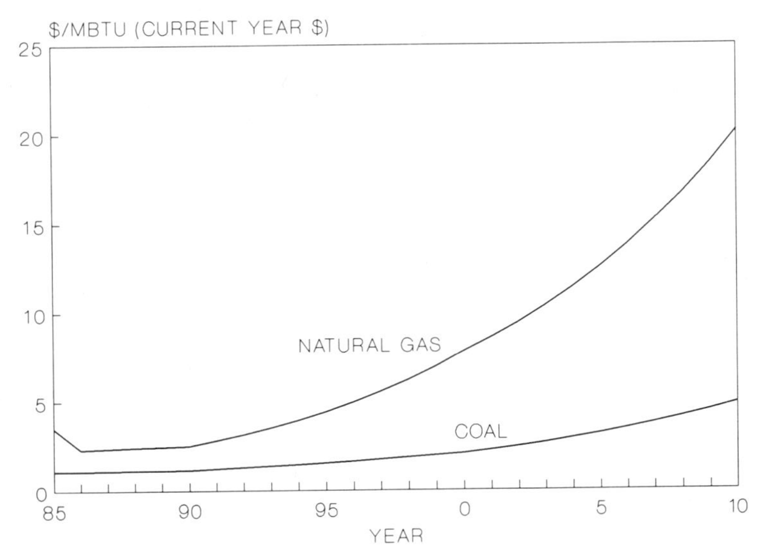

FIGURE 14.24 Fuel cost forecasts.

from year 1985. Column 2 presents the uniform series factor from the installation year through the study horizon year of 2012. Column 3 presents the present-worth factor from the capacity installation year to 1985. Thus, the investment charges of a 1997 100-MW coal-unit addition over the 1997 to 2012 time period and present worth discounted to 1985 is calculated as:

$$1.795 \text{ escalation factor} \cdot 1200 \text{ \$/kW} \cdot (.18 \text{ fixed-charge rate}) \cdot 7.823 \text{ uniform series factor} \cdot (100{,}000 \text{ kW}) \cdot .000001 \cdot .318 \text{ PW factor} = \$96.4\text{M}$$

SOLUTION. The data from Tables 14.15 and 14.16 are used in the dynamic programming solution technique.

The first step in the solution procedure is to begin with the final destination (year 2012) and proceed toward the starting point (year 1993) as in the stagecoach example. Since the final 10 years, from 2003 to 2012, constitute an extrapolation period with no capacity additions, the cumulative costs from 2003 to 2012 can be calculated by accumulating the production cost results of each column from Table 14.15. Table 14.17 presents these results.

The next step is to continue building the decision tree back toward the starting point (1993) using the dynamic programming procedure.

Consider, as an example, the case of a year-2002 coal or combined-cycle decision when three combined-cycle units and six coal units have been added through year 2001. In this case, the 2002 decision is characterized as in Table 14.18. The present-worth cost of each route from 2001 to 2002 is evaluated using the data of Tables 14.17, 14.16, and 14.15. A comparison of two alterna-

TABLE 14.15 Annual Present Worth Power System Production Cost

	Number of Combined-Cycle Units										
Year	10	9	8	7	6	5	4	3	2	1	0
1993										231.9005	231.4447
1994									230.5227	229.6054	229.0432
1995								229.6406	228.2345	227.2153	226.5684
1996							228.2352	226.4136	224.9134	223.8434	223.1258
1997						227.2710	225.0503	223.1011	221.5244	220.4077	219.6233
1998					226.7249	224.1259	221.7894	219.7349	218.0933	216.9303	216.0795
1999				226.6716	223.6589	220.9204	218.4491	216.3100	214.6118	213.4042	212.4947
2000			227.1229	223.6689	220.5415	217.6467	215.0489	212.8365	211.1123	209.8524	208.8790
2001		229.5942	225.8263	222.4371	219.3266	216.4726	213.8927	211.7181	210.0690	208.8370	207.8703
2002	232.3437	228.2225	224.5532	221.2087	218.1131	215.2976	212.7630	210.6495	209.0500	207.8314	206.8694
2003	229.6301	225.0249	220.8995	217.1231	213.6171	210.4102	207.5006	205.0244	203.0688	201.4989	200.1896
2004	227.0613	221.9726	217.3941	213.1907	209.2811	205.6915	202.4174	199.5903	197.2922	195.3862	193.7466
2005	224.6307	219.0591	214.0302	209.4046	205.0983	201.1344	197.5059	194.3396	191.7123	189.4849	187.5312
2006	222.3321	216.2780	210.8014	205.7585	201.0619	196.7321	192.7591	189.2650	186.3213	183.7869	181.5347
2007	220.1596	213.6233	207.7016	202.2460	197.1658	192.4783	188.1704	184.3597	181.1120	178.2843	175.7488
2008	218.1074	211.0893	204.7251	198.8613	193.4039	188.3667	183.7334	179.6167	176.0772	172.9695	170.1653
2009	216.1702	208.6705	201.8662	195.5988	189.7704	184.3914	179.4420	175.0299	171.2102	167.8355	164.7766
2010	214.3429	206.3616	199.1197	192.4531	186.2599	180.5467	175.2902	170.5930	166.5045	162.8754	159.5753
2011	212.6204	204.1577	196.4805	189.4191	182.8671	176.8273	171.2726	166.3002	161.9540	158.0824	154.5542
2012	210.9981	202.0539	193.9438	186.4918	179.5870	173.2280	167.3837	162.1458	157.5528	153.4504	149.7065

TABLE 14.16 Investment Charge Calculation
(USING 18% ANNUAL FIXED-CHARGE RATE)

Capacity Installation Year	(1) Investment Cost Escalation Index from 1985	(2) Uniform Series Factor to the Study-Horizon Year	(3) Present-Worth Factor to 1985	(4) Combined-Cycle Investment Cost Charges for 100-MW Unit (M $)	(5) Coal-Unit Investment Cost Charges for 100-MW Unit (M $)
1993	1.477	8.513	0.466	58.09	126.74
1994	1.551	8.364	0.424	54.48	118.87
1995	1.628	8.201	0.385	50.99	111.25
1996	1.710	8.021	0.350	47.60	103.86
1997	1.795	7.823	0.318	44.32	96.42
1998	1.885	7.606	0.289	41.12	89.73
1999	1.979	7.366	0.263	38.02	82.96
2000	2.078	7.103	0.239	34.99	76.36
2001	2.182	6.813	0.217	32.05	69.91
2002	2.292	6.495	0.197	29.15	63.61

tive routes shows that the coal alternative is less expensive in 2002 *if* the power system had three combined-cycle and six coal units in year 2001.

Consider another example, in the year 2002 when two combined-cycle units and seven coal units had been added through year 2001. Again, the cumulative present-worth cost from year 2003 to the destination year, 2012, is added to the year 2002 present-worth cost (investment plus production) to compute the cumulative present worth cost from 2002 to the destination. These data are presented in Table 14.19. In this case, it is more economical to add a coal unit in the year 2002 if the power system already had added two combined-cycle and seven coal units.

The data for Table 14.17 can be updated with the optimal cost data from the results from Tables 14.18 and 14.19. This augmented table is shown as Table 14.20. The optimal decision table showing which generation type is the most economical is shown as Table 14.21.

Consider one more example, but with decision year 2001. The objective is to calculate the optimal route to the destination (2012) from year 2001, given that two combined-cycle units (and six coal units) were added through the year 2000. The optimal route can be computed by using the cumulative cost to the destination from year 2002 (Table 14.20) and the annual production cost and investment cost of year 2001 (Tables 14.15 and 14.16). Table 14.22 presents the optimization calculation. The least cost addition in 2001 is a combined-cycle-type, given that two combined-cycle units were added through year 2000.

TABLE 14.17 Cumulative Present-Worth Power System Production Cost to the Destination, 2003–2012 (Million $)

	Number of Combined-Cycle Units										
Year	10	9	8	7	6	5	4	3	2	1	0
2003	2196.1	2128.3	2066.9	2010.5	1958.1	1909.8	1865.5	1826.3	1792.8	1763.6	1737.5

TABLE 14.18 2002 Optimization Process If Three Combined-Cycle (CC) and Six Coal Units Were Added Through Year 2001

	Cumulative Present-Worth Cost, 2002–2012	
	Add CC in 2002 (Four CC Total)	Add Coal in 2002 (Three CC Total)
Cumulative		
PW cost, 2003–2012	1865.5	1826.3
PW production cost, 2002	212.7	210.6
PW investment cost, 2002	29.2	63.6
Total cost (2002–2012)	2107.4	2100.5

The table containing the optimal cumulative cost to the destination (Table 14.20) and the table containing the optimal unit type decision (Table 14.21) can be updated as shown in Tables 14.23 and 14.24, respectively.

If this process were continued, the dynamic programming process would lead to an optimal cumulative present-worth cost table as shown in Table 14.25. The least-cost capacity plan has a present-worth cost of $4783.234M.

The optimal decision type corresponding to Table 14.25 is presented in Table 14.26; this optimal plan needs explanation. The optimal unit type to add in 1993 is a combined-cycle unit. Since a combined-cycle unit is added in 1993, the power system enters 1994 with one combined-cycle unit. To determine the optimal decision in 1994, use Table 14.26 entry at (row) year 1994 and (column) one combined-cycle. The table indicates that combined-cycle is again the optimal decision in 1994. Thus, the optimum power system enters 1995 with two combined-cycle units. The table entry in (row) year 1995 and (the column of) two combined-cycle units again indicates combined-cycle is the optimal decision. Similarly for 1996. In 1997, the optimum power system enters with four combined-cycle units. However, the optimal decision in 1997 is the addition of a coal unit. Thereafter, coal units are the optimal addition type.

TABLE 14.19 2002 Optimization Process if Two Combined-Cycle and Seven Coal Units Were Added Through Year 2001

	Cumulative Present-Worth Cost, 2002–2012	
	Add CC in 2002 (Three CC Total)	Add Coal in 2002 (Two CC Total)
Cumulative		
PW cost, 2003–2012	1826.3	1792.8
PW production cost, 2002	210.6	209.1
PW investment cost, 2002	29.2	63.6
Total cost (2002–2012)	2066.1	2065.5

TABLE 14.20 Optimal Cumulative Present-Worth Cost to the Destination (2012)

	Number of Combined-Cycle Units										
Year	10	9	8	7	6	5	4	3	2	1	0
2002								2100.5	2065.5		
2003	2196.1	2128.3	2066.9	2010.5	1958.1	1909.8	1865.5	1826.3	1792.8	1763.6	1737.5

TABLE 14.21 Optimal Unit Type Decision

	Number of Combined-Cycle Units										
Year	10	9	8	7	6	5	4	3	2	1	0
2002								Coal	Coal		

TABLE 14.22 2002 Optimization Process If Two Combined-Cycle and Six Coal Units Were Added Through Year 2001

	Cumulative Present-Worth Cost, 2002–2012	
	Add CC in 2002 (Three CC Total)	Add Coal in 2002 (Two CC Total)
Cumulative		
PW cost, 2002–2012	2100.5	2065.5
PW production cost, 2001	211.7	210.1
PW investment cost, 2001	32.1	69.9
Total cost (2001–2012)	2344.3	2345.5

The optimal generation addition plan adds four combined-cycle units in the early 1990s, and then adds six coal units in the late 1990s and early 2000s. This change in the generation type is driven principally by the differential fuel price escalation between natural gas and coal (reference to Figure 14.24). Combined-cycle units are added, while natural gas fuel has a low price, and the transition is made to coal units as the natural gas price rises in the 1990s.

It is interesting to note the sensitivity of the total power system expansion cost to changes in the percent of combined-cycle units. The optimal addition plan had 40% combined-cycle units. A plan with 50% combined-cycle units (installed in 1993, 1994, 1995, 1996, and 1997) has a higher system expansion cost. Figure 14.25 presents the variation of the expansion cost with the percent

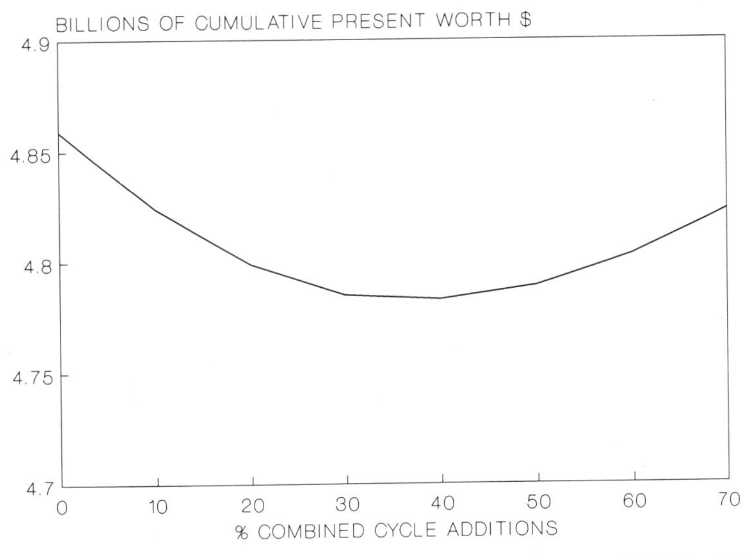

FIGURE 14.25 Sensitivity of expansion costs versus percent of STAG additions.

TABLE 14.23 Optimal Cumulative Present-Worth Cost to the Destination (2012)

	Number of Combined-Cycle Units										
Year	10	9	8	7	6	5	4	3	2	1	0
2001									2344.3		
2002								2100.5	2065.5		
2003	2196.1	2128.3	2066.9	2010.5	1958.1	1909.8	1865.5	1826.3	1792.8	1763.6	1737.3

TABLE 14.24 Optimal Unit Type Decision

	Number of Combined-Cycle Units										
Year	10	9	8	7	6	5	4	3	2	1	0
2001									CC		
2002								Coal	Coal		

TABLE 14.25 Optimal Cumulative Present-Worth Cost (Millions of 1985 $)

	Number of Combined-Cycle Units										
Year	10	9	8	7	6	5	4	3	2	1	0
1993											4783.234
1994										4493.241	4436.717
1995									4208.234	4152.628	4098.254
1996								3927.603	3873.403	3820.048	3768.794
1997							3651.763	3599.383	3547.529	3497.345	3449.649
1998						3386.863	3330.012	3280.107	3231.500	3184.921	3140.698
1999					3132.610	3073.001	3018.486	2970.636	2925.698	2882.638	2841.834
2000				2887.757	2825.989	2769.118	2717.075	2671.364	2630.002	2590.406	2552.438
2001			2650.876	2587.728	2529.087	2475.112	2425.666	2382.167	2344.295	2307.588	2271.896
2002		2420.131	2355.133	2295.374	2239.844	2188.722	2141.856	2100.532	2065.473	2031.014	2000.644
2003	2196.053	2128.291	2066.962	2010.547	1958.113	1909.806	1865.475	1826.264	1792.805	1763.654	1737.529

TABLE 14.26 Optimal Decision Matrix

	Number of CC Units										
Year	9	8	7	6	5	4	3	2	1	0	Year
93										CC	93
94									CC	CC	94
95								CC	CC	CC	95
96							CC	CC	CC	CC	96
97						Coal	CC	CC	CC	CC	97
98					Coal	Coal	Coal	CC	CC	CC	98
99				Coal	Coal	Coal	Coal	CC	CC	CC	99
00			Coal	Coal	Coal	Coal	Coal	CC	CC	CC	00
01		Coal	Coal	Coal	Coal	Coal	Coal	CC	CC	CC	01
02	Coal	Coal	Coal	Coal	Coal	Coal	Coal	Coal	CC	CC	02

Decison Pass
Annual Results Year = 1993 Alternative = Coal

of combined-cycle additions. Note that the minimum has a broad range. Changing the percent combined-cycle by 10% only changes the present-worth cost by approximately 0.1%.

The optimal capacity addition plan is influenced by the study-horizon year. A shorter study-horizon year leads to capacity addition plans that add less capital-intensive and more fuel-cost-intensive plant types (e.g., gas-turbine, combined-cycle plants). Table 14.27 shows the dynamic programming solution to the same example but with a study-horizon period of 2003, 9 years less than the prior example. This optimal capacity additions plans has seven combined-cycle units added from 1993 to 1999, and three coal units thereafter. While this plan is optimal over the 10-year period of 1993 to 2002, it is not at all optimal over the 1993–2012 time period, as shown in Figure 14.25. Thus, the proper selection of the planning horizon year is very important.

This example illustrates the application of dynamic programming to the solution of generation planning problems. It was simplified by having only two capacity addition types of the same megawatt size. Because of this, it was possible to characterize the library of annual production cost values by the number of combined-cycle units. When the unit sizes are dissimilar, the library of cases becomes larger. The dimensions of the library significantly increase with the number of capacity addition type alternatives. For some power system planning applications, it is sometimes useful to construct a library that does not contain all the route alternatives but interpolates within the library to obtain the annual production cost values for all routes.

In summary, the dynamic programming solution technique can be applied to optimal generation addition planning. Although the technique finds an exact solution, it does require a large amount of computational resources.

14.6.3 Approximate Technique: Annual Decision Decomposition

The objective of using approximate techniques in generation planning is to reduce the number of alternative plans that need to be evaluated by using insights and information about the specific nature of the electric-utility generation planning problem. While exact techniques such as dynamic programming provide an exact solution, they consume large amounts of computer execution costs.

The objective of the approximation technique is to develop a plan that is very close to the exact optimal answer but only at a fraction of the computational cost. In practice, many approximation techniques will converge to the exact answer. One very widely used approximation technique is the "annual decision decomposition" method. The objective of this technique is to decompose the generation addition decision into 20 yearly optimizations, rather than one large 20-year coupled-optimization search. By decomposing the problem into 20 yearly optimizations, each optimization can be performed significantly faster compared to one large optimization.

TABLE 14.27 Optimal Capacity Addition Plan: Study Horizon Year 2003
OPTIMAL DECISION MATRIX[a]

	Number of CC Units										
Year	9	8	7	6	5	4	3	2	1	0	Year
93										CC	93
94									CC	CC	94
95								CC	CC	CC	95
96							CC	CC	CC	CC	96
97						CC	CC	CC	CC	CC	97
98					CC	CC	CC	CC	CC	CC	98
99				CC	CC	CC	CC	CC	CC	CC	99
00			Coal	CC	CC	CC	CC	CC	CC	CC	00
01		Coal	Coal	Coal	CC	CC	CC	CC	CC	CC	01
02	Coal	Coal	Coal	Coal	Coal	Coal	CC	CC	CC	CC	02
03											03

[a]CC move to left; coal move down.

The annual decision decomposition technique can be applied in three steps.

Step 1. A simplification is first made. Generation addition decisions are considered as a series of yearly optimization problems. In each yearly problem, load growth is assumed not to increase in subsequent years. This step is illustrated in Figure 14.26. In 1995, the decision to be made is whether to add a steam (S), combined-cycle (CC), or gas-turbine (GT) unit as the alternative. In the decomposition procedure, there is no load growth subsequent to 1995; consequently, no capacity additions are needed subsequent to 1995. Therefore, the power plant capacity factors are constant through the 1995 to 2015 time-period for a 20-year planning study.

Step 2. Step 2 is illustrated in Figure 14.27. For each year, a reliability, production cost, and investment cost analysis is performed for the three decision alternatives: steam, combined-cycle, and gas turbine. The reliability evaluation computes how much capacity of each type (steam, combined-cycle, and gas turbine) is required to satisfy the loss of load probability or reserve margin requirements for the system in this year (i.e., 1995). The power system production cost analysis is performed according to the assumption in step 1, specifically, that there is no load growth subsequent to 1995. Since the capacity factors are constant, only one power system production cost simulation is necessary, because the other 19 annual simulations are identical except for fuel and O&M cost escalation. The 20-year production cost simulation can be calculated using levelized values of fuel and O&M costs. Thus, a significant 20-to-1 computer execution time savings its achieved because only one power system production cost simulation is performed, rather than 20 individual annual production simulations. Finally, an investment cost analysis is performed for those generation addition alternatives in 1995. The levelized annual production costs and the annual levelized investment

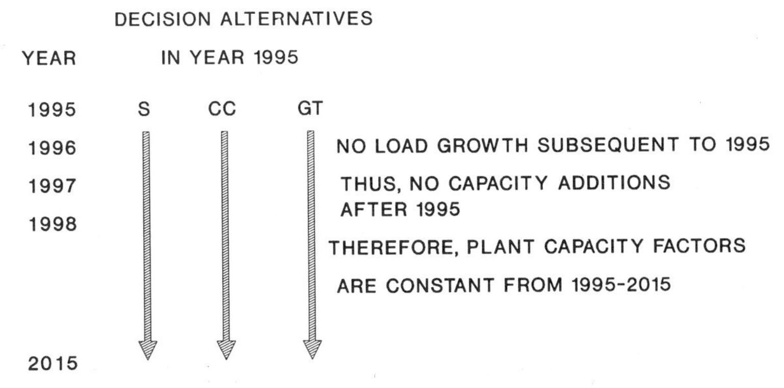

FIGURE 14.26 Decision alternatives.

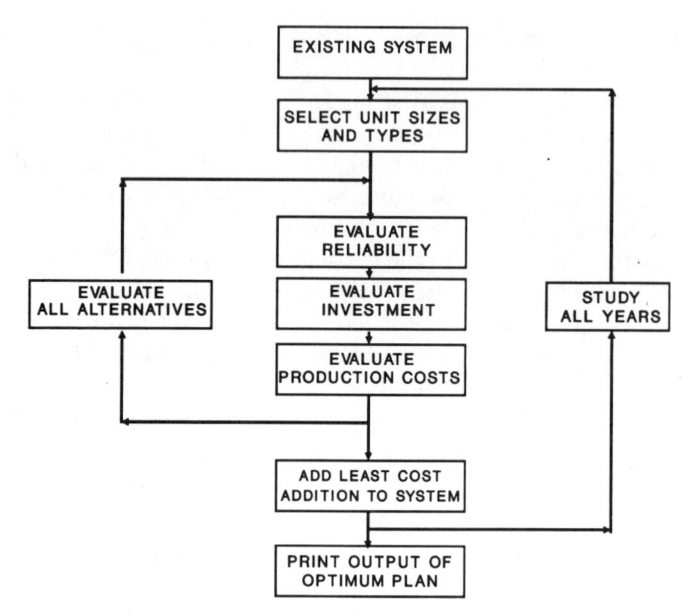

FIGURE 14.27 Generation planning methodology.

costs for each alternative are summed together to obtain a total cost for each alternative (steam plant, combined-cycle plant, and gas-turbine plant addition). The generation type that has the minimum cost is then added to the power system.

Step 3. The procedure then advances to the next year, 1996; using the load conditions of 1996 and the capacity including the optimal capacity addition of 1995, as determined in step 2. This procedure continues through time from 1995 out to 2014, each year adding the optimal type of plant to the power system. Step 2 is repeated 20 times for each study year of a 20-year study.

Example. The same example used in the dynamic programming procedure will illustrate this method.

In 1993, there are two alternatives, combined-cycle (CC) and coal-steam. First the combined-cycle alternative is evaluated on the basis of levelized (20-year, in this example) annual system expansion costs. A system production cost and investment cost analysis is performed. The evaluation summary is illustrated in Table 14.28 by principal plant type.

The fuel and O&M costs used in Table 14.28 are 20-year levelized values. The decision cost is the sum of the levelized annual fuel, O&M, and investment costs. The combined-cycle alternative has an evaluated cost of $821.6M/year.

Table 14.29 presents the results of the coal-unit addition alternative to the power system in 1993. The coal alternative has an evaluated cost of $832.9 M/year. The combined-cycle unit has a lower cost in 1993 and, therefore, is the unit type to add to the power system at that point.

TABLE 14.28 New Combined-Cycle Cost Evaluation in Year 1993 Using Levelized Costs

Type	Installed Capacity (MW)	Annual MWh Generation	Capacity Factor	Fuel Cost ($/MBtu)	System Fuel Cost (M $/Year)	System O&M Cost (M $/Year)	Incremental Plant Charge (M $/Year)	Total (M $/Year)
Hydro	41	323,244	.900	0.00	0	1	0	0.8
Nuclear	330	2,023,560	.700	1.47	31	27	0	57.9
Coal—new	635	4,450,080	.800	2.30	100	59	0	159.0
Coal—steam	2130	13,026,513	.698	2.30	299	181	0	480.5
CC—new	100	227,136	.259	8.31	15	2	15	31.8
CC—old	0	0	.000	8.31	0	0	0	0.0
O&G steam	569	870,982	.175	8.31	76	11	0	86.8
GT—new	0	0	.000	8.31	0	0	0	0.0
GT—old	171	30,671	.020	8.31	3	0	0	3.6
PSH	268	−146,730	.063	0.00	0	1	0	1.2
Emergency tie	0	468	.000	23.73	0	0	0	0.1
Generation	4244	2,605,922			525	282	15	821.6
Purchases	100	569,400			0			0.0
Total	4344	21,375,322			525	282	15	821.6

Decision Pass
Annual Results Year = 1993 Alternative = Combined-Cycle

TABLE 14.29 New Coal Unit Cost Evaluation in Year 1993 Using Levelized Costs

Type	Installed Capacity (MW)	Annual MWh Generation	Capacity Factor	Fuel Cost ($/MBtu)	System Fuel Cost (M $/Year)	System O&M Cost (M $/Year)	Incremental Plant Charge (M $/Year)	Total (M $/Year)
Hydro	41	323,244	.900	0.00	0	1	0	0.8
Nuclear	330	2,023,560	.700	1.47	31	27	0	57.9
Coal—new	735	550,880	.800	2.30	116	68	32	215.9
Coal—steam	2130	12,537,692	.672	2.30	288	177	0	465.1
CC—new	0	0	.000	8.31	0	0	0	0.0
CC—old	0	0	.0	8.31	0	0	0	0.0
O&G steam	569	884,641	.177	8.31	77	11	0	88.0
GT—new	0	0	.000	8.31	0	0	0	0.0
GT—old	171	31,859	.021	8.31	3	0	0	3.7
PSH	268	−146,730	.063	0.00	0	1	0	1.2
Emergency tie	0	778	.000	23.73	0	0	0	0.2
Generation	4244	20,805,922			516	285	32	832.9
Purchases	100	569,400			0			0.0
Total	4311	21,375,322			516	285	32	832.9

Decision Pass
Annual Results Year = 1993 Alternative = Coal

A final production simulation is made of 1993 with the combined-cycle optimal unit added and using actual 1993 (not levelized) values of fuel and O&M costs in documenting the final cost results. This evaluation is presented in Table 14.30.

The procedure is then repeated on an annual basis for each year through the study to year 2004.

The result of the annual decision decomposition method is an optimal or near-optimal expansion. The generation additions that are added to the system are identified in the printout in Table 14.31. In this table generation additions are shown year by year in the appropriate columns. The first column shows the year of the study and the subsequent seven columns present the different types of capacity (hydroelectric, nuclear, coal, combined-cycle, oil and gas steam, gas turbine, and pumped storage hydro). The numerical value presented in a row and column indicates the megawatt capacity of the addition that was optimally added in each year. This is an illustrative way of summarizing the results of the generation expansion plan. The annual decision decomposition method added five combined-cycle units prior to 1998 and added five coal units thereafter.

Table 14.32 presents a summary of the expansion costs by year. The annual costs are expressed in current-year millions of dollars per year. The annual and cumulative present worth costs are also calculated. The cumulative present worth cost is useful for comparison with other generation planning cases and for sensitivity analysis.

The results of the annual decision decomposition method and the dynamic programming method differ in this example. The decomposition method added a combined-cycle unit in 1997, whereas the dynamic programming method added a coal unit. Figure 14.25 illustrates that the expansion cost difference is only .1%. The decomposition method produced a very good near optimal answer. Generally, this method produces the exact result of a dynamic programming method. When it does not, the result is very nearly identical.

It should be noted that the decisions for the first 4 years of the expansion are the same in the two methods. As was indicated previously, a primary purpose of the planning process is to decide on the next couple of units in the context of a long-term system expansion. The process is then repeated each year to adjust for changing power system and economic conditions.

In summary, the annual decision decomposition method is an alternative approach to the dynamic programming solution, offering the opportunity to reduce calculational resources to solve the problem while yielding results that are the optimal or very nearly optimal.

14.7 SUMMARY

This chapter reviewed several widely used methodologies for generation planning. The levelized bus-bar method is the simplest, and often finds application

TABLE 14.30 Cost Calculation for Optimum Combined-Cycle Type Addition in Year 1993 Using Actual 1993 Costs

Type	Installed Capacity (MW)	Annual MWh Generation	Capacity Factor	Fuel Cost ($/MBtu)	System Fuel Cost (M $/Year)	System O&M Cost (M $/Year)	Incremental Plant Charge (M $/Year)	Total (M $/Year)
Hydro	41	323,244	.900	0.00	0	1	0	0.5
Nuclear	330	2,023,560	.700	1.03	22	19	0	40.7
Coal—new	635	4,450,080	.800	1.36	59	41	0	100.8
Coal—steam	2130	13,026,513	.698	1.36	178	127	0	304.9
CC—new	100	227,136	.259	3.51	6	2	15	22.6
CC—old	0	0	.000	3.51	0	0	0	0.0
O&G steam	569	870,982	.175	3.51	32	8	0	39.7
GT—new	0	0	.000	3.51	0	0	0	0.0
GT—old	171	30,671	.020	3.51	1	0	0	1.6
PSH	268	−146,730	.063	0.00	0	1	0	0.9
Emergency tie	0	468	.000	10.04	0	0	0	0.0
Generation	4244	20,805,922			299	198	15	511.7
Purchases	100	569,400			0			0.0
Total	4344	21,375,322			299	198	15	511.7

Decision Pass
Annual Results Year = 1993

TABLE 14.31 Optimal Generation Expansion Plan

	Plant Type						
Year	Hydro	Nuclear	Coal	Combined Cycle	O&G STM	GT	PSH
1993				100			
1994				100			
1995				100			
1996				100			
1997				100			
1998			100				
1999			100				
2000			100				
2001			100				
2002			100				
Total	0	0	500	500	0	0	0

as a discussion vehicle for interpreting the results of more detailed analysis methods. The screening curve and load-duration approaches are methods that can provide additional insight into the key economic and performance factors driving the optimal generation capacity decision. However, the workhorses of the utility industry are the detailed, power system reliability, production simulation, and investment cost methods. Automated generation planning methods are of great benefit in improving human productivity during the development of a comprehensive power system plan.

BIBLIOGRAPHY

Bensky, M. H., H. G. Stoll, R. S. Szczepanski, and R. E. Usher, "The Cost Benefits of Alternative Generation Reserve Levels," *Proceedings of the American Power Conference*, Vol. 40, 1978.

Dees, D. L., G. E. Haringa, H. G. Stoll, and R. S. Szczepanski, "Solving Today's Capital and Fuel Supply problems in the Selection of New Generation," *Proceedings of the American Power Conference*, Vol. 37, 1975.

Garver, L. L., "Factors Affecting the Planning of Interconnected Electric Utility Bulk Power Systems," *Factors Influencing Electric Utility Expansion*, Vol. II, Enver Masud, ed., CONF-770869, U.S. Dept. Energy, Division of Electric Energy Systems, Washington, DC, 20545, August 1977.

Garver, L. L., "The Electric Utilities," *Handbook of Operations Research, Models, and Applications*, Vol. II, Chapter II-6, Joseph J. Moder and Salah E. Elmaghraby, eds., Nostrand Reinhold, New York, 1978.

Garver, L. L., H. G. Stoll, and R. S. Szczepanski, "Impact of Uncertainty on Long Range Generation Planning," *Proceedings of the American Power Conference*, Vol. 38, 1976, pp. 1165–1174.

TABLE 14.32 Power System Expansion Costs

EXPANSION COST SUMMARY (ALL COSTS IN MILLIONS OF DOLLARS)

				Annual				(1985 $) Present Worth	
								Annual	Cumulative
Year	MW Peak Load	MW Installed Capacity	Reserve Margin	Fuel Cost	O&M Cost	Investment Cost	Total Cost	Total Cost	Total Cost
1993	3578	4244	1.214	299	198	15	512	217	217
1994	3674	4344	1.210	332	212	30	574	221	438
1995	3770	4444	1.205	369	226	46	642	225	663
1996	3855	4544	1.205	410	241	63	714	228	891
1997	3940	4644	1.204	456	257	81	794	230	1121
1998	4025	4744	1.203	497	277	122	895	236	1357
1999	4110	4844	1.203	540	299	164	1003	240	1597
2000	4195	4944	1.202	588	321	209	1118	243	1840
2001	4280	5044	1.202	649	346	256	1251	248	2088
2002	4365	5144	1.201	717	371	306	1394	251	2339

Hillier, F. S., and G. J. Lieberman, *Operations Research*, 2nd ed., Holden Day, San Francisco, 1974.

REFERENCE

Tag™-Technical Assessment Guide, Vol. 1: *Electricity Supply-1986*, Electric Power Research Institute, Palo Alto, Ca, 1986.

PROBLEMS

1 In 1990, the fuel costs and future escalation of three fuels are projected to be:

	Coal	Natural Gas	Distillate Oil
1990 fuel cost ($/MBtu)	2.0	4.50	5.50
Escalation (1991–2010)	6%/year	9%/year	8%/year
Escalation (2011–2020)	6%/year	5%/year	5%/year

The utility has a 10% present-worth discount rate.

a Compute the 20-year levelized fuel cost (1991–2010) for these three fuels.

b Compute the 30-year levelized fuel cost (1991–2020) for these three fuels.

2 A utility is considering the addition of coal steam, combined-cycle, and gas turbines. The characteristics of these are (all costs in 1990 $):

	Average Heat Rate (Btu/kWh)	1990 Capital Cost ($/kWh)	O&M Fuel Cost ($/MBtu)	Fixed O&M Cost ($/kW/Year)	Variable O&M Cost ($/MWh)
Coal steam	9800	1400	2.0	15	5
Combined-cycle (gas)	8200	650	3.5	9	3
Gas turbine (gas)	11800	350	3.5	1	5

The escalation of coal is projected to be 5%/year. Natural gas is projected to escalate at 8%/year. The utility has a 10%/year discount rate and an 18%/year fixed charge rate. Compute the 20-year levelized owning costs for these alternatives as a function of the unit capacity factor (a screening curve).

3 An existing power system has the following aggregate capacity at the beginning of year 1995:

Generating Unit Type	Rated Capacity (MW)	Average Heat Rate (MBtu/MWh)	Average Availability (%)	Fuel Cost in 1995 ($/MBtu)
Coal steam	2500	10.000	80%	3.0
Oil steam	2500	10.000	80%	7.0
Gas turbine (oil)	2222	12.000	90%	7.0
Combined-cycle (oil)	1176	8.000	85%	7.0
Nuclear steam	1250	10.000	80%	2.0
	9648			

Reliability studies indicate that the utility requires an additional 1000 MW of effective capacity in the year 1995. The utility is considering two options: (a) adding new combined-cycle units or (b) converting 2500 MW of oil-steam to 2500 MW of coal-fired generation. The utility would also add either combined-cycle or gas-turbine additions in conjunction with this alternative. The characteristics of these alternatives are:

Alternative	MW Size	Percent Availability	MW Effective Capacity	Capital Cost ($/kW)	Heat Rate (MBtu/MWh)
Oil-to-coal conversion of steam units	2500	80%	—	750	10.0
New combined-cycle unit	392	85%	350	800	8.0
New gas turbines	50	90%	45	400	12.0

The fuel price is forecast to escalate over the next 20 years as:

Fuel Type	Fuel Price Escalation Rate (% Year) 1996–2015
Coal	6
Oil	11
Nuclear	6

The utility uses a 10%/year present-worth discount rate and a 20% levelized annual fixed-charge rate. On the basis of a simplified 20-year levelized annual analysis of how the generation system would operate in 1995, determine which is the best alternative:

1. Combined-cycle additions.
2. Oil-to-coal conversion of oil-steam units:
   a. With combined-cycle units
   b. With gas-turbine units.

The annual load duration curve is:

Hours per Year	Load (MW)
1000	8000
4000	6000
3760	4000

a Compute the 20-year levelized operating costs ($/MWh) for each generating unit type.

b How many combined-cycle units and gas turbines are required to satisfy the 1000-MW effective capacity of new additions?

c Which of the options is most economic?

15

CAPACITY RESOURCE PLANNING

Chapter 14 presented the foundations of optimal generation planning. This chapter expands on these concepts by describing key driving factors that influence the optimal capacity plan.

Capacity resource planning involves integrated ''supply-side'' planning and ''demand-side'' planning. ''Supply-side'' planning involves the determination of least-cost generation, transmission, and distribution equipment to serve the customer load requirements. ''Demand-side'' planning involves the determination of programs to manage the customer load demands and achieve least-cost system operation. Chapter 14 discussed generation ''supply-side'' planning, and Chapters 7 and 8 discussed demand forecasting. This chapter discusses a methodology for integrated supply-side and demand-side resource planning.

The subjects of indifference value and marginal costing are also discussed along with the impacts of small (or incremental) improvements to existing equipment.

Generation capacity projects typically have high capital investments and long economic lives and are operated in an economic and business environment that may also be subject to uncertainties. Consequently, capacity projects have considerable business risk inherent in providing the forecasted benefits. Business risk considerations are presented in this chapter to illustrate optimal planning under uncertainty and risk.

15.1 SENSITIVITY OF THE OPTIMAL MIX OF ADDITIONS

The optimal mix of capacity additions is dependent on many factors. Among them are existing generation capacity composition, fuel costs, O&M costs, plant costs, target generation reserve level, load-demand profile, load management options, and economic parameters, including fixed-charge rate and present-worth rate.

The generation system example of Section 14.6 can be used to illustrate this sensitivity. At the beginning of 1993, the power system has the following capacity:

	MW
Hydroelectric	41
Nuclear	330
Coal steam	2765
Combined-cycle	0
Oil steam	569
Gas turbine	171
Pumped storage hydro	268
	4144

The annual load factor is 68%, the minimum reserve margin target is 20%, and the utility is investor-owned. Business parameters include a 10% present-worth rate, an 18% levelized fixed-charge rate, and a 20-year economic evaluation period.

Figure 15.1 presents the optimal system addition composition for several cases when coal steam, combined-cycle, and gas turbines are the optimization candidates. The base case has 600 MW of coal additions, 300 MW of combined-cycle additions, and 200 MW of gas-turbine additions.

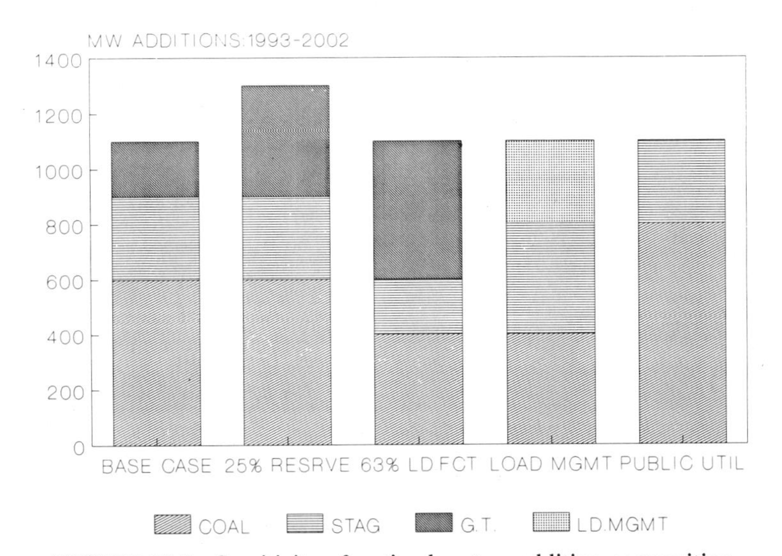

FIGURE 15.1 Sensitivity of optimal system addition composition.

If the reserve margin is increased from 20 to 25%, more capacity is required. Gas turbines are the most economical type of additional capacity to meet the peak capacity need of a higher reserve margin.

If the load factor (average-load demand/peak-load demand) is reduced from 68% to 63% while maintaining the same peak load, then the annual load *energy* is reduced. This implies that the system generation capacity factor is decreased. This leads to lower base-load (coal-steam) capacity additions and more gas-turbine additions.

A sensitivity case is performed in which 300 MW of load management is installed during the 1993–2002 time period. It is assumed that a *hypothetical* "load management" device operates for the three summer and three winter months by displacing the 6-hour period of peak load to the off-peak load periods for each day and is independent of the daily temperatures. It is also assumed that the load management devices reduce the peak load while maintaining total system energy (refer to Chapter 8 for more realistic load management device data). Therefore, it reduces the need for gas turbines. Because the load management additions were assumed to occur during the 1993–2002 time period, when only coal-unit additions were added in this example, load management displaces some of the coal-unit additions as well.

A sensitivity case was run in which the present-worth rate and fixed-charge rate were reduced to those typical of a publicly owned municipal or state-owned utility. A publicly owned utility has lower money costs and can have an 8% present-worth rate and a 13% fixed-charge rate. Lower money costs make capital-intensive plant types more economically attractive. Figure 15.1 illustrates that, in this case, gas-turbine additions are reduced to zero with a corresponding increase in capital-intensive coal plants.

15.2 INTEGRATED DEMAND–SUPPLY PLANNING

Automated generation planning simulation programs determine the optimal schedule of generation capacity additions. Input data describing conventional generating unit candidates can be characterized in terms of the generating unit size, plant cost, fuel cost, plant efficiency, plant reliability, O&M costs, and operation characteristics.

In some cases, a more "complex" generating unit candidate (such as a phased construction of a combined-cycle generating unit) cannot be characterized conveniently in terms of those parameters. In this case, a hybrid of the manual generation planning methodology and the automated generation planning methodology may be used. First, several alternative addition scenarios of the "complex" generation candidate are developed. Each scenario of the "complex" candidate addition schedule is then input as a fixed pattern of additions into the automated generation planning program. The other conventional candidate generation alternatives are then optimized based on this fixed pattern, using an automated generation planning program. The generation expansion cost of the power system is computed for each scenario. Finally, the optimal plan is determined as the scenario with the least cost.

In many cases, demand-side alternatives need evaluation as well. Demand-side alternatives may require special details that are not readily included in the planning program. Conservation programs and load management are such alternatives. A hybrid of manual and automated planning also may be used in these cases.

Studies involving demand-side alternatives require more generalized cost-objective criteria. A plan with the least total present-worth costs is *not* necessarily the best plan. An example illustrates this. Table 15.1 presents two conceptual alternative results. Both plans have the same starting-point costs in the year 1999. However, plan A is based primarily on generation additions, whereas plan B is based on significant demand conservation. The incremental megawatt-hour electricity demand growth as well as the incremental cost in year 2000 for plan B are both less. Although one might be tempted to conclude that plan B is better, the ratio of the incremental cost divided by the incremental megawatt-hour demand is higher for plan B. Similarly, the total cost per megawatt-hour is higher for plan B. Since plan A provides lower electricity costs to the consumer (7.0 $/MWh compared to 7.333 $/MWh, it is concluded that plan A is better.

When alternatives are compared that have different energy, the cost criteria must be the incremental cumulative present worth *$/MWh*. Only in the special case of equal MWh may the criteria be the incremental cumulative present worth *$ cost*.

TABLE 15.1 Demand-Side Alternatives—Example

	Plan A (Generation Additions)	Plan B (Demand Conservation)
	Starting Point	
Embedded MWH	1000	1000
Costs	5000	5000
Cost/MWh	5.0	5.0
	Incremental	
Incremental MWh growth	1000	500
Incremental cost	9000	6000
Incremental cost/MWh	9.0	12.0
	Total	
Total MWh	2000	1500
Total cost	14000	11000
Cost/MWh	7.0	7.333

15.2.1 Integrated Demand–Supply Planning Methodology

Integrated demand and supply planning is an essential utility planning process (Hirst, 1988). Demand-side alternatives influence the supply-side and the power system influences the economic merits of demand-side alternatives. In addition to utility economic issues in supply-side planning, there are customer economic issues in demand-side planning. A successful demand-side plan must meet the customer's economic criteria as well as the utility's. Figure 15.2 presents an integrated demand and supply planning methodology framework.

It is sometimes useful for preliminary screening analysis visualization to define the "cost" of an end-use alternative on a ¢/kWh basis. The procedure is to divide the annual investment charges by the annual kWh electricity saving. Candidate programs can then be sorted and arranged in a priority list beginning with the least ¢/kWh. The cumulative potential megawatt peak load reduction is plotted and contrasted with the ¢/kWh cost of supply alternatives as shown in Figure 15.3. This figure illustrates 165 MW of end-use programs are potentially economic.

The screening curve of Figure 15.3 is useful for visualization of priorities. However, it does not account for how the end-use program integrates with the utility load demand and supply. For example, an end-use program may change the load demand for only a few hours per year during the summer peak-load periods. The "cost" per kilowatt-hour may be high. However, the utility supply alternative for these peak hours has a higher cost. Thus, this end-use program is warranted, even though it would not appear beneficial based on a screening curve such as that shown in Figure 15.3. Similarly, some end-use programs with moderate "cost" per kilowatt-hour may influence the utility

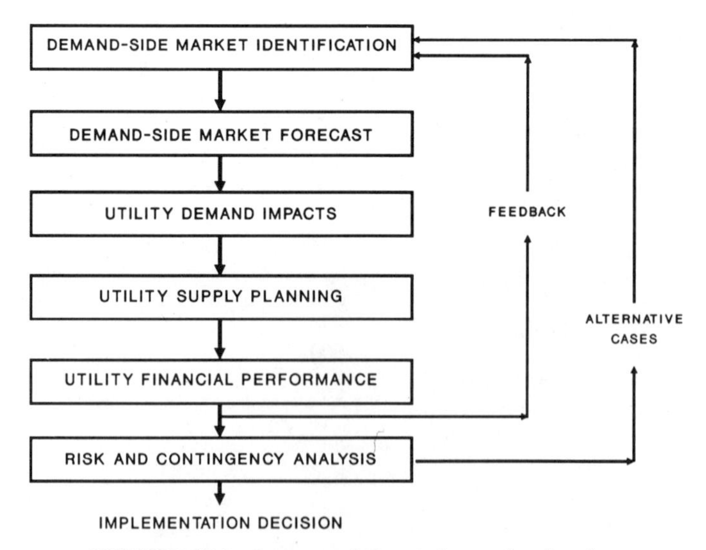

FIGURE 15.2 Integrated demand–supply planning.

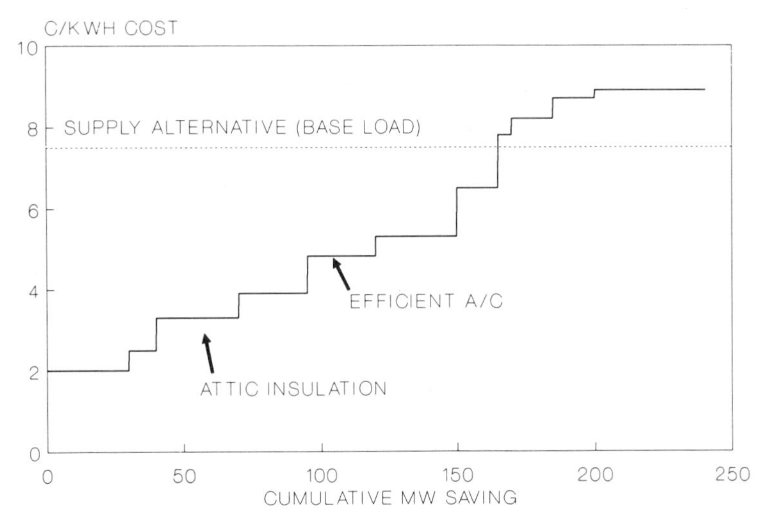

FIGURE 15.3 A demand-side screening curve useful for visualizing economic penetration.

load during periods of low utility replacement cost. While these programs appear beneficial based on a screening curve (Figure 15.3), they may be noneconomic when viewed from an integrated demand–supply analysis. Thus, an integrated demand–supply analysis is required.

The integrated demand–supply planning procedure of Figure 15.2 begins with a demand-side market identification of potential customers with specific demand-side programs. First, a candidate list of demand-side alternatives is prepared. This list might contain cycling of air conditioners, water heater control, home insulation incentive program, off-peak price discount, industrial cogeneration, or other options. Data is obtained from load research studies on the impact of each demand-side candidate on the utility (time-of-day demand, servicing requirements) as well as on the customer ("lifestyle" changes). A demand-side market forecast is then performed for each candidate on the basis of the key customer driving factors, including investment requirements, customer economic benefits, availability of financing, and lifestyle intrusion.

The utility impacts are evaluated in terms of overall load demand, supply impacts, and corporate financial performance. Because the electric utility influences the end-use consumer's decision (through electricity rates, incentive payments, and service reliability) and the end-use consumer directly influences the electric utility (by participating or not participating), a feedback interaction exists.

The final segment of this procedural structure involves implementation decisions based on corporate objectives, uncertainty, and confinement of business risks.

It is important to note that the mechanism for end-use penetration is different from that of capacity supply. In supply planning, the decision being addressed is when, how much and what type of capacity to add. In this case, the utility has complete control over the specification of the supply decision. In end-use planning, the utility must coordinate the decisions with a second party, the consumer. In end-use planning, a utility may find significant consumer inertia and, therefore, an evolutionary market penetration may result. The next several subsections amplify on the framework of Figure 15.2.

15.2.1.1 Demand-Side Market Identification. The application of end-use management involves a business relationship between the utility and the end-user (consumer). Both the utility and the end-user are impacted by this business relationship. The utility is impacted by several factors, including the manner in which the end-user consumes electricity (time-of-day; month; coincidence with generation, transmission, and distribution system load demands), servicing requirements, reliability of the end-use management device, and the dependability of the operation of the end-use management device.

This step identifies the key customer classes and their end-use energy consumption. Opportunities exist for conservation programs and electricity promotion programs. Conservation programs include efficient building insulation, lighting, space conditioning equipment, and hot water (Bulla, 1982).

Promotional programs can encourage consumers to switch from their current energy sources to electricity. The objective is to identify opportunities to increase societal economic efficiency through increased electricity usage. Promotional programs can also be aimed at increasing off-peak electricity consumption.

Customer segmentation is an important consideration in demand-side planning. It is needed because customer benefits of end-use management are not the same for all customers. Customers with electric heating may be more amenable to time-of-day (TOD) metering than nonelectric heating customers. Customers with lower family income may be more amenable to end-use management than higher-income families. Industries with a high electricity expense per unit of added value may have a greater incentive for end-use management than do industries with low electricity expense per unit of added value.

15.2.1.2 Demand-Side Market Forecast. Just as supply planning involves determining the optimal mix of alternative capacity additions (i.e., coal-steam, gas turbine, combined cycle, pumped-storage hydro, etc.), end-use planning involves determining the optimal choice of alternative end-use management candidates (controlling air conditioning and/or water heaters, TOD metering, conservation promotions, cogeneration participations, etc.). In supply planning, the key factors that lead to a mix of generation types are the significant seasonal and daily variations of load demands which, in turn, lead to initial cost versus operating cost trade-offs between alternative supply types. In end-use planning, the key factors that lead to a mix of end-use types are the significant seasonal and daily variations of load demands and the mix of customer classes on the system. Each end-use management candidate also has different utility impacts (i.e., air conditioning control impacts summer loads, while water heater control impacts loads for the entire year) along with different initial cost and operating cost characteristics.

The demand side market forecast segment of Figure 15.2 addresses issues of:

- The overall economic benefits of the end-use candidate to the electricity consumer, such as investment payback period. While economics may not be the sole driving force of end-use penetration, it almost always is a key determinant.
- The potential end-use penetration of each candidate. This is derived from many factors, including consumer economic benefits, promotional incentives, consumer life style, and advertising and marketing thrusts. Consumer surveys are one way to derive the potential penetration.
- Computation of the total hourly load changes as a result of end-use management strategy. This is performed by multiplying the number of end-use devices by the hourly kilowatt change per device, recognizing diversity.

15.2.1.2.1 Demand-Side Forecast Example. The following example illustrates a market forecast. The economic analysis has been simplified to illustrate the concepts.

Example. A summer peaking utility is considering instituting a residential efficient central air conditioner–heat-pump program. Consumers are expected to purchase 25,000 replacement and 20,000 new construction units per year, most with seasonal performance factors of (SPF) 7.5 for cooling and 2.4 for heating. The utility will offer a rebate incentive if the residential home owner installs an efficient unit with 9.0 SPF for cooling and 2.8 SPF for heating.

An efficient unit will save 1400 kWh/year of energy and 0.50-kW peak load during the summer. The more efficient unit costs the home owner an additional $450. Prior surveys have shown the home owner's purchase probability is proportional to the perceived investment payback.

Investment Payback (Years)	Purchase Probability (%)
1	95
2	90
3	50
4	20
5	5

The utility residential electric rate is 7¢/kWh, the marginal energy cost is 5¢/kWh; and the marginal generation, transmission, and distribution plant replacement capacity cost is $90/kW/year.

Calculate the expected megawatt load reduction per year if the utility provided a \$150 rebate incentive to those purchasing the efficient unit.

SOLUTION. The customer receives a \$98/year electricity savings (1400 kWh • \$0.07/kWh).

If the utility provided no rebate, the customer investment payback (ignoring discount factors) is \$450/\$98/year or 4.6 years. This has a purchase probability of 9%.

If the utility provided a \$150/kW rebate incentive, the customer investment payback is \$300/\$98/year, or 3.1 years. This has a purchase probability of 45%. Thus, the expected increase in the sale of efficient units is

$$(.47 - .09) \cdot (25{,}000 + 20{,}000 \text{ units/year}) = 17{,}100 \text{ units/year}$$

The expected megawatt peak load savings per year is 17,000 units/year • 0.5-kW units, or 8.55 MW.

The "cost" to the utility for this peak load "saving" may appear to be \$300/kW (\$150 rebate divided by a 0.5-kW benefit), thereby appearing very attractive relative to other resource alternatives. However, all the costs have not been included. From a utility cost perspective, the following calculations indicate the utility investment pay-back is 8.8 years:

Energy savings (1400 kWh • \$0.5/kWh)	\$ 70
Capacity savings (0.5 kW • \$90/kW/year)	\$ 45
Lost energy sales (1400 kW • \$0.07/kWh)	\$−98
Net savings	\$ 17
Investment	\$150
Pay-back years (\$150/\$17/year)	8.8 years

If the utility finances the investment and has a 15% annual fixed-charge rate for this investment, the annual investment charges are \$22.5/year, and the utility losses \$5.5/year per air conditioning (A/C) unit.

If a utility loses money serving one segment of customers, utility earnings decline. Conceptually, the overall utility rates will be readjusted upward at the next regulartory rate proceeding so that the utility achieves its regulated return on rate base. The net effect is that the one segment of customers become subsidized by the higher rate of the other utility customers.

In summary, if the utility were not to provide a rebate program, then only a very few high-efficiency units would be sold in the marketplace. On the other hand, if the utility provided a large rebate, many units would be sold but the utility would lose money.

What is lacking in both analysis alternatives is the overall assessment of whether the high-efficiency program is good or not good. The issue is whether overall society achieves increased economic efficiency through the program. The utility rebate and the utility revenue sales are only transfer payments from one sector to another and do not influence the overall societal economic benefits. Thus, the first step is to calculate the net societal economic benefits (see Chapter 5):

Incremental energy savings (1400 kWh • $0.05/kWh)	$ 70
Capacity savings (.5 kW • $90.kW/year)	$ 45
Net savings	$115
Investment	$450
Pay-back years	3.9 years

Thus, overall society would benefit with a 3.9-year investment pay-back.

Suppose that the utility rebate is adjusted so that the consumer receives a 3.9-year investment payback. Repeating the customer calculations, a utility rebate of $67 provides a consumer pay-back of 3.9 years. The $67 rebate also provides the utility a pay-back of 3.9 years (and a net saving of $7/year). This pay-back leads to a 23% purchase probability with 10,350 units purchased and a 5.2-MW peak-load reduction.

Suppose that the utility rebate were varied as shown in Figure 15.4. The utility net benefit is the product of the utility net benefit per A/C unit times the number of units purchased. If the rebate is small, then the consumer investment payback years are large and the purchase probability is small. However, the utility net benefit per A/C unit is large. As the utility rebate is increased, the purchase probability increases but the utility net benefit per A/C unit decreases. Thus, there is a rebate that yields maximum utility net benefits. In this example, the value is approximately $60. Thus, the utility has an economic incentive in this example to promote a rebate as part of the A/C program.

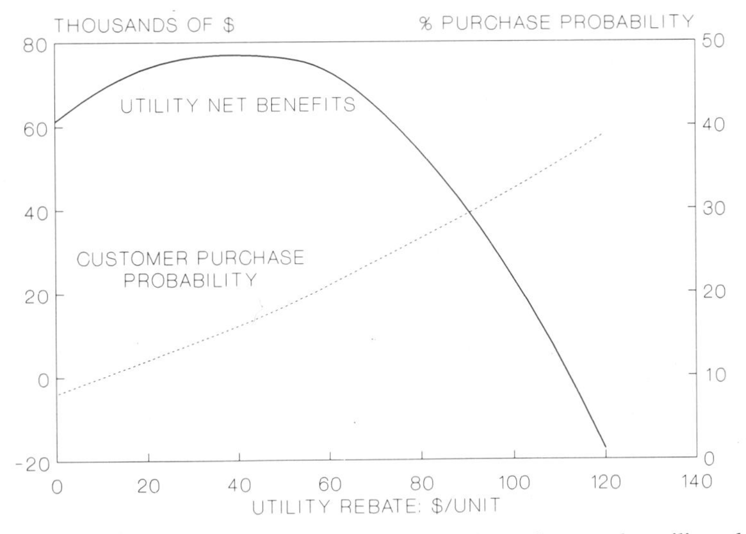

FIGURE 15.4 Utility economic net benefits are dependent on the utility rebate.

A consumer could argue the utility rebate should be $113. At this value the utility net benefit is zero. In this case all the utility costs are recovered.

From a societal perspective, each A/C unit purchased produces a net societal benefit. Thus, if more units are purchased, a greater total societal benefit is achieved. This would argue for a $113 utility rebate.

Thus the actual utility rebate determination should include all the elements of the issue and provide overall societal benefits, consumer benefits, and utility benefits. In order for the natural economic marketplace forces to work, each stakeholder must reap net benefits.

In this example, the marginal cost is greater than the average cost. Suppose that the marginal cost is less, with a utility marginal energy cost of 4¢/kWh and marginal capacity cost of $50/kW/year. The benefits to the three stakeholders with no rebate is:

	Consumer	Utility	Society
Energy savings	98	56	56
Capacity savings		25	25
Lost energy sales		−98	
Net Savings	98	−17	81
Investment	$450	0	450
Pay-back years	4.6 years		5.6 years
Investment charges	67.5		67.5
Net benefit	30.5	−17	13.5

In this case the overall society has a long pay-back, the utility loses money, and the consumer has a long pay-back. The overall societal benefit is positive, suggesting that the program leads to increased economic efficiency for overall society. However, the utility loses money and has no funds on which to offer a rebate other than to cross-subsidize the rebate from other customer's funds. But the consumer is receiving a net benefit greater than the society's net benefit, and thus no rebate is warranted.

In this case the utility might elect not to promote conservation and let the marketplace economics act undisturbed by a rebate; then the consumer purchase probability would be 9%.

Finally, suppose that the utility marginal energy cost is 3.03¢/kWh and the marginal plant replacement capacity cost is $50/kW/year. In this case, the overall societal benefits are:

Incremental energy savings (1400 kWh • $0.0303/kWh)	$42.5
Capacity savings (.5 kW • $50/kW/yr)	$25
Net savings	$67.5
Investment	$450
Pay-back years	6.7 years
Annual investment charges ($450 • .15 fixed-charge rate)	$67.5
Net societal benefit	$ 0

The net societal benefit is zero, based on a 15%/year carrying cost of the $450 additional investment. This corresponds to a 6.7 year investment pay-back. Thus, it would be concluded in this case that the efficient A/C program does not lead to overall increased societal economic efficiency. However, residential consumers would still view it as economic because their decisions are based on average residential rates, not marginal costs.

This example has utilized a simplified economic analysis procedure to illustrate the market forecasting concepts. The analysis assumed values for the power system replacement energy costs. However, these values will change depending on the penetration of end-use and supply-side programs. In addition, this analysis only examined one point in time. Power system replacement costs can markedly change through time, thus leading to changing demand-side economics. A detailed multiple-year power system simulation evaluation procedure using present worth and financial simulation analysis is necessary.

15.2.1.3 Utility Demand Impacts. The utility demand impacts segment of Figure 15.2 has three computation steps:

- Adjust load forecast by demand side management programs.
- Develop load model representations for the operations simulation model requirements (generation level, transmission level, and distribution level).
- Provide the corporate financial model with necessary load data for rate analysis and customer class segmentations.

15.2.1.4 Utility Supply Planning. For a specified end-use strategy, utility supply-side impacts may be determined by using conventional supply-side planning methods.

The four functional areas of utility supply impacts are generation, transmission, distribution, and others. The "others" function includes offices, buildings, and second businesses (such as a natural gas business). The second business may be influenced by electricity end-use alternatives. For example, the gas business of a combined electric-gas utility would be impacted by promoting electric heat pumps to displace natural gas and oil space heating.

Generation planning involves three functions: reliability, production simulation, and investment costing. In reliability evaluations it is important to account for the potential seasonal contribution of end-use management. For example, if an A/C control program results in 100 MW of load relief during the three summer months, the equivalent capacity credit (the amount of capacity that can be deferred, canceled, or sold) based on system reliability considerations can be approximately 100 MW for a strongly summer peaking utility, or it can be approximately 25 MW for a utility with equal summer and winter peaks. However, this assumes that end-use management is 100% reliable. Since some end-use management candidates are dependent on proper functioning of electronic and mechanical equipment, an appropriate equivalent forced outage rate should be used to evaluate reliability contributions of the end-use candidate.

Recent history and future projections of electric utility capital expenditures for plant and equipment typically have generation constituting 65% of the total, transmission 10%, and distribution 25%. Although average transmission contributions may be small, transmission costs from canceling or deferring a power plant due to end-use management penetration can contribute 5–20% to the overall benefits evaluation. While transmission planning typically involves time-consuming power flow, stability, and short-circuit analysis under normal and contingency cases, less detailed methods that correlate transmission expenditures to base-load generation capacity expenditures are reasonable alternatives for scoping analysis.

Distribution expenditures for long-range capital expenditure projections are typically computed by using correlative models. These models may be based on functions of residential and commercial load growth, housing and commercial completions, expected loading characteristics, and average lot sizes.

Demand-side management impacts on the distribution system can be important. One factor contributing to these impacts from end-use management is due to the residential (or distribution) peak that typically occurs 2–6 hours after the generation system peak. Some load management devices that reduce load during the generation system peak have an energy-pay-back period of several hours after the peak. Thus, the end-use device during the energy-pay-back period may add to the distribution peak.

15.2.1.5 Utility Financial Performance. Along with addressing traditional concerns such as providing reliable low-cost electricity, utilities must also provide service that is responsive to the financial well-being of the utility. To help meet this goal, demand-side management has several beneficial financial attributes. It can be economically attractive and not capital-intensive, and it may save capital by obviating the need for long lead-time generation and transmission capacity.

A corporate financial analysis can evaluate the overall cost of electricity, $/kWh, for each demand–supply scenario. Many corporate models have provisions for performing a cost-of-service analysis where the total costs of utility operations are allocated to the customer classes (and subclasses). The cost of service calculation can be based on marginal costs or on embedded cost concepts. After the costs of service are allocated, average customer rates are computed. This cost of service analysis forms the basis for evaluating detailed rate structure impacts (including energy charge, demand charge, and customer service charge).

The financial analysis of the integrated demand–supply plan can evaluate the utility benefits of many alternatives. These include incentive payments to consumers for end-use conservation, power purchase from industrial cogenerators, or promotional campaigns to encourage off-peak electricity sales.

In the overall framework structure presented in Figure 15.2, the integrated demand/supply planning procedure is decoupled into an optimal end-use planning procedure followed by an optimal supply planning procedure. The end-use planning procedure began with assumptions of electricity rates, incentive payments, and load management control policy. The feedback linkage provides the framework for adjusting the end-use assumptions to iteratively approach an ''optimum'' demand- and supply-side plan. Typically one or two iterations are required.

15.2.1.6 Risk and Contingency Analysis. Electric utility supply decisions are traditionally based on minimizing electric rates while maintaining adequate service reliability and financial performance. Other considerations entering into the decision process include short-term versus long-term cost minimization; technological risk; local economic considerations; and financial, ecological, and regulatory issues.

In the previous subtasks, the ''optimal'' end-use/supply plan was based on minimizing electric rates consistent with adequate service reliability and acceptable consumer lifestyle intrusion. The measurement of this plan in meeting other corporate objectives has not yet been addressed.

How are the benefits of the attractiveness of such an ''optimal'' plan judged? This judgment can be made only by comparing and contrasting candidate plans. As a result, another (or set of) ''near-optimum'' alternative candidate plan(s) needs to be generated for this comparison. This near-optimum alternative candidate plan(s) can be generated by using the same framework as illustrated in Figure 15.2. One approach is to exogenously specify one (or several) end-use penetration candidates or supply candidates and permit the framework to optimize the plan around the remaining unspecified candidates.

The major demand–supply planning decisions made by utilities involve enormous quantities of money, typically $1 billion for a large power plant. These large expenditure decisions may need to be made periodically. Utility executives need a clear perspective of the risks associated with a decision because of this immense capital expenditure. Many sensitivity and scenario cases are performed to examine and assess the risks and impacts of one uncertainty or another.

A methodology for evaluating the impacts of uncertainty would begin by reviewing the key business assumptions of the study. These assumptions are characterized in terms of their expected magnitude and probability of occurring.

For those events that are judged as having a significant magnitude-probability product, either a corrective action plan or a contingency plan is developed (events of low seriousness and low probability are pruned from further consideration). A corrective action plan is taken *now*, and modifies the optimal plan so that the magnitude impact of the potential event on the plan is modified. A contingency plan is developed so that if the event takes place, corrective actions can be taken at that time.

Corrective action plans result in a cost impact on the optimal plan even if the event does not occur. When the costs of implementing the corrective action plans are factored into the optimal plan, the plan may not necessarily by optimal. For this reason, it would be necessary to reevaluate the other candidate plans to identify whether a better candidate may be found on the basis of the costs of implementation.

An example of planning under uncertainty and risk is presented in Section 15.6.

15.2.1.7 Summary. The previous subsections illustrated a framework (Figure 15.2) for an integrated analysis of demand–supply planning. While the methodology has been presented in a general formulation, certain steps may be omitted depending on the specific application. In today's utility business environment, integrated demand–supply planning is a necessary ingredient for reliable least-cost electricity.

15.3 INDIFFERENCE VALUE CALCULATIONS

It is often desirable to calculate the utility indifference value of a technology device option. For example, consider calculating the value of the hypothetical load management option to the utility power system example of Section 15.1. The indifference value will be calculated for several megawatt target penetrations: 50, 100, 200, 300, and 400 MW. A base-case load management penetration is assumed to begin in 1997 with 100 MW. Increments of 100 MW are added each year until the target penetration amount is achieved. The generation planning evaluation was performed over the 1993 to 2007 time period.

Table 15.2 presents the results from the generation planning simulation program. The base case has a cost of \$8098.1M. The 100-MW load-management case has a cost of \$7990.1M. The potential savings due to load management is \$99M.

Investment charges on a 100-MW load-management addition in 1997 with a cost of \$100/kW (in 1993 \$) is then calculated. The \$100/kW value is arbi-

TABLE 15.2 Value of Load Management

	Base Case	50 MW	100 MW	200 MW	300 MW	400 MW
Present-worth cost of expansion plan excluding load management investment charges[a]	8089.1	8064.6	7990.1	7915.3	7861.8	6833.5
Potential savings[a]		24.5	99.0	173.8	227.3	255.6
Present-worth investment charges of load management at 100 \$/kW[a]		7.5	14.9	29.7	44.3	58.5
Value of load management, \$/kW (1993\$)		329	664	585	513	436

[a]Millions of 1993 \$.

trary, and is used only for computational convenience. The load management equipment escalates for 5 years (from 1993 to 1997) at 5%/year, or by a factor of 1.276. Investment charges are calculated over the study period, 1997–2007. The uniform series factor (11 years, 10%/year discount rate) is 6.5. The investment charges for 100-MW of load management costing $100/kW in 1993 are:

$$\$100/\text{kW} \cdot 1.276 \text{ escalation} \cdot 6.50 \text{ USF} \cdot .18 \text{ fixed-charge rate} \cdot 100{,}000 \text{ kW} = \$14.9\text{M}$$

The value of load management can then be calculated as that investment cost that equals the potential savings. In the 100-MW penetration case:

$$\$14.9\text{M} \cdot \$/\text{kW} \frac{\text{value of load management}}{100 \ \$/\text{kW}} = \$99.0 \text{ M}$$

$$\text{Value of load management} = 99.0 \cdot 100/14.9 = 664 \ \$/\text{kW}$$

Figure 15.5 presents the results for other megawatt penetration values. Note that the gross economic benefits of load management decreases with increasing penetration (for penetrations 100 MW or larger). This is the general case. The benefits of any one technology typically have diminishing returns with increasing penetration.

The graph of Figure 15.5 (or Table 15.2) may be used to evaluate the optimal load management penetration level. Assuming an implementation cost of $550/kW in 1993$, it is economical to implement a load management penetration having a value greater than $550/kW. In this example, the economic penetration level is between 200 and 300 MW.

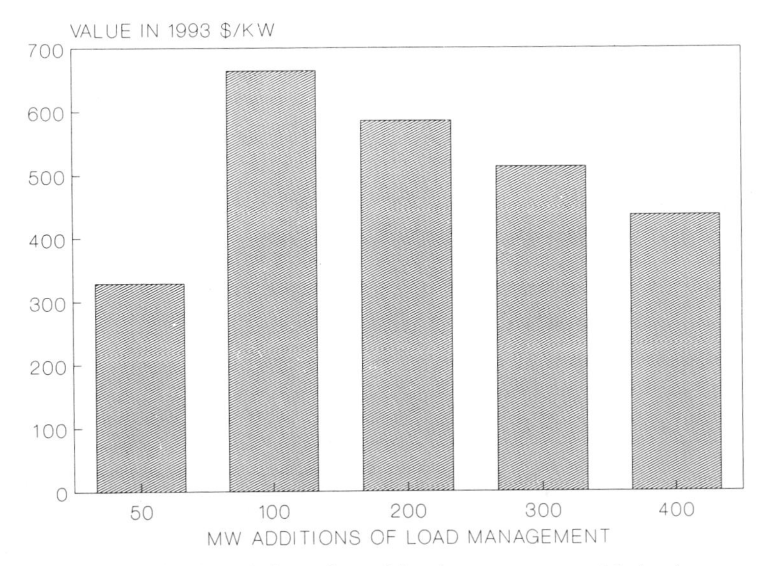

FIGURE 15.5 Sensitivity of the value of load managment with load management penetration.

The load management case of 50-MW penetration requires some clarification. It has a value of \$329/kW, which is much less than the 100-MW penetration case. In this example, the generation alternatives are sized at 100 MW. Displacing the addition of a new plant provides a potential savings of \$1770/kW for a coal unit or \$800/kW for a combined-cycle plant. In the case of a 50-MW load management penetration, this size is not large enough to displace a generating unit. Thus, the savings from the 50-MW load management are only a result of incremental fuel and O&M cost savings. The next section further amplifies this issue.

15.4 MARGINAL COSTING

It is often necessary to know the marginal cost of operating a power system. Marginal cost is the utility cost associated with providing one additional unit of energy (or power demand). Terms such as "marginal cost," "avoided cost," "incremental cost," "decremental cost," and "replacement cost" are often used interchangeably. Marginal cost data are used in electricity rate structures, generation planning, and power purchase planning. One segment of the U.S. Public Utilities Regulatory Policies Act (PURPA) requires that utilities pay up to avoided cost rates to cogenerators and small power producers for their produced electricity.

The marginal cost of power is typically calculated by subtracting the power system's total operation costs of two simulations, a reference case and a change case in which a small power purchase (or sale) is included. The marginal cost of power is dependent on the hourly time-of-day profile of the power purchase (or sale), the utility operating system, and the term of power purchase (or sale). The term of power purchase is important because the purchase may permit the utility to defer future capacity additions.

15.4.1 Short-Term Marginal Cost

"Short-term marginal cost" refers to the condition when a power purchase (or load-demand reduction) does not result in the immediate deferment of a capacity addition in a specified year. In this case, the utility has the same installed capacity either with or without the power purchase. The marginal cost calculated in this case is referred to as the marginal "energy" cost, because the cost contributions are the result of savings in energy generation only.

Marginal cost has an hourly time-of-day (TOD) profile. Figure 15.6 illustrates a conceptual daily load profile (lower curve) and the capacity types serving the load demand. Also shown is a second (incremental) load-demand profile.

The megawatt power difference between the base and the incremental load profiles is the amount of the power purchase. The type of generating unit serving the incremental purchase is the least expensive power type available to meet the incremental purchase load demand. For example, at hour 10, the incremental power type is oil-steam because nuclear and coal-unit capacity are

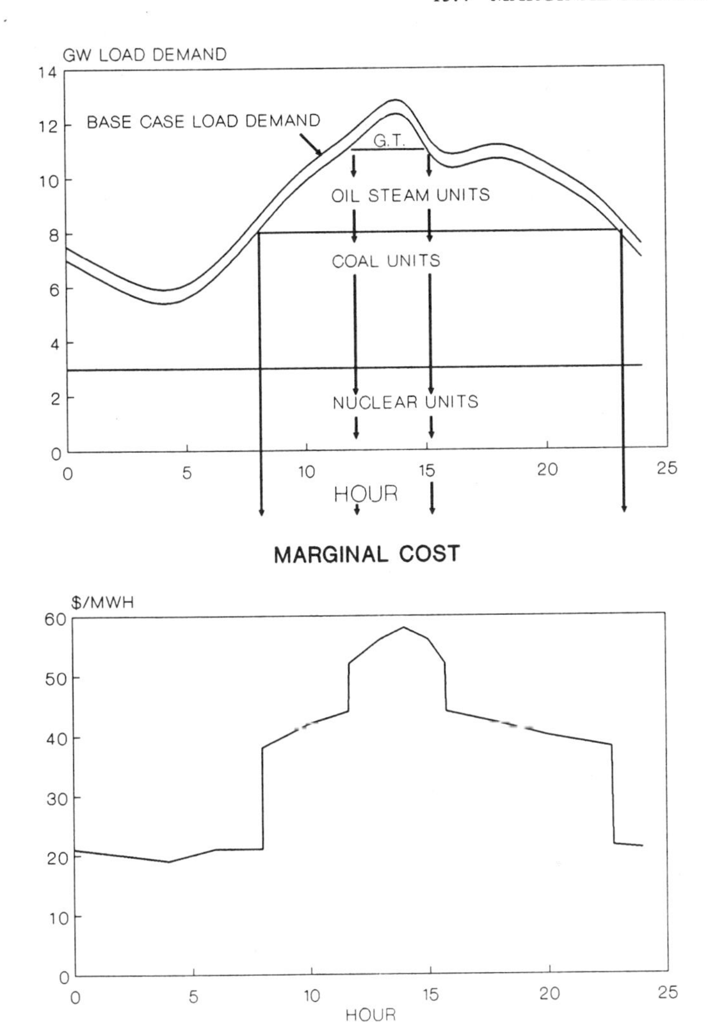

FIGURE 15.6 Marginal energy cost time-of-day characteristics.

already generating at maximum capability. Consequently, the oil-steam units must provide incremental power during this hourly period.

The bottom half of Figure 15.6 shows the incremental power cost based on a coal cost of $2MBtu and an oil–gas cost of $4/MBtu with steam-heat rates ranging from 9500 Btu/kWh to 11,000 Btu/kWh. The gas-turbine heat rates range from 12,000 to 14,000 Btu/kWh. The marginal costs of Figure 15.6 range from $20 to $60/MWh and are discontinuous as the incremental generation type changes. The daily average is $37/MWh. Note that the marginal cost should also include the variable O&M costs in addition to the fuel costs.

In some electricity rate structures, an average marginal cost over several hours may be used. Typically, on-peak and off-peak periods are defined. In this example, a time-period definition could be that the peak hour period is from hour 12 to hour 16, and the off-peak periods are from hour 0 to 12 and hour 16 to 24. Alternatively, three average time periods could be defined as the peak period from hour 12 to hour 16, the intermediate peak periods from hour 8 to hour 12 and from hour 16 to hour 22, and the off-peak periods from hour 0 to hour 8 and from hour 22 to hour 24.

15.4.2 Long-Term Marginal Cost

When a power purchase (or a load-demand reduction) can result in the deferment of a capacity addition in a specified year, then the marginal cost includes not only the short-term marginal cost (marginal energy cost) contribution but also the capacity displacement credit.

The calculation of the long-term marginal costs is dependent on the type of capacity that is displaced and the hourly TOD profile of the power purchase (or load-demand reduction).

Consider the case of a uniform marginal power purchase, or a load-demand decrease of 500 MW, as illustrated in Figure 15.7. The base-case capacity is also shown. Suppose that reliability studies indicate that a 500-MW coal unit could be canceled or deferred to some future year if a 500-MW load decrease is obtained (neglecting unit availability considerations for the moment).

Because the 500-MW marginal load decrease is exactly matched by a 500-MW coal capacity decrease, the energy generation by capacity type does not change markedly. A 500-MW uniform slice is removed from the base-load coal capacity and a 500-MW uniform slice is removed from the load. Consequently, the marginal energy cost change is small and is comprised of only the fuel and operating cost of the 500-MW coal unit, \$19/MWh in this example.

However, the capacity displacement credit is large. For a coal unit with a

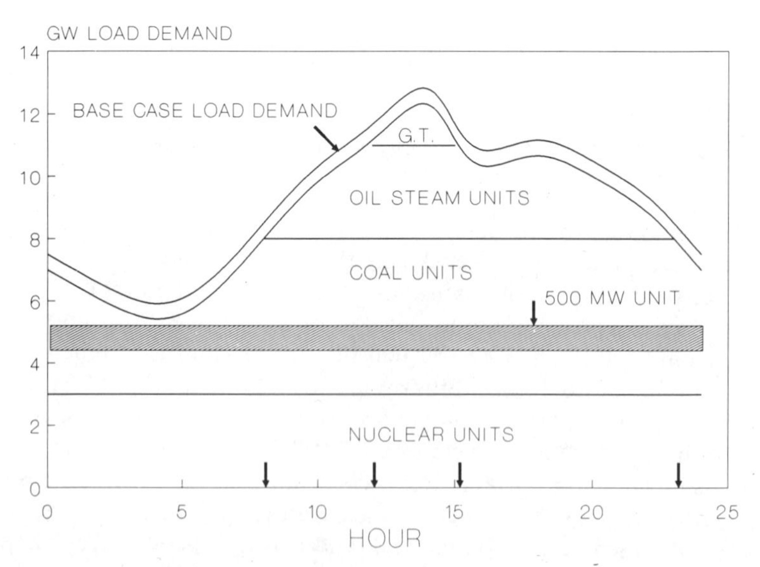

FIGURE 15.7 Impact of power purchases on long-term marginal cost.

capacity cost of $1500/kW, an 18% fixed charge rate, and 8760 hours/year of capacity displacement credit, the capacity credit is:

$$(\$1500/\text{kW})\ (.18\ \text{FCR}) \cdot 1000\ \text{kW/MW}/8760\ \text{hour/year} = \$30.8/\text{MWh}$$

Thus, the total long-term marginal cost is 30.8 + 19 = $49.8/MWh, (neglecting O&M costs).

Note that the long-term marginal cost ($49.8/MWh) is larger than the short-term average marginal energy cost of $37/MWh. This is the general rule.

Consider the same case of a 500-MW uniform marginal power purchase, but assume that 500 MW of gas-turbine capacity is displaced. The energy generation for the coal- and oil-steam unit type is the same as that calculated for the short-term marginal cost case. Long-term marginal energy generation for the gas-turbine type is influenced by the heat-rate difference between new gas turbines and the existing gas turbines, which will be assumed to be nearly zero. Thus, the long-term marginal energy cost averaged over the day is $37/MWh.

The capacity displacement value can also be calculated per kilowatt-hour of load change using a gas-turbine capital cost of $300/kW as:

$$\$300/\text{kW} \cdot .18\ \text{FCR} \cdot 1000\ \text{kW/MW}/8760\ \text{hour/year} = \$6/\text{MWh}$$

Thus, the long-term marginal cost is 37 + 6 = $43/MWh (neglecting O&M costs).

In this example, the gas-turbine displacement had the lower total marginal cost. This is dependent, however, on the specific utility power system characteristics. Thus, the long-term marginal cost depends on the type of capacity displaced by the incremental load demand or power purchase.

When gas-turbine capacity is displaced, the long-term marginal *energy* cost component is high, but the long-term marginal *capacity* cost component is low. When coal capacity is displaced, the long-term capacity cost component is high but the energy cost component is low. For a midrange capacity type, such as combined-cycle capacity, the *capacity* and *energy* components are in the medium cost range.

An example that illustrates these principles is for a 100-MW, 8760 hour/year marginal power purchase beginning in 1993 using the power system example of Section 15.1. The reference power system has two 100-MW gas turbine additions in 1993, followed by three combined-cycle units (one in each year from 1994 through 1996), followed by one coal unit per year thereafter.

The net long-term result of the marginal purchase is the cancellation of a base-load power plant. As shown in Table 15.3, the first effect of the marginal 100-MW purchase is to delay one gas turbine addition in 1993. In 1994, the marginal purchase results in the displacement of a combined-cycle unit and the replacement of a gas turbine. In 1997, a coal unit is displaced and a combined-cycle unit is replaced.

The composition of the marginal cost components is illustrated in Figure 15.8. In 1993, the principal marginal cost component is fuel cost. In 1994 and 1997, the investment charge component experiences an incremental change in 1994 and 1997 as a result of capacity displacement shifts to combined cycle

TABLE 15.3 Capacity Displacements for a 100-MW Power Purchase

Year	Reference Case	Incremental 100-MW Purchase	Change Incremental–Base
1993	2-GT	1 GT	− 1 GT
1994	1 CC	1 GT	1 GT − 1 CC
1995	1 CC	1 CC	
1996	1 CC	1 CC	
1997	1 Coal	1 CC	1 CC − 1 coal
1998	1 Coal	1 Coal	
1999	1 Coal	1 Coal	
2000	1 Coal	1 Coal	

and coal, respectively. The fuel cost component also responds to the capacity displacement shift.

15.4.3 Time-of-Day Marginal Cost

The examples presented previously were based on a marginal purchase (or load change) that occurs for 8760 hours/year. If a marginal power source operates for less than 8760/year (and during the peak and intermediate peak hours), then capacity and energy credit are different. The value per megawatt-hour of the fuel component is higher because marginal energy cost is averaged only over the peak and intermediate peak hours. The capacity credit component is also higher because capacity credits are divided by a lower megawatt-hour value.

When the marginal purchase occurs off-peak, the capacity credit component is typically zero and the only contribution is due to off-peak replacement

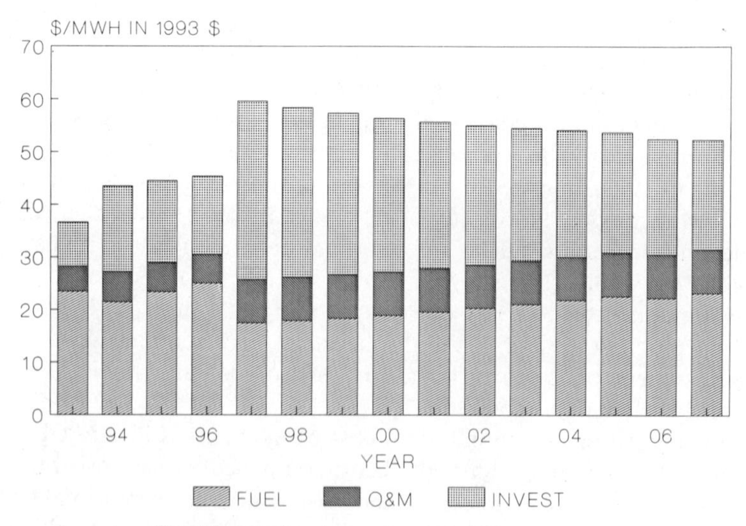

FIGURE 15.8 Annual marginal cost trend.

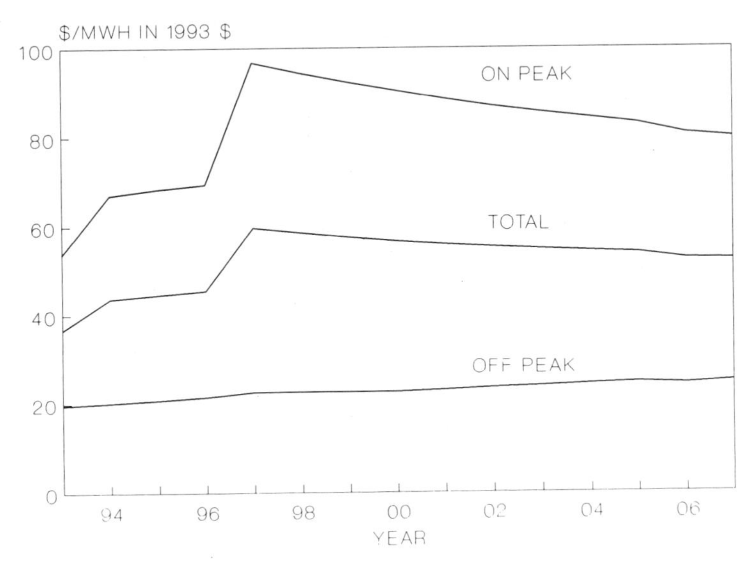

FIGURE 15.9 Marginal cost trend for 12-hour peak and 12-hour off-peak periods.

energy. Figure 15.9 illustrates the marginal cost trend for 12-hour peak and 12-hour off-peak periods, expressed in 1993 dollars. In this example, the ratio of peak to off-peak marginal cost is in the $\frac{3\text{–}5}{1}$ range.

Depending on the peak-period definition and the specific utility power system, peak marginal cost may include a contribution due only to displacing gas-turbine capacity. The off-peak contribution may include contributions from a change in the generation mix, such as changing from a combined-cycle unit to a coal unit. Figure 15.10 presents the case in which the peak period is defined

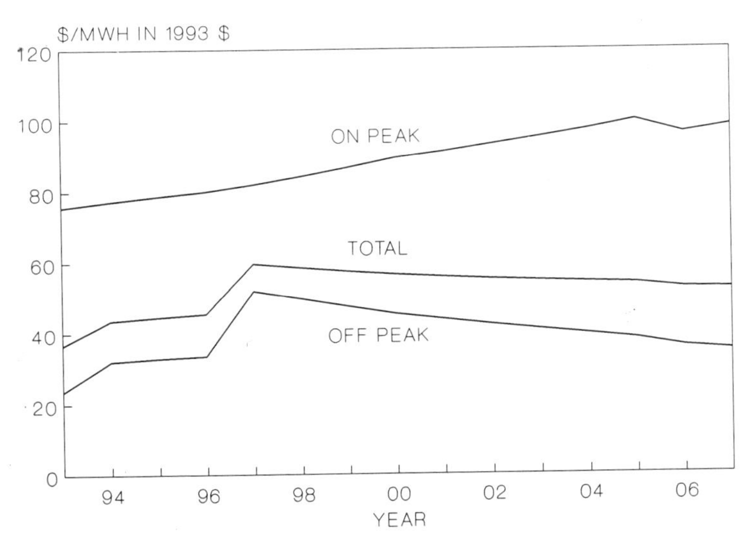

FIGURE 15.10 Marginal cost trend for 6-hour peak and 18-hour off-peak periods.

as 6 hours and the off-peak as 18 hours. The off-peak case resulted in a capacity mix change, and the peak case resulted in a gas-turbine displacement. Note, however, that the peak:off-peak price ratio is still in the $\frac{3-5}{1}$ range. These peak and off-peak marginal costs may be used in developing a TOD differentiated marginal cost rate.

While marginal costs may be expressed in \$/MWh, they may also be segmented into an energy cost component in \$/MWh and a capacity cost component \$/kW/year. The energy cost component typically includes fuel cost and variable O&M cost, and the capacity cost component typically includes investment cost changes and fixed O&M costs.

15.4.4 Intermediate-Range Marginal Costs

A power purchase (from an industrial cogenerator, e.g.) may provide reliable power, but the utility may have adequate capacity without the purchase to achieve their reliability target level. In this case, the power supplier (cogenerator) does not displace a utility capacity addition until perhaps several years into the future. From a utility avoided-cost viewpoint, the power supplier is not eligible to receive capacity credit for the power.

However, the power supplier does provide reliable power, and this results in the utility having a higher service reliability and a lower probability of customer power interruption. Because electricity consumers benefit from reduced power interruptions, the power supplier can be rewarded partial capacity credit for these benefits. The calculation of the reduced power interruptions and the consequential "societal" benefits were presented in Chapter 11, Section 11.2.

A graph of the total consumer cost of electricity is shown in Figure 15.11.

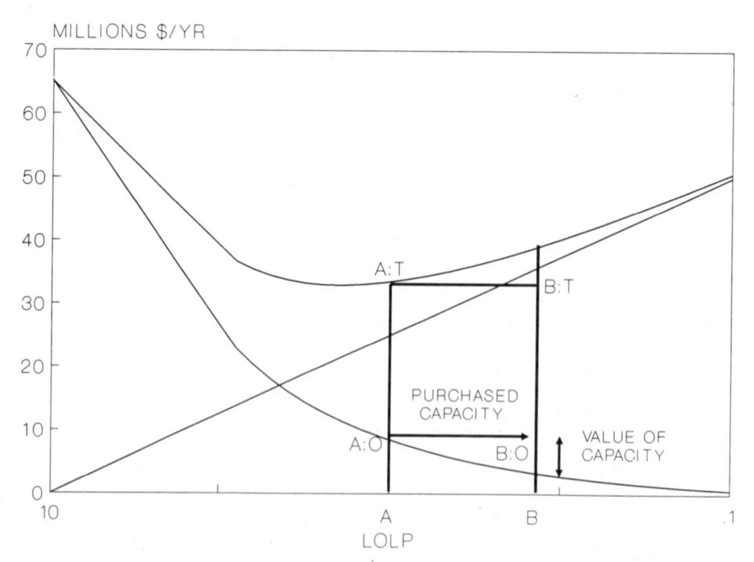

FIGURE 15.11 Total consumer cost of electricity.

The total cost is comprised of the sum of the utility cost of service (that recovered through electricity rates), and the consumer cost of outages when electricity is not supplied. The optimum value is shown at the LOLP level corresponding to the minimum total electricity cost.

Suppose that the utility has a higher reserve margin (or LOLP) than the optimum at point A in Figure 15.11. An independent power producer (or cogenerator) supplying an additional amount of capacity increases the utility reserve margin (or LOLP) to point B. The independent power producer decreases the consumer cost of outages from point A:0 to B:0. Thus, the independent power producer provides the consumer a value and can be rewarded with a payment. If the independent power producer received the total incremental value of the provided capacity, the value difference between points B:0 and A:0, the electricity rates would be adjusted upward by this same value. However, the total cost of electricity, point B:T, would be the same (as point A:T).

The value per KW of independently produced power can be calculated as:

$$\text{Value of produced power (in \$/kW/year)} =$$

$$-\frac{\partial \text{ consumer outage cost}}{\partial \text{ kW}} = -\frac{\partial \text{ consumer outages}}{\partial \text{ LOLP}} \cdot \frac{\partial \text{ LOLP}}{\partial \text{ kW}} \tag{15.1}$$

Using the outage component of Equation (11.3) in Chapter 11 gives

$$\text{Value of produced power} =$$

$$- M \cdot \text{equivalent peak hour} \cdot \text{outcost} \cdot \frac{\partial \text{ LOLP}}{\partial \text{ kW}}$$

$$= + M \cdot \text{equivalent peak hour} \cdot \text{outcost} \cdot \frac{1}{M} \text{LOLP} \tag{15.2}$$

where M = inverse logarithmic slope of the LOLP reliability curve versus MW

Equivalent peak hours = equivalent daily peak-load duration in hours

Outcost = consumer cost of unserved energy in \$/MWh

Substituting Equation (11.4) of Chapter 11:

$$\text{LOLP}_{\text{opt}} = \left(\frac{\text{annual capacity charges}}{\text{equivalent peak hour} \cdot \text{outcost}}\right) \tag{15.3}$$

into Equation (15.2) yields:

$$\text{Value of produced power (in \$/kW/year)} = \text{annual capacity charges} \cdot \left(\frac{\text{LOLP}}{\text{LOLP}_{opt}}\right) \quad (15.4)$$

where Annual capacity charges = annual cost of providing peaking capacity in \$/kW/year

LOLP_{opt} = optimum (least total cost) LOLP

Thus, the value of independently produced power (in \$/kW/year) is equal to the annual peaking capacity charge (in \$/kW/year) times the ratio of the current LOLP value divided by the optimum value. If the reliability in a given year is .3 days/year, the capitalized cost of peaking capacity is \$300/kW, and the utility reliability target is 1.0 day/year, then the avoided capacity cost credit is based on a capacity payment of (.3 days/day)/(1.0 days/day) • \$300/kW • .20 fixed-charge rate = \$18/kW/year.

This adjusted capacity cost credit component is added to the avoided energy cost component to compute the total marginal cost.

15.4.5 Summary

Marginal system costs are used in electricity rate design, generation planning, and power purchasing planning. Marginal costs may be calculated using power system planning programs based on daily, seasonal, and TOD bases. Figure 15.12 illustrates a conceptual trend of the real ¢/kWh marginal cost with time.

In the short-term and intermediate-term total marginal cost is comprised largely of marginal energy cost contributions. A small contribution arises from the marginal peaking capacity contribution, which increases with time accord-

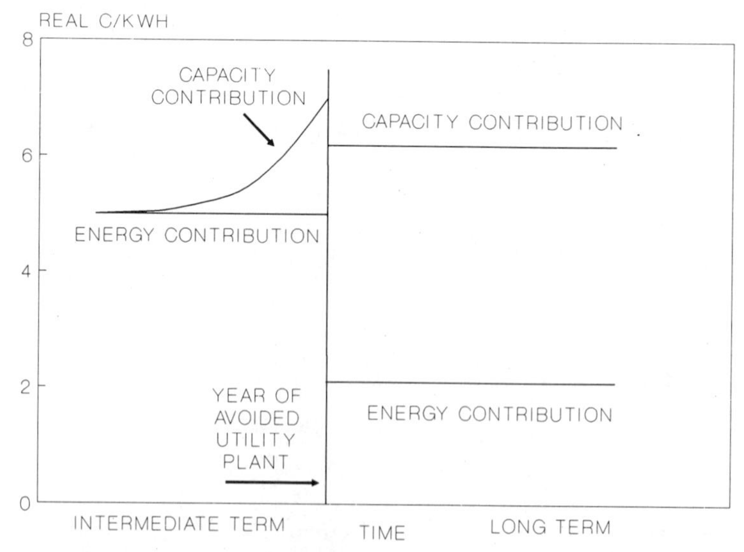

FIGURE 15.12 Conceptual trend of marginal cost versus time.

ing to Equation (15.4) until the date of the avoided utility generation plant. For years following the date of the avoided utility plant, the long-range marginal cost equations apply. In this case, energy cost contributions may be much smaller than capacity cost contributions if a capital-intensive base-load power plant is avoided.

The examples illustrated the long-term marginal cost calculations in which the utility avoided capacity were conventional power generation types, coal-steam gas turbine, and combined cycles. A utility also can construct a cogeneration plant in association with an industrial or other process steam user; hence the utility avoided capacity could also be a cogeneration plant.

15.5 SMALL IMPROVEMENT PROJECTS

Many utility economic studies require the evaluation of small improvement projects to the power system. Examples include evaluating the benefit of a new turbine rotor for a generating unit, deciding which manufacturer's product and price for a generating unit is the best, and evaluating whether to buy low-loss (but higher investment cost) distribution transformers. Hundreds of these types of evaluation may be performed each year in a utility (Felak and Stoll, 1987).

Any one of the small incremental improvement projects barely influences the overall utility economic operation. When accumulated, however, these projects can lead to noticeable economic improvements. Accounting for the incremental contribution of each project is an important economic evaluation.

Detailed methods can be used to evaluate small projects. However, in view of the small incremental changes and the number of project evaluations, it is more practical to use approximate techniques. Approximate techniques use power system sensitivity parameters derived from detailed evaluation methods. The incremental replacemental (marginal) energy cost in $/MWh is a key sensitivity parameter.

Example. A utility is purchasing a 400-MW coal-fired steam-generating unit and has received proposals from two manufacturing suppliers. The generating unit is to be installed in 1993 and will operate base load at 95% capacity factor when available. The characteristics of the two alternative proposals are:

	Alternative 1	Alternative 2
Unit size (MW)	400	410
Forced outage rate (%)	5	10
Scheduled outage rate (%)	10	10
Fuel cost (20-year levelized) ($/MBtu)	3.0	3.0
Net plant heat rate (Btu/kWh)	10,000	10,200
Plant cost (millions $)	560	540
Cost ($/kW)	1400	1350

Power system reliability (LOLP) studies found the M-slope to be 250 MW. Gas turbines may be used for adjusting reliability. Assume that gas turbines cost $300/kW and have a 5% forced outage rate. The power system incremental replacement energy cost (marginal energy cost) is $55/MWh (20-year levelized).

Determine the alternative with the lowest 20-year present-worth cost for these economic parameters:

Inflation	6%/year
Present-worth rate	10%/year
Levelized annual fixed-charge rate	$20%/year

SOLUTION. Use alternative 1 as a reference, and calculate the economic benefits of alternative 2 relative to alternative 1.

A. Reliability. The two alternatives differ in megawatt rating and reliability. Compute the additional cost so that both alternatives contribute the same to power system reliability. Compute the generating unit load-carrying capability using Garver's formula.

	Alternative 1	Alternative 2
M-slope	250	250
Unit size (MW)	400	410
Forced outage rate (%)	5	10
Load-carrying capability	354.9	323.1
Deficiency in load-carrying capability	Reference	31.8 MW

The deficiency in load-carrying capability can be evaluated in terms of 31.8 MW of additional load-carrying capacity of gas turbines to provide backup to the alternative 2 generating units.

Because gas turbines have a 5% forced outage rate, alternative 2 would need to add 31.8 MW/(1 − .05) = 33.5 MW of gas turbines. These gas turbines would cost $300/kW to install.

For equal system reliability, the alternatives are:

	Alternative 1	Alternative 2
Capacity additions	400 MW coal	410 MW coal 33.5 MW gas turbine
Cost of gas turbines	(Reference)	$10.05M

B. Production Cost. *Evaluate production cost differences.* The average power produced by each of these two generating units is:

	Alternative 1	Alternative 2
	Replacement Energy Cost	
Forced outage rate	.05	.10
Scheduled outage rate	.10	.10
Maximum availability	(.95) • .9 = .855	.9 • .9 = .81
Energy output/year	400 • 8760 • .855 • .95	410 • 8760 • .81 • .95
	2,846,124 MWh	2,763,736 MWh
Energy difference	Reference	82,388 MWh
Annual levelized replacement Energy Cost ($year @ /MWh $55)		$4,531 M/year
	Fuel Cost	
Heat rate (Btu/kWh)	10,000	10,200
Fuel cost ($/MBtu) (levelized)	3.0	3.0
Energy output (MWh)	2,846,124	2,763,736
Annual fuel cost	$85,383,720M/year	$84,570,322M/year
Annual levelized fuel cost-difference($/year)	Base	$ – 0.813M/year

The annual system production cost differences are summarized as:

	Alternative 1	Alternative 2
Replacement energy differences	Base	$4.531M/year
Fuel cost differences	Base	$ – 0.813M/year
Annual production cost difference (levelized)	Base	$3.718M/year

The annual levelized production cost can be translated into an equivalent capitalized value by dividing by the annual levelized fixed charge rate of 20%.

	Alternative 1	Alternative 2
Capitalized equivalent Production cost differences Over 20 years	Base	$18.59M

C. Total Costs. The equipment capital cost and the capitalized equivalent production cost can now be added together to obtain the total cost.

	Alternative 1	Alternative 2
Coal plant cost	$560M	$540M
Gas-turbine reliability costs gas turbine at $300/kW	(Reference)	$10.05M
Capitalized equivalent production cost difference	(Reference)	$18.59M
Total evaluated costs	$560M	$568.64M

Alternative 1 is the economical choice, since it has a lower total 20-year present-worth evaluated cost. It is interesting to note that the initial choice may have been alternative 2 because it has a higher megawatt rating and a lower plant investment. However, the lower forced outage rate and better heat rate of alternative 1 leads to a lower total economic evaluation.

Generating unit upgrade (and life extension) projects may also be evaluated using an approximate incremental project evaluation methodology (McDonough, Nair, and Stoll, 1987). Upgrade projects can reduce forced outages, maintenance outages, maintenance expenses, and improve plant heat rate. The first step in these evaluations is to compute the economic value of a one-day reduction in forced and maintenance outages and a 1-Btu improvement in heat rate for the upgrade generating unit candidate.

A one-day reduction in a maintenance outage leads to a net replacement energy saving. Net replacement energy saving is the difference between the incremental production energy cost and the upgrade candidate generating unit production cost.

The benefit of a one-day maintenance outage saving is computed as:

$$\begin{aligned}
&\text{Maintenance day benefit (\$/day/year)} \\
&= (\text{MW rating}) \cdot 24 \text{ hours/day} \cdot (\text{capacity factor}) \\
&\cdot [(\text{system replacement energy cost: \$/MWh}) \\
&- (\text{unit fuel cost: \$/MBtu}) \cdot (\text{unit heat rate: MBtu/MWh})] \qquad (15.5)
\end{aligned}$$

(This is also referred to as "daily replacement energy benefit".)

The system replacement energy cost used in Equation (15.5) is the average value during the time of day that the upgrade candidate operates, and during the season when maintenance is performed. The capacity factor is the average unit value during the season when maintenance is performed.

A one-day reduction in a forced outage leads to a replacement energy saving (as in the maintenance outage case) *and* a replacement capacity saving. The replacement capacity saving is credited because the forced outage reduction leads to fewer capacity purchases and reduces future capacity needs. The replacement capacity saving is computed as:

$$\text{Daily capacity benefit (\$/day/year)} = \text{MW rating} \cdot \frac{\text{annual replacement capacity cost \$/MW/year}}{\text{365 days}} \qquad (15.6)$$

A forced outage is random, and could occur throughout the year (365 days). Annual replacement capacity cost is multiplied by the probability of a forced outage (1/365) contributing to a replacement capacity saving. The annual replacement capacity cost in \$/MW/year is the annual cost of the emergency capacity power purchases or the annual capacity cost charges of a new (peaking-type) capacity. Thus, the value of a one-day forced outage rate saving is:

$$\text{Forced day benefit (\$/day/year)} = \text{daily replacement energy benefit} + \text{daily capacity benefit} \qquad (15.7)$$

Note that the daily replacement energy benefit [Equation (15.5)] for a forced outage may be numerically different from the maintenance day benefit, because the forced outage calculation must use capacity factor and system replacement energy cost values based on annual operating year average values, whereas the maintenance day benefit is calculated by using values typical during the period of planned maintenance.

A heat-rate improvement may result in several benefits. If the generating unit fuel input is limiting on the unit (boiler-limited), then improving unit efficiency permits the unit to generate *more* MW power for the *same* fuel input. If the generating unit is output-limited (turbine-generator-limited), then improving the unit efficiency permits the unit to generate the *same* megawatt power but use *less* fuel. The case that applies is generating-unit-specific and dependent on plant design margins.

Consider the case of an output-limited generating unit (fuel savings case). The annual value of a 1-Btu heat-rate improvement is calculated as:

$$\begin{aligned} &\text{Fuel saving benefit (\$/Btu/year)} \\ &= \text{(8760 hours)} \cdot \text{(annual capacity factor)} \\ &\cdot \text{(fuel cost: \$/MBtu)} \cdot \text{(MW rating)} \cdot (.001) \end{aligned} \qquad (15.8)$$

Consider the case of a fuel-limited generating unit. For the *same* fuel, additional "free" megawatt power output is produced. The additional megawatt may be used to reduce the power system's most expensive generating unit. The megawatt increase is proportional to the heat rate improvement. The megawatt increase corresponding to a one-Btu improvement is:

$$\Delta\text{MW} = \text{(MW rating)} \cdot \frac{\text{1 Btu}}{\text{unit heat rate: Btu/kWh}} \qquad (15.9)$$

Thus, the annual replacement energy cost saving per Btu is:

Replacement energy cost benefit ($/Btu/year)

$$= \frac{\text{MW rating}}{\text{unit heat rate: Btu/kWh}}$$

$$\cdot \text{(8760 hours)} \cdot \text{(annual capacity factor)}$$

$$\cdot \text{(system replacement energy cost: \$/MWh)} \qquad (15.10)$$

In addition, there is a replacement capacity saving credit (in the fuel-limited case) because of the additional megawatts.

Replacement capacity cost benefit ($/Btu/year)

$$= \frac{\text{MW rating}}{\text{unit heat rate: Btu/kWh}}$$

$$\cdot \text{(annual replacement capacity cost: \$/MW/year)} \qquad (15.11)$$

Equations (15.8) through (15.11) can be combined into one equation as:

Heat-rate saving benefit ($/Btu/year)

$$= \text{fuel saving benefit}$$

$$+ \text{fuel limit} \cdot \text{(replacement energy cost benefit}$$

$$- \text{fuel saving benefit} + \text{replacement capacity cost benefit)}$$

$$= \text{(8760 hours)} \cdot \text{(annual capacity factor)} \cdot \text{(fuel cost: \$/MBtu)}$$

$$\cdot \text{(MW rating)} \cdot .001$$

$$+ \text{fuel limit} \cdot \left\{ \frac{\text{MW rating}}{\text{unit heat rate: Btu/kWh}} \right.$$

$$\cdot \text{(8760 hours)} \cdot \text{(annual capacity factor)}$$

$$\cdot \text{[(system replacement energy cost: \$/MWh)}$$

$$- \text{(fuel cost: \$/MBtu)} \cdot \text{(unit heat rate: MBtu/MWh)]}$$

$$+ \frac{\text{MW rating}}{\text{unit heat rate: Btu/kWh}}$$

$$\left. \cdot \text{(annual replacement capacity cost: \$/MW/hour)} \right\} \qquad (15.12)$$

where Fuel limit = 0 if fuel input is not limited but power output is limited (turbine-generator-limited)

= 1 if fuel input is limited but power output is not limited (boiler-limited)

Example. A 200-MW coal-fired generating unit operating at 75% capacity factor has an asphalt insulated generator stator. In Chapter 9, the probability of a stator girth crack failure was evaluated at 32.2% over the next 10 years. For the utility to upgrade, the investment pay-back period must be less than 4 years. Should the utility upgrade the stator in 1990 using a mica-based insulation material (with practically zero failure probability)? The case parameters are:

MW rating	200 MW
Unit capacity factor	75%
Fuel cost (1990 $/MBtu)	2.0
Unit heat rate (Btu/kWh)	10,000
Replacement energy cost (1990 $/MWh)	40.0
Replacement capacity cost (1990 $/kW/year)	60.0
Annual stator failure rate (assume uniform/year)	3.22%/year
Outage repair time duration if failed	84 days
Repair costs if failed (1990$)	$200,000
Upgrade investment cost (1990$)	$800,000

The forced outage rate cost per day is calculated using Equations (15.5), (15.6), and (15.7).

$$\begin{aligned}\text{Maintenance day benefit} &= 200 \cdot 24 \cdot .75 \cdot [55 - (2.0 \cdot 10.0)] \\ &= \$126{,}000/\text{day/year}\end{aligned}$$

$$\begin{aligned}\text{Capacity day benefit} &= 200 \cdot 60{,}000/365 \\ &= \$32{,}877/\text{day/year}\end{aligned}$$

$$\begin{aligned}\text{Forced day benefit} &= 126{,}000 + 32{,}877 \\ &= \$158{,}877/\text{day/year}\end{aligned}$$

The annual expected probability weighted benefit per year is:

$$\begin{aligned}\text{Expected annual benefit} &= .0322 \cdot (\$158{,}877/\text{day} \cdot 84 \text{ days} \\ &\quad + 200{,}000) \\ &= \$436{,}170/\text{year}\end{aligned}$$

The investment pay-back (not including cost escalation) is (800,000/436,170) = 1.83 years. Thus, the utility should implement the generator upgrade.

In summary, this section has illustrated economic analysis procedures for use in evaluating small improvement projects.

15.6 PLANNING UNDER UNCERTAINTY

Capacity addition plans are dependent on many forecast parameters, including load demand, fuel pricing, plant cost, technology availability, environmental and regulatory requirements, and financial forecasts. Unfortunately, the forecast of these parameters is subject to uncertainty. Each has a likelihood of assuming a range of values in the future. A plan that is least-cost under a reference set of forecast parameters may not be so under an alternative set of forecast parameters. A key objective in planning is to recognize these uncertainties and to develop a plan that can be adapted to changing business conditions and be least-cost under the probability-weighted range of forecast parameters.

Planning under uncertainty is a dynamic year-by-year process. The planning process begins with parameters that are fact today. One can have only a perception of what these parameters might be in the future.

The analysis procedure generally involves identifying the potential uncertain (or risky) events and assigning a probability to the event. A planning analysis is performed that accounts for any associated corrective action plans by the utility to evaluate the impacts of the events. The impacts may then be probability-weighted, and a composite utility impact value can be computed. This process may be repeated by examining alternative or contingency plans.

15.6.1 Load Growth Uncertainty

Load-demand growth is one of the key forecast parameters that is subject to uncertainty. Load growth is influenced by many factors, including the national economy, the local economy, energy prices, and conservation. The forecast of future load growth over the midterm (5–10 years) and long term (10–30 years) is subject to broad uncertainty.

History reveals that the load growth during the 1960s averaged 7%/year on a national basis, 4.5%/year during the 1970s, and is forecast to grow at 2.5%/year through the year 2000. Utilities reported to the North American Electric Reliability Council (NERC) a 90% confidence band of load growth from 0 to 4.5%/year over a 10-year future forecast period (NERC, 1986). This broad load growth uncertainty places a significant added burden of responsibility on the utility capacity planner.

It is difficult for utilities to adapt to load-demand changes because of the long lead times associated with constructing capacity. For example, coal-fired power plants have lead times of 6 years or more, and load management programs require several years to initiate customer participation.

A load uncertainty analysis procedure is presented in Figure 15.13. Data on the load forecast and load uncertainties are translated into load growth scenarios. Load growth scenarios are a finite sample of potential load growth outcomes to be studied. Other data input is the future planned additions and

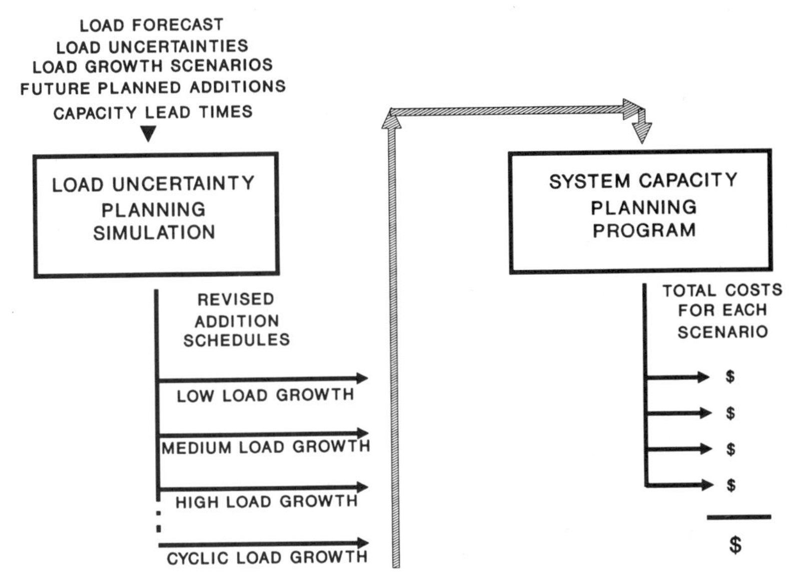

FIGURE 15.13 Load uncertainty analysis procedure.

capacity lead times. The load uncertainty planning simulation evaluates how corrective action plans for future planned capacity additions would change in response to each load-growth scenario.

The corrective action plans for future capacity addition schedules are evaluated through the use of a generation planning simulation program to evaluate the power system fuel, O&M, and investment charges. The result is a present-worth expansion cost for each load-growth scenario. These costs can then be weighted by the probability of each scenario to obtain an "expected" evaluated cost for the plan.

The procedure illustrated in Figure 15.13 would be repeated for other strategy plans of future capacity additions. Examples of strategy plans include lower/higher reserve margins, more/less demand-side capacity additions, or more/less short lead-time generation capacity. The plan with the lowest probability-weighted evaluated cost is the optimal plan (assuming no other considerations besides load uncertainty).

15.6.2 Load Growth Uncertainty Example

To illustrate the application of the methodology for planning with load uncertainty, the same example power system of Section 15.1 is used. The utility plans for a 20% reserve margin target. Three load growth scenarios are to be studied. It is assumed that the utility planner has developed a plan for the base-case forecast.

	Low	Base Case	High
Probability (%)	20	60	20
MW growth trend each year (MW/year)	43	85	127

The utility's forecast load growth is adapted to new growth trends as they become evident. However, because load growth may experience short-term fluctuations due to weather, national and local economic business cycles, and other factors, any adjustment from the forecast load growth to reflect recent historical load growth trends requires several years of data. If long-term load growth were to decrease in 1994 to 43 MW/year for the next 10 years, a utility would not immediately revise the forecast from 85 to 43 MW/year, but would reduce the forecast to some intermediate value until more data are available to substantiate the long-term change.

Figure 15.14 illustrates model results of the load growth revision process. In 1994, a new trend emerges. In each succeeding year, a new forecast is made on the basis of new data. One mathematical model of this process is:

$$\text{New load growth forecast (years)} = (1 - \alpha) \cdot (\text{previous load growth forecast}) + \alpha \cdot (\text{true load growth rate}) \tag{15.13}$$

where α is a parameter between 0 and 1 that reflects the "inertia" associated with changing the load forecast to follow a new trend. If $\alpha = 1$, then a new trend is used with no delay. In this example, a value of $\alpha = .40$ is used. This

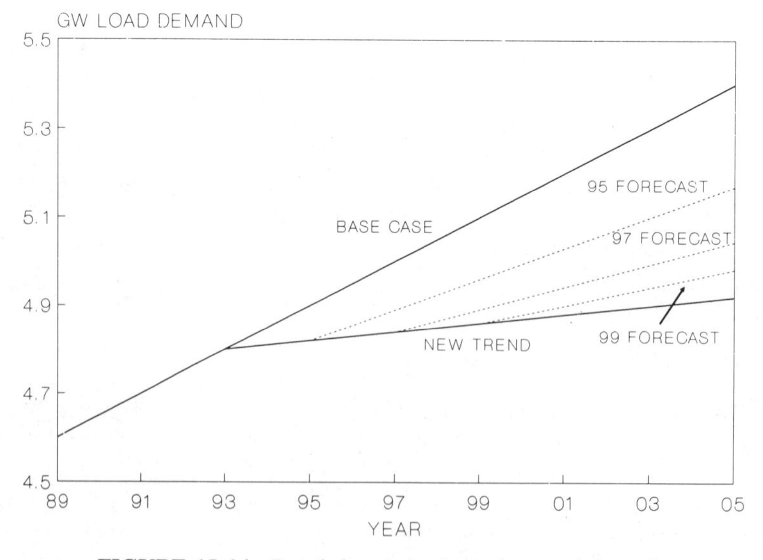

FIGURE 15.14 Load-demand adaptation model result.

model results in an exponential response characteristic to an incremental change in the long-term load growth trend.

For this example, the generation capacity lead times are:

Type	Construction Lead Time
Coal	6 years
Combined-cycle	4 years
Gas-turbine	2 years

The construction expenditure pattern [excluding "allowance for funds used during construction" (AFDC) and escalation] for the generating units is:

Construction Year	Coal	Combined Cycle	Gas Turbine
1	5%	10%	20%
2	10%	30%	80%
3	15%	30%	
4	25%	30%	
5	30%		
6	15%		
Total	100%	100%	100%

The construction expenditure pattern data for the generating units are required in the event that a unit is delayed. A construction delay can result when the load growth is low and less generation capacity is needed than originally planned. Construction delays result in the accumulation of AFDC in the "construction work in progress" (CWIP) plant account and, therefore, higher escalated costs for project completion.

Three scenarios must be evaluated: base-case load growth, low load growth, and high load growth. The base-case load growth scenario was examined in Section 15.1. The starting year is 1993 with the load growth given.

15.6.2.1 High Load Growth Scenario. In the high load-growth scenario, the "actual" load demand in year 1994 is 3716 MW. However, the forecast value for 1994 is 3674 MW, based on the last previous forecast. Consequently, the utility adjusts its 1994 forecast to account for this forecast error. It is assumed that the load forecast is revised using an $\alpha = .4$ [in Equation (15.13)]. Table 15.4 presents the new load growth forecast.

The new 1994 load growth forecast (last column of Table 15.4) is the basis by which a utility planner *in 1994* would revise the capacity needs. The original capacity plan and the new 1994 load forecast produce reserve margin values less than the target value of 20% as shown in the first two columns of Table 15.5.

TABLE 15.4 1994 Load Forecast Model

Year	Old Forecast	High Load Growth Scenario	New 1994 Forecast
1994	3674	3716	3716
1995	3770	3854	3829
1996	3855	3981	3931
1997	3940	4108	4032
1998	4025	4235	4134
1999	4110	4362	4236
2000	4195	4489	4338
2001	4280	4616	4440
2002	4365	4743	4541
2003	4450	4870	4643
2004	4535	4997	4745
2005	4620	5124	4847
2006	4705	5251	4949
2007	4790	5378	5050

TABLE 15.5 1994 Planning Revision to the Capacity Addition Plan

	Old Additions Forecast			Revised Additions Forecast	
	(1)		(2)	(3)	(4)
Year	Additions (MW)	Type	Reserve Margin with New Load Forecast (%)	New Forecast Additions	Reserve Margin
1994	100	CC	20	100 CC	20%
1995	100	CC	19	100 CC	19%
1996	100	CC	18	100 CC + 100 GT	21%
1997	100	Coal	18	100 Coal	20%
1998	100	Coal	17	100 Coal	20%
1999	100	Coal	17	100 Coal + 100 CC	21%
2000	100	Coal	16	100 Coal	21%
2001	100	Coal	16	100 Coal	20%
2002	100	Coal	15	100 Coal	20%
2003	100	Coal	15	100 Coal + 100 coal	22%
2004	100	Coal	15	100 Coal	21%
2005	100	Coal	14	100 Coal	21%
2006	100	Coal	14	100 Coal	20%
2007	100	Coal	14	100 Coal	20%

The addition plan needs revision to assure a 20% reserve margin. Columns 3 and 4 of Table 15.5 present the calculations. The shortest generation capacity lead time is 2 years for gas turbines. Consequently, the earliest capacity addition is a 1996 gas-turbine unit addition. Even with the 1996 gas-turbine addition, capacity is also needed in 1999. Since the coal unit lead time is 6 years, a combined-cycle unit is added. In 2003, another 100-MW capacity addition is required, which can be achieved with a coal-unit addition.

In summary, in 1994, a utility planner would recognize the need for more capacity additions. Because of lead time requirements, the optimal economic type of capacity additions may not be added. Instead, the corrective action plan in response to higher load growth must add capacity additions that meet the lead-time requirements.

The same process is repeated for the other forecast years, 1995 through year 2007. Figure 15.15 illustrates how the load forecast model predicts the utility planner's load forecast for the year 2007 under a high load growth scenario. Most of the forecast adaptation occurs within 3 years after an abrupt long-term load demand change. The final generation addition plan is presented in Table 15.6. Note that the final corrective additional plan includes an additional 800 MW, of which 300 MW was identified by the year 1994. While the preferred economic choice would have been to add coal units, lead-time requirements (at the time the capacity need was discovered) did not permit addition of the optimal economic generation capacity.

The revised addition plan is now input into a power system generation capacity planning model in which the system expansion costs are calculated. The

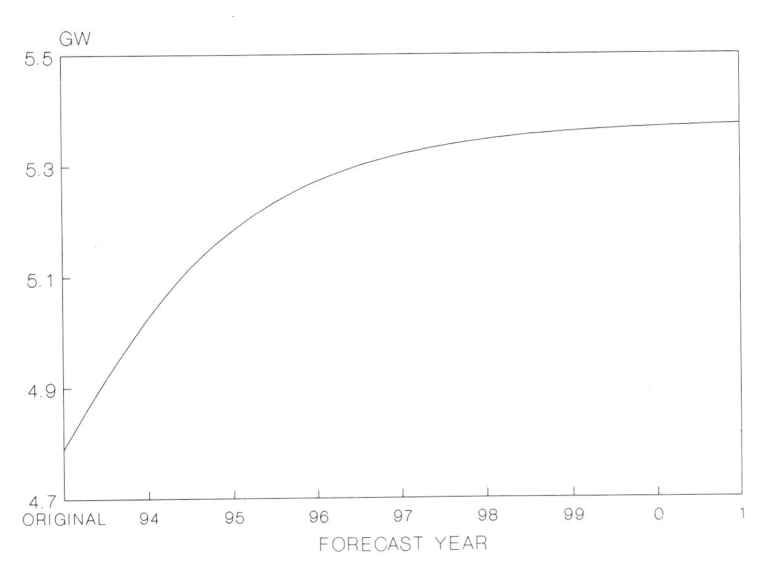

FIGURE 15.15 Year 2007 load demand projections by forecast year.

TABLE 15.6 Final Capacity Addition Plan—High Load Growth Scenario

Year	Additions
1994	100 CC
1995	100 CC
1996	100 CC + 100 GT
1997	100 Coal + 100 GT
1998	100 Coal
1999	100 Coal + 100 CC
2000	100 Coal + 100 CC
2001	100 Coal + 100 CC
2002	100 Coal
2003	100 Coal + 100 coal
2004	100 Coal
2005	100 Coal + 100 coal
2006	100 Coal
2007	100 Coal + 100 coal

results from this evaluation yield a 20-year present-worth expansion cost of $9118.3M.

15.6.2.2 Low Load Growth Scenario. The low load growth scenario evaluation analysis is similar to the high load growth scenario in that a year-by-year utility planning perspective is taken in which a revised forecast is made each year and generation additions are adapted to assure that reserve margin (or reliability) is within the target range of 20%. In the low load growth scenario, generating units are delayed to avoid having excess reserve margins. In this example, it will be assumed that generating units with 40% or more completed construction would not be delayed.

Table 15.7 presents the resulting capacity additions corresponding to the low load growth scenario. Column 1 presents the original planned additions and column 2, the revised capacity addition plan in response to the low load growth scenario. The original installation year, years delayed, and the years of construction at the time of delay are also shown. These last two items are needed when calculating the final plant cost due to the accumulation of AFDC during the construction delay.

Because of construction delays for completing the coal power plants, AFDC accumulates together with price escalation, resulting in higher plant costs. The plant costs of the delayed units are calculated and presented in Table 15.8.

These plant costs and the revised capacity addition schedule (Table 15.7) are input into a power system generation capacity planning model and the system expansion costs calculated. The results from this evaluation yield a 20-year present-worth cost of $7002.4M.

The composite expansion cost of the three load growth scenarios can be

TABLE 15.7 Capacity Additions—Low Load—Growth Scenario

Year	(1) Original Capacity Addition Planned (MW)	(2) Final Revised Capacity Addition Planned (MW)	(3) Original Installation Year	(4) Years Delayed	(5) Years of Construction When Delayed
1994	100 CC	100 CC			
1995	100 CC	100 CC			
1996	100 CC	100 CC			
1997	100 Coal				
1998	100 Coal				
1999	100 Coal				
2000	100 Coal	100 Coal	1997	3	3
2001	100 Coal				
2002	100 Coal	100 Coal	1998	4	2
2003	100 Coal				
2004	100 Coal	100 Coal	1999	5	1
2005	100 Coal				
2006	100 Coal	100 Coal	2000	6	0
2007	100 Coal				

calculated by weighting the expansion cost with the probability of the scenario. The *levelized* cost per kilowatt-hour over the time period is calculated as \$44.2/MWh.

15.6.2.3 Optimal Reserve Margin Example. The previous example was calculated on the basis of a 20% reserve margin policy. The analysis can be repeated to explore the sensitivity of the total costs to alternative generation reserve margin targets in an environment with load uncertainty. The analysis is repeated for reserve margins of 15 and 25%. Figure 15.16 illustrates the results of this analysis.

When the target reserve margin is low (e.g., 15%), consumer outage costs are high. This is especially true in the event of a high load growth scenario.

TABLE 15.8 Plant Costs of Delayed Plant Additions

Unit	Year Delayed	Year of Construction When Delayed	Plant Cost of Delayed Unit: Expressed in % Plant Cost Increase in Final Installation Year Dollars
2000 Coal	3	3	4.9
2002 Coal	4	2	3.4
2004 Coal	5	1	1.5
2006 Coal	6	0	0

FIGURE 15.16 Expansion costs versus reserve margin under uncertainty.

When the target reserve margin is high (i.e., 25%), then consumer costs increase slightly as a result of higher investment charges on the amount of capacity of the higher reserve margin. Thus, there is an optimum value in this power system near the 20% value.

15.6.3 Fuel Cost Uncertainty

Fuel cost, especially petroleum and natural gas cost, is another key forecast uncertainty. Oil prices quadrupled in 1974, doubled in 1979, and plummeted by one-half from 1982 to 1986. Natural gas prices followed a similar course and are projected to maintain a price parity with oil. Coal prices had been under a severe cost ceiling because of competition from low oil and gas prices prior to 1974, but the higher oil and gas prices of the mid-1970s permitted coal prices to escalate. Coal prices have been stable through the 1980s in response to the oil and natural gas price trends.

Most projections of the future price of oil indicate a rapidly escalating price during the 1990s as steadily increasing world demand causes excess world supply capacity to shrink. The low oil and gas prices of the mid 1980s may only exacerbate future price escalation because of the significantly reduced exploration for new fields during the 1980s.

Future fuel prices are very uncertain. In view of historical fuel price changes, future fuel prices are likely to be highly cyclical in nature in contrast to gradual price increases.

The long lead times and long equipment lives of conventional power equipment offer limited flexibility in developing corrective action plans in response to changes in the fuel price environment. One generation concept that has inherent flexibility is the progressive generation concept, as illustrated in Figure 15.17. "Progressive generation" refers to a phased approach to capacity additions. Gas turbines are installed in the first phase to meet short-term peaking capacity needs. In Phase 2, a heat-recovery steam generator (HRSG) and steam turbine are added to obtain the high efficiency of a combined-cycle plant. Finally, the Phase 3 addition of a coal gasifier permits the use of the lower cost coal fuel.

The phased addition approach provides opportunity for adapting to changes. If the price of oil and gas is low, then the Phase 3 addition of a gasifier may be postponed. Should the load growth rate increase, then gas turbines may be installed that can be converted later to combined cycle and integrated gasification combined cycle (IGCC). This flexibility provides value to a utility.

Fuel price uncertainty is typically analyzed on a discrete scenario basis similar to load growth uncertainty procedures.

15.6.3.1 Fuel Price Uncertainty Example. The value of fuel flexibility in the progressive generation concept will be illustrated by a generation planning example that uses the same power system data as in Section 15.1 (Heiges, 1985).

Two generation plans will be contrasted, a progressive generation plan and a plan with only conventional coal unit additions. The capital cost and operating cost parameters for the conventional coal plant and the IGCC plant are assumed to be equal in this example. The fuel price is uncertain, with a 20% probability of a 100% increment in 1997, a 20% probability of a 100% decre-

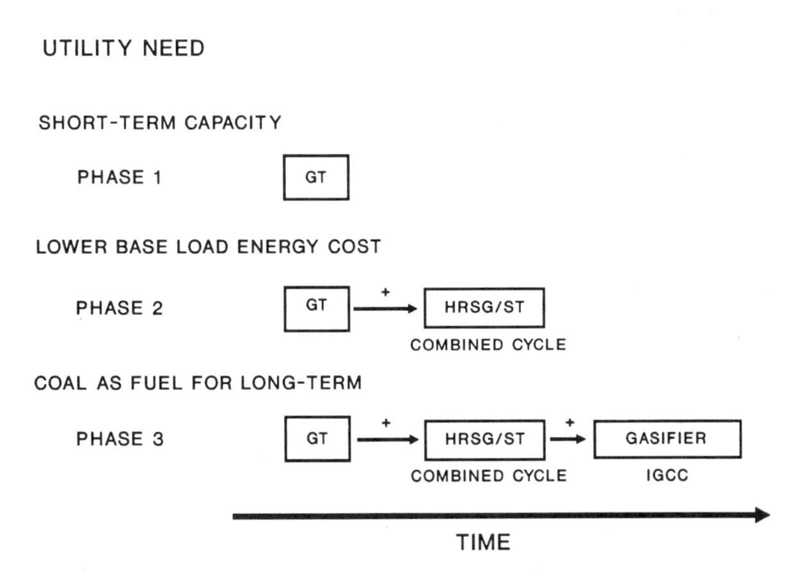

FIGURE 15.17 Progressive generation concept.

ment in 1997, and a 60% probability of the base-case fuel price escalation. The base case has a 6%/year real-price increase for oil and natural gas and a 1.5%/year real-price increase for coal. These 1997 step-change fuel price scenarios are presented for purposes of illustration; a more complete analysis might include other years and other fuel price excursions.

The *base-case* generation addition plans are contrasted in Table 15.9. The conventional coal technology plan uses gas turbine additions in 1993, combined-cycle additions in 1994 through 1996, and conventional coal technology through the year 2000. Coal technology is elected in the late 1990s because of the high real-price escalation of oil and natural gas fuels.

The progressive generation plan also uses gas turbines in 1993 and combined-cycle additions in the mid-1990s. However, because of the capability for future conversion from combined-cycle to coal-based IGCC, a combined-cycle unit is installed in 1997. This unit is converted to an IGCC unit in 2001. Another combined-cycle unit is converted in 2004, and a combined-cycle unit is added in 2006 to balance the generation mix.

The progressive generation plan has a small economic advantage compared to the conventional coal plant plan of $16M (cumulative present worth 1993 $). The $16M benefit has a break-even plant cost value of $27/kW of IGCC. Thus, if the IGCC plant cost $27/kW more than the conventional coal plant, the two plans would have identical cumulative present-worth expansion-plan costs. The IGCC benefit is derived from the capability to convert the

TABLE 15.9 MW Addition Plans: Base-Case Fuel Price

	Conventional Coal Technology Plan			Progressive Generation Plan		
Year	Coal	Combined-Cycle	Gas-Turbine	IGCC	Combined-Cycle	Gas-Turbine
1993			200			200
1994		100			100	
1995		100			100	
1996		100			100	
1997	100				100	
1998	100			100		
1999	100			100		
2000	100			100		
2001	100			200	−100 (conversion)	
2002	100			100		
2003	100			100		
2004	100			200	−100 (conversion)	
2005	100			100		
2006	100				100	
2007	100			100		
Total	1200	300	200	1200	300	200

combined-cycle plant to a coal-fueled IGCC at a later time period when oil and natural gas fuel prices are high.

15.6.3.2 High Fuel Price. Table 15.10 contrasts the conventional coal technology plan with the progressive generation plan for the case of the high fuel cost scenario. The construction lead time for conversion from a combined-cycle unit to an IGCC unit is assumed to be 4 years. Thus, the first conversion of a combined-cycle could be completed in year 2001, based on a 1997 fuel price excursion.

The progressive generation plan proceeds to convert combined-cycle units to coal-fueled IGCC in the 2001–2004 time period. Simple-cycle gas turbines are also converted, first to combined cycle and ultimately to IGCC.

The progressive generation case has a slight economic penalty compared to the conventional coal plan in the case of a high fuel price. The IGCC addition plan has a negative \$4.0M cumulative present worth (1993 \$) benefit, or a break-even capital cost of −\$7/kW of IGCC plant cost.

15.6.3.3 Low Fuel Price. Table 15.11 contrasts the conventional coal technology plan with the progresssive generation plan for the case of the low fuel cost scenario.

It is assumed that a conventional coal plant has a 6-year construction time and the combined-cycle unit conversion to IGCC has a 4-year construction

TABLE 15.10 MW Addition Plans: High Fuel Price Case

	Conventional Coal Technology Plan			Progressive Generation Plan		
Year	Coal	Combined-Cycle	Gas-Turbine	IGCC	Combined-Cycle	Gas-Turbine
1993			200			200
1994		100			100	
1995		100			100	
1996		100			100	
1997	100				100	
1998	100			100		
1999	100			100		
2000	100			100		
2001	100			200	−100 (Conversion)	
2002	100			200	−100 (Conversion)	
2003	100			200	−100 (Conversion)	
2004	100			200	−100 (Conversion)	
2005	100				200	−100 (Conversion)
2006	100			200	−100 (Conversion)	
2007	100				200	−100 (Conversion)
Total	1100	300	200	1300	300	0

TABLE 15.11 MW Addition Plans: Low Fuel Price Case

	Conventional Coal Technology Plan			Progressive Generation Plan		
Year	Coal	Combined-Cycle	Gas-Turbine	IGCC	Combined-Cycle	Gas-Turbine
1993			200			200
1994		100			100	
1995		100			100	
1996		100			100	
1997	100				100	
1998	100			100		
1999	100			100		
2000	100			100		
2001	100					100
2002			100			100
2003			100			100
2004			100			100
2005			100			100
2006			100			100
2007			100			100
Total	500	300	800	300	400	900

time. It is also assumed that a coal plant or an IGCC conversion with less than 10% construction in 1997 would be canceled as a result of the 100% fuel price decrease.

With the advent of low fuel prices, the conventional coal plant plan adds gas turbines from the earliest opportunity in the year 2002. The progressive generation plan adds gas turbines beginning in year 2000.

The progressive generation plan offers a significant benefit in the low fuel price scenario. One contributing factor is the extra combined-cycle unit installed in 1998 that would have been converted to an IGCC if fuel prices were medium or high. However, in the low price scenario this unit is not converted, resulting in savings in conversion investment charges. Another contributing factor is the flexibility of the shorter construction cycle for the modular conversion phases. The progressive generation plan has a total coumulative present-worth savings of \$216M. This savings has an equivalent break-even capital cost value of \$365/kW of IGCC.

15.6.3.4 Fuel Cost Uncertainty—Summary. Table 15.12 presents a summary of the cumulative present-worth benefits of the progressive generation plan compared to the conventional coal generation plan. The progressive generation plan has an expected benefit of \$52M over the 1997–2007 time period.

The benefits also can be presented on a break-even capital cost basis, as shown in Table 15.13. The progressive generation plan has an \$88/kW expected advantage.

TABLE 15.12 Cumulative Present-Worth Benefits of Progressive Generation

	Fuel Price Scenario			Expected (Probability-Weighted)
	Low	Medium	High	
Probability	20%	60%	20%	
IGCC benefit (Millions of present-worth $)	216	+16	−4	$52M

This example has illustrated a methodology for analyzing fuel cost uncertainty. The example concluded an $88/kW benefit to progressive generation, a value that is dependent on the power system and various scenarios. The progressive generation plan also has benefits derived from flexibility in an uncertain load growth environment. The combined fuel and load growth benefits have been studied by Ellert and Heiges (1985).

15.7 SUMMARY

This chapter has expanded on the foundations of optimum capacity planning.

The sensitivity of the optional mix of generation additions was examined. An increase in the reserve margin target or a decrease in load factor results in an increase in peaking capacity mix. Decreasing the costs of money and financing costs leads to an increase in more capital-intensive plant types.

An integrated demand and supply planning methodology framework was presented. The cost criteria objective for combined demand and supply planning is the least-cost cumulative present-worth incremental energy cost (in $/MWh).

The marginal cost of operating a power system is used in electricity rate analysis, generation planning and power purchase planning. Marginal costs can be segmented into short-, intermediate-, and long-term costs.

Planning under uncertain business conditions is an essential planning requirement. A key ingredient is to assess uncertainties and then develop a plan that can be adapted to changing business conditions and can thereby result in least-cost utility operation under a realm of possible business environment outcomes.

TABLE 15.13 Break-even Capital Cost Benefits of Progressive Generation

	Fuel Price Scenario			Expected (Probability-Weighted)
	Low	Medium	High	
Probability (%)	20	60	20	
IGCC benefit ($/kW)	365	27	−7	88 $/kW

REFERENCES

Bulla, D. N., "Marketing Energy Conservation," 1982 Conference on Load Management—Public Utility Commission of Texas, 1982.

Ellert, F. J., and H. H. Heiges, "Economic Benefits of Progressive Generation," Associated Edison Illuminating Companies, April 1985.

Felak, R. P., and H. G. Stoll, "The Economics of System-Wide Power Plant Upgrades," Proceedings of the American Power Conference (April 1987).

Heiges, H. H., "Progressive Generation (PROGEN)—The Economic Answer to Load Growth and Fuel Price Uncertainty," *Frontiers of Power*, Oklahoma State University, 1985.

Hirst, E., "Creating Viable Conservation/Load Management Programs," *Journal of Energy*, Vol. 13, No. 1, January 1988.

McDonough, C. M., N. K. Nair, and H. G. Stoll, "Applying Failure Probability Methods to Power Plant Upgrade/Life Extension Economic Evaluations," 14th INTER-RAM Conference for the Electric Power Industry (May 1987, Toronto).

NERC, "1986 Electricity Supply and Demand," North American Electric Reliability Council, October 1986, Princeton, NJ.

PROBLEMS

1 A utility is considering the replacement of a high-pressure turbine rotor on a 35-year-old generating unit. Engineering estimates that over the next 10 years the current rotor has the following potential failure modes:

Event	Expected Frequency over Next 10 Years	Additional Maintenance Outage Days per Event	Additional Forced Outage Days per Event	Repair Cost per Event (Million $)
Inspection every 3 years	3.0	21		0.04
Cracks near turbine blades	0.2		180	0.10
Surface cracks	0.5		7	0.005

A new replacement turbine rotor is expected to cost $3.0M, have a 25-Btu/kWh efficiency advantage, and not require any inspections during the first 10 years. The unit has a capability of 200 MW, is expected to operate at 50% capacity factor, and has a fuel cost of 2.0 $/MBtu and a heat rate of 10,500 Btu/kWh. The utility power system replacement energy cost averages $25/MWh, and the capacity credit cost is $90/kW/year. What is the investment pay-back for a replacement rotor?

2 Using the data of Problem 1, compute the investment payback if the system replacement energy cost is 35 $/MWh.

16

BULK POWER TRANSMISSION PLANNING

In the history of North American electric power supply, utilities have grown from isolated systems that supply their customers with power from local resources to highly interconnected systems of generating stations and high-voltage transmission lines that encompass large geographic areas and many utilities. Today, there are four major interconnected areas in the contiguous United States and Canada: the Eastern Interconnection, the Western Systems Coordinating Council, the Electric Reliability Council of Texas, and the Hydro-Quebec system.

16.1 INTERCONNECTED BULK POWER SYSTEMS

Within each interconnected area, individual generating systems are connected together by alternating-current (AC) transmission lines. All parties are in synchronous operation, and AC frequency is nearly the same at all points in a given interconnected area. A disturbance that occurs at any location in the interconnected area is felt at all other points in the power system. Transfer of power from one point to another within an interconnected area occurs over many of the transmission lines in the network, not just on the desired contract path.

The transmission facilities of interconnected electric power systems perform several functions simultaneously:

- They provide multiple paths between various generation sources and their loads.

- They provide for power transfers from one geographic area to another to achieve overall system operating economies.
- They interconnect the bulk power facilities of individual utilities so that they can better withstand major system disturbances without interruption of service.

This multiplicity of functions provided by bulk power transmission systems requires close cooperation in system planning and operation among all members of an interconnected area. In addition, the four major interconnected areas have the capability of interchanging power over direct-current (DC) transmission lines, which extends the requirements for coordinated planning and operation.

The North American Electric Reliability Council (NERC) was founded by the electric utility industry in 1968 to promote the reliability of bulk power supply in the electric utility systems of North America. It is concerned with the ability of these systems to supply the aggregate electrical power and energy requirements of consumers at all times. This includes taking into account scheduled and unscheduled outages of system components and the ability of the bulk power electric system to withstand sudden disturbances such as electrical short-circuits and/or unanticipated loss of system components.

This chapter presents the fundamentals of bulk power transmission lines and associated equipment in the steady state. Transfer limits, the concept of surge impedance loading, and steady-state line loadability are discussed. The steady-state AC and DC network equations are derived and solution methods presented. The chapter concludes with an AC transmission planning example. The dynamic aspects of system performance and resulting transmission implications are covered in Chapter 17. Also presented are methods for upgrading transmission capability by compensating lines with series and/or shunt capacitance, adding phase-shifting transformers to enhance power-flow control, uprating transmission voltage level, and converting AC lines to DC.

This material is presented with the assumption that the reader is familiar with basic concepts of power system analysis, including balanced three-phase systems (including Eaton and Cohen, 1983; Gross, 1979; Miller, 1982; Neuenswander, 1971; Stevenson, 1962; and Weedy, 1979). Familiarity with the symmetrical component method of analyzing unbalanced three-phase systems is assumed in Chapter 17.

16.2 TRANSMISSION PLANNING METHODOLOGY OVERVIEW

In transmission planning, there is a tradeoff between the level of detail and the number of possibilities or alternatives studied. Figure 16.1 illustrates a basic approach to transmission planning.

When a transmission planning study is begun, the objective is to evaluate a broad range of alternatives, including the uncertainty of future transmission

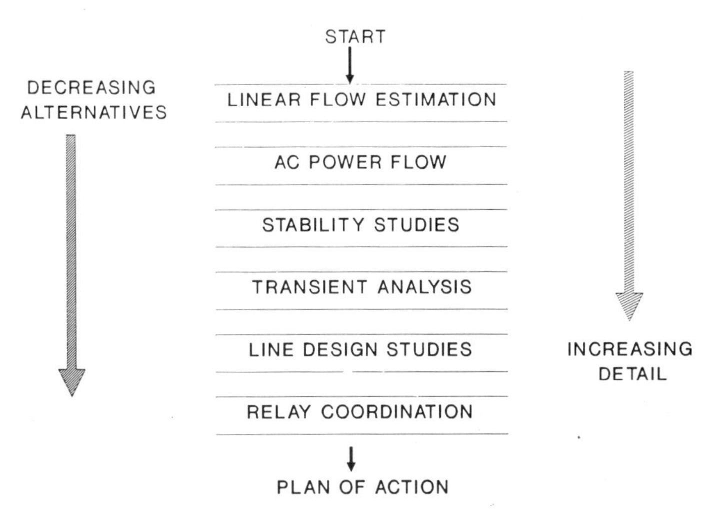

FIGURE 16.1 A basic approach to transmission planning.

configurations of the entire interconnected area. The degree of modeling fidelity is reduced to permit the study to be performed within a reasonable amount of time and at a reasonable cost. As the analysis is conducted and the planning solution focuses on a few alternatives, modeling fidelity may be increased. After the transmission concept plan is confirmed, additional detail is required for the design and coordination studies.

At the beginning of the analysis procedure is the scoping phase. The analysis method could include a simple linear flow estimation of the transmission network, which may be adequate for understanding alternatives. This may be followed by more detailed studies using AC power flow and transient stability analysis to further reduce the alternatives and to focus on the most promising candidates. Finally, line design studies and relay coordination may be used to complete the plan of action.

Although the computer plays a key role in transmission planning, the human planner is even more essential. The planner must select the important factors to study: the load forecast, generation expansion, voltages, and route preference. The planner uses a computer program to synthesize a conceptual network expansion that performs generation dispatches, projects line outage cases, identifies capacity shortages, and selects circuit additions.

16.3 REVIEW OF POWER SYSTEM FUNDAMENTALS

16.3.1 Per Phase Analysis

Balanced three-phase power systems may be analyzed by computing the results for one phase and extrapolating the results to three-phase symmetry.

The single-phase representation is generally based on the line-to-line voltage and the phase current. The voltage and power relationships are:

$$V_{\mathrm{LN}} = \frac{V_{\mathrm{LL}}}{\sqrt{3}}$$

where V_{LN} is the line-to-neutral voltage and V_{LL} is the line-to-line voltage and

$$P_{1\Phi} = V \cdot I^*$$

where $P_{1\Phi}$ is the single-phase power flow, I is the line current, and * denotes the complex conjugate. The three-phase power is

$$P_{3\Phi} = 3\ P_{1\Phi} = \sqrt{3}\ V_{\mathrm{LL}} I^* = 3\ V_{\mathrm{LN}} I^*$$

Three-phase circuit diagrams are typically reduced to a single-phase diagram (referred to as a "one-line diagram").

16.3.2 Per Unit System

Solutions to network problems are expressed in volts, amperes, volt-amperes, ohms (Ω), and angles. It is very convenient, especially in view of the many voltage levels involved, to express these parameters in terms of a per unit system using an absolute base as a reference.

The per unit (pu) parameters are then:

$$Z_{\Omega} = Z_{\mathrm{pu}} \cdot Z_{\mathrm{base}} \tag{16.1}$$

$$I_A = I_{\mathrm{pu}} \cdot I_{\mathrm{base}} \tag{16.2}$$

$$V_{\mathrm{kV}} = V_{\mathrm{pu}} \cdot \mathrm{kV}_{\mathrm{base}} \tag{16.3}$$

$$S_{\mathrm{MVA}} = S_{\mathrm{pu}} \cdot \mathrm{MVA}_{\mathrm{base}} \tag{16.4}$$

where Z = impedance, I = line current, V = line-to-line voltage, S = three-phase power (P, Q, or S), pu denotes per unit, and base denotes reference base.

The choice of the base values is related by Kirchhoff's laws:

$$\mathrm{KV}_{\mathrm{base}} = I_{\mathrm{base}} \frac{Z_{\mathrm{base}}}{1000} \tag{16.5}$$

$$\mathrm{MVA}_{\mathrm{base}} = \mathrm{kV}_{\mathrm{base}} \frac{I_{\mathrm{base}}}{1000} \tag{16.6}$$

Because of the values related in Equations (16.5) and (16.6), only two of the base parameters of Equations (16.1) to (16.4) are independent. The other two base parameters are determined by Equations (16.5) and (16.6). Typically, the MVA_{base} and KV_{base} are chosen.

The Z_{base} is then given by:

$$Z_{base} = \frac{kV_{base} \cdot (1000)}{I_{base}} = \frac{kV_{base} \cdot kV_{base}}{MVA_{base}} \tag{16.7}$$

Often the impedance parameters of a generating unit or transmission line are given on one base while the transmission planning problem is solved on a different base. In this case, Equation (16.7) can be written as

$$Z_{\Omega} = Z_{pu-1} \cdot Z_{base-1} = Z_{pu-2} \cdot Z_{base-2} \tag{16.8}$$

Solving for Z_{pu-2} yields:

$$Z_{pu-2} = Z_{pu-1}\frac{Z_{base-1}}{Z_{base-2}} = \left[\frac{(kV_{base-1})^2}{(kV_{base-2})^2} \cdot \frac{MVA_{base-2}}{MVA_{base-1}}\right] \cdot Z_{pu-1} \tag{16.9}$$

Example. A 500-kV transmission line has the following parameters:

$$Z = R + jX = .022 + j.55\ \Omega \text{ per mile}$$

$$B_{charging} = 7.28 \times 10^{-6}\ \mho \text{ per mile}$$

Compute the per unit impedance and line charging admittance using a 500-kV base and a 100-MVA base.

The impedance base on a 500-kV base is:

$$Z_{base} = \frac{(500)^2}{100} = 2500\ \Omega$$

$$R_{pu} = \frac{.022}{2500} = .0000088 \text{ per mile}$$

$$X_{pu} = \frac{.55}{2500} = .00022 \text{ per mile}$$

$$B_{charging\ pu} = (7.28 \times 10^{-6}) \cdot 2500 = .0182 \text{ per mile}$$

On a network where several voltage levels are present because of transformers, it is necessary to define a kilovolt base for each voltage level of the network. If the nominal voltage levels are chosen as the base and transformers are set at nominal tap position, then 1 pu on one side is equivalent to 1 pu on the other side.

16.4 REVIEW OF LONG AC TRANSMISSION LINES

Analysis of long transmission lines (length > 100 miles) requires special analysis in order to account for phase shifts and wave phenomena. Figure 16.2 shows a model of a differential section of transmission line with an impedance of Z Ω/mile and an admittance of Y mhos/mile. The voltage equation [Equation (16.10)] states that the differential change in voltage is equal to current times impedance times the differential change in X distance. The current equation states that differential change in current is equal to voltage times admittance times the differential distance X. (Note the X is measured positively from the load to the source.)

$$dV = Iz\,dX \qquad \text{and} \qquad dI = Vy\,dX \tag{16.10}$$

Let dX = differential distance (miles), z = impedance per mile ($R + jX$), and y = admittance per mile ($G + jB$). Equation (16.10) can be written as:

$$\frac{dV}{dX} = zI \qquad \text{and} \qquad \frac{dI}{dX} = yV \tag{16.11}$$

Differentiating these equations again yields:

$$\frac{d^2V}{dX^2} = z\frac{dI}{dX} \qquad \text{and} \qquad \frac{d^2I}{dX^2} = y\frac{dV}{dX} \tag{16.12}$$

Substituting for dI/dX and dV/dX from Equation (16.11) yields:

$$\frac{d^2V}{dX^2} = zyV \qquad \text{and} \qquad \frac{d^2I}{dX^2} = yzI \tag{16.13}$$

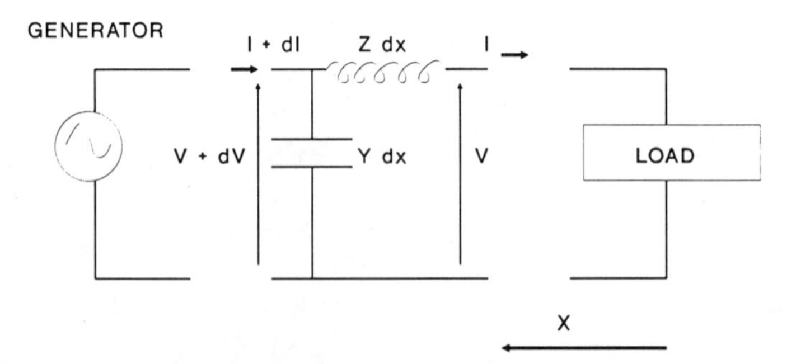

FIGURE 16.2 A model of a differential section of transmission line with an impedance of Z Ω per mile and an admittance of Y ℧ per mile.

It is useful to define the characteristic impedance of the line as:

$$Z_c = \sqrt{z/y} \tag{16.14}$$

and propagation constant as:

$$\gamma = \sqrt{yz} = j \cdot [-j\gamma] \tag{16.15}$$

On a transmission line, z is strongly inductive $(+jX)$ and the admittance is strongly capacitive $(+jB)$. Hence the propagation constant has a large imaginary component and usually a small real component.

The solution to Equation (16.13) is:

$$\begin{aligned} V &= V_R \cosh \gamma X + I_R \cdot Z_c \sinh \gamma X \\ &= V_R \cos(-j\gamma X) + jI_R \cdot Z_c \sin(-j\gamma X) \end{aligned} \tag{16.16}$$

$$\begin{aligned} I &= I_R \cosh \gamma X + \frac{V_R}{Z_c} \sinh \gamma X \\ &= I_R \cos(-j\gamma X) + j\frac{V_R}{Z_c} \sin(-j\gamma X) \end{aligned} \tag{16.17}$$

where V_R is the complex voltage at $X = 0$ (load end) and I_R is the complex current at $X = 0$ (load end).

Substitution of Equations (16.16) and (16.17) into Equations (16.13) or (16.11) will verify the solution (Stevenson, 1962).

It is useful to construct a π equivalent circuit of a long transmission line as shown in Figure 16.3.

The equation for the equivalent circuit is:

$$V_S = \left(\frac{Z_{line} Y_{line}}{2} + 1\right) V_R + Z_{line} I_R \tag{16.18}$$

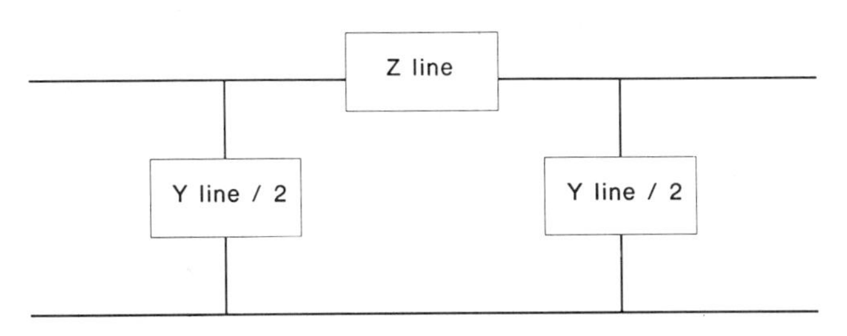

FIGURE 16.3 Review of a π-equivalent circuit of a long transmission line.

The equivalent circuit must give the same V_S as the long transmission line. Thus, the corresponding terms of Equation (16.18) must be equal to the terms of Equation (16.16). This yields:

$$\begin{aligned} Z_{\text{line}} &= Z_c \sinh \gamma \ell \\ &= \sqrt{\frac{z}{y}} \sinh \gamma \ell \\ &= z\ell \frac{\sinh \gamma \ell}{\gamma \ell} = Z\left(\frac{\sinh \gamma \ell}{\gamma \ell}\right) \end{aligned} \tag{16.19}$$

where $z\ell = Z =$ total line impedance.

The coefficient for V_R must also match in Equations (16.18) and (16.16). Consequently:

$$\frac{Z_{\text{line}} Y_{\text{line}}}{2} + 1 = \cosh \gamma \ell \tag{16.20}$$

$$\frac{Z_c \sinh \gamma \ell \; Y_{\text{line}}}{2} + 1 = \cosh \gamma \ell \tag{16.21}$$

Rewriting these equations yields:

$$\begin{aligned} \frac{Y_{\text{line}}}{2} &= \frac{1}{Z_c} \frac{\cosh \gamma \ell - 1}{\sinh \gamma \ell} \\ &= \frac{1}{Z_c} \tanh \frac{\gamma \ell}{2} \\ &= \sqrt{\frac{y}{z}} \tanh \frac{\gamma \ell}{2} \\ &= \frac{y\ell}{2} \frac{\tanh(\gamma \ell/2)}{\gamma \ell/2} = \frac{Y}{2} \frac{\tanh(\gamma \ell/2)}{\gamma \ell/2} \end{aligned} \tag{16.22}$$

where $y\ell = Y =$ total line admittance.

The equivalent parameters corresponding to the π-equivalent line of Figure 16.3 are:

$$Z_{\text{line}} = Z\left[\frac{\sinh \gamma \ell}{\gamma \ell}\right] \tag{16.23}$$

$$\frac{Y_{\text{line}}}{2} = \frac{Y}{2}\left[\frac{\tanh(\gamma \ell/2)}{\gamma \ell/2}\right] \tag{16.24}$$

The bracketed hyperbolic terms are small adjustment factors necessary for long-transmission-line analysis.

Series resistance and conductive admittance are typically small. If they are assumed to be zero, then the characteristic impedance becomes equal to the square root of L divided by C, where L is the inductance of the transmission line and C is the capacitance. The propagation constant then become j times 2π times the frequency (e.g., 60 Hz) times the square root of the product of L time C.

Table 16.1 presents an example that illustrates these adjustment factors for long lines.

From the example, for a 345-kV line at 200 miles, the impedance of the line is 3% lower and the admittance is 1% higher by including these adjustment factors than by not including them. This can have an influence on the loading of a long transmission line and, therefore, it is necessary to include these adjustments for transmission line lengths greater than 100 miles.

16.5 SURGE IMPEDANCE LOADING

Figure 16.4 presents a π diagram of a transmission line that is terminated with its characteristic impedance. When line losses are neglected, the characteristic impedance of a line is referred to as its "surge impedance."

Surge impedance loading is defined as the power delivered by a line to a purely resistive load equal to its surge impedance. Surge impedance loading can be calculated by using Equation (16.25) to calculate the line current.

$$I_{\text{line}} = \frac{V_{\text{line}}}{\sqrt{3}\, Z_c} \tag{16.25}$$

TABLE 16.1 Adjustment Factors for Long Lines

Example: 345-kV line (neglect losses)

$z = j\ .0005$ pu/mile (100-MVA base)

$y = j\ .0080$ pu/mile (100-MVA base)

$Z_c = \sqrt{.0005/.0080} = .25$ pu

Base impedance $= (345)^2/100 = 1190$

$Z_c = .2500 \cdot 1190 = 298\ \Omega$

$\gamma = j(\sqrt{(.0005 \cdot .0080)} = j\ .002$ mile $^{-1}$

Adjustment Factors	Line Length in Miles 100	200	400
$(\sinh \gamma\ell)/\gamma\ell$ factor	0.9933	0.9735	0.8966
$(\tanh \gamma\ell/2)/\gamma\ell/2$ factor	1.0033	1.0135	1.0569

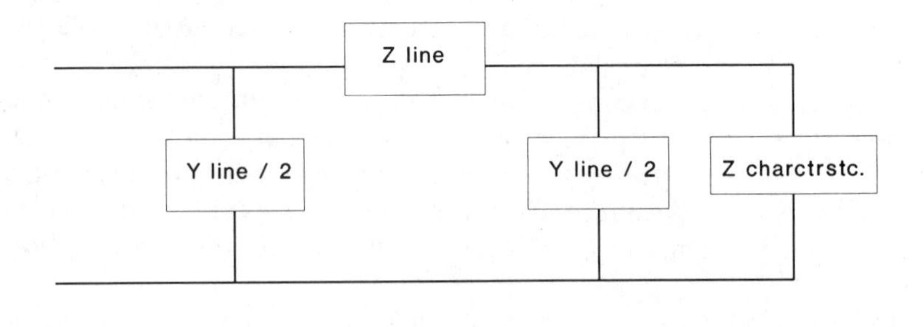

FIGURE 16.4 Surge impedance loading.

where I_{line} = line current on one phase of a three-phase system
V_{line} = line-to-line voltage on a three-phase system
Z_c = characteristic impedance of the line with no losses (surge impedance)

The power through the load is the product of current times voltage because Z_c is purely resistive. Surge impedance loading (SIL) for all three power phases is:

$$\text{SIL} = \sqrt{3}\left[\frac{V_{line}}{\sqrt{3}\,Z_c}\right] V_{line} = \frac{V_{line}^2}{Z_c} \tag{16.26}$$

For a 345-kV line with a Z_c = 298 Ω, the surge impedance loading is:

$$\text{SIL} = \frac{(345)^2}{298} = 399 \text{ MW}$$

Surge impedance loading for typical voltage levels is presented in Table 16.2. Also shown is the maximum thermal rating from Table 11.5. (Thermal limits are discussed in Section 16.8.1.) The maximum thermal rating is approximately three times the surge impedance value but varies with the specific conductor size and line design.

When a lossless line is terminated in its characteristic impedance [$V_R = I_R \cdot Z_c$ in Equation (16.16)], the voltage magnitude is constant at all points along the line. In Equation (16.16), then:

$$V = V_R\,[\cos(-j\gamma X) + j\,\sin(-j\gamma X)] = V_R e^{\gamma X}$$

which is a phasor of magnitude V_R with a phase shift down the transmission line.

When line loading is less than surge impedance loading, $R > Z_c$, then the voltage magnitude increases toward the load. When line loading is greater than

TABLE 16.2 Typical Transmission-Line Loadings

Voltage level: kV	138	230	345	500	765
Surge impedance loading: MVA	64	132	390	910	2210
Maximum thermal: MVA	196	396	1187	2756	7017
Ratio: maximum thermal/SIL	3.06	3.0	3.04	3.03	3.18

surge impedance loading, $R < Z_c$, then the voltage magnitude on a line decreases toward the load.

16.6 STEADY-STATE TRANSMISSION SYSTEM MODELS

Transmission lines are generally represented by a π circuit as shown in Figure 16.5. For short- and medium-length lines, the π equivalent values can be computed by using per mile quantities times the length of the line in miles. For lines longer than 50–100 miles and for cables, the π equivalent values must be computed according to the distributed parameter theory (as computed in Section 16.4).

Transformers are usually represented as a series impedance plus an ideal tap ratio, as shown in Figure 16.6. For some purposes, a shunt branch representing the magnetizing reactance is also included.

Generators are typically connected to the high-voltage transmission network through a generator step-up (GSU) transformer. The generator voltage regulator is configured to regulate the high voltage side of the GSU. Thus, for steady-state studies, the generator is represented as producing active power to the generator terminal bus and regulating the GSU high-side bus voltage by providing leading or lagging reactive power. The bus voltage will be held to the specified value, provided the required reactive power is within the capability of the generator. Figure 16.7 presents a typical generator reactive capability curve. Three limit lines are present. Near rated generator MVA, the magnitude of the stator current limits reactive capability. This is expressed by:

$$I_{\text{limit}} = \frac{\sqrt{P^2 + Q^2}}{V}$$

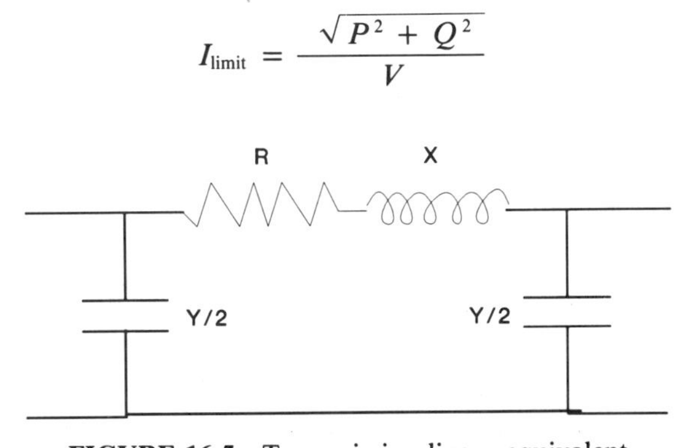

FIGURE 16.5 Transmission line π equivalent.

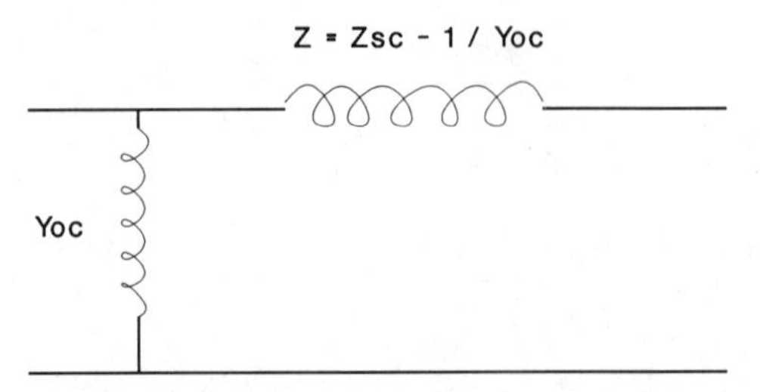

FIGURE 16.6 Transformer equivalent.

Reactive output is also limited by a field-heating limit for lagging reactive power ("overexcited" condition) and stator-core end-heating limit for leading reactive power. In most cases, the generator is sized such that when the turbine produces full net-power output, the generator can produce full net-power output at the rated power factor (typically 85–90%). Thus, 85–90% active generator power is equivalent to 100% turbine power. The generating unit capability curve of Figure 16.7 can be approximated by a rectangle:

$$0 \leq P \leq .85,$$

$$-.40 \leq Q \leq .55$$

in steady-state power-flow analysis.

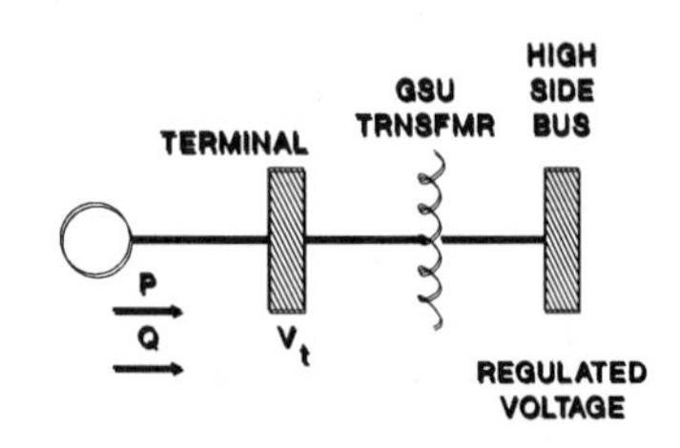

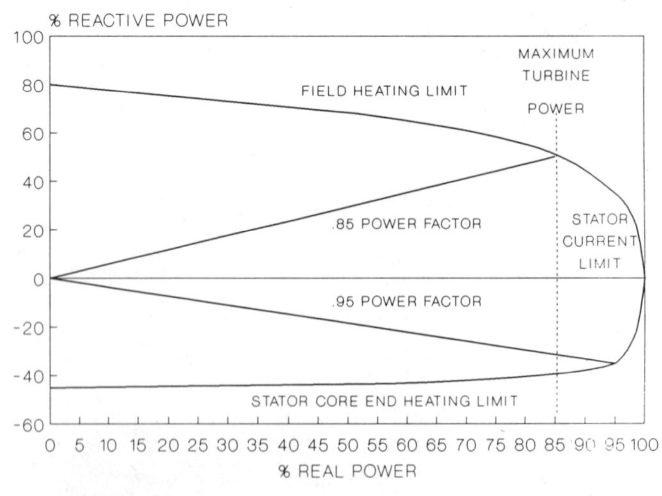

FIGURE 16.7 Typical generator reactive capability curve.

16.7 TYPICAL TRANSMISSION PARAMETERS

Figure 16.8 illustrates the trend of transmission voltages during this century. Transmission voltages have historically doubled every 20 years. However, 765 kV appears to be a plateau in this curve. That rating may remain constant for several years because of a variety of factors, including a reduced rate of load growth and environmental considerations.

Typical transmission-line characteristics per mile for various voltage levels are presented in Table 16.3. These parameters are given on a 100-MVA base and the voltage base of the line kilovolt level. While line reactance in ohms per mile decreases only slightly at higher voltage levels, the per unit values exhibit a significant decrease with voltage level. Line resistance in ohms per mile decreases with voltage level as a result of increased conductor bundle size. When expressed on a per unit base, line resistance decreases very rapidly with voltage level, since the base ohms vary as the square of the voltage level.

Table 16.4 presents typical transformer reactances. Although only one value is shown, transformers have a broad range of values, depending on their design. Note that reactance is based on the transformer base. For power system analysis, the values must be expressed on a common system base.

16.8 OVERVIEW OF TRANSMISSION-LINE LOADABILITY

Transmission-line power loading is generally governed by four considerations:

- Thermal limits due to I^2R heating on the line. Limits are set to avoid damaging the conductor due to annealing and to avoid excessive sag,

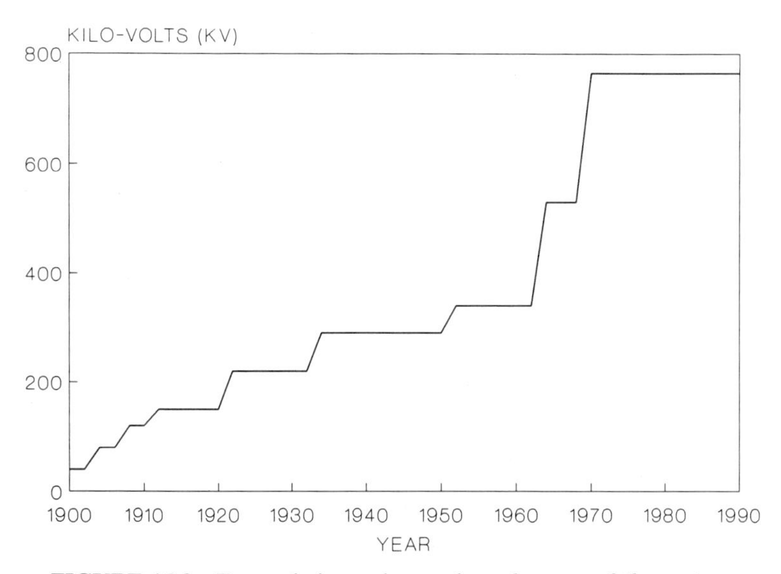

FIGURE 16.8 Transmission voltages since the turn of the century.

TABLE 16.3 Typical Transmission-Line Characteristics (100-MVA Base)

kV Level	Reactance per Mile (pu)	Resistance per Mile (pu)	Capacitance per Mile (pu)	Z_{base} (Ω)
230	.00150	.00020	.0030	529
345	.00050	.000050	.00865	1190
500	.00024	.000096	.0180	2500
765	.000091	.0000035	.0470	5852

which would encroach on the required clearance. Severe overloading can cause transmission lines to stretch or break.

- Voltage limits at the receiving end. The acceptable target is typically 95–98% for normal conditions. Voltage collapse must also be avoided for maximum credible contingencies.
- Stability limits, in which the power angle between the sending and receiving buses is limited by the dynamics of regulating systems.
- Transient stability limits, based on system response to large disturbances.

On transmission lines that are less than 50 miles, the maximum loading is usually limited by thermal considerations. Lines 50–200 miles long are often limited by voltage considerations unless shunt compensation is provided. Long lines are likely to be limited by stability considerations, which is the subject of Chapter 17.

The next section discusses thermal limits, and the subsequent section presents an overview of steady-state stability limits. Voltage considerations are discussed in Section 16.9.

16.8.1 Transmission-Line Thermal Limits

The thermal limits of a line can be determined if the heat balance of the line is known. (See "Current Temperature Characteristics of Aluminum Conductors," *ALCOA Conductor Engineering Handbook*.)

$$q_c + q_r = q_s + I^2R \tag{16.27}$$

TABLE 16.4 Typical Transformer Reactances (per Unit on Transformer Base)

High-Side kV Level	Reactance
230	.11
345	.13
500	.15
765	.17

where q_c = convected heat loss to the surrounding air in W/foot
q_r = radiated heat loss from the line in W/foot
q_s = solar heat input from the sun in W/foot
I^2R = heat input to the line from resistance heating of the line where I is in amperes and R is in Ω/foot at 60 Hz

This equation can be solved to obtain the maximum current capability of the line:

$$I_{max} = \left(\frac{q_c + q_r - q_s}{R}\right)^{1/2} \tag{16.28}$$

The convected heat-loss component for a given line configuration is typically calculated on the basis of an empirical correlation of heat loss with conductor diameter, wind speed, conductor (line) temperature, ambient temperature, and air density. It is typical practice to assume a wind velocity of 2 feet/second, 100°C conductor, and 40°C ambient. Different conditions (typically winter) lead to higher maximum current limits.

Radiated heat loss depends on the fourth power of the conductor temperature and the ambient temperature. Conductor emissivity for aluminum can range from .23 for a new conductor to .90 for a conductor well blackened by years of service. The heat gain from the sun depends on the sun's altitude, line direction, and conductor solar absorption coefficient.

Typical contributions of these three heat terms for a 345-kV line with 100°C conductor, 40°C ambient, 1.108-inch diameter conductor, resistance of .0290 Ω/1000 feet are:

	Sea Level	10,000-Foot Altitude
Convective heat loss	25.03 W/foot	20.68 W/foot
Radiative heat loss (.5 emissivity)	7.41 W/foot	7.41 W/foot
Solar heat gain (.5 absorption coefficient)	4.30 W/foot	5.37 W/foot
Maximum current	993 A	885 A

When applied at higher altitudes, a conductor has a lower maximum current because convective heat loss is less, as a result of decreased air density, and solar gain is larger, as a result of higher solar intensity. These factors are often offset by a decreased maximum ambient temperature.

For a given set of wind, temperature, altitude, and emissivity conditions, maximum current can be correlated with conductor diameter. For a 40°C ambient, 100°C conductor at sea level in a 2 feet/second wind, the maximum current for a multiple layer (aluminum conductor-steel reinforced) ACSR conductor is:

$$I_{max} = 870D^{1.3} \tag{16.29}$$

where D is the conductor diameter in inches.

Table 16.5 presents typical ACSR transmission-line characteristics for several voltage classes at the previously stated conditions.

In addition to these steady-state thermal limits, one or more short-term (10 minutes to 1 hour) limits are often computed for use during emergency conditions. The short-term limits are applicable for examining the adequacy of the transmission system during emergency or contingency conditions. The short-term limit may be approximately 20% higher than the normal rating (see Davidson et al., 1969). The short-term rating is calculated using a time dynamic model of a transmission-line segment. Higher short-term currents are permissible if the maximum temperature of the conductor does not exceed its limit.

16.8.2 Steady-State Transmission-Line Stability Overview

The steady-state stability limit can be approximated by assuming the transmission line is represented as illustrated in Figure 16.9. Further assume that the load (receiving) bus can be regulated to maintain a one-per-unit voltage level.

The complex current is equal to the difference between the sending and receiving voltages divided by the impedance:

$$\overline{I}_L = \frac{\overline{E}_G - \overline{E}_L}{jX} \tag{16.30}$$

The complex load power is the product of the voltage times the conjugate of the current. Assuming that the load is at a zero reference angle and the generator at angle δ yields:

$$\text{Complex power} = \overline{E}_L \overline{I}_L^* = \frac{(E_L)(E_G)}{X} \underline{/90^\circ - \delta} - \frac{(E_L)^2}{X} \underline{/90^\circ} \tag{16.31}$$

where * indicates the complex conjugate operation.

The active power can be expressed as the real part of the complex power:

$$P = \frac{(E_L)(E_G)}{X} \cos(90^\circ - \delta) - \frac{(E_L)^2}{X} \cos(90^\circ)$$

$$P = \frac{(E_L)(E_G)}{X} \sin \delta \tag{16.32}$$

TABLE 16.5 Typical Steady-State Thermal Transmission-Line Limits

Voltage Level: kV	69	115	230	345	500	765
Number of conductors	1	1	1	2	3	4
Conductor diameter (inch)	.72	1.10	1.10	1.108	1.165	1.382
Maximum current/conductor (A)	568	985	985	994	1061	1324
Bundle maximum current (A)	568	985	985	1988	3183	5296
Maximum MVA	68	196	396	1187	2756	7017

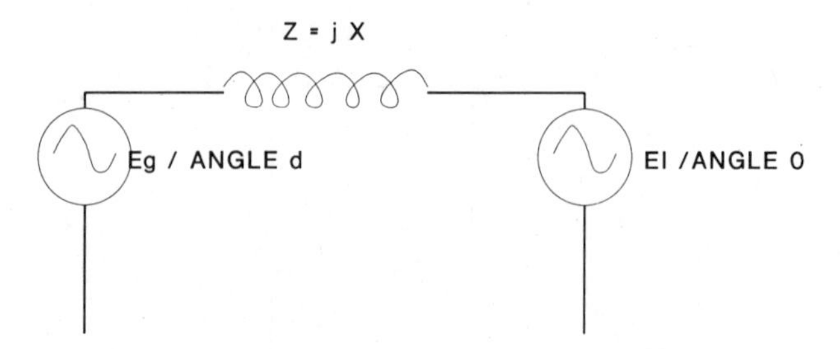

FIGURE 16.9 Steady-state stability limit.

The reactance of the line is the product of the length of the line in miles times the reactance per mile (neglecting long-line effects). Thus, the power that can be transmitted is inversely proportional to the length of the line.

Note also from Equation (16.32) that the sin(δ) is largest when δ is 90°. If the load and generation buses are at one per unit voltage, then the power transfer follows the power angle curve as illustrated in Figure 16.10. If transmission lines were operated at a 90° angle, then there would not be any margin available in the event of a generation or transmission-line outage. Consequently, transmission lines must be operated with stability margin.

For transmission planning, a 30% steady-state stability margin for normal transmission-line loading is typically used in the "normal state" case, where all generator units and all transmission lines are in service.

Because transmission lines are interconnected in a power system network, power system network considerations also influence transmission-line loading limits. Subsequent sections expand on these considerations.

16.9 TRANSMISSION-LINE LOADABILITY

In the previous section, the steady-state stability loading of a transmission line was introduced under the assumption that the receiving voltage was capable of being regulated to 1.0-pu unit. Voltage regulation requires that there be a variable reactive power source such as a generating unit, synchronous condenser or static var system.

When the receiving bus is not firmly supported, then line loadability is further reduced (St. Clair, 1953). This section describes these cases in further detail (Dunlop et al., 1979). Figure 16.11 illustrates the line.

The equation governing the flow to the receiving system is:

$$P_2 + jQ_2 = E_2([E_1 - E_2]\, y_{12})^* - E_2(E_2 y_{22})^* \tag{16.33}$$

Assume that the line is lossless, then:

$$y_{12} = \frac{-j}{X_{\text{line}}} \tag{16.34}$$

POWER ANGLE CURVE

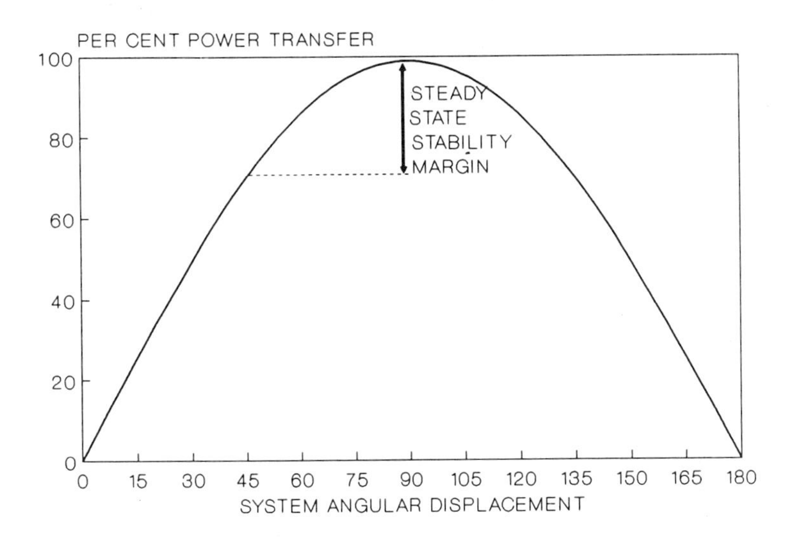

STEADY STATE STABILITY MARGIN CURVE

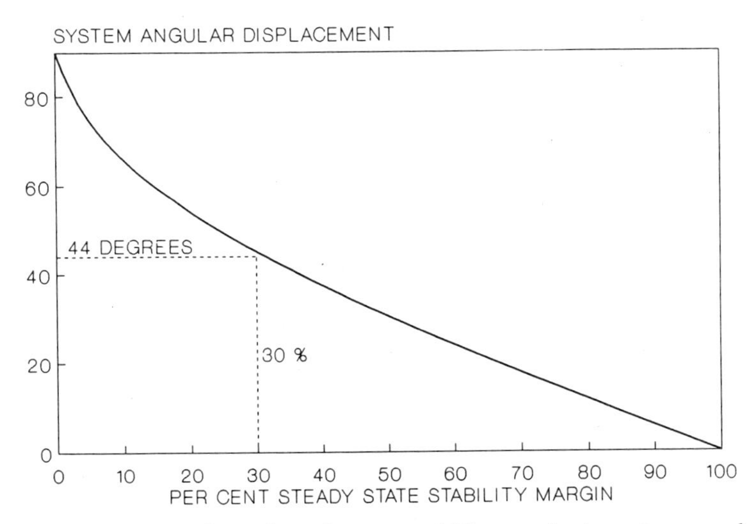

FIGURE 16.10 Conversion of steady-state stability margin to system angular displacement.

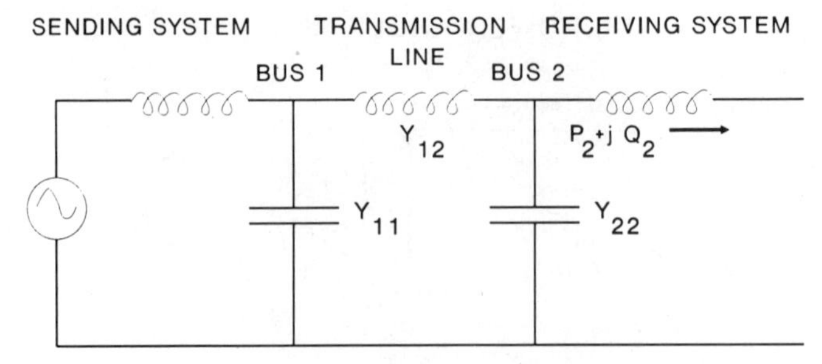

FIGURE 16.11 Line loadability.

$$y_{22} = \frac{jB_{\text{line}}}{2} \tag{16.35}$$

$$P_2 + jQ_2 = [|E_1|\,|E_2|\sin(\theta_1 - \theta_2) + j|E_1|\,|E_2|\cos(\theta_1 - \theta_2) - |E_2|^2] \cdot \frac{1}{X_{\text{line}}} + j\,\frac{B_{\text{line}}}{2}\,|E_2|^2 \tag{16.36}$$

Taking the real and imaginary parts of (16.36) yields:

$$P_2 = \frac{V_1 V_2}{X} \sin \delta \tag{16.37}$$

$$Q_2 = \frac{V_2}{X} [V_1 \cos \delta - AV_2] \tag{16.38}$$

where B = line capacitance
X = reactance of line
$V_2 = |E_2|$
$V_1 = |E_1|$
$\delta = \theta_1 - \theta_2$
$A = [1 - \frac{BX}{2}]$

Letting $\cos(\phi)$ be the power factor of the receiving load yields:

$$\cos \phi = \frac{P_2}{\sqrt{P_2^2 + Q_2^2}} \tag{16.39}$$

$$\tan \phi = \frac{Q_2}{P_2} \tag{16.40}$$

Substituting into Equations (16.37) and (16.38) yields:

$$P_2 = \frac{V_1 V_2 \sin \delta}{X} = \frac{1}{\tan \phi} \frac{V_2}{X} [V_1 \cos \delta - AV_2] \tag{16.41}$$

or

$$V_1 \sin \delta = \frac{1}{\tan \phi} [V_1 \cos \delta - AV_2] \tag{16.42}$$

$$V_2 = \frac{1}{A} [V_1 (\cos \delta - \sin \delta \tan \phi)] \tag{16.43}$$

The power transferred is given by Equation (16.37):

$$P_2 = \frac{V_1^2}{A \cdot X} \sin \delta [\cos \delta - \sin \delta \tan \phi] \tag{16.44}$$

Source voltage is typically set at the high end of the source transformer tap or regulator setting, typically 1.05-pu voltage. The load voltage under loading could drop to .95 pu. This is acceptable because at .95 voltage, the transformer taps would be at the low end of their range setting and would be able to provide rated voltage on the secondary. The bus load demand would not change for receiver voltages in the .95–1.05 range. A typical transmission system design criterion is to maintain bus voltage levels not lower then .95 pu during normal steady-state operation.

Figure 16.12 illustrates the power flow and receiving voltage on a 345-kV line 100 miles long when the receiving load has a 10° lagging power-factor angle. Voltage and power flow are dependent on the power angle, δ, across the transmission line. Power flow is expressed in per unit of surge impedance loading.

The power flow has the sinusoidal character as illustrated in Section 16.6.2. Receiving load-bus voltage decreases as more power is transferred across the line. Line loadability is limited when the receiving load voltage decreases to the .95 voltage limit, which occurs at a power angle of 20°. The power transferred is 1.8 times the surge impedance loading.

Figure 16.13 illustrates a 345-kV line 300 miles long. The receiving load voltage intersects with the .95 limit at an angle of 35°. The power flow is 1.07 times the surge impedance loading.

Note that at very low power flows (typically during low load periods), or small power angles, the voltage on the receiving load bus is greater than 1.05 as a result of line charging capacitance. However, only a small amount of lagging vars are required for compensation. On a practical basis, this can be controlled in several ways, including: operating generating units on the receiv-

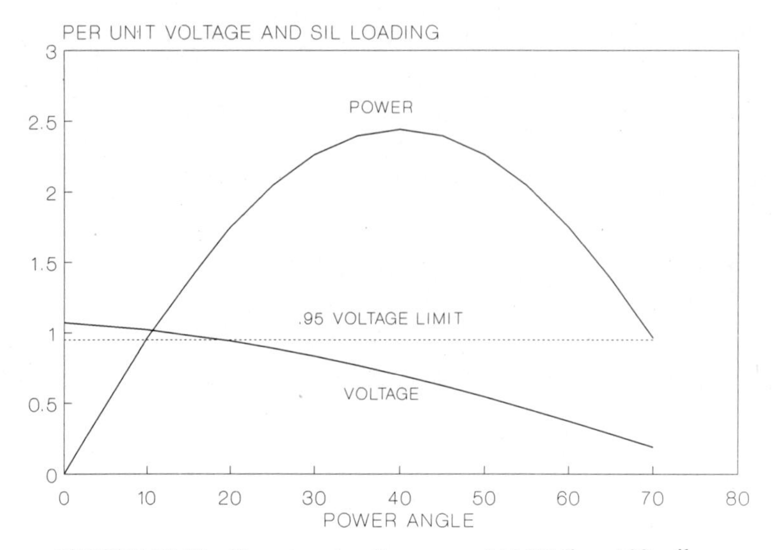

FIGURE 16.12 Power and voltage on a 345-kV line, 100 miles.

ing bus to absorb vars to counteract the capacitive vars, by installing shunt reactors on the line, or by switching off some transmission lines to increase loading on those remaining connected.

Figure 16.14 presents the case of a 500-mile, 345-kV line. The .95 voltage limit is reached at a power angle of 65°. However, this angle is greater than the normal steady-state stability limit angle of 44°, based on a 30% steady-

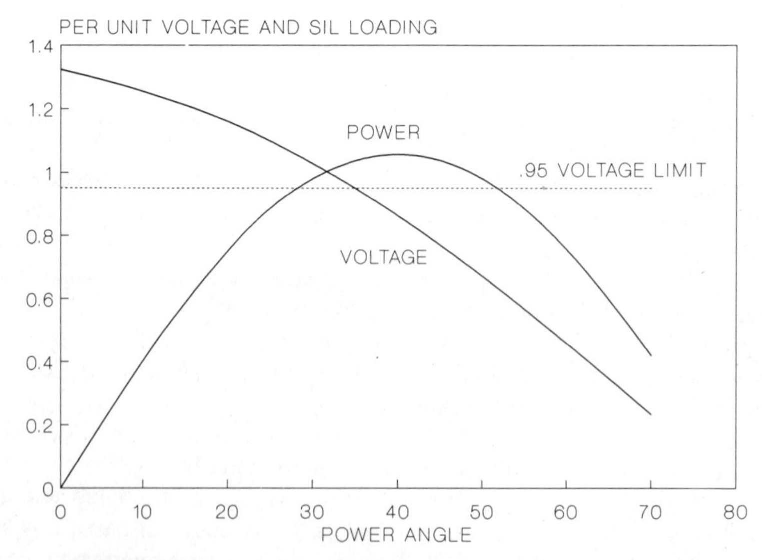

FIGURE 16.13 Power and voltage on a 345-kV line, 300 miles.

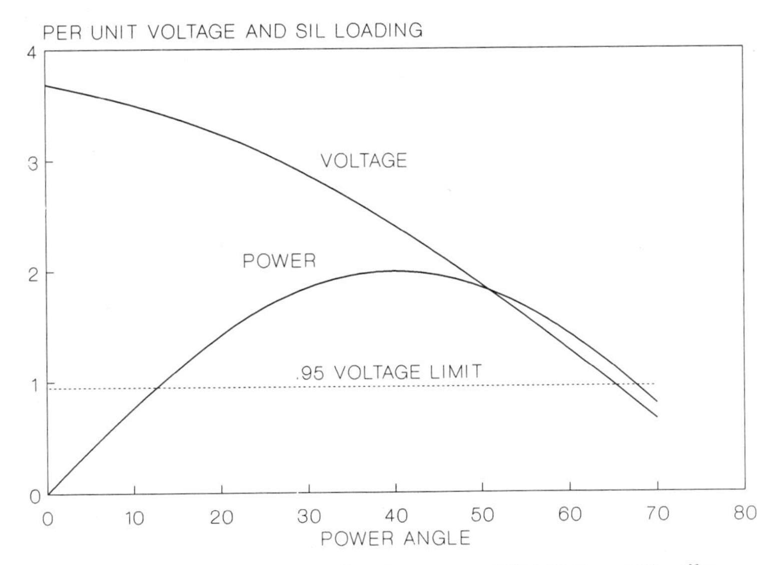

FIGURE 16.14 Power and voltage on a 345-kV line, 500 miles.

state stability margin. For long line distances, the steady-state stability angle is the limiting constraint.

For medium-length lines (less than 300 miles) voltage consideration is a limiting constraint. Figure 16.15 illustrates these constraints for a 200-mile, 345-kV line. The receiver load voltage at maximum power is .78 pu and maximum power is 1.34 SIL. Imposing a 30% steady-state voltage stability margin would

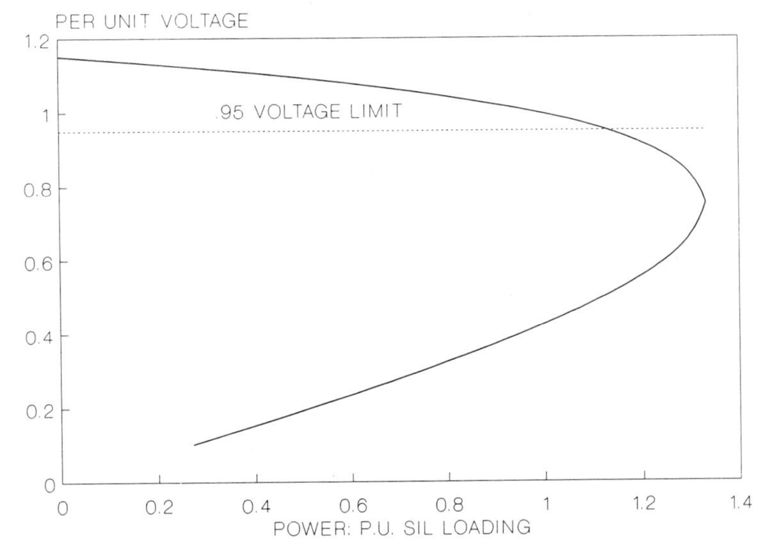

FIGURE 16.15 Voltage stability 345-kV, 200 miles.

lead to a line loadability rating of 1.1 SIL. Coincidently, this is also in the range of the .95 voltage constraint limit.

These line loadability considerations are generalized in Figure 16.16 (Dunlop et al., 1969). Line loadability is expressed in terms of a surge impedance loading multiplier and applies for any voltage class. For short lines (less than 50 miles) line loadability is limited by thermal characteristics of the line or by terminal capability. Lines 50–300 miles long are limited by voltage conditions, whereas long lines are limited by angular stability. This line loadability curve serves as a guide in loading transmission lines. Actual line loadability, however, is dependent on the network. When network conditions are different from those used to derive the line loadability guide shown in Figure 16.16, then a different line loadability is appropriate. If required, shunt and series compensation (discussed in Chapter 17) and intermediate voltage support allow even long lines to be loaded near their thermal limits.

16.10 TRANSMISSION OUTAGE CASCADING AND ISLANDING

Transmission outage cascading is a significant concern in transmission planning. Transmission cascading generally begins with a transmission-line outage. The outage of one line in a transmission system then leads to overloading of another line. This line could be tripped out of service by a relay that recognizes this overloaded condition. In this event, two lines are out of service and can cause further overloads on other lines. This process of a one-line outage causing one or more other lines to fail is called a "cascading outage." Cascading

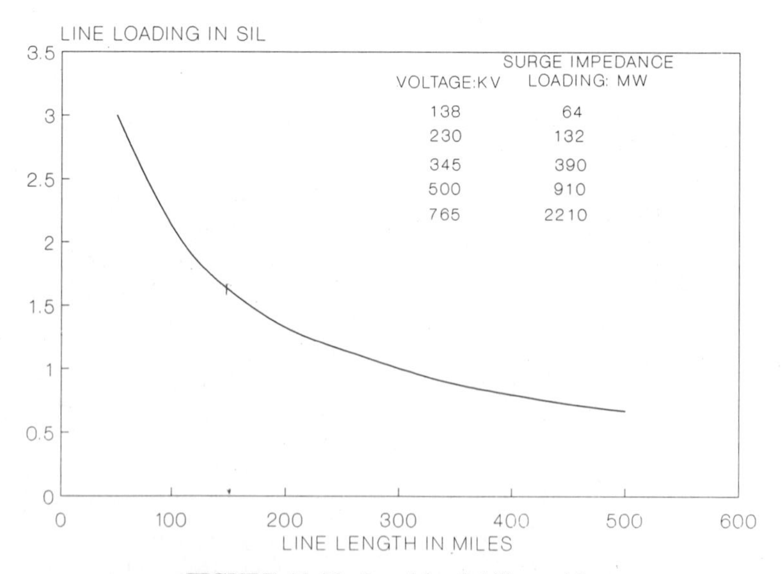

FIGURE 16.16 Load loadability guide.

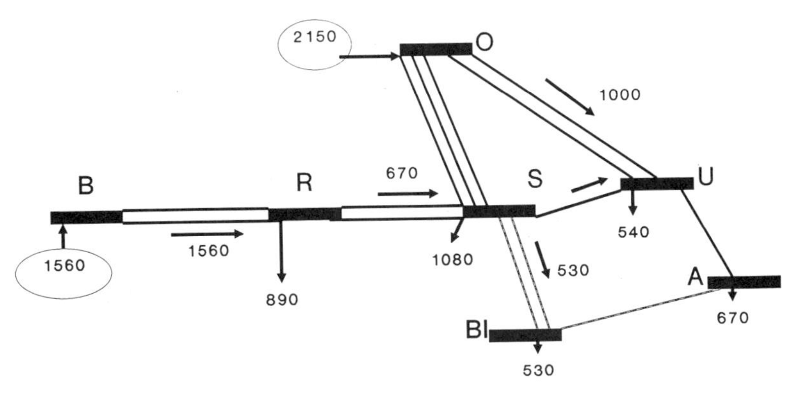

FIGURE 16.17 Hypothetical transmission system—base case.

sometimes continues until large amounts of load are shed. The power system may no longer be interconnected and becomes instead a group of isolated "islands."

A hypothetical transmission system is shown in Figures 16.17 through 16.19. In this example, the lines from *B* to *R* are carrying 1560 MW; the right-of-way is rated for 1560 MW. The load at *R* is 890 MW; consequently, the power lines then transmit 670 MW on to *S*. The balance of the network to the east of *S* is dominated by the generation in O and the loads on the other buses.

Consider what happens in the event of a line outage. Suppose that one of the B to R lines trips out as shown in Figure 16.18. Then 1560 MW is being forced to flow from *B* to *R* along a right-of-way that is now only rated for 780 MW. Most likely, this line would not be able to support the 1560-MW load and, after several consequential events, the line would trip. As a result of generation overspeed the generation at *B* would then trip out. The result, shown in Figure 16.19, is a net generation in the system of 2150 MW and a

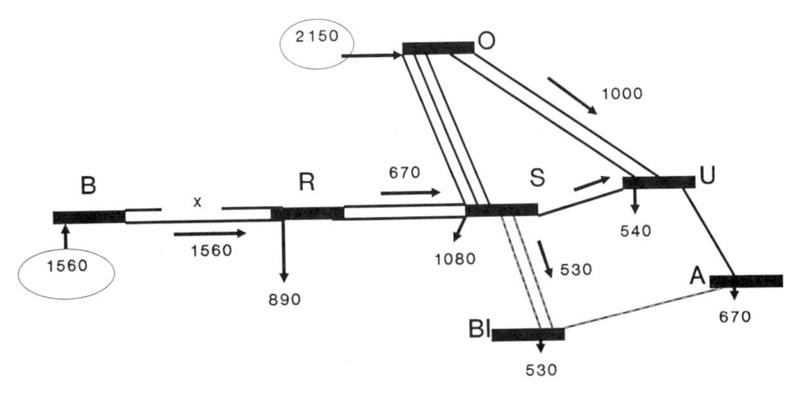

FIGURE 16.18 Transmission system with one line tripped.

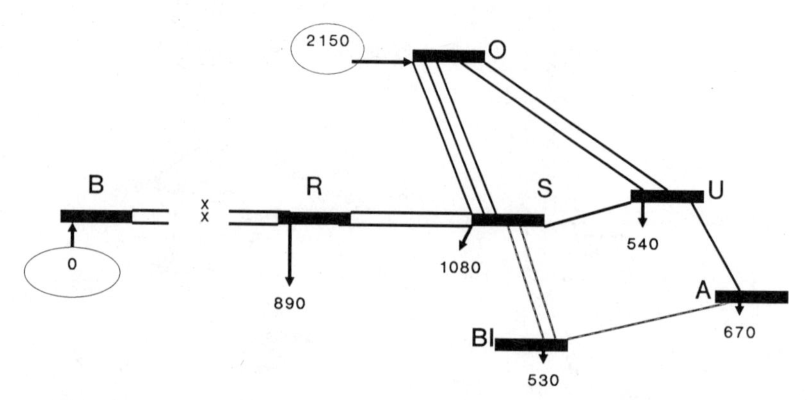

FIGURE 16.19 Transmission system following second line trip-out.

load of 3710 MW, resulting in a deficiency of 1560 MW, or 42%. Because of low frequency, the generation-load imbalance would lead to load shedding by the automatic relays. The result would be an islanded power system capable of serving only 58% of the load.

In this simple illustration, one line outage cascaded to a two-line outage and an islanded power system with 42% of its load shed.

The transmission-line outage cascading phenomenon is an important element in system planning. This phenomenon is usually taken into consideration by designing the transmission network to be capable of withstanding the outage of all single-contingency and some multiple-contingency events.

16.11 RADIAL TRANSMISSION NETWORKS

One of the simplest transmission networks to analyze is a radial network. Much basic insight can be gained by first examining a radial network before examining a general interconnected network.

Transmission cost parameters as presented in Table 16.6 are important fac-

TABLE 16.6 AC Transmission Cost Parameters
TYPICAL 1990 $ (SPARSE, MIDWESTERN UNITED STATES)

kV Voltage Level	Line Costs per Mile (Millions $)	Terminal Costs (Millions $)
138	.21	.42
161	.24	.48
230	.30	.84
345	.48	1.44
500	.72	2.40
765	1.32	4.20

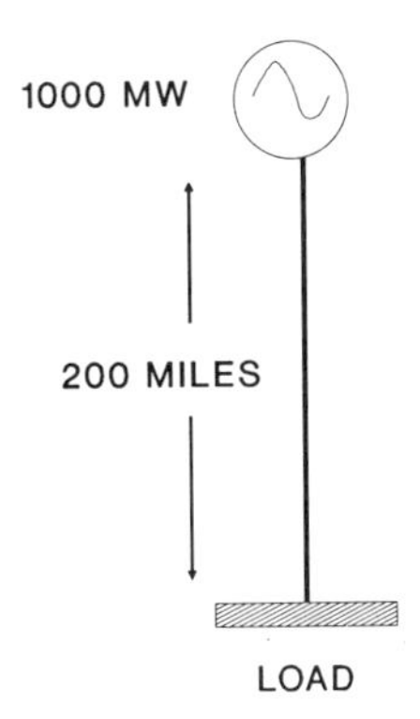

FIGURE 16.20 Radial line transmission example.

tors in transmission planning. This table presents typical data on the average cost per mile of transmission lines for the average line cost in a sparse midwestern United States type of utility. Costs vary due to line design and single versus double circuit towers. It also shows the termination cost, including transformers, circuit breakers, and protection equipment. Termination cost can be higher if there is a voltage transformation at the bus. A substation with transformation has a cost in the range of $12/kVA. For example, a 345-kV line could cost $500,000/mile and be terminated with equipment costing $1.5M. The cost in other regions of the United States can be markedly higher depending on the cost of land, labor productivity, and labor rate cost.

Consider an example of a radial line, illustrated in Figure 16.20. What would be the least expensive voltage level (either a 345- or 500-kV transmission line) to transmit 1000 MW of power over a 200-mile radial right-of-way to a load center? Table 16.7 summarizes the data.

Using the line-loadability guide of Figure 16.16, the loadability of a 200-mile-long 500-kV line is 910 MW times a 1.3 multiplier, or 1183 MW. The line loadability of a 345-kV line is 390 MW times 1.3, or 507 MW. The line cost is $720,000/mile at 500 kV, the terminal cost is $2.4M per terminal. For this analysis, assume zero losses. The total line cost is then $148.8M for the 500-kV line and $98.88M for the 345-kV line. Figure 16.21 presents the preliminary

TABLE 16.7 Data for the Radial Line Example

	500 kV	345 kV
Length (miles)	200	200
MW circuit capability	1183	507
Line cost (millions $/mile)	.72	.48
Terminal cost (millions $/terminal)	2.4	1.44
Losses	0	0
Total line cost	$148.8M	$98.88M

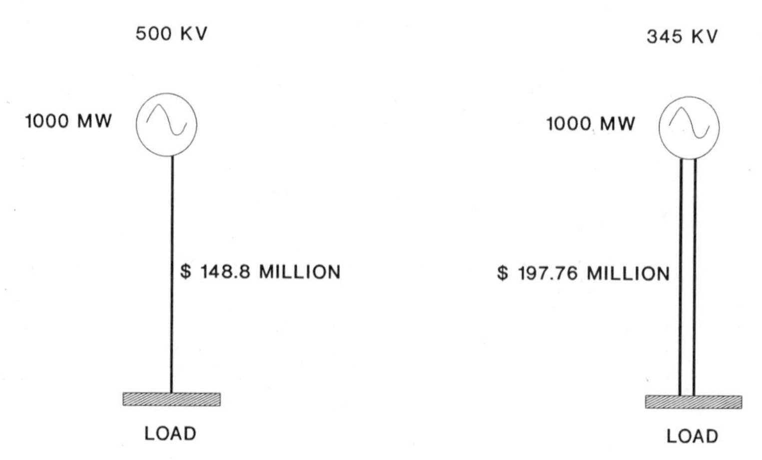

FIGURE 16.21 Radial line example calculation.

conclusions from this study. The load can be served by one 500-kV line that would cost $148.8M. The cost of the 345-kV design would be $197.76M because two lines are required to serve this power.

This solution provides power from the generation unit to the load center. However, if a line were to trip, then power would be lost. Examine this radial line again, but under the condition of a single-line contingency. Figure 16.22 illustrates a single-line contingency design for both the 500-kV and the 345-kV line. The total cost of the 500 kV line is $297M, the cost of the 345-kV line is $296M. For this example, in the single-line contingency planning criterion, both line voltage levels have approximately the same cost. In this case, the choice would have to be made based on other factors, such as losses and compatibility with existing voltage levels in the system.

The single-line contingency is typically the most widely used transmission

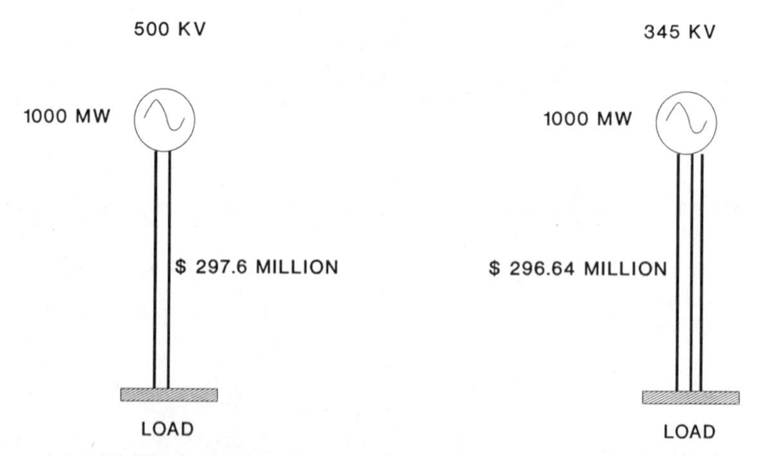

FIGURE 16.22 Single-line contingency case of radial line example.

planning criteria. However, there are cases for which it is prudent to plan for a right-of-way contingency in areas where the entire right-of-way could outage, such as over steep mountains where there may be a significant snow pack or hazardous winds, or in tornado-prone areas.

16.12 INFLUENCE OF LINE LOSSES

In the example discussed in the previous section, it was assumed that there were no line losses. This is an unreasonable assumption, however, because line losses can have significant economic value.

Consider estimating the line losses for a 200-mile 345-kV line. The resistance is .01 pu on a 100-MVA base, based on the typical values listed in Table 16.1. For a power flow of 1.0 surge impedance loading (SIL), the power flow is 390 MW or 3.90 pu of 100 MVA base. Since the voltage is near 1.0 pu, the current is 3.90 pu if the power factor of the load is near unity. The power losses are then estimated as:

$$\text{Line losses} = I^2R = (3.9)^2\ .01 = .152 \text{ pu}$$

The ratio of the line losses to the power flow is .152/3.9 = .039 or 3.9%.

At this point, reexamine the example problem of the previous section and include line losses. The line-loss evaluation is presented in Table 16.8.

The current in each line is first calculated. The I^2R losses are then evalu-

TABLE 16.8 Line Losses for the Example in Section 16.8 (Single Line Contingency Case)

	Voltage Plan	
	500 kV	345 kV
Number of lines	2	3
Power flow/line (MW)	500	333
Power flow/line: per unit	5.0	3.33
Losses: per unit/line	.05	.111
Losses: MW/line (MW)	5.0	11.1
Hours/year	8,760	8,760
Transmission utilization factor	.7	.7
Annual energy losses (MWh)	61,320	204,195
Power system value of losses		
Capitalized replacement energy ($/MWh)	200	200
Capitalized replacement capacity ($/kW)	300	300
Total cost of losses (capitalized)		
Replacement energy	$12.26M	$40.83M
Replacement capacity	$ 1.50M	$ 3.33M
Total	$13.76M	$44.16M

ated. These losses would occur for 8760 hours/year except when the generating units are on outage or when the power is reduced during periods of low system load demand.

A transmission utilization factor is introduced to account for the annual line loadings. Since losses are proportional to the square of the power flow, the transmission utilization factor used for the loss calculation is calculated as:

$$\frac{\int_0^{8760} \text{load}^2(t)\, dt}{8760 \cdot (\text{peak load})^2}$$

Annual energy losses can then be evaluated as the product of megawatt losses times 8760 times the transmission utilization factor.

The cost (or value) of losses is comprised of two components, replacement energy and demand or capacity cost. Replacement energy results from a marginal generating unit replacing energy that was consumed in the losses. This is calculated using a typical levelized value of $40/MWh and capitalized by dividing by a .20 annual levelized fixed-charge rate. Demand or capacity cost is the second component of the cost of losses. A value of $300/kW is used to reflect the incremental cost of peaking capacity (or load management) plus transmission capacity cost. The total capitalized cost of losses is $13.7M for the 500-kV plan and $44M for the 345-kV plan.

When the cost of losses is added to the transmission investment cost, the total evaluated cost of the plan then becomes:

	Voltage Plan	
	500 kV	345 kV
Investment cost (million $)	297.6	296.6
Capitalized cost of losses (million $)	13.8	44.2
	$311.4M	$340.8M

Thus, the 500-kV plan is shown to have a lower cost.

In summary, since the evaluated cost of losses for new transmission lines can represent 5–15% of the total line cost, transmission losses are an important planning consideration.

16.13 STEADY-STATE NETWORK POWER-FLOW EQUATIONS

The radial transmission line examples provided an introduction to the cost concepts necessary for planning. In general, however, transmission systems are configured in a network that is more difficult to analyze.

A transmission system is described by a network of buses and branches. The high-voltage network elements (69 or 138 kV and above) are represented in the transmission network model. A bus is typically a substation. The substation may be a connecting point for several transmission lines, a connecting point for low-voltage distribution circuit loads, or a connecting point for a generating unit. A branch is an element (of nonnegligible impedance) connecting two buses, typically a transmission line or transformer.

The transmission network is generally specified as having specified active and reactive load demands and generation output at each bus. (Variants to this are discussed later.) The objective is to determine the power flow on each transmission line, the voltage level of each bus, and relative phase angle of each bus.

One requirement is the network must have a solution. The network conditions may be specified such that no steady-state solution exists. For example, this may result when the power flows on one or more transmission lines are too high or when the reactive power supply is constrained. If this is the case, then the solution method will not converge to an answer.

The network power flow equations are coupled nonlinear equations that are solved using iterative solution methods. To start these methods requires an initial estimate of the solution. If the initial estimate is not "close" to the solution, the solution method may not converge. Thus, it is required that the initial estimate be reasonably close to the final solution.

This section reviews the AC power flow equations of a network and describes methods for their solution.

The power flow entering a transmission network branch element in Figure 16.23 is given by:

$$P_{KM} + jQ_{KM} = E_K I^*_{KM} + E_K I^*_{KK}$$
$$P_{KM} + jQ_{KM} = E_K[E_K - E_M)y_{KM}]^* + E_K\,[E_K y_{KK}]^* \tag{16.45}$$

where y_{KM} = admittance of the circuit branch element connecting bus K to bus M

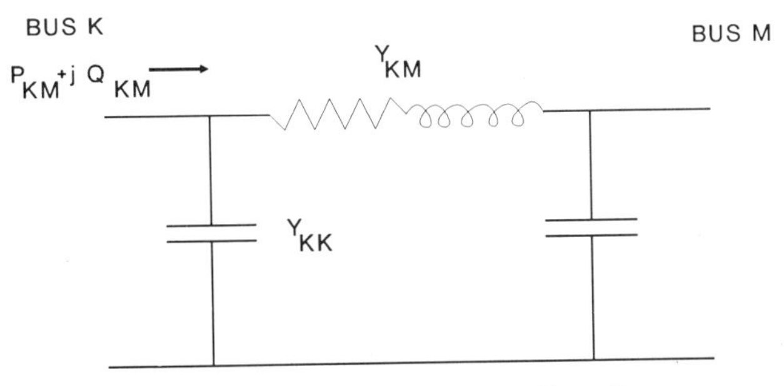

FIGURE 16.23 Network power-flow diagram.

$*$ = superscript asterisk denotes complex conjugate
y_{KK} = admittance of the circuit branch element connecting bus K to ground

If $P_K + jQ_K$ is the total injected power (generation minus load) into bus K, then this power must equal the power into all the transmission branch elements connected to the bus. Thus:

$$P_K + jQ_K = \sum_{\substack{M \\ M \neq K}} E_K[(E_K - E_M)y_{KM}]^* + E_K(E_K\, y_{KK})^* \tag{16.46}$$

where M denotes all buses connected to K and $*$ denotes complex conjugate.
This can be rewritten by taking the conjugate of (16.46) as:

$$\frac{P_K - jQ_K}{E_K^*} = \sum_{\substack{M \\ M \neq K}} Y_{MK}E_M + Y_{KK}E_K \tag{16.47}$$

where

$$Y_{MK} = -y_{KM} \tag{16.48}$$

$$Y_{KK} = \sum_{\substack{M \\ M \neq K}} y_{KM} + y_{KK} \tag{16.49}$$

Given the injected complex power at each bus, the voltage at each bus can be determined by iterative solution of Equation (16.47) for all buses in the system.

However, not all buses have a specified P and Q. Table 16.9 presents a summary of the principal bus types.

The solution approach is to first delete the reference bus from Equation (16.47). One approach, referred to as the "Gauss–Seidel method," iterates toward a solution by first estimating the voltage values. Equation (16.47) is then used to improve upon the estimate. This can be expressed as:

$$E_K^{(i)} = \frac{P_K - jQ_K}{Y_{KK}E_K^{(i-1)*}} - \frac{1}{Y_{KK}} \sum_{M<K} Y_{MK}E_M^{(i)} + \sum_{M>K} Y_{MK}E_M^{(i-1)} \tag{16.50}$$

where superscript (i) denotes iteration number i and superscript $(i-1)$ denotes the previous iteration. Equation (16.50) solves Equation (16.47) by solving for E_K using the best available voltage estimates. The equation for E_M for $M<K$ would have been solved prior to bus K. Hence, the voltage for iteration i is available. The E_M for $M>K$ have not been determined, so Equation (16.50) must use the E_M from the previous iteration.

TABLE 16.9 Summary of Principal Bus Types

Bus Type	P	Q	$\|E\|$	θ	Comments
Reference bus			✓	✓	The reference bus is typically chosen as a bus with a generating unit; the voltage is specified (e.g., 1.05) and the angle reference is specified ($\theta = 0$); the power output of the reference bus will "swing" to meet the net requirements of the power system
Load	✓	✓			General load bus
Fixed impedance load					The shunt impedance of the load-to-ground is specified
Generator	✓ if $Q < Q_{\max^+}$ $Q > Q_{\max^-}$		✓		Generator is specified by active power output and bus voltage (e.g., 1.05); however, if the reactive output from the generator is greater than the unit can supply, then the generator is a fixed P and Q source and the voltage level will vary (this may be used to represent a generating unit, synchronous condenser, switched capacitors, and static var sources)
	✓ if $Q = Q_{\max^-}$ or $Q = Q_{\max^+}$	✓			

Equation (16.50) is really two equations, one for the active power and another for the reactive power. One equation is used for the real component of the voltage, and the second equation is used for the imaginary component of the voltage. These two components can then be vectorially combined to yield voltage magnitude and angle.

The Gauss–Seidel method was one of the first AC network solution techniques. The method has good global convergence characteristics but converges at a slow rate. Thus this method requires a large number of iterations before converging to a solution. In practice the Gauss–Seidel method may be used for a few iterations to obtain a good initial starting point for a more rapidly converging method.

Equation (16.46) can also be written in polar form as:

$$P_K + jQ_K = \sum_{\substack{M \\ M \neq K}} |E_K| e^{j\theta_K} [|E_K| e^{-j\theta_K} - |E_M| e^{-j\theta_M}] y^*_{KM} + |E_K|^2 \, y^*_{KK} \tag{16.51}$$

Let

$$Y_{MK} = - y_{KM} \qquad \text{for} \qquad M \neq K \tag{16.52}$$

$$Y_{KK} = \sum_{\substack{M \\ M \neq K}} y_{KM} + y_{KK} \tag{16.53}$$

and

$$Y_{MK} = G_{MK} + jB_{MK} \qquad \text{for all} \qquad M, K \tag{16.54}$$

Equation (16.51) then yields:

$$P_K + jQ_K = \sum_M |E_K| \, |E_M| \, (G_{MK} - jB_{MK}) e^{j(\theta_K - \theta_M)} \tag{16.55}$$

$$= \sum_M |E_K| \, |E_M| \cdot [G_{MK} \cos(\theta_K - \theta_M) + B_{MK} \sin(\theta_K - \theta_M)$$

$$+ j \, G_{MK} \sin(\theta_K - \theta_M) - jB_{MK} \cos(\theta_K - \theta_M)] \tag{16.56}$$

One approach to solving Equation (16.56) is to expand the equation in a Taylor series in terms of $\Delta\theta_M$ and $\Delta E_M / E_M$.

To expand Equation (16.56) is a Taylor series, the derivative terms are required.

$$H_{KM} = \frac{\partial P_K}{\partial \theta_M} = |E_K| \, |E_M| \, [G_{MK} \sin(\theta_K - \theta_M) - B_{MK} \cos(\theta_K - \theta_M)] \tag{16.57}$$

$$H_{KK} = \frac{\partial P_K}{\partial \theta_K} = - |E_K|^2 \, B_{KK} - Q_K \tag{16.58}$$

$$N_{KM} = \frac{\partial P_K}{\partial |E_M| / |E_M|}$$
$$= |E_K| \, |E_M| \, [G_{MK} \cos(\theta_K - \theta_M) + B_{MK} \sin(\theta_K - \theta_M)] \tag{16.59}$$

$$N_{KK} = \frac{\partial P_K}{\partial |E_K| / |E_K|} = |E_K|^2 \, G_{KK} + P_K \tag{16.60}$$

$$J_{KM} = \frac{\partial Q_K}{\partial \theta_M} = - |E_K| \, |E_M| \, [G_{MK} \cos(\theta_K - \theta_M) + B_{MK} \sin(\theta_K - \theta_M)] \tag{16.61}$$

$$J_{KK} = \frac{\partial Q_K}{\partial \theta_K} = -|E_K|^2 G_{KK} + P_K \tag{16.62}$$

$$L_{KM} = \frac{\partial Q_K}{\partial |E_M|/|E_M|} = |E_K|\,|E_M|\,[G_{MK} \sin(\theta_K - \theta_M) - B_{MK} \cos(\theta_K - \theta_M)] \tag{16.63}$$

$$L_{KK} = \frac{\partial Q_K}{\partial |E_K|/|E_K|} = -|E_K|^2 B_{KK} + Q_K \tag{16.64}$$

One solution approach is to solve the set of equations iteratively:

$$P_K + jQ_K = \overline{P}_K^{(i-1)} + j\overline{Q}_K^{(i-1)} + \sum_M H_{KM}\,\Delta\theta_M^{(i)} + N_{KM}\frac{|\Delta E_M^{(i)}|}{|E_M^{(i)}|} + j\left(\sum_M J_{KM}\,\Delta\theta_M^{(i)} + L_{KM}\frac{|\Delta E_M^{(i)}|}{|E_M^{(i)}|}\right) \tag{16.65}$$

where $\overline{P}_K^{(i-1)}$ = evaluation of active component of right-hand side of Equation (16.56)) after $i-1$ iterations

$\overline{Q}_K^{(i-1)}$ = evaluation of reactive component of right-hand side of Equation (16.56) after $i-1$ iterations

$\Delta\theta_M^{(i)}$ = a small correction term for angle θ_M

$|\Delta E_M^{(i)}|/|E_M^{(i)}|$ = a small correction term for relative voltage magnitude

Rearranging the terms in Equation (16.65) yields a left side of:

$$P_K + jQ_K - \overline{P}_K^{(i-1)} - j\overline{Q}_K^{(i-1)}$$

$$= \text{bus active power mismatch} + j\ \text{bus reactive power mismatch}$$
$$= \Delta P + j\Delta Q \tag{16.66}$$

The bus mismatch is the difference between the actual bus power and the calculated bus power based on the solved voltage terms of the right side of Equation (16.56).

The equations expressed in matrix notation are:

$$\begin{bmatrix} \Delta P \\ \\ \Delta Q \end{bmatrix} = \begin{bmatrix} H & N \\ \\ J & L \end{bmatrix} \begin{bmatrix} \Delta\theta \\ \\ \dfrac{|\Delta E|}{|E|} \end{bmatrix} \tag{16.67}$$

These equations could be solved for $\Delta\theta$ and $|\Delta E|/|E|$. Then θ and E could be updated by:

$$\theta_M^{(i)} = \Delta\theta_M + \theta_M^{(i-1)} \tag{16.68}$$

$$|E^{(i)}| = |E^{(i-1)}| + \frac{|\Delta E|}{|E|} \cdot |E^{(i-1)}| \tag{16.69}$$

The updated values of θ and E are substituted into the right side of Equation (16.56), which is then used to compute a new bus mismatch. Equations (16.57) through (16.64) can also be updated, but this is usually only done after several iterations in order to reduce computer time.

After several iterations, the above procedure (termed the "Newton–Raphson power-flow solution method") converges to the answer.

The Newton–Raphson method works well and was the workhorse power flow solution technique until the advent of the "fast decoupled technique" in 1973 (see Stott and Alsac, 1973).

The fast dccoupled technique begins with Equation (16.67) and makes several simplifying assumptions.

$$\cos(\theta_K - \theta_M) \approx 1 \tag{16.70}$$

$$G_{KM} \sin(\theta_K - \theta_M) << B_{KM} \tag{16.71}$$

$$Q_K << B_{KK}E_K \tag{16.72}$$

$$\text{Neglect coupling matrices } J \text{ and } N \tag{16.73}$$

After inserting Equations (16.70) through (16.73) into Equations (16.57) to (16.64), the resulting equations take the form:

$$[\Delta P] = [|E|\ B'\ |E|]\ \Delta\theta \tag{16.74}$$

$$[\Delta Q] = [|E|\ B''\ |E|]\ \frac{|\Delta E|}{|E|} \tag{16.75}$$

Several other simplifications are made:

- Remove terms from B' that largely influence var flows, that is, shunt reactances and off-nominal in-phase transformer taps. Neglecting the series resistances in calculating the B' matrix yields slightly improved convergence (Stott and Alsac, 1973). This leads to the DC power-flow matrix.
- Remove the angle shifting influences in B'' of phase shifting transformers.
- Take the left-hand voltage $|E_K|$ out of the matrix and move it to the other side. The right-hand voltage, $|E_M|$, is replaced by unity in B'.

The final fast decoupled power flow equations become:

$$\left[\frac{\Delta P}{|E|}\right] = [B'][\Delta\theta] \tag{16.76}$$

$$\left[\frac{\Delta Q}{|E|}\right] = [B''][\Delta E] \tag{16.77}$$

Equations (16.76) and (16.77) are then used iteratively to solve the power flow. The B' and B'' matrices are constant for every iteration. Since they do not change, they can be inverted at the beginning of the procedure and saved. The procedure is:

Step 0 Invert the B' and B'' matrices.
Step 1 Calculate $\Delta P/|E|$, using Equations (16.56) and (16.66).
Step 2 Solve for $\Delta\theta$, using Equation (16.76), and update θ, using Equation (16.68).
Step 3 Calculate $\Delta Q/|E|$, using Equations (16.56) and (16.66).
Step 4 Solve for ΔE, using Equation (16.77), and update $|E|$, using Equation (16.69).
Step 5 Test for convergence.

$$\max_M |\Delta P_M| \leq \text{convergence test}$$

$$\max_M |\Delta Q_M| \leq \text{convergence test}$$

If not converged, proceed to step 1.

The fast decoupled technique can converge in several iterations in a "well-behaved" transmission network which does not require any automatic transformer tap changes, or generator reactive limits. The fast decoupled technique is a very fast technique, and has become the workhorse method of the utility industry.

Example. A power system is comprised of three buses connected by 345-kV lines 200 miles long as shown in Figure 16.24. Compute the power flow and bus voltages, using the fast decoupled technique. Assume that the voltage at reference bus 1 is held at 1.05. Assume that the transmission line parameters are:

Line admittance: $y_{12} = y_{23} = -j10 + 1$; $y_{13} - j20 + 2$

Line admittance-to-ground (line charging): $b_{13} = -j3.4$; $b_{12} = b_{23} = j1.7$

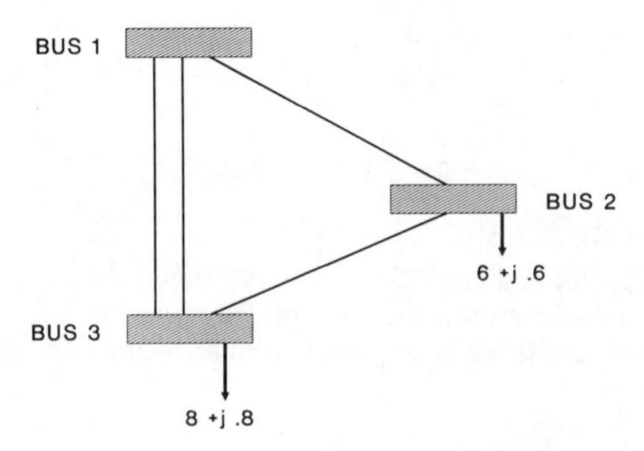

FIGURE 16.24 Sample power system network of 345-kV lines, 200 miles long.

The Y matrix is based on Equations (16.48) and (16.49):

$$[Y] = \begin{bmatrix} -j27.45+3 & j10-1 & j20-2 \\ j10-1 & -j18.3+2 & j10-1 \\ j20-2 & j10-1 & -j27.45 \end{bmatrix} \tag{16.78}$$

(columns numbered 1, 2, 3)

Note that half of the line charging admittance, b, is associated with each line end node.

SOLUTION. The B' matrix for the fast decoupled technique is evaluated at $\theta = 0$ and has the shunt reactance (line charging) admittance terms removed:

$$[B'] = \begin{bmatrix} 30 & -10 & -20 \\ -10 & 20 & -10 \\ -20 & -10 & 30 \end{bmatrix} \tag{16.79}$$

(columns numbered 1, 2, 3)

The B'' matrix is the negative reactive segment of the Y matrix:

$$[B''] = \begin{bmatrix} 27.45 & -10 & -20 \\ -10 & 18.3 & -10 \\ -20 & -10 & 27.45 \end{bmatrix} \tag{16.80}$$

(columns numbered 1, 2, 3)

Since bus 1 is the reference bus, column 1 and row 1 can be deleted from the matrices. The B' matrix can be inverted to:

$$(B')^{-1} = \begin{bmatrix} .06 & .02 \\ .02 & .04 \end{bmatrix} \tag{16.81}$$

The B'' matrix can be inverted to:

$$(B'')^{-1} = \begin{bmatrix} .0682 & .0249 \\ .0249 & .0454 \end{bmatrix} \tag{16.82}$$

Assume that $|E_2| = |E_3| = 1.0$ and $\theta_2 = \theta_3 = 0$. The power flows on the lines can then be calculated using Equation (16.46) as shown in Table 16.10. On the basis of the calculated line flows, the bus net power can be calculated and the bus mismatch (error) evaluated as presented in Table 16.11. Substituting the active power mismatch into Equation (16.76) gives:

$$[\Delta\theta] = [B']^{-1} [\frac{\Delta P}{|E|}] \tag{16.83}$$

$$= \begin{bmatrix} .06 & .02 \\ .02 & .04 \end{bmatrix} \begin{bmatrix} -5.95 \\ -7.90 \end{bmatrix} = \begin{bmatrix} -.515 \\ -.435 \end{bmatrix} \begin{matrix} \text{radians} \\ \text{radians} \end{matrix} = \begin{bmatrix} -29.52° \\ -24.92° \end{bmatrix}$$

The power flows on the lines can be calculated by using these new angle values as shown in Table 16.12. The bus net power can also be evaluated as shown in Table 16.13.

TABLE 16.10 Calculated Line Flows

Start Bus No.	End Bus No.	P	Q	Start Bus No.	End Bus No.	P	Q
1	2	0.1	−0.4	2	1	−0.0	−1.4
2	3	0.0	−0.9	3	2	0.0	−0.9
3	1	−0.1	−2.7	1	3	0.1	−0.8

TABLE 16.11 Bus Net Power

	Specified		Calculated		Error	
Bus No.	P	Q	P	Q	P	Q
1	0.00	0.00	0.16	−1.24	−0.16	1.24
2	−6.00	−0.60	−0.05	−2.20	−5.95	1.60
3	−8.00	−0.80	−0.10	−3.55	−7.90	2.75

TABLE 16.12 Calculated Line Flows

Start Bus No.	End Bus No.	P	Q	Start Bus No.	End Bus No.	P	Q
1	2	5.2	.7	2	1	−5.0	.3
2	3	0.8	−.9	3	2	0.8	−.7
3	1	−8.7	−.1	1	3	9.1	.5

TABLE 16.13 Bus Net Power

	Specified		Calculated		Error	
Bus No.	P	Q	P	Q	P	Q
1	0.00	0.00	14.31	1.18	−14.31	−1.18
2	−6.00	−0.60	−5.75	−0.58	−0.25	−0.02
3	−8.00	−0.80	−7.89	−0.79	−0.11	−0.01

The voltage can then be calculated by using Equation (16.77)

$$[\Delta E] = [B'']^{-1}\left[\frac{\Delta Q}{|E|}\right] = \begin{bmatrix} .0682 & .0249 \\ .0249 & .0454 \end{bmatrix}\begin{bmatrix} -.39 \\ -.04 \end{bmatrix} = \begin{bmatrix} -.028 \\ -.012 \end{bmatrix} \quad (16.84)$$

The angles and voltages calculated after one iteration are:

$$\theta = \begin{bmatrix} -29.52 \\ -24.92 \end{bmatrix} \text{ degrees;} \qquad E = \begin{bmatrix} .972 \\ .988 \end{bmatrix}$$

A second iteration is then performed. The results from subsequent iterations are illustrated in Figures 16.25 through 16.27.

Figure 16.25 shows the maximum percent bus power mismatch after each iteration. In this figure a half-iteration corresponds to a $P{:}\Delta\theta$ evaluation. A

FIGURE 16.25 Maximum percent bus power mismatch.

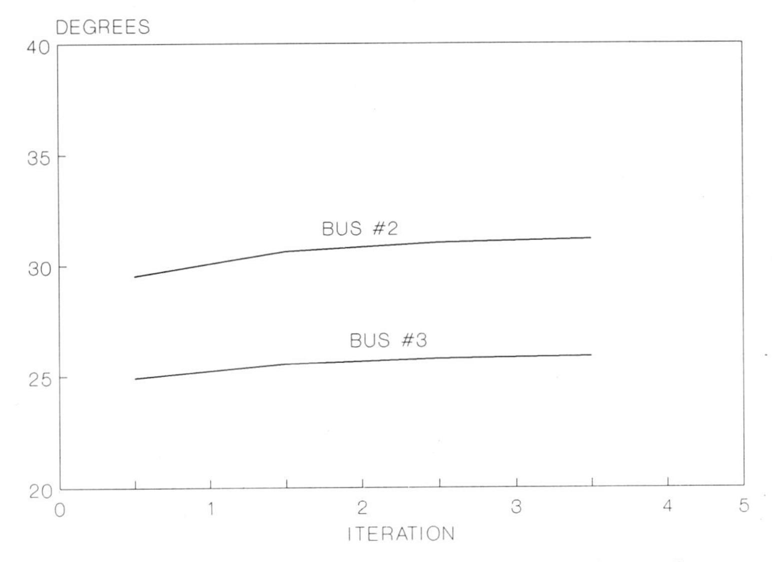

FIGURE 16.26 Fast decoupled convergence rate bus angle.

$Q{:}\Delta E$ evaluation completes a full iteration. During the $P{:}\Delta\theta$ half-iteration, the active power mismatch is driven toward zero, while the reactive power mismatch increases slightly. Similarly during the $Q{:}\Delta E$ half-iteration, the reactive mismatch is driven toward zero. The overall convergence is very rapid, requiring only three iterations.

The bus angle and voltages are presented in Figures 16.26 and 16.27. Note

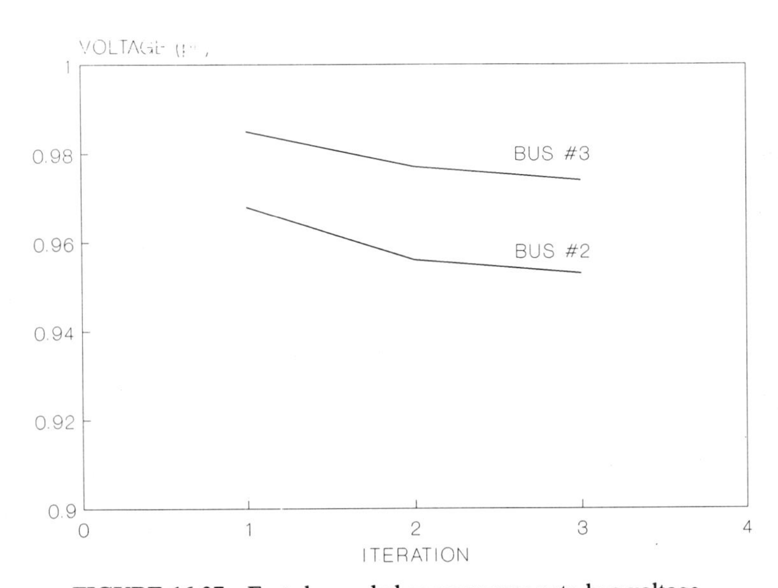

FIGURE 16.27 Fast decoupled convergence rate bus voltage.

that the first iteration provides a very realistic estimate of these values. The first half-iteration of the fast decoupled technique is the often used "DC power-flow" approximation.

In summary, the fast decoupled technique is a rapidly converging method and is widely used for transmission system analysis. Note that for distribution systems, with low X/R ratios, these simplifying assumptions are not valid, and the performance of the Fast Decoupled method may not be as good.

16.14 HVDC TRANSMISSION

The discussion thus far has focused on AC transmission. While AC transmission comprises the majority of the world's electric networks, high-voltage direct-current (HVDC) transmission is assuming increasing importance. As of 1987, more than 50 GW of HVDC systems were installed, committed, or actively being studied (Fink and Beaty, 1987).

The choice of HVDC transmission may be based on several factors. Among them are:

- *Nonsynchronous connection.* Systems connected by HVDC do not have to maintain synchronism. Thus, power transfers between large independent AC systems, such as the eastern and western U.S. systems, Hydro-Quebec system, and Texas system, can be effected without installing significant transmission capacity to maintain synchronous operation.
- *Cables.* For underground or underwater cables, HVDC transmission is much less costly than AC, with a break-even distance of approximately 25 miles. At high voltage and power levels, DC may be the only choice.
- *Performance Benefits.* Because DC power flow is highly adjustable, an HVDC system can play the same role as a phase-angle regulating transformer in forcing flow patterns in the network that are different from normal flows as determined by Kirchhoff's laws. In addition, rapid modulation of DC power can be used to stabilize AC power-flow oscillations. HVDC controls can be used to regulate AC bus voltages in the same manner as a synchronous condenser or static var control system.
- *Economy.* Despite the cost of AC–DC conversion equipment, HVDC transmission lines are less costly than AC lines for long distances, as a result of the fewer conductors (two versus three) and smaller transmission towers. The break-even point is usually between 350 and 500 miles. Since the direct cost difference is usually small, selection of DC is usually motivated by one or more of the first three factors.

Figure 16.28 shows the major elements of a two-terminal HVDC system. The rectifier, shown on the left, has an AC input and produces DC to the DC line. An AC filter is included to filter out the higher-order harmonics generated

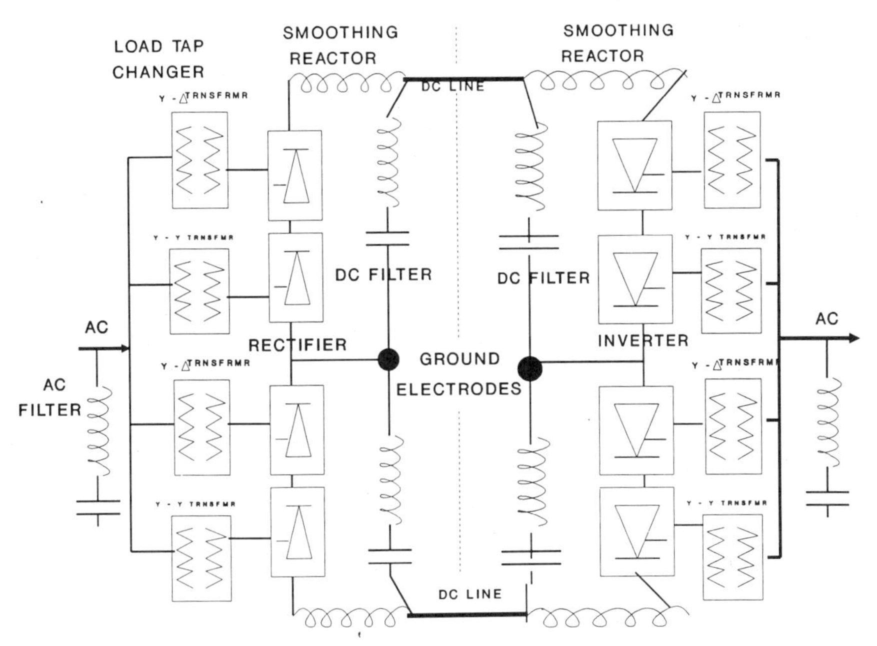

FIGURE 16.28 Elements of an HVDC system.

by the converters. The figure shows a "12-pulse converter" with four-thyristor converter groups. A thyristor is a solid-state switching device that carries current only in one direction and then not until the "gate" is momentarily energized by a current pulse. The gate timing provides a method to control HVDC power flow. The thyristor converters are fed from a load tap-changer transformer. The use of both Y-Δ and a Y-Y transformers to feed each thyristor converter group reduces the harmonic content of the AC as well as the HVDC system. A smoothing reactor smooths the ripple from the DC voltage. The inverter has a DC input and produces AC output. It has the same hardware component configuration as the rectifier, except that the polarity of the thyristors in the circuit is reversed. In most systems, either converter can operate as rectifier or inverter to permit power transfer in either direction. The AC and DC filters are required to smooth the ripple (harmonics) in the AC and DC currents. This prevents interference with communication circuits as well as other problems.

HVDC systems can be classified as back-to-back systems or point-to-point systems. Back-to-back systems have the rectifier and inverter in the same substation and find application in connecting two asynchronous power systems. In point-to-point systems, the rectifier and inverter are located a considerable distance from each other (hundreds of miles) and are connected by a DC transmission line or cable. Point-to-point HVDC systems are normally bipolar, that is, one positive and one negative pole, but can operate as monopolar if necessary, either with a metallic return via the second pole's conductor or, for short periods, with earth return.

16.14.1 Conceptual HVDC Converter Operation

The operation of the converter can be conceptually viewed by first examining a simplified three-pulse bridge as shown in Figure 16.29. Assume that the transformer inductance is zero. Figure 16.30 illustrates the phase to neutral voltage waveforms and the line currents. One line-to-line voltage, V_{ba}, is also shown.

During the time interval from -120° to 0°, phase A voltage is positive. The current flow is through phase A, through the thyristor valve, and through the smoothing reactor to the load.

At the time of 0°, the voltage on phase B is greater than that on phase A, and line-to-line voltage B to A is positive. Because voltage on phase B is greater than phase A, thyristor A is shut off and thyristor B conducts. This causes current to flow on phase B through to the load.

A simplified but more realistic HVDC terminal model is shown in Figure 16.31. This model includes the effect of the commutation inductance, which consists largely of transformer leakage inductance. During the time period from -120° to 0°, phase A is conducting.

At an angle of 0°, voltage B to A becomes positive and begins to drive current through thyristor B and reduce current in thyristor A. However, the phase currents cannot change instantaneously through the transformer inductance. The smoothing reactor (filter) also acts to maintain a constant load current. Therefore, the sum of phase A and phase B current is approximately constant. The result is that thyristor A and B both conduct for a period of time. This results in a phase-to-phase short-circuit on the transformer.

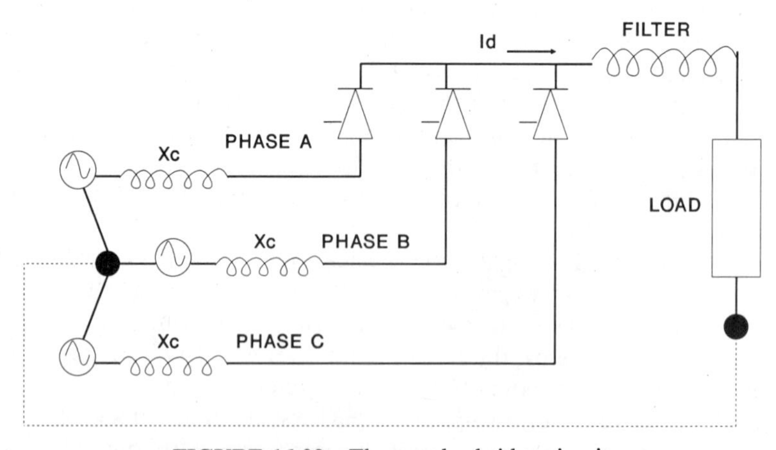

FIGURE 16.29 Three-pulse bridge circuit.

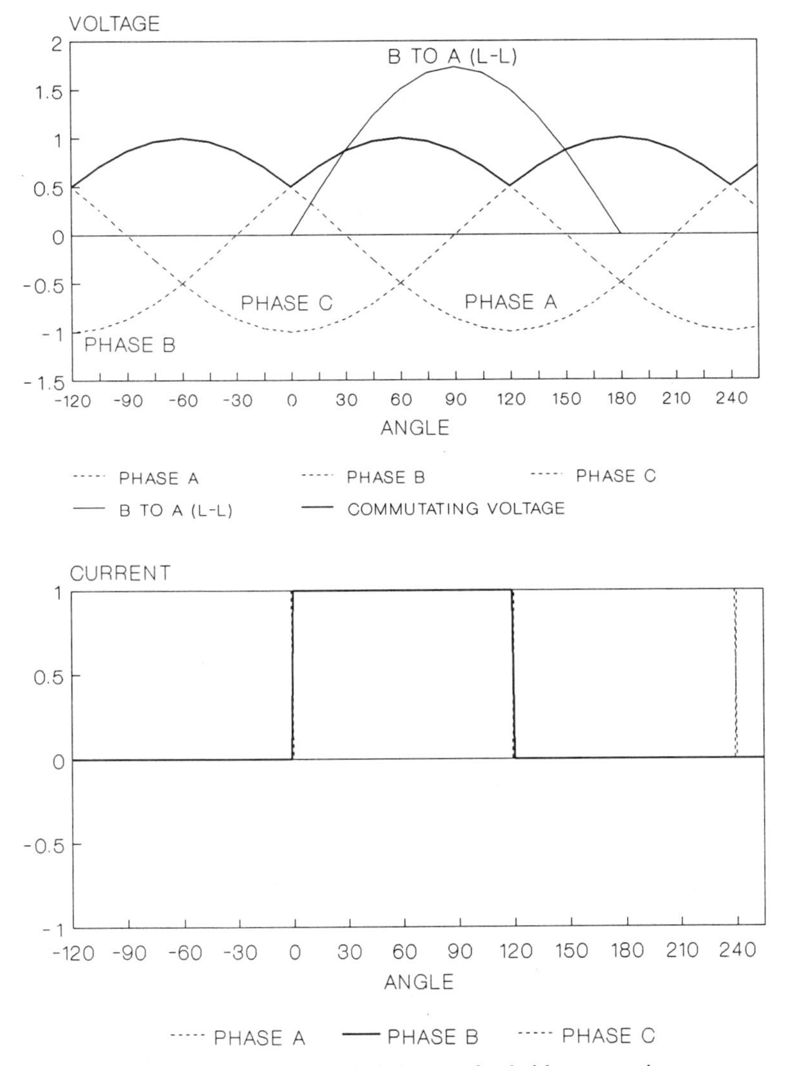

FIGURE 16.30 Simplified three-pulse bridge operation.

During this overlap time period, the bridge equations for phase A and phase B can be written by a voltage loop equation:

$$V_A - L_c \frac{d}{dt} i_A = V_B - L_c \frac{d}{dt} i_B \tag{16.85}$$

where V = phase voltage to ground
L_c = commutating reactance
i = instantaneous line current

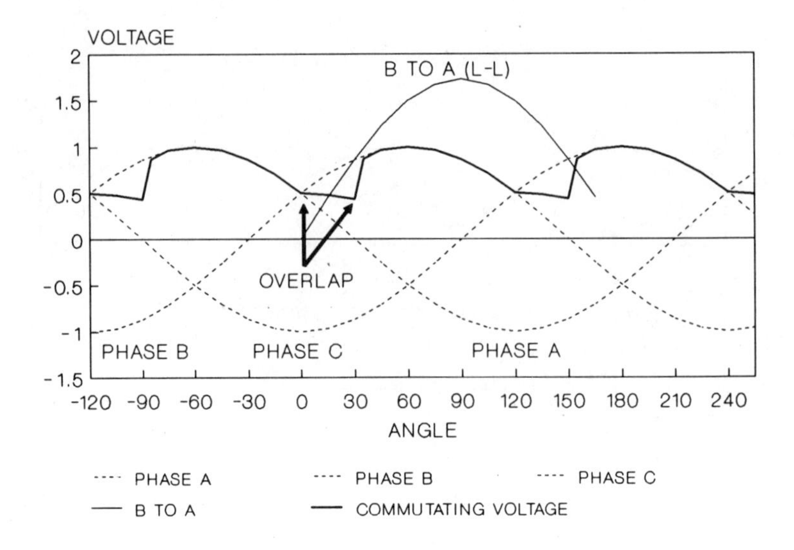

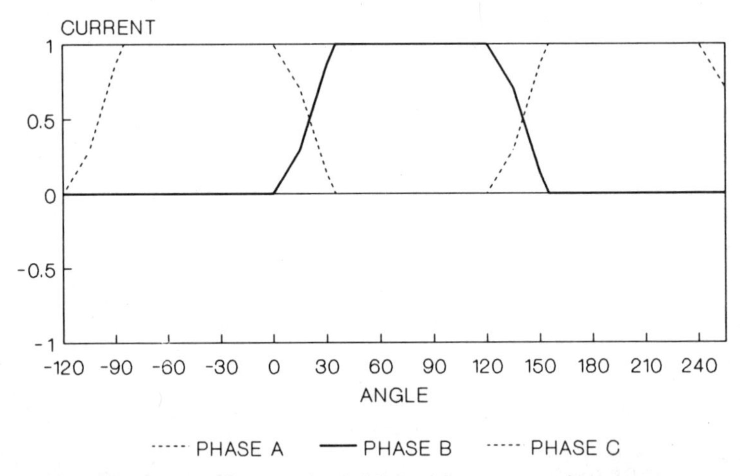

FIGURE 16.31 Three-pulse bridge with commutating reactance.

Because the smoothing reactor tends to maintain a constant current:

$$i_A + i_B = i_D \text{ (constant)} \tag{16.86}$$

These equations can be written as:

$$V_{BA} = V_B - V_A = 2L_c \frac{di_B}{dt} \tag{16.87}$$

The phase currents are:

$$i_B(t) = \frac{1}{2L_c} \int_0^t V_{BA}(t)\, dt \tag{16.88}$$

$$i_A(t) = i_D - \frac{1}{2L_c} \int_0^t V_{BA}(t)\, dt \tag{16.89}$$

The phase B current increases while phase A current decreases. When the phase A current becomes zero, thyristor A stops conducting. The voltage across the commutating reactance (termed the cummutating voltage) during the overlap period is:

$$V_{\text{commutating}} = V_B(t) - L_c \frac{di_B(t)}{dt} \tag{16.90}$$

The commutating voltage has a ''notch'' decrease from the ideal phase voltage during the overlap period. This effect contributes to a lower average DC voltage.

In the analysis hereto, the conduction through the thyristor began at the time at which the voltage of phase B equaled that of phase A. The B thyristor gate was pulsed at this instant to initiate conduction.

Consider when the thyristor bridge is fired at a *30° gate delay* as shown in Figure 16.32. In this case, the average commutating voltage (over a 360° period) is less than the case when the gate is fired at 0° (Figure 16.31). Varying the gate delay is a principal means of controlling DC power flow.

Consider now the case when the gate delay is long, approximately 85°, as shown in Figure 16.33. The commutation delay time is shorter in this case because the B-to-A line-to-line voltage is larger at the commutation time and, therefore, more strongly driving the phase current commutation.

In this example, the commutation voltage has both positive and negative swings and averages to approximately zero. Thus, no active power can be transferred to the load.

Note also that the phase currents are 90° lagging (reactive) the phase voltage. Therefore, when the delay angle is near 90°, the HVDC converter consumes vars from the AC system and is not transmitting active power.

Consider the case of delaying the gate time to 140°, as shown in Figure 16.34. In this case, the commutation still takes place but the commutation voltage is negative. The product of the phase current and the phase voltage is negative, implying that the converter absorbs power from the load. By reversing the terminal polarity, this configuration serves as an inverter, converting DC into AC power.

Three angles are shown on Figure 16.34:

α = gate delay
μ = commutation overlap angle

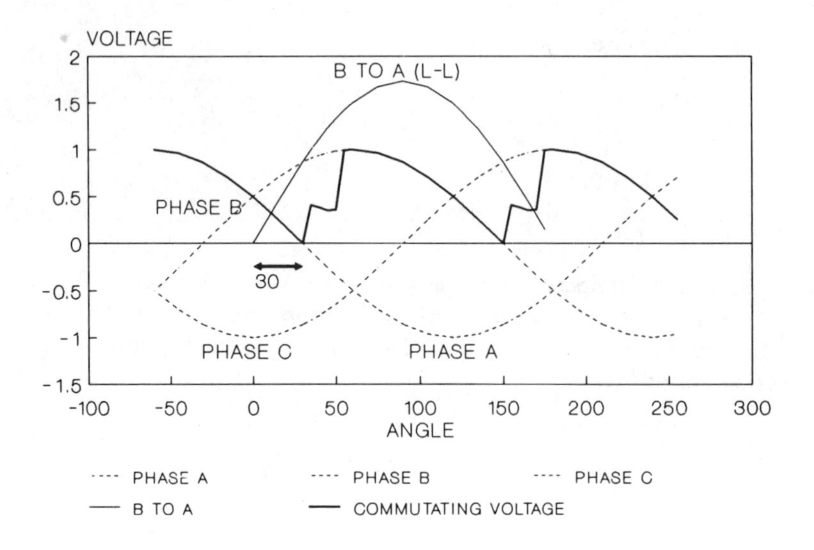

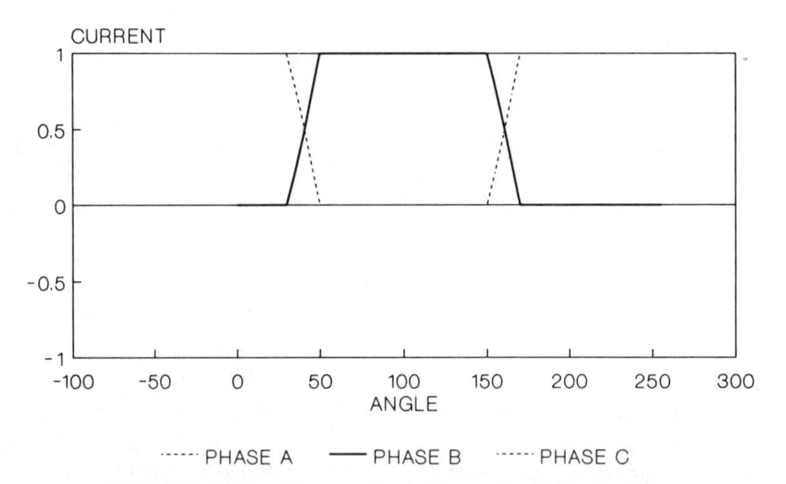

FIGURE 16.32 Three-pulse bridge with 30° gate delay.

γ = extinction angle, from the end of commutation to the end of the positive cycle of the commutating voltage V_{BA}

The extinction angle must be maintained as positive.

The process for inversion and rectification are "mirror" symmetrical, as noted by a comparison of Figures 16.32 and 16.34. This can be noted by reversing the sign of the inverter voltage and by reversing the inverter X-axis direction. When these two operations are completed, the inverter and rectifier have the same wave shapes. Thus, the angle α for the rectifier and angle γ for the inverter have the same physical interpretation.

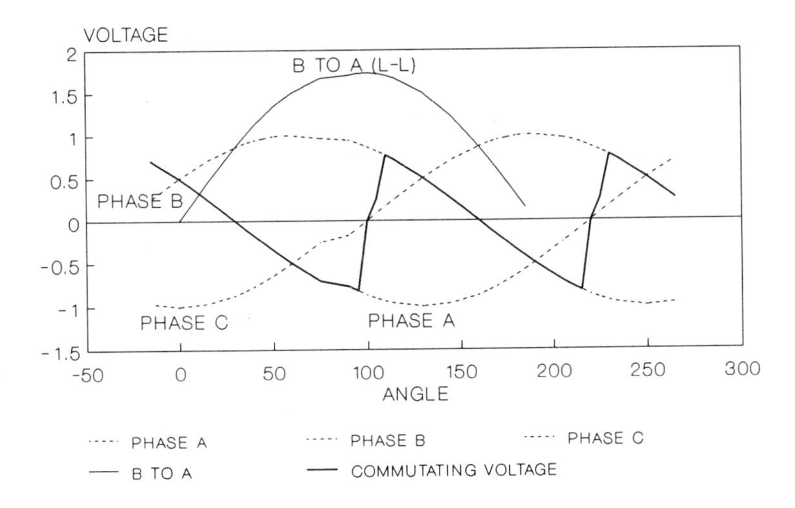

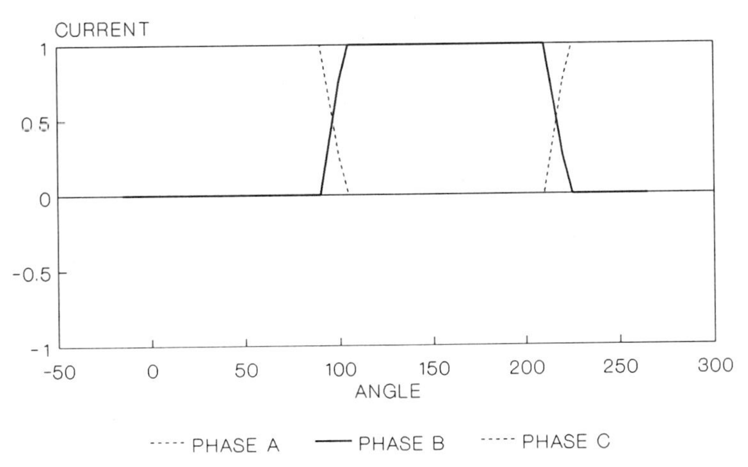

FIGURE 16.33 Three-pulse bridge with 85° gate delay.

The three-pulse bridge demonstrates the principles, but increased effectiveness results from using a two-sided bridge. This is illustrated conceptually in Figure 16.35. A second set of thyristors is inserted between the negative DC node and the AC phases. Thus, while the positive bridge side conducts through one phase (e.g., phase A), another phase, phase C, conducts on the negative side (e.g., phase B or C).

The commutating voltage waveform is the sum of the positive and negative bridge sides as shown in Figure 16.36. Note that the commutation voltage is proportional to the line-to-line voltage and that the voltage profile is smoother because of six pulses (commutations) per cycle. The line currents have both

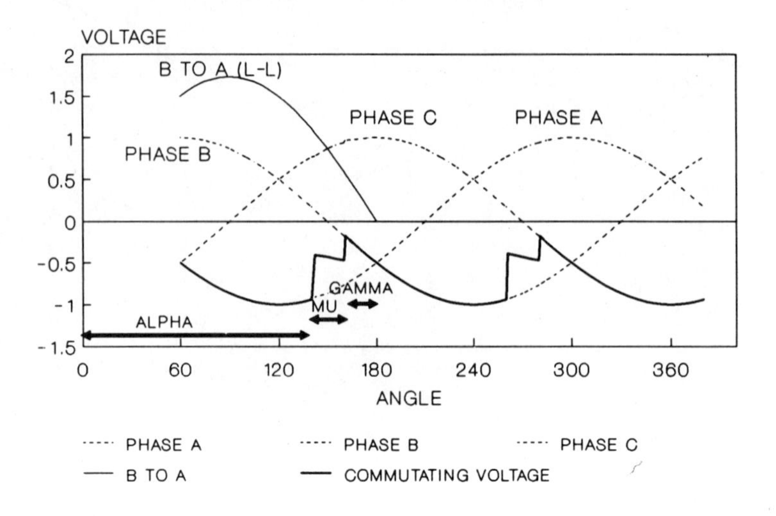

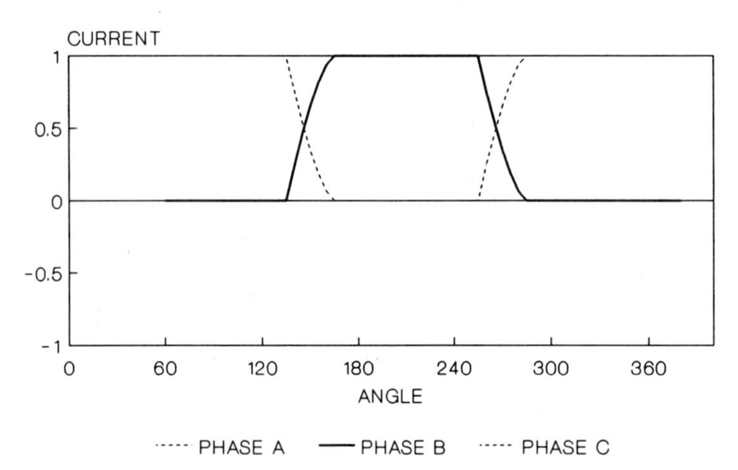

FIGURE 16.34 Three-pulse bridge with 140° gate display.

positive and negative swings. The smoothed voltage profile and the cyclic line currents lead to reduced AC and DC filtering requirements.

16.14.2 Converter Equations

Converter equations can be derived by analysis of the single six-pulse converter bridge.

Neglecting the commutation (overlap) notch for the moment, the average direct voltage for rectifier operation can be obtained by integrating the average direct voltage for rectifier operation over a 60° segment of the voltage wave-

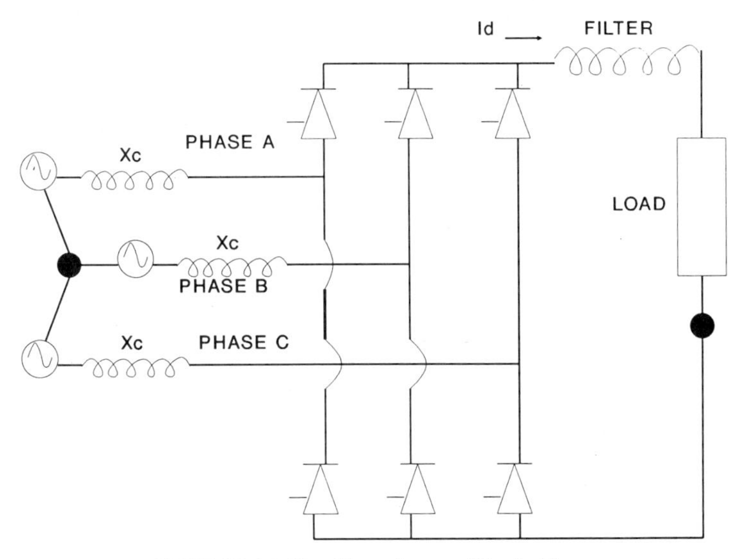

FIGURE 16.35 Six-pulse rectifier bridge.

form, as indicated in Figure 16.37, and dividing the result by $\pi/3$ radians (60°):

$$V_{\mathrm{dr}} = \frac{\text{area } a\text{–}b\text{–}c\text{–}d}{\pi/3} = \frac{\sqrt{2}}{\pi/3} V_{\mathrm{Vr}} \int_{\alpha - 30^\circ}^{\alpha + 30^\circ} \cos\theta \, d\theta \tag{16.91}$$

where V_{V_r} is the root-mean-square (RMS) voltage (kV) applied to the converter valve, and α is the gate delay angle. The RMS voltage applied to the converter can be expressed as:

$$V_{\mathrm{Vr}} = t_{\mathrm{r}} \, V_{\mathrm{ar}} \, V_{\mathrm{Br}} \tag{16.92}$$

where subscript r denotes rectifier and subscript i denotes inverter, and:

V_{ar}, V_{ai} = voltage magnitude (pu) at the "commutating bus"
V_{Br}, V_{Bi} = base voltage (kV) on transformer secondaries
t_{r}, t_{i} = transformer tap ratio $\left[\dfrac{\text{converter side (pu)}}{\text{AC bus side (pu)}}\right]$

The $\sqrt{2}$ in Equation (16.91) is required to obtain the peak value from the RMS value. Integration of Equation (16.91) and Equation (16.92) gives:

$$V_{\mathrm{dr}} = \frac{3\sqrt{2}}{\pi} t_{\mathrm{r}} \, V_{\mathrm{ar}} \, V_{\mathrm{Br}} \cos\alpha \tag{16.93}$$

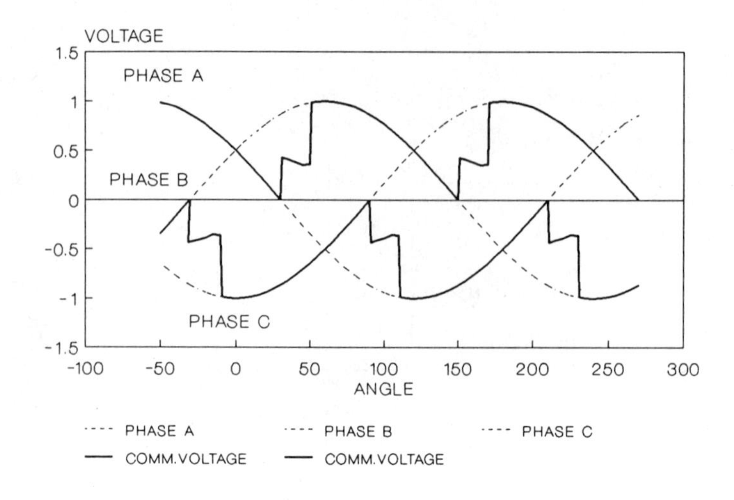

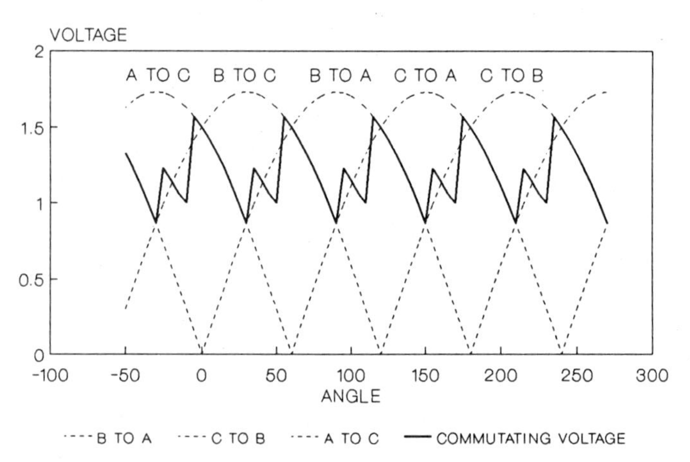

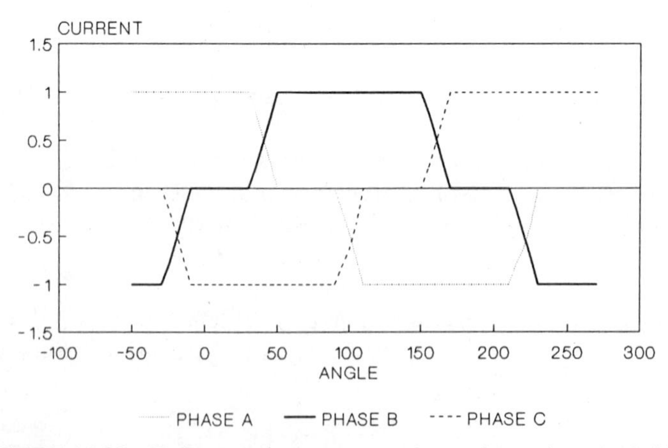

FIGURE 16.36 Voltage and current waveforms for a six-pulse bridge.

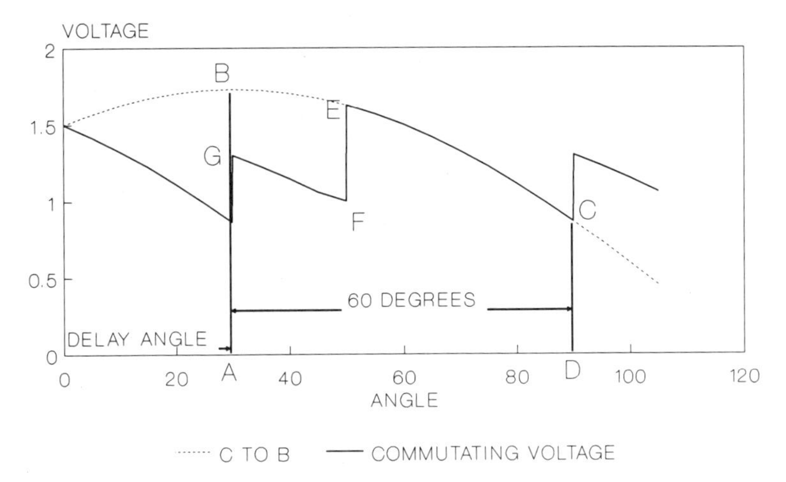

FIGURE 16.37 Analysis of six-pulse converter bridge.

For operation in the inverter mode ($\alpha > 90°$), it was shown that the waveforms are the same as for rectifier operation, except that the extinction angle (γ) replaces α and the polarity is reversed. The relation between α and γ depends on the overlap angle (μ) since

$$\alpha + \gamma + \mu = 180° \tag{16.94}$$

If γ becomes zero, a commutation failure will occur, resulting in a short-circuit within the converter and a failure to invert. A minimum γ, typically about 12°, is maintained under normal conditions to prevent this. Both the minimum α and γ are typically set at 15–18° to provide control range in both directions.

The effect of the commutation overlap on the direct voltage can be obtained by integrating the voltage drop in the commutating reactance over each 60° segment of the waveform, area *GBEF* in Figure 16.37. From Equation (16.90), we obtain:

$$V_X = \frac{\int_0^{\pi/3} V\,dt}{\pi/3} = \frac{\omega L_c \int_0^{I_d} di}{\pi/3} \tag{16.95}$$

$$V_X = \frac{3}{\pi} I_d X_c \tag{16.96}$$

This voltage drop is subtracted from the value obtained in Equation (16.93) to give the final value for the average direct voltage.

$$V_{dor} = \frac{3\sqrt{2}}{\pi} t_r V_{ar} V_{Br}$$

$$V_{dr} = V_{dor} \cos \alpha - \frac{3}{\pi} X_{cr} I_d$$

where V_{dor} = rated no load voltage at zero gate delay.

Figure 16.38 illustrates the overall volt-ampere relationship of V_d and I_d with α and γ. The rectifier typically operates with a minimum delay angle of 18°. As current is increased, voltage decreases. The inverter has the same overall characteristics but has a negative voltage. The inverter typically operates with a minimum extinction angle of 10–20°. The rectifier delay angle is typically regulated to control the current on the DC line, while the inverter delay angle is typically regulated to control the voltage. The example in Section 16.14.4 will illustrate this.

The active power for a single pole is obtained simply as the product of the direct current and voltage:

$$P_r = V_{dr} I_{dr} \tag{16.97}$$

The corresponding AC power is given by:

$$P = \sqrt{3}\, t \cdot V_{ar} \cdot V_{Br} \cdot I_{Ar} \cdot \cos \phi \tag{16.98}$$

where I_A is the AC current in kiloamperes, on the converter side of the transformer. Therefore:

$$\cos \phi = \frac{V_{dr} I_d}{\sqrt{3}\, t \cdot V_{ar} \cdot V_{Br} \cdot I_{Ar}} \tag{16.99}$$

and V_{dor} is defined as:

$$V_{dor} = \frac{3\sqrt{2}}{\pi} t \cdot V_{ar} \cdot V_{Br} \tag{16.100}$$

With perfect filtering, only the AC fundamental current component enters or leaves the DC system. The magnitude of the current can be evaluated by analyzing the Fourier series of the phase A current.

Assume an ideal case of no commutation overlap. The phase current is equal to unity at 0° through 120°, is 0 from 120° to 180°, is negative from 180° to 300°, and is 0 from 300° to 360°. The Fourier coefficient (Kreyszig, 1962) for the fundamental component is (after shifting the X axis by 30° for symmetry):

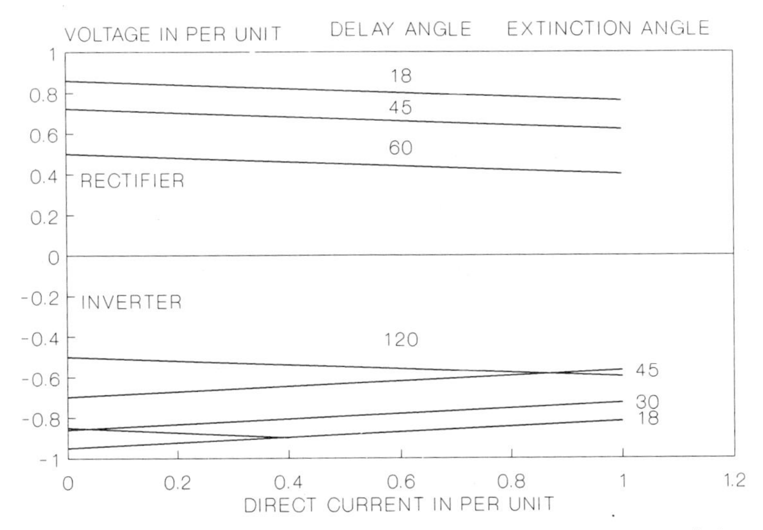

FIGURE 16.38 Rectifier and inverter voltage current characteristics.

$$\text{Fourier coefficient} = \frac{1}{\pi}\int_{30^\circ}^{150^\circ} 1 \sin\theta \, d\theta + \frac{1}{\pi}\int_{210^\circ}^{330^\circ} -1 \sin\theta \, d\theta$$

$$= \frac{4\sqrt{3}}{2\pi} = \frac{2\sqrt{3}}{\pi}$$

In this case the fundamental component of the phase A current is:

$$I_A = \frac{2\sqrt{3}}{\pi}\left(\frac{1}{\sqrt{2}}\right) I_d = \frac{\sqrt{6}}{\pi} \cdot I_d$$

where $\sqrt{2}$ is inserted to convert the phase current to a RMS value.

A more detailed Fourier analysis of the current waveform (Adamson and Hingorani, 1960) leads to:

$$I_A = \frac{\sqrt{6}}{\pi} \cdot k \cdot I_d \tag{16.101}$$

where k is a factor very close to unity and includes the effects of the commutation overlap.

The value of k is given by:

$$[k = \frac{\sqrt{[\cos 2\alpha - \cos 2(\alpha + \mu)]^2 + [2\mu + \sin 2\alpha - \sin 2(\alpha + \mu)]^2}}{4[\cos\alpha - \cos(\alpha + \mu)]} \tag{16.102}$$

for values of $\mu \leq 60°$, which is the case for HVDC systems. This value is about .995 for normal operation and is usually taken to be 1. Using the k value of 1. gives the equations for reactive power in Equations (16.121) and (16.125).

Substituting Equations (16.100) and (16.101) into Equation (16.99) yields:

$$\cos \phi = \frac{V_d \cdot I_d}{k \cdot V_{dor} \cdot I_d} = \frac{V_d}{k V_{dor}} \tag{16.103}$$

The equation for the inverter, the DC line, and the rectifier is

$$\begin{aligned} V_{dr} &= V_{di} + R_d I_d \\ &= V_{dor} \cos \alpha - \frac{3}{\pi} X_{cr} I_d \\ &= V_{doi} \cos \gamma - \frac{3}{\pi} X_{ci} I_d + R_d I_d \end{aligned}$$

The current is then solved as:

$$I_d = \frac{V_{dor} \cos \alpha - V_{doi} \cos \gamma}{R_d - 3/\pi\,(X_{ci} - X_{cr})}$$

Because the current is dependent on the difference between the rectifier and inverter voltages, a small change in the transformer tap settings or gate angles can produce a large change in DC current (and thereby DC power transmission).

16.14.3 Per Unit Definitions

The commutating reactance is normally given as a per unit value of the converter transformer rating. Assuming that the transformer rating is consistent with the DC system requirements, the impedance base value is given by:

$$X_{cB} = \frac{\pi}{6} \frac{V_{doN}}{I_{dN}} \tag{16.104}$$

where I_{dN} = rated direct current (kA), V_{doN} = rated ideal direct voltage (kV), and the subscript N denotes the normalized base value. This assumes a transformer design based on current flow at rated DC value for two-thirds of each cycle, as shown in Figure 16.37, so the RMS current is:

$$I_{LN} = \left[\frac{1}{3}(0) + \frac{2}{3} I_{dN}^2\right]^{\frac{1}{2}} \tag{16.105}$$

$$I_{LN} = \frac{\sqrt{2}}{\sqrt{3}} I_{dN} \tag{16.106}$$

The volt-ampere rating of the transformer is, therefore:

$$MVA_N = \sqrt{3}\, V_{VN} I_{LN} = \sqrt{3}\, \frac{V_{doN}}{(3\sqrt{2}\ /\pi)} \frac{\sqrt{2}}{\sqrt{3}} I_{dN} \tag{16.107}$$

or

$$MVA_N = \frac{\pi}{3} I_{dN} V_{doN} \tag{16.108}$$

The reactance base is then given by:

$$X_{cB} = \frac{(V_{VN})^2}{MVA_N} = \frac{[\pi/3\sqrt{2})\ V_{doN}]^2}{(\pi/3)\ I_{dN}\ V_{doN}} \tag{16.109}$$

$$X_{cB} = \frac{\pi}{6} \frac{V_{doN}}{I_{dN}} \tag{16.110}$$

The value of V_{doN} can be related to the rated direct voltage V_{dN} as follows:

$$V_{dN} = V_{doN} \cos\alpha - \frac{3}{\pi} X_c I_{dN} \tag{16.111}$$

or

$$V_{dN} = V_{doN} \cos\alpha - \frac{3}{\pi} \overline{X}_c \cdot X_{cB} I_{dN} \tag{16.112}$$

$$V_{dN} = V_{doN} \cos\alpha - \frac{3}{\pi} \overline{X}_c \left(\frac{\pi}{6} \frac{V_{doN}}{I_{dN}} \right) I_{dN} \tag{16.113}$$

$$V_{dN} = V_{doN} \left(\cos\alpha - \frac{\overline{X}_c}{2}\right) \tag{16.114}$$

For typical values $\alpha = 18°$ and $\overline{X}_c = .13$ pu:

$$V_{doN} = 1.13\ V_{dN} \tag{16.115}$$

Thus, the formula for the impedance base can be written as:

$$X_{cB} \approx 1.13 \frac{\pi}{6} \frac{V_{dN}}{I_{dN}} \tag{16.116}$$

For many analysis purposes, the simplified representation of a two-terminal HVDC system shown in Figure 16.39 can be used, assuming completely balanced operation. The buses indicated by V_{ar} and V_{ai} are termed the "commutating buses." They are the closest points where sinusoidal voltage waveforms, unaffected by converter ripple, can be assumed. Often, this is the bus at the high side of the converter transformer. The DC voltage source is proportional to the cosine of the delay and extinction angle. The rectifier commutating reactance appears to the DC equations as a resistance, and the inverter commutating reactance appears to the DC equations as a negative resistance.

The basic equations for the HVDC system are summarized by component as follows:

DC line:

$$V_{dr} = V_{di} + R_d I_d \tag{16.117}$$

Rectifiers:

$$V_{dor} = \frac{3\sqrt{2}}{\pi} t_r V_{ar} V_{Br} \tag{16.118}$$

$$V_{dr} = V_{dor} \cos \alpha - \frac{3}{\pi} X_{cr} I_d \tag{16.119}$$

$$P_r = N_P V_{dr} I_d \tag{16.120}$$

$$Q_r \approx P_r \tan \left[\cos^{-1} \frac{V_{dr}}{V_{dor}}\right] \tag{16.121}$$

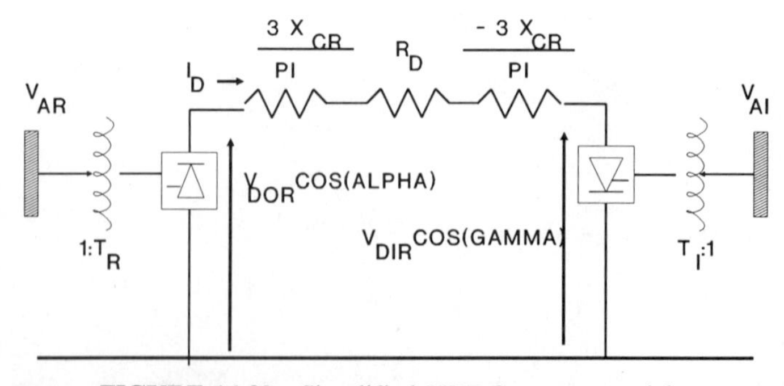

FIGURE 16.39 Simplified HVDC system model.

Inverters:

$$V_{doi} = \frac{3\sqrt{2}}{\pi} t_i V_{ai} V_{Bi} \tag{16.122}$$

$$V_{di} = V_{doi} \cos \gamma - \frac{3}{\pi} X_{ci} I_d \tag{16.123}$$

$$P_i = N_P V_{di} I_d \tag{16.124}$$

$$Q_i \approx P_i \tan [\cos^{-1} \frac{V_{di}}{V_{doi}}] \tag{16.125}$$

It is also necessary to determine a base for the AC voltage on the secondary (converter) side of the transformer, as:

$$V_B = \frac{V_{dN} + (3/\pi) X_c I_{dN}}{(3\sqrt{2}/\pi) \cos \alpha_o} \tag{16.126}$$

where

α_0	= normal firing angle (replace with γ_0 for inverter).
V_{ar}, V_{ai}	= voltage magnitude (pu) at "commutating bus" (AC high-voltage side)
X_{cr}, X_{ci}	= commutating reactance (Ω) per pole
V_{dr}, V_{di}	= direct voltage (kV) per pole
V_{Br}, V_{Bi}	= base voltage (kV) on transformer secondaries (converter side)
I_d	= direct current (kA)
R_d	= dc line resistance (Ω) per pole
α	= firing angle of rectifier
γ	= extinction angle of inverter
P_r	= active power (MW) from AC system to rectifier
P_i	= active power (MW) from inverter to AC system
Q_r, Q_i	= reactive power (Mvar) drawn from AC system
t_r, t_i	= transformer tap ratio $\left[\frac{\text{converter side (pu)}}{\text{AC bus side (pu)}}\right]$
N_P	= number of poles (one or two)

These equations are written so that all quantities are positive for normal operation. The converter angle of one converter, usually the inverter, is used to control the direct voltage (V_{di}). The other converter controls the direct current (I_d). Transformer taps are used to keep the converter angles within range during steady-state operation for varying AC bus voltages and DC levels.

Typical parameters for HVDC systems are given in Table 16.14. The line

TABLE 16.14 Typical HVDC System Parameters

Number of Poles N_P	Rated Power (MW)	Rated Voltage (kV/Pole) V_{dN}	Rated Current (kA) I_{dN}	Line Length (Miles)	Line Resistance (Ω/Mile) R_d	Costs 1990 $: Line Costs per Mile (Million $)	Costs 1990 $: Terminal Costs (Million $)
2	2000	±500	2.0	850	.02	.45	100
2	2000	±450	2.22	1000	.018		
2	3150	±600	2.625	500	.014		
1	200	82	2.44	–	–		
2	1000	145	3.45	–	–		

Commutating reactance (X_c)	.12–.18 pu
Normal converter angle (α, γ)	16–18° (±1–2°)
Minimum rectifier angle (α_{min})	8°
Minimum inverter angle (γ_{min})	12°
Transformer tap range (t)	±10%
Number of transformer tap steps	16 or 32

costs of a HVDC system are in the range of two-thirds of the comparable AC line costs. The terminal costs are in the range of $50/kW.

For multiterminal HVDC systems, the previous converter equations still apply, except that there may be more than one rectifier and/or inverter. The DC line equation is replaced by an admittance (conductance) matrix relating the direct voltages and currents at all the terminals. All but one of the converters control the direct current inflow or outflow at their nodes; the remaining converter controls the direct voltage.

16.14.4 Example

A bipolar ±500-kV HVDC line is 400 miles long with a line resistance of 0.015 Ω/mile. The converters are rated for 2000 A with commutating reactance of .13 pu. Given the following conditions:

$$V_{ar} = 1.03 \text{ pu}$$
$$I_d = 2000 \text{ A}$$
$$\alpha = \alpha_o = 18°$$
$$V_{ai} = 1.05 \text{ pu}$$
$$V_{di} = 488 \text{ kV}$$
$$\gamma = 18°$$

determine the transformer per unit tap ratios required to obtain these conditions and the resulting P and Q at each converter.

Compute base values:

$$V_{dN} = 500 \text{ kV}$$

$$I_{dN} = 2 \text{ kA}$$

$$X_{cBr} = X_{cBi} = 1.13 \frac{\pi}{6} \frac{500}{2} = 147.9 \ \Omega$$

$$X_{cr} = X_{ci} = (.13)(147.9) = 19.2 \ \Omega$$

$$V_{Br} = V_{Bi} = \frac{500 + (3/\pi)(19.2)(2)}{(3\sqrt{2}/\pi) \cos 18°} = 418.0 \text{ kV}$$

At inverter:

$$V_{di} = V_{doi} \cos \gamma - \frac{3}{\pi} X_{ci} I_d$$

$$V_{doi} = \frac{488 + (3/\pi)(19.2)(2.)}{\cos 18°} = 551.7 \text{ kV}$$

$$V_{doi} = \frac{3\sqrt{2}}{\pi} t_i V_{ai} V_{Bi}$$

$$t_i = \frac{551.7}{(1.35)(1.05)(418.0)} = .931 \text{ pu}$$

$$P_i = N_p V_{di} I_d = (2.)(488)(2) = 1952 \text{ MW}$$

$$Q_i = P_i \tan [\cos^{-1} \frac{V_{di}}{V_{doi}}] = 1952 \tan [\cos^{-1} \frac{488}{551.7}]$$

$$Q_i = 1029 \text{ Mvar}$$

At rectifier:

$$V_{dr} = V_{di} + R_d I_d = 488 + (400)(.015)(.2)$$

$$V_{dr} = 500 \text{ kV}$$

$$V_{dor} = \frac{500 + (3/\pi)(19.2)(2.)}{\cos 18°} = 564.3 \text{ kV}$$

$$t_r = \frac{564.3}{(1.35)(1.03)(418.0)} = .971 \text{ pu}$$

$$P_r = (2.)(500)(2.) = 2000 \text{ MW}$$

$$Q_r = 2000 \tan [\cos^{-1} \frac{500}{564.3}] = 1046 \text{ Mvar}$$

Thus, the losses in transmitting the power are 48 MW or 2.4%. The losses may be reduced by increasing the DC conductor size. The selection of conductor size is an economic tradeoff between the increased cost of the transmission line and the value of decreased losses. Losses of 0.5–1%/100 miles are typical.

If the scheduled power were to be reduced to 1000 MW with 500 kV at the rectifier, then repeating the above calculations yields:

$$P_{rectifier} = 1000 \text{ MW}$$
$$Q_{rectifier} = 430 \text{ Mvar}$$

$$\begin{aligned} P_{\text{inverter}} &= 988 \text{ MW} \\ Q_{\text{inverter}} &= 430 \text{ Mvar} \\ \\ \text{Losses} &= 12 \text{ MW} \end{aligned}$$

If the power were to be suddenly reduced to 1000 MW, then the firing angle of the rectifier would need to be changed to reduce the current while the inverter angle is adjusted to maintain voltage. This is necessary because the transformer tap adjustment is too slow to respond to sudden power-flow commands. Repeating the calculations yields:

$$\begin{aligned} P_{\text{rectifier}} &= 1000 \text{ MW} \\ Q_{\text{rectifier}} &= 522 \text{ Mvar} \\ \alpha_{\text{rectifier}} &= 23.2^\circ \\ \\ P_{\text{inverter}} &= 988 \text{ MW} \\ Q_{\text{inverter}} &= 492 \text{ Mvar} \\ \gamma_{\text{inverter}} &= 21.8^\circ \\ \\ \text{Losses} &= 12 \text{ MW} \end{aligned}$$

Note that the losses are the same in the last two cases but reactive power requirements are higher for transient adjustments than for steady-state transformer tap adjustments.

Incorporation of HVDC systems into power-flow calculations requires that these DC system equations be solved simultaneously with the normal AC power-flow equations. This can be done by either (1) converting the DC equations to a consistent per unit base and adding them to the AC equation set or (2) solving the AC and DC equations separately and iteratively.

The normal constraints for power-flow calculations are similar to the previous example. For these constraints, the direct voltage at the inverter, direct current, and desired firing angles are given, together with the AC bus voltages as determined by the AC network solution. The required tap ratios and P, Q values are computed. The latter are added to the P and Q injections at the AC buses for the next iteration of the AC solution, which may result in a change in bus voltages. It may also be necessary to change the firing angles from the desired values if (1) transformer tap limits are exceeded or (2) discrete tap steps are modeled. For a more detailed discussion of the incorporation of HVDC systems into power flow, see Arrillaga et al. (1983).

16.14.5 HVDC Transmission Planning Summary

In planning the addition of an HVDC system, engineering studies must be performed in a number of areas to ensure successful integration of the AC and DC systems. These analyses include:

- Steady-state performance studies to specify required transformer tap ranges and assess the impact of different power transfer levels and directions and impacts on AC system conditions.
- Reactive compensation studies to determine the reactive power requirements of the converters and appropriate methods to regulate AC voltage under varying operating conditions.
- Studies of measures to limit dynamic overvoltage when DC power transfer is suddenly interrupted.
- Filter specification studies to limit any harmful effects of ripple generated by the converters.
- Insulation coordination studies among the components of the DC and AC systems for determining anticipated overvoltages due to lightning, switching transients, and other cases.
- Dynamic performance studies to assess the impact on system stability and specify HVDC controls to improve AC system dynamic performance. This topic is discussed in more detail in Section 17.13.

16.15 HORIZON-YEAR TRANSMISSION PLANNING

In transmission planning, it is generally useful to plan the network on a horizon-year basis. The horizon year may be 10 or 20 years into the future.

An example will illustrate the reasoning for a horizon-year approach. A 5000-MW power system utilizes a 345-kV transmission line voltage "backbone." The total system annual load growth might be 125 MW/year. For this system, a 300-MW generation addition every other year (i.e., on alternate-years basis) would be adequate to meet capacity addition requirements. Total system load growth, however, is comprised of very small load growth increments at each load bus. A system represented by 100 load buses would have, on the average, an annual load bus growth of only 1.25 MW. A new 345-kV transmission line has a surge impedance loading (SIL) capability of 400 MW. Since it would not be obvious where a new transmission line is needed, the placement decision is dependent on the future transmission network requirements.

An alternative approach is to first examine the transmission system in a horizon year. According to the insights gained though the horizon-year perspective, a yearly transmission plan can be developed that builds toward the horizon year. Figure 16.40 illustrates this concept.

The transmission network must be designed to meet reliability or adequacy criteria. When viewed from a steady-state condition (no transients), typical criteria are:

1. The network in its "normal state" must not have any lines overloaded beyond the thermal line limit or terminal equipment rating. Typically this limit is most applicable to short transmission lines.

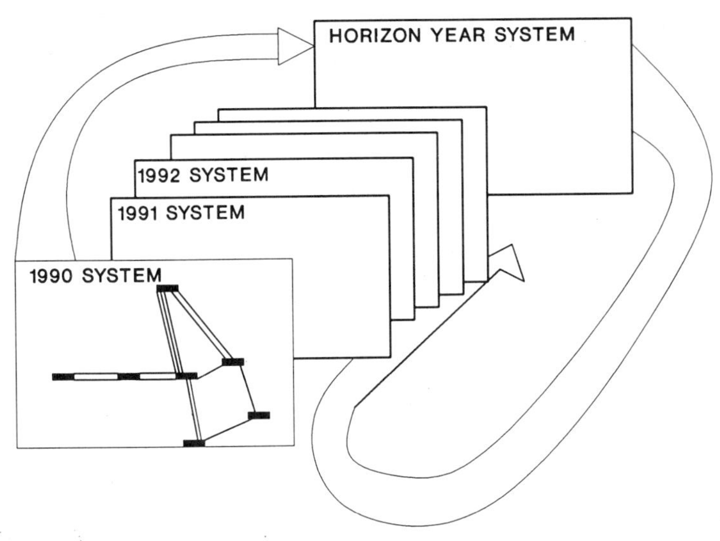

FIGURE 16.40 The horizon-year approach to transmission expansion planning.

2. The network in its "normal state" must have no voltages less than V_{min} nor greater than V_{max}, where V_{min} is typically 95–98% and V_{max} is 103–105%.
3. The network in its "normal state" must not have any uncompensated transmission line loaded beyond an angle of 45°.
4. The network must be capable of supplying full load during any single-line (three-phase) outage (single-line contingency) with:
   a. No bus voltage less than, CV_{min}, where CV_{min} is in the range of 90–95%.
   b. No transmission line loaded beyond its emergency line-rating limit (typically a 10-minute overload rating based on thermal line limits).
   c. No uncompensated transmission line loaded beyond an angle of 60°.
5. The network must be capable of supplying full load subject to the criteria of rule 4. for any single outage of a generating unit, transformer or other power system equipment.

In some special cases, a transmission right-of-way may be subject to outage due to high winds and bad weather. In this case, the transmission design criteria are modified to include weather-related phenomena as one of the criteria.

Another criterion often used is a security constraint with a line on maintenance. If any transmission line or element is out on maintenance, the transmission system must be capable of meeting the "normal state" criteria by a redispatch of generating units. The redispatch of generating units permits the line loadings to be changed such that the network is secure with regard to any single-line outage. This can usually be accomplished if the previous criteria are achieved. However, there can be special cases where this constraint would require additional transmission equipment to be added.

The transmission network design must be adequate for all load demands throughout the year. Although the peak-load hour is usually the most limiting, other hours with slightly lower loads can be more limiting. Consider a power system in which the most efficient generation is far from the load center and peaking generation is located at the load center. During peak-load periods, peaking generation is operating and providing var and voltage support at the load center. For lower load demands, peaking generation may not be running and may not be on line. In this case, there is no var or voltage support at the load center, and the voltage at the load center could drop below the voltage design criteria. Several load conditions may need to be examined.

If the transmission system is inadequate, then additional transmission equipment is required. The equipment may include new transmission lines, additional compensation, and additional terminal equipment. The objective is to add equipment that minimizes the present worth of transmission system cost while meeting the transmission design criteria. There are many alternative ways to achieve the transmission design criteria. The minimum present-worth alternative is found by evaluating the present worth of the annual cost of each alternative and selecting the lowest cost alternative.

Figure 16.41 illustrates a typical transmission planning approach based on

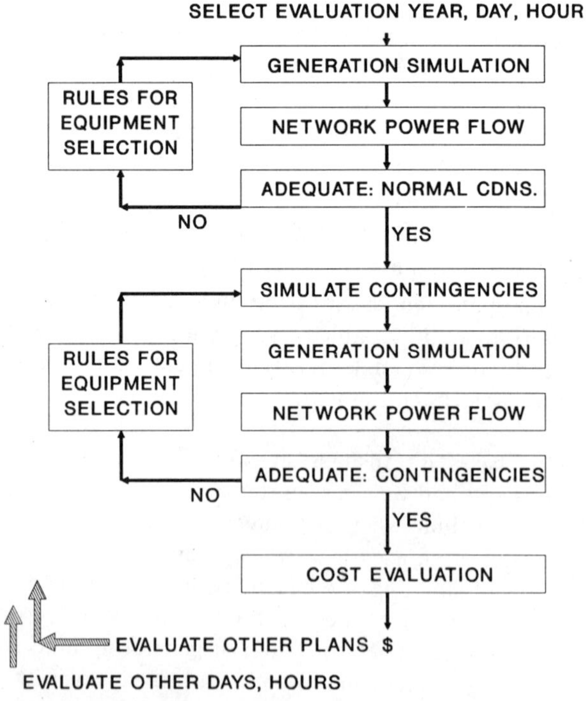

FIGURE 16.41 Typical transmission planning approach.

steady-state design criteria. The system is evaluated on the basis of the specified year, month, day, and hour. These parameters determine the system load demand as well as bus power (active, reactive) demands. The generation system is simulated to determine which generating units are operating and at what power level. Following this, the network power flow is calculated. If the system is not adequate, then a set of transmission selection rules must be adopted to add new equipment that satisfies the transmission design criteria. After the system is designed to be adequate under "normal" conditions, the system must be tested and modified, if necessary, to be adequate for contingency situations. The transmission system is then adequate for the year, and the cost of any new equipment added this year is evaluated.

This calculation would be repeated each year as the transmission network is designed toward the "horizon-year" network. The present worth of this alternative is computed according to the cost of each annual equipment addition and the cost of power system losses.

The present-worth costs of alternative transmission designs derived from using alternative rules of equipment selection are compared. The alternative with the least cost is selected.

In summary, this section illustrates a steady-state transmission planning analysis based on a horizon-year focus. While a steady-state transmission system analysis is useful, a transient transmission system analysis should be performed. Transient considerations are discussed in Chapter 17.

16.16 TRANSMISSION PLANNING EXAMPLE

The example presented in this section is a horizon-year transmission plan. The transmission network is hypothetical, but the bus locations correspond to portions of an actual system.

Example. The existing network is graphically illustrated in Figure 16.42. All the lines are 345 kV except those to BI, which are 230 kV. The specific transmission-line data are listed in Table 16.15.

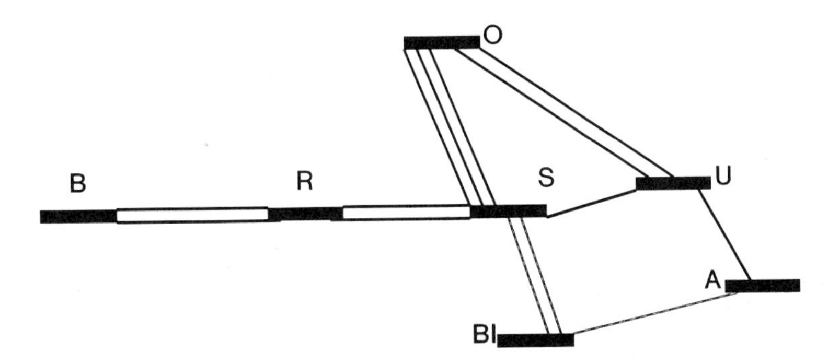

FIGURE 16.42 Hypothetical transmission network.

TABLE 16.15 Transmission Network

Origin Name	Destination Name	R-O-W Origin Number	R-O-W Destination Number	Distance (Miles)	kV Level	Number of Circuits
A	U	1	2	75	345	1
A	BI	1	4	100	230	1
U	S	2	3	50	345	1
U	O	2	7	75	345	2
S	BI	3	4	100	230	2
O	S	7	3	50	345	3
S	R	3	5	75	345	2
R	B	5	6	100	345	2

In the horizon year, year 2000, the total system load is forecast to be 8000 MW. The loads at each bus, in per unit of peak load, are listed in Table 16.16.

The generation capacity in the year 2000 is 10000 MW, comprised of the capacity types at each bus as listed in Table 16.17.

The characteristics of each generating unit are based on fuel type. Table 16.18 presents these data based on projected costs in year 2000.

Design the network to achieve a 95% minimum bus voltage in the "normal state" with no transmission line angle greater than 45°; a 92% minimum bus voltage and a maximum line angle of 60% for all single-line contingencies.

SOLUTION. The power system is simulated for the peak hour in the year 2000 (the July–September time period). The dispatch of these units is shown in Table 16.19 (units were derated by the forced outage rates).

The network has considerable generation (75%) at B and O and a significant (57%) amount of load at U, S, A, and BI. On the basis of these parame-

TABLE 16.16 Per Unit Load Table

Location Number	Name	Per Unit Load
1	A	.19200
2	U	.09700
3	S	.19300
4	BI	.09500
5	R	.16000
6	B	.26300
7	O	.00000
8	SA	.00000
9	S	.00000
10	E	.00000

TABLE 16.17 Generating Units

Plant No.	Unit Name	Location No.	Location Name	Service Year	MW Size	Heat Rate (MBtu/MWh)	Fuel Type	Dispatch Penalty Factor	Forced Outage Rate
1	A-Stm	1	A	51	400	10.000	1	1.00	0.100
2	A-Gt	8	SA	68	300	12.000	2	1.00	0.080
3	BI-Gt	4	BI	70	200	12.000	2	1.00	0.080
4	O-Stm	7	O	65	1200	10.000	1	1.00	0.100
5	O-Nuc	7	O	78	1600	10.000	3	1.00	0.150
6	O-Coal	7	O	82	1300	10.000	4	1.00	0.150
7	R-Stm	5	R	74	600	10.000	1	1.00	0.100
8	B-Coal	6	B	76	3400	10.000	4	1.00	0.150
9	B-Hyd	6	B	58	1000	10.000	6	1.00	0.050

ters, the transmission network is moving a large amount of power from B to buses 200–400 miles east.

The power flow for this existing transmission network did not converge. The nonconverged solution suggested a problem at A. A second line was added from U to A. The resulting power flow is presented in Figure 16.43. This figure presents the voltage and angle at each bus. The lowest voltage is .856 pu at BI. The largest angle across any line is 31° from S to BI.

The next step is to improve the bus voltage levels by adding lines. A line was added from B to R to improve the voltage along the network as illustrated in Figure 16.44.

A line was added from S to BI and S to U to improve voltage to the 95% level. The final network satisfying the design goals under "normal" conditions is presented in Figure 16.45.

Now the network must be tested for single-line outage contingencies. The procedure is to outage one line at a time, compute the power flow, check for acceptable design limits, and add any necessary lines to ensure that the network will be adequate. Table 16.20 summarizes the contingency cases.

The final network is presented in Figure 16.46. The voltage and angle values

TABLE 16.18 Generation Types

Type No.	Type Name	Fuel Cost ($/MBtu)	O&M Cost $/kW/year	O&M Cost $/MWh	Minimum Power (pu)	Fuel at Min. Power (pu)	Minimum Down Hours
1	Resid-Oil	10.000	10.0	2.0	0.250	0.280	6
2	Dist-Oil	14.000	8.0	6.0	0.900	0.900	0
3	Nuclear	2.000	30.0	10.0	0.300	0.340	60
4	Coal-East	4.000	25.0	15.0	0.400	0.440	30
5	Coal-Low SO_2	4.500	25.0	10.0	0.400	0.440	30
6	Hydro	0.000	12.0	4.0	0.001	0.001	0

TABLE 16.19 Hourly Economic Dispatch July-September 2000

Priority No.	Unit No.	Unit Name	MW Size	Weekday—Hour 1 Available MW Capacity	Weekday—Hour 1 Dispatch MW Output	Fuel Cost ($/Hour)	Fuel Type
1	9	B-Hyd	1000	950	950	0	6
2	5	O-Nuc	1600	1360	1360	27,200	3
3	8	B-Coal	3400	2890	2890	115,600	4
4	6	O-Coal	1300	883	883	35,312	4
5	1	A-Stm	400	360	360	36,000	1
6	7	R-Stm	400	360	360	36,000	1
7	4	O-Stm	1200	863	783	78,639	1
8	2	A-Gt	300	276	248	41,731	2
9	3	BI-Gt	200	184	166	27,821	2
10	10	Emergency	4000	4000	0	3,600	7
					8000	401,903	

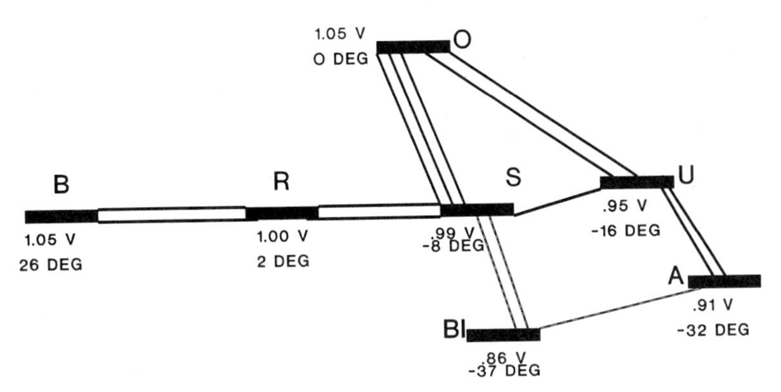

FIGURE 16.43 Hypothetical transmission network with a problem at "A" resolved by adding a second line from U to A.

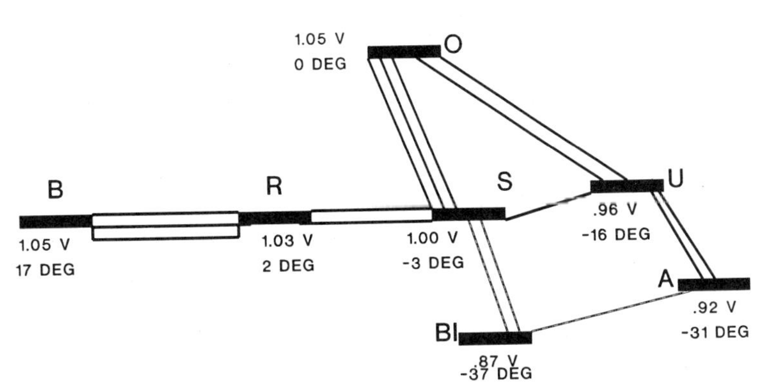

FIGURE 16.44 Hypothetical transmission network with a second line added to improve voltage along the network.

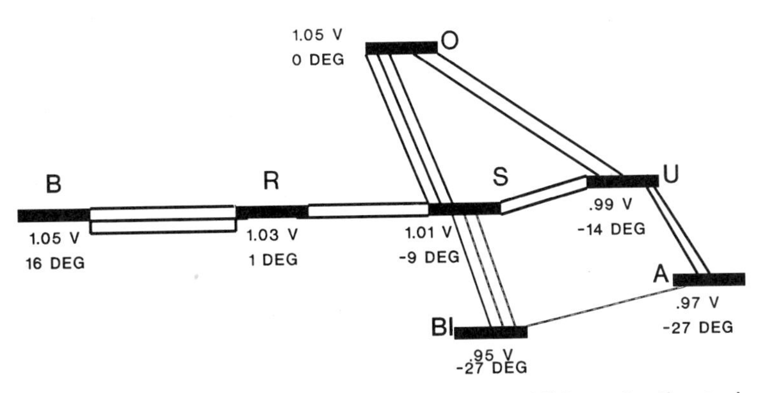

FIGURE 16.45 Hypothetical transmission network—addition of a line to improve voltage and satisfy design goals.

TABLE 16.20 Outage Contingency Cases

Line Contingency Case: Outage Line		Lowest Bus Voltage		
From	To	Value	Location	Solution Remedy
B	R	.945	BI	None
S	R	.938	BI	None
O	S	.930	A	None
S	BI	.887	BI	Add line: S–BI
A	BI	.967	A	None
U	O	.932	A	None
U	S	.961	A	None
A	U	.809	A	Add line: A–U

are those corresponding to the "normal" state. This network will pass all single-line contingency tests.

In summary, a one-horizon-year transmission plan was developed. A yearly plan can now be constructed by analyzing each year and building toward the horizon-year plan.

Many alternative horizon-year plans and annual year implementations can be developed by using variations in line loading due to compensation, intermediate voltage support, different voltage levels, alternative routing, and other factors. The transient stability of this network also needs examination and is discussed further in Chapter 17. The alternative to be selected is a judgment based on many factors, including minimum cost, flexibility to alternative forecasts of load growth, capacity addition, and other considerations, including reliability of the power system and environmental feasibility.

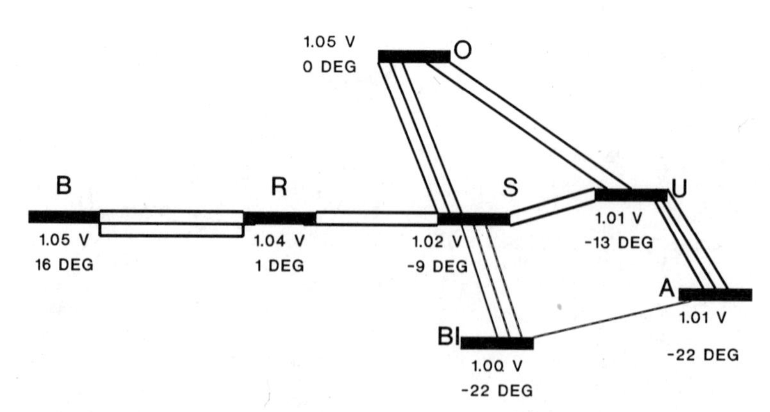

FIGURE 16.46 Hypothetical transmission network—designed to pass all single-line contingencty tests.

REFERENCES

Adamson, C., and N. G. Hingorani, *HVDC Power Transmission*, Garraway Ltd., London, 1960.

ALCOA, "Current Temperature Characteristics of Aluminum Conductors," *ALCOA Conductor Engineering Handbook*, 1958, Aluminum Company of America.

Arrillaga, J., C. P. Arnold, and B. J. Harker, *Computer Modeling of Electrical Power Systems*, Wiley, New York, 1983.

Davidson, G. A., "Short-Time Thermal Ratings for Bare Overhead Conductors," *IEEE Transactions on Power Apparatus and Systems*, Vol. PAS-88, 1969, pp. 194–199.

Dunlop, R. D., R. Gutman, and P. P. Marchenko, "Analytical Development of Loadability Characteristics for EHV and UHV Transmission Lines," *IEEE Transactions on Power Apparatus and Systems*, Vol. PAS-98, 1979, pp. 606–617.

Eaton, J. R., and E. Cohen, *Electric Power Transmission Systems*, Prentice-Hall, Englewood Cliffs, NJ, 1983.

Fink, D. G., and H. W. Beaty, "Direct Current Power Transmission," in *Standard Handbook for Electrical Engineers*, 12th ed., McGraw-Hill, New York, 1987, Chapter 15.

Gross, C. A., *Power System Analysis*, Wiley, New York, 1979.

Kreyszig, E., *Advanced Engineering Mathematics*, Wiley, New York, 1962.

Miller, T. J. E., *Reactive Power Control in Electric Systems*, Wiley, New York, 1982.

Neuenswander, J. R., *Modern Power Systems*, International Text Book Company, 1971.

St. Clair, H. P., "Practical Concepts in Capability and Performance of Transmission Lines," *AIEE Transactions on Power Apparatus and Systems*, Vol. PAS-72, 1953, pp. 1152–1157.

Stevenson, W. D., Jr., *Elements of Power Systems Analysis*, McGraw-Hill, New York, 1962.

Stott, B., and O. Alsac, "Fast Decoupled Load Flow," *IEEE Transactions on Power Apparatus and Systems,* Vol. PAS-74, 1974, pp. 859-869.

Weedy, B. M., *Electric Power Systems*, Wiley, New York, 1979.

PROBLEMS

1 A remote 500-MW generating unit is to be installed in 1995 and is located 150 miles from a load center. The generating unit is expected to be base-loaded. On the basis of the line loadability guide of Figure 16.16 and the economic data of Table 16.6, determine which line design voltage, 230 kV, 345 kV, or 500 kV, is least-cost. Use a single-outage contingency criterion and neglect losses.

2 A second unit addition of 500 MW is to be installed in year 2000. Using the data of Problem 1, determine which line design voltage and the number and timing of the line additions is least-cost. Use a 5%/year annual cost escalation rate, a 10%/year present-worth rate, and an 18%/year levelized fixed-charge rate. Using a present-worth cost analysis over the 1995 to 2015 time horizon.

3 A power system consists of three buses connected by 345-kV lines, 200 miles long. The bus loads (100-MVA base) are:

$$\begin{aligned} \text{Load}_1 &= 0 \\ \text{Load}_2 &= 6 + j.6 \\ \text{Load}_3 &= 6 + j.6 \end{aligned}$$

The line admittance is:

$$\begin{aligned} y_{12} &= -j20 + 2 \\ y_{23} = y_{13} &= -j10 + 1 \end{aligned}$$

The line admittance to ground is:

$$\begin{aligned} b_{12} &= j3.4 \\ b_{23} = b_{13} &= j1.7 \end{aligned}$$

A generator is located at bus 1 and holds the voltage at 1.05. Solve for the power flows using the fast decoupled technique.

4 In the power system shown in Figure 16.24 the bus loads grow by year 2010 to:

$$\begin{aligned} \text{Load bus}_2 &= 7 + j.7 \\ \text{Load bus}_3 &= 9 + j.9 \end{aligned}$$

Using the data of the example near the end of Section 16.13 (but with the new loads), design a least-cost transmission system in this horizon year 2010 to permit the network to have bus voltages not less than .95 under normal conditions. Use 345-kV lines and cost data of Table 16.6.

5 Continue with Problem 4, and design the system to have bus voltages not less than .92 under single-line outage contingencies.

6 Assume that the network of Problems 4 and 5 is operated for all 8760 hours/year. Can the addition of another line be cost-justified on the basis of reduced line losses? Assume a $40/MWh levelized generation cost and an 18% levelized annual fixed-charge rate.

17

POWER SYSTEM STABILITY

Successful power system operation requires reliable and uninterrupted service to the consumer. One requirement for reliable service is that all generating units must remain synchronous, except between areas connected by only DC connections. If generating units lose synchronism, large voltage and power fluctuations may occur, transmission lines may be tripped by protective relays, isolation of generation capacity and load centers may occur, and consumer loads may be shed.

It is assumed here that the reader has familiarity with synchronous machine theory and the symmetrical component method for evaluating unbalanced three-phase operation. These topics are reviewed only briefly.

17.1 SYNCHRONOUS OPERATION AND STABILITY

The power system must be designed to remain synchronous, both under normal conditions and during abnormal system disturbances. Abnormal system disturbances include lightning strikes on transmission lines, generating unit trips, transmission-line faults, and transmission system terminal faults.

Generating units (synchronous machines) do not easily lose synchronism during system disturbances. A generating unit has synchronizing torques that act to maintain synchronism between the connected generating units.

When a power system is subjected to a disturbance, each generating unit's terminal voltage, rotor speed, and rotor angle (relative to other synchronous machines) change. The rotor will accelerate and decelerate in response to net mechanical input from the turbine and net electrical output from the generator. The net electrical output from the generator is dependent on the transmission system condition prior to, during, and after the disturbance, as well as the machine's rotor angle and magnetic flux.

If, following the disturbance, generator oscillations damp out to a satisfactory new state, then the power system is considered to be stable with regard to the disturbance.

A synchronous machine is affected by many factors during a transient power system disturbance:

- As long as oscillations continue, currents are induced in the amortisseur circuits (i.e., damper windings) or iron of the rotor and produce damping. These effects, referred to as "subtransient" effects, typically have time constants of less than .1 second.
- Currents are induced in the field windings. These circuits have time constants of several seconds and are referred to as "transient" effects.
- The machine's field voltage responds to changes in terminal voltage through the excitation system. The excitation system typically reaches its ceiling voltage in .1 to .5 seconds, depending on the design type.
- The machine input power from the turbine is constant until the speed governor senses a change in speed or until some external signal calls for a power change. The turbine power feedback response requires several seconds. Large sustained power excursions from turbines require several minutes to accomplish. Early-valve actuation (EVA) systems are sometimes employed to reduce power quickly after a nearby fault.

The transmission system plays a key factor in power system stability. A transmission system with large power-flow capability permits generating unit changes to occur in synchronism. A transmission system with limited power-flow capability can, during a power system disturbance, restrict power-flow changes among generating units and lead to power system instability. One means of enhancing power system stability is to strengthen the transmission system network. This is expensive, however, so other less expensive alternatives must first be considered.

Power system stability includes several aspects, as follows:

- "Steady-state stability" refers to the viability of a particular system operating condition, without any severe disturbance applied. Generally this type of stability is characterized by a steady-state power-transfer limit beyond which synchronism cannot be maintained. An estimate of this limit can be determined simply by analysis of the peak of the power-angle curve for a two-machine equivalent of the system.
- Some power systems experience growing oscillations of system variables at high power-transfer levels. These oscillations, typically in the 0.2–2-Hz range, lead eventually to loss of synchronism or to line tripping and cascading outages due to large power swings. This type of instability

has often been referred to as "dynamic instability." This term is now deprecated, since all stability problems are dynamic in nature. The term "oscillatory steady-state instability" is preferred. Since it involves only small perturbations from an operating condition, the oscillatory steady-state stability of a power system can be studied by linearizing the system equations about the operating point. Linear system stability theory and methods, such as modal (eigenvalue) analysis, can be used to determine the steady-state stability for various operating conditions, control system settings, and other system modifications.

- "Voltage instability" is a special case of steady-state instability that is characterized by the collapse of voltage in a load area as a result of a small increase in power transfer or weakening of the transmission system. It is influenced by the voltage sensitivity of the load. Some loads, such as most motor loads, which are either insensitive to voltage changes or may increase with a voltage reduction, aggravate this condition. Analysis of this phenomenon requires accurate modeling of nonlinear load–voltage relationships.
- "Transient stability" refers to the system's ability to withstand large disturbances, such as those due to transmission system faults (short-circuits), followed by loss of one or more transmission circuits. Because of the large variations in system variables, it is necessary to include nonlinear effects in analyzing transient stability. These nonlinearities include the inherent sinusoidal relationship between synchronous machine torque and angle, saturation and limits in machines and control systems, and nonlinear relationships between load active and reactive power and the impressed voltage and frequency. Analysis of transient stability is normally performed by detailed step-by-step numerical integration (dynamic simulation) of power system equations, during and after the test disturbance. Other methods, based on Lyapunov stability theory, are also under development.
- "Subsynchronous reasonance" (SSR) refers to oscillations in the 10–60-Hz range due to interactions between turbine-generator rotor torsional resonant frequencies and either the natural frequencies of series-compensated transmission lines or of high-speed control systems, such as those of HVDC systems. Damaging stresses can be produced in turbine generators as a result of these oscillations. The SSR phenomenon can result in a "steady-state" instability in which the oscillations grow without bound. More often, however, the oscillations simply exhibit poor damping and result in large rotor torques and poorly damped fluctuations. This may follow a large disturbance, such as a fault or a switching operation. These problems can be analyzed by methods similar to those used for steady-state and transient stability, but with more detailed modeling of transmission and rotor torsional dynamics.
- "Long-term stability" refers to the system's ability to remain intact after a disturbance causes an imbalance of generation and load, often accom-

panied by significant loss of transmission capability (Dunlop et al., 1975). Generation–load imbalance results in frequency excursions and excessive power flows on transmission elements, and may lead to cascade tripping of lines and generators (Liuni, Schulz, and Turner, 1975). This problem is not a true "stability" problem and is sometimes referred to as a "viability crisis." It is "long-term" with respect to the time scale of tens of seconds involved with other stability problems. This is due to slow response times and long time constants for power plant equipment to frequency deviations and generation–load imbalance. Detailed analysis of this problem is difficult because of the wide range of equipment and operating conditions involved. Approximate analysis can be used to design corrective measures, such as automatic underfrequency load shedding.

All the previously stated stability phenomena must be evaluated when planning a system. On a particular system at a particular stage in its evolution, certain stability problems will be more limiting than others and will need to be studied in more detail. Limiting stability phenomena can change as a system evolves, or even for different parts of the same system or for different operating conditions. Therefore, all types of instability should be considered when a new situation is considered.

17.2 TRANSMISSION EQUIPMENT OUTAGES

Transmission lines experience outages of approximately 1.5 outages per 100 miles per year, which varies depending on line design, geography, isokeraunic (lightning activity) level, and location (*IEEE,* 1967). Figure 17.1 illustrates the composition of these outages for 230–345-kV lines in the United States.

Sixty percent of all outages are temporary line-to-ground faults, as illustrated in Figure 17.2. Ten percent of the outages are caused by permanent line-to-ground faults. Ten percent of the outages are not associated with any fault but are temporary and caused by undesired relay settings, personnel error, and terminal equipment failures.

Lightning, a major cause of line outages, accounts for .6 outages/100 miles/year, or 42% of the outages. Ninety percent of lightning outages are associated with line-to-ground faults as shown in Figure 17.3.

Transformers, relays, lightning arresters, circuit breakers, and other equipment also have outages.

The mean time-to-repair following a forced outage of transmission equipment ranges from a few hours to several days, with an average of approximately one day. In addition to forced outages, equipment is periodically maintained and inspected. This requires the transmission-line section to be removed from service for typically 3 days/year. A large power transmission system will typically have one or more transmission-line sections out of service every day. These normal outages precondition the transmission network toward a

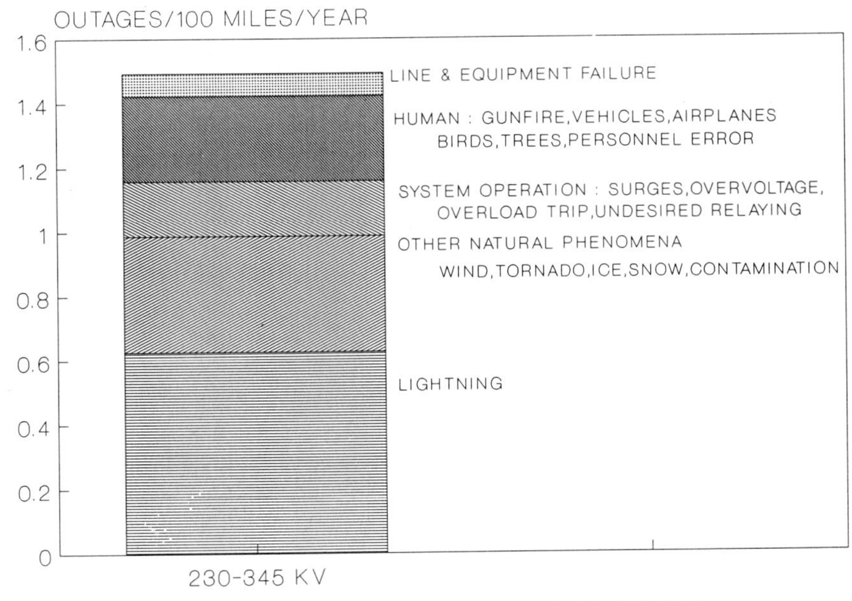

FIGURE 17.1 Outages per 100 miles of 230–345-kV lines.

weakened state. Generating units are typically dispatched to mitigate the potential effects of additional transmission forced outages associated with this preconditioned state.

The distinction between temporary outages (< 0.5 second) and permanent outages is important for relay protection of the transmission network. Tempo-

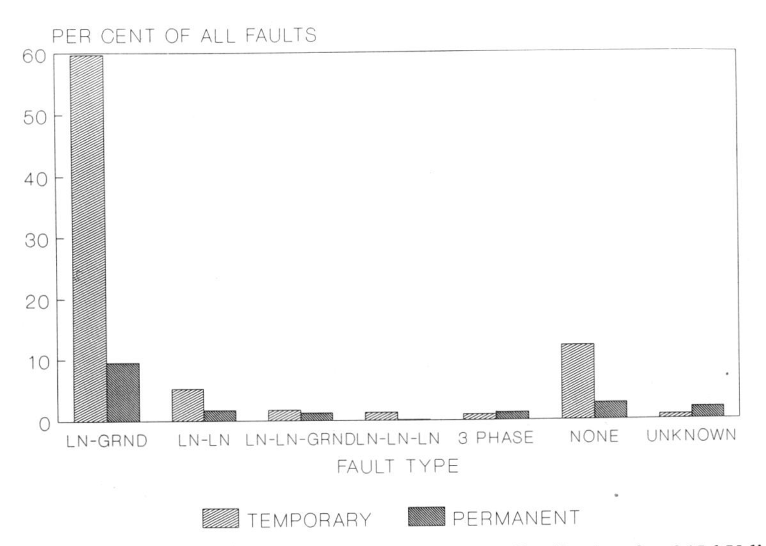

FIGURE 17.2 Temporary and permanent outages, distribution for 345-kV lines.

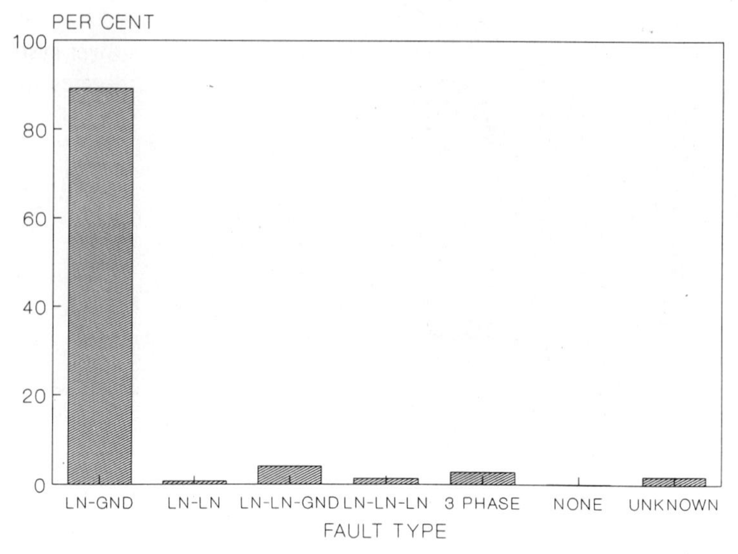

FIGURE 17.3 Distribution of lightning outages of 345-kV lines.

rary outages may be quickly cleared by deenergizing the transmission line long enough for the temporary fault to clear itself, typically within 15–30 cycles. The line may then be reenergized and returned to service. Line-to-ground faults are permanent only 10% of the time (Figure 17.4).

The probability of permanent faults increases with the fault severity as shown in Figure 17.4. A three-phase fault (including ground) has a 55% probability of being permanent, although the incidence of a three-phase fault is very

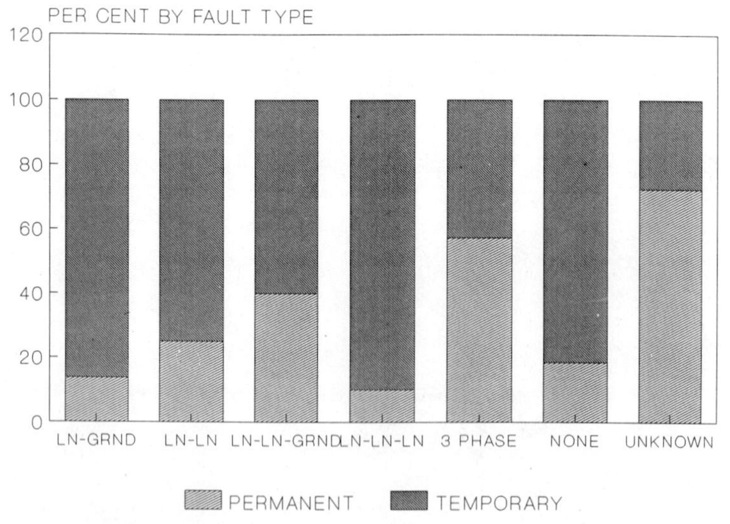

FIGURE 17.4 Permanent and temporary outages, distribution for 345-kV lines.

low (.03 outages per 100 miles per year, as shown in Figure 17.2). Since attempted reclosure on permanent faults is more damaging to power system stability than no reclosure attempt at all, automatic reclosure strategies are sometimes based on detecting the fault type.

17.3 TRANSMISSION STABILITY PLANNING CRITERIA

Conceptually, transmission system outage statistics and generation system outage statistics could be employed to enumerate all the possible combination states of generation and transmission equipment out of service and identify the annual probability of each state. A stability analysis could then be performed for each state. This analysis would evaluate the power system's capability to withstand outages associated with that state. If the system were not stable, a long-term power system dynamics analysis could assess the extent of the load lost. The expected number of system failures and expected megawatt-hours of load lost could be computed by weighting the results by the probability of each state. The result would be a measure of overall generation–transmission system reliability.

Unfortunately, this evaluation is extremely computer-intensive and has not been practical except for analysis of small sections of the transmission system.

Instead, a deterministic approach is used in which several contingency types are examined. The contingencies must be chosen to severely test the network, yet be probable enough to warrant investigation. Contingency types include:

- A permanent three-phase fault that is detected by protective equipment and cleared in the primary relaying time (typically three to five cycles) by initiating circuit breakers to remove the faulted equipment from the network. A fault location near a major generating unit is typically chosen.
- A single-phase fault in combination with a stuck breaker. In this case the faulted equipment is not removed from the network by the primary (local) breaker. The fault must be cleared by remote breaker backup (at a more distant location) in which two or more transmission lines may be lost. A remote breaker backup is designed to have a time delay and clears the fault in backup relay time (typically 10–15 cycles).
- A sudden trip of a generating unit, usually one of the larger units, or a major transmission tie from a neighboring system.
- Fast reclosing on a permanent fault.
- Multiple single-phase lightning strikes on double-circuited transmission lines.
- Loss of a transmission substation connecting several lines.

If the transmission network is not stable for the specified contingencies, then the network must be improved. The improvements can include adding new transmission circuits; adding equipment such as series or shunt capacitors, static–var controls or braking resistors; improving relay and breaker interruption times; or improving the transient response characteristics of generating units by installing high-response exciters or early valve actuation equipment. Power system stabilizers should be considered if the problem is oscillatory steady-state instability.

17.4 TRANSIENT STABILITY

When a severe disturbance is placed on a power system, a large mismatch between turbine input power, generator output power and load-demand power occurs. During the transient response period, synchronous machines accelerate and decelerate with respect to each other. The accelerating and decelerating torques associated with the transient result in oscillations among the machines with frequencies typically in the range of 0.2 Hz to several Hz.

The first swing transient stability of a system to a disturbance can be determined from the first cycle of rotor angle swing response. The "first swing" typically has a period of 0.5 second to several seconds. Figure 17.5 illustrates stable and unstable machine swing curves.

A transient stability study focuses on the first few seconds of time following a disturbance. In analyzing a large interconnected system with weak interconnections between areas, the oscillation period for interarea oscillations may be as much as several seconds. Conversely, a single machine oscillating with respect to the rest of the system will have a period of about 1 second or less.

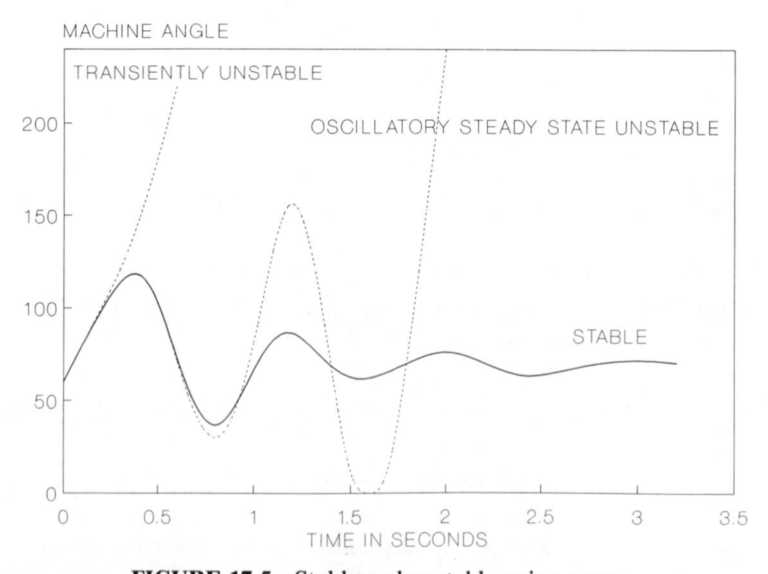

FIGURE 17.5 Stable and unstable swing curves.

Systems often exhibit multimodal behavior where more than one mode of oscillation, for example, a local machine mode and an interarea mode, are significantly perturbed. In this case, loss of synchronism may not occur on the first swing but on a later swing, because of the way in which the modes add and subtract. In this case and in the case of growing oscillations due to an "oscillatory steady-state instability," as much as 20–30 seconds of time simulation may be required to evaluate system stability. More modeling detail is also required to represent these phenomena.

The result from "first-swing" or "multiple-swing" transient analysis may indicate whether a power system is stable or unstable. It does not evaluate the potential extent of power interruptions if the power system is unstable. A "long-term dynamic" analysis proceeds much further in time and evaluates the impact of later automatic relay and operator actions on consequential cascading equipment outage and final stable power system operation. Sometimes, the final stable power system operation consists of several nonconnected "islanded" power systems operating with reduced loads. Accurate analysis of long-term dynamics requires detailed models of the prime-mover dynamics, relay settings, systemwide automatic generation control (AGC) and operator actions. While methods have been developed for such detailed analysis, (see Concordia et al., 1966) they are *not* widely used at present because of the lack of production-grade programs and the difficulty of obtaining all of the required data. Simplified models can be used to predict system frequency excursions and evaluate load-shedding schemes to restore generation–load balance.

Due to the nature of transient stability analysis, several simplifying assumptions can be made. The following simplifications are made in nearly all large-scale stability studies:

1. Only fundamental frequency effects are examined.
2. Unbalanced faults, such as line-to-ground, are analyzed in terms of the positive sequence symmetrical component. Equivalent negative and zero sequence stubs are inserted into the positive sequence network.
3. The impact of machine speed and frequency variations on reactances and voltages are neglected.

Prior to the advent of large computers, most transient stability calculations were performed using simplified, "classical" assumptions. The additional assumptions used in classical stability analysis are:

1. Mechanical torque (or power) is assumed to be constant during the first swing transient, and speed-governor effects are ignored.
2. Amortisseur effects are neglected.
3. The rotor flux linkages (e'_q) are assumed to be constant over the first swing cycle. In actuality, the generator is demagnetized by fault currents; however, the voltage regulator reinforces the field voltage to increase the

field flux in 0.1–0.5 seconds, depending on the exciter. This is a good assumption for a fast-response exciter, but is optimistic for slower exciters. The simplest "classical" model of a machine neglects transient saliency and assumes the voltage behind transient reactance, e'_q is constant.

4. Loads are represented as constant admittances.

17.5 REVIEW OF SYNCHRONOUS MACHINE REPRESENTATION

The complete presentation of synchronous machine theory and applications can be a complete textbook in itself. Several textbooks are noted in the references. This section presents a review of the equations used in power system stability analysis.

Synchronous machine equations are derived by writing flux linkage and voltage equations for the three stator core windings (for a three-phase machine), direct-axis field winding, and direct and quadrature axis amortisseur windings.

Amortisseur circuits are formed by a bar or cage winding embedded in the pole faces of a salient pole rotor. On round rotor generators typical of high speed 1800- and 3600-rpm (revolutions per minute) machines, the field core iron of the rotor forms amortisseur circuits. Amortisseur circuits help damp out oscillations of the rotor.

The stator equations are then simplified by a Park transformation, which translates the stator equations to a reference axis rotating with the rotor. The stator equations can then be described by direct-axis, quadrature-axis, and zero-axis equations.

The steady-state equations using per unit machine parameters are then derived for a transient model of the synchronous machine.

The terminal conditions of the generator as expressed in direct and quadrature axis components are:

$$e_d = E_t \sin \delta \tag{17.1}$$

$$e_q = E_t \cos \delta \tag{17.2}$$

$$i_d = I_t \sin(\delta + \alpha) \tag{17.3}$$

$$i_q = I_t \cos(\delta + \alpha) \tag{17.4}$$

where α = power factor angle between terminal voltage and terminal current
e_d = direct-axis stator voltage
e_q = quadrature axis stator voltage
E_t = terminal voltage
δ = angle between the rotor quadrature axis and terminal voltage

I_t = terminal current
i_d = direct-axis stator current
i_q = quadrature axis stator current

A stator voltage proportional to the field current E_I is defined as

$$E_I = X_{ad} i_{fd} \tag{17.5}$$

where X_{ad} is the mutual reactance from the field to the direct axis and i_{fd} is field current. In per unit notation, E_I is defined as 1.0 pu at rated open-circuit conditions.

An equation for the stator voltage proportional to the field flux linkages, e'_q, is also written.

$$e'_q = E1 - (X_d - X'_d) \cdot i_d \tag{17.6}$$

where i_d is the stator direct axis current, X_d is the direct axis synchronous reactance, and X'_d is the transient reactance.

This equation states that the stator voltage proportional to the field flux linkages is equal to the voltage produced by the field current alone less the demagnetizing influence of the direct axis stator current. At rated open-circuit conditions, e'_q is unity in per unit notation.

The stator voltage equations then reduce to:

$$e_d = X_q i_q - R_a i_d \tag{17.7}$$

$$e_q = E_I - X_d i_d - R_a i_q \tag{17.8}$$

where X_q is the quadrature axis synchronous reactance and R_a is the stator resistance. It is useful to define a fictitious voltage, E_q, lying on the machine quadrature axis as

$$E_q = E_I - (X_d - X_q) i_d \tag{17.9}$$

Substitution of Equation (17.9) into Equation (17.8), multiplication of Equation (17.8) by j, and adding to Equation (17.7) yields the phasor equation:

$$\overline{E}_t = j \cdot E_q - jX_q \cdot \overline{I}_t - R_a \overline{I}_t \tag{17.10}$$

where superscript $^-$ indicates a phasor quantity.

Thus, Equation (17.10) relates generator terminal conditions to the quadrature axis voltage, E_q. Thus, E_q and the angle δ are defined from a phasor diagram in terms of the generator terminal conditions.

Then, i_d and i_q are calculated by using Equations (17.3) and (17.4) followed by E_I, using Equation (17.5), and finally e'_q, using Equation (17.6).

In a "classical" model used in transient analysis, the field flux linkages, e'_q, do not change over the short time duration of a transient. This is only an approximation. In reality, a fast excitation system can drive the field current and increase the field flux linkages. Similarly, a slow excitation system may lead to a decrease of the field flux linkages.

Assume constant flux linkages and the machine is connected to an infinite bus power system of voltage E_s through an equivalent inductance, X_e (Figure 17.6). The machine voltage equations are (neglecting stator resistance):

$$e_d = X_q i_q = E_t \sin \delta = -X_e i_q + E_s \sin \delta_s \quad (17.11)$$

$$e_q = -X'_d i_d + e'_q = E_t \cos \delta = +X_e i_d + E_s \cos \delta_s \quad (17.12)$$

where δ_s is the angle between E_s and the quadrature axis and δ is the angle between the terminal voltage and the quadrature axis. Thus:

$$i_d = \frac{e'_q - E_s \cos \delta_s}{X'_d + X_e} \quad (17.13)$$

$$i_q = \frac{E_s \sin \delta_s}{X_q + X_e} \quad (17.14)$$

The electrical power output is:

$$P_{elec} = e_d i_d + e_q i_q \quad (17.15)$$

which can be reduced to:

$$P_{elec} = \frac{e'_q E_s \sin \delta_s}{(X'_d + X_e)} + \frac{E_s^2 (X'_d - X_q)}{2(X'_d + X_e)(X_q + X_e)} \sin 2\delta_s \quad (17.16)$$

In Equation (17.16), the first term is the largest contributor. The second term is a reluctance torque contribution associated with the transient saliency (X'_d

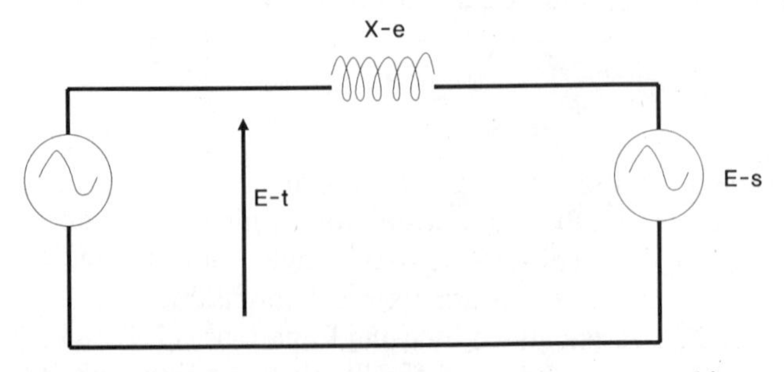

FIGURE 17.6 Transient model of synchronous machine.

$- X_q$). Because X_q is larger than X'_d, the term has a negative contribution for angles less than 90°. For angles greater than 90°, it contributes a positive torque which enhances stability.

Alternatively, if saliency is neglected (i.e., $X'_d = X_d$) and it is assumed that the voltage behind the transient reactance e' is constant, then the electrical power output is:

$$P_{elec} = \frac{e' E_s \sin \delta_s}{(X'_d + X_e)} \tag{17.17}$$

Example 17.1. A machine is connected through a step-up transformer ($X = .10$, nominal taps) and two parallel transmission lines to an infinite bus. The two transmission lines are 200 miles long and operate at 230 kV, with a .30-pu impedance per line (100-MVA base). The synchronous machine has the following parameters (on a 100-MVA base):

$$X_d = 1.0; \qquad X'_d = .15; \qquad X_q = .95$$

Assume that resistances are zero. The machine delivers 2.0-pu active power, with a terminal voltage of 1.05. The infinite bus voltage is 1.0.

Compute the transient power-transfer capability as a function of the power angle, δ_s assuming constant flux linkages.

SOLUTION. The first step is to compute the initial operating conditions of the machine.

The infinite bus reference angle is chosen to be zero. The angle of the generator terminals is calculated from the power transfer equation:

$$2.0 = \frac{(1.05)(1.0) \sin \delta}{.1 + .3/2}$$

$$\sin \delta = \frac{(2.0)(.25)}{(1.05)(1.0)} \qquad \text{or} \qquad \delta = 28.4°$$

The current can be found as the solution of the voltage-drop equation across the transmission line:

$$1.05 \angle 28.4° = j.25\, I + 1.0\angle 0$$

$$I = \frac{.924 + j.50 - 1.0}{j.25} = 1.998 + j.305 = 2.02 \angle 8.69°$$

The steady-state synchronous machine diagram can now be completed as shown in Figure 17.7.

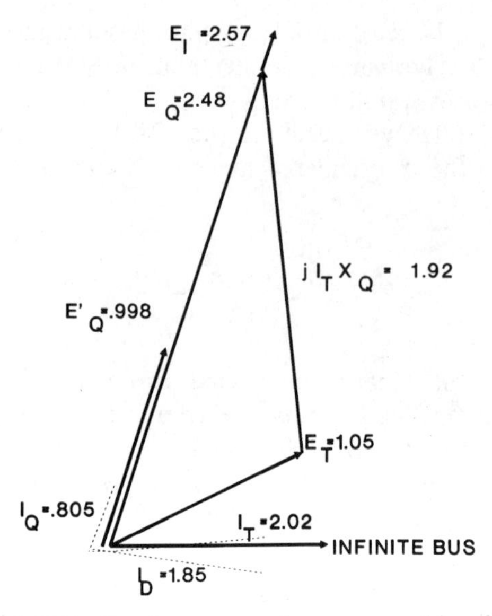

FIGURE 17.7 Steady-state synchronous machine diagram.

The voltage, E_q, behind X_q and the step-up transformer, can be calcuated as:

$$\begin{aligned} E_q &= 1.05 \angle 28.4° + (j.95)\, 2.02 \angle 8.69° \\ &= .924 + j.50 + j1.90 - .290 = .634 + j2.40 = 2.479 \angle 75.2° \end{aligned}$$

The E_q vector defines the q axis. Thus, the terminal current can be projected onto the q axis as:

$$i_q = I_t \cos(75.2 - 8.69) = 2.02 \cos(66.5) = .805$$

Consequently, the direct axis current is:

$$i_d = 2.02 \sin(66.5) = 1.852$$

The field current, E_I, is then calculated as:

$$\begin{aligned} |E_I| &= |E_q| + i_d(X_d - X_q) \\ &= 2.479 + 1.852(1.0 - .95) = 2.572 \end{aligned}$$

The field flux linkage, e'_q, is calculated as:

$$|e'_q| = |E_I| - i_d(X_d - X'_d) = 2.572 - 1.852(1.0 - .15) = .998$$

Alternatively, the "classical" model of constant voltage behind transient reactance can be used. The voltage behind the direct transient reactance, e', is found by:

$$\begin{aligned} e' &= E_t + jX'_d I_t \\ &= (.924 + j.50) + (j.15) \cdot (1.998 - j.305) \\ &= .878 + j.80 = 1.187 \angle 42.3^\circ \end{aligned}$$

The power equation for both the constant flux linkge assumption and voltage behind transient reactance is shown in Figure 17.8. For this example, because saliency leads to slightly higher power transfers at larger angles, inclusion of saliency in the machine models leads to slightly enhanced stability results.

17.6 SWING EQUATION

The generator swing equation describes the angular motion of a generating unit in response to the net torque acting on it. The swing equation is the principal equation used in assessing "first swing" power system stability.

The total kinetic energy of a rotating body is:

$$KE = \frac{1}{2} I \omega_{mech}^2 \tag{17.18}$$

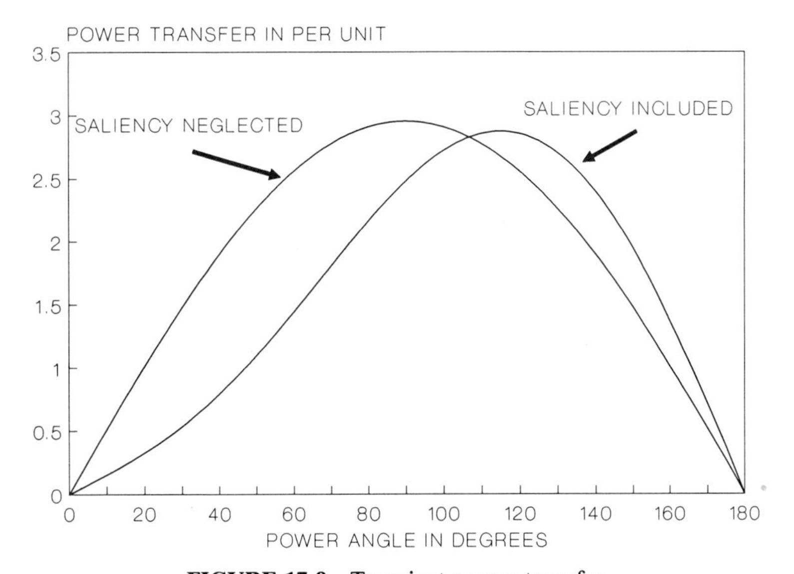

FIGURE 17.8 Transient power transfer.

where I is the moment of inertia of the generating unit, including turbine and generator, and ω_{mech} is the mechanical angular velocity of the rotor shaft in radians per second.

It is convenient to define the machine inertia constant, H, as the kinetic energy of the machine at rated speed divided by the MVA rating of the machine:

$$H = \frac{\frac{1}{2} I \omega_0^2}{\text{MVA}} \tag{17.19}$$

where H has units of MW seconds/MVA or seconds and ω_o is the rated mechanical speed. Typical machine inertia coefficients H tend to decrease slowly with increasing generating unit size. Table 17.1 presents typical inertia constants. On steam turbines, the length of the last stage blades is a significant contributing factor to the overall machine H constant.

The kinetic energy of the generating unit per MVA then becomes:

$$\frac{KE}{\text{MVA}} = H \left(\frac{\omega_{mech}}{\omega_0} \right)^2 \tag{17.20}$$

The torque equation states that the net torque acting on a rotating body is equal to the moment of inertia times the angular acceleration:

$$T_{net} = I \frac{d}{dt} (\omega_{mech}) \tag{17.21}$$

TABLE 17.1 Typical Machine Inertia Constants

Unit Type (rpm)	MVA Rating	H Coefficient (Seconds)
Nuclear		
1800	1400	6.0
Fossil steam		
3600	1400	3.3
	600	3.4
	100	3.5
	600 (high pressure)	1.6
1800	600 (low pressure)	5.0
Hydroelectric		
200–400	50	3.2
80–120	50	2.8

Substituting for I gives:

$$T_{net} = \frac{2H(\text{MVA})}{\omega_0^2} \frac{d}{dt} (\omega_{mech}) \tag{17.22}$$

This equation may be expressed in per unit notation by multiplying by ω_0 /MVA. Then:

$$T_{net} = \frac{2H}{\omega_0} \frac{d}{dt} (\omega_{mech}) \qquad \text{in per unit notation} \tag{17.23}$$

Since electrical speeds are more convenient to analyze for a system of machines operating at different speeds, the mechanical speed may be converted to electrical speed by multiplying by the number of poles. Note that the poles cancel in the numerator and denominator. Thus,

$$T_{net} = \frac{2H}{\omega_s} \frac{d}{dt} (\omega) \qquad \text{in per unit notation} \tag{17.24}$$

where ω_s = synchronous electrical speed
ω = electrical speed
T_{net} = net torque output in per unit notation

This equation, called the "swing" equation, describes how the machine angular frequency responds to the net torque acting on the unit.

The swing equation can be multiplied by ω. Since power is torque times ω, and ω is near unity in per unit notation, then per unit torque and power are almost numerically equal. Thus, approximately:

$$P_{net} \approx \frac{2H}{\omega_s} \frac{d\omega}{dt} \qquad \text{in per unit notation} \tag{17.25}$$

The inertia coefficient H is usually specified on the machine-base MVA. For system studies in which many units are connected to the power system network, the inertia constant H may need to be converted to the system MVA base. Thus:

$$H_{\text{system base}} = H_{\text{machine base}} \cdot \left(\frac{\text{MVA}_{\text{machine base}}}{\text{MVA}_{\text{system base}}} \right)$$

17.7 ONE MACHINE CONNECTED TO AN INFINITE BUS

Many stability concepts are illustrated by studying a generating unit connected through a transmission line to an infinite bus. The insights gained in studying this simple system can be translated to more realistic networks. The next section begins with the fundamental principle of the "equal area stability criterion." Further sections expand on this, including discussions of fault types, stability enhancement through faster excitation systems, early control valve actuation, speed-governor controls, and system changes, such as adding series and shunt compensation.

17.7.1 Equal-Area Criterion

One consequence of the swing curve is the stability principle of the "equal-area criterion."

Consider the case of one generating unit connected to an infinite bus through a transmission line. The generating unit is initially producing power, and is at an angle, δ_o, with the infinite bus. This condition is illustrated in Figure 17.9. The mechanical input to the generating unit is the same as the electrical output.

Suppose that the mechanical torque (or power) is increased at time $t = 0$. The machine begins to accelerate because the mechanical power input is greater than the electrical power output. The acceleration is given by:

$$\frac{d^2\delta}{dt^2} = \frac{\pi f}{H}(P_{\text{mech}} - P_{\text{elec}}) \tag{17.26}$$

This equation, when multiplied by $2(d\delta/dt)$, leads to:

$$\frac{d}{dt}\left(\frac{d\delta}{dt}\right)^2 = \frac{2\pi f}{H}(P_{\text{mech}} - P_{\text{elec}})\frac{d\delta}{dt} \tag{17.27}$$

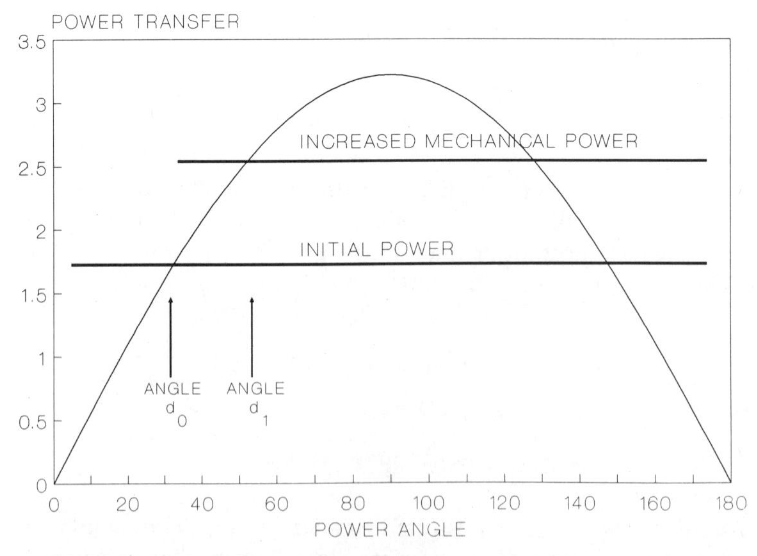

FIGURE 17.9 One machine supplying power—initial production.

Multiplying by dt and rearranging yields:

$$d\left[\frac{H}{2\pi f}\left(\frac{d\delta}{dt}\right)^2\right] = \left[P_{\text{mech}} - P_{\text{elec}}\right] d\delta \tag{17.28}$$

The left-hand term represents the incremental change in machine kinetic energy. The right-hand term represents the incremental change in accelerating energy.

The machine begins to accelerate at $t = 0^+$ because the mechanical input is greater than the electrical output. The rotor continues to accelerate until angle δ_1 is reached, as illustrated in Figure 17.10. At this point, the mechanical power input equals the electrical output. The rotor continues to advance past angle δ_1, as a result of the rotor momentum. At this point the rotor begins to decelerate because the electrical power output exceeds the mechanical input. The rotor continues to decelerate until it reaches angle δ_2. At this time, area B equals area A. Area A is referred to as the "accelerating energy" and area B, the "decelerating energy."

This phenomenon can be expressed mathematically by integrating Equation (17.28). This yields:

$$\frac{H}{2\pi f}\left(\frac{d\delta}{dt}\right)^2 = \int_0^{\delta} (P_{\text{mech}} - P_{\text{elec}})\, d\delta \tag{17.29}$$

Of interest is the angle δ_2, at which $(d\delta/dt) = 0$. This represents the point at which the machine angular speed is equal to the synchronous speed, and provides an indication of stability on the first swing.

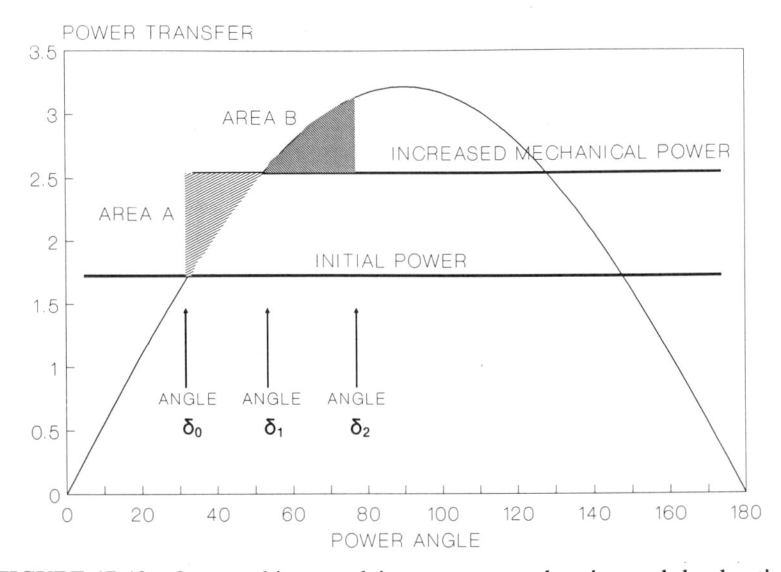

FIGURE 17.10 One machine supplying power—acceleration and deceleration.

Thus, first swing stability criterion is

$$\int_0^\delta (P_{mech} - P_{elec})\, d\delta = 0 \tag{17.30}$$

The equal area rule can also be applied to a faulted power system. Figure 17.11 illustrates this for the case of a generating unit operating at an initial angle δ_o. At time $t = 0^+$, a falut occurs, which reduces the amount of power that the generating unit can transfer to the infinite bus. The fault is removed after the power angle has increased to δ_{cc}. Area A represents the accelerating energy on the generating unit.

When the faulted line has been removed, the generating unit can transfer power along the "postfault" power curve. Since the electrical power is greater than the mechanical power, the generating unit decelerates. Area B represents the decelerating energy. Area A equals area B at angle δ_p, which is the largest angular displacement of the rotor relative to the infinite bus. Because the generator electrical output is larger than the mechanical turbine input at angle δ_p, the rotor decelerates and the angle swings back toward zero. Ultimately, because of the damping of resistance, winding loss, and control systems, the rotor will return to angle δ_f.

Of interest in power system analysis is the critical clearing angle and critical clearing time for fault removal in order for the power system to remain stable.

A formula for the maximum critical clearing angle can be derived. Let:

P_{mech} = mechanical turbine power input
P_{elec} = electrical power transfer

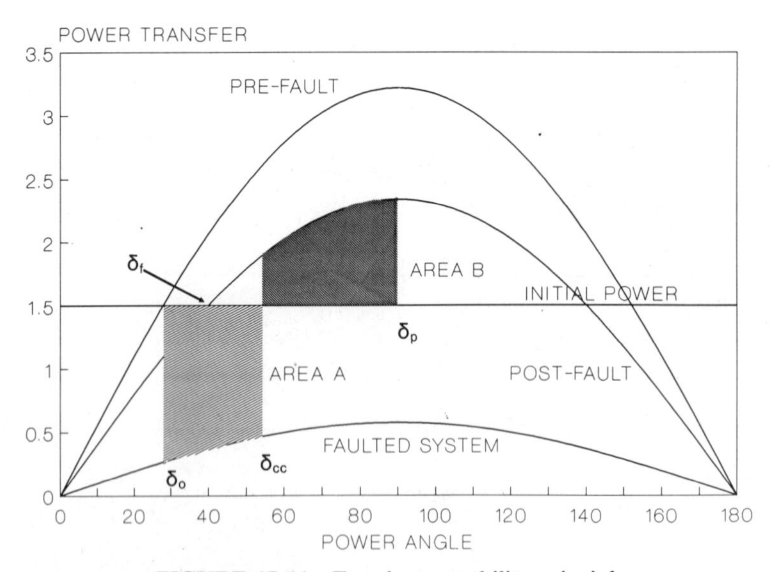

FIGURE 17.11 Equal-area stability principle.

$= P_{\text{pre}} \sin \delta$ prior to the fault
$= P_{\text{fault}} \sin \delta$ during the fault
$= P_{\text{post}} \sin \delta$ postfault
δ_{cc} = critical clearing angle

Critical clearing occurs when the swing angle of area B assumes its largest value. This occurs when the

$$P_{\text{post}} \sin \delta_{pp} = P_{\text{mech}} \tag{17.31}$$

$$\delta_{pp} = \arcsin \frac{P_{\text{mech}}}{P_{\text{post}}} \tag{17.32}$$

For the equal area rule to be satisfied, area A must equal area B. Performing the integration yields:

$$\begin{aligned} \text{Area A} &= \text{area B} \\ (\delta_{cc} - \delta_0)\, P_{\text{mech}} &- P_{\text{fault}}\,(\cos \delta_0 - \cos \delta_{cc}) \\ &= P_{\text{post}}(\cos \delta_{cc} - \cos \delta_{pp}) - P_{\text{mech}}(\delta_{pp} - \delta_{cc}) \end{aligned} \tag{17.33}$$

Solving for the critical clearing angle yields:

$$\cos(\delta_{cc}) = \frac{P_{\text{post}} \cos \delta_{pp} - P_{\text{fault}} \cos\delta_0 + P_{\text{mech}}(\delta_{pp} - \delta_0)}{P_{\text{post}} - P_{\text{fault}}} \tag{17.34}$$

The critical clearing time may be found by integrating the swing equation over the fault-time interval.

$$\frac{H}{\pi f} \frac{d^2\delta}{dt^2} = P_{\text{mech}} - P_{\text{fault}} \sin \delta \tag{17.35}$$

Expanding electrical power transfer during the fault in a Taylor series about an angle, δ, leads to:

$$P_{\text{fault}} \sin \delta = P_{\text{fault}} \sin \delta_0 + P_{\text{fault}}\,(\cos \delta_0) \cdot (\delta - \delta_0) \tag{17.36}$$

This leads to:

$$\frac{d^2\delta}{dt^2} + \omega_n^2\, \delta = B \tag{17.37}$$

where

$$\omega_n = \text{swing frequency} = \sqrt{\frac{(P_{\text{fault}} \cos \delta_0)\, \pi f}{H}} \tag{17.38}$$

$$B = \frac{\pi f}{H} \{P_{mech} - [P_{fault} \sin \delta_0 - P_{fault} (\cos \delta_0) \delta_0]\} \tag{17.39}$$

Taking the Laplace transform yields:

$$s^2\delta - s\delta_0 + \omega_n^2 \delta = \frac{B}{s} \tag{17.40}$$

or

$$\delta = \frac{B}{s(s^2 + \omega_n^2)} + \frac{s\,\delta_0}{s^2 + \omega_n^2} \tag{17.41}$$

$$\delta(t) = \frac{B}{\omega_n} \int_0^t \sin \omega_n t + \delta_0 \cos \omega_n t$$

$$= \frac{-B}{\omega_n^2} [\cos \omega_n t - 1] + \delta_0 \cos \omega_n t \tag{17.42}$$

Thus, the generator swing is sinusoidal with time (given the approximation used in expanding the electrical power-transfer function).

This may be expanded for small fault-time intervals to:

$$\delta(t) = \frac{B}{\omega_n^2} \frac{\omega_n^2\, t^2}{2} + [\frac{\omega_n^2\, t^2}{2}]\, \delta_0 + \delta_0 \tag{17.43}$$

$$2[\delta(t) - \delta_0] = \frac{\pi f}{H} [P_{mech} - (P_{fault} \sin \delta_0)]t^2 \tag{17.44}$$

This approximation yields:

$$t_{cc} = \sqrt{\frac{2(\delta_{cc} - \delta_0)}{\pi f/H\,(P_{mech} - P_{fault} \sin \delta_0)}} \tag{17.45}$$

Often, the clearing time is known and the objective is to determine the maximum first swing angle and stability margin.

When the clearing angle is given, the maximum first swing angle, δ_p, may be calculated by rearranging Equation (17.33) as the solution of:

$$P_{post} (\cos \delta_p) + P_{mech} (\delta_p)$$

$$= P_{mech}\, \delta_0 + P_{fault} (\cos \delta_0 - \cos \delta_c) + P_{post} \cos \delta_c \tag{17.46}$$

The stability margin can then be defined as the unused portion of the decelerating energy divided by the total decelerating energy.

$$\text{Stability margin} = \frac{\int_{\delta p}^{\delta pp}(-P_{\text{mech}} + P_{\text{post}} \sin \delta)d\delta}{\int_{\delta c}^{\delta pp}(-P_{\text{mech}} + P_{\text{post}} \sin \delta)d\delta}$$

$$= \frac{P_{\text{post}}(\cos \delta_{\text{p}} - \cos \delta_{\text{pp}}) - P_{\text{mech}}(\delta_{\text{pp}} - \delta_{\text{p}})}{P_{\text{post}}(\cos \delta_{\text{c}} - \cos \delta_{\text{pp}}) - P_{\text{mech}}(\delta_{\text{pp}} - \delta_{c})} \quad (17.47)$$

Example 17.2. Consider the case of a permanent three-phase fault on a transmission line. A generating unit ($X'_{\text{d}} = .15$, step-up transformer $X = .10$ and $H = 4$ seconds) is operating at 1.5-pu power (100-MVA base) and delivering power over two parallel 200-mile-long 230-kV lines ($X = .3$ pu) to an infinite bus. A solid three-phase fault occurs on one of the lines at $t = 0$ at the high side of the generator step-up transformer. The fault is removed by circuit breakers. Determine the maximum critical clearing angle for the system to be stable. Use a machine model of constant voltage behind the transient reactance with the terminal voltage initially regulated at the high side of the generator step-up transformer at 1.05 (prior to the fault).

SOLUTION

Part A. The impedance diagram prior to the fault is shown in Figure 17.12. The clearing angle can be calculated from the swing-angle approximation. The power delivered prior to the fault is:

$$1.5 = \frac{(1.05)(1.0)}{.15} \sin \delta; \qquad \delta = 12.4^\circ$$

The current can be found as a solution of the voltage-drop equation:

$$1.05 \angle 12.4^\circ = j.15I + 1.0; \qquad I = 1.51 \angle -6.5^\circ$$

The voltage behind the transient reactance is:

$$e' = 1.225 \angle 29.3^\circ$$

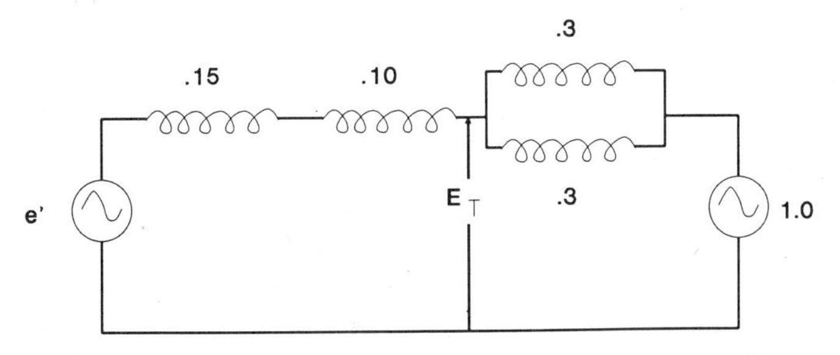

FIGURE 17.12 Three-phase fault example—impedance diagram.

Thus, the power-flow equation prior to the fault is given by:

$$P_{elec} = P_{pre}(\sin \delta) = \frac{(1.225)(1.0) \sin \delta}{.40} = 3.063 \sin \delta$$

During the fault:

$$P_{elec} = P_{fault}(\sin \delta) = 0$$

After the breakers remove the fault, the power transfer is given by:

$$P_{elec} = P_{post}(\sin \delta) = \frac{(1.225)(1.0)}{.55} \sin \delta = 2.228 \sin \delta$$

The largest postfault angle is:

$$\delta_{pp} = \arcsin \frac{1.5}{2.228} = 132.3°$$

Substituting into Equation (17.34) yields:

$$\cos \delta_{cc} = \frac{2.228 \cos (132.3°) - 0 + 1.5 \cdot (132.3° - 29.3°) \cdot (\pi/180)}{2.228}$$

$$\cos \delta_{cc} = .537$$

$$\delta_{cc} = 57.5°$$

Or the breaker must clear the fault in:

$$57.5° - 29.3° = 28.2°$$

Part B. The critical clearing time is from Equation (17.45):

$$t_{cc} = \sqrt{\frac{2(57.5 - 29.3) \cdot \pi/180}{(\pi\ 60/4)(1.5)}} = \sqrt{\frac{.98}{70.68}}$$

$$= .118 \text{ seconds or } 7.08 \text{ cycles}$$

This is within the interrupt time range of typical primary breakers.

Part C. Calculate the maximum swing angle and stability margin if the breakers clear the fault in five cycles, .083 seconds. From Equation (17.44):

$$\delta_c = \delta_0 + \frac{\pi f}{2H}\,[P_{\text{mech}} - P_{\text{fault}} \sin \delta_0]\, t_c^2$$

$$= 29.3 \cdot \frac{\pi}{180} + \frac{\pi 60}{2(4)}\,(1.5)(.083)^2$$

$$= .511 + .2435 = .7549 \text{ or } 43.3^\circ$$

The maximum first swing angle can be calculated as the solution of Equation (17.46):

$$2.228 \cos \delta_p + 1.5\, \delta_p = 1.5(29.3 \cdot \frac{\pi}{180}) + 2.228 \cos(43.3^\circ)$$

By iteration, the solution is $\delta_p = 87.6^\circ$ and the stability margin is 53.4%.

Figure 17.13 illustrates the stability margin for several values of the breaker clearing times in cycles. The stability margin decreases rapidly for increasing clearing time. As the clearing time increases, two phenomena occur. The clearing angle increases initially proportionally to the square of the clearing time. This causes the accelerating energy to increase. The decelerating energy must balance, but the decelerating area (proportional to cos δ_p) increases very slowly with increasing swing angles.

Machines with higher inertia ($H = 6$) are more stable because they accelerate more slowly, thereby providing less angular change for a given clearing time.

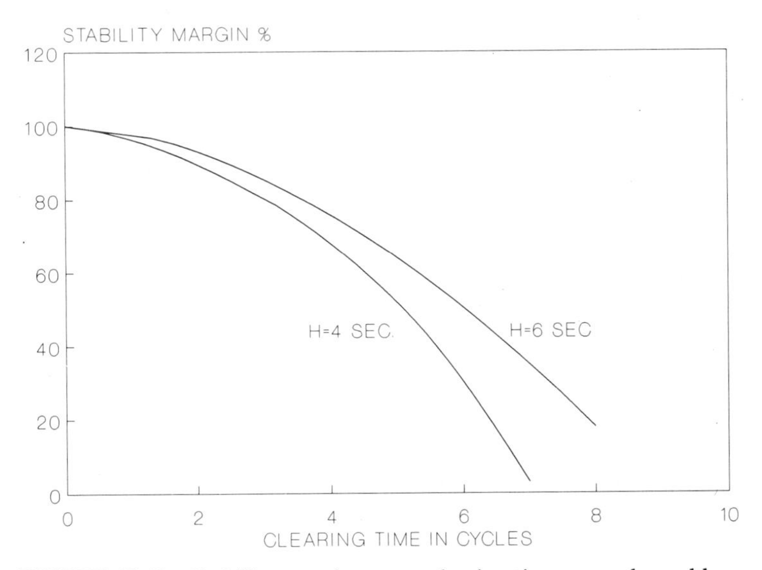

FIGURE 17.13 Stability margin versus clearing time example problem.

17.7.2 Unsymmetrical Faults

Most of the faults that occur on power systems are unsymmetrical faults. The method of symmetrical components is used to evaluate their impact (see Gross, 1979, or Stevenson, 1962).

The symmetrical-component method translates an unbalanced three-phase system into three sequence networks and provides procedures for connecting the sequence networks at the point of the fault. The connection diagrams for several types of faults are shown in Figure 17.14 (Kimbark, 1962).

The power output of synchronous machines is in only the positive sequence. No voltage sources exist in the negative and zero sequence components. If resistance is neglected, then the negative and zero-sequence component contribution to the power transfer is zero. Therefore, the positive sequence diagram is of primary interest.

A short-circuit can be represented in the positive sequence network by connecting a shunt impedance, Z_F, at the point of the fault. The value of Z_F depends on the type of fault, the negative sequence impedence, Z_2; and zero-sequence impedence, Z_o, as viewed from the point of the fault.

From an inspection of the sequence connection diagrams in Figure 17.14, the shunt fault impedance values may be determined, as shown in Table 17.2.

Example 17.3. Consider the parameters of the prior example, except that a single line-to-ground, a line-to-line, or a line-to-line-to-ground fault occurs instead of a three-phase fault. Compute the stability margin for each fault type.

The first step is to calculate the equivalent Thevenin impedance of the symmetrical sequence diagrams at the fault location.

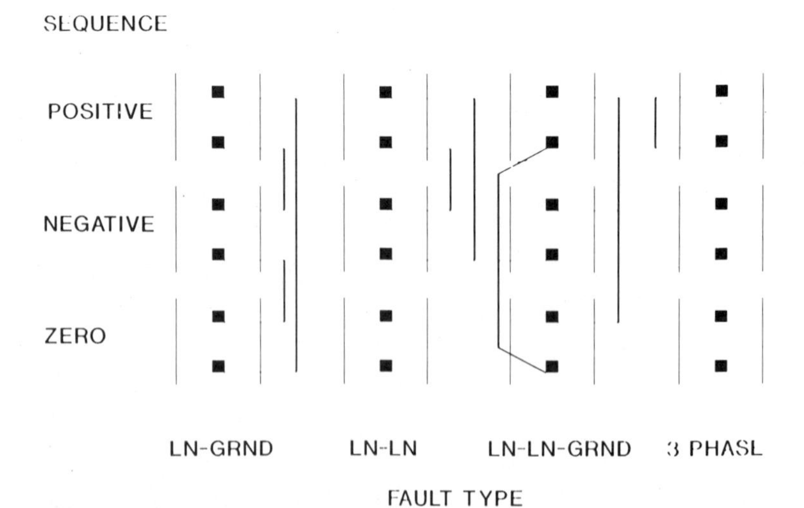

FIGURE 17.14 Connection diagrams for unbalanced faults.

TABLE 17.2 Fault Shunt Impedance Values

Type of Short-Circuit	Fault Shunt Impedance
Line-to-ground	$Z_0 + Z_2$
Line-to-line	Z_2
Line-to-line-ground	$Z_0 Z_2/(Z_0 + Z_2)$
Three-phase	0

Negative-Sequence Diagrams. The transformer and transmission-line reactances are the same for positive or negative sequences. The synchronous machine negative-sequence reactance is approximately equal to the subtransient reactance, $X''_d = .10$, in this example. The sequence diagram during the fault is shown in Figure 17.15*a*.

Zero-Sequence Diagram. The transmission-line reactance is assumed to be twice the positive sequence value. The synchronous machine reactance is approximately equal to the leakage reactance, .03 in this example. Since the step-up transformer is assumed to be connected in a ΔY arrangement, no zero-sequence currents can flow into the generator. Figure 17.15*b* presents the zero-sequence diagram.

Using symmetrical component theory, the sequence diagrams are connected in series for a line-to-ground fault (Figure 17.15*c*).

A star-mesh transformation can be used to compute the transfer impedance (Stevenson, 1962):

$$X_{TR} = X_A + X_B + \frac{X_A X_B}{X_F}$$

$$= .25 + .15 + \frac{(.25)(.15)}{.1607} = .633$$

The resulting sequence diagram is shown in Figure 17.15*d*.

The power transferred during the line-to-ground fault is:

$$P_{\text{fault}} = \frac{(1.225)(1.0)}{.633} \sin \delta$$

Consider the line-to-line fault case. In this case, only the negative sequence diagram is in the shunt path (Figure 17.15*e*). The power transferred is:

$$P_{\text{fault}} = \frac{1.225(1.0)}{.8375} \sin \delta$$

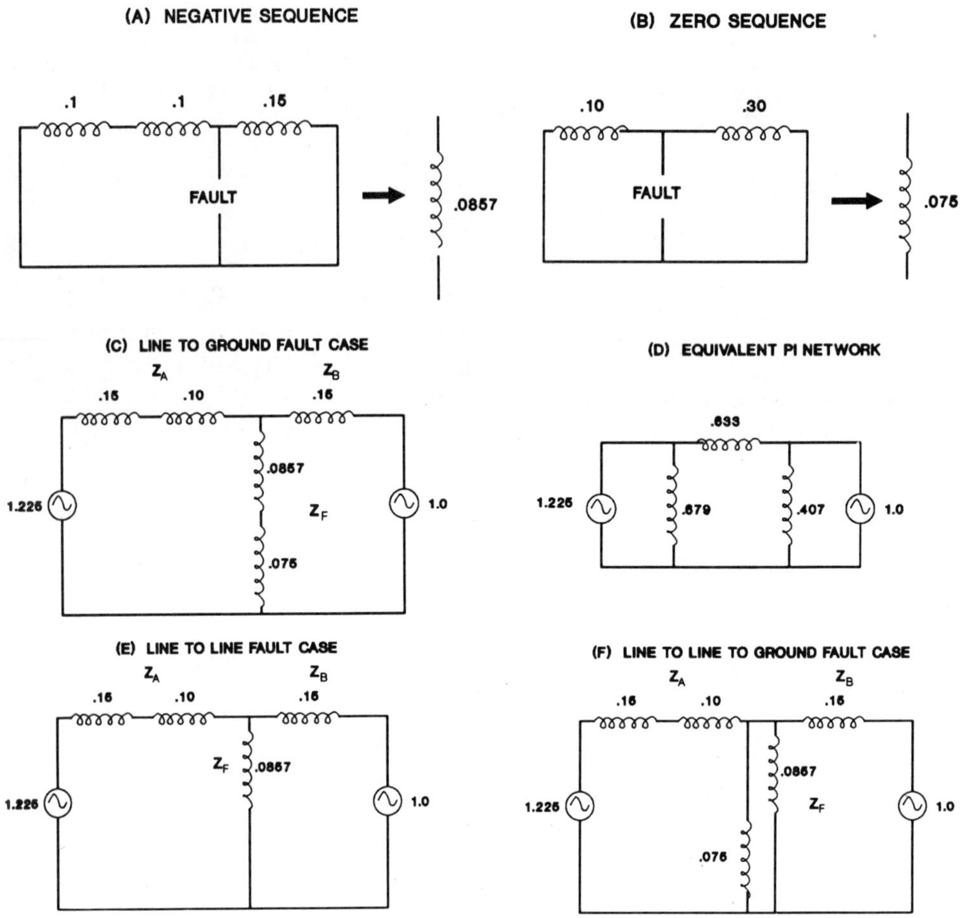

FIGURE 17.15 (*a*) Negative sequence; (*b*) zero sequence; (*c*) line-to-ground fault; (*d*) line-to-ground fault-transfer impedance; (*e*) line-to-line fault; (*f*) line-to-line ground fault.

Consider the line-to-line-ground fault case. In this case, the sequence diagram has the zero and negative sequence shunt paths in parallel (Figure 17.15*a*). The power transferred is:

$$P_{\text{fault}} = \frac{(1.225)(1.0)}{1.337} \sin \delta$$

Figure 17.16 presents the results of the stability margin calculations for several fault types as a function of breaker clearing time. The most common fault type, single line-to-ground, is least stressful (most stability margin), while the three-phase fault has the least margin. In some cases, a single line-to-ground

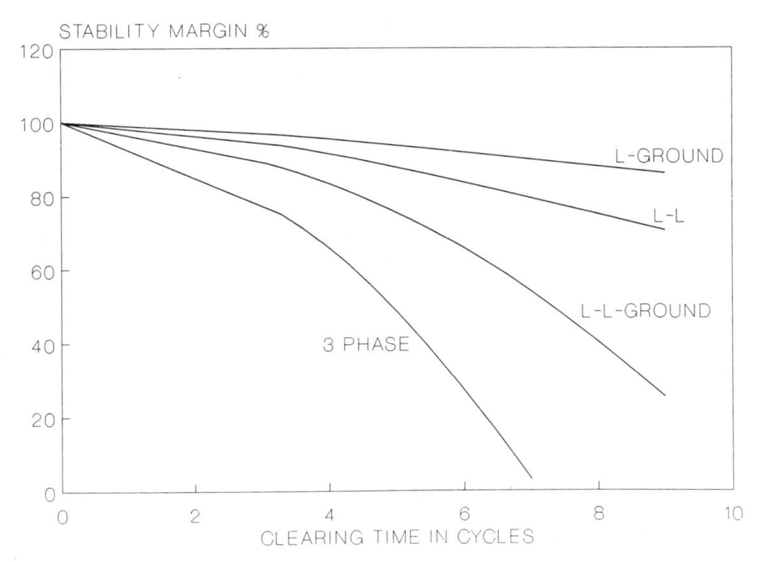

FIGURE 17.16 Stability margin by fault type.

fault may not cause instability even when left connected, and the critical clearing time is infinite.

17.7.3 High-Speed Reclosing

Most faults on transmission lines are temporary and may be cleared by deenergizing the line, waiting for the fault path to be deionized, and then reenergizing the line. Lightning is the most common temporary fault, but other causes include swinging wires and temporary contact with foreign objects.

The benefits of high-speed reclosing can be conceptually illustrated by use of the equal-area stability method. Figure 17.17 illustrates a case where the power transfer is unstable with respect to the fault. Area A, the acceleration energy, is larger than the deceleration energy, area B.

Consider now the case with high-speed reclosing (Figure 17.18). The fault is cleared at the same time as that in Figure 17.17. However, several cycles later, the breakers are reclosed on the faulted line, which is assumed to have had only a temporary fault. In this case, the line is reenergized and power can be transmitted over the prefault transfer capability. The deceleration energy, area B, is markedly enhanced and the system is stable for this example.

While Figure 17.18 illustrated a case where fast reclosing was successful, an unsuccessful reclosing (on a permanent fault or on a fault not cleared by the reclosure period) is more detrimental to stability than no reclosure at all. This can be observed in Figure 17.18 by noting that an unsuccessful reclosing would contribute to accelerating energy.

High-speed reclosing is typically applied after 20–30 cycles (0.5 second)

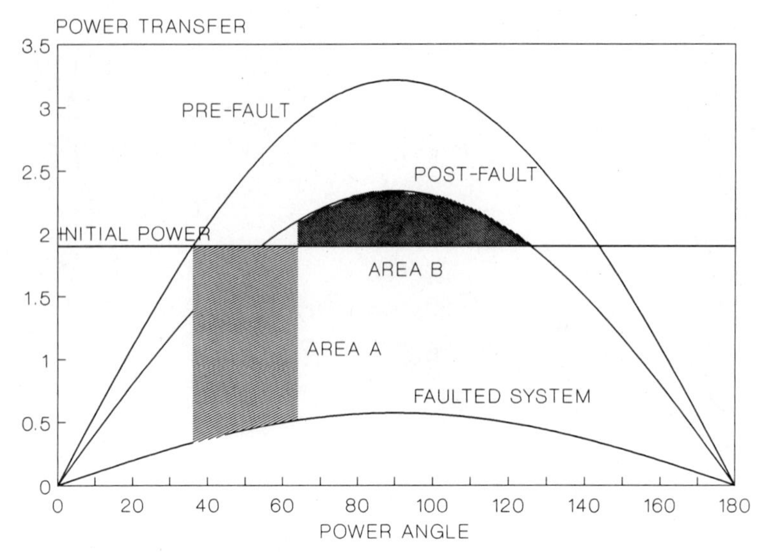

FIGURE 17.17 Unstable power transfer.

after the fault is removed. Lower voltages require less time delay (20 cycles for 220-kV lines). This has been found to give ample time for deionization of most temporary faults on EHV lines. If the first reclosure is unsuccessful, many utilities automatically try a second reclosure after additional time delay.

High-speed reclosing near generating units has potentially harmful effects on turbine-generator shafts because of the step change in the applied torques

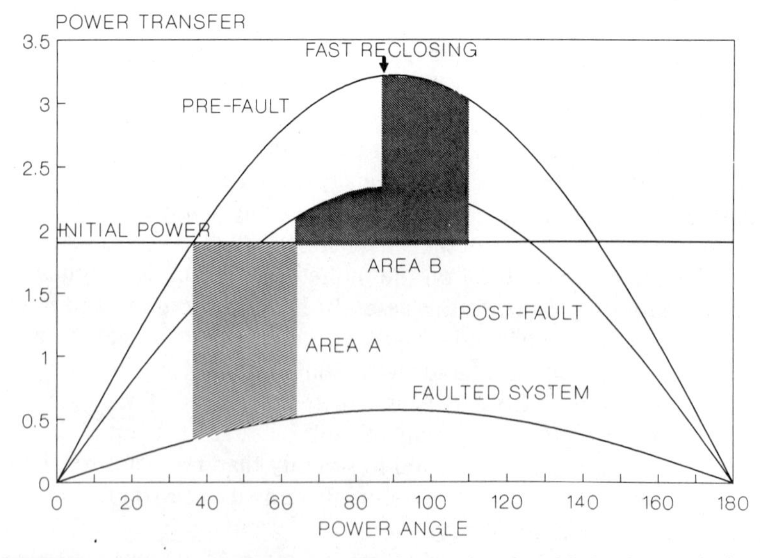

FIGURE 17.18 Unstable power transfer rendered stable by fast reclosing.

(Berdy et al., 1982). Successful reclosing can lead to shaft torques that are three times higher than steady-state values. Unsuccessful reclosure on a permanent fault can lead to shaft torques that are six times the steady-state value. These cyclical fault torques can significantly reduce shaft life. For faults near generating units, turbine-generator manufacturers may recommend the following procedures for high-speed reclosing:

1. That it be used only on less severe fault types (i.e., line-to-ground).
2. That sequential reclosing of breakers be used in which the breaker at the far end of a line be closed first to determine whether the fault persists prior to reclosing the breaker near the generating unit.
3. That reclosure be delayed for 10 seconds (600 cycles) to permit initial turbine-generator torque transients to damp.

The decision to use high-speed reclosing to augment power transfer capability is dependent on the probability of successful high-speed reclosing. High-speed reclosing should be avoided when there is a high probability of permanent fault and, therefore, is not recommended on transmission cables since nearly all cable faults are permanent.

Some utilities sense the fault type. If the fault is a line-to-ground, high-speed reclosing is actuated in view of the high probability of only a temporary fault. In this case, however, "single-pole switching" is even more advantageous, as discussed in the next section.

High-speed reclosure is especially beneficial for single-circuit lines interconnecting two utility systems. Without high-speed reclosure, the power-transfer limit is zero. High-speed reclosure permits a power transfer except for infrequent permanent faults. High-speed reclosing on double-circuit lines is beneficial in cases of faults involving both circuits simultaneously or nearly simultaneous multiple lightning strikes. In general, high-speed reclosing may be used to improve the postfault condition and not as an essential element of stability.

17.7.4 Single-Pole Switching

Thus far, it has been assumed that circuit breakers would deenergize all three phases of a faulted line, regardless of fault type (referred to as "three-pole" or "group" operation). However, more than 70% of all faults are of the line-to-ground type, and 85% of these are temporary. For these cases, it is possible to deenergize and clear only the faulted phase while maintaining power transmission on the nonfaulted phases. This mode of operation is called "single-pole switching." The circuit breakers and relays must be designed for single-pole operation.

Figures 17.19 and Figure 17.20 contrast the stability impact from the operation of three-pole switching and single-pole switching for the case of a temporary line-to-ground fault on a single-circuit line using high-speed reclosing. For three-pole switching (Figure 17.19), the line-to-ground fault is cleared at 60°,

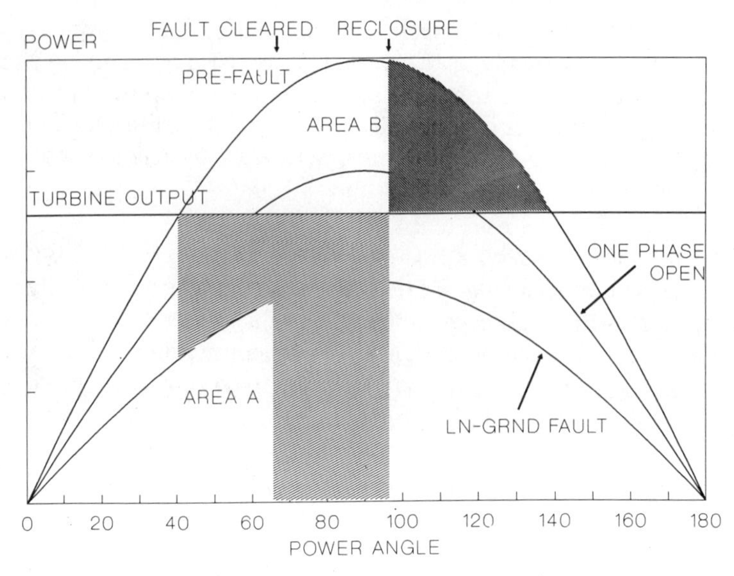

FIGURE 17.19 Three-pole operation.

at which time the breakers deenergize the line and the power transfer is zero. The power transfer remains at zero until a successful reclosure, at 97° in this example. The acceleration energy (area A) is much larger than the deceleration energy (area B). Thus, with three-pole switching, the power system is unstable.

Figure 17.20 illustrates the single-pole switching operation for the fault. The fault is cleared at 60°, at which time the breakers open only one phase of the

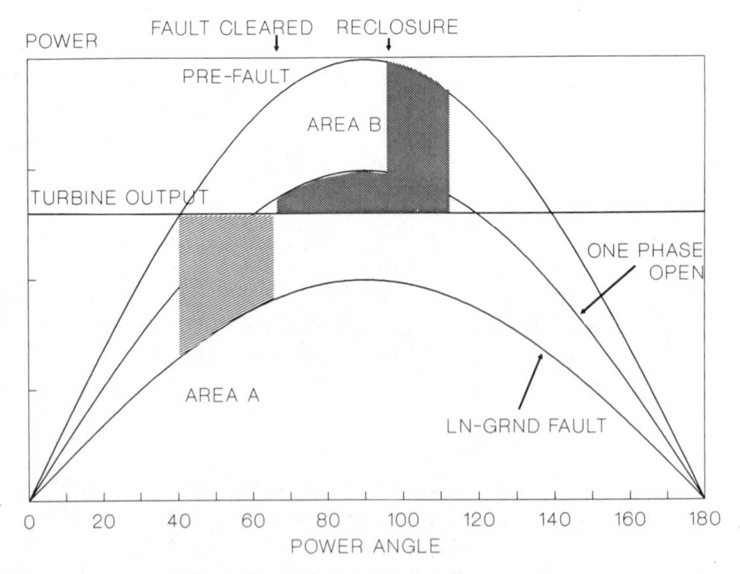

FIGURE 17.20 Single-pole operation.

line. The acceleration energy (area A) is small. At 100°, the line is reclosed. Thus, the deceleration energy (area B) is equal to area A and the power system is stable.

Because of capacitive coupling between the open phase and the connected phases, reclosing must be exercised with care. During a temporary fault, capacitive coupling prolongs the temporary arc current, called "secondary arc current." The secondary arc current requires longer waiting periods for single-pole reclosure than is required for three-pole switching. To shorten this time, the faulted conductor can be grounded when the circuit breakers are open to deenergize the faulted conductor.

Single-pole switching is analyzed using symmetrical components. The connection diagrams for several types of open lines are shown in Figure 17.21. Note that the sequence networks are connected at the points on each side of the open line.

Since the positive sequence network is of primary interest, the open conductor can be represented as a *series* impedance, Z_c, in the positive sequence network. The series impedance Z_c is expressed in terms of the positive and negative sequence impedances.

Type of Open Line	Series Impedance Z_c in Positive Sequence Network
One conductor open	$Z_o Z_2/(Z_o + Z_2)$
Two conductors open	$Z_o + Z_2$

Example 17.4. The power system of Examples 17.2 and 17.3 is delivering 1.7-pu power and experiences a line-to-ground fault. Calculate the stability

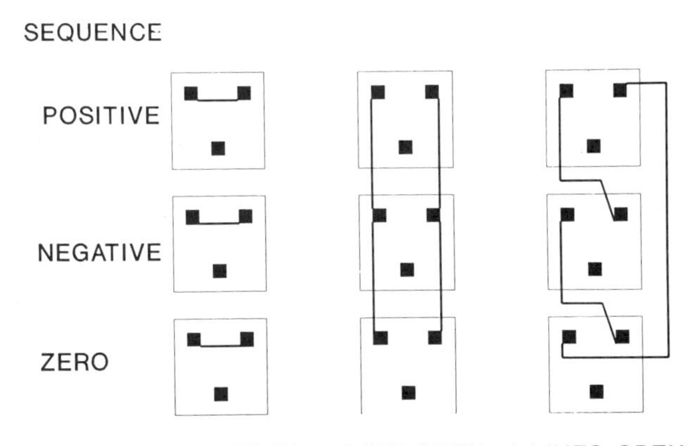

FIGURE 17.21 Sequence connection diagram.

margin for the case of single-pole switching of the faulted line to clear the fault.

SOLUTION. The power delivered prior to the fault is:

$$1.7 = \frac{(1.05)(1.0)}{.15} \sin \delta; \qquad \delta = 14.1°$$

The current can be found as a solution of the voltage-drop equation:

$$1.05 \angle 14.1° = j.15I + 1.0; \qquad I = 1.71 \angle -4.7°$$

The voltage behind transient reactance is:

$$\begin{aligned} e' &= E_{HS} + j(X'_d + X_t)I_t = (1.018 + j.2558) + j.25 \cdot (1.705 - j.12) \\ &= 1.250 \angle 33.0° \end{aligned}$$

The power delivered prior to the fault is:

$$P = \frac{1.250 \cdot 1.0}{.40} \sin \delta = 3.125 \sin \delta$$

From Example 17.3, the power delivered during the line-to-ground fault is:

$$P = \frac{(1.250)(1.0)}{.633} \sin \delta = 1.974 \sin \delta$$

The power delivered after the fault is cleared, and one phase is open, is computed from symmetrical component analysis. The Thevenin equivalent impedances of the negative and zero sequence networks are evaluated by looking into the networks at the location of the open lines.

The negative-sequence diagram is shown in Figure 17.22*a*; the zero-sequence diagram is shown in Figure 17.22*b*. The series impedance stub inserted into the positive-sequence network is:

$$Z_F = \frac{(.42)(.686)}{(.42 + .686)} = .261$$

The equivalent positive sequence network is shown in Figure 17.22*c*.

The network is then reduced to an equivalent impedance. The power transfer is described by

$$P_{post} = \frac{(1.250)(1.0)}{.445} \sin \delta = 2.81 \sin \delta$$

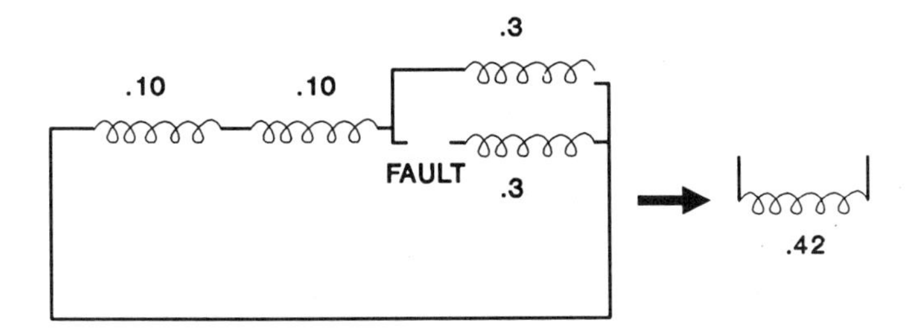

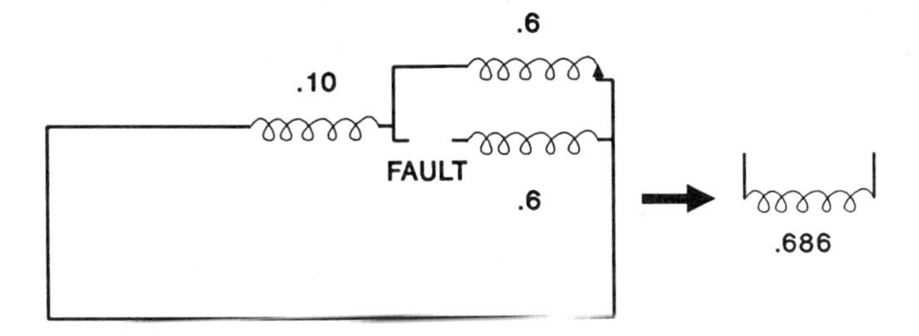

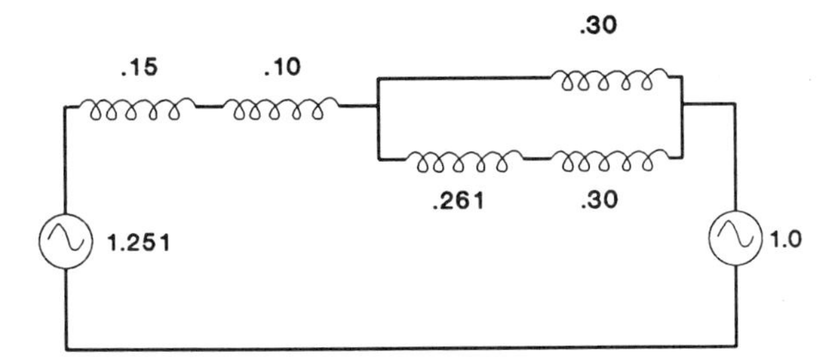

FIGURE 17.22 (*a*) Negative sequence; (*b*) zero sequence; (*c*) equivalent positive sequence.

The transient equations can now be integrated as in Example 17.2 and the stability margin determined. Figure 17.23 contrasts the stability margin results for several fault type cases. Note that for temporary faults, single-pole switching of double-circuit lines leads to improved stability margins.

In summary, for temporary faults, single-pole switching provides improved system reliability by improving the system stability margin, especially on single-circuit ties between systems.

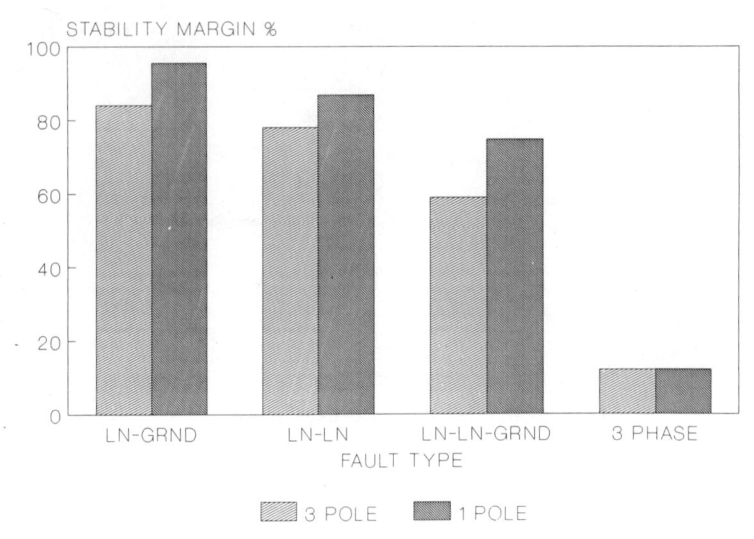

FIGURE 17.23 Contrast of stability margin from three-pole and one-pole switching.

17.7.5 Early Valve Actuation

In the previous analysis, it has been assumed that the mechanical turbine torque (power) is constant during and after the fault. If the mechanical turbine power could be rapidly reduced during a fault, then stability could be enhanced. One approach is to rapidly close the steam intercept valve on a steam generating unit. The intercept valve is located prior to the reheat (intermediate pressure) section of the turbine, and controls approximately 60–70% of the turbine power output, as shown in Figure 17.24. Rapidly closing the intercept valve is referred to as "early valve actuation" (EVA).

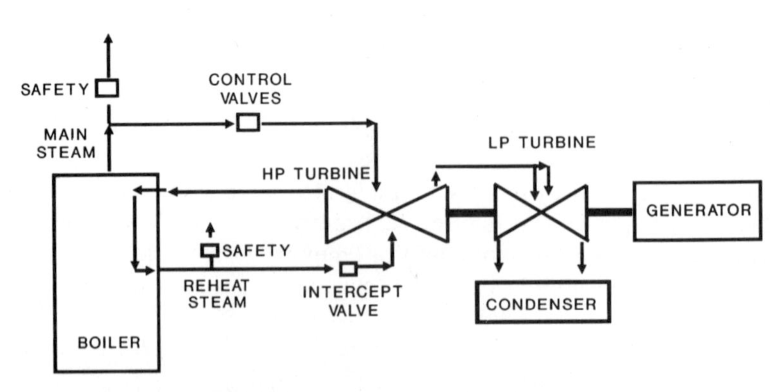

FIGURE 17.24 Steam plant layout.

The intercept valve is chosen because it can be closed and then reopened after one second without causing a large increase in the boiler pressure (Brown, 1973). Closing the control valve would lead to a larger boiler pressure transient and could result in boiler safety valve operation, which might require the plant to temporarily shutdown for reseating of these valves.

Figure 17.25 illustrates a typical EVA intercept valve and turbine power response for a tandem-compound turbine-generator (Brown, 1973). The intercept valve is closed in 0.1–0.2 second following the close signal. After one second, the valve is slowly opened. The turbine power output decreases in response to the valve closing. Approximately one second is required to achieve 40% power output due to the time constant associated with the steam already in the turbine, which is continuing to produce power. After the intercept valve slowly opens, turbine power increases rapidly from stored energy in the piping and boiler as well as the nonlinear flow–valve position characteristics.

Figure 17.26 presents a power-angle plot for an EVA example. Because of early valve actuation, the deceleration energy (area B) is equal to the acceleration energy (area A). Without early valve actuation, the deceleration area would be less than area A and the system would be unstable. Thus, early valve actuation improves the stability margin of the power system.

The control logic for early valve actuation typically uses plant measurements rather than transmission equipment relays. Actuation depends on fulfilling two requirements:

1. The accelerating power (difference between power input and power output) exceeds a constant setting.
2. The rate of change of power exceeds a constant setting.

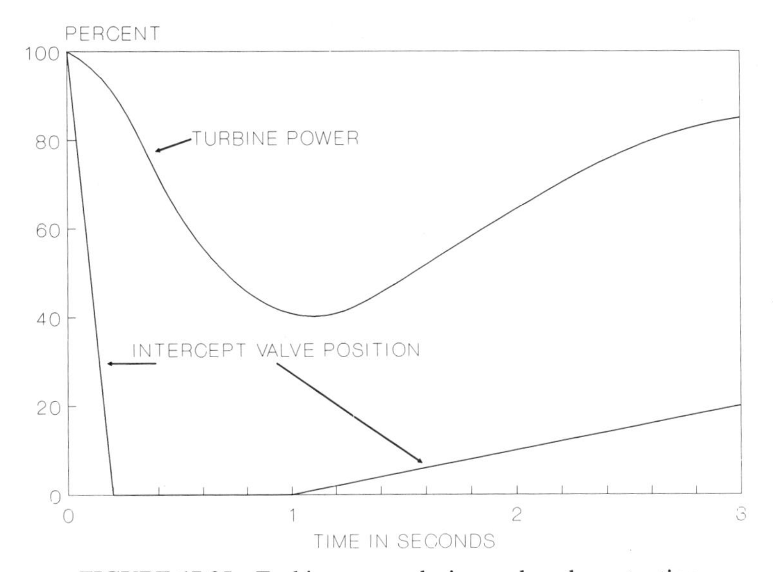

FIGURE 17.25 Turbine power during early valve actuation.

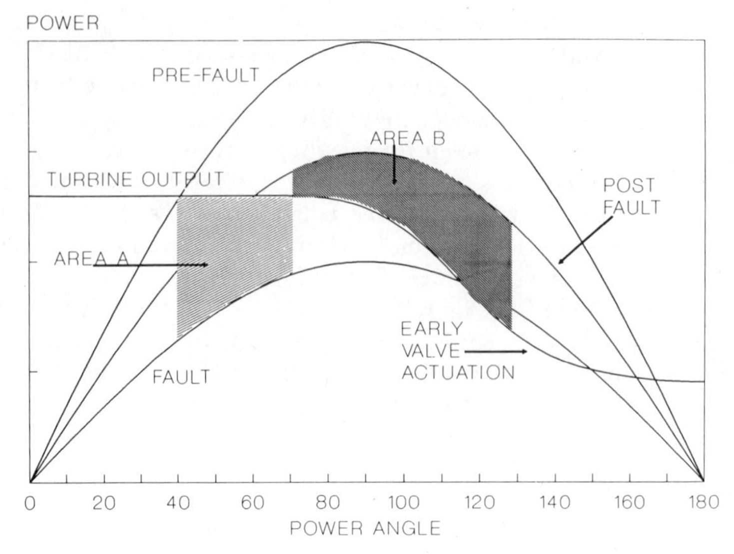

FIGURE 17.26 Early valve actuation.

These two settings are set to recognize the fault severity level and to initiate early valve actuation only for severe fault types.

While early valve actuation provides stability enhancement benefits, it places a very severe duty on the steam valve, which is quite massive (2–4 tons) and which must close in approximately 0.1 second. The tradeoff must be made between improved stability and increased valve maintenance cost and potential valve replacement. Consequently, EVA is typically initiated only for severe fault cases.

17.8 EXCITATION SYSTEM IMPACT ON STABILITY

The excitation system supplies power to the field circuit of a synchronous generator. The resulting field current produces field flux. Heretofore, field flux linkages have been assumed to be constant. For first swing stability analysis, this is an optimistic approximation over the first 0.5 seconds for low-response exciters. Newer static exciters with high-response capability can enhance stability by providing higher flux–linkage rate changes. The gain of high-response exciters must be coordinated with the power system characteristics in order to provide adequate damping of rotor oscillations. Frequently power system stabilizers are applied to develop the maximum capability of high-response exciters to provide both higher synchronizing torque and greater damping of the power system swings.

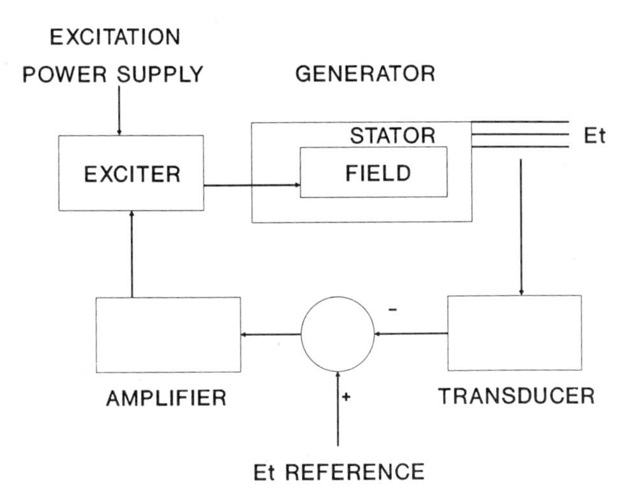

FIGURE 17.27 Simplified excitation system diagram.

The excitation system (Figure 17.27) consists of three key components:

- A transducer, which converts terminal voltage (typically the high side of the generator step-up transformer) to a DC signal proportional to terminal voltage.
- An automatic voltage regulator, which compares the terminal voltage to the terminal reference value, generates an error signal, and amplifies this signal to control the exciter.
- An exciter, which develops the power to excite the field windings.

The synchronous machine equation for the field is:

$$\frac{d\psi_{fd}}{dt} = e_{fd} - R_{fd}\, i_{fd} \tag{17.48}$$

where ψ_{fd} is the field flux linkages, e_{fd} is the field voltage, and i_{fd} is the field current.

Substituting for the machine voltage definitions yields:

$$\frac{X_{ffd}}{X_{ad}} \frac{de'_q}{dt} = \frac{R_{fd}}{X_{ad}} E_{fd} - \frac{R_{fd}}{X_{ad}} E_I \tag{17.49}$$

where E_I = voltage proportion to field current ($E_I = X_{ad} \cdot i_{fd}$)
E_{fd} = voltage proportional to the field voltage ($E_{fd} = X_{ad} \cdot e_{fd}/R_{fd}$)
e'_q = voltage proportional to the field flux linkages ($e'_q = X_{ad}/X_{ffd} \cdot \psi_{fd}$)

X_{ad} = mutual inductance from the field to the stator direct axis
X_{ffd} = field self inductance
R_{fd} = field resistance

The open-circuit field time constant is defined as:

$$T'_{do} = \frac{X_{ffd}}{R_{fd}} \tag{17.50}$$

Thus, the field voltage equation becomes:

$$\frac{E_{fd} - E_I}{T'_{do}} = \frac{de'_q}{dt} \tag{17.51}$$

The field flux linkage equation also gives:

$$E_I = e'_q + (X_d - X'_d)\, i_d \tag{17.52}$$

Equations (17.51) and (17.52) in conjunction with the direct and quadrature axis steady-state machine equations, Equations (17.11) and (17.12), describe the synchronous generator.

The excitation system components have time-response characteristics. The Laplace transform description of a conventional rotating exciter system is shown in Figure 17.28. Table 17.3 presents a typical set of parameters for a low-response excitation system. The excitation system has an open-loop gain at zero frequency ($s = 0$) of K_A/K_E. The major time constant governing the system is the exciter. Output from the amplifier is generally limited to 130–250% pu, based on the generator at rated power conditions. A saturation function may be used to account for saturation effects in the exciter. The stabilizing loop feeds the time derivative of the field voltage back into the amplifier to provide stabilization.

When the amplifier output is not at ceiling value, the excitation system gain

TABLE 17.3 Typical Low-Response-Ratio Excitation System Parameters

Response ratio	.5
T_R	0.01 second
K_A	200
T_A	0.05 second
K_E	1.0
T_E	0.50 second
K_F	0.03 second
T_F	1.0 second
V_{max}	1.30 pu of rated output

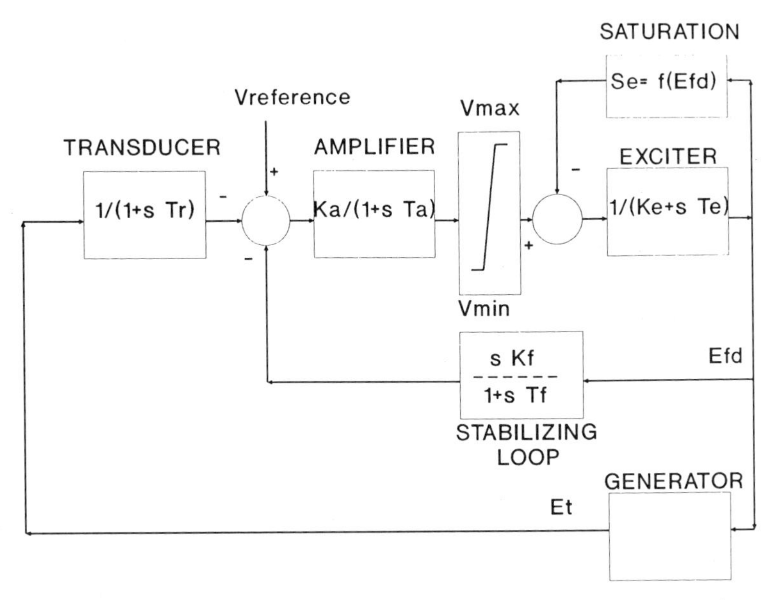

FIGURE 17.28 Laplace Transform description of a conventional rotating exciter system.

for small signals may be analyzed with frequency-domain techniques. The transducer time constant is small, and can normally be neglected. The rest of the system can be represented as follows (see Figure 17.29):

$$\frac{\Delta E_{\mathrm{fd}}}{\Delta E_{\mathrm{t}}} = \frac{G(s)}{1 + G(s)\,H(s)} = \text{gain} \tag{17.53}$$

where

$$G(s) = \frac{K_{\mathrm{A}}}{(1 + sT_{\mathrm{A}})(K_{\mathrm{E}} + sT_{\mathrm{E}})}$$

$$H(s) = \frac{K_{\mathrm{F}}s}{1 + sT_{\mathrm{F}}} \tag{17.54}$$

The excitation system gain can be written as:

$$\text{Gain} = \frac{K_{\mathrm{A}}(1 + sT_{\mathrm{F}})}{(1 + sT_{\mathrm{F}})(K_{\mathrm{E}} + sT_{\mathrm{E}})(1 + sT_{\mathrm{A}}) + sK_{\mathrm{F}}K_{\mathrm{A}}} \tag{17.55}$$

At low frequencies ($s = 0$), the gain is equal to $K_{\mathrm{A}}/K_{\mathrm{E}}$. The feedback loop reduces this gain to $T_{\mathrm{F}}/K_{\mathrm{F}}$ for midrange frequencies greater than $\omega = 1/\mathrm{T_F}$ radian, termed "transient gain reduction." This can be shown for $1/T_{\mathrm{F}} \leq \omega$

$\leq 1/T_E$, noting that $1/T_A < 1/T_F$ and K_A is very large. The gain can be expressed as:

$$\text{Gain}\left(\frac{1}{T_F} \leq \omega \leq \frac{1}{T_E}\right) = \frac{K_A j\omega T_F}{(j\omega T_F)(K_E)(j\omega T_A) + j\omega K_F K_A}$$

$$\approx \frac{K_A j\omega T_F}{j\omega K_F K_A} \approx \frac{T_F}{K_F} \tag{17.56}$$

For large ω, $G(s)$ becomes small; consequently:

$$1 + G(s)\,H(s) = 1 \quad \text{for large} \quad \omega \tag{17.57}$$

and

$$\text{Gain (large } \omega) = G(s) \tag{17.58}$$

A plot of the logarithm of system gain versus frequency is shown in Figure 17.30 for the case of $K_A = 200$, $T_A = .05$, $K_E = 1$, $T_E = .5$, $K_F = .03$, and $T_F = 1$.

The conventional excitation system has very high gain at low frequencies, thus providing good steady-state voltage regulation. At midfrequencies, $0.1 < \omega < 10$, the gain is reduced to provide more stable machine operation in the range of the system swing frequencies. The dominant high-frequency break point (or pole) is at $\omega = 10$, which implies a time constant of 0.1 second.

The small signal gain analysis is applicable only if the amplifier is not at the ceiling value. During medium and large transients (faults, line outages, etc.), the excitation system gain quickly causes the amplifier to reach the ceiling value. After reaching the ceiling value, the excitation system response is then governed by the time lag of the exciter itself, which has a time constant of 0.5–1.0 second. This significantly slows the overall excitation system time-response characteristics.

The modern solid-state high ceiling, high-initial-response excitation system

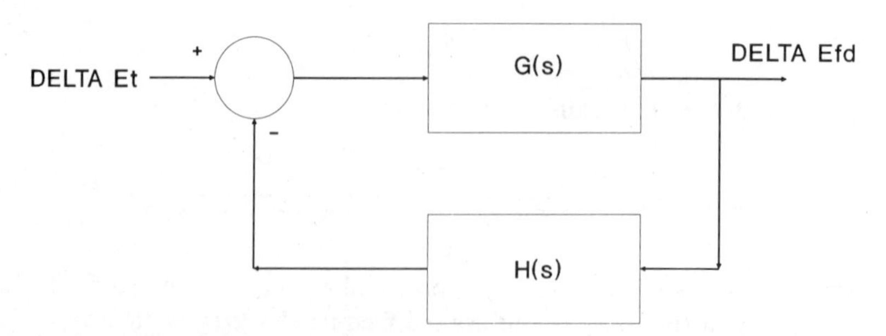

FIGURE 17.29 Excitation system gain for small signals.

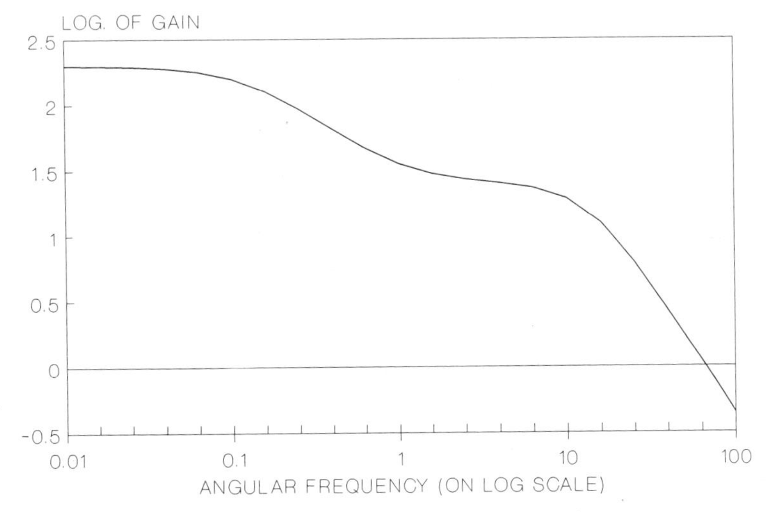

FIGURE 17.30 Excitation system gain.

can be represented by the same block diagram as the conventional excitation system of Figure 17.30, with the exception that the exciter time constant and saturation block can be eliminated because of the very small time constants of the solid-state device. The lack of an exciter time constant leads to faster excitation system response. Typical high-response ratio excitation system parameters are given in Table 17.4.

Solid-state exciters require a source of power supply. Some designs use supply from the generator terminal output through a transformer (referred to as "potential source supply"). Unfortunately, during a fault, the terminal voltage typically drops, which then reduces the excitation system power supply capability during the critical time period when it should be near ceiling values. A "compound source" design that uses both terminal voltage and terminal current for the power supply can solve this problem. Other designs that employ a rotating ac generator mounted on the generator shaft.

TABLE 17.4 Typical High-Response Ratio Excitation System Parameters

Response ratio	6
T_R	0.01 second
K_A	200
T_A	0.05 second
K_F	0.05 second
T_F	1.0 second
V_{max}	2.50 pu of rated output

17.8.1 Excitation System Influence

As an illustration of the influence of the excitation system on power system stability, consider the data of Example 17.2 but using a salient machine representation with $X_d = 1$, $X'_d = .15$, $X_q = .95$, power transfer $P = 1.45$, and three-phase fault-clearing time = 0.08 seconds. Assuming a classical model of constant field flux linkages, the swing curve may be calculated by numerically integrating the swing equation. The resulting swing curve is shown in Figure 17.31. Note that the machine first accelerates from 60° (with reference to the quadrature axis) to 120° at the maximum swing. The machine then swings back to 35° at 0.8 second. It continues to oscillate undamped with a period of approximately one second. The terminal voltage varies due to the voltage drop across the transient reactance.

Undamped oscillation is unacceptable for stable power system operation. In this example, several sources of damping have been omitted, including the effects of amortisseur windings in the machine, resistance in the generator and transmission network, damping due to the characteristics of the electrical loads, and damping provided by the generator field.

The excitation system is a powerful means to introduce damping into power system oscillations. When a fault occurs, machine terminal voltage drops. This causes the excitation system to respond by increasing voltage, which causes more electrical power to be delivered. More delivered power slows the rate of machine acceleration. Because of time lags in the excitation system, the net effect on the oscillating system is to add or subtract a damping torque.

As an example, consider the results of a simplified excitation system, which is represented by the data of Table 17.5. The excitation system has a gain of 400, a transient gain of 33, a 0.5-second exciter time constant, and the genera-

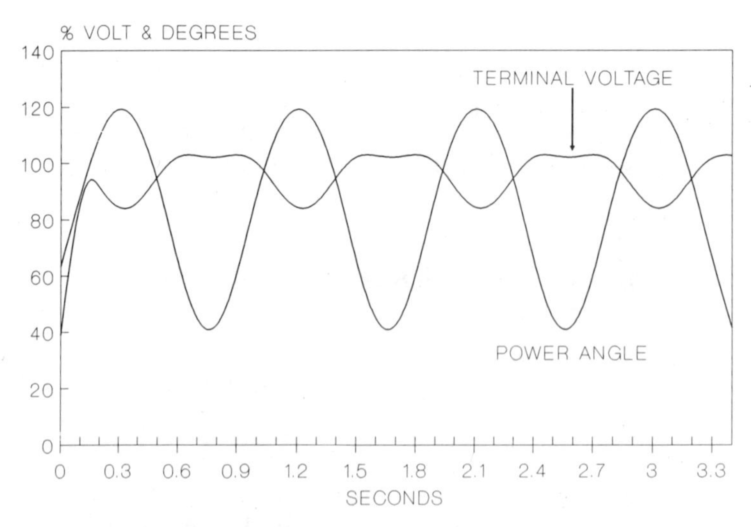

FIGURE 17.31 Swing curve example—constant E'_q.

TABLE 17.5 Exciter Characteristics for Example

T_R	$= 0$
K_A	$= 400$
T_A	$= 0.05$
K_E	$= 1$
T_E	$= 0.5$
K_F	$= 0.03$
T_F	$= 1$
V_{max}	$= 1.3$

tor field has a 5-second time constant. This excitation system is considered to have a slow-response characteristic typical of machines constructed during the 1950s.

The results from the numerical integration of the exciter, generator and swing equations are shown in Figure 17.32. Both the power angle and the terminal voltage swing are shown. The fault on the system from $t = 0$ to $t = 0.08$ seconds leads to a high fault current that partially demagnetizes the field and causes the terminal voltage to decrease to 40%. After the fault is removed, the terminal voltage returns to 101% but continues to fall during the first swing as more power is delivered from the generator. After the machine stops accelerating at time = 0.4 seconds, the generator decelerates and the angle returns to 40° at time = 0.8 seconds. The terminal voltage then rises to 105% as a result of the lower power output.

The cycle continues with a period of one second. However, note that the oscillations are dampened. After 3 seconds, the oscillations are half the initial

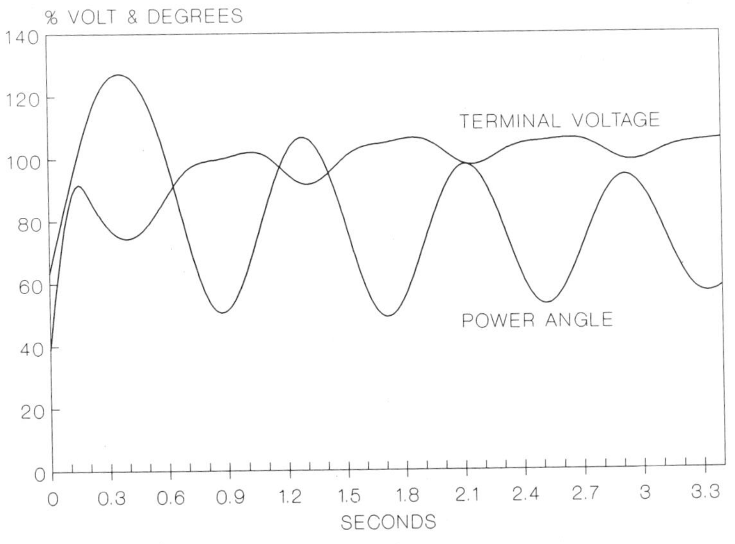

FIGURE 17.32 Swing curve example—slow exciter.

magnitude. The excitation system has contributed a strong damping contribution to the machine oscillations.

A subsequent section will further analyze the damping contributions of the excitation system and supplementary control signals.

17.8.2 Excitation System Response Ratio

The excitation system can significantly influence the transient performance of a synchronous machine. One measure of the effectiveness is the "excitation system voltage response ratio." The response rato is the slope of an equivalent line that encompasses the same voltage time area as the actual exciter response, as shown in Figure 17.33. The curves *AB, AC,* and *AD* (straight line) have the same voltage time area and hence the same response ratio. The time period is 0.5 second, over which the time integration is performed. The 0.5-second time interval corresponds to the first swing cycle during which a strong excitation source is required.

Conventional rotating exciters typically have response ratios from 0.5 to 2.0. Modern solid-state exciters have a high initial-response characteristic similar to curve *AB* of Figure 17.33. Table 17.6 presents a comparison of the characteristics of these two types of exciter.

17.8.3 Excitation System Damping

The excitation system is a contributor to improving first-swing stability by providing a strong synchronizing torque. A high-response ratio exciter im-

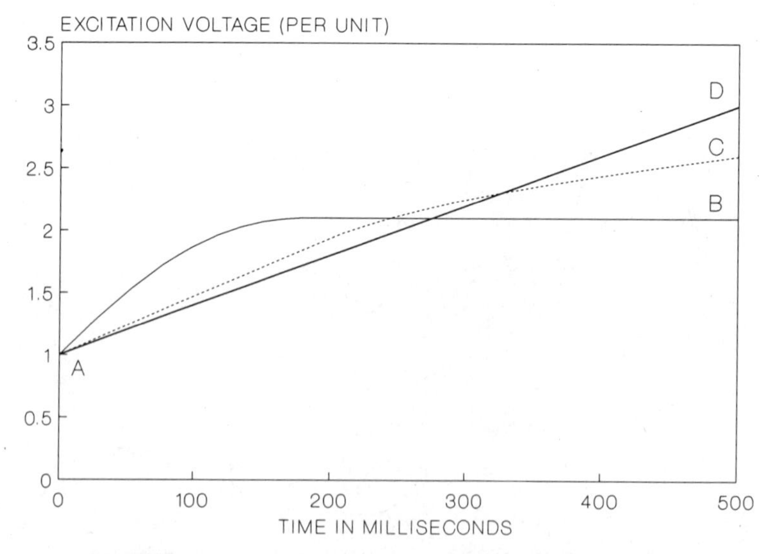

FIGURE 17.33 Excitation system response ratio example.

TABLE 17.6 Exciter Response Characteristics Comparison

Excitation System Voltage Response Ratio	Ceiling Voltage (% of Rated Load Field Voltage)	
	Conventional Exciter	High-initial-Response Exciter
0.5	140	
2.0	190	160
3.5		200
6.0		260

proves the first-swing stability margin. In addition, one of the functions of the excitation system is to provide damping of the machine oscillations following a transient.

Figure 17.34 presents the swing curve for the example of the previous section but with a high-response exciter with 250-pu ceiling voltage and an excitation system time constant of 0.05 second (high initial response). The high-response system reduces the first swing from 129° for a 0.5 response-rate excitation system to a 119° for the high-initial-response system with a 6.0 response rate. While the high-response-rate system contributes a strong synchronizing torque term, it contributes a negative damping, which results in the machine loosing stability during the fourth cycle.

An analysis of damping effects can be evaluated by using a frequency-domain analysis of the feedback control system of the synchronous machine and its interconnection to the power system.

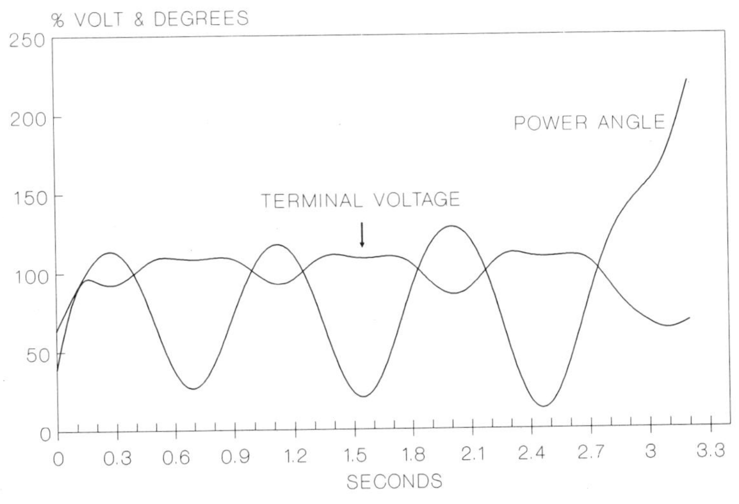

FIGURE 17.34 Swing curve example—fast exciter.

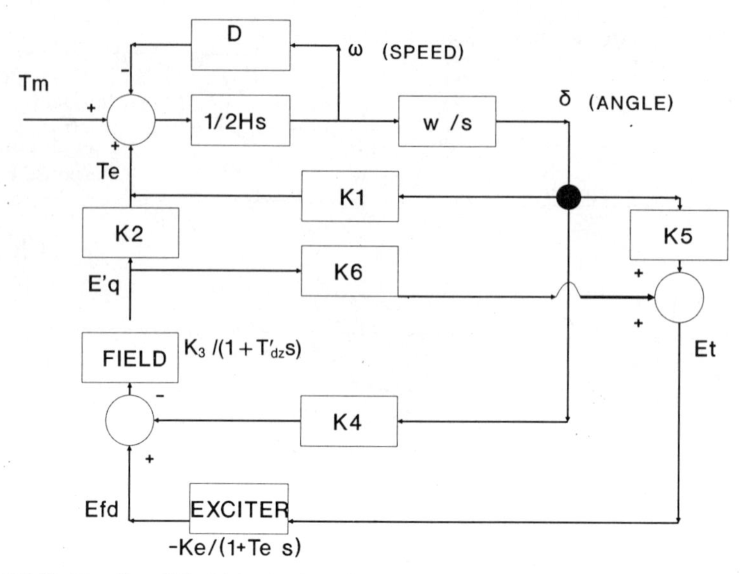

FIGURE 17.35 Simplified block diagram of a complex synchronous machine system.

17.8.4 Excitation System Dynamics

The excitation system is one feedback element of a complex synchronous machine system. Figure 17.35 illustrates a simplified block diagram of this system applicable for small changes of variables (de Mello and Concordia, 1968).

The upper-left corner of the diagram describes how a difference between mechanical, electrical, and speed damping torque lead to a net torque that causes a machine to change angular speed, ω, and angle (swing equation). The angle change also causes the terminal voltage and, in turn, the field voltage to change. The terminal voltage is dependent on the angle and field flux linkage voltage, e'_q. The excitation system and field have time constants that introduce time lags. The values of these coefficients are first derived, followed by an analysis of their impact on damping.

Figure 17.36 presents a phasor diagram of a synchronous machine. The machine equation for the field is:

$$T'_{do} \cdot s \cdot e'_q = E_{fd} - E_I \tag{17.59}$$

$$E_I = e'_q + (X_d - X'_d) \cdot i_d \tag{17.60}$$

where s is the Laplace transform operator.

By the phasor diagram:

$$E_{bus} \cdot \cos\delta + i_d \cdot (X_e + X'_d) = e'_q \tag{17.61}$$

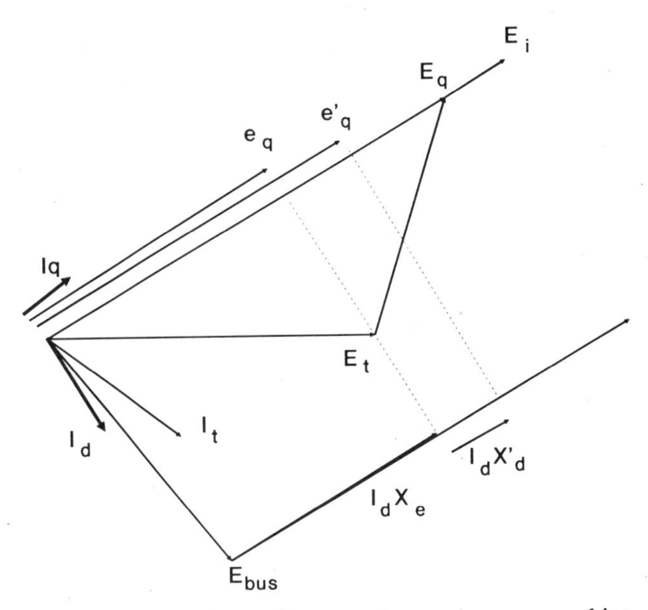

FIGURE 17.36 Phasor diagram of a synchronous machine.

Substituting for E_I into Equation (17.59) and i_d in Equation (17.60) yields:

$$(1 + sT'_{do}) \cdot e'_q = E_{fd} - \left(\frac{X_d - X'_d}{X_e + X'_d}\right) \cdot (e'_q - E_{bus} \cos \delta) \quad (17.62)$$

Define K_3 and T'_{dz} as:

$$K_3 = \frac{X_e + X'_d}{X_e + X_d} \quad (17.63)$$

$$T'_{dz} = K_3 \cdot T'_{do} \quad (17.64)$$

The constant T'_{dz} is the effective field time constant under load. The equation then reduces to:

$$(1 + sT'_{dz}) \cdot e'_q = K_3 \cdot E_{fd} + K_3 \cdot \left(\frac{X_d - X'_d}{X_e + X'_d}\right) \cdot E_{bus} \cdot \cos \delta \quad (17.65)$$

Consider a small change to E_{fd} and δ. By partial differentiation:

$$\Delta e'_q = \left(\frac{K_3}{1 + T'_{dz}\, s}\right) \cdot (\Delta E_{fd} \cdot K_4\, \Delta\delta) \quad (17.66)$$

where

$$K_4 = (\frac{X_d - X'_d}{X_e + X'_d}) \cdot E_{bus} \cdot \sin \delta_0 \tag{17.67}$$

This equation identifies two feedback blocks in the block diagram of Figure 17.35.

The relationship of the terminal voltage to the machine angle and flux linkage, e'_q, can be derived from the phasor diagram (Figure 17.36):

$$e_q = e'_q - i_d \cdot X'_d = e'_q - (\frac{e'_q - E_{bus} \cdot \cos \delta}{X_e + X'_d}) \cdot X'_d \tag{17.68}$$

$$e_d = E_{bus} \cdot \sin \delta - i_q \cdot X_e = X_q i_q \tag{17.69}$$

$$i_q = \frac{E_{bus} \sin \delta}{X_e + X_q} \tag{17.70}$$

Hence:

$$e_d = E_{bus} \cdot \sin \delta \cdot (1 - \frac{X_e}{X_e + X_q}) \tag{17.71}$$

Since $E_t^2 = e_d^2 + e_q^2$, by partial differentiation:

$$\Delta E_t = \frac{e_d}{E_t} \Delta e_d + \frac{e_q}{E_t} \Delta e_q = [\frac{e_d}{E_t} \frac{\partial e_d}{\partial e'_q} + \frac{e_q}{E_t} \frac{\partial e_q}{\partial e'_q}] \Delta e'_q + [\frac{e_d}{E_t} \frac{\partial e_d}{\partial \delta} + \frac{e_q}{E_t} \frac{\partial e_q}{\partial \delta}] \Delta \delta \tag{17.72}$$

But

$$\frac{\partial e_q}{\partial e'_q} = \frac{X_e}{X_e + X'_d} \tag{17.73}$$

$$\frac{\partial e_d}{\partial e'_q} = 0 \tag{17.74}$$

$$\frac{\partial e_q}{\partial \delta} = \frac{-X'_d}{X_e + X'_d} \cdot E_{bus} \cdot \sin \delta_0 \tag{17.75}$$

$$\frac{\partial e_d}{\partial \delta} = E_{bus} \cdot \cos \delta \cdot \frac{X_q}{X_e + X_q} \tag{17.76}$$

Thus, the coefficients K_5 and K_6 are defined as:

$$K_5 = \frac{e_{do}}{E_{to}} \cdot \left(\frac{X_q}{X_e + X_q} \right) \cdot E_{bus} \cdot \cos \delta_0 - \frac{e_{qo}}{E_{to}} \cdot \left(\frac{X'_d}{X_e + X'_d} \right) \cdot E_{bus} \cdot \sin \delta_0 \quad (17.77)$$

$$K_6 = \frac{e_{qo}}{E_{to}} \cdot \left(\frac{X_e}{X_e + X'_d} \right) \quad (17.78)$$

Thus, the terminal voltage change equation is:

$$\Delta E_t = K_5 \, \Delta\delta + K_6 \, \Delta e'_q \quad (17.79)$$

The electrical torque equation is:

$$T_e = \frac{e'_q \, E_{bus}}{(X'_d + X_e)} \cdot \sin \delta + \frac{E_{bus}^2 \, (X'_d - X_q)}{2(X'_d + X_e)(X_q + X_e)} \cdot \sin 2\delta \quad (17.80)$$

Define K_1 and K_2 as:

$$K_1 = \frac{\partial T_e}{\partial \delta} = \frac{e'_{qo} \, E_{bus}}{X'_d + X_e} \cdot \cos \delta_0 + \frac{E_{bus}^2 \cdot (X'_d - X_q) \cdot \cos 2\delta_0}{(X'_d + X_e) \cdot (X_q + X_e)} \quad (17.81)$$

$$K_2 = \frac{\partial T_e}{\partial e'_q} = \frac{E_{bus}}{X'_d + X_e} \cdot \sin \delta_0 \quad (17.82)$$

Then

$$\Delta T_e = K_1 \, \Delta\delta + K_2 \, \Delta e'_q \quad (17.83)$$

A damping torque proportional to the speed deviation from synchronous speed is also included. This term typically includes the windage losses and amortisseur winding effects, if these are not explicitly represented in the generator model. The damping torque is of the form:

$$\Delta T_{damping} = D \, \Delta\omega \quad (17.84)$$

The block diagram of the incremental changes to the synchronous machine is complete, as shown in Figure 17.35. The upper part of the diagram represents the acceleration equations. The difference in mechanical and electrical torque is integrated twice to calculate the change in angle, $\Delta\delta$. The synchronizing coefficient, K_1, feeds back an electrical torque.

The excitation system model has been reduced to a simple model of a gain, K_E, and a single time constant, $1 + s \cdot T_E$. Note the negative sign; a decrease in terminal voltage causes an increase in the field voltage.

It is useful to envision that all variables are varing sinusoidally with small amplitudes; hence $s = j\omega$. A phase diagram can then be drawn in which the variable, $\Delta\delta$, is on the X axis and the angular velocity, $\Delta\omega$ (the deviative of $\Delta\delta$, $j\omega\Delta\delta$), is on the y axis. Variables that are in phase with $\Delta\delta$ produce synchronizing torques, and variables that are in phase with $\Delta\omega$ produce damping torques, as shown in Figure 17.37.

The damping contributions to the synchronous machine system are now examined.

The middle of Figure 17.35 contains the field flux linkage equation, the bottom of the figure contains the excitation system equations.

Consider first the effect of field flux linkages on stability. In this case, consider no exciter action, or $\Delta E_{fd} = 0$. The change in electrical torque due to the field, T_{ef}, is:

$$\frac{\Delta T_{ef}}{\Delta\delta} = \frac{-K_2K_3K_4}{1 + j\omega T'_{dz}} = \frac{K_2K_3K_4[-1 + j\omega T'_{dz}]}{1 + (\omega T'_{dz})^2} \tag{17.85}$$

Since the real part is negative and the imaginary part is positive, this field torque contribution provides a negative synchronizing coefficient and a positive damping coefficient. The negative synchronizing coefficient is due to field demagnetization by the armature current.

The total electrical torque coefficient includes K_1, which leads to:

$$\frac{\Delta T_{ef}}{\Delta\delta} = K_1 - \frac{K_2K_3K_4}{1 + (\omega T'_{dz})^2} + \frac{j\omega T'_{dz}K_2K_3K_4}{1 + (\omega T'_{dz})^2} \tag{17.86}$$

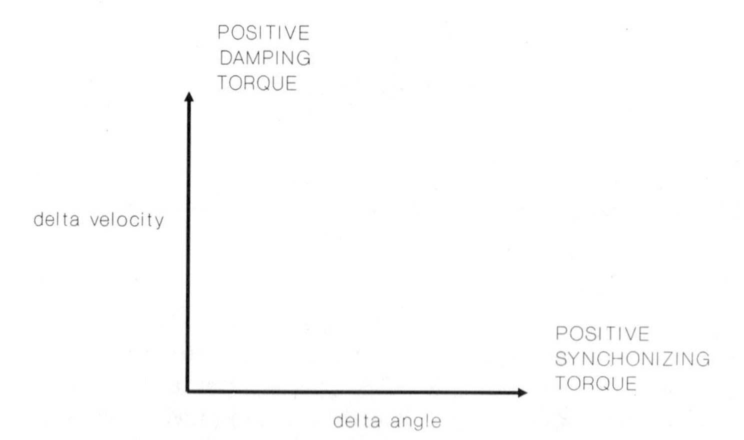

FIGURE 17.37 Damping and synchronizing torques based on variables.

Thus, the electrical synchronizing coefficient for no exciter action may be positive or negative, depending on the loading of the machine. For a heavily loaded unit, the synchronizing coefficient may be negative, which results in an unstable machine. However, the damping coefficient of the field is always positive, which contributes to damping of the oscillations.

In the "classical model" in which e'_q is constant, $K_4 = 0$, and there is no damping of oscillations as was shown in Figure 17.31.

Consider now the case of an active exciter. The electrical field torque is:

$$\Delta T_{ef} = K_2 \, \Delta e'_q \tag{17.87}$$

$$\Delta e'_q = \frac{K_3}{1 + T'_{dz} s} (\Delta E_{fd} - K_4 \, \Delta\delta) \tag{17.88}$$

$$\Delta E_{fd} = \frac{-K_E \, \Delta E_t}{1 + T_E s} = \frac{-K_E}{1 + T_E s} (K_5 \, \Delta\delta + K_6 \, \Delta e'_q) \tag{17.89}$$

Thus,

$$\Delta e'_q = \frac{K_3}{1 + T'_{dz} s} \left(\frac{-K_E K_5}{1 + T_E s} \Delta\delta - \frac{K_E K_6}{1 + T_E s} \Delta e'_q - K_4 \, \Delta\delta \right) \tag{17.90}$$

$$\Delta e'_q (1 + T'_{dz} s) + \frac{K_3 K_E K_6}{1 + T_E s} \Delta e'_q = - \left(\frac{K_3 K_E K_5}{1 + T_E s} + K_3 K_4 \right) \Delta\delta \tag{17.91}$$

Multiplying by $(1 + T_E s) / K_3$ yields:

$$\frac{\Delta T_{ef}}{\Delta\delta} = \frac{-K_2[(K_E K_5 + K_4) + K_4 T_E s]}{\dfrac{T_E T'_{dz}}{K_3} s^2 + \left(\dfrac{T_E}{K_3} + \dfrac{T'_{dz}}{K_3} \right) s + \dfrac{1}{K_3} + K_E K_6} \tag{17.92}$$

Set $s = j\omega$, and separate the real and imaginary parts of this equation by multiplying by the conjugate of the denominator:

$$\left(\frac{1}{K_3} + K_E K_6 - \omega^2 \frac{T_E T'_{dz}}{K_3} \right) - j\omega \left(\frac{T_E}{K_3} + \frac{T'_{dz}}{K_3} \right)$$

The denominator is then:

Denominator D =

$$\left(\frac{1}{K_3} + K_E K_6 - \omega^2 \frac{T_E T'_{dz}}{K_3} \right)^2 + \omega^2 \left(\frac{T_E}{K_3} + \frac{T'_{dz}}{K_3} \right)^2 \tag{17.93}$$

$$\text{Real numerator RN} = -K_2[(K_E K_5 + K_4)\left(\frac{1}{K_3} + K_E K_6 - \omega^2 \frac{T_E T'_{dz}}{K_3}\right) + \omega^2 K_4 T_E\left(\frac{T_E}{K_3} + \frac{T'_{dz}}{K_3}\right)] \tag{17.94}$$

$$\text{Imaginary numerator IN} = -K_2[\omega K_4 T_E\left(\frac{1}{K_3} + K_E K_6 - \omega^2 \frac{T_E T'_{dz}}{K_3}\right) - \omega\left(\frac{T_E}{K_3} + \frac{T'_{dz}}{K_3}\right)(K_E K_5 + K_4)] \tag{17.95}$$

Thus, the electrical field torque is:

$$\frac{\Delta T_{ef}}{\Delta \delta} = \frac{\text{RN}}{\text{D}} + j\frac{\text{IN}}{\text{D}} \tag{17.96}$$

Consider the special case of a high-gain exciter (K_E large) that is very fast ($T_E = 0$). In this case for $\omega < 1$:

$$\text{Real}\ \frac{\Delta T_{ef}}{\Delta \delta} = \frac{-K_2 K_E K_5 K_E K_6}{(K_E K_6)^2} = \frac{-K_2 K_5}{K_6} \tag{17.97}$$

In the case of heavy line loading, $K_5 < 0$. Thus, a fast high-gain exciter provides a positive synchronizing torque coefficient.

Examination of the imaginary part of the field torque contribution indicates that:

$$\text{Imag}\ \frac{\Delta T_{ef}}{\Delta \delta} = \frac{+K_2 K_E K_5 T'_{do}\omega}{\text{D}} \tag{17.98}$$

which is negative damping when $K_5 < 0$.

For a heavily loaded transmission system, a fast, high-gain, high-response exciter provides good synchronizing torque but negative damping.

The K coefficients for the example of Section 17.8.1 are shown in Table 17.7 for several conditions. During light load conditions, the K_5 coefficient is positive. For one line in service, K_5 is negative. During a power-swing transient, the delivered power can reach 2.0 as the machine swings through a large power angle (100°). In this case, the K_5 coefficient is strongly negative.

The field torque for several cases is tabulated in Table 17.8, again using the example of Section 17.8.1 (evaluated at the system swing frequency). Using data typical of a saturated amplifier at ceiling in which the dominant time constant is the exciter's time constant, the conventional exciter provides positive damping over the broad range of unit loading. The synchronizing coefficient is slightly negative, but not large enough to counteract the K_1 synchronizing power coefficient. The fast, high-gain exciter, however, provides good

TABLE 17.7 *K* Coefficients for Example in Section 17.8.1

Conditions	2 Lines in Service (Light Load)	2 Lines in Service	1 Line in Service	1 Line in Service at High Power Transient
X_e	0.25	0.25	0.4	0.4
Power	0.7	1.45	1.45	2.0
K_1	1.54	2.07	1.42	1.02
K_2	1.46	2.20	1.74	1.82
K_3	0.32	0.32	0.39	0.39
K_4	1.24	1.87	1.48	1.54
K_5	0.09	+.001	−.07	−0.22
K_6	0.56	0.466	0.55	0.54

damping at low-power output but negative damping at high-power output. At high-power output (when stability is a problem and a strong synchronizing coefficient is very beneficial), the fast, high-gain exciter provides a strong synchronizing coefficient. Unfortunately, negative damping at high-power output is a problem. Thus, some supplemental damping is beneficial for high-gain, high-initial-response excitation systems on heavily loaded power systems.

17.8.5 Supplementary Stabilizing Systems (Power-System Stabilizers)

The objective of the high-response excitation system is to drive the field voltage quickly to ceiling value to enhance transient stability. For heavily loaded transmission lines, this type of system yields a positive synchronizing coefficient but a negative damping coefficient. Therefore, addition of a supplementary stabilizing signal, often called a "power system stabilizer" (PSS) into the excitation system is beneficial.

TABLE 17.8 Field Torque Coefficients

High-Response Exciter Gain	2 Lines in Service Light Load 0.7		2 Lines in Service High Load 1.45		1 Line in Service High Load 1.45		1 Line During Transient 2.0	
	Synch	Damp	Synch	Damp	Synch	Damp	Synch	Damp
$K_E = 0$	−.003	.04	−.005	.08	−.004	.06	−.006	+.08
$K_E = 10$	−.002	.07	−.01	.09	−.02	.04	−.03	−.03
$K_E = 25$	−.01	.11	−.02	.09	−.02	−.01	−.005	−.20
$K_E = 50$	−.07	.16	−.04	.08	+.003	−.09	+.166	−.39
			Conventional Exciter					
$K_E = 25$, $T_E = .5$	.01	+.05	−.006	+.09	−.02	+.06	−.09	+.06

One approach to improving damping is to derive a signal based on machine speed. Because machine speed leads the machine angle, δ, by 90°, the machine-speed signal lies on the positive damping axis. Consider a feedback loop as shown in Figure 17.38 involving a transfer function $G(s)$ that modulates the input signal to the exciter. The goal is to have $G(s)$ times the excitation system transfer function be nearly without phase shift. Because the speed signal, ω, lies on the damping axis, the resulting ΔT_{ef} contribution would be pure damping torque.

The incremental torque equation of Equation (17.92) with $K_4 = 0$ [since it is not in the $G(s)$ feedback loop] is written as:

$$\frac{\Delta T_{ef}}{\Delta E_t} = \frac{-K_2 K_E}{[(1/K_3) + K_E K_6] + [(T_E/K_3) + (T'_{dz}/K_3)]s + (T_E T'_{dz}/K_3)s^2} \tag{17.99}$$

which is factored into:

$$\frac{\Delta T_{ef}}{\Delta E_t} = \frac{K}{(1 + T_1 s)(1 + T_2 s)} \tag{17.100}$$

Thus, the transfer function of the change in electrical torque to a change in angular speed is given by:

$$\frac{\Delta T_{ef}}{\Delta \omega} = G(s)\frac{K}{(1 + T_1 s)(1 + T_2 s)} \tag{17.101}$$

For $\Delta T_{ef}/\Delta \omega$ to be a pure gain, then $G(S)$ should have a functional form:

$$G(s) = K(1 + T_1 s)(1 + T_2 s) \tag{17.102}$$

This time response is impossible to create because the order of the numerator is larger than the denominator. However, it can be approximated with a lead–lag network transfer function.

Instead, a transfer function of the following form is selected:

$$G(s) = \frac{sK_S\,(1 + sT'_{Q1})(1 + sT'_{Q2})}{(1 + sT_Q)(1 + sT_{Q1})(1 + sT_{Q2})} \tag{17.103}$$

This form has a washout term (the $s/(1 + sT_Q)$ factor) so that at low frequencies the gain is zero and will not respond to the steady-state drift of angular frequency. The remaining terms are lead–lag pairs to compensate the excitation system network.

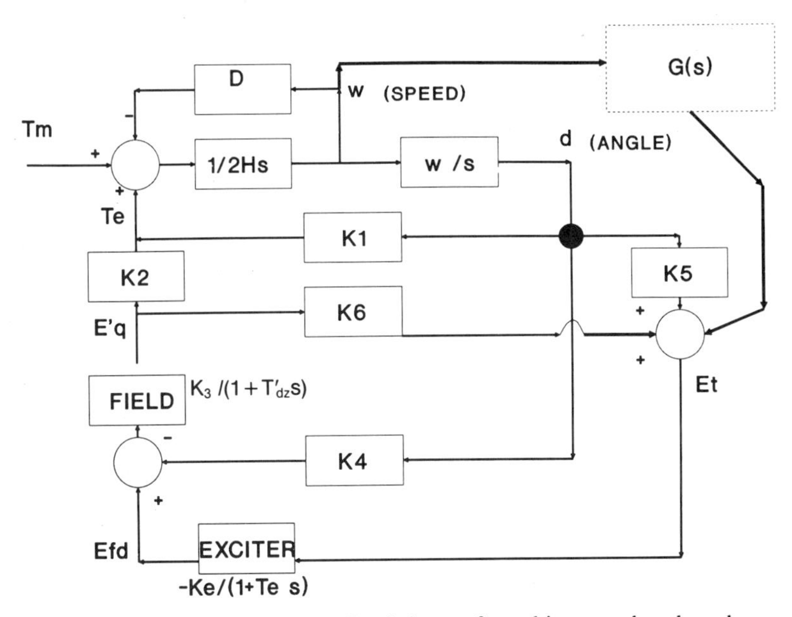

FIGURE 17.38 Signal feedback loop of machine speed and angle.

Before compensating the network with damping, calculate how much damping is required. Consider the swing equation:

$$\frac{2H}{\omega}\frac{d^2\delta}{dt^2} + \frac{D}{\omega}\frac{d\delta}{dt} + K_1\delta = 0 \tag{17.104}$$

The Laplace transform roots of this equation are:

$$s = -\frac{D}{4H} \pm \sqrt{\left(\frac{D}{4H}\right)^2 - (K_1\,\omega/2H)} \tag{17.105}$$

The natural frequency of oscillation, excluding damping is:

$$\omega_N = \sqrt{\frac{K_1\omega}{2H}} \tag{17.106}$$

When $(D/4H)^2 > K_1\omega/2H$, then the oscillations are overdamped, with no complex roots. An overdamped system is very sluggish. Critical damping occurs when $(D/4H)^2 = K_1\omega/2H$. The damping ratio is defined as:

$$\zeta = \frac{\text{total damping}}{\text{critical damping}} = \frac{D/4H}{\sqrt{K_1\omega/2H}} = \frac{D}{4H\omega_N} \tag{17.107}$$

A typically desired damping ratio is .5. Thus,

$$.5 = \frac{D}{4H\,\omega_N} \tag{17.108}$$

and

$$D = 2H\,\omega_N \tag{17.109}$$

The gain of the stabilizing transfer function at the machine's natural frequency is calculated as:

$$2H\,\omega_N + D_{\text{other}} = \frac{\Delta T_{\text{ef}}}{\Delta E_t}(\omega_N) \cdot |G(\omega_N)| \tag{17.110}$$

where D_{other} are all of the other sources of positive and negative damping of the machine system. In this example, all other sources are set equal to zero. In addition, the phase angle of the stabilizer should negatively match that of the excitation system.

The stabilizer must act over a broad range of operating and transmission conditions. Consequently, an exact determination of the coefficients must be made with this consideration (Larsen and Swann, 1981).

Figure 17.39 presents the required stabilizer gain for several operating conditions as a function of angular frequency using the example data of Section 17.8.1. The natural system swing frequency is in the range of 8 radians/sec-

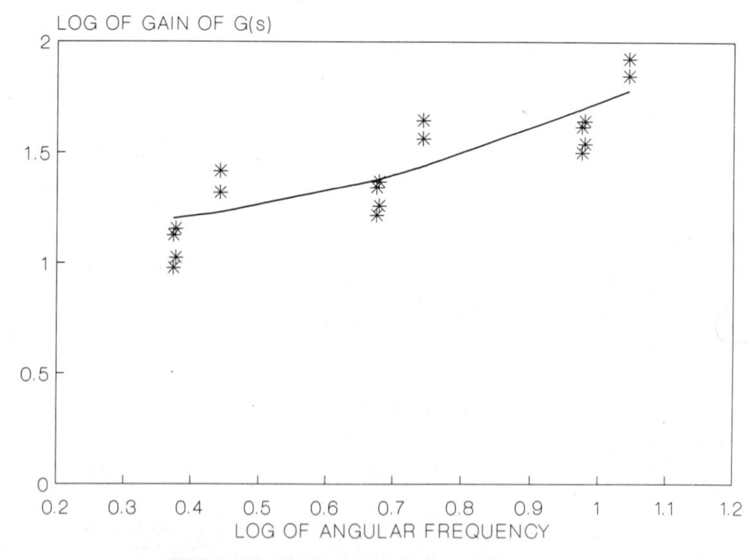

FIGURE 17.39 Required stabilizer gain.

ond. Points are plotted near $\omega = 5$ corresponding to power loadings of .7, 1.45, and 1.8 pu, and with one and two lines in service. Points are also plotted for these same conditions at twice and half this frequency. Also shown is a proposed feedback function, $G(S)$. The values used in this example are:

K_S = 40
T_Q = 3 seconds
T'_{Q1} = .2 second
T_{Q1} = 0.05 second
T'_{Q2} = 0.2 second
T_{Q2} = 0.05 second

These values are typical of those recommended by deMello and Concordia (1968) and of those used in practice. The gain, K_S, is the principal variable used to set the stabilizer gain to match that required.

The required stabilizer phase angle is presented in Figure 17.40, as well as the proposed stabilizer phase change. The T'_{Q1} and T'_{Q2} coefficients are the principal variables used to establish the stabilizer angle to match that required.

Figure 17.41 illustrates the resulting change in machine-swing stability using the power system stabilizer. The machine's first cycle swing is to 123°, compared to 130° for the slow excitation system, and thereafter the swing becomes damped in two cycles. The stabilizer achieves a remarkable damping of the oscillations and improves the stability margin.

Note the PSS signal in Figure 17.41. The stabilizer provides a signal to the

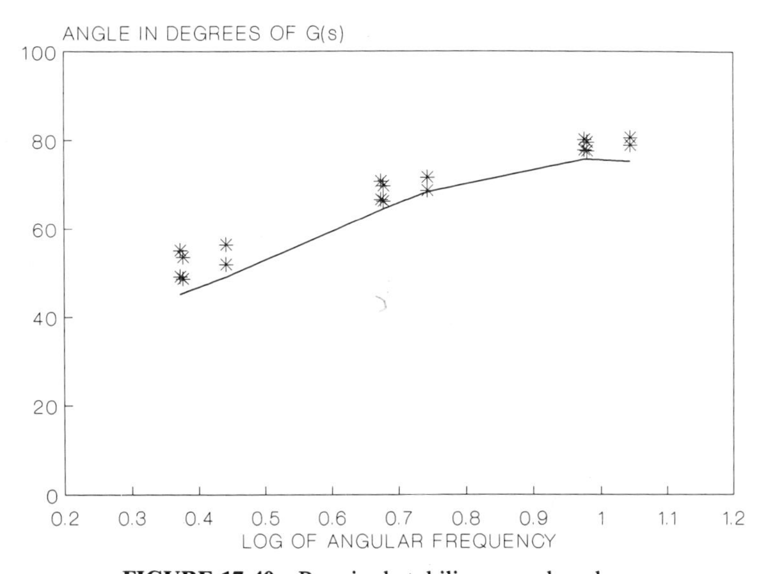

FIGURE 17.40 Required stabilizer angular phase.

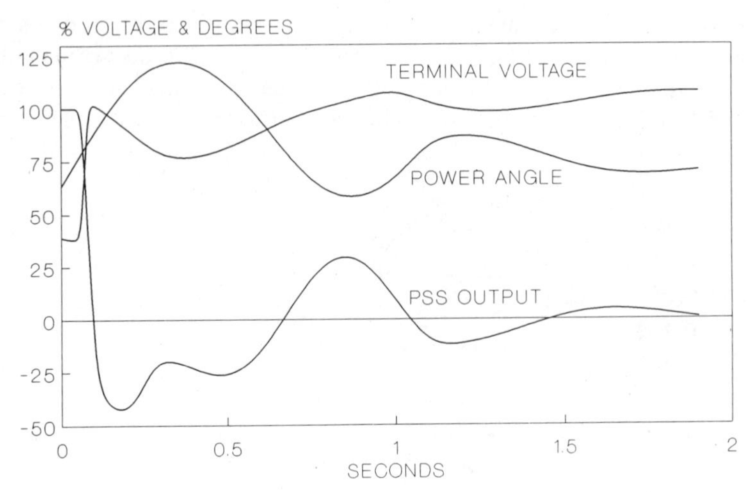

FIGURE 17.41 Swing curve example with stabilizer—high-response-rate excitation system.

excitation system during the first 0.08 second of the fault to increase the field. After the fault has been removed, the stabilizer begins to direct the excitation system to reduce voltage because the rate of speed change is decreasing. During the first negative swing, the stabilizer directs the exciter to increase the field. This is in contrast to the terminal voltage signal to the exciter to decrease field voltage on the negative swing. Stabilizer actions are taken in lead of the time required for them to act on the generator. Stabilizer anticipates excitation system lag times by using machine speed, machine acceleration, and rate of change of machine acceleration. Other input signals, such as change in electrical power, can also be used.

In summary, the power system stabilizer (PSS), using external signals, can provide strong damping torques. Its parameters must be tuned to the power system in which it operates.

A high-response-ratio excitation system with a complementary PSS is an important contribution to improving power system stability and permitting larger power transfer across the transmission system network. A 6.0 versus 0.5 response-ratio excitation system can increase power transfer in the range of 0–5% depending on the transmission network.

17.9 PRIME MOVER DAMPING

Heretofore it has been assumed that the torque input to the synchronous generator is constant during the transient. However, the mechanical torque equation can be written as:

$$T_M = \frac{P_M}{\omega} \tag{17.111}$$

Differentiating, we then obtain:

$$\Delta T_{\mathrm{M}} = \left(\frac{\partial T_{\mathrm{M}}}{\partial P_{\mathrm{M}}}\right) \Delta P_{\mathrm{M}} + \left(\frac{\partial T_{\mathrm{M}}}{\partial \omega_{\mathrm{M}}}\right) \Delta\omega \tag{17.112}$$

$$\Delta T_{\mathrm{M}} = \frac{1}{\omega} \Delta P_{\mathrm{M}} - \frac{P_{\mathrm{M}}}{\omega^2} \Delta\omega \tag{17.113}$$

Since ω is approximately one per unit, it follows that:

$$\Delta T_{\mathrm{M}} = \Delta P_{\mathrm{M}} - P_{\mathrm{M}} \Delta\omega \tag{17.114}$$

The change in mechanical torque input is composed of two terms: a change in power input from the turbine and a change in the shaft speed.

The first term, a change in power input from the turbine, will be analyzed in greater detail in a subsequent section on prime mover analysis. This term accounts for power changes as the control valves are changed to increase or decrease the power level.

The second term, a change in shaft speed, arises because the prime mover power output is constant and, as shaft speed increases, the torque must decrease according to Equation (17.111). This leads to a small damping contribution to the swing equation because the torque is proportional to $\Delta\omega$. This damping is often referred to as "steam damping" in steam turbines.

Figure 17.42 illustrates the effect of this damping using the same data as in Section 17.8.1 with an exciter with e'_q = constant. With no "steam damping," or constant *torque* input, the swing curve exhibits no damping. With "steam damping," or constant *power,* the swing oscillations are slightly damped.

Prime mover damping is a contributing source of damping for all types of generating units, but varies depending on the type.

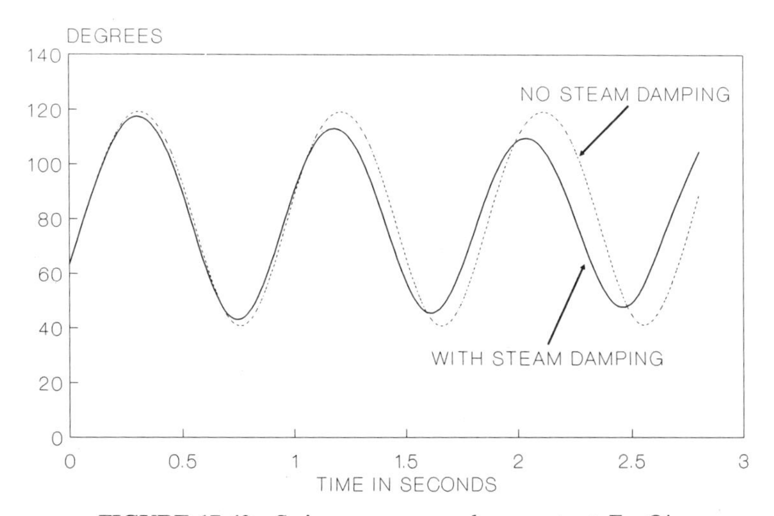

FIGURE 17.42 Swing curve example—constant E - Q'.

17.10 PRIME MOVER RESPONSE

The stability assessment has previously been based on the ability of the generation–transmission system to withstand the first few seconds of generator power angle oscillations in synchronism following faults. Except for early valving action, the nominal response of the prime movers (turbines) is usually too slow to materially influence transient stability, thereby justifying the assumption of constant prime mover mechanical power.

When studying large interconnected power systems and power flows across interconnecting transmission lines, the time period of interarea oscillations is longer (1–5 seconds) than the intraarea machine inertia oscillations (0.5–1.0 second). In this case, response and governing characteristics of the prime movers are important considerations in the stability assessment (deMello, Ewart, and Temoshok, 1966).

The governor system measures the speed (frequency) of a generating unit and feeds an error control signal to the control valves to increase or decrease power to reset the speed (frequency) of the unit. Figure 17.43 shows a control system for a typical steam unit. The frequency error is multiplied by a gain, K_G, and integrated ($1/s$) to force the speed error to zero.

To permit many units to share load, a "droop" characteristic is included as a feedback loop with gain, R. The governor system block diagram is converted to a transfer function diagram in Figure 17.44 (Wood and Wollenberg, 1984, Chapter 9). The change in valve position is equal to the speed error divided by the speed droop R and multiplied by a time-constant function. The term $1/K_G R$ is defined as the governor time constant:

$$\frac{1}{K_G R} = T_G \qquad \text{(governor time constant)} \tag{17.115}$$

Typically, R, the speed droop, is set to .05 and T_G ranges from 0.1 to 0.2 second.

A steam-power plant block diagram for a drum-type boiler and single-reheat turbine generator is shown in Figure 17.45. The boiler steam drum and

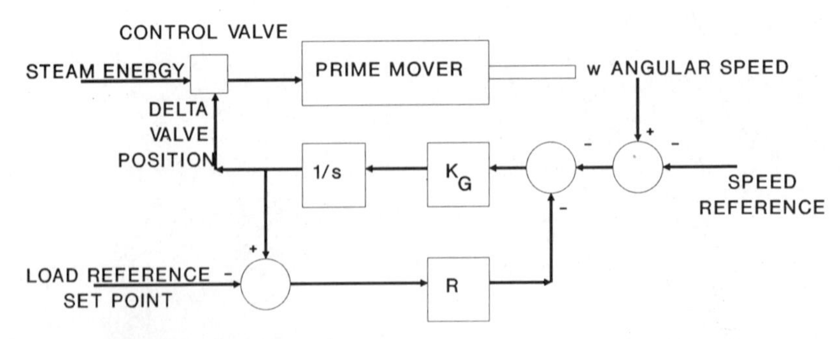

FIGURE 17.43 Speed-control system for a typical steam unit.

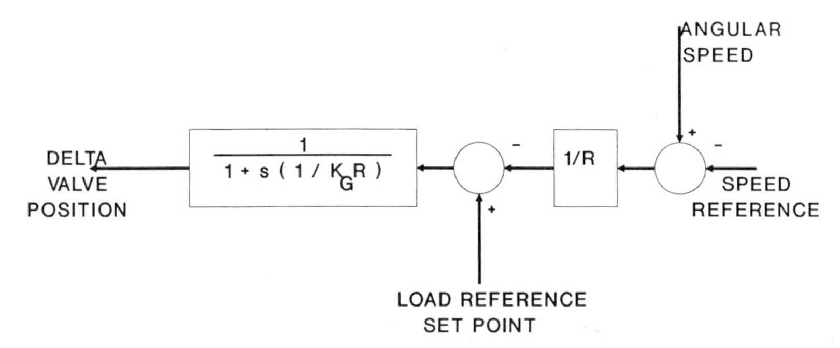

FIGURE 17.44 Transfer function diagram.

superheater constitute a large pressurized volume of energy with storage times of 60–120 seconds. One approximation for short stability studies less than 30 seconds is to assume constant steam drum pressure.

The high-pressure (HP) steam turbine and control-valve steam chest act as another storage volume for steam. This volume is very small, in the range of 0.05–0.4 second. Opening a control valve results in a quick power response in the HP turbine section because of this small steam volume.

The reheater and cold and hot reheat legs make up another steam storage volume in the range of 5–10 seconds. This volume influences the time response of the intermediate pressure (IP) and low pressure (LP) turbines. Both IP and LP turbines produce 60–70% of the total power of a generating unit.

The tandem-compound unit can be represented by the Laplace transform linear block diagram of Figure 17.46 (IEEE, "Dynamic Models for Steam and Hydro Turbines," 1973). The control valve change causes steam to flow through the HP turbine with a time constant. The fraction of power produced by the HP turbine is added to the total mechanical power of the generating

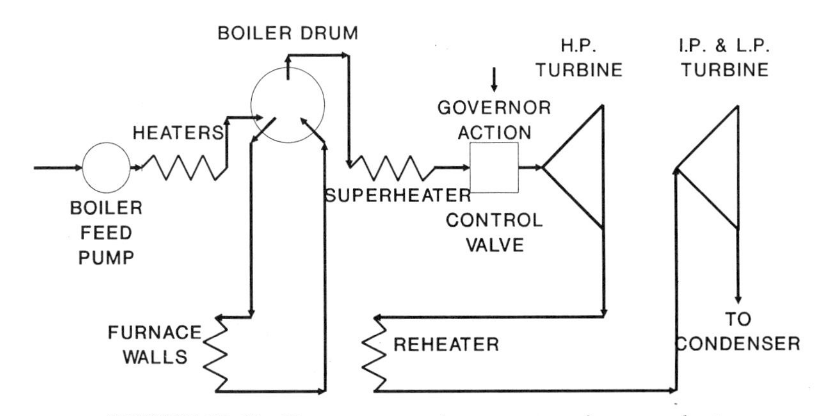

FIGURE 17.45 Drum-type tandem-compound power plant.

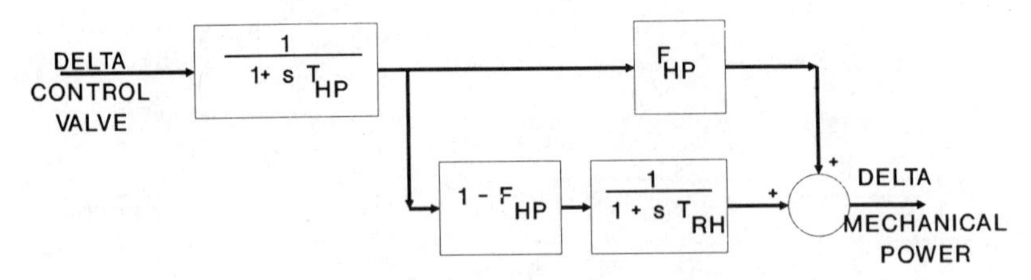

FIGURE 17.46 Steam plant linear block diagram.

unit. The IP/LP fraction (1. – HP fraction) is further delayed by the reheat time constant.

The transfer function is:

$$\Delta P_{\text{mech}} = \left(\frac{1}{1 + sT_{\text{HP}}}\right)\left[F_{\text{HP}} + \frac{(1 - F_{\text{HP}})}{1 + sT_{\text{RH}}}\right] \Delta \text{ control valve} \quad (17.116)$$

or

$$\Delta P_{\text{mech}} = \frac{(1 + F_{\text{HP}}\, T_{\text{RH}} s)}{(1 + sT_{\text{HP}})(1 + sT_{\text{RH}})} \Delta \text{ control valve} \quad (17.117)$$

A typical response-time trend of a single reheat turbine to a 10% change in the control valve flow is shown in Figure 17.47. The power increase reaches 3% after 0.2 seconds and then responds more slowly to the remaining IP and LP power with a time lag of 5 seconds.

This model can be expanded to include other power plant turbine configurations such as nonreheat, double-reheat, and cross-compound units.

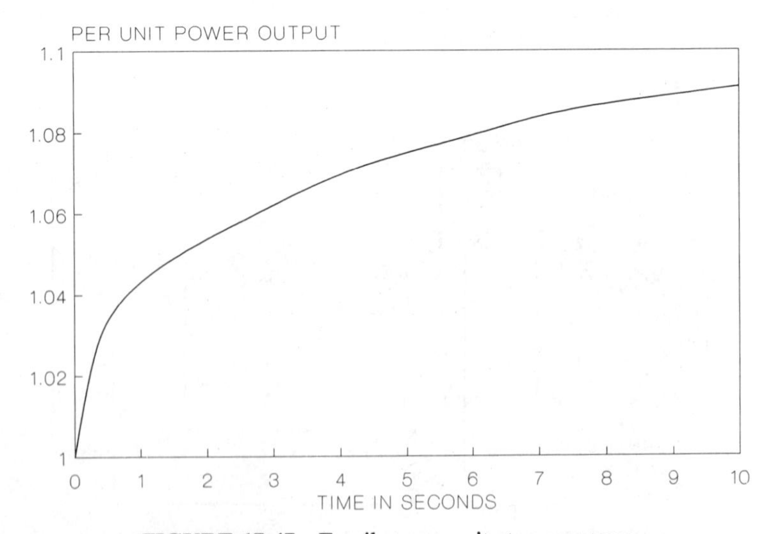

FIGURE 17.47 Fossil-steam unit step response.

For analysis times greater than 20–30 seconds for large (10%) step changes in valve position, the boiler pressure decreases and power output response is affected. Gas- and oil-fired steam units are more responsive than coal-fired steam units. Figure 17.48 illustrates this effect for a 10% step change in power requirements (Concordia et al., 1966).

The graph of Figure 17.48 is based on a turbine-lead–boiler-follow mode of control. In this case, the boiler controls sense the pressure decrease and steam flow increase, thereby increasing the firing rate of the boiler. This control mode unfortunately leads to deviation in boiler pressures and temperatures and has led to the adoption of control modes referred to as "coordinated" or "integrated" boiler–turbine control modes (IEEE, "MW Response of Fossil Fueled Steam Units," 1973).

In this mode, the turbine control valve also receives a boiler control signal based on boiler pressure. The boiler also receives a control signal based on the megawatt power output deviation. In this way, the turbine control valve opening is controlled by boiler conditions as well, resulting in less transient temperature and pressure changes. This coordinated control mode is especially needed in the case of once-through-boilers, where there is no large boiler steam drum volume.

The response characteristics have other limitations. During power increase excursions, the opening of turbine valves may be limited to allow no more than a 10% drop in boiler pressure to prevent carry-over of water from the boiler to the turbine, which could severely damage the turbine. Combustion controls may not stand large transients due to transient mismatches in fuel and air and the potential for explosions.

Hydroelectric unit governors are similar to steam units with the addition

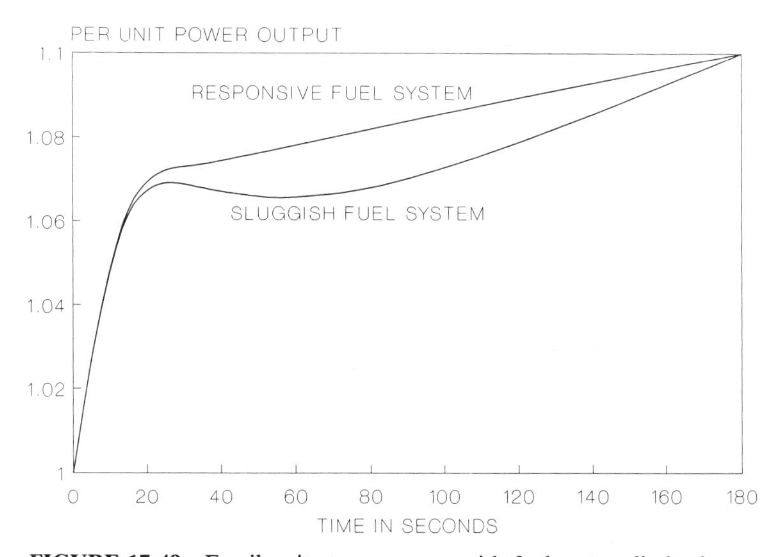

FIGURE 17.48 Fossil unit step response with fuel system limitations.

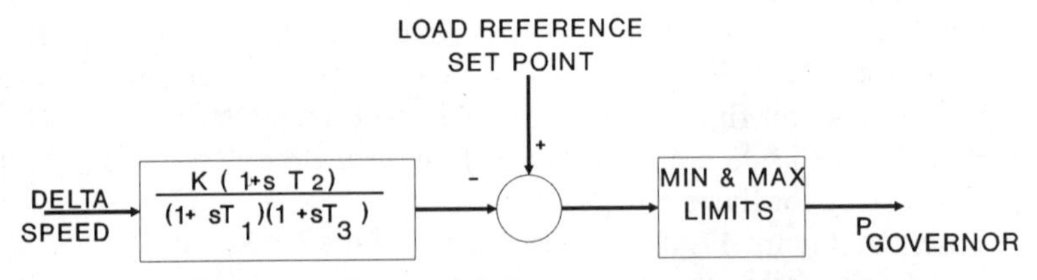

FIGURE 17.49 Typical hydro unit governor model.

of transient droop compensation to achieve stable performance. The transfer function of a typical hydroelectric-governor can be represented as shown in Figure 17.49. Typical values are T_1 = 15–25 seconds, T_2 = 2–4 seconds, and T_3 = 0.5–1 second.

A linearized hydraulic turbine model is shown in Figure 17.50. The water time constant, T_W, has values in the range of 0.5–2 seconds, depending on the water head potential, penstock length, and water velocity (IEEE, "Dynamic Models for Steam and Hyrdro Turbines," 1973). The hydro unit time response proceeds initially in the opposite direction from that desired, as shown in Figure 17.51 for a 10% step change in control valve. By slowing down the control valve (gate) a more stable response characteristic is obtained, as illustrated in Figure 17.52 for a ramp change in the control valve.

17.11 INFLUENCE OF ELECTRICAL LOADS ON STABILITY

In "classical" stability analysis, loads are characterized as being of constant impedance. An AC power-flow analysis establishes the pretransient conditions, and the impedance values are computed from the bus power and voltage. For more accurate analysis, loads are represented as components of constant impedance, constant current, and constant power and are frequency sensitive.

Loads are typically represented by formulas of the form:

$$P = P_o\,[a_0 + a_1 \frac{V}{V_o} + a_2(\frac{V}{V_o})^2]\,(1 + L_P\,\Delta f) \tag{17.118}$$

$$Q = Q_o[b_0 + b_1 \frac{V}{V_o} + b_2(\frac{V}{V_o})^2](1 + L_Q\,\Delta f) \tag{17.119}$$

P GOVERNOR

$$\frac{1 - sT_W}{1 + .5sT_W}$$

P MECHANICAL

FIGURE 17.50 Typical hydro turbine model.

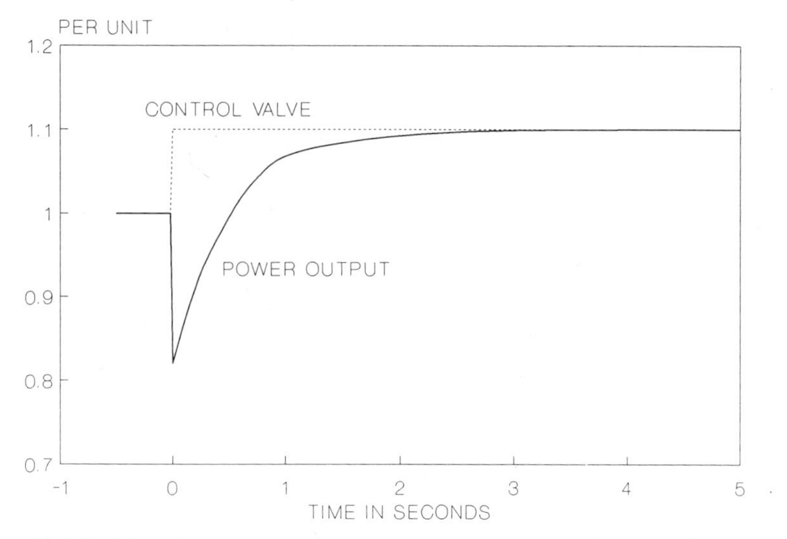

FIGURE 17.51 Hydro unit time response—control valve step change.

where a_0, b_0 = fraction of load represented by constant power/var
a_1, b_1 = fraction of load represented by constant current
a_2, b_2 = fraction of load represented by constant impedance
L_p, L_Q = linearized frequency sensitivity coefficients
V_o, P_o, Q_o = initial values from power-flow solution

A constant-impedance load tends to decrease voltage oscillations on a power system. When voltage dips during a transient, the load current decreases proportionally, thereby tending to reduce the severity of the voltage dip.

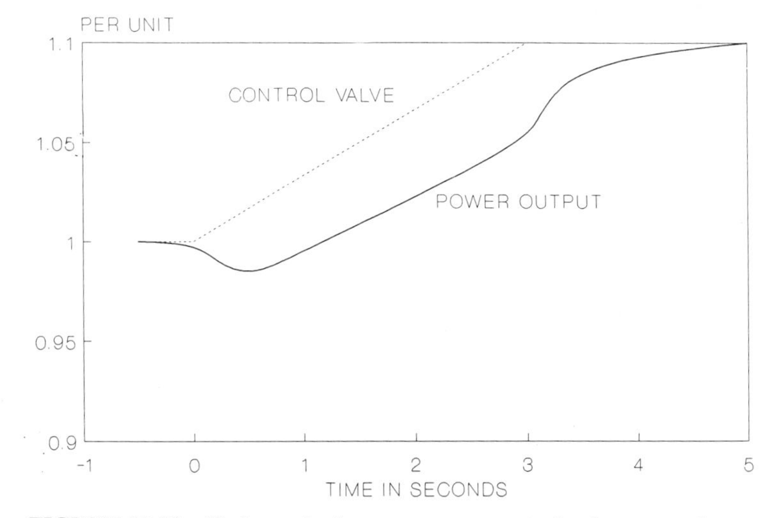

FIGURE 17.52 Hydro unit time response—control valve ramp change.

A constant power load tends to increase voltage oscillations. When voltage dips during a transient, the load current *increases* proportionally, thereby causing the voltage to dip even further. This is especially pronounced for reactive loads, since they have a greater influence on system voltage levels.

Simulation studies may be used to assess the effect of load representation on system stability. One study (Gentile et al., 1981) analyzed a radial line connecting two generators as shown in Figure 17.53. Loads on buses 1 and 2 were fixed, while the load characteristics on bus 3 were varied.

Two sets of studies were performed, remote generation studies in which generator 1 had a 3100-MVA rating (operated at 2700 MW), and generator 2 had an 800 MVA rating (operated at 119 MW). The load characteristics of load bus 3 were then varied. This set of studies corresponds to a situation in which most of the generation is remote from the principal load center. The other set of studies interchanged the generating units (unit 2 operated at 2700 MW and unit 1 at 63 MW). This set corresponded to a situation in which most of the generation is local to the principal load demands.

The fault position was at load center 2. The maximum swing angle between the two generators was taken as an indication of system stability. The maximum swing angle was calculated for a series of fault clearing times.

Figure 17.54 illustrates the stability effect for variations in the active power component of load 3 with the reactive component fixed at constant impedance.

For faults that are cleared in the primary clearing time, 0.08–0.10 second, the load modeling differences are small. However, for backup relay clearing times, 0.15 second, the effect for remote generation is significant. The constant impedance load model is more stable than a constant current load model. Similarly, but not indicated on the figure, a constant power load model is even less stable. The results can be explained physically. When the fault occurs, it interrupts power transfer from the large generator to the remote load. Generator 1 then speeds up and generator 2 slows down under increased load demand. A load model with a greater reduction in load demand with voltage, such as a constant impedance model, will lessen the deceleration of generator 2 with respect to generator 1.

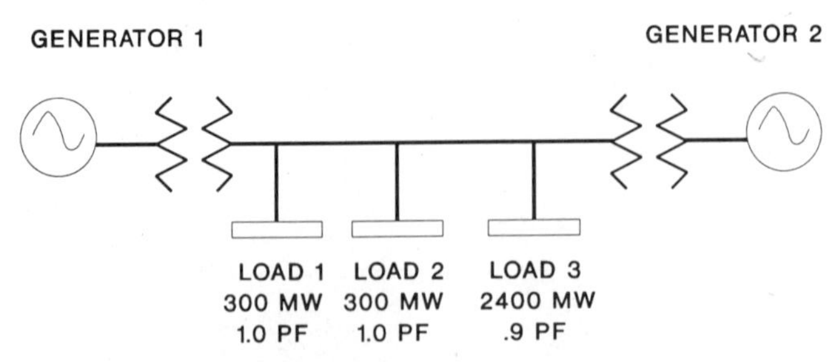

FIGURE 17.53 Radial line connection in simulation study.

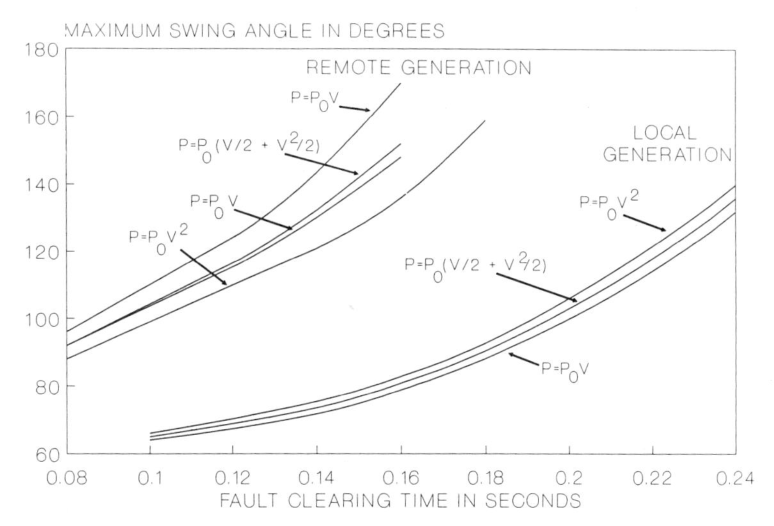

FIGURE 17.54 Effect of stability on changing active power load models with reactive power loads modeled as constant impedance.

The case of local generation shows a much smaller, but opposite, effect. When a fault occurs, generator 2 accelerated. A load model in which the load held most constant is most stable.

The results for the reactive power load characteristics were treated by representing the active power with a constant current. Figure 17.55 presents the reactive power model's impact on stability. Reactive power results for the remote generation case show smaller differences and results opposite to those of

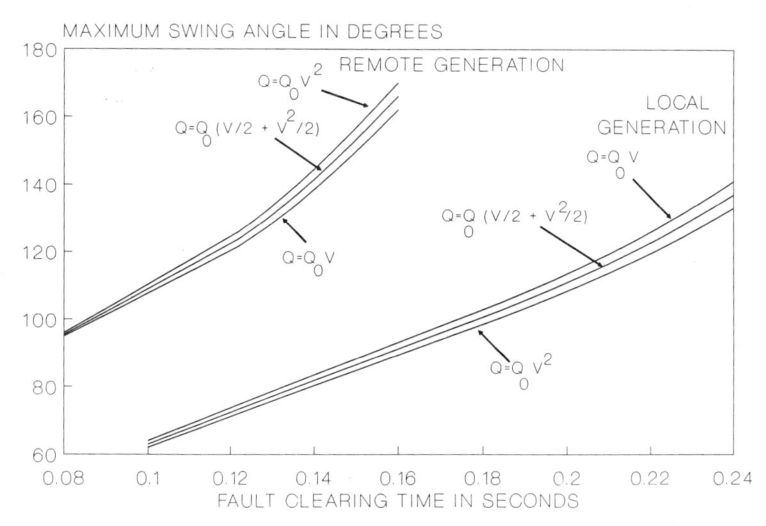

FIGURE 17.55 Effect on stability on changing reactive power load models with active power loads modeled as constant current.

the active power case. A fault on the system causes generator 2 to accelerate. Acceleration is lessened by decreased load at load 3. Loads with more constant reactive power requirements, specifically, constant current, will hold the voltage and load down at load 3.

The frequency sensitivity of the loads also affects stability. The more the active power of the load decreases with decreasing frequency, the more stable the power system becomes. Positive load-frequency coefficients provide a positive damping torque, which helps to damp oscillations.

Another load model formulation involves an exponential voltage characteristic:

$$\frac{P}{P_0} = P_A \left(\frac{V}{V_0}\right)^{KPV1} \cdot (1 + KPF1 \cdot \Delta \text{ frequency}) + (1 - P_A)\left(\frac{V}{V_0}\right)^{KPV2} \tag{17.120}$$

$$\frac{Q}{P_0} = Q_A\left(\frac{V}{V_0}\right)^{KQV1} (1 + KQF1) + \left(\frac{Q_0}{P_0} - Q_A\right)\left(\frac{V}{V_0}\right)^{KQV2} (1 + KQF2 \cdot \Delta \text{ frequency}) \tag{17.121}$$

The active power consists of one frequency-dependent term (motors, fluorescent lights, etc.) and one non-frequency-dependent term (resistive devices). The reactive power component is comprised of two components. The first component represents the loads contributed by uncompensated loads; the second represents the net reactive contribution from feeder and transformer reactances and capacitive compensation. Both Q_0 and P_0 are the initial values from the power flow.

It is possible to synthesize the load model data by aggregating individual end-use devices based on the load composition at each bus and individual device characteristics. Typical values for polynomial coefficients by customer class are presented in Table 17.9 (Price et al., 1987). Figures 17.56 and 17.57 illustrate this typical load characteristic. Utility-specific values should be used in stability studies.

In view of the large quantity of electricity consumed in induction motors by all end users on a load bus, it may also be desirable to represent induction model load contributions by a dynamic induction motor model. The induction motor's torque equations and electrical circuit equations are modeled.

In summary, proper load modeling can have an important influence on the stability of a power system. Generally, constant-impedance loads are the most stable and constant-power loads the least stable for remote generation. The load model has less of an influence for local generation.

TABLE 17.9 Typical Polynomial Load Model Coefficients

Customer Class	Active Power (P)			Reactive Power (Q)		
	Constant Power	Constant Current	Constant Impedance	Constant Power	Constant Current	Constant Impedance
Residential	−0.08	0.67	0.41	0.05	−2	2.5
Commercial	0	1	0	1.25	−4.5	4.25
Industrial	0.83	0.24	−0.07	0	0	1

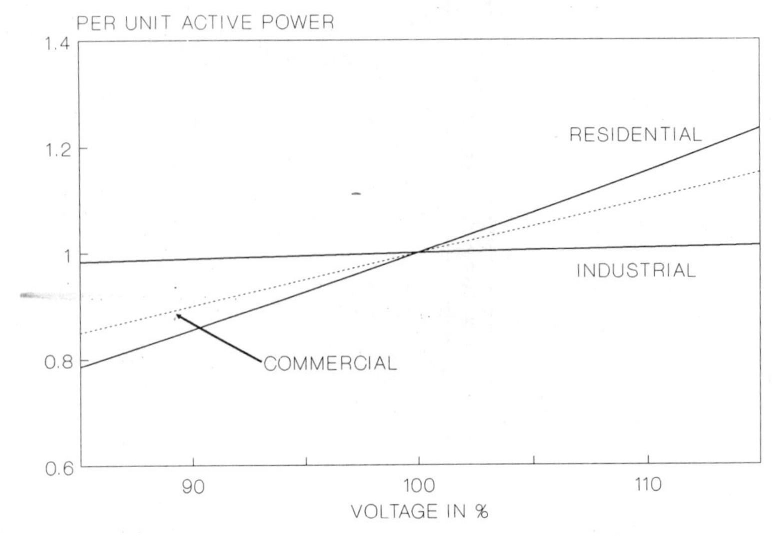

FIGURE 17.56 Active power voltage dependence.

17.12 TRANSMISSION-LINE COMPENSATION

Transmission lines may be compensated with shunt capacitance or series capacitance to improve power-transfer capability (Miller, 1982).

17.12.1 Shunt Compensation

The objective of applying shunt compensation to power transmission lines is to improve power transfer capability by adding reactive vars to the transmission

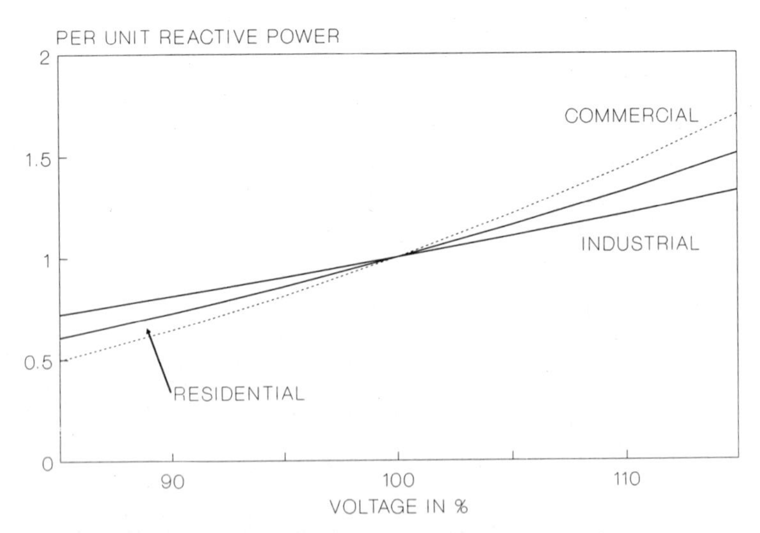

FIGURE 17.57 Reactive power voltage dependence.

network to maintain a desired voltage level during normal and fault conditions (Hauth and Moran, 1978). While fixed or switched shunt capacitors are sometimes adequate, actively controlled shunt compensation is required for maximum benefit. Synchronous condensers had been traditionally used for this purpose, but more recently static–var support systems have proved economical and reliable.

Shunt compensation must be capable of being switched rapidly to accommodate and relieve system transients and to prevent overvoltages.

The static–var system (SVS) is typically configured as illustrated in Figure 17.58. The SVS is essentially a capacitor and a variable inductor. The inductor value is varied by a thyristor switch where the conduction angle is regulated by the voltage magnitude on the HVAC bus. The thyristor switch is an off–on device only, it regulates the inductor current flow by conducting from 0° to 360° of each power cycle. If the voltage magnitude decreases, then the regulator forces the thyristor switch to conduct for a smaller portion of each cycle, thereby reducing the inductance and yielding more capacitive vars. Because the thyristor chops the inductor current cycle, it introduces harmonics which are then removed in the harmonic filter.

One method to visualize the shunt compensation benefit to system power transfer is to analyze a simple transmission line transferring power between two voltage sources as shown in Figure 17.59. An SVS system, installed at the midpoint of the transmission line, is represented as an inductive admittance, $-jB$. For symmetry, the shunt compensation system is shown in two half-sections.

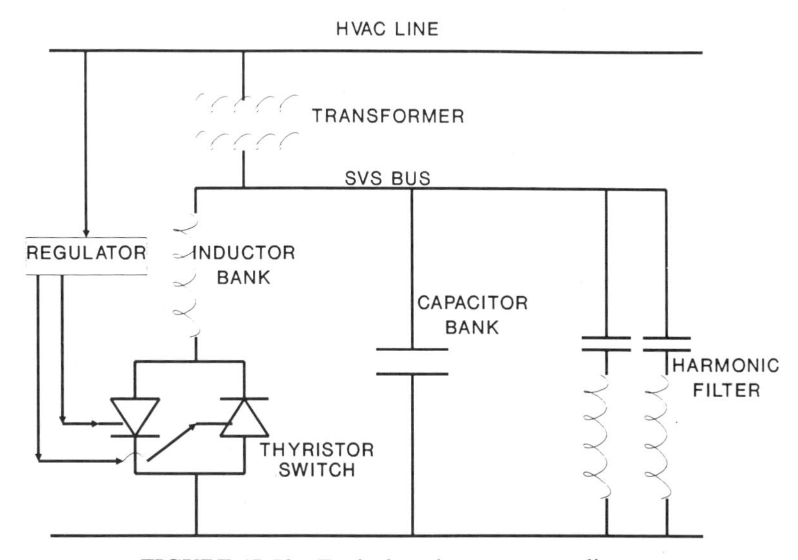

FIGURE 17.58 Typical static–var system diagram.

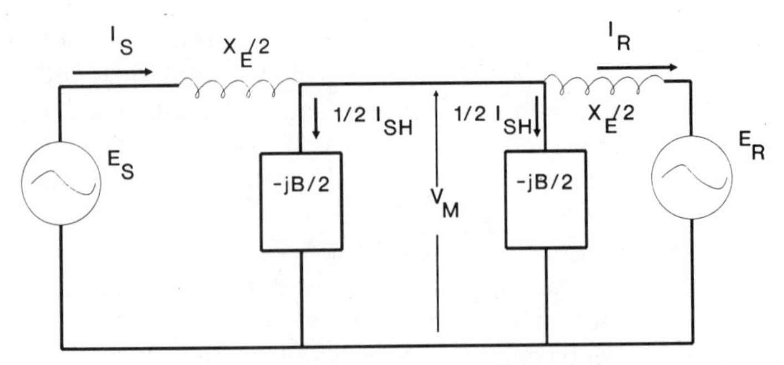

FIGURE 17.59 Simplified transmission-line power transfer.

The midpoint voltage, $\overline{V}_{\mathrm{m}}$, is calculated as:

$$\overline{V}_{\mathrm{m}} = \overline{E}_{\mathrm{S}} - j\,\frac{X_{\mathrm{e}}}{2}\,\overline{I}_{\mathrm{S}} = \overline{E}_{\mathrm{R}} + j\,\frac{X_{\mathrm{e}}}{2}\,\overline{I}_{\mathrm{R}} \tag{17.122}$$

The shunt current is:

$$\overline{I}_{\mathrm{R}} - \overline{I}_{\mathrm{S}} = \overline{I}_{\mathrm{SH}} = -jB\,\overline{V}_{\mathrm{m}} \tag{17.123}$$

The current at the midpoint may be expressed as:

$$\overline{I}_{\mathrm{S}} = \overline{I}_{\mathrm{m}} + \frac{\overline{I}_{\mathrm{SH}}}{2} \tag{17.124}$$

$$\overline{I}_{\mathrm{R}} = \overline{I}_{\mathrm{m}} - \frac{\overline{I}_{\mathrm{SH}}}{2} \tag{17.125}$$

Adding the two equations yields:

$$\overline{I}_{\mathrm{m}} = \frac{\overline{I}_{\mathrm{S}} + \overline{I}_{\mathrm{R}}}{2} \tag{17.126}$$

Substituting into Equation (17.122) yields:

$$\overline{I}_{\mathrm{m}} = \frac{\overline{E}_{\mathrm{S}} - \overline{E}_{\mathrm{R}}}{jX_{\mathrm{e}}} \tag{17.127}$$

Adding the two equations for V_{m} gives:

$$2\,\overline{V}_{\mathrm{m}} = \overline{E}_{\mathrm{R}} + \overline{E}_{\mathrm{S}} + \frac{jX_{\mathrm{e}}}{2}\,(\overline{I}_{\mathrm{R}} - \overline{I}_{\mathrm{S}}) \tag{17.128}$$

$$2\,\overline{V}_{\mathrm{m}} = \overline{E}_{\mathrm{R}} + \overline{E}_{\mathrm{S}} + \frac{jX_{\mathrm{e}}}{2}\, jB\, \overline{V}_{\mathrm{m}} \tag{17.129}$$

Solving for V_{m} yields:

$$\overline{V}_{\mathrm{m}} = \frac{\overline{E}_{\mathrm{R}} + \overline{E}_{\mathrm{S}}}{2[1 + (X_{\mathrm{e}}B/4)]} \tag{17.130}$$

The complex power at the midpoint can then be calculated as:

$$P + jQ = V_{\mathrm{m}}\, I^{*}_{\mathrm{m}} \tag{17.131}$$

$$P + jQ = \left\{ \frac{\overline{E}_{\mathrm{R}} + \overline{E}_{\mathrm{S}}}{2\,[1 + (X_{\mathrm{e}}B/4)]} \right\} \left[\frac{\overline{E}^{*}_{\mathrm{R}} - \overline{E}^{*}_{\mathrm{S}}}{jX_{\mathrm{e}}} \right] \tag{17.132}$$

$$P + jQ = \frac{\overline{E}^{2}_{\mathrm{R}} - \overline{E}^{2}_{\mathrm{S}} + \overline{E}^{*}_{\mathrm{R}}\overline{E}_{\mathrm{S}} - \overline{E}_{\mathrm{R}}\overline{E}^{*}_{\mathrm{S}}}{2[1 + (X_{\mathrm{e}}B/4)]jX_{\mathrm{e}}} \tag{17.133}$$

where $*$ denotes the complex conjugate operation. Let

$$\overline{E}_{\mathrm{R}} = Ee^{-j(\delta/2)} \qquad \text{and} \qquad \overline{E}_{\mathrm{s}} = Ee^{j(\delta/2)}$$

Let E_{m} be the reference angle and $|E_{\mathrm{R}}| = |E_{\mathrm{S}}| = |E|$. Then:

$$P + jQ = \frac{E^{2}\, e^{j\delta} - E^{2}\, e^{-j\delta}}{2[1 + (X_{\mathrm{e}}B/4)]jX_{\mathrm{e}}} \tag{17.134}$$

The active power is:

$$P = \frac{E^{2} \sin \delta}{X_{\mathrm{e}}\, [1 + (X_{\mathrm{e}}B/4)]} \tag{17.135}$$

The midpoint voltage is:

$$V_{\mathrm{m}} = \frac{E \cos (\delta/2)}{1 + (X_{\mathrm{e}}B/4)} \tag{17.136}$$

Consider the case where the midpoint voltage is controlled to equal the sending voltage. Then, by Equation (17.136), the shunt admittance must be adjusted so that:

$$\left(1 + \frac{X_{\mathrm{e}}B}{4}\right) = \cos\left(\frac{\delta}{2}\right) \tag{17.137}$$

Substituting into (17.136) and using the double-angle sine rule yields:

$$P = \frac{E^2}{X_e} \frac{\sin \delta}{\cos (\delta/2)} = \frac{2E^2}{X_e} \sin \frac{\delta}{2} \tag{17.138}$$

The amount of shunt compensation necessary to achieve midpoint voltage support is calculated from Equation (17.137) as:

$$B = \frac{4}{X_e} \left(\cos \frac{\delta}{2} - 1\right) = \frac{4\, P_{base}}{E^2} \left(\cos \frac{\delta}{2} - 1\right) \tag{17.139}$$

where $P_{base} = E^2/X_e$. The value of B includes the external shunt compensation as well as the π equivalent line capacitance.

Figure 17.60 illustrates the locus of operating points of Equation (17.138), in per unit of P_{base}. Note that the potential power transfer is double that of an uncompensated line ($B = 0$).

Figure 17.60 also illustrates power-transfer curves of Equation (17.135) for several B values. In this figure, the maximum B value is taken as 2.0. The power-transfer characteristic follows the locus of operating points up to where the required B equals B_{max}. For angles greater than 120°, the power transfer characteristic follows the B_{max} curve.

Shunt compensation can be a contributor to improved power system stability. Figure 17.61 illustrates an equal-area example without shunt compensation. In this example, area B is less than area A; therefore, this system is unstable. Figure 17.62 presents the same example but with controlled shunt compensation. In this case, area B easily equals area A and the power system is stable.

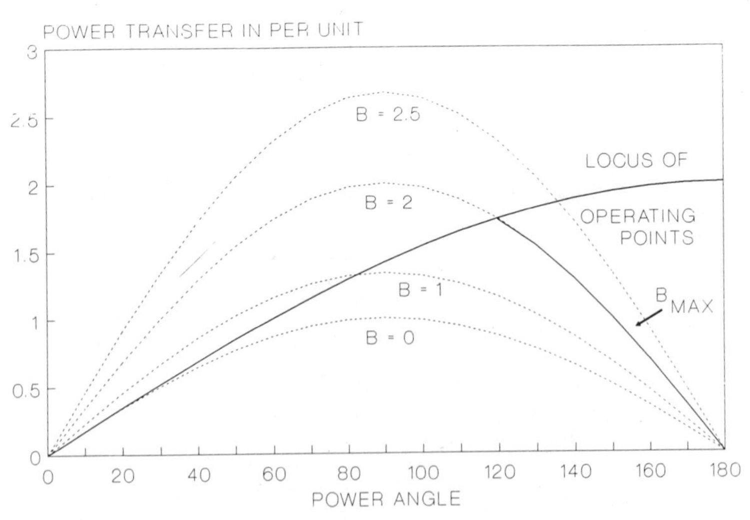

FIGURE 17.60 Power-transfer capability with shunt compensation.

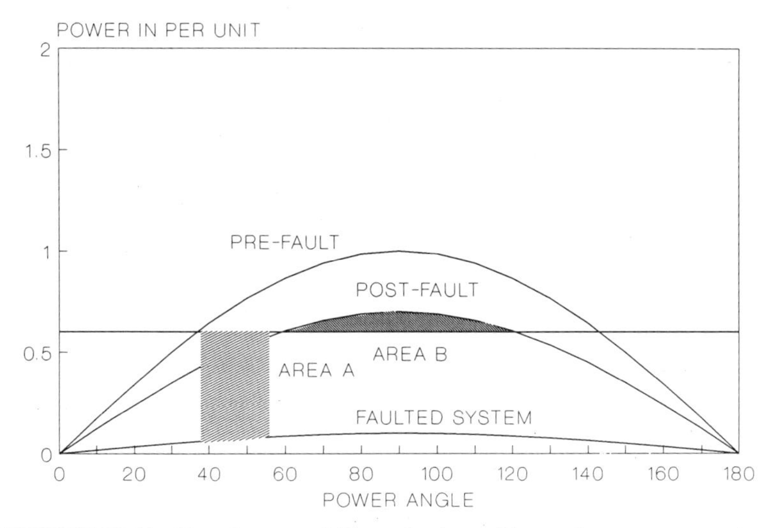

FIGURE 17.61 Equal-area stability criteria—without shunt compensation.

The dynamic performance of a static–var system can be modeled as shown in Figure 17.63, in which K_{SVS} is typically 20 to 100 and T_{SVS} is 0.03 seconds. The static–var system quickly responds to voltage swings and provides a very strong positive synchronizing coefficient. The static–var system can also have supplemental signals applied to the voltage-control junction to enhance damping characteristics (just as supplementary signals are applied to enhance damping on generator excitation systems). The supplementary signals can be driven by system frequency change or rate of change of line current.

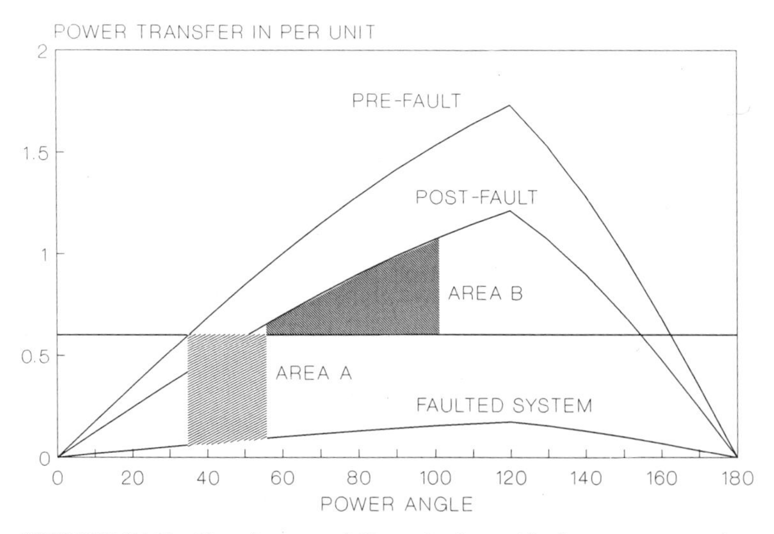

FIGURE 17.62 Equal-area stability criteria—with shunt compensation.

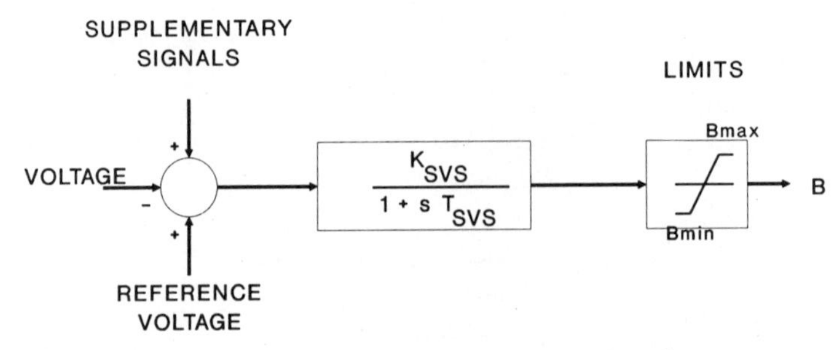

FIGURE 17.63 Dynamic performance model of static–var system.

A static–var system can be an economical method for improving power transfer for medium- and long-length lines where high-speed voltage control and/or transient stability are limiting.

17.12.2 Series Compensation

The objective of series compensation for transmission lines is to improve power transfer by adding series capacitance to the network. This reduces the effective reactance of the transmisson line.

One way to visualize the series capacitance benefit to power transfer is to examine a simple example as shown in Figure 17.64. The transmission line has a series capacitor installed at the line midpoint. This example neglects the shunt capacitance of the line.

Shunt reactors are typically applied with series capacitors to compensate for the line shunt capacitance, to reduce the reactive burden at the line terminals (generators) and to better maintain the line voltage profile at light load.

The power transfer can be readily calculated as:

$$P = \frac{E_S\, E_R}{X_e - X_c} \sin \delta \tag{17.140}$$

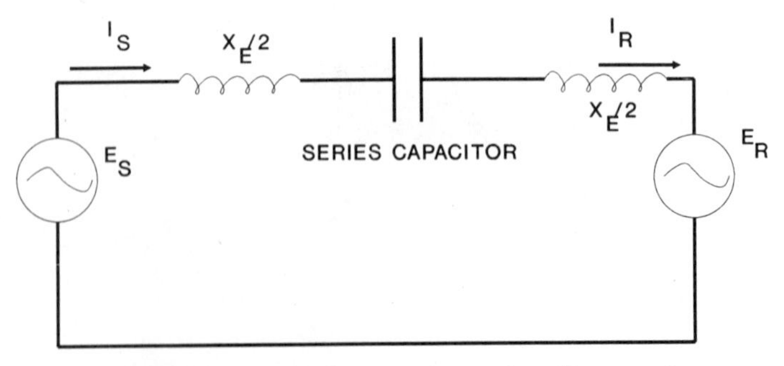

FIGURE 17.64 Series capacitance benefit example.

Thus, power transfer can be significantly increased by series compensating the line.

During normal operation, the current through the capacitors typically leads to capacitor voltages that are 30–40% of rated capability. During faults, however, the capacitors could be damaged unless they are bypassed.

One approach to protecting series capacitors is to use a spark-gap technique as shown in Figure 17.65. The main gap is set to spark over at a preset level to protect the capacitors during a fault. Current flowing in the spark-gap circuit then causes the bypass breaker to close, which extinguishes the spark-gap current. When the bypass current returns to normal after the fault has been removed, the bypass breaker opens, typically with a three- to five-cycle opening delay.

Modern protective equipment can use a nonlinear resistor, such as a zinc oxide varistor, to limit capacitor voltage. A varistor has a volt-ampere characteristic with an abrupt knee, below which no current flows, and above which conduction occurs. A varistor protective system will maintain voltage across the capacitor during the fault and will remove itself almost instantaneously after the fault. To protect the varistor from extreme fault duty, a spark gap and bypass breaker are also added.

In transient simulations, the effect of breaker delay, during which time capacitors are bypassed, has an important influence on the power-transfer stability limit. Figure 17.66 illustrates the same equal-area case of Figure 17.61, except that the lines are 50% series compensated. Area A is the acceleration energy. The fault is cleared at 45°. After the fault is cleared, electrical power transfer is governed by the uncompensated line until the bypass breaker is opened. After this time, the postfault is governed by the compensated line. The stability margin of Figure 17.66 would be considerably larger if the bypass breaker time were shortened. Varistor protection systems can be a beneficial enhancement to system stability by reducing this delay time.

A series capacitor in series with the line inductance forms a resonant circuit whose oscillation frequence is:

$$f_c = \frac{1}{2\,\pi\,\sqrt{LC}} = 60\text{ Hz} \cdot \sqrt{\frac{X_c}{X_e}}$$

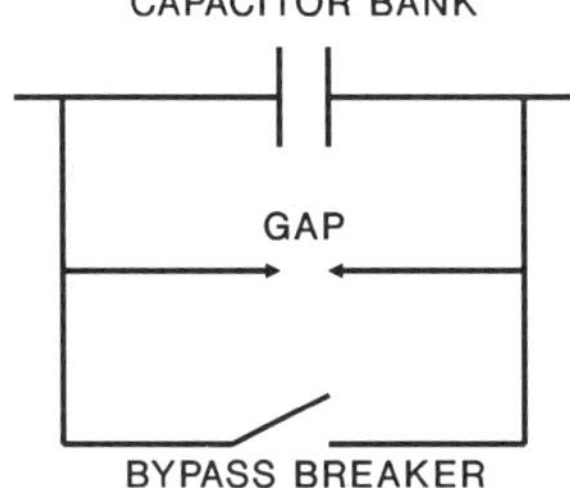

FIGURE 17.65 Spark-gap protection technique.

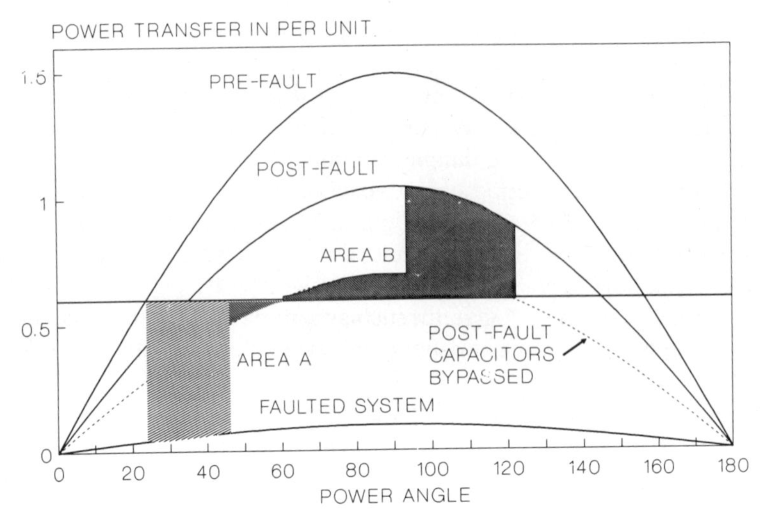

FIGURE 17.66 Equal-area stability criteria—with 50% series compensation.

Since X_c typically represents 25–70% of the line inductance, the oscillation frequency is subsynchronous, less than 60 Hz. This resonance can interact with the torsional modes of any turbine generators near the series compensated line. The interactions can lead to increased fatigue life consumption of the turbine-generator shafts and, in very severe cases, can lead to the breaking of shafts (Bowler, 1976). The likelihood of problems is larger for highly compensated lines. Corrective action for this subsynchronous resonance includes installation of filter circuits, use of excitation control to increase damping of the oscillations, and use of static var compensator and controls to damp the oscillations.

In summary, series capacitive compensation is a cost-effective method for improving power transfer for medium and long transmission lines but must be applied with due regard to potential subsynchronous resonance effects. In deciding between shunt and series compensation, the following points may be helpful:

1. If there are stability difficulties on a particular line, shunt compensation can be installed without disturbing the power flow. However, if series capacitors are used, the whole network flow is changed.
2. If there is a need to increase transfer over a particular line, a series capacitor is appropriate, sized to preserve the same flows on the parallel lines.
3. For long transmission lines from a generating plant:
   a. To a load area without generation—use shunt.
   b. To an infinite bus system—use series.
   c. Break-even may be when generation at load area equals remote generation.

17.13 IMPACT OF HVDC ON SYSTEM STABILITY

As discussed in Section 16.14, high-voltage direct-current (HVDC) systems, because of their flexibility of control, can be used to improve system stability. HVDC controls can rapidly change or modulate the firing angles of the converters, which will, in turn, cause rapid changes or modulation of the direct current and voltage. Resulting changes in the ac active and reactive power levels can be used to improve transient stability, damp oscillations, and regulate AC bus voltages.

Contingencies on the DC system, such as sudden loss of DC power transfer, can also adversely affect AC system stability. Such contingencies must therefore be studied.

The HVDC control system design must be tailored to the particular AC/DC system configuration. System characteristics that will influence control design include:

1. Whether the DC system connects isolated ac systems or parallels AC transmission circuits.
2. Whether there is a significant amount of generation near the rectifier or inverter end or both.
3. The relative strength (short circuit ratio) of the AC systems at each end.

Detailed dynamic simulations of the combined AC/DC system are required during the planning of an HVDC system. The modeling required for such analysis is beyond the scope of this book. (Arrillaga, et al., 1983) Effects that must be modeled include:

1. *Power Modulation Control.* The DC power is often modulated to increase the damping of oscillations and to improve transient stability. The modulation controller receives signals such as AC system frequency or flow on a key AC transmission corridor and generates a signal that is added to the signals that control direct current and voltage.
2. *DC System Controller.* This system produces the current and voltage orders based on setpoints, modulation signals, and other factors. It includes the following:
   a. *Power Controller.* It is often desired to regulate power flow rather than current. The power controller converts a power setpoint into a current order, considering the direct voltage level.
   b. *Telemetry.* When power control is used, the resulting current order must be transmitted to the other terminal, with some telemetry delay, for use in case of a mode shift.
   c. *Block and Restart.* Conduction is temporarily blocked in the converters as a result of depressed AC voltages, such as a nearby fault. In order to improve system stability, DC conduction is restarted as quickly as possible, but not so quickly as to cause voltage collapse.

3. *Mode-Shifting Logic.* In Section 16.17, the basic control mode of most HVDC systems was described. The inverter controls the direct voltage (or holds constant extinction angle) and the rectifier controls direct current. However, if the AC bus voltage at the rectifier drops too low, because of a nearby fault or voltage swing, for example, a viable operating point may no longer exist for this control mode. A change in control mode (mode shift) then occurs, in which the rectifier holds constant firing angle and the inverter controls direct current at a slightly reduced level until the AC voltage recovers.
4. *Converter Equations.* The converter equations are as described in Section 16.14.
5. *DC Network.* Depending on the degree of detail required, the dc network may be modeled as:
   a. Algebraic relationships between direct currents and voltages.
   b. Differential equations including smoothing reactor and line inductance and capacitance.

Detailed studies may require modeling of the time constants and other characteristics of the firing angle control circuits and other details of the controllers, not represented in a basic model described above.

17.14 SYSTEM STABILITY ANALYSIS

The stability analysis of a power system involves the analysis of many generating units coupled through transmission network equations to the end-use consumer loads, as illutrated in Figure 17.67. Each generating unit is represented by an excitation system, synchronous generator, swing equation, governor, turbine, energy source, and controls. Depending on the stability problem being analyzed, parts of the generating unit representation may not be needed in the analysis.

When studying a large power system, not all generating units swing with the same oscillation frequency. Groups of units that are electrically well connected tend to swing together, while units not well connected tend to have slower oscillations with respect to each other.

Large-scale transient stability computer programs are used to solve generating unit dynamic equations; static transmission network equations; and static and dynamic equations for loads, SVS, HVDC systems, and other devices. The number of dynamic equations per generating unit can become large.

If an excitation system and transient machine model are used to represent each generating unit, then 11 or more coupled, dynamic (time-derivative) equations must be solved. Because of the range of time constants in these equations, short integration time steps (about 0.01 second) are required. This leads to a significant computational burden when representing hundreds of generating units.

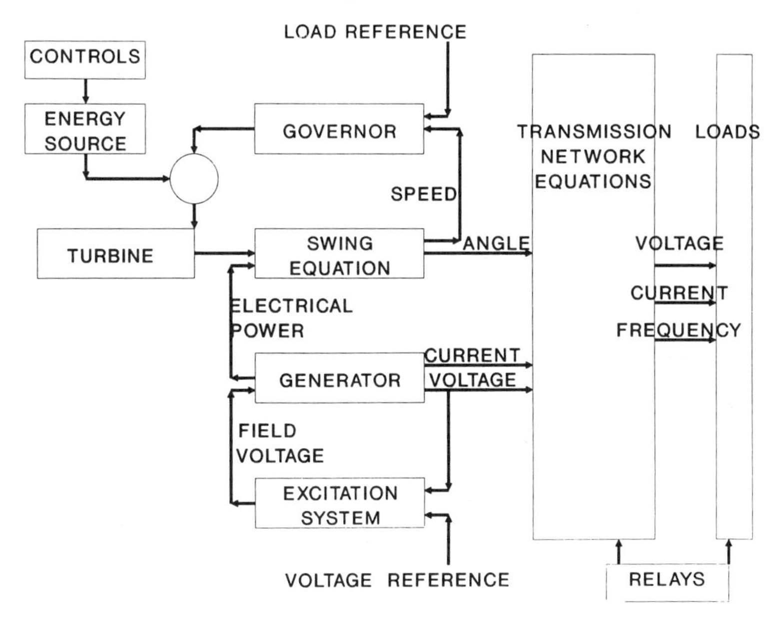

FIGURE 17.67 Simplified schematic of many generating units coupled through the transmission network to end-use consumer load.

The network solution also requires considerable computer time since it must be computed for every time-integration step and usually requires iterative solution. Transient stability analysis consumes significant computer resources.

17.14.1 Transient Stability System Example

Many North American transmission facilities provide for large electric power transfers over long distances from widely separated geographic areas to achieve overall system operating economies. The final transmission stability analysis for many projects must include relevant plans for a large and diverse geographic area, and must address the plans of other utilities as well.

A simplified representation of an example system is shown in Figure 17.68. This example system is comprised of 105 GW of capacity, representative of a large region of the United States. The region may be conceptually viewed as four connected areas, each with a substantial generation and transmission infrastructure. While the actual network is comprised of more than 2000 major transmission lines, 400 generating plants and 1500 load buses, Figure 17.68 illustrates only the major interconnection distances and normal power flows. Areas 1, 3, and 4 have low-cost coal, nuclear, and hydroelectric generation. These areas sell power to region 2, which is largely comprised of higher-cost oil and natural gas units, although it has some nuclear and hydroelectric generation as well.

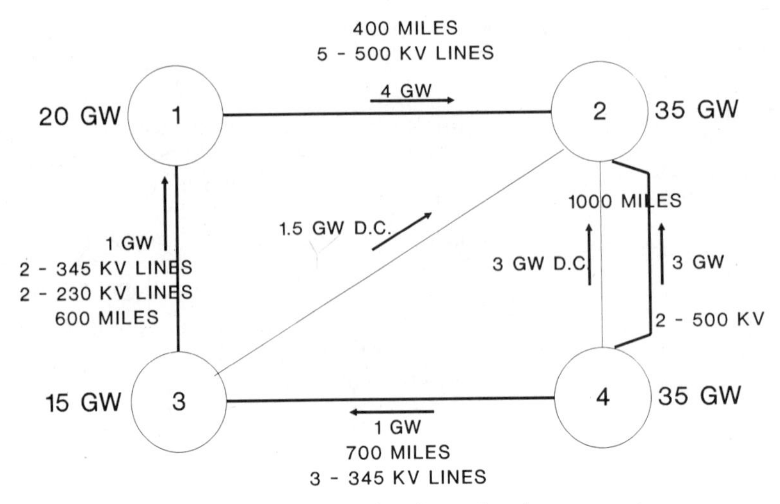

FIGURE 17.68 Schematic of a regional power system.

This regional system includes series compensated lines, static var compensation, synchronous condensers, phase-shifting transformers, and power system stabilizers. In addition, several HVDC systems are included within the ac network. Power transfer into area 2 exceeds 10 GW, roughly 10% of the total region generation. The large power transfer across this system results in unique system steady-state and transient performance characteristics that are typical of such long-distance power transfers.

A critical disturbance on the example system is a trip-out of some of the heavily loaded transmission interconnections between areas. This places the burden of transferring the excess generation in the sending area across the remaining lines to the receiving area. A double-circuit outage of two 500-kV lines connecting area 4 to area 2 is evaluated.

When this double circuit has an outage, the sending area generation accelerates. Because of the high level of precontingency line loadings, the remaining transmission system is not able to transfer the extra power. The sending and receiving areas lose synchronism. One means to maintain synchronism (and stability) is to trip out some generation in the sending areas. This reduces sending area generation acceleration since the generation in the sending areas is now more nearly matched to the load. Enough generation can be tripped to keep the sending and receiving systems stable during the subsequent power swing.

The tripping of generation, however, ultimately results in a frequency drop that may initiate automatic and undesirable underfrequency load shedding. When less sending generation is dropped, the system becomes unstable with angle and voltage characteristics as shown in Figures 17.69 and 17.70. The sending and receiving areas separate relative to one another with a subsequent voltage collapse in the generation-deficient receiving area. The separation time

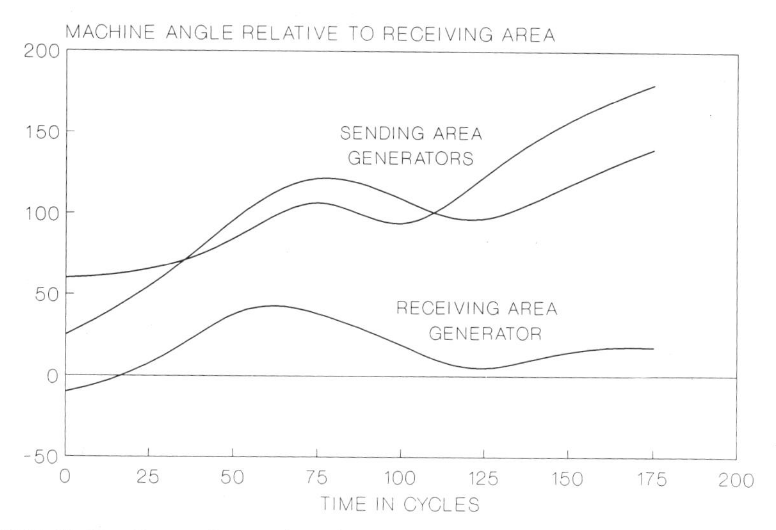

FIGURE 17.69 Generation angle swing curves during a double-circuit fault of 3-GW transfer from area 4 to area 2.

is approximately 2 seconds, and is typical of power transfers over large distances that have weak transmission linkages.

This large-region stability problem may be remedied by several approaches, including adding additional transmission lines and/or enhancing the performance of existing lines. Adding new lines is capital intensive, whereas enhancing the performance of existing lines can be less capital intensive. An example will

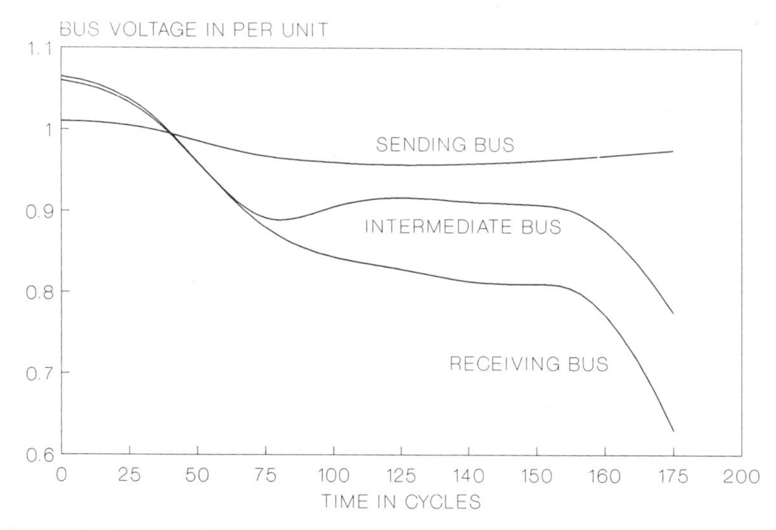

FIGURE 17.70 Bus voltages during a double-circuit fault of 3-GW transfer from area 4 to area 2.

be developed illustrating transmission-line enhancement using static–var and series compensation methods; but first, let us explore the principles of using shunt and series compensation with a simplified example.

From the standpoint of long-distance power transfer between widely separated geographic areas, there are many intermediate buses with local loads and some with generation. To maintain acceptable voltage variations at these intermediate locations, it is necessary to hold as constant a transmission voltage profile across the system as possible. This can be done by supplementing the natural transmission line charging with a variety of shunt var supply devices. Heavy maximum line loading is only really practical if some series compensation is used. Otherwise, the angular separation across the transmission system between the sending and receiving systems becomes extremely large. This large angular separation would appear across all the lines interconnecting the distinct geographic areas.

In large regional power networks that involve many distinct utility operating companies, some utilities are electrically in parallel to the flow path from the sender to the receiver area such as utilities in area 3. In the example, 1 GW of power (loop flow) flows through area 3 because it is in parallel with sender area 4 to receiver area 2. There may be utilities in area 3 that do not wish to participate in a scheduled power transfer. The loop flow may increase their own line losses for power transfer within their utility and may reduce their own power transfer capability within the system.

If power lines are only compensated with shunt var devices, then all intermediate systems, including those that might not wish to participate in the scheduled power transfer, experience large power transfers. As part of the regional system, these intermediate systems would also have to install supplementary var devices and participate equally in heavy circuit loadings. For this example, if the maximum system line angular separation were limited to 70°, only those systems that were series (series and shunt) compensated would experience power transfer in excess of one per unit surge impedance loading (SIL). Therefore, supplementary var compensation would be limited to only those systems participating in higher circuit loadings.

Long distance transmission lines are segmented many times and connected to other lines at intermediate buses. A typical configuration of series and shunt compensation and the resulting line voltage and angle profiles are shown in Figure 17.71, which illustrates two equal segments of 150 miles each. Two power-loading cases are shown, one at 1.0 SIL and another at 2.5 SIL. Using power flow calculations, the range of the switched capacitors, static–var control, and series capacitors was determined to provide acceptable voltage profiles and angular separation for light and heavy loadings. Note that shunt reactors are added at the series capacitor locations to control voltage during light load periods.

The relationship between line loading, series compensation and the required amount of net supplementary var supply is generalized in Figure 17.72 for the case of a 600-mile transmission line with three equally spaced, intermediate

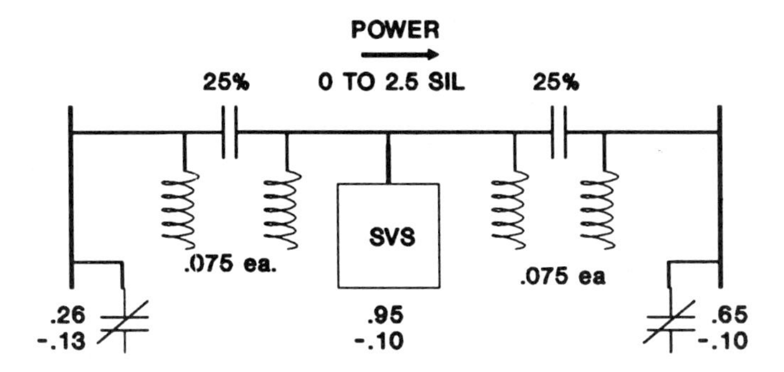

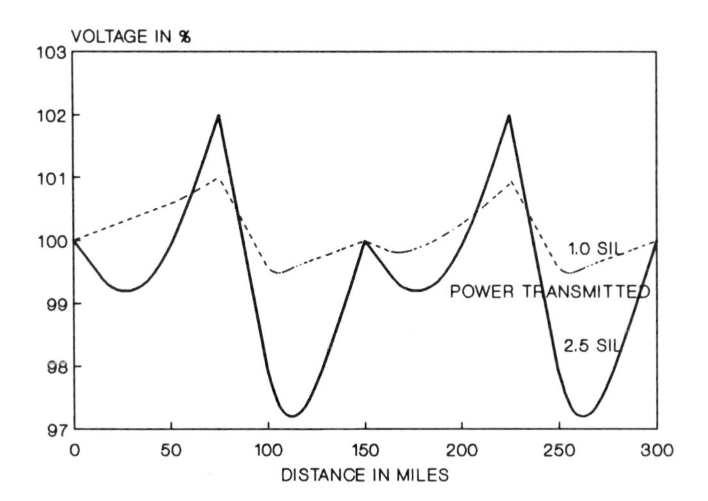

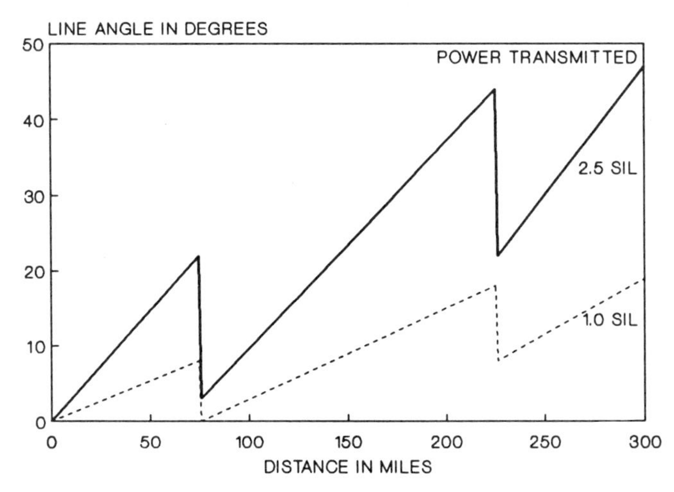

FIGURE 17.71 Example of a near-optimally compensated transmission line.

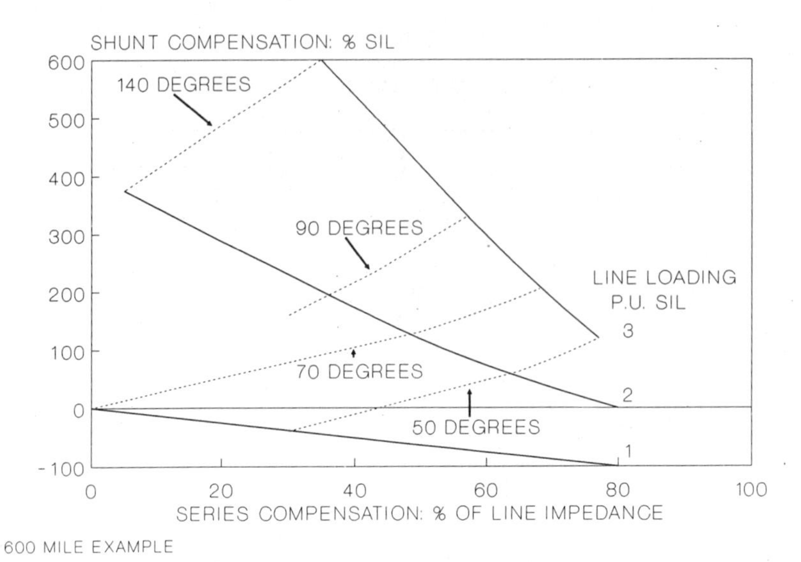

FIGURE 17.72 Required shunt and series compensation to hold flat line voltage for 600-mile line.

switching stations. For example, if it is necessary to hold a flat voltage profile at a maximum power transfer per transmission line of 2.5 per unit of SIL, then a 50% series-compensated line needs net supplementary var requirements of 250% of SIL with a line angle of approximately 85°. If significant variation is expected in line loading, as would normally be the case, then to maintain flat voltage, the bulk of the supplementary var supplies must be either switchable or of the SVS type.

Using the foregoing principles of series and shunt compensation, an appropriate distribution and sizing of the individual static var systems (SVS) was determined for the example system. This resulted in a stable case with voltage and angle characteristics as shown in Figures 17.73 and 17.74. Note the large initial angular separation across the system (105°) and the large separation (210°) at the point of maximum swing. This is typical of and unique to the case of long-distance power transmission across extensive systems with many points of intermediate voltage support by local generation and static var systems. The systems swing well beyond 180° and return to new steady-state operating points with angular separations considerably in excess of 90°. Special supplementary controls may be required to insure adequate damping of subsequent oscillations and that minimum voltages during swings are held to 85% or higher. The output response of several of the static var systems (SVS) during the disturbance is illustrated in Figure 17.75. Note that these SVS are idling prior to the disturbance. To enhance system stability, the SVS must have the additional capability to hold 85–90% voltage during the power-swing conditions.

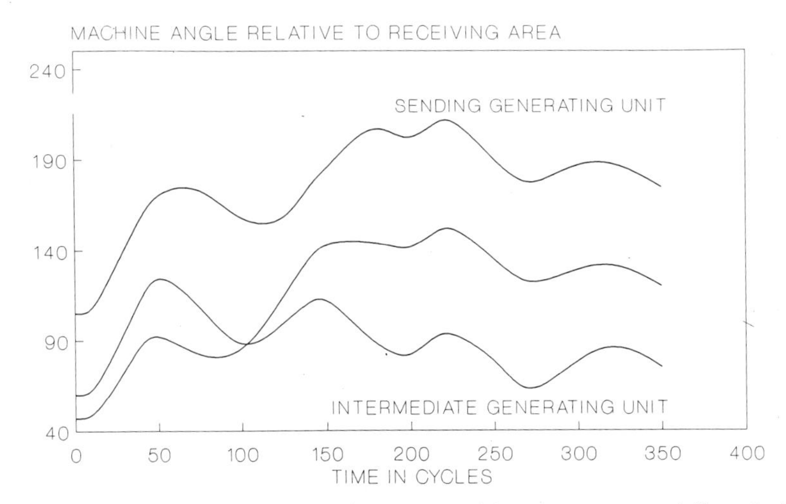

FIGURE 17.73 Generation angle swing curves with var compensated lines during a double-circuit fault of 3-GW transfer from area 4 to area 2.

The ratio between the required static var control kilovolt-ampere size and the initial megawatt flow on the ties that are being tripped is not fixed, but increases as the system's initial angular separation increases. That is, each additional megawatt that is transferred requires progressively more reactive power, as shown in Figure 17.76. This incremental ratio of Mvar/MW is indicative of an approach to a practical stability limit. The incremental ratio was determined to vary from two to three for the example system.

Another means of improving the power transfer stability limit is to use a

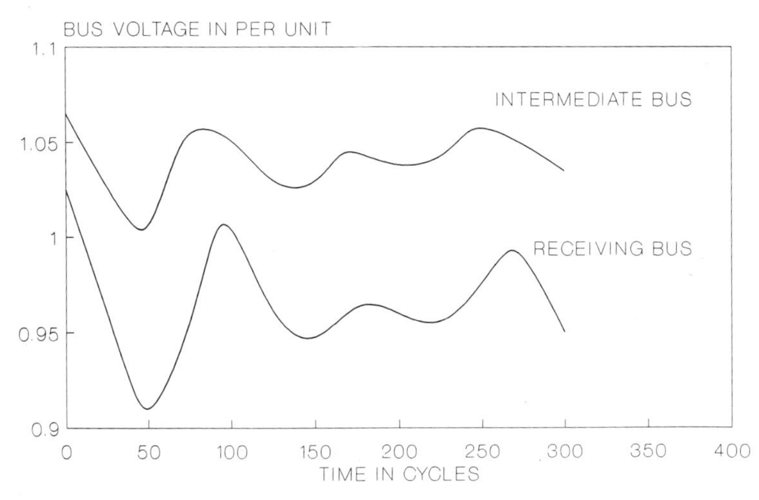

FIGURE 17.74 Bus voltage with var compensated lines during a double-circuit fault of 3-GW transfer from area 4 to area 2.

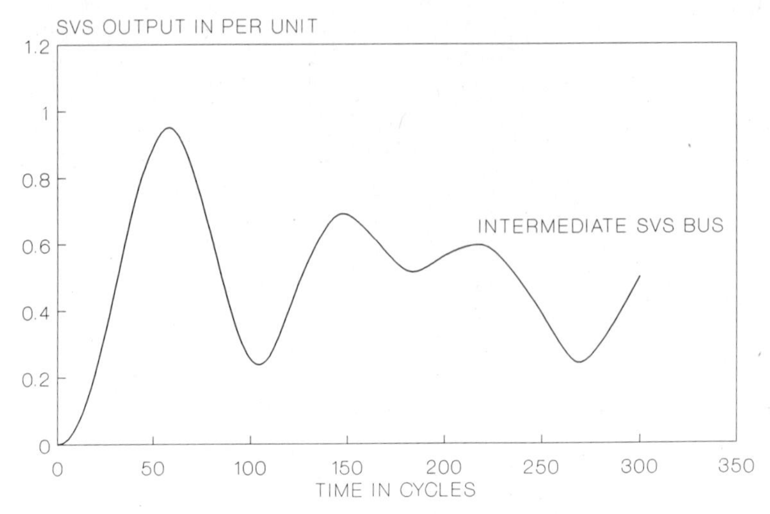

FIGURE 17.75 Static–var control response during a double-circuit fault of the 3-GW transfer from area 4 to area 2.

combination of phase-shifting transformers and series capacitors on the transmission lines between areas 1 and 3. A phase-shifting transformer can impede or aid power flow by inserting a positive or negative angle through the transformer. The circuit representation of a phase-shifting transformer is a constant phase-angle shift in series with a simple inductor.

Transmission lines of this example were equipped with both phase shifters and series compensation. The phase-shifting transformers were adjusted to maintain the same initial flow conditions as in the original case; otherwise, a

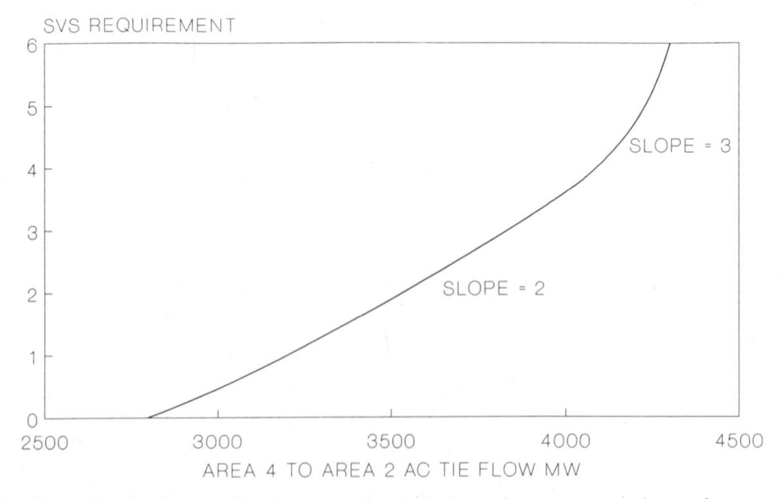

FIGURE 17.76 Relationship between increased static–var control requirement and increased initial tie flow.

much larger power transfer would occur on the series-compensated lines. The series compensation permits the lines to carry much larger power flows during a transient and, thereby, to greatly enhance system stability. The result is that, for the same initial disturbance, the static–var system requirements in this example are reduced by half for the equivalent system transient performance. Thus, phase-shifting transformers and series compensation can be used to effectively complement each other.

17.15 SUMMARY

The stability aspect of transmission planning is normally a multistep process that starts with steady-state and pre- and postdisturbance power-flow calculations. This is expanded on by "classical" stability cases (simplified equipment and system representations that include relay operations). Finally, a detailed stability analysis for selected critical cases verifies satisfactory system performance for cost optimization of the various complex and often interrelated options under consideration.

This final step, although very time consuming and expensive, is essential. Required modifications are to be expected based on the results obtained with initial simplified approaches such as equal area criteria and "classical" system representation. In the final analysis, it has been demonstrated many times that systems do, in fact, perform essentially as predicted by careful and detailed stability analyses. Therefore, the results of transmission stability analyses can be relied on with confidence.

REFERENCES

Arrillaga, J., C. P. Arnold, and B. J. Harker, *Computer Modeling of Electrical Power Systems,* Wiley, New York, 1983.

Berdy, J., P. G. Brown, C. A. Mathews, D. N. Walker, and S. B. Wilkinson, "Automatic High Speed Reclosing Near Large Generating Stations," CIGRE 34-02 (1982).

Bowler, C. E. J., "Understanding Subsynchronous Resonance," IEEE Publication 76 CH, 066-0-PWR, Symposium on Analysis and Control of Subsynchronous Resonance, 1976.

Brown, P. G., "Early Valve Actuation as an Aid to System Stability," Pennsylvania Electric Association, Relay Committee Meeting (1973).

Concordia, C., F. P. deMello, L. K. Kirchmayer, and R. P. Schulz, "Effect of Prime-Mover Response and Governing Characteristics on System Dynamic Performance," American Power Conference, Vol. 27 (1966).

deMello, F. P., D. N. Ewart, and M. Temoshok, "Turbine Energy Controls Aid in Power System Performance," American Power Conference (1966).

deMello, F. P., and C. Concordia, "Concepts of Synchronous Machine Stability as Affected by Excitation System," IEEE *Transactions on Power Apparatus and Systems,* Vol. PAS-88, 1969.

Dunlop, R. D., D. N. Ewart, and R. P. Schulz, "Use of Digital Computer Simulations to Assess Long-Term Power System Dynamic Response," *IEEE Transactions on Power Apparatus and Systems,* Vol. PAS-94, No. 3, 1975, pp. 850–857.

Gentile, T. J., S. Ihara, A. Murdoch, and N. W. Simons, *Determining Load Characteristics for Transient Performance: Volume 2—Load Model Guidelines,* EL-850, Volume 2, Research Project (EPRI) 849-1, March 1981.

Gross, C. A., *Power System Analysis,* Wiley, New York, 1979.

Hauth, R. L., and R. J. Moran, "The Performance of Thyristor Controlled Static Var Systems in HVAC Applications, Part 1," IEEE Tutorial Course Text 78 EH0135-4-PWR, July 1978.

IEEE, "Dynamic Models for Steam and Hydro Turbines in Power System Studies," IEEE Committee Report, *IEEE Transactions on Power Apparatus and Systems,* Vol. PAS-93 November 1973, pp. 1904–1915.

IEEE, "Extra High Voltage Line Outages," *IEEE Transactions on Power Apparatus and Systems,* Vol. PAS-86, No. 5, May 1967, pp. 547–562.

IEEE, "MW Response of Fossil Fueled Steam Units," IEEE Working Group on Power Plant Response to Load Changes, *IEEE Transactions on Power Apparatus and Systems,* Vol. PAS-92, March 1973, pp. 455–463.

Kimbark, E. W., *Power System Stability,* Wiley, New York, 1962.

Larsen, E. V., and D. A. Swann, "Applying Power System Stabilizers," *IEEE Transactions on Power Apparatus and Systems,* Vol. PAS-100, June 1981, pp. 3017–3046.

Luini, J. F., R. P. Schulz, and A. E. Turner, "A Digital Computer Program for Analyzing Long-Term Dynamic Response of Power Systems," IEEE PICA Conference Proceedings, 1975.

Miller, T. J. E., *Reactive Power Control in Electric Systems,* Wiley, New York, 1982.

Price, W. W., K. A. Wirgau, A. Murdoch, J. Mitsche, E. Vaahedi, and M. A. El-Kady, "Load Modeling for Power Flow and Transient Stability Studies," *IEEE Transactions on Power Systems,* Vol. 3, February 1988, pp. 180–187.

Stevenson, W. D., *Elements of Power System Analysis,* McGraw-Hill, New York, 1962.

Wood, A., and B. Wollenberg, *Power Generation Operation and Control,* Wiley, New York, 1984.

PROBLEMS

1 A generation unit ($X'_d = .15$, step-up transformer $X = .10$, $H = 3$ seconds) operating at 1.00-pu power (100-MVA base) is delivering power over two parallel 400-mile-long transmission lines ($X = .6$ per line) to infinite bus with 1.0-pu voltage. The generation voltage is regulated at 1.05 at the high side of the ΔY generator step-up transformer. A solid three-phase fault occurs at the high side (transmission-line side) of the generator step-up transformer. The fault is removed by circuit breakers. Use a machine model of constant voltage behind transient reactance. Calculate (a) the

largest postfault angle, (b) the critical clearing angle, and (c) the critical clearing time.

2 Use the same data as Problem 1, but with a three-phase fault midway (200 miles) down the transmission line. Calculate (a) the largest postfault angle, (b) the critical clearing angle, and (c) the critical clearing time.

3 Use the same data as Problem 1, but with a simultaneous common single line-to-ground fault on each line at the high side of the step-up transformer. Calculate the same fault consequence as in Problem 1.

4 Using the same data as in Problem 3, a common line-to-ground fault is cleared by a single pole operation on the faulted phases with successful reclosure at 30 cycles. Calculate (a) the largest post-fault angle, (b) the critical clearing angle, and (c) the critical clearing time.

5 For the conditions of Problem 1, calculate the *K* coefficients (Section 17.8.4) for the system conditions of both lines in service, and one line in service.

6 Using the results of Problem 5, calculate the field torque coefficients corresponding to conventional excitation system ($K_E = 25$, $T_E = 0.5$ second) and high-response excitation system ($K_E = 100$, $T_E = 0.05$ second) for two lines in service and one line in service.

7 Determine the required power system stabilizer gain and phase angle (driven by the machine speed of Figure 17.38) that would stabilize the high response excitation system of Problem 6 at half, twice, and the natural frequency of oscillation. Use a .5 damping ratio.

18

PREPARING FOR THE NEXT CENTURY

This book has presented the foundation for electric utility system least-cost planning. References at the end of each chapter provide additional technical depth in many of the subjects. Application of this foundation requires the conscientious collection of technical performance data, objective appraisals of costs and other economic data, dedication to performing accurate power system evaluations, and patient and creative thought in developing better ways to serve the customer of electricity.

The most important ingredient to successful utility planning, however, is creative thought in developing better ways to aggressively serve the customer. The customer has the need for energy, and electricity is one method of serving this energy need. The customer requires efficient, economic, convenient, and high-quality service, and the requirements of that service can be met by electrical energy.

Political, social, environmental, and regulatory forces influence utility planning. These forces establish the boundaries and level of the business playing field. These boundaries are constantly changing, creating obstacles or opportunities. The turbulent era of the 1970s and 1980s nurtured these influences. Thus, a new utility industry is emerging where aggressive business creativity and flexibility are necessary success elements for the future.

A partial list of potential new directions for the power industry includes:

- Increased emphasis by consumers, regulatory agencies, and investors on low-cost, high-quality electric service.
- A focus on utility projects with contained business risks. Prudency disallowances by regulatory agencies during the 1980s have and may focus utilities toward policies of containing business risks and sharing remaining

risks with consumers, equipment manufacturers, and project developers. Planning strategies that are flexible to accommodate uncertainties will be important.

- Partial deregulation and market competition of certain electricity services may allow the potential for greater economic efficiency, thereby passing some of the savings on to the consumer. The 1978 Public Utility Regulatory Policies Act (PURPA) is viewed as a first step toward deregulation of the power generation industry. The number of cogeneration and independent power projects undertaken since the early 1980s is evidence of changes in the generation of electricity.
- A trend may evolve toward unbundling of electricity services into separate generation, transmission, and distribution functions. Unbundling may increase deregulation and competition in generation and transmission services.
- Increased economic efficiencies may be obtained from utility mergers, acquisitions, and operating partnerships. In the late 1980s, 20% of the investor-owned utilities generated 60% of the electrical production in the United States. Further concentration within the utility industry may occur if operational efficiencies can be realized and if regulatory agencies approve.
- If generation services are unbundled and deregulated, competition for the construction of new generation plants may result. Potential competitors for owning and operating new plants include service-area utilities as well as utilities from outside the service area, cogenerators, and independent (nonutility) power producers.
- Increased emphasis on cost reduction will promote more power from low-cost generating units being transmitted to distant areas of higher generation cost. Transmission networks were originally designed to transport power from local generation locations to local load. New transmission capacity has been and will continue to be built where low-cost power is available and contractual wheeling agreements can be successfully negotiated to facilitate the transfer of electrical energy over greater distances.
- If the generation function is deregulated, then access to transmission services becomes important to providing a way for generated electricity to be obtained by the end-user. Transmission may continue to be regulated, but with an open-access policy. Deregulation of transmission to a "common carrier" status may be a possible future outcome.
- A more competitive utility power industry with increased energy wheeling may raise state and local concerns. Just as states have wanted to preserve their own environmental quality and conserve natural resources, interstate transmission lines and generating units sited in one state but not significantly benefiting that state, may be perceived as undesirable.
- Increased social and environmental concerns may determine the fuels used in new and existing generating units. Issues, including acid rain,

waste disposal, and greenhouse effect, may become dominant forces in the utility planning and operational decisions.

- And finally, but not necessarily all inclusively, the pendulum of change for deregulation of the utility industry may swing back toward containing regulation. In regulated industries when excess supply exists (as it has in many areas of the United States utility industry in the late 1980s), an incentive exists to encourage competition, which tends to reduce prices. However, when there is a shortage of supply (potentially the utility industry in the 1990s and into the next century) and producers raise prices in response to the natural econmic principles of supply/demand pricing, then incentives may arise to increase regulation for political and social reasons.

The electric industry has evolved dramatically during the more than 100 years of its existence. Every industry can typically be characterized by a life cycle from infancy through maturity to possible demise or revitalization. The power industry has been characterized by early infancy (1880s through 1900s), teenage rapid technological change (1900–1920), early adulthood business consolidation (1920–1940), and mature evolution (1940–1970). The 1970s and 1980s have been an era of turbulence and reaction to a changing social, political, and economic world. The power industry will continue to see technological change, but the key issue for the 1990s is the development of utility business strategies for serving more knowledgeable and mature end-use customer demands with high-quality and least-cost service in the rapidly changing business environment of the 1990s. The utility power industry has an exciting future as it evolves to meet these needs.

APPENDIX COMPOUND INTEREST FORMULAS

2.0 Percent/Year Discount Rate

	Single Payment		Uniform Annual Series			
	Compound Amount Factor	Present Worth Factor	Sinking Fund Factor	Compound Amount Factor	Capital Recovery Factor	Uniform Series Factor
Years	Find F Given P	Find P Given F	Find R Given F	Find F Given R	Find R Given P	Find P Given R
1	1.02	0.98039	1.00000	1.000	1.02000	0.980
2	1.04	0.96117	0.49505	2.020	0.51505	1.942
3	1.06	0.94232	0.32676	3.060	0.34676	2.884
4	1.08	0.92385	0.24262	4.122	0.26262	3.808
5	1.10	0.90573	0.19216	5.204	0.21216	4.713
6	1.13	0.88797	0.15853	6.308	0.17853	5.601
7	1.15	0.87056	0.13451	7.434	0.15451	6.472
8	1.17	0.85349	0.11651	8.583	0.13651	7.325
9	1.20	0.83676	0.10252	9.755	0.12252	8.162
10	1.22	0.82035	0.09133	10.950	0.11133	8.983
11	1.24	0.80426	0.08218	12.169	0.10218	9.787
12	1.27	0.78849	0.07456	13.412	0.09456	10.575
13	1.29	0.77303	0.06812	14.680	0.08812	11.348
14	1.32	0.75788	0.06260	15.974	0.08260	12.106
15	1.35	0.74302	0.05783	17.293	0.07783	12.849
16	1.37	0.72845	0.05365	18.639	0.07365	13.578
17	1.40	0.71416	0.04997	20.012	0.06997	14.292
18	1.43	0.70016	0.04670	21.412	0.06670	14.992
19	1.46	0.68643	0.04378	22.840	0.06378	15.678
20	1.49	0.67297	0.04116	24.297	0.06116	16.351
21	1.52	0.65978	0.03878	25.783	0.05878	17.011
22	1.55	0.64684	0.03663	27.299	0.05663	17.658
23	1.58	0.63416	0.03467	28.845	0.05467	18.292
24	1.61	0.62172	0.03287	30.422	0.05287	18.914
25	1.64	0.60953	0.03122	32.030	0.05122	19.523
26	1.67	0.59758	0.02970	33.671	0.04970	20.121
27	1.71	0.58586	0.02829	35.344	0.04829	20.707
28	1.74	0.57438	0.02699	37.051	0.04699	21.281
29	1.78	0.56311	0.02578	38.792	0.04578	21.844
30	1.81	0.55207	0.02465	40.568	0.04465	22.396
31	1.85	0.54125	0.02360	42.379	0.04360	22.938
32	1.88	0.53063	0.02261	44.227	0.04261	23.468
33	1.92	0.52023	0.02169	46.111	0.04169	23.989
34	1.96	0.51003	0.02082	48.034	0.04082	24.499
35	2.00	0.50003	0.02000	49.994	0.04000	24.999

(continued)

2.0 Percent/Year Discount Rate (Continued)

	Single Payment		Uniform Annual Series			
	Compound Amount Factor	Present Worth Factor	Sinking Fund Factor	Compound Amount Factor	Capital Recovery Factor	Uniform Series Factor
Years	Find F Given P	Find P Given F	Find R Given F	Find F Given R	Find R Given P	Find P Given R
36	2.04	0.49022	0.01923	51.994	0.03923	25.489
37	2.08	0.48061	0.01851	54.034	0.03851	25.969
38	2.12	0.47119	0.01782	56.115	0.03782	26.441
39	2.16	0.46195	0.01717	58.237	0.03717	26.903
40	2.21	0.45289	0.01656	60.402	0.03656	27.355
41	2.25	0.44401	0.01597	62.610	0.03597	27.799
42	2.30	0.43531	0.01542	64.862	0.03542	28.235
43	2.34	0.42677	0.01489	67.159	0.03489	28.662
44	2.39	0.41840	0.01439	69.502	0.03439	29.080
45	2.44	0.41020	0.01391	71.892	0.03391	29.490

4.0 Percent/Year Discount Rate

	Single Payment		Uniform Annual Series			
	Compound Amount Factor	Present Worth Factor	Sinking Fund Factor	Compound Amount Factor	Capital Recovery Factor	Uniform Series Factor
Years	Find F Given P	Find P Given F	Find R Given F	Find F Given R	Find R Given P	Find P Given R
1	1.04	0.96154	1.00000	1.000	1.04000	0.962
2	1.08	0.92456	0.49020	2.040	0.53020	1.886
3	1.12	0.88900	0.32035	3.122	0.36035	2.775
4	1.17	0.85480	0.23549	4.246	0.27549	3.630
5	1.22	0.82193	0.18463	5.416	0.22463	4.452
6	1.27	0.79031	0.15076	6.633	0.19076	5.242
7	1.32	0.75992	0.12661	7.898	0.16661	6.002
8	1.37	0.73069	0.10853	9.214	0.14853	6.733
9	1.42	0.70259	0.09449	10.583	0.13449	7.435
10	1.48	0.67556	0.08329	12.006	0.12329	8.111
11	1.54	0.64958	0.07415	13.486	0.11415	8.760
12	1.60	0.62460	0.06655	15.026	0.10655	9.385
13	1.67	0.60057	0.06014	16.627	0.10014	9.986
14	1.73	0.57748	0.05467	18.292	0.09467	10.563
15	1.80	0.55526	0.04994	20.024	0.08994	11.118
16	1.87	0.53391	0.04582	21.825	0.08582	11.652
17	1.95	0.51337	0.04220	23.697	0.08220	12.166

4.0 Percent/Year Discount Rate (Continued)

	Single Payment		Uniform Annual Series			
	Compound Amount Factor	Present Worth Factor	Sinking Fund Factor	Compound Amount Factor	Capital Recovery Factor	Uniform Series Factor
Years	Find F Given P	Find P Given F	Find R Given F	Find F Given R	Find R Given P	Find P Given R
18	2.03	0.49363	0.03899	25.645	0.07899	12.659
19	2.11	0.47464	0.03614	27.671	0.07614	13.134
20	2.19	0.45639	0.03358	29.778	0.07358	13.590
21	2.28	0.43883	0.03128	31.969	0.07128	14.029
22	2.37	0.42196	0.02920	34.248	0.06920	14.451
23	2.46	0.40573	0.02731	36.618	0.06731	14.857
24	2.56	0.39012	0.02559	39.083	0.06559	15.247
25	2.67	0.37512	0.02401	41.646	0.06401	15.622
26	2.77	0.36069	0.02257	44.312	0.06257	15.983
27	2.88	0.34682	0.02124	47.084	0.06124	16.330
28	3.00	0.33348	0.02001	49.968	0.06001	16.663
29	3.12	0.32065	0.01888	52.966	0.05888	16.984
30	3.24	0.30832	0.01783	56.085	0.05783	17.292
31	3.37	0.29646	0.01686	59.328	0.05686	17.588
32	3.51	0.28506	0.01595	62.701	0.05595	17.874
33	3.65	0.27409	0.01510	66.209	0.05510	18.148
34	3.79	0.26355	0.01431	69.858	0.05431	18.411
35	3.95	0.25342	0.01358	73.652	0.05358	18.665
36	4.10	0.24367	0.01289	77.598	0.05289	18.908
37	4.27	0.23430	0.01224	81.702	0.05224	19.143
38	4.44	0.22529	0.01163	85.970	0.05163	19.368
39	4.62	0.21662	0.01106	90.409	0.05106	19.584
40	4.80	0.20829	0.01052	95.025	0.05052	19.793
41	4.99	0.20028	0.01002	99.826	0.05002	19.993
42	5.19	0.19258	0.00954	104.819	0.04954	20.186
43	5.40	0.18517	0.00909	110.012	0.04909	20.371
44	5.62	0.17805	0.00866	115.413	0.04866	20.549
45	5.84	0.17120	0.00826	121.029	0.04826	20.720

6.0 Percent/Year Discount Rate

	Single Payment		Uniform Annual Series			
	Compound Amount Factor	Present Worth Factor	Sinking Fund Factor	Compound Amount Factor	Capital Recovery Factor	Uniform Series Factor
Years	Find F Given P	Find P Given F	Find R Given F	Find F Given R	Find R Given P	Find P Given R
1	1.06	0.94340	1.00000	1.000	1.06000	0.943
2	1.12	0.89000	0.48544	2.060	0.54544	1.833
3	1.19	0.83962	0.31411	3.184	0.37411	2.673
4	1.26	0.79209	0.22859	4.375	0.28859	3.465
5	1.34	0.74726	0.17740	5.637	0.23740	4.212
6	1.42	0.70496	0.14336	6.975	0.20336	4.917
7	1.50	0.66506	0.11914	8.394	0.17914	5.582
8	1.59	0.62741	0.10104	9.897	0.16104	6.210
9	1.69	0.59190	0.08702	11.491	0.14702	6.802
10	1.79	0.55840	0.07587	13.181	0.13587	7.360
11	1.90	0.52679	0.06679	14.972	0.12679	7.887
12	2.01	0.49697	0.05928	16.870	0.11928	8.384
13	2.13	0.46884	0.05296	18.882	0.11296	8.853
14	2.26	0.44230	0.04758	21.015	0.10758	9.295
15	2.40	0.41727	0.04296	23.276	0.10296	9.712
16	2.54	0.39365	0.03895	25.672	0.09895	10.106
17	2.69	0.37136	0.03544	28.213	0.09544	10.477
18	2.85	0.35034	0.03236	30.906	0.09236	10.828
19	3.03	0.33051	0.02962	33.760	0.08962	11.158
20	3.21	0.31181	0.02718	36.786	0.08718	11.470
21	3.40	0.29416	0.02500	39.993	0.08500	11.764
22	3.60	0.27751	0.02305	43.392	0.08305	12.042
23	3.82	0.26180	0.02128	46.996	0.08128	12.303
24	4.05	0.24698	0.01968	50.815	0.07968	12.550
25	4.29	0.23300	0.01823	54.864	0.07823	12.783
26	4.55	0.21981	0.01690	59.156	0.07690	13.003
27	4.82	0.20737	0.01570	63.706	0.07570	13.211
28	5.11	0.19563	0.01459	68.528	0.07459	13.406
29	5.42	0.18456	0.01358	73.640	0.07358	13.591
30	5.74	0.17411	0.01265	79.058	0.07265	13.765
31	6.09	0.16426	0.01179	84.802	0.07179	13.929
32	6.45	0.15496	0.01100	90.890	0.07100	14.084
33	6.84	0.14619	0.01027	97.343	0.07027	14.230
34	7.25	0.13791	0.00960	104.184	0.06960	14.368
35	7.69	0.13011	0.00897	111.435	0.06897	14.498
36	8.15	0.12274	0.00839	119.121	0.06839	14.621
37	8.64	0.11579	0.00786	127.268	0.06786	14.737
38	9.15	0.10924	0.00736	135.904	0.06736	14.846

6.0 Percent/Year Discount Rate (Continued)

	Single Payment		Uniform Annual Series			
	Compound Amount Factor	Present Worth Factor	Sinking Fund Factor	Compound Amount Factor	Capital Recovery Factor	Uniform Series Factor
Years	Find F Given P	Find P Given F	Find R Given F	Find F Given R	Find R Given P	Find P Given R
39	9.70	0.10306	0.00689	145.058	0.06689	14.949
40	10.29	0.09722	0.00646	154.762	0.06646	15.046
41	10.90	0.09172	0.00606	165.047	0.06606	15.138
42	11.56	0.08653	0.00568	175.950	0.06568	15.225
43	12.25	0.08163	0.00533	187.507	0.06533	15.306
44	12.99	0.07701	0.00501	199.758	0.06501	15.383
45	13.76	0.07265	0.00470	212.743	0.06470	15.456

8.0 Percent/Year Discount Rate

	Single Payment		Uniform Annual Series			
	Compound Amount Factor	Present Worth Factor	Sinking Fund Factor	Compound Amount Factor	Capital Recovery Factor	Uniform Series Factor
Years	Find F Given P	Find P Given F	Find R Given F	Find F Given R	Find R Given P	Find P Given R
1	1.08	0.92593	1.00000	1.000	1.08000	0.926
2	1.17	0.85734	0.48077	2.080	0.56077	1.783
3	1.26	0.79383	0.30803	3.246	0.38803	2.577
4	1.36	0.73503	0.22192	4.506	0.30192	3.312
5	1.47	0.68058	0.17046	5.867	0.25046	3.993
6	1.59	0.63017	0.13632	7.336	0.21632	4.623
7	1.71	0.58349	0.11207	8.923	0.19207	5.206
8	1.85	0.54027	0.09401	10.637	0.17401	5.747
9	2.00	0.50025	0.08008	12.488	0.16008	6.247
10	2.16	0.46319	0.06903	14.487	0.14903	6.710
11	2.33	0.42888	0.06008	16.645	0.14008	7.139
12	2.52	0.39711	0.05269	18.977	0.13269	7.536
13	2.72	0.36770	0.04652	21.495	0.12652	7.904
14	2.94	0.34046	0.04130	24.215	0.12130	8.244
15	3.17	0.31524	0.03683	27.152	0.11683	8.559
16	3.43	0.29189	0.03298	30.324	0.11298	8.851
17	3.70	0.27027	0.02963	33.750	0.10963	9.122
18	4.00	0.25025	0.02670	37.450	0.10670	9.372
19	4.32	0.23171	0.02413	41.446	0.10413	9.604
20	4.66	0.21455	0.02185	45.762	0.10185	9.818

(continued)

8.0 Percent/Year Discount Rate (Continued)

	Single Payment		Uniform Annual Series			
	Compound Amount Factor	Present Worth Factor	Sinking Fund Factor	Compound Amount Factor	Capital Recovery Factor	Uniform Series Factor
Years	Find F Given P	Find P Given F	Find R Given F	Find F Given R	Find R Given P	Find P Given R
21	5.03	0.19866	0.01983	50.423	0.09983	10.017
22	5.44	0.18394	0.01803	55.457	0.09803	10.201
23	5.87	0.17032	0.01642	60.893	0.09642	10.371
24	6.34	0.15770	0.01498	66.765	0.09498	10.529
25	6.85	0.14602	0.01368	73.106	0.09368	10.675
26	7.40	0.13520	0.01251	79.955	0.09251	10.810
27	7.99	0.12519	0.01145	87.351	0.09145	10.935
28	8.63	0.11591	0.01049	95.339	0.09049	11.051
29	9.32	0.10733	0.00962	103.966	0.08962	11.158
30	10.06	0.09938	0.00883	113.283	0.08883	11.258
31	10.87	0.09202	0.00811	123.346	0.08811	11.350
32	11.74	0.08520	0.00745	134.214	0.08745	11.435
33	12.68	0.07889	0.00685	145.951	0.08685	11.514
34	13.69	0.07305	0.00630	158.627	0.08630	11.587
35	14.79	0.06763	0.00580	172.317	0.08580	11.655
36	15.97	0.06262	0.00534	187.102	0.08534	11.717
37	17.25	0.05799	0.00492	203.071	0.08492	11.775
38	18.63	0.05369	0.00454	220.316	0.08454	11.829
39	20.12	0.04971	0.00419	238.942	0.08419	11.879
40	21.72	0.04603	0.00386	259.057	0.08386	11.925
41	23.46	0.04262	0.00356	280.781	0.08356	11.967
42	25.34	0.03946	0.00329	304.244	0.08329	12.007
43	27.37	0.03654	0.00303	329.583	0.08303	12.043
44	29.56	0.03383	0.00280	356.950	0.08280	12.077
45	31.92	0.03133	0.00259	386.506	0.08259	12.108

10.0 Percent/Year Discount Rate

	Single Payment		Uniform Annual Series			
	Compound Amount Factor	Present Worth Factor	Sinking Fund Factor	Compound Amount Factor	Capital Recovery Factor	Uniform Series Factor
Years	Find F Given P	Find P Given F	Find R Given F	Find F Given R	Find R Given P	Find P Given R
1	1.10	0.90909	1.00000	1.000	1.10000	0.909
2	1.21	0.82645	0.47619	2.100	0.57619	1.736
3	1.33	0.75131	0.30211	3.310	0.40211	2.487
4	1.46	0.68301	0.21547	4.641	0.31547	3.170
5	1.61	0.62092	0.16380	6.105	0.26380	3.791
6	1.77	0.56447	0.12961	7.716	0.22961	4.355
7	1.95	0.51316	0.10541	9.487	0.20541	4.868
8	2.14	0.46651	0.08744	11.436	0.18744	5.335
9	2.36	0.42410	0.07364	13.579	0.17364	5.759
10	2.59	0.38554	0.06275	15.937	0.16275	6.145
11	2.85	0.35049	0.05396	18.531	0.15396	6.495
12	3.14	0.31863	0.04676	21.384	0.14676	6.814
13	3.45	0.28966	0.04078	24.523	0.14078	7.103
14	3.80	0.26333	0.03575	27.975	0.13575	7.367
15	4.18	0.23939	0.03147	31.772	0.13147	7.606
16	4.59	0.21763	0.02782	35.950	0.12782	7.824
17	5.05	0.19784	0.02466	40.545	0.12466	8.022
18	5.56	0.17986	0.02193	45.599	0.12193	8.201
19	6.12	0.16351	0.01955	51.159	0.11955	8.365
20	6.73	0.14864	0.01746	57.275	0.11746	8.514
21	7.40	0.13513	0.01562	64.003	0.11562	8.649
22	8.14	0.12285	0.01401	71.403	0.11401	8.772
23	8.95	0.11168	0.01257	79.543	0.11257	8.883
24	9.85	0.10153	0.01130	88.497	0.11130	8.985
25	10.83	0.09230	0.01017	98.347	0.11017	9.077
26	11.92	0.08391	0.00916	109.182	0.10916	9.161
27	13.11	0.07628	0.00826	121.100	0.10826	9.237
28	14.42	0.06934	0.00745	134.210	0.10745	9.307
29	15.86	0.06304	0.00673	148.631	0.10673	9.370
30	17.45	0.05731	0.00608	164.494	0.10608	9.427
31	19.19	0.05210	0.00550	181.944	0.10550	9.479
32	21.11	0.04736	0.00497	201.138	0.10497	9.526
33	23.23	0.04306	0.00450	222.252	0.10450	9.569
34	25.55	0.03914	0.00407	245.477	0.10407	9.609
35	28.10	0.03558	0.00369	271.025	0.10369	9.644
36	30.91	0.03235	0.00334	299.127	0.10334	9.677
37	34.00	0.02941	0.00303	330.040	0.10303	9.706
38	37.40	0.02673	0.00275	364.044	0.10275	9.733

(*continued*)

10.0 Percent/Year Discount Rate (Continued)

	Single Payment		Uniform Annual Series			
	Compound Amount Factor	Present Worth Factor	Sinking Fund Factor	Compound Amount Factor	Capital Recovery Factor	Uniform Series Factor
Years	Find F Given P	Find P Given F	Find R Given F	Find F Given R	Find R Given P	Find P Given R
39	41.14	0.02430	0.00249	401.448	0.10249	9.757
40	45.26	0.02209	0.00226	442.593	0.10226	9.779
41	49.79	0.02009	0.00205	487.852	0.10205	9.799
42	54.76	0.01826	0.00186	537.637	0.10186	9.817
43	60.24	0.01660	0.00169	592.401	0.10169	9.834
44	66.26	0.01509	0.00153	652.641	0.10153	9.849
45	72.89	0.01372	0.00139	718.905	0.10139	8.863

12.0 Percent/Year Discount Rate

	Single Payment		Uniform Annual Series			
	Compound Amount Factor	Present Worth Factor	Sinking Fund Factor	Compound Amount Factor	Capital Recovery Factor	Uniform Series Factor
Years	Find F Given P	Find P Given F	Find R Given F	Find F Given R	Find R Given P	Find P Given R
1	1.12	0.89286	1.00000	1.000	1.12000	0.893
2	1.25	0.79719	0.47170	2.120	0.59170	1.690
3	1.40	0.71178	0.29635	3.374	0.41635	2.402
4	1.57	0.63552	0.20923	4.779	0.32923	3.037
5	1.76	0.56743	0.15741	6.353	0.27741	3.605
6	1.97	0.50663	0.12323	8.115	0.24323	4.111
7	2.21	0.45235	0.09912	10.089	0.21912	4.564
8	2.48	0.40388	0.08130	12.300	0.20130	4.968
9	2.77	0.36061	0.06768	14.776	0.18768	5.328
10	3.11	0.32197	0.05698	17.549	0.17698	5.650
11	3.48	0.28748	0.04842	20.655	0.16842	5.938
12	3.90	0.25668	0.04144	24.133	0.16144	6.194
13	4.36	0.22917	0.03568	28.029	0.15568	6.424
14	4.89	0.20462	0.03087	32.393	0.15087	6.628
15	5.47	0.18270	0.02682	37.280	0.14682	6.811
16	6.13	0.16312	0.02339	42.753	0.14339	6.974
17	6.87	0.14564	0.02046	48.884	0.14046	7.120
18	7.69	0.13004	0.01794	55.750	0.13794	7.250
19	8.61	0.11611	0.01576	63.440	0.13576	7.366
20	9.65	0.10367	0.01388	72.052	0.13388	7.469

12.0 Percent/Year Discount Rate (Continued)

	Single Payment		Uniform Annual Series			
	Compound Amount Factor	Present Worth Factor	Sinking Fund Factor	Compound Amount Factor	Capital Recovery Factor	Uniform Series Factor
Years	Find F Given P	Find P Given F	Find R Given F	Find F Given R	Find R Given P	Find P Given R
21	10.80	0.09256	0.01224	81.699	0.13224	7.562
22	12.10	0.08264	0.01081	92.503	0.13081	7.645
23	13.55	0.07379	0.00956	104.603	0.12956	7.718
24	15.18	0.06588	0.00846	118.155	0.12846	7.784
25	17.00	0.05882	0.00750	133.334	0.12750	7.843
26	19.04	0.05252	0.00665	150.334	0.12665	7.896
27	21.32	0.04689	0.00590	169.374	0.12590	7.943
28	23.88	0.04187	0.00524	190.699	0.12524	7.984
29	26.75	0.03738	0.00466	214.583	0.12466	8.022
30	29.96	0.03338	0.00414	241.333	0.12414	8.055
31	33.56	0.02980	0.00369	271.293	0.12369	8.085
32	37.58	0.02661	0.00328	304.848	0.12328	8.112
33	42.09	0.02376	0.00292	342.429	0.12292	8.135
34	47.14	0.02121	0.00260	384.521	0.12260	8.157
35	52.80	0.01894	0.00232	431.664	0.12232	8.176
36	59.14	0.01691	0.00206	484.463	0.12206	8.192
37	66.23	0.01510	0.00184	543.599	0.12184	8.208
38	74.18	0.01348	0.00164	609.831	0.12164	8.221
39	83.08	0.01204	0.00146	684.010	0.12146	8.233
40	93.05	0.01075	0.00130	767.091	0.12130	8.244
41	104.22	0.00960	0.00116	860.142	0.12116	8.253
42	116.72	0.00857	0.00104	964.359	0.12104	8.262
43	130.73	0.00765	0.00092	1081.083	0.12093	8.270
44	146.42	0.00683	0.00083	1211.813	0.12083	8.276
45	163.99	0.00610	0.00074	1358.230	0.12074	8.283

14.0 Percent/Year Discount Rate

	Single Payment		Uniform Annual Series			
	Compound Amount Factor	Present Worth Factor	Sinking Fund Factor	Compound Amount Factor	Capital Recovery Factor	Uniform Series Factor
Years	Find F Given P	Find P Given F	Find R Given F	Find F Given R	Find R Given P	Find P Given R
1	1.14	0.87719	1.00000	1.000	1.14000	0.877
2	1.30	0.76947	0.46729	2.140	0.60729	1.647
3	1.48	0.67497	0.29073	3.440	0.43073	2.322
4	1.69	0.59208	0.20320	4.921	0.34320	2.914
5	1.93	0.51937	0.15128	6.610	0.29128	3.433
6	2.19	0.45559	0.11716	8.536	0.25716	3.889
7	2.50	0.39964	0.09319	10.730	0.23319	4.288
8	2.85	0.35056	0.07557	13.233	0.21557	4.639
9	3.25	0.30751	0.06217	16.085	0.20217	4.946
10	3.71	0.26974	0.05171	19.337	0.19171	5.216
11	4.23	0.23662	0.04339	23.045	0.18339	5.453
12	4.82	0.20756	0.03667	27.271	0.17667	5.660
13	5.49	0.18207	0.03116	32.089	0.17116	5.842
14	6.26	0.15971	0.02661	37.581	0.16661	6.002
15	7.14	0.14010	0.02281	43.842	0.16281	6.142
16	8.14	0.12289	0.01962	50.980	0.15962	6.265
17	9.28	0.10780	0.01692	59.118	0.15692	6.373
18	10.58	0.09456	0.01462	68.394	0.15462	6.467
19	12.06	0.08295	0.01266	78.969	0.15266	6.550
20	13.74	0.07276	0.01099	91.025	0.15099	6.623
21	15.67	0.06383	0.00954	104.768	0.14954	6.687
22	17.86	0.05599	0.00830	120.436	0.14830	6.743
23	20.36	0.04911	0.00723	138.297	0.14723	6.792
24	23.21	0.04308	0.00630	158.659	0.14630	6.835
25	26.46	0.03779	0.00550	181.871	0.14550	6.873
26	30.17	0.03315	0.00480	208.333	0.14480	6.906
27	34.39	0.02908	0.00419	238.499	0.14419	6.935
28	39.20	0.02551	0.00366	272.889	0.14366	6.961
29	44.69	0.02237	0.00320	312.094	0.14320	6.983
30	50.95	0.01963	0.00280	356.787	0.14280	7.003
31	58.08	0.01722	0.00245	407.737	0.14245	7.020
32	66.21	0.01510	0.00215	465.820	0.14215	7.035
33	75.48	0.01325	0.00188	532.035	0.14188	7.048
34	86.05	0.01162	0.00165	607.520	0.14165	7.060
35	98.10	0.01019	0.00144	693.573	0.14144	7.070
36	111.83	0.00894	0.00126	791.673	0.14126	7.079
37	127.49	0.00784	0.00111	903.507	0.14111	7.087
38	145.34	0.00688	0.00097	1030.998	0.14097	7.094

14.0 Percent/Year Discount Rate (Continued)

	Single Payment		Uniform Annual Series			
	Compound Amount Factor	Present Worth Factor	Sinking Fund Factor	Compound Amount Factor	Capital Recovery Factor	Uniform Series Factor
Years	Find F Given P	Find P Given F	Find R Given F	Find F Given R	Find R Given P	Find P Given R
39	165.69	0.00604	0.00085	1176.338	0.14085	7.100
40	188.88	0.00529	0.00075	1342.025	0.14075	7.105
41	215.33	0.00464	0.00065	1530.909	0.14065	7.110
42	245.47	0.00407	0.00057	1746.236	0.14057	7.114
43	279.84	0.00357	0.00050	1991.709	0.14050	7.117
44	319.02	0.00313	0.00044	2271.548	0.14044	7.120
45	363.68	0.00275	0.00039	2590.565	0.14039	7.123

16.0 Percent/Year Discount Rate

	Single Payment		Uniform Annual Series			
	Compound Amount Factor	Present Worth Factor	Sinking Fund Factor	Compound Amount Factor	Capital Recovery Factor	Uniform Series Factor
Years	Find F Given P	Find P Given F	Find R Given F	Find F Given R	Find R Given P	Find P Given R
1	1.16	0.86207	1.00000	1.000	1.16000	0.862
2	1.35	0.74316	0.46296	2.160	0.62296	1.605
3	1.56	0.64066	0.28526	3.506	0.44526	2.246
4	1.81	0.55229	0.19738	5.066	0.35738	2.798
5	2.10	0.47611	0.14541	6.877	0.30541	3.274
6	2.44	0.41044	0.11139	8.977	0.27139	3.685
7	2.83	0.35383	0.08761	11.414	0.24761	4.039
8	3.28	0.30503	0.07022	14.240	0.23022	4.344
9	3.80	0.26295	0.05708	17.518	0.21708	4.607
10	4.41	0.22668	0.04690	21.321	0.20690	4.833
11	5.12	0.19542	0.03886	25.733	0.19886	5.029
12	5.94	0.16846	0.03241	30.850	0.19241	5.197
13	6.89	0.14523	0.02718	36.786	0.18718	5.342
14	7.99	0.12520	0.02290	43.672	0.18290	5.468
15	9.27	0.10793	0.01936	51.659	0.17936	5.575
16	10.75	0.09304	0.01641	60.925	0.17641	5.668
17	12.47	0.08021	0.01395	71.673	0.17395	5.749
18	14.46	0.06914	0.01188	84.141	0.17188	5.818
19	16.78	0.05961	0.01014	98.603	0.17014	5.877
20	19.46	0.05139	0.00867	115.380	0.16867	5.929

(continued)

16.0 Percent/Year Discount Rate (Continued)

	Single Payment		Uniform Annual Series			
	Compound Amount Factor	Present Worth Factor	Sinking Fund Factor	Compound Amount Factor	Capital Recovery Factor	Uniform Series Factor
Years	Find F Given P	Find P Given F	Find R Given F	Find F Given R	Find R Given P	Find P Given R
21	22.57	0.04430	0.00742	134.840	0.16742	5.973
22	26.19	0.03819	0.00635	157.415	0.16635	6.011
23	30.38	0.03292	0.00545	183.601	0.16545	6.044
24	35.24	0.02838	0.00467	213.977	0.16467	6.073
25	40.87	0.02447	0.00401	249.214	0.16401	6.097
26	47.41	0.02109	0.00345	290.088	0.16345	6.118
27	55.00	0.01818	0.00296	337.502	0.16296	6.136
28	63.80	0.01567	0.00255	392.502	0.16255	6.152
29	74.01	0.01351	0.00219	456.303	0.16219	6.166
30	85.85	0.01165	0.00189	530.311	0.16189	6.177
31	99.59	0.01004	0.00162	616.161	0.16162	6.187
32	115.52	0.00866	0.00140	715.747	0.16140	6.196
33	134.00	0.00746	0.00120	831.266	0.16120	6.203
34	155.44	0.00643	0.00104	965.269	0.16104	6.210
35	180.31	0.00555	0.00089	1120.711	0.16089	6.215
36	209.16	0.00478	0.00077	1301.025	0.16077	6.220
37	242.63	0.00412	0.00066	1510.189	0.16066	6.224
38	281.45	0.00355	0.00057	1752.819	0.16057	6.228
39	326.48	0.00306	0.00049	2034.270	0.16049	6.231
40	378.72	0.00264	0.00042	2360.754	0.16042	6.233
41	439.32	0.00228	0.00037	2739.474	0.16037	6.236
42	509.61	0.00196	0.00031	3178.790	0.16031	6.238
43	591.14	0.00169	0.00027	3688.396	0.16027	6.239
44	685.73	0.00146	0.00023	4279.539	0.16023	6.241
45	795.44	0.00126	0.00020	4965.265	0.16020	6.242

18.0 Percent/Year Discount Rate

	Single Payment		Uniform Annual Series			
	Compound Amount Factor	Present Worth Factor	Sinking Fund Factor	Compound Amount Factor	Capital Recovery Factor	Uniform Series Factor
Years	Find F Given P	Find P Given F	Find R Given F	Find F Given R	Find R Given P	Find P Given R
1	1.18	0.84746	1.00000	1.000	1.18000	0.847
2	1.39	0.71818	0.45872	2.180	0.63872	1.566
3	1.64	0.60863	0.27992	3.572	0.45992	2.174
4	1.94	0.51579	0.19174	5.215	0.37174	2.690
5	2.29	0.43711	0.13978	7.154	0.31978	3.127
6	2.70	0.37043	0.10591	9.442	0.28591	3.498
7	3.19	0.31393	0.08236	12.142	0.26236	3.812
8	3.76	0.26604	0.06524	15.327	0.24524	4.078
9	4.44	0.22546	0.05239	19.086	0.23239	4.303
10	5.23	0.19106	0.04251	23.521	0.22251	4.494
11	6.18	0.16192	0.03478	28.755	0.21478	4.656
12	7.29	0.13722	0.02863	34.931	0.20863	4.793
13	8.60	0.11629	0.02369	42.219	0.20369	4.910
14	10.15	0.09855	0.01968	50.818	0.19968	5.008
15	11.97	0.08352	0.01640	60.965	0.19640	5.092
16	14.13	0.07078	0.01371	72.939	0.19371	5.162
17	16.67	0.05998	0.01149	87.068	0.19149	5.222
18	19.67	0.05083	0.00964	103.740	0.18964	5.273
19	23.21	0.04308	0.00810	123.413	0.18810	5.316
20	27.39	0.03651	0.00682	146.628	0.18682	5.353
21	32.32	0.03094	0.00575	174.021	0.18575	5.384
22	38.14	0.02622	0.00485	206.345	0.18485	5.410
23	45.01	0.02222	0.00409	244.487	0.18409	5.432
24	53.11	0.01883	0.00345	289.494	0.18345	5.451
25	62.67	0.01596	0.00292	342.603	0.18292	5.467
26	73.95	0.01352	0.00247	405.272	0.18247	5.480
27	87.26	0.01146	0.00209	479.221	0.18209	5.492
28	102.97	0.00971	0.00177	566.480	0.18177	5.502
29	121.50	0.00823	0.00149	669.447	0.18149	5.510
30	143.37	0.00697	0.00126	790.947	0.18126	5.517
31	169.18	0.00591	0.00107	934.318	0.18107	5.523
32	199.63	0.00501	0.00091	1103.495	0.18091	5.528
33	235.56	0.00425	0.00077	1303.124	0.18077	5.532
34	277.96	0.00360	0.00065	1538.686	0.18065	5.536
35	328.00	0.00305	0.00055	1816.649	0.18055	5.539
36	387.04	0.00258	0.00047	2144.646	0.18047	5.541
37	456.70	0.00219	0.00039	2531.682	0.18039	5.543
38	538.91	0.00186	0.00033	2988.385	0.18033	5.545

(*continued*)

18.0 Percent/Year Discount Rate (Continued)

	Single Payment		Uniform Annual Series			
	Compound Amount Factor	Present Worth Factor	Sinking Fund Factor	Compound Amount Factor	Capital Recovery Factor	Uniform Series Factor
Years	Find F Given P	Find P Given F	Find R Given F	Find F Given R	Find R Given P	Find P Given R
39	635.91	0.00157	0.00028	3527.294	0.18028	5.547
40	750.38	0.00133	0.00024	4163.206	0.18024	5.548
41	885.44	0.00113	0.00020	4913.583	0.18020	5.549
42	1044.82	0.00096	0.00017	5799.028	0.18017	5.550
43	1232.89	0.00081	0.00015	6843.853	0.18015	5.551
44	1454.81	0.00069	0.00012	8076.746	0.18012	5.552
45	1716.68	0.00058	0.00010	9531.559	0.18010	5.552

20.0 Percent/Year Discount Rate

	Single Payment		Uniform Annual Series			
	Compound Amount Factor	Present Worth Factor	Sinking Fund Factor	Compound Amount Factor	Capital Recovery Factor	Uniform Series Factor
Years	Find F Given P	Find P Given F	Find R Given F	Find F Given R	Find R Given P	Find P Given R
1	1.20	0.83333	1.00000	1.000	1.20000	0.833
2	1.44	0.69444	0.45455	2.200	0.65455	1.528
3	1.73	0.57870	0.27473	3.640	0.47473	2.106
4	2.07	0.48225	0.18629	5.368	0.38629	2.589
5	2.49	0.40188	0.13438	7.442	0.33438	2.991
6	2.99	0.33490	0.10071	9.930	0.30071	3.326
7	3.58	0.27908	0.07742	12.916	0.27742	3.605
8	4.30	0.23257	0.06061	16.499	0.26061	3.837
9	5.16	0.19381	0.04808	20.799	0.24808	4.031
10	6.19	0.16151	0.03852	25.959	0.23852	4.192
11	7.43	0.13459	0.03110	32.150	0.23110	4.327
12	8.92	0.11216	0.02526	39.581	0.22526	4.439
13	10.70	0.09346	0.02062	48.497	0.22062	4.533
14	12.84	0.07789	0.01689	59.196	0.21689	4.611
15	15.41	0.06491	0.01388	72.035	0.21388	4.675
16	18.49	0.05409	0.01144	87.442	0.21144	4.730
17	22.19	0.04507	0.00944	105.931	0.20944	4.775
18	26.62	0.03756	0.00781	128.117	0.20781	4.812
19	31.95	0.03130	0.00646	154.740	0.20646	4.843
20	38.34	0.02608	0.00536	186.688	0.20536	4.870

20.0 Percent/Year Discount Rate (Continued)

	Single Payment		Uniform Annual Series			
	Compound Amount Factor	Present Worth Factor	Sinking Fund Factor	Compound Amount Factor	Capital Recovery Factor	Uniform Series Factor
Years	Find F Given P	Find P Given F	Find R Given F	Find F Given R	Find R Given P	Find P Given R
21	46.01	0.02174	0.00444	225.026	0.20444	4.891
22	55.21	0.01811	0.00369	271.031	0.20369	4.909
23	66.25	0.01509	0.00307	326.237	0.20307	4.925
24	79.50	0.01258	0.00255	392.484	0.20255	4.937
25	95.40	0.01048	0.00212	471.981	0.20212	4.948
26	114.48	0.00874	0.00176	567.377	0.20176	4.956
27	137.37	0.00728	0.00147	681.853	0.20147	4.964
28	164.84	0.00607	0.00122	819.223	0.20122	4.970
29	197.81	0.00506	0.00102	984.068	0.20102	4.975
30	237.38	0.00421	0.00085	1181.882	0.20085	4.979
31	284.85	0.00351	0.00070	1419.258	0.20070	4.982
32	341.82	0.00293	0.00059	1704.110	0.20059	4.985
33	410.19	0.00244	0.00049	2045.932	0.20049	4.988
34	492.22	0.00203	0.00041	2456.118	0.20041	4.990
35	590.67	0.00169	0.00034	2948.342	0.20034	4.992
36	708.80	0.00141	0.00028	3539.010	0.20028	4.993
37	850.56	0.00118	0.00024	4247.812	0.20024	4.994
38	1020.67	0.00098	0.00020	5098.374	0.20020	4.995
39	1224.81	0.00082	0.00016	6119.050	0.20016	4.996
40	1469.77	0.00068	0.00014	7343.859	0.20014	4.997
41	1763.73	0.00057	0.00011	8813.630	0.20011	4.997
42	2116.47	0.00047	0.00009	10577.360	0.20009	4.998
43	2539.77	0.00039	0.00008	12693.830	0.20008	4.998
44	3047.72	0.00033	0.00007	15233.590	0.20007	4.998
45	3657.26	0.00027	0.00005	18281.310	0.20005	4.999

INDEX